Norepinephrine

Frontiers of Clinical Neuroscience

Series Editors James H. Wood, M.D.
Division of Neurosurgery
Emory University School of Medicine
Atlanta, Georgia

Benjamin Rix Brooks, M.D.
Departments of Neurology and Medical Microbiology
University of Wisconsin School of Medicine
Madison, Wisconsin

VOLUME 1 **Neurobiology of Mood Disorders**
Robert M. Post, M.D., and James C. Ballenger, M.D.

VOLUME 2 **Norepinephrine**
Michael G. Ziegler, M.D., and C. Raymond Lake, M.D., Ph.D.

VOLUME 3 **Bacterial Infections of the Central Nervous System**
H. Richard Winn, M.D., and William M. Scheld, M.D.

VOLUME 4 **Evoked Potentials**
Roger Q. Cracco, M.D., and Ivan Bodis Wollner, M.D.

VOLUME 5 **Neurobiology of Brain Tumors**
Michael Salcman, M.D.

Frontiers of Clinical Neuroscience
VOLUME 2

NOREPINEPHRINE

Edited by
MICHAEL G. ZIEGLER, M.D.
Assistant Professor
Department of Medicine
University of California, San Diego
Medical Center
University Hospital
San Diego, California

C. RAYMOND LAKE, M.D., Ph.D.
Professor
Departments of Psychiatry and Pharmacology
Uniformed Services University of the Health Sciences
Hebert School of Medicine
Bethesda, Maryland

WILLIAMS & WILKINS
Baltimore/London

Editor: James L. Sangston
Associate Editor: Jonathan W. Pine, Jr.
Copy Editor: Andrea Clemente
Design: Joanne Janowiak
Illustration Planning: Reginald Stanley
Production: Carol L. Eckhart

Made in the United States of America

Library of Congress Cataloging in Publication Data

Main entry under title:

Norepinephrine.

(Frontiers of clinical neuroscience; v. 2)
Includes index.
1. Noradrenalin—Physiological effect—Addresses, essays, lectures. 2. Catecholamines—Physiological effect—Addresses, essays, lectures. I. Ziegler, Michael G. II. Lake, C. Raymond. III. Series.
[DNLM: 1. Norepinephrine. W1 FR946DM v. 2 / WK 725 N834]
QP572.N6N67 1984 615′.78 83-14809
ISBN 0-683-09381-9

Composed and printed at the
Waverly Press, Inc.

Foreword

In addition to being important in their own right, catecholamines have played a remarkable role in the development of many of our concepts of how chemicals mediate intercellular communication in the neurone and endocrine systems. Epinephrine was the first hormone to be identified, the concept of chemical neurotransmission was conceived to explain the similarities in the action of epinephrine and sympathetic nerve stimulation, and the term "receptor substance" was introduced to explain the actions of epinephrine. These events all occurred at about the turn of the century. The role of adenyl cyclase actuation in initiating responses to epinephrine led to the concepts of second messengers for hormone actions. Norepinephrine was identified first as the neurotransmitter released from sympathetic nerve endings, then as a neurotransmitter in brain. The unfolding of the processes for synthesis, storage, release, action, and termination of action and metabolism of norepinephrine provided a model for understanding other neurotransmitters. The demonstration that many psychoactive drugs affect these processes, that drugs which block adrenergic receptors are useful in treatment of hypertension, anxiety, etc., and that the catecholamines themselves have therapeutic benefits has evoked interest in catecholamines by psychiatrists, internists, anesthesiologists, neurologists, and scientists interested in how the nervous system functions.

The field of catecholamine research provides a classic example of the productive interaction of basic research and clinical investigations. Not only have fundamental discoveries yielded useful clinical applications, but frequently, laboratory studies have been conducted in response to clinical observations. It is clearly fitting that a series on Frontiers of Clinical Neuroscience single out norepinephrine for one of its earliest subjects.

In this volume, Drs. Ziegler and Lake have brought together a group of investigators who have been active in various aspects of research on norepinephrine and have organized a book which presents the state-of-the-art in bridging basic and clinical studies of norepinephrine. The research methods used, the fundamental bases of understanding the role of norepinephrine in bodily functions, and current views of the role of catecholamines in relation to various clinical states are considered in detail. It is indeed a source of pride that a number of the authors have been or currently are my colleagues at the National Institute of Health. It is reasonable to expect that the high level of sophistication which has been reached in catecholamine research will contribute to our understanding of disease processes and provide new exciting and important diagnostic procedures and therapeutic approaches.

I. J. Kopin, M.D.

Preface

Hormone, neurotransmitter and neuromodulator, norepinephrine (NE) plays an important role in the three most prominent diseases of Western man: hypertension, heart disease, and mental illness. It is so central to body functions that enormous efforts have been expended in developing techniques to measure NE, its receptors, and its effects. These attempts have been so successful that we can now measure NE and its receptors in nearly every tissue.

These new methods have advanced our understanding of noradrenergic systems enough to provide a tremendous increase in knowledge—or in confusion. This book assembles these recent advances in a manner that allows the reader to grasp the current state of knowledge, to re-evaluate old ideas, and assimilate new techniques.

This volume chronicles the many recent advances that have radically altered our views of NE. Assays are now so sensitive that they can measure changes in plasma NE that result from what were once considered trivial alterations in posture or emotion. The structures of synaptic clefts where NE is released are not all identical; differences exist which expose receptors to a 100-fold range of NE concentrations. NE neurons in the brain have a well-ordered anatomy that was unknown just a few years ago, and NE co-exists with other neurotransmitters in the same cell. Blood NE levels are determined as much by individual rates of clearance and prior caloric intake as they are by neuronal release.

When all these pieces of information are added together, they vastly improve our understanding of NE. A more cohesive and comprehensive picture emerges, one that not only provides answers to old problems, but that leads us to the most important new questions. This book can serve as a reference for the reader who has specific questions, while those who read through the text will be rewarded with a new understanding of NE and of how the body uses neurohumoral transmission.

MICHAEL G. ZIEGLER, M.D.

Acknowledgments

Thanks and appreciation are extended to Laurie Taylor-Donald, Kathy Moriarty, Cindy Scher, Michael Batenjany, and Carole Ziegler for their expertise and assistance in preparing this volume. We commend the editorial and production staff at Williams & Wilkins for their patience, skill, and support, and we thank the contributing authors for their excellent chapters.

Contributors

Gary Aston-Jones, Ph.D.
Assistant Professor, Center for Neurobehavioral Science, SUNY at Binghamton, Binghamton, New York

James E. Barrett, Ph.D.
Associate Professor, Department of Psychiatry, Uniformed Services University of the Health Sciences, Bethesda, Maryland

Richard A. Becker, M.D.
Clinical Assistant Professor, Department of Medicine, University of Texas Medical School at San Antonio, and Consultant, U.S. Army Institute of Surgical Research, Brooke Army Medical Center, Fort Sam Houston, San Antonio, Texas

John A. Bevan, M.D.
Professor, Department of Pharmacology, Brain Research Institute, School of Medicine, Center for the Health Sciences, University of California, Los Angeles, California

Floyd E. Bloom, M.D.
Director, Arthur Vining Davis Center for Behavioral Neurobiology, The Salk Institute, San Diego, California

Robert M. Bowker, V.M.D., Ph.D.
Research Scientist and Assistant Professor, Department of Anatomy and Cell Biology, Columbia University and the New York Psychiatric Institute, New York, New York

Nguyen Thanh Buu, Ph.D.
Assistant Professor, Clinical Research Institute, University of Montreal, Montreal, Quebec, Canada

Bart Chernow, M.D.
Director of Research, Critical Care Medicine Branch, Internal Medicine Service, Naval Hospital, and Associate Professor, Department of Medicine, Uniformed Services University of the Health Sciences, Bethesda, Maryland

Alison E. Cole, Ph.D.
Post Doctoral Fellow, Neurophysiology Lab, National Institute of Neurological and Communicative Diseases and Stroke, National Institutes of Health, Bethesda, Maryland

Harry T. Colfer, M.D.
Instructor of Internal Medicine, University of Michigan Medical Center, Ann Arbor, Michigan

Richard J. Contrada, A.B.
Research Associate, Department of Psychology, SUNY at Stony Brook, Stony Brook, New York

Joe Dan Coulter, Ph.D.
Associate Professor, Marine Biomedical Institute, University of Texas Medical Branch, Galveston, Texas

Phillip E. Cryer, M.D.
Associate Professor, Department of Medicine, Division of Metabolism, Washington University School of Medicine, Saint Louis, Missouri

Lenard R. Durrett, M.D., Ph.D.
Post-Doctoral Fellow, Department of Anesthesiology, Division of Cardiology, University of Texas Medical Branch, Galveston, Texas

Csaba Farsang, M.D.
Associate Professor, Second Department Medicine, Semmelweis Medical University, Budapest, Hungary

Giora Feuerstein, M.D.
Assistant Professor, Neurobiological Research Institute, Uniformed Services University of the Health Sciences, Bethesda, Maryland

Chris K. Finton, M.D.
Fellow, Cardiovascular Disease Branch, Internal Medicine Service, Naval Hospital, Bethesda, Maryland

Steven L. Foote, Ph.D.
Staff Scientist, Arthur Vining Davis Center for Behavioral Neurobiology, The Salk Institute, San Diego, California

Joel P. Gallagher, Ph.D.
Associate Professor, Department of Pharmacology and Toxicology, University of Texas Medical Branch, Galveston, Texas

David G. Glass, Ph.D.
Vice Provost for Research Graduate Studies, SUNY at Stony Brook, Stony Brook, New York

David S. Goldstein, M.D., Ph.D.
Medical Staff Fellow, National Heart, Lung and Blood Institute, National Institutes of Health, Bethesda, Maryland

Robert E. Goldstein, M.D.
Associate Professor, Departments of Medicine and Physiology, Uniformed Services University of the Health Sciences, Bethesda, Maryland

Hans-Georg Güllner, M.D.
Visiting Scientist, Hypotension-Endocrine Branch, National Heart, Lung, and Blood Institute, National Institutes of Health, Bethesda, Maryland

Wyatt R. Hume, D.D.S., Ph.D.
Associate Professor, Section of Oral Biology, School of Dentistry, Center for the Health Sciences, University of California, Los Angeles, California

Hans Ibsen, M.D.
Assistant Professor, Rigshospitalet, University Hospital, Copenhagen, Denmark

Paul A. Insel, M.D.
Assistant Professor, Division of Pharmacology, Department of Medicine, University of California, San Diego, La Jolla, California

Stevo Julius, M.D., Sc.D.
Professor, Department of Internal Medicine, Division of Hypertension, University of Michigan Medical School, Ann Arbor, Michigan

Harry R. Keiser, M.D.
Clinical Director, National Heart, Lung, and Blood Institute, National Institutes of Health, Bethesda, Maryland

Henry Khachaturian, Ph.D.
Post-Doctoral Fellow, Mental Health Research Institute, University of Michigan, Ann Arbor, Michigan

Peter T. Kissinger, Ph.D.
Professor, Department of Chemistry, Purdue University, West Lafayette, Indiana

Walter Kobinger, M.D.
Professor, Ernst Boehringer Institute of Drug Research, Department of Pharmacology, Vienna, Austria

Otto Kuchel, M.D., Sc.D.
Professor, Clinical Research Institute, University of Montreal, Montreal, Quebec, Canada

George Kunos, M.D., Ph.D.
Professor, Departments of Medicine, Pharmacology and Therapeutics, McGill University, Montreal, Quebec, Canada

C. Raymond Lake, M.D., Ph.D.
Professor, Departments of Psychiatry and Pharmacology, Uniformed Services University of the Health Sciences, Bethesda, Maryland

Lewis Landsberg, M.D.
Associate Professor, Department of Medicine, Harvard Medical School, Department of Medicine, Beth Israel Hospital, Boston, Massachusetts

Pauline Lerner, Ph.D.
Chemist, Office of Nutrition and Food Sciences, Bureau of Foods, Food and Drug Administration, Washington, D.C.

Gertrude D. Maengwyn-Davies, Ph.D.
Professor Emeritus, Department of Pharmacology, Physiology, and Biophysics, Georgetown University School of Medicine and Dentistry, Washington, D.C.

Lee Frank Major, M.D.
Associate Professor, Department of Psychiatry, Upstate Medical Center, Clinical Campus, SUNY at Binghamton, Binghamton, New York

Ali Jose Milano, M.D.
Professor, Departmento de Farmacolgica, Universidad de Los Andes, Venezuela, South America

Jonathan Moss, M.D., Ph.D.
Associate Professor, Department of Anesthesia, Harvard University School of Medicine, Massachusetts Eye and Ear Infirmary, Boston, Massachusetts

Harvey J. Motulsky, M.D.
Assistant Research Pharmacologist, Division of Pharmacology, Department of Medicine, University of California, San Diego, La Jolla, California

Sheila M. Muldoon, M.D.
Associate Professor, Department of Anesthesiology and Physiology, Uniformed Services University of the Health Sciences, Bethesda, Maryland

John T. O'Brian, M.D.
Head, Endocrine Branch, Naval Hospital, Bethesda, Maryland

Sol I. Rafjer, M.D.
Assistant Professor, Department of Medicine, Section of Cardiology, Committee on Clinical Pharmacology, University of Chicago, Chicago, Illinois

Thomas G. Rainey, M.D.
Assistant Professor, Department of Medicine, Uniformed Services University of the Health Sciences, Head, Critical Care Medicine Department, Naval Hospital, Bethesda, Maryland

Maria Dolores Ramirez-Gonzalez, Ph.D.
Department of Pharmacology and Therapeutics, McGill University, Montreal, Quebec, Canada

John L. Reid, M.D.
Regius Professor of Materia Medica, Department of Materia Medica, University of Glasgow, Stobhill General Hospital, Glasgow, Scotland

Michael F. Roizen, M.D.
Associate Professor, Department of Anesthesia, University of California School of Medicine, San Francisco, California

Peter C. Rubin, M.D.
Wellcome Senior Fellow, Department of Materia Medica, University of Glasgow, Glasgow Scotland

Patricia Shinnick-Gallagher, Ph.D.
Associate Professor, Department of Pharmacology and Toxicology, University of Texas Medical Branch, Galveston, Texas

Ronald E. Shoup, Ph.D.
Research Director, Bioanalytical Systems, Inc., West Lafayette, Indiana

Matthew B. Stern, M.D.
Assistant Professor, Department of Neurology, University of Pennsylvania and Graduate Hospital, Philadelphia, Pennsylvania

Richard Earl Tessel, Ph.D.
Associate Professor, Department of Pharmacology and Toxicology, University of Kansas, Lawrence, Kansas

Nguyen Bich Thoa, Ph.D.
The Bio-Medical Group, Bethesda, Maryland

George M. Vaughan, M.D.
Major, Medical Corps, and Chief, Internal Medicine Branch, Department of Army, U.S. Army Instructor of Surgical Research, Brooke Army Medical Center, Fort Sam Houston, San Antonio, Texas

O. Humberto Viveros, Ph.D.
Group Leader, Department of Medicinal Biochemistry, Wellcome Research Labs, Research Triangle Park, North Carolina

J. Michael Walker, Ph.D.
Post-Doctoral Fellow, Mental Health Research Institute, University of Michigan, Ann Arbor, Michigan

B. Gunnar Wallin, M.D.
University Hospital, Department of Clinical Neurophysiology, Uppsala, Sweden

Stanley J. Watson, M.D., Ph.D.
Associate Research Scientist and Associate Professor, Department of Psychiatry, Mental Health Research Institute, University of Michigan, Ann Arbor, Michigan

Karin N. Westlund, Ph.D.
Post-Doctoral Fellow, Department of Anatomy, University of Texas Medical Branch, Galveston, Texas

Steven P. Wilson, Ph.D.
Assistant Medical Research Professor, Department of Pharmacology, Duke University Medical Center, Durham, North Carolina

James H. Wood, M.D.
Director, Cerebral Blood Flow Laboratories, Assistant Professor, Division of Neurosurgery, Emory University Clinic, Atlanta, Georgia

James B. Young, M.D.
Assistant Professor, Harvard Medical School, Department of Medicine, Beth Israel Hospital, Boston, Massachusetts

Michael G. Ziegler, M.D.
Assistant Professor, Department of Medicine, University of California, San Diego Medical Center, University Hospital, San Diego, California

Contents

SECTION 1

Introduction and Measurements

CHAPTER ONE

The Sympathetic Nervous System in Man: Its Evaluation and the Measurement of Plasma NE

C. RAYMOND LAKE, M.D., Ph.D., BART CHERNOW, M.D., GIORA FEUERSTEIN, M.D., DAVID S. GOLDSTEIN, M.D., Ph.D., and MICHAEL G. ZIEGLER, M.D.

Abbreviations used in Chapter 1 are: SNS, sympathetic nervous system; NE, norepinephrine; E, epinephrine; DA, dopamine; CA, catecholamines; DBH, dopamine-β-hydroxylase; dopa, 3,4-dihydroxyphenylalanine; NTS, nucleus tractus solitarius; 5-HT, serotonin; PNMT, phenylethanolamine-N-methyltransferase; MAO, monoamine oxidase; MAO-I, monoamine oxidase inhibitor; COMT, catechol-o-methyltransferase; TH, tyrosine hydroxylase; ATP, adenosine triphosphate; cAMP, cyclic adenosine monophosphate; NM, normetanephrine; MHPG, 3-methoxy-4-hydroxyphenylglycol; M, metanephrine; VMA, vanillylmandelic acid or 3-methoxy-4-hydroxymandelic acid; HPLC, high pressure liquid chromatography; CV, coefficient of variation; EDTA, disodium-ethylenediamine-tetra-acetate; EGTA, ethyleneglycol-bis-(β-amino-ethyl-ether)N,N′-tetra-acetic acid.

*The authors wish to acknowledge the contributions of the following individuals: Kathy Moriarty, Laurie Taylor-Donald, Dr. Sheryle Alagana, Rhona Bosin, Audrey Reid, Christelle Coen, Rosemarie Quirk, Cindy Scher, Duke Owen, Marilyn R. Stefanye, and Dominic S. D'Angona, Jr.

HISTORY

The chemical structure of epinephrine (E) was reported in 1897 (1). Langley (87) and his student, Elliot (35), suggested a relationship between E, or an E-like compound, and the sympathetic nervous system (SNS). Cannon, in 1914, demonstrated the sympatho-adrenal medullary system's "emergency function" of mediating the organism's physiological preparation for threatening situations (19). In the late 1940s von Euler (42) and Peart (115) confirmed that norepinephrine (NE), the precursor of E, was the primary neurotransmitter of the SNS, and, since adrenalectomy decreases urinary E but not NE, concluded that the SNS is the primary site of origin of NE, whereas E originates from the adrenal glands (43). NE was first measured in plasma by a fluorometric method in 1950 (95). Since then there have been many substantial improvements in catecholamine (CA) measurement and technology which are discussed in detail in Chapters 2 and 3.

ANATOMY

Central Autonomic Connections

Central autonomic and somatic centers overlap extensively; no clear autonomic or somatic centers are known. Autonomic reflexes exist at various levels of the cerebrospinal axis:

1. Spinal cord reflexes are clearly demonstrable in spinal animals (animals whose spinal cords have been cut at the C-1 or C-2 level) and involve responses of sweat glands and visceral organs, e.g., gastrointestinal and genitourinary activities. However, spinal animals lack reflex sympathetic responses to blood volume or pressure changes. Furthermore, they must be maintained on artificial respiration since all of the motor respiratory nerves (phrenic, intercostal) originate from lower spinal areas. However, parasympathetic regulation of cardiac functions (through the vagus) is preserved. Spinal animal preparations are often used to examine the role of the vagus in modulating cardiac functions since cardiac sympathetic reflexes are eliminated. Further details of spinal noradrenergic pathways are given in Chapter 5.

2. The medulla oblongata and the pons include vital centers that integrate control of blood pressure. Elimination of the medulla oblongata (high spinal cord cuts) results in shock with blood pressure falling to 40 to 60 mm Hg, but midcollicular cuts affect neither basal blood pressure or the baroreflex mechanism. It is currently recognized that only certain nuclei in the medulla oblongata are necessary for cardiovascular regulation; adrenergic elements within the medulla are important in these functions (22, 99). Destruction of discrete NE cell clusters within the tractus solitarius or the A-1 neurons results in hypertension (127) through activation of the SNS and hypothalamic release of vasopressin (12). The nucleus tractus solitarius (NTS) is a major area for relay of chemo- and baroreceptor information (14, 114). The impulses generated at the baroreceptive areas (carotid sinus, left and right atrium, aortic arch, and large veins and arteries in the thoracic-cardiopulmonary area) are transferred by neurons located in the nodosal ganglia of the ninth and tenth cranial nerves into the solitarius tract. The tonic input of these afferent pathways inhibits peripheral sympathetic discharge and increases vagal tone. Two drugs thought to act on adrenergic elements in the NTS to produce hypotension are clonidine (a central α-2-adrenoreceptor agonist) and α-methyl dopa (31) (Chapters 7 and 19).

The medulla oblongata also contains the two major motor nuclei of the vagal outflow to the heart: the dorsal motor nucleus of the vagus, which lies in close proximity to the NTS, and the nucleus ambiguus at the ventrolateral side of the medulla oblongata. The medulla also contains the essential respiratory neurons which maintain respiratory automaticity. These neurons are more diffuse but are found mainly in the reticular formation and the parabrachial nucleus.

3. The hypothalamus is the main integration site of the various functions regulated by the autonomic nervous system. The various nuclei within the hypothalamus are involved in regulation of temperature (anterior hypothalamus), intravascular volume (supraoptic and paraventricular nuclei), blood pressure, reaction to stress, emotions, sleep, and sexual reflexes. The posterolateral nuclei are especially involved in sympathetic connections, and, therefore, electrical stimulation in these sites can cause massive activation and discharge of the sympathoadrenal axis (99).

4. Although little is known of thalamic and striatal levels of autonomic integration, lesions in this region result in autonomic disturbances.

5. The cortex is the highest level of sympathetic integration. Cardiovascular, gastrointestinal, and other autonomic functions can be successfully modulated through conscious, intentional efforts (100). Cortical areas associated with hippocampal nuclei (amygdala) and the olfactory bulb are responsible for control of emotional and/or stress-induced responses to visceral functions (96). Limbic system disorders in NE neurotransmission may result in symptoms characteristic of some subtypes of schizophrenia (81, 145) and major affective disorders (123). For additional details of CA CNS anatomy the reader is referred to Chapters 5, 6, 7, and 27.

Peripheral Autonomic Connections

The efferent component (motor side) of the autonomic nervous system consists of two parts: (1) sympathetic and (2) parasympathetic. Chapter 5 includes a detailed discussion of the organization of the descending noradrenergic pathways in the spinal cord.

Sympathetic cell bodies lie in the intermediolateral horn of the spinal cord from the eighth cervical to the second or third lumbar segment. The axons of these nerve cells pass through the ventral nerve root of the same spinal segment, but after a short distance, leave the spinal nerve and form a short branch called the white ramus, composed of myelin-coated nerves, which terminates in a sympathetic ganglion located near the spinal nerve. There are three groups of sympathetic ganglia: (1) paravertebral ganglia as described, which lie on either side of the vertebral column to form the lateral chains; (2) prevertebral ganglia, which lie at the ventral side of the bony spinal column; and (3) terminal ganglia which lie close to the organs supplied by the postganglionic fibers. Most of the preganglionic neurons terminate in the lateral paravertebral ganglia at the level of exit, but some ascend in the chain to synapse in the third or fourth ganglion above the level of exit; caudal extensions are even greater since the lateral chain reaches the sacral level despite the termination of the preganglionic outflow in the upper lumbar region. A single ganglion receives fibers from many spinal segments (convergence). Also, single preganglionic fibers emanating from a spinal segment may branch and innervate several sympathetic ganglia (divergence). The 22 paravertebral ganglia are connected to each other by nerve trunks; each ganglion also has a second connection to the spinal nerve root by a grey ramus composed of nonmyelinated fibers. This bundle arises from nerve cells within the ganglion and carries postganglionic fibers back to the spinal nerve from which connections are distributed distally to sweat glands, pilomotor muscle, and blood vessels of skeletal muscles and skin. Two major sympathetic ganglia, the celiac and mesenteric, are formed from both preganglionic and postganglionic fibers.

At the upper part of the paravertebral chain is the large stellate ganglion, composed primarily of postganglionic neuron fibers since most of the preganglionic fibers are short. The paravertebral sympathetic chain ends with the superior cervical ganglion, which, along with the middle and inferior ganglia, innervate intra- and extracranial structures.

Postganglionic fibers emanating from sympathetic ganglia reach all of the organs in the head, neck, thorax, and abdomen. The trunk and limbs are supplied by sympathetic fibers in spinal nerves. The glands and smooth muscles in the abdomen and pelvis are supplied by postganglionic fibers arising from the prevertebral ganglia. NE is the neurotransmitter released from almost all postganglionic SNS fibers, while acetylcholine is released by the preganglionic fibers and the fibers to the adrenal medulla.

The adrenal medulla is composed of chromaffin tissue which is embryologically homologous to sympathetic ganglia. These chromaffin cells are innervated by preganglionic sympathetic fibers originating from the fifth through ninth spinal segments and grouped in the greater splanchnic nerve. The chromaffin cells differ from sympathetic postganglionic fibers primarily by having the enzyme phenylethanolamine-N-methyltransferase (PNMT) which converts NE to E, the major hormone secreted from the adrenal medulla. A normal adrenal gland appears to contribute less than 2% to circulating NE levels but is responsible for almost all of the E (15). The chromaffin cells also differ from sympathetic nerve endings in their resistance to 6-hydroxy-dopamine, and their somewhat different mechanisms for CA release and uptake (151).

The termination of the postganglionic SNS fibers at smooth muscles or glandular cells is characterized by multiple branchings called a terminal effector plexus or reticulum (Fig. 1.1). This plexus, typically located at the adventitio-medial junction, surrounds the muscle of the media "like a mesh stocking" (9, 13, 147). Varicosities containing vesicle packages of the neurotransmitter NE and the enzyme dopamine-β-hydroxylase (DBH) occur repeatedly along the multiple divisions of the axons. Each plexus may extend over an area of 0.2 to 0.5 mm in diameter. Each adrenergic neuron gives rise to approximately 10 cm of terminal axon and 26,000 varicosities containing 260 pg of NE (9). The distance between the varicosity and the muscle cell varies from 100 Å in arterioles and small arteries to up to 20,000 Å in large vessels (Chapter 4). Myogenic propagation functions as the major mechanism for transmission of activation in small vessels. In large vessels the rate of the contractile response is slower and is regulated primarily by the rate of NE diffusion through the vessel

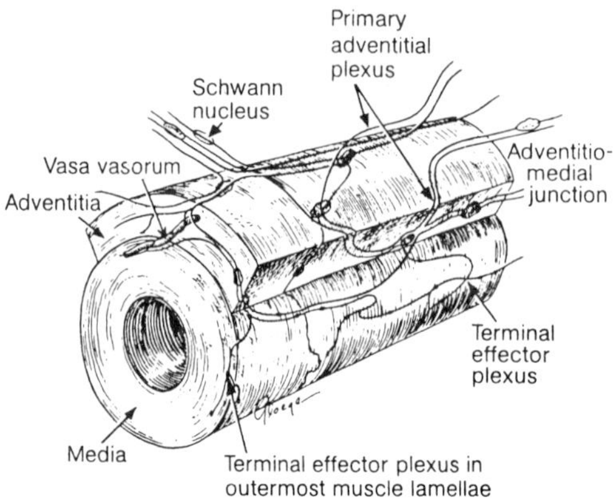

Figure 1.1. Schematic arrangement of a noradrenergic neuroeffector apparatus. Postganglionic sympathetic axons branch into bundles called primary plexuses located in the loose adventitial layers of blood vessels. The fibers from the primary plexus extend to the adventitiomedial junction and into the outermost layers of the smooth muscular media forming terminal effector plexuses that have the characteristic beaded appearance. These beads are varicosities that form the actual neuroeffector junctions across which NE transmits SNS information. (Reproduced with permission from M. A. Verity (147).)

wall (9, 93, 113) (Fig. 1.2). The reader is referred to Chapter 4 for details of this synapse.

BIOSYNTHESIS OF CATECHOLAMINES: THE ROLE OF ENZYMES

The CA are synthesized from the essential amino acid tyrosine by several enzymatically catalyzed steps (Fig. 1.3) occurring in the postganglionic SNS neurons, in the adrenal medullary cells, and in certain neurons in the brain (6, 11, 50, 73, 105). The first two steps, from tyrosine to 3,4-dihydroxyphenylalanine (dopa) and from dopa to dopamine (DA), take place in the cytoplasm of these cells. DA then enters the storage vesicles, which are about 400 Å in diameter, where it is converted to NE by DBH (Fig. 1.4). In those neurons utilizing DA as their neurotransmitter, no DBH is present. In the adrenal medulla, the NE may leave the vesicles in order to be methylated by PNMT to form E in the cytoplasm; the E then reenters another group of vesicles where it is stored until released from the medullary cells. E comprises about 80% of the CA in these cells and NE about 20%.

Unlike the CA, the enzymes are synthesized entirely in the neuronal cell bodies. Tyrosine hydroxylase (TH), a soluble enzyme found in the nerve cell body, axons, and terminals, has a molecular weight of about 192,000 and requires molecular oxygen, ferrous ions, and tetrahydropteridine as cofactors (104). NE and DA, which compete for the pteridine cofactor, inhibit TH, allowing for intraneuronal negative feedback regulation of NE formation (105) (Fig. 1.4). The conversion of tyrosine to dopa is the slowest reaction of the series and appears to be the rate-limiting step (91, 141) (Figs. 1.3 and 1.4).

Dopa is converted to DA through the action of aromatic amino acid decarboxylase (dopa decarboxylase), which has a molecular weight

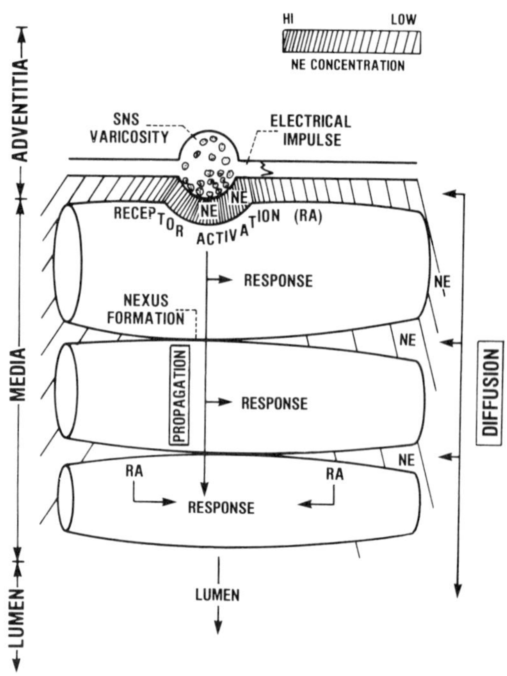

Figure 1.2. Extension of response beyond the immediate neuroeffector junction. NE released from SNS varicosities in the adventitiomedial junction of a blood vessel first activates directly innervated smooth muscle cells in the outermost layers of the media. Deeper muscle layers are recruited by one or both of two mechanisms: (1) diffusion of NE which decreases in concentration as it reaches and activates nonjunctional receptors; (2) propagation of the response between adjacent smooth muscle cells through low-resistance connections called nexus formations. *RA*, receptor activation. (Modified from B. Ljung (93).)

of about 109,000. Dopa decarboxylase is nonspecific and can therefore decarboxylate a variety of L-aromatic amino acids, in addition to dopa. It requires, and is tightly bound to, its cofactor, pyridoxal phosphate (94).

DBH is highly concentrated in NE-containing vesicles of the SNS terminal varicosities but can also be found in the SNS nerve cell bodies and axons. It is a mixed function oxidase which hydroxylates the β-carbon of DA to form NE. A relatively nonspecific enzyme with a molecular weight of about 290,000, DBH contains copper and requires ascorbic acid or an oxygen acceptor. Each DBH molecule has four identical subunits of about 77,000 molecular weight, each consisting of two polypeptide chains (50, 105). Since it is contained in the storage vesicles and released with NE, its levels in plasma have been investigated as an index of SNS activity (50, 74, 77, 152). A more in-depth discussion of DBH is contained under the heading Dopamine-β-Hydroxylase in this chapter and in Chapter 8.

PNMT, the enzyme that converts NE to E, is highly localized in the cytoplasm of the chromaffin cells of the adrenal medulla and in certain neurons in the CNS. It has a molecular weight of around 30,000 (4, 27). PNMT catalyzes the transfer of a methyl group from S-adenosylmethionine to NE or to other β-hydroxylated phenylethylamines. This property of the enzyme has been used in the development of radioenzymatic assays for DBH and NE (76, 150, 152) (Chapter 2).

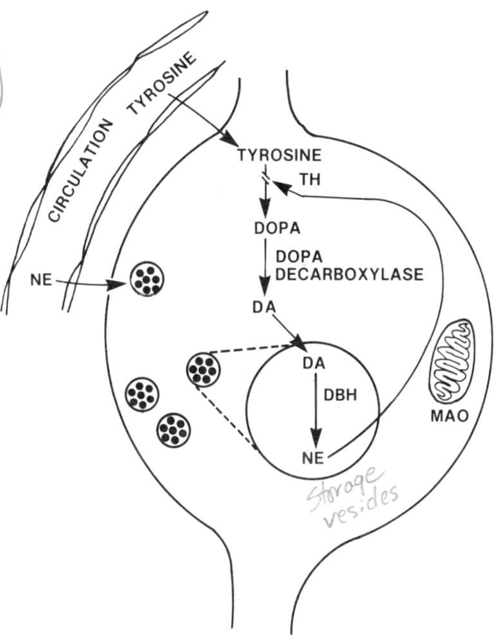

Figure 1.4. Uptake and biosynthesis of NE within a SNS varicosity. Tyrosine and NE can be taken up into postganglionic SNS nerves. In the cytoplasm tyrosine is metabolized to DA which, along with cytoplasmic NE, can be actively taken up into storage vesicles where DBH converts the DA to NE. (Modified from J. Axelrod and I. J. Kopin (5).)

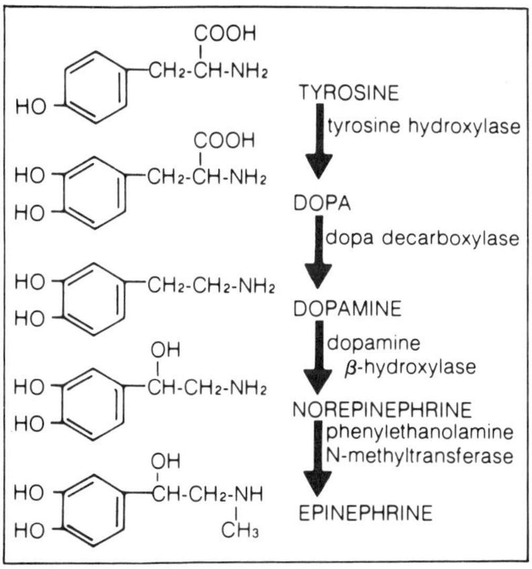

Figure 1.3. Biochemical synthesis of the catecholamines.

Since synthesis of NE from tyrosine occurs in the terminal varicosities, and the protein-synthesizing apparatus of the SNS neuron is localized in the cell body, these enzymes must be transported distally along the axons. TH and DBH are transported (probably via the microtubular system) at a rate of 1 to 10 mm/hour. DBH appears to be synthesized at a rate of 5% of the total amount in a ganglion per hour.

REGULATION OF CATECHOLAMINE METABOLISM

TH, DBH, and PNMT levels appear to be genetically linked and controlled by a single regulatory locus. During prolonged SNS activity, production of these enzymes is increased to accelerate the rate of NE and E synthesis. NE and DA in the neuronal cytoplasm inhibit TH (Fig. 1.4; Fig. 1.5, step 20). With acute SNS activation the amount of extraneuronally re-

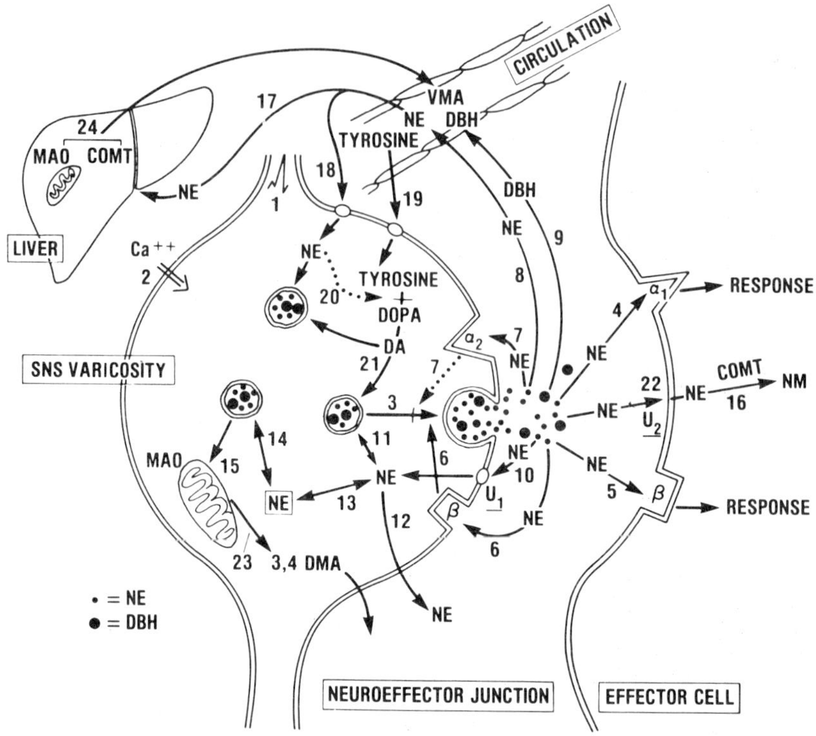

Figure 1.5. Norepinephrine release and metabolism in the SNS varicosity. Step *1*, arrival of action potential *2*, influx of calcium ions; *3*, migration of vesicles to varicosity membrane at neuroeffector junction; *4*, exocytosis and extrusion of varicosity contents (NE and DBH) into neuroeffector junction; NE acts at α-1-receptor on effector cell; *5*, NE acts at β-receptor on effector cell; *6*, NE acts at presynaptic β-receptor; *7*, NE acts at presynaptic α-2-receptor to downregulate further NE release; *8*, NE diffuses out of the neuroeffector junction into the circulation; *9*, DBH diffuses into the circulation; *10*, NE is actively taken back up into varicosity (uptake-1); *11*, intraneuronal NE is taken up into vesicle; *12*, intraneuronal NE diffuses out of varicosity; *13*, intraneuronal NE is stored in another NE pool not initially available for release; *14*, this pool actively exchanges NE with the vesicular pool; *15*, NE can leak out of the vesicular pool and be oxidized by intracellular MAO associated with mitochondria; *16*, *17*, extraneuronal NE is initially metabolized by COMT in the effector cells, in the liver, and in red cells; *18*, *19*, circulating NE and its precursor, tyrosine, are actively taken up by SNS varicosities; *20*, intraneuronal NE inhibits TH (tyrosine hydroxylase); *21*, intraneuronal DA is actively taken up by the vesicles; *22*, released NE is removed from the neuroeffector junction by uptake-2 (a nonspecific large capacity process); *23*, MAO oxidizes intraneuronal NE to 3,4-dihydroxymandelic acid (3,4-DMA); *24*, NE is usually metabolized by both MAO and COMT to 3-methoxy,4-hydroxymandelic acid (VMA); (Modified from J. Axelrod and R. Weinshilboum (6).)

leased NE increases, thereby reducing NE leakage from the storage vesicles into the cytoplasm and decreasing TH inhibition (Fig. 1.5, steps 15 and 20). In addition, TH molecules already present are activated by changes in their kinetic properties, *i.e.*, an increase in the number of active sites per molecule (73).

The glucocorticoids of the adrenal cortex are needed to maintain TH, DBH, and PNMT activities of the adrenal medulla and may be involved in the early differentiation of nerve cells to chromaffin-type cells. The intra-adrenal portal vascular system carries the glucocorticoids from the cortex to the medulla where they stimulate the synthesis of PNMT.

Norepinephrine Storage and Release

NE, ATP, DBH, and chromogranins (a class of proteins involved in CA binding) are stored in the osmophilic storage vesicles in the SNS terminal varicosities (Figs. 1.4 and 1.5) (5, 50, 73, 105). Cytoplasmic β-hydroxylated or catechol compounds including NE, E, DA, tyramine, and amphetamines are taken up actively by the vesicles (Fig. 1.5, steps 11, 14, and 21). ATP and magnesium are required for this somewhat selective, stereospecific, temperature-dependent process which is blocked by reserpine and many hydroxylated phenylethylamines capable of displacing NE, such as tyramine. This displacement increases cytoplasmic catabolism (Fig. 1.5, steps 15 and 23) and diffusion of NE through the axoplasmic membrane, producing effects similar to those of normally released NE (Fig. 1.5, step 12). Indirectly acting sympathomimetic drugs (*i.e.*, tyramine) may also inhibit intraneuronal monoamine oxidase (MAO). Several drugs, such as the tricyclic antidepressants and cocaine, block uptake-1 of NE across the axoplasmic membrane (Fig. 1.5, steps 10 and 18).

At least two pools of NE exist: one is readily released upon acute nerve stimulation, while the second is maintained in reserve in the vesicles (Fig. 1.5, step 14). NE can be concentrated in the storage vesicles against a 200-fold gradient, resulting in a concentration 10,000-fold greater than plasma.

NE is released from the postganglionic terminal varicosities in response to a depolarizing action potential (Fig. 1.5, step 1). The influx of calcium ions that follows an action potential (Fig. 1.5, step 2; Fig. 1.6, step 2) causes fusion of the vesicular membrane to the prejunctional membrane (Fig. 1.5, step 3; Fig. 1.6, steps 1 to 5) (73). The vesicular contents, including NE, DBH, ATP, and chromogranins are then extruded into the junctional cleft by a process called exocytosis (Figs. 1.5 and 1.6). Microtubules and microfilaments seem to be involved in the release process, since colchicine and cytochalasin-B inhibit exocytosis (Fig. 1.6,

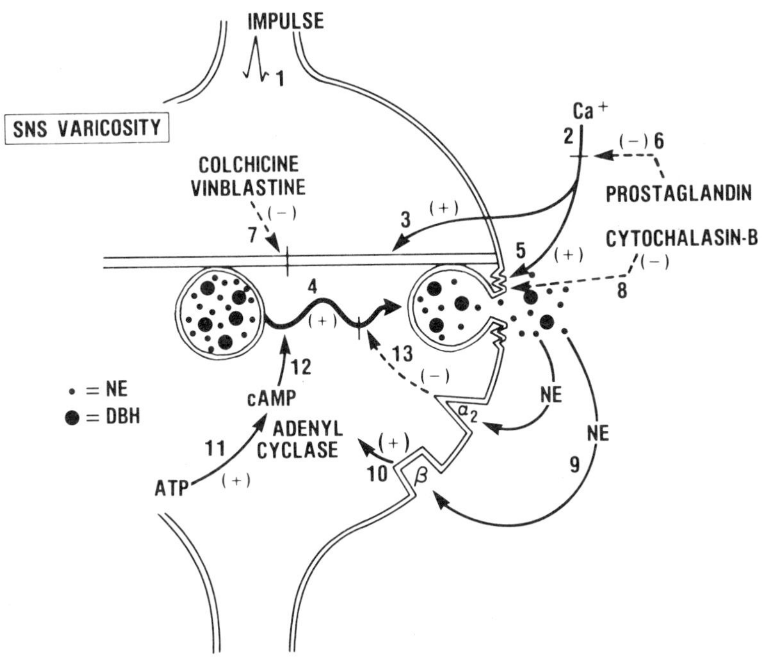

Figure 1.6. Details of postulated mechanisms of exocytosis in the SNS varicosity. Step *1*, arrival of action potential at varicosity. *2–4*, calcium influx enhances migration of NE-containing vesicles along microtubules serving as cytoskeletons to guide vesicles to the membrane adjacent to the effector cell; calcium may cause the microtubules to contract. *5*, calcium may also cause contraction of microfilaments in the varicosity membrane to enlarge the opening during exocytosis. *6*, prostaglandins inhibit stimulation-induced release, possibly by inhibiting the influx of calcium ions. *7*, vinblastine and colchicine inhibit stimulation-induced release by interruption of the microtubular system. *8*, cytochalasin-B inhibits NE release, possibly by interfering with calcium-induced contraction of the microfilaments. *9–12*, junctional NE can activate presynaptic β-receptors which enhance adenylate cyclase activity, increasing the formation of cAMP from ATP. cAMP is necessary for exocytosis. *13*, presynaptic α-2-receptor activation downregulates further NE release. (Modified from I. J. Kopin (73).)

steps 7 and 8). The microtubules may function as cytoskeletons that direct the vesicles to the proper place on the prejunctional membrane (Fig. 1.6, steps 3 and 4). Calcium may cause microfilaments in the prejunctional membrane to contract and enlarge the opening, allowing exocytosis to occur (Fig. 1.6, step 5). Prostaglandin E2 inhibits exocytosis, presumably by interfering with the role of calcium (Fig. 1.6, step 6). Phenoxybenzamine and phentolamine, α-receptor antagonists, increase the release of NE by blocking the prejunctional α-2-receptors where activation downregulates NE release (133) (Fig. 1.5, step 7 and Fig. 1.6, step 13). Prejunctional β-receptor activation enhances further NE release by a positive feedback mechanism, probably mediated through cyclic adenosine monophosphate (cAMP) (Fig. 1.6, steps 9 to 12) (142). The postjunctional effector cell receptors may be involved in neuronal "long-loop" negative feedback mechanisms to further modulate NE release. The reader is referred to Chapters 4, 18, and 19 for detailed information on the NE synapse and receptors.

Catabolism and Inactivation of Norepinephrine

The action of NE may be terminated by at least four mechanisms: (1) active reuptake by the prejunctional SNS neuron (uptake-1) (68) (Fig. 1.5, step 10); (2) dilution and diffusion out of the junctional cleft (Fig. 1.2; Fig. 1.5, step 8); (3) enzymatic degradation by MAO, catechol-O-methyltransferase (COMT) and phenolsulfotransferase (Fig. 1.5, steps 15, 16, 24 and Fig. 1.7); or (4) a high capacity, low affinity process by which NE is taken up by the effector and other cells and subsequently catabolized (uptake-2) (68) (Fig. 1.5, step 22). It is clear that uptake-1 is the major mechanism of inactivation of endogenous NE since its inhibition prolongs the actions of NE substantially more than does the inhibition of MAO and COMT (6). Reuptake is a rapid and economical mechanism for terminating NE action; not only can

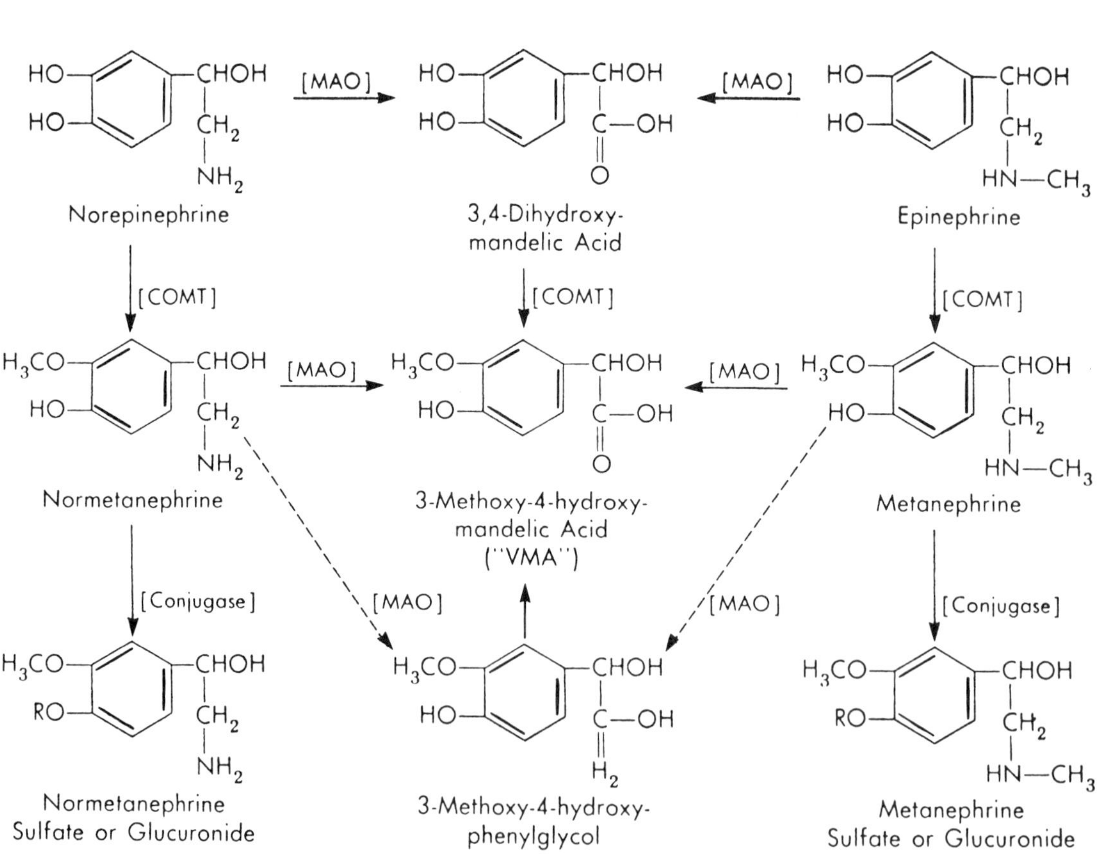

Figure 1.7. Catabolic breakdown of catecholamines. (Modified from I. J. Kopin (72).)

NE be taken up through the axoplasmic membrane, but it can also be taken up into a storage granule from which it can be released again (5, 72).

The synaptic cleft width also determines the dominant inactivation process. In larger vessels where cleft width is greatest, diffusion of NE away from the smooth muscle cells is important (Fig. 1.2). Neuronal reuptake predominates in smaller vessels where reaction time is considerably faster (8, 11) (Chapter 4).

MAO is primarily associated with mitochondria and is present in SNS varicosities (Fig. 1.5, step 15) (105). Through deamination, it inactivates NE that leaks out of, or that is displaced from, the storage vesicles, or excess NE that is taken up by the prejunctional membrane.

COMT requires magnesium ions and S-adenosylmethionine as a methyl donor (105). It methylates a wide variety of catechols including NE, E, and DA and is utilized in the COMT CA radioenzymatic assay (Chapter 2). Although present in most tissues, COMT is most plentiful in the liver and kidneys (Fig. 1.5, step 24). Methylation, rather than oxidation, is generally the first degradative step for circulating CA released from the adrenal medulla, for NE released from sympathetic nerve endings that diffuses into the general circulation, and for exogenously administered CA. More than two-thirds of exogenously administered, tritium-labeled NE is O-methylated by liver COMT to normetanephrine (NM) before being deaminated by MAO; less than one-fourth of NE is directly deaminated (5, 71, 72).

Deamination is generally the route of intraneuronal metabolism. Intraneuronal CA are initially deaminated by MAO to the corresponding aldehyde (3,4-dihydroxyphenylmandel-aldehyde) which is rapidly oxidized to 3,4-dihydroxymandelic acid by aldehyde oxidase (Fig. 1.5, steps 15 and 23; Fig. 1.7) (105). Most of the product of the initial action of both enzymes, MAO and COMT, is predominantly acted upon by the other catabolic enzyme to form 3-methoxy-4-hydroxymandelic acid (vanillylmandelic acid, VMA) (Fig. 1.5, step 24; Fig. 1.7). VMA accounts for more than 60% of the total urinary catabolites of endogenous CA (13). Substantial NM and metanephrine (M) are reduced to 3-methoxy-4-hydroxyphenylethylglycol (MHPG) which can subsequently be conjugated, converted to VMA (13), or excreted into the urine (Fig. 1.7). Conjugation with glucuronide or sulfate is common. (See Chapter 17 for information on the conjugation of CA.) The average 24-hour urinary excretion of CA and CA metabolites in humans are: (1) NE, from about 25 to 100 μg; (2) E, 2 to 8 μg; (3) NM, 100 to 400 μg; (4) M from adrenal E, 100 to 200 μg; (5) VMA, 2 to 4 mg and (6) MHPG, 1 to 2 mg (105) (Fig. 1.7).

When E is infused in subjects receiving the β-blocking drug propranolol, plasma E levels are three times as high as in subjects infused with E without propranolol. While propranolol alone increases NE 4-fold, phentolamine, an α-blocking drug, only doubles endogenous plasma NE levels. Thus Cryer *et al.* conclude that the primary clearance mechanism of NE and E in man involves β- rather than α-receptors (29). Propranolol even more effectively blocks clearance of isoproterenol (unpublished observations), possibly indicating interference with uptake-2 mechanisms. However, phenoxybenzamine, an α-receptor antagonist, inhibits NE uptake-1 and -2 *in vitro* (68). Another possible explanation for the β-mediated decrease in NE clearance is postulated by Brown *et al.* (16): rather than affecting uptake or metabolism, β-antagonists such as propranolol may decrease heart rate and cardiac output, causing a decrease in the perfusion of organs that extract NE from the circulation. Consistent with these data are the findings that the lungs extract primarily NE (56, 140).

EFFECTS OF PERIPHERALLY RELEASED NOREPINEPHRINE

The effects of junctionally released NE are determined by numerous factors. Response of the postjunctional effector cell depends on the target organ, its predominant cell-surface receptor type, synaptic cleft width, and nerve density (8, 9) (Chapter 4).

Ahlquist (2) proposed the terms "α-" and "β-receptors" for sites on smooth muscle cells. This model has been expanded to four NE receptor types: α-1, α-2, β-1 and β-2. Stimulation of cardiac β-1-receptors increases activity of the heart (Chapters 18, 19, 20), while β-2-receptor stimulation relaxes smooth muscles in blood vessels and bronchioles (Table 1.1). The prejunctional β-receptor facilitates increased NE release, as discussed previously (Fig. 1.6, steps 9 to 12). Adrenergic receptors in man have been reviewed recently (45, 101) (Chapters 18 and 19), and some of their effects are summarized in Table 1.1.

Table 1.1
α and β-Adrenergic Responses

ORGAN	EFFECT	α	β_1	β_1 and β_2	β_2
Blood Vessels	a) Constricts arterioles & venules	+	0		0
	b) Dilates arterioles (esp in skeletal mm)	0	0		+
Heart	a) ↑ rate of S-A node discharge	0	+		0
	b) ↑ contractility	0	+		0
	c) ↑ conduction velocity	0	+		0
Lungs	a) Dilates bronchial mm.	0	0		+
GI	a) ↓ peristalsis	+		+	
Metabolic Effects	a) Lipolysis	0		+	
	b) ↓ insulin release	+		0	
	c) ↑ insulin release	0		+	
	d) ↑ formation of c-AMP	0		+	
	e) ↑ fasting blood sugar	±		+	
	f) ↑ renin secretion			+	

In the peripheral SNS, α-1-receptors predominate on postjunctional effector sites on smooth muscle and gland cells. Some of the major α-adrenergic responses include: contraction of most smooth muscles (especially vascular), relaxation of intestinal smooth muscle, inhibition of insulin secretion, stimulation of salivary potassium ions and water secretion, aggregation of platelets, and stimulation of glycogen breakdown in the liver (45) (Table 1.1). α-2-Receptors, located on the prejunctional neuronal membrane, act to downregulate NE release (Fig. 1.5, step 7; Fig. 1.6, step 13). Extrajunctional α-receptors are also located on smooth muscle cells, in the brain and on platelets. Intracellular "second messenger" mechanisms are postulated for each of the four adrenergic receptors. Adenylate cyclase is activated by β-receptors, increasing cellular cAMP. This process is inhibited by α-2-receptors. α-1-Receptor activation increases intracellular calcium levels and phosphoinositol hydrolysis (101) (Chapters 18 and 19).

Additional factors at the neuroeffector junction modulate adjustments in the amount of NE available to activate α- and β-receptors (138). A decrease in pH and hyperosmolarity, as well as moderate increases in potassium ions, adenosine, and adenine nucleotides, depress NE release. These metabolic changes produce vasodilation which mediates local hyperemia by "disconnecting the blood vessels from sympathetic control" (138). Neurotransmitters such as acetylcholine, histamine, and serotonin (5-HT), activate specific receptors on the prejunctional SNS axons, thereby reducing NE release, while angiotensin II augments NE release via local action on the SNS varicosities. A decrease in local temperature decreases NE release but increases postjunctional α-receptor affinity for NE. E can activate the prejunctional β-2-receptors to augment release of NE (142). In order for this to occur, circulating levels of E have to be high enough to allow diffusion from the circulation to SNS neuroeffector junctions (86). This may occur at times of significant stress and major generalized SNS activation when plasma levels reach concentrations of 1 to 2 ng/ml (139).

A major function of the peripheral SNS is the maintenance of blood pressure during stresses such as standing up (as demonstrated in Figs. 1.8 and 1.9 which show significant increases in CA with sitting, tilting, and standing). Normally, standing causes large increases in hydrostatic pressure in venous capacitance vessels without appreciable changes in blood pressure. As much as 800 ml of blood can collect in the veins of the legs upon standing, immediately reducing venous return to the right heart (Fig. 1.10). Right ventricular stroke volume decreases, and within one or two beats, left ventricular output is reduced, proportionately lowering systemic arterial pressure. The initial response may begin in the right atrium from which sensory receptors re-

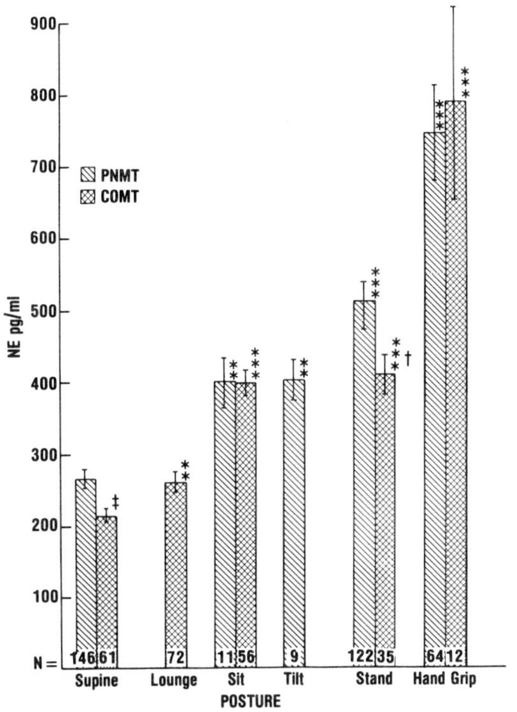

Figure 1.8. Plasma NE: effects of assay methodology, posture, and exercise. Means ± SEM. N shown at bottom of each bar. †, $p < .05$ from same posture PNMT method, 2-tail *t*-test; ‡, $p < .01$ from same posture PNMT method, 2-tail *t*-test; **, $p < .01$ from supine, same assay method, paired *t*-test; ***, $p < .001$ from supine, same assay method, paired *t*-test.

lay information via afferent fibers to the NTS in the medullary vasomotor center. These afferent fibers are joined by similar ones from the left atrium, but the major afferents are from the baroreceptors in the carotid arteries and along the thoracic aorta (137). Low blood pressure diminishes stretch of the baroreceptors and decreases afferent activity (14). This activates the SNS, resulting in generalized vasoconstriction, renin release, and cardiac stimulation, both directly and by increasing circulating E from the adrenal glands; vagal tone is decreased to additionally enhance cardiac activity (Fig. 1.10). For details of the peripheral SNS activation with assumption of an upright posture, see Plasma Levels of Norepinephrine in this chapter and Figures 1.8 and 1.9.

All of these rapidly produced phenomena are especially critical to a person quietly standing in place. Muscular contraction in the legs, while walking, acts to efficiently augment venous return and reduces venous pressure in the ankles to about 20 mm Hg from about 80 mm Hg when standing still (122). Without these complex hemodynamic reflex changes, we could not stand upright.

NE and E have some role in most body functions (many to be discussed in this text), especially in the responses to external stimuli interpreted as threatening or stressful (19, 48, 50, 146). Peripheral E is currently known to function only as a hormone; NE can also act as a hormone when circulating levels approach 2 ng/ml (139) during, for example, strenuous physical exercise (34, 47) and sexual intercourse (153). In man there is frequent dissociation between the effects of different stimuli on SNS and adrenal medullary CA release. Physical exercise and postural change tend to increase NE more than E, while hypoglycemia, caffeine, and psychological stressors release relatively more E than NE (34, 129) (Chapter 21).

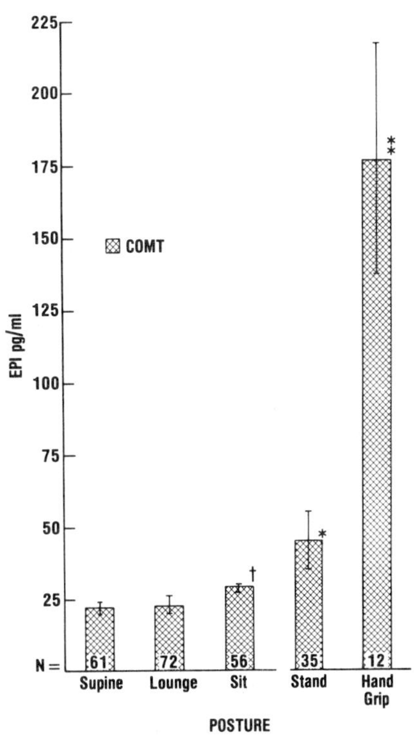

Figure 1.9. Plasma E: effects of posture and exercise. All samples measured by COMT method. Means ± SEM. *N* shown at bottom of each bar. †, $p < .05$ from lounge, 2 tail *t*-test; *, $p < .05$ from supine, 2 tail *t*-test.

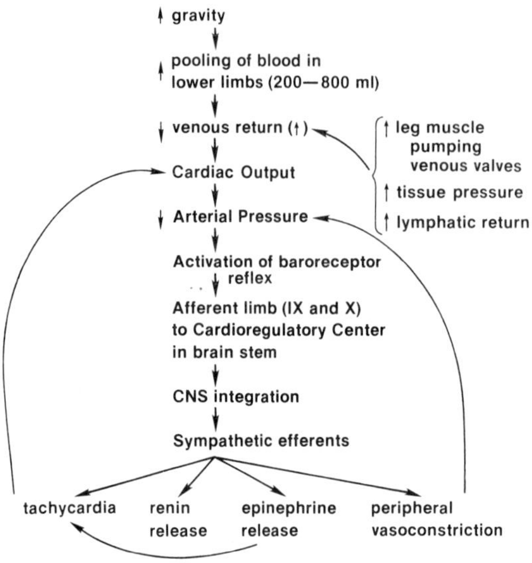

Figure 1.10. Physiological mechanisms involving SNS activation upon standing-up. Gravity causes blood to pool in the legs causing decreased venous return, decreased cardiac output, decreased arterial pressure and activation of the baroreceptor reflex. The SNS is activated to increase heart rate, renin release, E secretion, and peripheral vasoconstriction.

INDICES OF SYMPATHETIC NERVOUS FUNCTION

Urinary Catecholamines and Catecholamine Metabolites

Techniques to measure NE and E in 24-hour samples of urine, which are magnitudes greater than NE and E levels in plasma, have been available for several decades (44). However, only a fraction of the released E and NE is excreted as the free amine. The 24-hour urinary content of NE, E, and their metabolites are given under Catabolism and Inactivation of Norepinephrine in this chapter. Recently noted was a high correlation in depressed patients between urinary NE and its three primary metabolites, NM, MHPG, and VMA (92). Because this correlation is probably present in normal subjects as well (W. Z. Potter, personal communication), the measurement of the excretion rate of any of the three metabolites or urinary NE may constitute an index of the general production rate of NE over hours or days.

Measurements of free NE, E, their conjugates, and their metabolites can account for most of the expected 24-hour release of the amines. Overnight and 24-hour urinary NE levels correlate with supine plasma NE (but not urinary E with plasma E) under basal conditions and after either volume expansion or contraction (66, 136). For a discussion of the use of urinary CA measurements in the diagnosis of pheochromocytoma, the reader is referred to Chapter 32. Prolonged or chronic stress can be assessed by monitoring urinary CA. However, attention must be given to many variables that can distort results such as diet, caffeine and alcohol intake, urinary pH, tobacco smoking, psychosocial environment, time of day, age, weight, posture, and degree of physical activity (48) (Table 1.2).

Numerous studies in man have been conducted to evaluate the sympathoadrenal medullary response to a variety of psychosocial stressors common to our daily lives (48). In these studies urine is collected only over the period of the stress, usually for 1 to 3 hours but occasionally for up to 24 hours. Subjects void before beginning the procedure, which must be long enough to allow adequate urine formation to enable urination at the end of the session. A measured amount of water can be given just before the procedure to ensure an adequate urine volume. Experimental sessions are matched with control sessions of the same length of time.

Until recently, the urinary metabolite, MHPG, was considered to be an index of central NE metabolism. It was thought that a high proportion of urinary MHPG was derived from brain catabolism of NE. However, it is now apparent that about half of the free MHPG that reaches the circulation from various tissues, including the brain, is converted to VMA. Adrenal medullary and presumably CNS E also contribute to urinary MHPG (Fig. 1.7). Because only about 20% of urinary MHPG is from the brain, urinary MHPG levels are not a reliable index of brain NE metabolism (13).

Dopamine-β-Hydroxylase

Dopamine-β-hydroxylase (DBH) has been suggested as an index of SNS activity for three reasons: (1) it is responsible for the formation of NE from DA (90, 105); (2) it is released with CA from the adrenal medulla (148) and from stimulated sympathetic nerves in perfused organs (33, 53, 55) and (3) it is present in plasma of humans and other animals (150, 151, 152). In animals, DBH levels are increased with

Table 1.2
Some Factors That Can Influence Human Plasma NE Values

1. Environment of subject (external stimuli)
2. Tobacco smoking
3. Caffeine consumption
4. Medications, including:
   Diuretics (hydrochlorothiazide, furosemide)
   Vasodilators (nitroglycerine, diazoxide, hydralazine)
   β-Antagonists (propranolol)
   Central α-agonists (clonidine)
   Depleters of CA (reserpine)
   Antidepressants (MAO-I)
   Marijuana
   CNS stimulants (amphetamine)
   Anorexiants (fenfluramine)
   DA agonists (bromocriptine)
   Neuroleptics (chlorpromazine)
5. Diet (carbohydrate intake)
6. Sodium intake (low salt diet)
7. Volume depletion
8. Time of day
9. Age
10. Hospitalized *vs.* outpatient status
11. Familiarity with the clinical environment (blood drawing)
12. Venipuncture *vs.* indwelling cannula
13. Anxiety and psychological stress
14. Posture
15. Exercise
16. Physical conditioning
17. Various illnesses (depression, mania, thyroid disease, congestive heart failure, pulmonary and renal hypertension, etc.)
18. Blood sample processing
19. Assay methodology

stress and decreased after chemical destruction of the sympathetic nerve endings, but not after adrenalectomy (151). Thus, DBH in plasma appears to come primarily from sympathetic nerve endings.

Levels of DBH in human plasma are generally increased by situations that increase sympathetic neuronal activity, *i.e.*, exercise (49, 119, 154), immersion of the hand in cold water (154), and insulin-induced hypoglycemia (111). However, there are cases in which increases in enzyme levels are not observed after stress (109). Furthermore, during the cold pressor test, levels of the other large protein molecules increase in parallel with DBH (143). Basal levels of DBH vary widely between normal individuals, and the degree of DBH elevation with stress is usually relatively small (10 to 25%) (77). Genetic regulation accounts for most of the wide variability in circulating DBH levels (152) (Fig. 1.11). Human plasma NE levels are less variable (Fig. 1.12). Thus, some investigators find DBH to be a useful index of sympathetic function (3, 54, 98, 120, 132, 135), but others do not (64, 65, 67, 77, 88, 102, 110, 126).

We measured DBH and NE in over 350 plasma samples from 174 subjects while resting supine, standing, and exercising (77). Although significant increments in both NE and DBH occur with postural change, the further significant increase in plasma levels of NE during exertion is not attended by any change in DBH levels (Fig. 1.13). The relatively small increment in DBH and other proteins upon standing may relate to an increase in lymph flow into the circulation that is not further enhanced with isometric hand exercise (77) (Fig. 1.13), or it may simply reflect hemoconcentration. There is no significant correlation between basal levels of DBH and NE, nor is there any correlation in their increments after

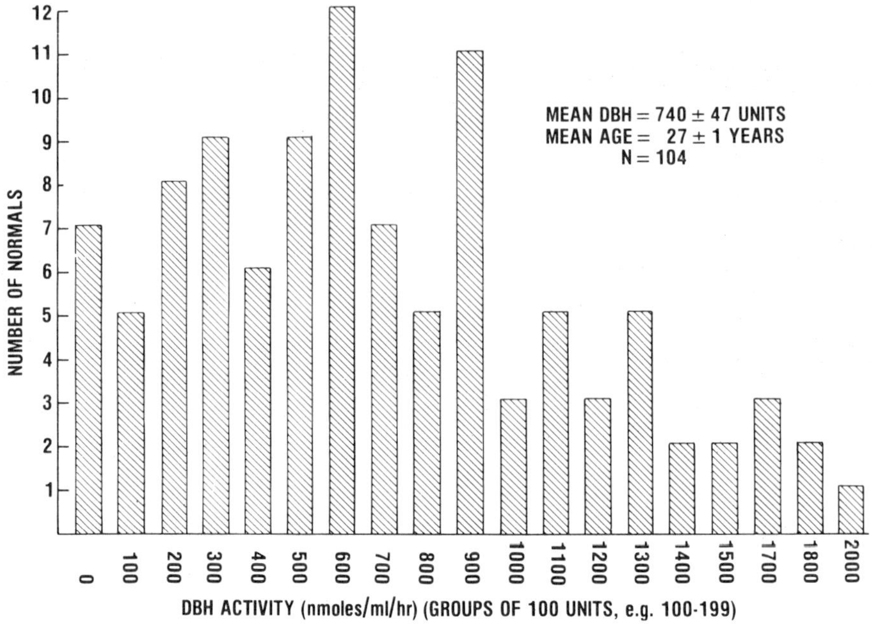

Figure 1.11. Nomograph for supine DBH in normotensive control subjects. Means ± SEM. nmoles/ml/hr = nanomoles of phenylethylamine converted to phenylethanolamine per milliliter of plasma per hour of incubation time.

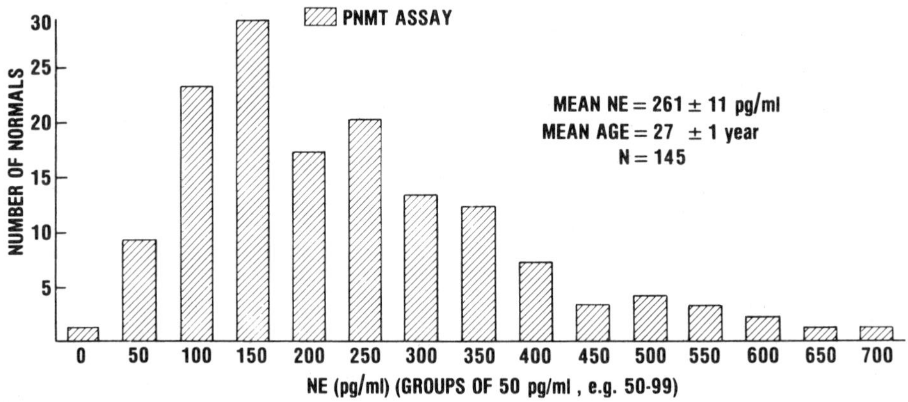

Figure 1.12. Nomograph for supine NE in normotensive control subjects. Samples assayed by the PNMT method. Means ± SEM.

standing. We therefore conclude that plasma DBH levels are a poor index of *acute* sympathetic neuronal activity (77). Chapter 8 deals with DBH in CSF.

Pressor Response Tests

Infusion of vasoactive drugs can yield information about SNS function (57). For instance, tyramine is taken up into SNS vesicles and causes release of NE, resulting in correlated blood pressure and plasma NE responses (10, 63). Pressor and NE responses to tyramine in patients with essential hypertension and normotensive controls are similar (10). The tyramine pressor test also proves useful in evaluating releasable stores of NE in patients with orthostatic hypotensive syndromes (121). Infusion of NE itself helps in the identification of patients with denervation hypersensitivity (121), but interpretation of pressor responses

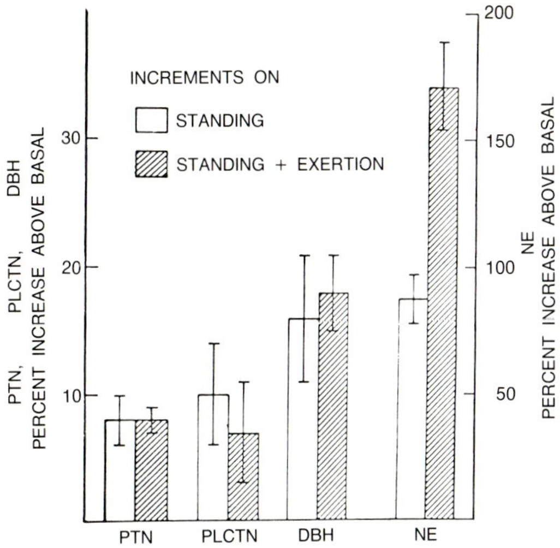

Figure 1.13. Increments in protein (*PTN*), prolactin (*PLCTN*), DBH and NE in plasma produced by standing and exertion. Means ± SEM. The results are percentage increases above supine resting values of the various substances. Means ± SEM. N = 37. (Reproduced with permission from C. R. Lake *et al.* (77).)

to noradrenergic receptor agonists must take into account not only receptor activity but also smooth muscle contraction, the wall to lumen ratio, and circulatory reflex controls. See Chapter 25 for additional information.

Norepinephrine Turnover in Plasma: Rate of Release and Clearance

Circulating NE levels are determined by sympathetically mediated NE release and the action of several removal and degradative processes, as discussed under the headings Norepinephrine Storage and Release and Catabolism and Inactivation of Norepinephrine. Therefore, without knowledge of the contribution of these removal processes, the relationship between circulating NE levels and sympathetic neural activity may be difficult to delineate. One can assess overall clearance of NE from the bloodstream by injecting NE and measuring concentrations over time (16, 126) (Chapter 14). This approach, however, may be flawed because the hypertensive effect of injected NE in pharmacological doses may inhibit endogenous release. When plasma NE levels of about 1500 pg/ml are attained, NE acts as a hormone to cause substantial physiological changes (139). Recently, investigators have used labeled tracer amounts of NE in man and have described the kinetics of plasma disappearance of the injected NE (36–41, 160). This involves the infusion of tritiated levo-NE of high specific activity (24 to 28 Ci/mmole) in supine resting subjects until steady-state plateau concentrations are reached, followed by determination of endogenous and tritiated NE levels and the clearance rate of tritiated NE (36, 60). The tracer amounts of NE elevate plasma levels by only about 3 pg/ml. In the steady-state, the apparent clearance of NE is equal to the infusion rate divided by the plasma concentration of the labeled NE. Similarly, the apparent release rate of NE is equal to the infusion rate divided by the steady-state specific activity of plasma NE. One group calculates the rate at which released NE reaches the circulation at 360 ng/m/min, and the NE clearance rate at 1.32 L/m/min. The loss of tritiated NE from the plasma is reported to be biexponential with the half-life of the rapid removal phase equal to 2 minutes, and the half-life of the slower phase equal to 34 minutes (40, 41). Since the first rapid decline was prolonged by desipramine, Esler and co-workers (40) suggested that this portion of the kinetics of NE disappearance reflects uptake-1 activity. Consideration of NE removal processes is important for a true understanding of the meaning of circulating NE levels in any disease state.

The method described above is applicable only to physiologically stable conditions where endogenous plasma NE levels are constant and is not useful in evaluating responsivity of the SNS. Junctional clearance mechanisms and the actual rate of release of NE at the nerve terminal are not directly measured. NE clearance from the plasma and the apparent rate of release into the plasma are evaluated. A 3-fold range in plasma clearance of infused NE in normal subjects is reported (46). Since the technique accounts for variable clearance mechanisms from plasma, it can provide more detailed information about sympathetic activity than plasma NE measurements alone.

Plasma Levels of Norepinephrine

Plasma NE levels can accurately reflect SNS activity only when several criteria are met and attention is given to numerous factors that can influence plasma CA levels (Table 1.2). A central question is whether NE concentrations released or in the junctional gap are proportional to the amount of circulating NE (see

Summary). Only a small portion of the released NE escapes enzymatic catabolism or reuptake and diffuses out of the junctional cleft into the bloodstream as noted above (Figs. 1.2 and 1.5). As discussed under the preceding heading, in some cases changes in the rate of NE release rate and clearance can obscure the relevance of plasma NE levels to SNS activity. This situation is most obvious if both rate of release and clearance are altered in the same direction. Esler *et al.* (37) report four orthostatic hypotension patients with decreased release and clearance but with normal resting plasma NE. Patients with this disorder also fail to normally increase NE plasma levels upon standing, indicating a significant defect of the SNS response to stress (157, 159). Clearly valuable information can be gained from the tritiated NE infusion technique, but identification of such a SNS deficit can be made by evaluating the response of plasma NE levels to orthostasis or stress (80, 157, 159).

For plasma NE levels to constitute an index of SNS activity, it must be assumed that in any individual, NE release, diffusion into the circulation, and reuptake, extraction, or clearance are in a "steady-state" at the time of blood sampling. When a person lies supine for 20 minutes after venipuncture, NE levels do not change for at least an additional 3 hours (Table 1.3) (76, 80). When a person stands for 5 minutes NE levels nearly double and remain constant after another 5 minutes of standing (Fig. 1.8) (76, 80).

External environmental stimuli should be carefully monitored and controlled to avoid psychological stress (Table 1.2). An adjustable hospital bed and relaxing background music

Table 1.3
Effect of Venipuncture and Time on Plasma NE and Pulse[a]

Time Post-venipuncture	NE (pg/ml)	Pulse Rate (beats/min)
0 min	335 ± 35	80 ± 4
20 min	262 ± 21	74 ± 3
Paired change	−73 ± 26[b]	−6 ± 2[b]
20 min	251 ± 60	62 ± 2
3 hr	280 ± 76	63 ± 3
Paired change	+29 ± 39 NS[c]	+1 ± 2 NS[c]

[a] Reprinted with permission from C. R. Lake *et al.* (76).
[b] $p < 0.01$, paired Student's *t*-test, 2-tailed.
[c] NS, $p > 0.05$, paired *t*-test not significant.

Table 1.4
Effect of Age on Plasma NE[a]

Age Range	Age Mean	N	Supine NE (pg/ml)
10–19	15 ± 0.5	24	199 ± 22
20–29	24 ± 0.4	51	271 ± 20
30–39	34 ± 1.0	10	222 ± 28
40–49	45 ± 0.6	19	308 ± 35
50–59	54 ± 0.6	15	336 ± 66
60–70	63 ± 0.7	10	378 ± 51
11–70	33 ± 1	130	274 ± 14

[a] Means ± SEM. Analysis of variance for supine NE by age decade. $F\ (1{,}129) = 11.8\ p < 0.001$.

in an otherwise quiet room is better than the treatment room table. The subject should not have ingested medication for several weeks with special attention paid to medications with long half-lives such as the neuroleptics. Drugs which elevate plasma NE levels include some neuroleptics such as chlorpromazine (but not haloperidol) (21), CNS stimulants such as amphetamine (131) and caffeine (128), nicotine (28, 124), diuretics (78, 79), vasodilators (23, 108), tricyclic antidepressants (40), and marijuana (52). Drugs known to lower plasma NE levels in humans include reserpine (23), guanethidine (23), fenfluramine (24, 82), MAO inhibitors (MAO-Is) (103, 118), and clonidine (25). Carbohydrate intake is also important (Chapter 15) (30, 85). A low-salt diet elevates NE levels, while saline loading or a high-salt diet has minimal effects (51, 78, 106, 130).

NE concentrations in CSF (158) and in plasma (69) are higher when subjects are awake. Subjects should be awake during blood drawing. Another group reports minimal 24-hour variation in NE but large E changes in samples analyzed by a fluorometric technique (134). Levin *et al.* (89) find an ultradian rhythm in plasma NE with a periodicity of 107 minutes. Higher daytime plasma NE may simply reflect increased mental and physical activity, including standing and walking.

Since plasma NE increases with age ($r = .54$; $p < .01$), age-matched control groups are important in experimental designs (Chapter 24) (58, 61, 80, 156). In Table 1.4, 129 normal subjects are grouped by age decade. The mean supine NE levels in this group demonstrate a significant increase with age. The higher plasma NE levels in older subjects may be due to decreased clearance of NE (41) or increased

Table 1.5
The Effects of Hospitalization on Plasma NE Levels in Essential Hypertensive (HES) and Normotensive (NT) Subjects

	SUPINE			STAND		
	NT (age) (N)		HES (age) (N)	NT (N)		HES (N)
CLINIC VISIT	338 ± 22 (41 ± 2) (64)	NS[2] →	348 ± 16 (42 ± 1) (131)	627 ± 50 (55)	NS →	660 ± 33 (128)
	↓ 0.01[3]		↓ 0.001	↓ 0.01		↓ 0.001
HOSPITALIZED	253 ± 24 (40 ± 3) (29)	NS →	248 ± 19 (42 ± 2) (44)	458 ± 40 (29)	NS →	425 ± 27 (44)

[1]NE in pg/ml; values are means ± SEM

[2]Student (2 tailed) "t" test, NS is not significant ($p > 0.05$)

[3]Student t-test, Clinic vs. Hospitalized.

sympathetic neuronal discharge rate (Chapter 28).

When comparing two groups with regard to plasma NE, both should be studied in similar medical environments. Patients with mild to moderate essential hypertension and normal volunteers were admitted to the Clinical Center at NIH where (after at least 2 days of inpatient status) their levels of plasma NE were compared to those of hypertensive and normal subjects studied in the clinic on an outpatient basis. NE levels of the inpatient populations were significantly lower in both supine and standing postures (Table 1.5). This finding may be analogous to the blood pressure decrease often observed in hospitalized hypertensive patients. Familiarity with the medical environment, venipuncture, and blood drawing procedures are expected to produce less initial anxiety upon venipuncture and a more rapid attainment of a "basal" plasma NE level.

Since the anxiety, fear, and pain some subjects associate with venipuncture can be acutely stressful, we had 15 laboratory-naive subjects lie supine for insertion of an indwelling needle in an antecubital vein. We withdrew blood immediately upon venipuncture, after 20 minutes, and after 3 hours (76). Blood samples drawn immediately after venipuncture contained a mean of 335 pg/ml NE but 20 minutes later contained only 262 pg/ml NE (Table 1.3). During the same period, subjects' pulse rates decreased from 80 to 74 beats/min. The decreases in both NE levels and heart rate are significant ($p < .01$, paired t-test), and the reduction in heart rate correlates with the 22% decrease in NE ($r = .58$, $p < .05$). Monitoring heart rate under these circumstances helps to evaluate subject relaxation. After lying quietly for 3 hours, the subjects' NE levels are no different than after 20 minutes (76). Robertson *et al.* (129) report resting supine NE levels of 201 pg/ml in blood obtained via an indwelling cannula but note a nonsignificant 13% increase in NE in blood drawn by acute venipuncture. Subjects in this study may have been familiar with the procedure of venipuncture.

Posture and exercise are major determinants of plasma CA levels (Figs. 1.8 and 1.9). Standing from the supine position activates the baroreceptor reflex and the SNS, changes reflected in a significant increase in circulating levels of NE and possibly E (Figs. 1.8 to 1.10). In this lab, using the COMT assay, mean supine NE levels of 215 ± 11 pg/ml (N = 61) increase to 411 ± 29 (N = 35) after standing in place for 5 minutes, an increase of about 91% (Fig. 1.8). The COMT data from the literature from 19 studies, excluding ours, of 430 normal subjects are similar: supine, 228 pg/ml; standing, 413 pg/ml (Table 1.6). Results of this lab's

Table 1.6
Plasma Catecholamines in Normotensive Controls: Results from a Literature Survey[a]

	PNMT	COMT	HPLC
Supine			
NE (pg/ml)	272 ± 20	228 ± 30	240 ± 13
E (pg/ml)		48 ± 8	45 ± 3
No. subjects	304	430	66
Age	38	39	28
No. studies/ no. labs	8/7	19/15	2/2
Standing			
NE	467 ± 40	413 ± 47	533 ± 42
E		60 ± 7	
No. subjects	243	213	7
No. studies/ no. labs	5/4	8/6	1/1

[a] Adapted from D. S. Goldstein (58, 59, 61). Excludes data from this laboratory. All means ± SEM weighted for size of the study.

PNMT assay of 146 supine normal subjects substantiate these data; we find an increase in NE of 226 ± pg/ml (in the 122 subjects who stood) from a supine value of 261 ± 11, an 87% increase (Fig. 1.8). Table 1.6 shows a mean supine NE by the PNMT assay from 304 normal subjects from the literature of 272 ± 20 that increases to 467 ± 40 pg/ml upon standing.

Using an adjustable lounge chair is a convenient and common means to passively change a subject's posture from semisupine (chair fully reclined) to seated. Plasma NE levels (as measured by the COMT assay) from subjects semisupine in a lounge chair (N = 72) are significantly greater ($p < .01$) than levels in fully supine subjects (Fig. 1.8). When the lounge chair is moved to the upright sitting position, the SNS is significantly activated, as demonstrated by an increase in NE from 261 ± 13 to 400 ± 19 pg/ml, or about a 60% increase (Fig. 1.8). Although adjustments in the position of the lounge chair can substantially activate the SNS, the increase in plasma NE caused by moving from a semisupine to a sitting position is significantly less than the increase in NE induced by standing from a supine posture ($p < .01$). (The lounge chair subjects had their blood sampled under the direction of Dr. David G. Glass (Chapter 21)).

Sitting up causes an increase in NE similar to that caused by a 60° head-up tilt (Fig. 1.8). Ten minutes after tilting, NE is 405 ± 30 pg/ml, approximately the level obtained from patients seated in a lounge chair (400 ± 19), on the edge of a bed, or in a regular chair (401 ± 37) (Fig. 1.8).

Isometric hand grip in standing subjects is a convenient means of further activating the SNS (75, 76). A hand dynamometer is preferable, but a rolled-up partially inflated blood pressure cuff may be substituted. Maintaining 30% of one's maximal ability for 5 minutes while remaining standing increases plasma NE levels in 64 subjects to 749 ± 67 pg/ml as measured by the PNMT technique in our lab (76, 80) (Fig. 1.8). By the COMT technique in 12 normal subjects, the hand grip raised NE levels to 792 ± 137, similar to values obtained by the PNMT assay (Fig. 1.8). Subjects can decrease the amount of SNS activation for a given work load by training in a physical fitness program (116). However, physical training can double the maximum NE level a subject can attain during his peak exertion level of isotonic exercise (unpublished observation). Supine resting NE is not changed by training (116).

The easiest and quickest technique used to evaluate SNS function is to: (1) take a supine resting blood sample only after the subject is relaxed, *i.e.*, 15 to 20 minutes after insertion of the needle or catheter of an indwelling heparin lock; then (2) draw a sample after the subject has stood quietly in place for 5 minutes; and optimally (3) obtain blood after 5 minutes of isometric hand grip (76). These data do not determine release or clearance rates, but quantify the net sum of spillover and clearance in the basal supine posture and test the ability of the SNS and baroreceptor reflex arc to respond normally to standardized postural and exercise stresses.

In this laboratory, E increases from 23 ± 2 pg/ml (N = 61) to 45 ± 5 after the 5-minute stand, which is a 100% increase (Fig. 1.9). Using Student's *t*-test for paired data from the 35 subjects who had both supine and standing samples drawn, plasma E increased by 18 ± 6 pg/ml ($p < .01$) (Fig. 1.9). In contrast, most studies from the literature show a nonsignificant increase in plasma E upon standing (Table 1.6). However, our mean supine resting E value is significantly lower than the literature's (23 ± 2 pg/ml *vs.* 48 ± 8; $p < .01$), but our mean standing E is similar, thus explaining our finding of a significant increment in E with standing (compare Table 1.6 with Fig. 1.9). Effort is taken to ensure that our subjects are maximally relaxed and absolutely supine before blood is sampled for CA assay. Supine E levels should indicate subject relaxation prior to blood sampling, although supine resting E levels are near the limits of sensitivity of the COMT assay. The correlation between the NE and E increments upon standing is nonsignificant, although the percent increases are similar. Isometric hand grip increases plasma E to 178 ± 40 pg/ml, significantly above standing and supine values (Fig. 1.9). Plasma E levels of semisupine lounge chair subjects are similar to those of fully supine individuals. Sitting up in the lounge chair significantly elevates plasma E by 8 ± 2 pg/ml ($p < .001$, paired t-test).

Different organs vary widely in regard to their contribution to or clearance of CA from the circulation (15). It appears that the lungs extract or inactivate about 25% of NE but do not substantially clear E and DA (56, 140). Patients with pulmonary hypertension lose this metabolic NE clearance function (140, 155). In dogs, the kidney is a major source of circulating NE, since renal venous blood contains almost double the NE levels of renal arterial blood (112, 125). Renal arterial stenosis or chronic bilateral renal parenchymal disease may increase plasma NE levels in patients with secondary hypertension (15, 84, 97). Variability in individual organ contribution to or clearance of circulating NE must be considered in light of numerous disease states and medications that can alter the function of such organs. Several other chapters deal extensively with related topics.

Some reports claim that blood CA levels deteriorate at room temperature within a few minutes (20), while others conclude that they are stable for hours (117). We find that CA are stable in blood handled with the following precautions. Tubes containing whole blood should be chilled immediately on ice. Heparin, acid-citrate-dextrose, disodium-ethylenediaminetetraacetic acid (EDTA), or ethyleneglycol-bis-(beta-amino-ethyl-ether) N, N′-tetra-acetic acid (EGTA) can be used in the initial collection tubes to prevent clotting. The red cells should be separated in a refrigerated centrifuge and the plasma transferred to polypropylene storage tubes and frozen within 60 minutes of the time of drawing. Antioxidants such as glutathione or ascorbic acid should be added to the storage tubes to decrease oxidation of the plasma CA. The plasma should be stored at −70°C until assayed for CA. Plasma NE stored at this temperature does not deteriorate significantly for at least a year. If the plasma proteins are precipitated with 0.3 N perchloric acid prior to freezing, the NE is stable at only −20°C.

Methods for measuring NE, discussed in detail in Chapters 2 and 3, include two radioenzymatic assays, commonly known as the PNMT and COMT techniques, and a high pressure liquid chromatographic (HPLC) technique. A recent paper reviews new developments in CA determinations (70). Numerous laboratories are currently able to measure CA; Table 1.6 indicates mean values of NE and E of about 800 normal subjects while supine and standing measured by various laboratories, excluding our own, using the three techniques. The PNMT assay usually gives slightly higher plasma NE values than the COMT assay, although the literature indicates there is little difference (Table 1.6). Data from our laboratory show that NE values form 146 volunteers measured by the PNMT technique (267 ± 13 pg/ml and 513 ± 28, supine and standing (N = 122), respectively), were significantly higher than NE values of 61 volunteers measured by the COMT technique (215 ± 1 and 411 ± 29, supine and standing N = 35) (Fig. 1.8). None of the subjects were assayed by both techniques, the most accurate way of examining assay differences. The HPLC technique gives a mean supine NE value in between but not significantly different from either radioenzymatic assay (Table 1.6).

The PNMT assay coefficient of variation (CV) in this laboratory on previously aliquoted blood bank plasma with a mean NE of 260 pg/ml through 12 different assays on different days (interassay CV) is 8% and for the same samples assayed 16 times in the same run, 6% (intraassay CV). For our COMT assay, a different pool of blood bank plasma was used with mean NE and E content of 382 and 102 pg/ml, respectively. The interassay CV for NE for 38 assays is 7% and the intraassay CV (20 samples), 5%; for E the CVs are 11 and 10%, respectively. Thus samples for any one study must be assayed by the same technique and preferably in the same batch.

To conduct accurate longitudinal studies, an individual's plasma NE should remain constant over time. This laboratory recently measured supine CA levels by the COMT assay in normal volunteers sampled at five time points: baseline (0 minutes), 5 minutes, 1 hour, 1 day, and 1 week. Standing and hand grip levels were assayed only at baseline and 1 week. As shown in Table 1.7, there are no differences in either NE or E values between any of the time points. In a separate group, blood was drawn from each subject for NE measurement twice, 2 weeks apart. Results demonstrate no mean differences in either the supine or standing values (76). Although plasma NE increases with age, the slope is not steep, and the largest increase is between the 4th and 5th decades (the 30s *vs.* the 40s) (Table 1.4) (156). Thus, one's circulating level of NE in the resting, supine (basal) state appears to be stable over months to years and is certainly so over days to weeks. Not only are basal levels consistent, but standing and quantitative isometric hand grip cause a consistent elevation in the same individual over time. Similarly, E values are consistent over time (Table 1.7). These data allow an individual's SNS function to be studied longitudinally.

SUMMARY

The evidence that plasma NE reflects SNS activity in man is substantial but largely indirect. In the mid-1800s, Bernard (7) found vasodilation after section of the sympathetic chain, and Brown-Sequard (17) reported vasoconstriction when the sympathetic chain was stimulated. Patients with Raynáud's disease who have had surgical sympathectomy have a decrease in venous levels of NE in the treated limb (107). Diabetic patients with autonomic neuropathy (157) and patients with SNS disorders, such as familial dysautonomia (157), idiopathic orthostatic hypotension (159), and Shy-Drager syndrome (159) demonstrate low plasma NE levels or an absence of NE release upon standing. A variety of stimuli that cause SNS activation in animals, such as exercise and environmental stress, also increase NE in man (59,34,76,80). Amphetamine and tyramine, drugs known to release NE from sympathetic nerve endings and to prevent its reuptake in animals, cause increases in circulating NE levels in man (118,131). The most direct evidence linking plasma NE levels to SNS activity is reported by Wallin *et al.* (18,144,149) who made direct electrical recordings of multiunit sympathetic activity from muscle branches of the peroneal nerve in 22 healthy human subjects (Chapter 28). Plasma NE correlated positively with levels of electrical sympathetic activity ($r = .81$, $p < .01$), suggesting to the authors that the overflow of NE from sympathetic junctions in muscle contributes significantly to plasma NE levels at rest. These data indicate that SNS electrical activity is reflected by plasma NE levels.

Measurement of supine plasma NE levels facilitates assessment of SNS tone; calculation of the increment of NE caused by standing for 5 minutes makes possible evaluation of the ability of the SNS to respond to orthostatic change. These techniques already have some application in clinical medicine as diagnostic and research tools. For example, three neuro-

Table 1.7
CA Stability with COMT Assay[a]

	0 min	5 min	1 hr	1 day	1 wk
Norepinephrine					
Supine	200 ± 9	200 ± 12	193 ± 21	243 ± 27	186 ± 16
(N)	(55)	(41)	(20)	(6)	(19)
Stand	409 ± 29				424 ± 45
(N)	(28)				(13)
Squeeze	1027 ± 149				1024 ± 200
(N)	(12)				(6)
Epinephrine					
Supine	18 ± 1	19 ± 2	26 ± 5	20 ± 4	16 ± 2
(N)	(55)	(41)	(10)	(6)	(19)
Stand	36 ± 4				31 ± 55
(N)	(28)				(13)
Squeeze	157 ± 38				109 ± 27
(N)	(12)				(6)

Means ± SEM. Paired Student's *t*-tests revealed no significant differences at different time points.

logical syndromes having severe orthostatic hypotension as a common symptom can be differentiated biochemically through this technique (121,157,159). Patients with very low supine NE levels (indicating peripheral SNS damage) are unlikely to respond to tyramine plus MAO-Is; they require directly acting sympathomimetic agents or volume expanders. Patients suffering from Shy-Drager syndrome should respond to indirectly acting sympathomimetic agents such as tyramine, since their peripheral SNS nerve endings appear intact, and their supine NE levels are normal (159). A third category of patients with marked tachycardia and normal NE levels may respond to selective β-blockers (121).

There continues to be hope of subdividing essential hypertensive patients, some of whom may have CA hyperactivity, along etiological lines (Chapters 19, 24, 25, 27). If a subgroup of patients with essential hypertension are discovered through this SNS evaluation technique to have enhanced CA neurotransmission early in their development of elevated blood pressure, pharmacological therapy geared to correcting the SNS malfunction can potentially resolve the hypertension (32). One group finds a better hypotensive response to a β-blocker in those essential hypertensive patients with elevated NE levels (32).

The causes of the functional psychoses, schizophrenia, and the major affective disorders are currently unknown, but abnormalities in CA neurotransmission are substantially implicated (81,83). Subsequent chapters will detail clinical implications of current CA research.

References

1. Abel JJ, Crawford AC: On the blood pressure raising constituent of the suprarenal capsule. *Bull Johns Hopkins Hosp* 8:151–157, 1897.
2. Ahlquist RP: A study of the adrenotropic receptors. *Am J Physiol* 153:586–600, 1948.
3. Alexander N, McClaskey J, Maronde RF: Elevated plasma dopamine-beta-hydroxylase activity in rats with neurogenic hypertension. *Life Sci* 18:655–662, 1976.
4. Axelrod J: Purification and properties of phenylethanolamine-N-methyltransferase. *J Biol Chem* 237:1657–1660, 1962.
5. Axelrod J, Kopin IJ: The uptake, storage, release and metabolism of noradrenaline in sympathetic nerves. *Prog Brain Res* 31:21–32, 1969.
6. Axelrod J, Weinshilboum R: Catecholamines. *N Engl J Med* 287:237–242, 1972.
7. Bernard C: Experimental research on the vascular and heat producing nerves of the great sympathetic. *Extrait des Comptes Rendues des Seances de l'Academie des Sciences.* LV:1–32, 1862.
8. Bevan JA: Some functional consequences of variation in adrenergic synaptic cleft width and in nerve density and distribution. *Fed Proc* 36:2439–2443, 1977.
9. Bevan JA: Some bases of differences in vascular response to sympathetic activity: Variations on a theme. *Circ Res* 45:161–171, 1979.
10. Bianchetti MG, Minder I, Beretta-Piccoli C, Meier A, Weidmann P: Effects of tyramine on blood pressure and plasma catecholamines in normal and hypertensive subjects. *Klin Wochenschr* 60:465–470, 1982.
11. Blaschko H: The activity of 1(−)dopa decarboxylase. *J Physiol* 101:337–349, 1942.
12. Blessing WW, Sved AF, Reis DJ: Destruction of noradrenergic neurons in rabbit brainstem elevates plasma vasopressin causing hypertension. *Science* 217:661–663, 1982.
13. Blombery PA, Kopin IJ, Gordon EK, Markey SP, Ebert MH: Conversion of MHPG to vanillylmandelic acid: Implications for the importance of urinary MHPG. *Arch Gen Psychiatry* 37:1095–1098, 1980.
14. Brown AM: Receptors under pressure: An update on baroreceptors. *Circ Res* 46:1–10, 1980.
15. Brown MJ, Jenner DA, Allison DJ, Dollery CT: Variations in individual organ release of noradrenaline measured by an improved radioenzymatic technique; limitations of peripheral venous measurements in the assessment of sympathetic nervous activity. *Clin Sci* 61:585–590, 1981.
16. Brown MJ, Lhoste FJM, Zamboulis C, Ind PW, Jenner DA, Dollery CT: Estimation of sympathetic activity in essential hypertension. *Clin Pharmacol Ther* 31:16–22, 1982.
17. Brown-Sequard CE: On the results of the section and the galvanization (electrical stimulation) of the great sympathetic nerve at the neck. In Thunot E: *Gazette Medicale de Paris.* Paris, 1854, pp 1–9.
18. Burke D, Sundlof G, Wallin BG: Postural effects on muscle nerve sympathetic activity in man. *J Physiol* 272:399–414, 1977.
19. Cannon WB: *Bodily Changes in Pain, Hunger, Fear, and Rape.* Boston, Branford, 1914.
20. Carruthers M, Conway N, Taggart P, Bates D, Somerville W: Validity of plasma-catecholamine estimations. *Lancet* ii:62–67, 1970.
21. Castellani S, Ziegler MG, van Kammen DP, Alexander PE, Siris SG, Lake CR: Plasma norepinephrine and dopamine-beta-hydroxylase activity in schizophrenia. *Arch Gen Psychiatry* 39:1145–1149, 1982.
22. Chalmers JP, Blessing WW, West MJ, Howe PRC, Costa M, Furness JB: Importance of new catecholamine pathways in control of blood pressure. *Hypertension* 3:393–416, 1981.
23. Chernow B, Lake CR, Coleman MD, Ziegler MG: Effect of antihypertensive therapy on sympathetic nervous system activity in patients with essential hypertension. *Fed Proc*, 43:72–77, 1984.
24. Chernow B, Lake CR, Cook D, Coyle J, Hughes

P, Coleman M, Ziegler MG: The effect of fenfluramine on sympathetic nervous system activity. *Int J Clin Pharmacol Res*, 1983.
25. Chernow B, Lake CR, Ziegler MG, Zaloga GP, Coleman MD: The effect of clonidine on sympathetic nervous system activity in essential hypertension. *Int J Clin Pharmacol Res* 3:9–15, 1983.
26. Christensen NJ: Plasma catecholamines in long-term diabetics with and without neuropathy and in hypophysectomized subjects. *J Clin Invest* 51:779–787, 1972.
27. Connett RT, Kirshner N: Purification and properties of bovine phenylethanolamine N-methyltransferase. *J Biol Chem* 245:329–334, 1970.
28. Cryer PE, Haymond MW, Santiago JV, Shah DD: Norepinephrine and epinephrine release and adrenergic mediation of smoking-associated hemodynamic and metabolic events. *N Engl J Med* 295:573–577, 1976.
29. Cryer PE, Rizza RA, Haymond MW, Gerich JE: Epinephrine and norepinephrine are cleared through beta-adrenergic, but not alpha-adrenergic mechanisms in man. *Metabolism* 29(Suppl 1):1114–1118, 1980.
30. Davidson L, Vandongen R, Beilin LJ: Effect of eating bananas on plasma free and sulfate-conjugated catecholamines. *Life Sci* 29:1773–1778, 1981.
31. Davies DS, Reid JL (eds): *Central Action of Drugs in Blood Pressure Regulation*. Baltimore, University Park Press, 1975.
32. deChamplain J, Cousineau D, Lapointe L: Evidence supporting an increased sympathetic tone and reactivity in a subgroup of patients with essential hypertension. *Clin Exp Hypertens* 2:359–377, 1980.
33. dePotter WP, Schaepdryver AI, Moerman ES, Smith AD: Evidence for the release of vesicle proteins together with noradrenaline upon stimulation of the splenic nerve. *J Physiol (Lond)* 204:102P–104P, 1969.
34. Dimsdale JE, Moss J: Plasma catecholamines in stress and exercise. *JAMA* 243:340–342, 1980.
35. Elliot TR: The action of adrenaline. *J Physiol* 32:401–467, 1905.
36. Esler M, Jackman G, Bobik A, Kelleher, Jennings G, Leonard P, Skews H, Korner P: Determination of norepinephrine apparent release rate and clearance in humans. *Life Sci* 25:1461–1470, 1979.
37. Esler M, Jackman G, Kelleher D, Skews H, Jennings G, Bobik A, Korner P: Norepinephrine kinetics in patients with idiopathic autonomic insufficiency. *Circ Res* 46(Suppl 1):47–48, 1980.
38. Esler M, Jackman G, Leonard P, Bobik A, Skews H, Jennings G, Kelleher D, Korner P: Determination of noradrenaline uptake, spillover to plasma and plasma concentration in patients with essential hypertension. *Clin Sci* 59:311s–313s, 1980.
39. Esler M, Leonard P, Kelleher D, Jackman G, Bobik A, Skews H, Jennings G, Korner P: Assessment of neuronal uptake of noradrenaline in humans: Defective uptake in some patients with essential hypertension. *Clin Exp Pharmacol Physiol* 7:535–539, 1980.
40. Esler M, Jackman G, Leonard P, Skews H, Bobik A, Korner P: Effect of norepinephrine uptake blockers on norepinephrine kinetics. *Clin Pharmacol Ther* 29:12–20, 1981.
41. Esler M, Leonard P, O'Dea K, Jackman G, Jennings G, Korner P: Biochemical quantification of sympathetic nervous activity in humans using radiotracer methodology: Fallibility of plasma noradrenaline measurements. *J Cardiovasc Pharmacol* 4:S152–S157, 1982.
42. Euler US von: Identification of the sympathomimetic ergone in adrenergic nerves of cattle (Sympathin N) with laevo-noradrenaline. *Acta Physiol Scand* 16:63–74, 1948.
43. Euler US von, Fraksson C, Hellstrom J: Adrenaline and noradrenaline output in urine after unilateral and bilateral adrenalectomy in man. *Acta Physiol Scand* 31:1–5, 1954.
44. Euler US von, Lishajko F: Improved technique for the fluorometric estimation of catecholamines. *Acta Physiol Scand* 51:348–355, 1961.
45. Exton JH: Mechanisms involved in alpha-adrenergic effects of catecholamines. In Kunos G: *Adrenoreceptors and Catecholamine Action*, New York, Wiley & Sons, 1981, vol 1, pp 117–129.
46. FitzGerald GA, Dollery CT: Biochemical indices of sympathetic function in man. *Trends Pharmacol Sci* 1:84–87, 1979.
47. Francis GS, Goldsmith SR, Ziesche SM, Cohn JN: Response of plasma norepinephrine and epinephrine to dynamic exercise in patients with congestive heart failure. *Am J Cardiol* 49:1152–1156, 1982.
48. Frankenhaeuser M: Experimental approaches to the study of catecholamines and emotion. In Levi L: *Emotions—Their Parameters and Measurement*. New York, Raven Press, 1975, pp 209–234.
49. Frewin DB, Downey JA, Levitt M: The effect of heat, cold, and exercise on plasma dopamine-beta-hydroxylase activity in man. *Can J Physiol Pharmacol* 51:986–989, 1973.
50. Friedhoff AJ (ed): *Catecholamines and Behavior*, New York, Plenum Press, 1975, vols 1, 2.
51. Fujita T, Henry WL, Bartter FC, Lake CR, DeLea CS: Factors influencing blood pressure in salt sensitive patients with hypertension. *Am J Med* 69:334–344, 1980.
52. Gash A, Karliner JS, Janowsky D, Lake CR: Effects of smoking marijuana on left ventricular performance and plasma norepinephrine in normal men. *Ann Intern Med* 89:448–452, 1978.
53. Geffen LB, Livett BG, Rush RA: Immunological chemical localization of chromagranins in sheep sympathetic nerves. *J Physiol (Lond)* 204:58P–59P, 1969.
54. Geffen LB, Rush RA, Louis WJ, Doyle AE: Plasma dopamine-beta-hydroxylase and noradrenaline amounts in essential hypertension. *Clin Sci* 44:617–620, 1973.
55. Gewirtz GP, Kopin IJ: Release of dopamine-beta-hydroxylase with norepinephrine by splenic nerve stimulation. *Nature* 227:406–407, 1970.
56. Gillis CN, Greene LH, Cronau LH, Hammond GL: Pulmonary extraction of 5-hydroxytrypta-

mine and norepinephrine before and after cardiopulmonary bypass in man. *Circ Res* 30:666–674, 1972.

57. Ghose K: Assessment of peripheral adrenergic activity and its interactions with drugs in man. *Eur J Clin Pharmacol* 17:233–238, 1980.
58. Goldstein DS: Plasma norepinephrine in essential hypertension: A study of the studies. *Hypertension* 3:48–52, 1981.
59. Goldstein DS: Plasma norepinephrine during stress in essential hypertension. *Hypertension* 3:551–556, 1981.
60. Goldstein DS, Horwitz D, Keiser HR, Polinsky RJ, Kopin IJ: Plasma 3-H-L-norepinephrine, 14 c-d norepinephrine 3H-d,l-isoproterenol kinetics in hypertension. *J Clin Invest*, in press.
61. Goldstein DS, Lake CR: Plasma norepinephrine and epinephrine levels in essential hypertension. *Fed Proc* 43:57–61, 1984.
62. Goldstein DS, Lake CR, Chernow B, Ziegler MG, Coleman MD, Taylor AA, Mitchell JR, Kopin IJ, Keiser HR: Age-dependence of hypertensive-normotensive differences in plasma norepinephrine. *Hypertension* 5:100–104, 1983.
63. Goldstein DS, Nurnberger J Jr, Gershon ES, Simmons S, Polinsky R, Keiser HR: Effects of injected sympathomimetic amines on plasma CA and circulating variables in man. *Life Sci* 32:1057–1063, 1983.
64. Grant C, Routh JI, Lawton W, Whitte DL: The effects of therapy for mild hypertension on circulating levels of dopamine-beta-hydroxylase. *Clin Chim Acta* 69:333–340, 1976.
65. Gutterberg T, Borud O, Stromme JH: Dopamine-beta-hydroxylase activity in serum following acute myocardial infarction: An evaluation of this parameter for routine use as an index of sympathetic activity. *Clin Chim Acta* 69:61–66, 1976.
66. Henry DP, Luft FC, Weinberger MH, Fineberg NS, Grim CE: Norepinephrine in urine and plasma following provocative maneuvers in normal and hypertensive subjects. *Hypertension* 2:20–28, 1980.
67. Horwitz D, Alexander RW, Lovenberg W, Keiser HR: Human serum dopamine-beta-hydroxylase: Relationship to hypertension and sympathetic activity. *Circ Res* 32:594–599, 1973.
68. Iversen LL: Uptake processes for biogenic amines. In Iversen LL, Iversen SD, Snyder SH: *Biochemistry of Biogenic Amines. Handbook of Psychopharmacology.* New York, Plenum Press, 1975, vol 3, pp 381–442.
69. Kato T, Hashimoto Y, Nagatsu T, Shinoda T, Okada T, Takeuchi T, Umezawa H: 24-hour rhythm of human plasma noradrenaline and the effect of fusaric acid, a dopamine beta-hydroxylase inhibitor. *Neuropsychobiology* 6:61–65, 1980.
70. Kleine TO: New developments for the determination of neurotransmitters and their metabolites in urine, blood plasma, cerebrospinal fluid and nervous tissue. *J Clin Chem Clin Biochem* 19:1085–1094, 1981.
71. Kopin IJ: Technique for the study of alternate metabolic pathways: Epinephrine metabolism in man. *Science* 131:1372–1374, 1960.
72. Kopin IJ: Metabolic degradation of catecholamines and relative importance of different pathways under physiological conditions and after the administration of drugs. In Blaschko H, Muscholl E: *Catecholamines. Handbook of Experimental Pharmacology.* New York, Springer and Verlag, 1972, vol 33, pp 270–272.
73. Kopin IJ: Catecholamine metabolism (and the biochemical assessment of sympathetic activity). *Clin Endocrinol Metab* 6:525–549, 1977.
74. Kopin IJ, Lake CR, Ziegler MG: Plasma levels of norepinephrine. *Ann Intern Med* 88:671–680, 1978.
75. Kozlowski S, Brezezinsha Z, Nazar K, Kozlowski W, Franczyk M: Plasma catecholamines during sustained isometric exercise. *Clin Sci Mol Med* 45:723–731, 1973.
76. Lake CR, Ziegler MG, Kopin IJ: Use of plasma norepinephrine for evaluation of sympathetic neuronal function in man. *Life Sci* 18:1315–1326, 1976.
77. Lake CR, Ziegler MG, Coleman MD, Kopin IJ: Lack of correlation of plasma norepinephrine and dopamine-beta-hydroxylase in hypertensive and normotensive subjects. *Circ Res* 41:865–869, 1977.
78. Lake CR, Ziegler MG: Effect of acute volume alterations on norepinephrine and dopamine-beta-hydroxylase in normotensive and hypertensive subjects. *Circulation* 57:774–778, 1978.
79. Lake CR, Ziegler MG, Coleman MD, Kopin IJ: Sustained sympathetic hyperactivity in hydrochlorothiazide-treated hypertensives. *Clin Pharmacol Ther* 26:428–432, 1979.
80. Lake CR: Relationship of sympathetic nervous system tone and blood pressure. *Nephron* 23:84–90, 1979.
81. Lake CR, Sternberg DE, van Kammen DP, Ballanger JC, Ziegler MG, Post RM, Kopin IJ, Bunney WE: Schizophrenia: Elevated cerebrospinal fluid norepinephrine. *Science* 207:331–333, 1980.
82. Lake CR, Ziegler MG, Coleman MD, Kopin IJ: Fenfluramine potentiation of antihypertensive effects of thiazides. *Clin Pharmacol Ther* 28:22–27, 1980.
83. Lake CR, Pickar D, Ziegler MG, Lipper S, Slater S, Murphy DL: High plasma norepinephrine levels in patients with major affective disorders. *Am J Psychiatry* 139:1315–1318, 1982.
84. Lake CR, Chernow B, Goldstein DS, Glass D, Coleman M, Ziegler MG: Plasma catecholamine levels in normal subjects and in patients with secondary hypertension. *Fed Proc* 43:52–56, 1984.
85. Landsberg L, Young JB: Fasting, feeding and regulation of the sympathetic nerve cells. *N Engl J Med* 298:1295–1301, 1978.
86. Langer SZ: The role of alpha and beta presynaptic receptors in the regulation of noradrenaline release elicited by nerve stimulation. *Clin Sci Mol Med* 51(Suppl 3):423–426, 1976.
87. Langley JN: The difference of behavior of central and peripheral pilomotor nerve cells. *J Physiol* 27:224–236, 1901.

88. Leon AS, Thomas PE, Sernatinger E, Canlas A: Serum dopamine-beta-hydroxylase activity as an index of sympathetic activity. *J Clin Pharmacol* 14:354–362, 1974.
89. Levin BE, Rappaport M, Natelson BH: Ultradian variations of plasma noradrenaline in humans. *Life Sci* 25:621–628, 1979.
90. Levin EY, Levenberg B, Kaufman G: The enzymatic conversion of 3,4-dihydroxyphenylethylamine to norepinephrine. *J Biol Chem* 235:2080–2086, 1960.
91. Levitt M, Spector S, Sjoerdsma A, Udenfriend S: Elucidation of the rate-limiting step in norepinephrine biosynthesis in the perfused guinea-pig heart. *J Pharmacol Exp Ther* 148:1–8, 1965.
92. Linnoila M, Karoum F, Potter WZ: High correlations of norepinephrine and its major metabolite excretion rates. *Arch Gen Psychiatry* 39:521–523, 1982.
93. Ljung B: Physiological patterns of neuroeffector control mechanisms. In Bevan JA, Burnstock G, Maxwell RA, Nedergaard CA: *Vascular Neuroeffector Mechanisms*. Basel, Karger, 1976, pp 143–155.
94. Lovenberg W, Weissbach H, Udenfriend S: Aromatic-L-amino acid decarboxylase. *J Biol Chem* 237:89–93, 1962.
95. Lund A: Simultaneous fluorometric determinations of adrenaline and noradrenaline in blood. *Acta Pharmacol Toxicol (Kbh)* 6:137–146, 1950.
96. MacLean PD: The limbic brain in relation to the psychoses. In Black P: *Physiological Correlates of Emotion*. New York, Academic Press, 1970, pp 130–146.
97. Masuyama Y, Nishimura T, Nishio I, Kuchi M: Noradrenaline clearance in hypertensive patients with various degrees of renal function. *Clin Sci* 57:213s–215s, 1979.
98. Mathias CJ, Smith AD, Frankel HL, Spalding JMK: Dopamine-beta-hydroxylase release during hypertension from sympathetic nervous overactivity in man. *Cardiovasc Res* 10:176–181, 1976.
99. Meyer P, Schmitt H (eds): *Nervous System and Hypertension*. Paris and New York, Wiley-Flammarion, 1979.
100. Miller NE: Biofeedback and visceral learning. *Ann Rev Psychol* 29:373–404, 1978.
101. Motulsky HJ, Insel PA: Adrenergic receptors in man. *N Engl J Med* 307:18–29, 1982.
102. Mueller RA, Millward DK, Woods JW: Circulating catecholamines, plasma renin and dopamine-beta-hydroxylase activity with postural stress. *Pharmacol Biochem Behav* 2:757–761, 1974.
103. Murphy DL, Lake CR, Slater S, Lipper S, Shiling D, de la Vega E, Ziegler MG: Psychoactive drug effects on plasma norepinephrine and plasma dopamine beta-hydroxylase in man. In Usdin E, Kopin IJ, Barchas J: *Catecholamines: Basic and Clinical Frontiers*. New York, Pergamon Press, 1979, vol 1, pp 918–920.
104. Nagatsu T, Levitt M, Udenfriend S: Tyrosine hydroxylase: The initial step in norepinephrine biosynthesis. *J Biol Chem* 239:2910–2917, 1964.
105. Nagatsu T: *Biochemistry of Catecholamines*. Tokyo, University Park Press, 1973.
106. Nicholls MG, Kiowski W, Zweifler AJ, Julius S, Schork MA, Greenhouse J: Plasma norepinephrine variations with dietary sodium intake. *Hypertension* 2:29–32, 1980.
107. Nielsen SL, Christensen NJ, Olsen N, Lassen NA: Raynaud's phenomenon: Peripheral catecholamine concentration and effect of sympathectomy. *Acta Chim Scand* 502:57–62, 1980.
108. Ogasawara B, Ogawa K, Sassa H: Effects of nitroglycerin ointment on plasma norepinephrine and cyclic nucleotides in congestive heart failure. *J Cardiovasc Pharmacol* 3:867–875, 1981.
109. Ogihara T, Nugent CA: Serum dopamine-beta-hydroxylase in three forms of acute stress. *Life Sci* 15:923–930, 1974.
110. Ogihara T, Nugent CA, Shen SW, Goldfein S: Serum dopamine-beta-hydroxylase activity in parents and children. *J Clin Lab Med* 85:566–573, 1975.
111. Okada F, Yamashita I, Suwa N, Kunita H, Hata S: Elevation of plasma dopamine-beta-hydroxylase activity during insulin-induced hypoglycemia in man. *Experientia* 31:70–71, 1975.
112. Oliver JA, Pinto J, Sciacca RR, Cannon PJ: Basal norepinephrine overflow into the renal vein: Effect of renal nerve stimulation. *Am J Physiol* 239(Renal Fluid Electrolyte Physiol 8):F371–F377, 1980.
113. Owman C: Neurogenic vasodilation mediated by the autonomic nervous system. *Triangle* 18:89–99, 1979.
114. Palkovits M, Mezey E, Zaboszky L: Neuroanatomical evidences for direct neural connections between the brainstem baroreceptor centers and the forebrain areas involved in the neural regulation of the blood pressure. In Meyer P, Schmitt H: *Nervous System and Hypertension*. New York and Paris, Wiley-Flammarion, 1979.
115. Peart WS: The nature of splenic sympathin. *J Physiol* 108:491–501, 1949.
116. Peronnet F, Cleroux J, Perrault H, Cousineau D, de Champlain J, Nadeau R: Plasma norepinephrine response to exercise before and after training in humans. *J Appl Physiol* 51:812–815, 1981.
117. Pettersson J, Hussi E, Janne J: Stability of human plasma catecholamines. *Scand J Clin Lab Invest* 40:297–303, 1980.
118. Pickar D, Lake CR, Cohen RM, Jimerson DC, Murphy DL: Alterations in noradrenergic function during clorgyline treatment. *Commun Psychopharmacol* 4:379–386, 1980.
119. Plantz G, Palm D: Acute enhancement of dopamine-beta-hydroxylase activity in human plasma after maximal work load. *Eur J Clin Pharmacol* 5:255–258, 1973.
120. Plantz G, Wiethold G, Appel E, Bohmer D, Palm D, Grobecker H: Correlation between increased dopamine-beta-hydroxylase activity and catecholamine concentration in plasma; determination of acute changes in sympathetic activity in man. *Eur J Clin Pharmacol* 8:181–188, 1975.
121. Polinsky RJ, Kopin IJ, Ebert MH, Weise V: Pharmacologic distinction of different orthostatic

hypotension syndromes. *Neurology* 31:1–7, 1981.

122. Pollack A, Wood EH: Venous pressure in the saphenous vein at the ankle in man during exercise and changes in posture. *J Appl Physiol* 1:649–650, 1948.
123. Post RM, Lake CR, Jimerson DC, Bunney WE, Wood JH, Ziegler MG, Goodwin FK: Cerebrospinal fluid norepinephrine in affective illness. *Am J Psychiatry* 135:907–912, 1978.
124. Quigley ME, Sheehan KL, Wilkes MM, Yen SSC: Effects of maternal smoking on circulating catecholamine levels and fetal heart rates. *Am J Obstet Gynecol* 133:685–690, 1979.
125. Reid AA, Lake CR, Moore J, Gagnon JA, Owen WW, Copley B: Localized renal norepinephrine (NE) in the canine: A possible mechanism for renal hemodynamic regulation. FASEB Abstract, p. 1724, 1982.
126. Reid JL, Kopin IJ: Significance of plasma dopamine-beta-hydroxylase activity as an index of sympathetic neuronal function. *Proc Natl Acad Sci USA* 71:4392–4394, 1974.
127. Reis DJ, Joh TH, Nathan MA, Renaud B, Snyder DW, Talman WT: Nucleus tractus solitarii: Catecholaminergic innervation in normal and abnormal control of arterial pressure. In Meyer P, Schmitt H: *Nervous System and Hypertension*. New York, Wiley-Flammarion, 1979, pp 147–164.
128. Robertson D, Frolich JC, Carr RK, Watson JT, Hollifield JW, Shand DG, Oates JA: Effects of caffeine on plasma renin activity, catecholamines and blood pressure. *N Engl J Med* 298:181–186, 1978.
129. Robertson D, Johnson GA, Robertson RM, Nies AS, Shand DG, Oates JA: Comparative assessment of stimuli that release neuronal and adrenalmedullary catecholamines in man. *Circulation* 59:637–643, 1979.
130. Romoff MS, Keusch G, Campese VM, Wang M, Friedler RM, Weidmann P, Massry SG: Effect of sodium intake on plasma catecholamines in normal subjects. *J Clin Endocrinol Metab* 48:26–31, 1979.
131. Rosenblatt JE, Lake CR, van Kammen DP, Ziegler MG, Bunney WE: Interactions of amphetamine, pimozide and lithium on plasma norepinephrine and dopamine-beta-hydroxylase in schizophrenic patients. *Psychiatry Res* 1:45–52, 1979.
132. Rysanek K, Kuhn E, Spankova H, Tovarek J, Voznicek J: The role of dopamine-beta-hydroxylase as an indicator of the sympathetic nerve tonus. *Act Nerv Super (Praha)* 17:256–257, 1975.
133. Saeed M, Sommer O, Holtz J, Bassenge E: Alpha-adrenoceptor blockade by phentolamine causes beta-adrenergic vasodilation by increased catecholamine release due to presynaptic alpha-blockade. *J Cardiovasc Pharmacol* 4:44–52, 1982.
134. Sauerbier I, Mayersbach H von: Circadian variation of catecholamines in human blood. *Horm Metab Res* 9:529–530, 1977.
135. Schanberg SM, Kirshner N: Serum dopamine-beta-hydroxylase as an indicator of sympathetic activity and primary hypertension. *Biochem Pharmacol* 25:617–621, 1976.
136. Schiffl H, Weidmann P, Meier A, Ziegler WH: Relationship between catecholamines and urinary catecholamine excretion rates in normal subjects and certain disease states. *Klin Wochenschr* 59:837–844, 1981.
137. Shepherd JT, Vanhoutte PM: *The Human Cardiovascular System: Facts and Concepts*. New York, Raven Press, 1979.
138. Shepherd JT, Vanhoutte PM: Local modulation of adrenergic neurotransmission. *Circulation* 64:655–666, 1981.
139. Silverberg AL, Shah SD, Haymond MW, Cryer PE: Norepinephrine: Hormone and neurotransmitter in man. *Am J Physiol* 234:E252–E256, 1978.
140. Sole MJ, Drobac M, Schwartz L, Hussain MN, Vaughan-Neil EF: The extraction of circulating catecholamines by the lungs in normal man and patients with pulmonary hypertension. *Circulation* 60:160–163, 1979.
141. Spector S, Sjoerdsma A, Zaltzman-Nirenberg P, Levitt M, Udenfriend S: NE synthesis from tyrosine-C14 in isolated perfused guinea pig heart. *Science* 139:1299–1301, 1963.
142. Starke K, Taube HD, Borowski E: Presynaptic receptor systems in catecholaminergic transmission. *Biochem Pharmacol* 26:259–268, 1977.
143. Stone RA, Kirshner N, Gunnells JC, Robinson RR: Changes of plasma dopamine-beta-hydroxylase activity and other plasma constituents during the cold pressor test. *Life Sci* 14:1797–1809, 1974.
144. Sundlof G, Wallin GB: The variability of muscle nerve sympathetic activity in resting recumbent man. *J Physiol* 272:383–397, 1977.
145. Torrey EF, Peterson MR: Schizophrenia and the limbic system. *Lancet* ii:942–946, 1974.
146. Usdin E, Kvetnansky R, Kopin IJ: Catecholamines and stress: Recent advances. In *Developments in Neuroscience*, New York, Elsevier-North Holland, 1980, vol 8.
147. Verity MA: Morphologic studies of the vascular neuroeffector apparatus. In Bevan JA, Furchgott RF, Maxwell RA, Somlyo AP: *Physiology and Pharmacology of Vascular Neuroeffector Systems*. Basel, Karger, 1971, pp 2–12.
148. Viveros OH, Arqueros L, Kirshner N: Release of catecholamines and dopamine-beta-oxidase from the adrenal medulla. *Life Sci* 7:609–618, 1968.
149. Wallin BG, Sundlof G, Eriksson B-M, Dominiak P, Grobecker H, Lindblad LE: Plasma noradrenaline correlates to sympathetic muscle nerve activity in normotensive man. *Acta Physiol Scand* 111:69–73, 1981.
150. Weinshilboum RM, Axelrod J: Serum dopamine-beta-hydroxylase activity. *Circ Res* 28:307–315, 1971.
151. Weinshilboum RM, Axelrod J: Serum dopamine-beta-hydroxylase: Decrease after chemical sympathectomy. *Science* 173:931–934, 1971.
152. Weinshilboum RM: Serum dopamine-beta-hydroxylase. *Pharmacol Rev* 30:133–166, 1979.
153. Wiedeking MC, Ziegler MG, Lake CR: Plasma noradrenaline and dopamine-beta-hydroxylase during sexual activity. *J Psychiatr Res* 15:139–145, 1979.

154. Wooten GF, Cardon PV: Plasma dopamine-beta-hydroxylase activity. *Arch Neurol* 128:103–106, 1973.
155. Zaloga GP, Chernow B, Fletcher JR, Hughes P, Rainey TG, Lake CR: Increased circulating plasma norepinephrine concentrations in *non*-cardiac causes of pulmonary hypertension. *Crit Care Med*, in press, 1984.
156. Ziegler MG, Lake CR, Kopin IJ: Plasma noradrenaline increases with age. *Nature* 261:333–335, 1976.
157. Ziegler MG, Lake CR, Kopin IJ: Deficient sympathetic nervous response in familial dysautonomia. *N Engl J Med* 294:630–633, 1976.
158. Ziegler MG, Lake CR, Wood JH, Ebert MH: Circadian rhythm in cerebrospinal fluid noradrenaline of man and monkey. *Nature* 264:656–658, 1976.
159. Ziegler MG, Lake CR, Kopin IJ: The sympathetic nervous system defect in primary orthostatic hypotension. *N Engl J Med* 296:293–297, 1977.

CHAPTER TWO

Radioenzymatic Assay of Catecholamines

LENARD R. DURRETT, M.D., Ph.D., ALI J. MILANO, M.D., and MICHAEL G. ZIEGLER, M.D.

INTRODUCTION

Assay techniques sufficiently sensitive to measure the level of catecholamines present in most biological fluids have only recently become available. However, there are now several techniques with sufficient sensitivity and reliability to be useful in the measurement of low levels of norepinephrine (NE). These include gas chromatography mass spectroscopy (41), high pressure liquid chromatography with electrochemical detection, a phenylethanolamine-*N*-methyltransferase (PNMT)-based radioenzymatic technique (41), and a catechol-*O*-methyltransferase (COMT)- based radioenzymatic technique. Of these methods, the COMT-based technique has become the most popular because the major equipment required to perform this method is commonly available in biochemistry laboratories. The method has greater absolute sensitivity than other techniques (although other techniques may be made more sensitive by preconcentration of the fluid to be measured), and the COMT measures NE, epinephrine (E), dopamine (DA), and several other catecholamines. Disadvantages of the COMT technique include its expense, the time involved to perform the assay, and its complexity.

Catechol-*O*-Methyltransferase (COMT)

The COMT technique to measure catecholamines involves the incubation of the catecholamine-containing fluid with the methyl donor [^{3}H]methyl-S-adenosylmethionine ([^{3}H] SAM), the enzyme COMT which catalyzes transfer of the [^{3}H]methyl group to the catecholamines, and buffers, salts, and a chelator. COMT catalyzes the formation of metabolites which are purified by solvent extraction and separated by chromatography. The radiolabeled metabolites can then be derivatized or counted directly to quantitate the amount of catecholamine present in the initial sample.

HISTORICAL DEVELOPMENT OF ASSAYS

The COMT assay has evolved over many years with several variations. The story of the evolution of this assay is an interesting tale with many useful lessons for anyone who either wishes to perform the assay or to interpret results obtained by the assay. For many years monoamine oxidase was considered to be the primary enzyme in the metabolism of catecholamines. However, studies reported in 1958 by Axelrod (1) and others (23) demonstrated that COMT was quantitatively a more important enzyme in the metabolism of catecholamines. In 1958, Axelrod and Tomchick (1) published the basic information upon which all further COMT assays have been based, including methods for purification of the enzyme from rat liver, stimulation of enzymatic activity by divalent cations, and demonstration of the wide range of catechols methylated by this enzyme. The commercial availability of radiolabeled SAM allowed the development of a more sensitive assay which used scintillation spectrometry rather than fluorescence assay to detect levels of the metabolites, providing a great increase in potential sensitivity of the assay.

In 1968, Engelman *et al.* (16) developed a double isotope derivative method for labeling and measuring catecholamines in biological materials. This method consisted of three basic steps: (1) incubation of the biological fluid with COMT and $^{14}CH_3$-SAM in a phosphate-buffered solution (pH 8.5); (2) extraction of the ^{14}C-derivatives by passage over a cation exchange resin; (3) conversion of radiolabeled metabolites to vanillin and extraction into toluene before liquid scintillation counting. ^3H-NE was added after incubation to account for recovery in the extraction process. Oxidation of β-hydroxylated metabolites to vanillin allowed for better extraction and decreased interference by DA (33). The sensitivity of this assay (160 pg) was much greater than other currently available fluorescent or bioassay techniques. The practical application of the assay was demonstrated even in this early report by showing its use in localizing a pheochromocytoma in a human patient. This method did not, however, distinguish between E and NE, and a further increase in sensitivity was necessary to provide an accurate assay of catecholamine levels in normal patients.

Nikodejevic *et al.* (28) characterized the conditions for methylation by COMT *in vitro*. They found the optimal buffer to be Tris at a pH of approximately 9.0. Mg^{2+} is also required as a cofactor (1 μM). They eliminated the cation exchange resin step and replaced it with a solvent extraction into toluene isoamyl alcohol. This assay gave an overall recovery of 10 to 20% and was linear from 50 to 300 pmoles.

Epinephrine and NE are similar molecules, and early assays did not separate their metabolites. Engleman and Portnoy (15) revised their original assay to include separation of metanephrine and normetanephrine by thin layer chromatography (TLC) before conversion to vanillin. They found less than 1% crossover between E and NE metabolites. Their average overall recovery was 19%, sensitivity was still only 250 pg, and samples of 5 to 10 ml of plasma were required. Siggers *et al.* (35) also used a double isotope dilution technique but then separated the O-methylated products by electrophoresis and an ether extraction. Coyle and Henry (9) separated the β-hydroxylated catechols E and NE from DA by oxidizing them to vanillin. This technique could measure DA and NE in brain tissue extracts at a sensitivity of 100 pg, but did not separate E and NE (30).

Up to this point, double-isotope methods were the standard procedure used for radioenzymatic assay of catecholamines. These assays were highly specific when compared to previous methods but, due to low specific activity of [^{14}C]SAM, were not very sensitive and therefore required large quantities (5 to 10 ml) of sample fluid. Passon and Peuler (31) combined and modified previous procedures to produce a single isotope assay with a 10-fold increase in sensitivity, which required less than 1 ml of plasma and gave separate determinations of E and NE in one sample. Sensitivity was significantly increased by using [^3H] SAM, which has a higher specific activity than [^{14}C]SAM (17). Following extraction with 3:2 toluene isoamyl alcohol, the products were evaporated to dryness, then taken up into ethanol and streaked on thin layer chromatography plates. The spots representing metanephrine (M) and normetanephrine (NM) were scraped and the metabolites eluted and converted to vanillin before counting. Internal standards were included to correct for conversion and recovery of catecholamine metabolites. This technique was later adapted by Moerman *et al.* (25) and used for measuring serum levels of E and NE in human subjects at rest and during exercise. A commercially available assay ("Cat-a-Kit," Upjohn, Kalamazoo, Mich.) was later produced based on modifications of Passon and Peuler's (31) technique. Cuello *et al.* (12) introduced the use of descending paper chromatography for separation of 3-methoxy-tyramine (3MT) from NM. This greatly improved the specificity for determination of DA but substantially prolonged assay time due to the 1- to 2-day development period required.

Throughout all this development of the COMT assay, it retained some serious drawbacks. Calcium ion is ubiquitous in tissues, but Ca^{2+} is a potent inhibitor of COMT enzymatic activity. This is partly overcome by use of the Ca^{2+} binding agent ethyleneglycoltetraacetic acid (EGTA), but EGTA can also bind Mg^{2+}, which stimulates COMT activity. Many tissues and fluids have extremely low levels of catecholamines, but they could not be concentrated by alumina chromatography prior to assay (as is commonly done in other assays) because Al^{3+} inhibits COMT activity. Gauchy *et al.* (19) overcame both these problems by concentrating samples on alumina and then eluting with alcohol to avoid Al^{3+} contamination. Unfortunately, the process is so laborious that it has not been widely adopted, and problems of inhibition of COMT by calcium and aluminum are common to most COMT assays.

Development of thin layer chromatography techniques resulted in further increases in specificity and sensitivity (13, 3). DaPrada and Zürcher (13) reported the selective measurement of E, NE, and DA from the same sample fluid. They achieved excellent separation of these compounds with a TLC system containing a ratio of chloroform to ethanol to ethylamine of 16:3:3 (see Table 4 of ref. 13). The sensitivity was also increased by extraction of the incubation mixture with ethyl ether containing tetraphenylborate. Ben-Jonathon and Porter (3) used two-dimensional chromatography to separate the 3-O-methylated derivatives and got only 1 to 2% crossover. They reported a sensitivity of 10 to 30 pg and could measure all three endogenous catecholamines simultaneously. Further alterations of Passon and Peuler's assay were made by changing the incubation technique and extraction or TLC methods (32, 36). Hortnagle *et al.* (20) increased sensitivity by using a smaller amount of [^{3}H] SAM with a higher specific activity and increasing the efficiency of the extraction procedure. By saturating the aqueous phase with K_2HPO_4 at pH of 10.5, the extraction of normetanephrine (NM) was increased from 28 to 70%, and blank counts were decreased.

Up to this point, most COMT assays relied upon solvent extraction procedures to separate the labeled catechol metabolites from the residual [^{3}H]SAM and other metabolites. Saller and Zigmond (34), however, developed a simplified extraction procedure. They used phosphotungstic acid to precipitate the residual [^{3}H]SAM out of the incubation solution. The reaction products were then separated by TLC. This gave a sensitivity of 15 to 25 pg, not as high as other techniques employing solvent extraction (13, 14), but much simpler. Over 100 tubes could be assayed in a single day, and additional extractions or TLC steps could be added to increase sensitivity. This technique would still be quite useful in assay of tissues such as brain or adrenal which contain high levels of certain catecholamines.

COMT ASSAY TECHNIQUES

The use of COMT-based radioenzymatic catecholamine assay procedures is now widespread and accepted as a sensitive and reliable technique for catecholamine determination. Many groups have reported slight modifications on Passon and Peuler's (31) or DaPrada and Zürcher's (13) assays, but most currently used techniques involve the basic scheme of:

1. Incubation with partially purified COMT and [^{3}H]SAM.
2. Solvent extraction.
3. Separation of metabolites by TLC, and
4. Vanillin conversion.

These assays usually give sensitivities below 10 pg for NE and 10 to 20 pg for E. Dopamine sensitivity is somewhat higher (50 to 100 pg) due to higher blank counts. The assay may also be used to measure other metabolites in the catecholamine pathway, such as DOPA (43), DOPAC (18), DOMA, or DOPEG (37), as well as certain catechol drugs (14, 26, 2).

Preparation and Handling of Samples

Collection, handling, and preparation of the biological samples are all critical steps in obtaining reliable data. Many errors can be made before assay which alter the results. Concentrations of NE in the blood fluctuate widely during normal daily activities due to alterations in tone of the sympathetic nervous system and rapid metabolism by extraneuronal enzymes. Plasma NE levels are often used as an indication of sympathetic neuronal activity (8, 11, 24). Standing, for example, increases peripheral vascular resistance and plasma catecholamine levels (25, 8, 11). Norepinephrine values usually double upon standing and may increase 3-fold with only moderate exercise (42). It is important that collection of blood samples be carried out in a manner not altering sympathetic tone. Emotional stress will increase NE levels in both human subjects (40) and experimental animals (31, 6). The painful stimulus of venipuncture can increase blood catecholamine levels (31, 6). It is best if experimental subjects have an indwelling intravenous catheter which can be left in place to allow blood to be drawn painlessly. Usually, if subjects remain recumbent for 20 to 30 minutes after venipuncture, plasma NE values can return to resting levels (25, 40). When working with experimental animals, it is important to consider the stress involved in manipulation when measuring blood catecholamines. Dogs can be conditioned over several months so that manipulation and venipuncture result in minimal stress (31, 27). In studies using rats as experimental animals, the use of restraint or venipuncture can cause a 10- to 50-fold elevation of plasma NE, so chronic indwelling catheters must be used to obtain reliable basal NE levels (38). Various experimental procedures may also alter plasma catecholamine levels. Plasma levels of NE and E increase approximately 3-fold in anesthetized cats upon thoracotomy (L. R. Durrett and M. G. Ziegler, unpublished results).

Choice of Subjects

The selection of subjects in human studies requiring measurement of catecholamines is also an important factor. Resting NE levels are affected by many variables. Several groups have reported differences in resting NE values among subjects divided according to age (24, 5, 21), sex (21), and blood pressure (7). Many experimental protocols include use of laboratory personnel or students as control subjects. However, P. C. Rubin and J. L. Reid note in Chapter 14 that male laboratory staff has significantly lower plasma NE levels compared to a similar group of normotensive outpatients. This suggests that being accustomed to the experimental environment also affects plasma NE levels. The choice of appropriately selected controls is, therefore, a critical factor in this type of experiment. Caffeine, nicotine, and alcohol or withdrawal from these drugs in people accustomed to them can also increase plasma NE levels.

Collection

Blood samples should be collected in plastic or glass tubes containing an anticoagulant. Norepinephrine is more stable in the presence of an antioxidant, acid, or cold temperature. Although NE is stable at 4°C in blood (24) for about 1 hour or less, it is very labile in cerebrospinal fluid (see Chapter 8). Following separation, and if kept for longer than 2 weeks, plasma should be frozen immediately and stored at −70°C or mixed with antioxidants or acid.

COMT Preparation

Since the original partial purification of the enzyme by Axelrod and Tomchick (1), most investigators have used similar but slightly modified techniques of isolation. We use the following technique based on the original Axelrod and Tomchick method, modified by Nickodejevic *et al.* (29).

Fresh rat livers are placed on ice and kept cold throughout the procedure. The livers are homogenized in 4 volumes of cold 1.15% KCl in a blender and centrifuged for 30 minutes at 14,000 × g. Fat is removed by wiping the tops of the centrifuge tubes with tissue or filtering through cheesecloth. The pH is lowered to 5.3 by the gradual addition (over 10 minutes) of cold 1 N acetic acid with gentle stirring. This suspension is then centrifuged at 14,000 × g for 10 minutes, the supernatant saved, and its pH adjusted to 6.8 with 0.5 M sodium phosphate buffer (pH 7.0). Ammonium sulfate (17.6 gm/100 ml of supernatant) is added slowly with gentle stirring over a period of 60 minutes. This suspension is then centrifuged for 10 minutes at 14,000 × g. The supernatant is then removed, and additional ammonium sulfate (16.2 gm/100 ml of supernatant) is added as in step 3. The solution is centrifuged and the supernatant discarded. The protein sediment from this 55% $(NH_4)_2SO_4$ step is resuspended in 100 ml of cold 50% ammonium sulfate in 0.05 M phosphate buffer (pH 7.0) and stirred gently for 15 minutes. This suspension is centrifuged at 14,600 × g for 10 minutes, the supernatant discarded, and the previous step repeated in 48% ammonium sulfate solutions. The final sediment is suspended in 33% ammonium sulfate, centrifuged, and the supernatant retained. More ammonium sulfate (24.3 gm/100 ml) is added to the pooled supernatant solution. This suspension is centrifuged for 10 minutes at 14,000 × g and the supernatant discarded. The pellet is dissolved in 1 mM phosphate buffer (pH 7.0).

This solution is dialyzed exhaustively against three changes of 0.005 M $NaHPO_4$, pH 7.4, containing freshly added 0.1 mM dithiothreitol over 16 hours. Enzyme activity is titrated by assaying the conversion of dihydroxyphenylacetic acid to [^{3}H]homovanillic acid (HVA) in the assay mix detailed later in this paper, extracting into toluene after acidifying the assay mixture, and counting the toluene layer. We use 1.5 times the amount of enzyme which gives maximal O-methylation of dihydroxyphenylacetic acid in the COMT assay for catecholamines. The enzyme is brought to a concentration of 10 mM reduced glutathione to preserve activity and 1 mM benzylhydroxylamine to inhibit dopa-decarboxylase activity. Addition of the decarboxylase inhibitor to the enzyme prevents conversion of dopa to dopamine with much less inhibitor than if it is added to the incubation mix.

ASSAY PROCEDURE AND PREPARATION

Prior to incubation, plasma and tissue samples should be centrifuged to remove any clots or precipitates. Aliquots of 100 μl of each sample are pipetted into 100 × 13 mm polypropylene tubes. Blanks contain water (100 μl H_2O) and standards contain samples with 1 ng of E, NE, and 200 pg DA added. Each 100 μl of incubation mix contains: 85 μl of a solution containing 90 mM $MgCl_2$, 20 mM EGTA and 200 mM Tris (pH 8.3), 5 μl[^{3}H]SAM, and 10 μl COMT and 0.06 mg reduced glutathione. Following addition of the mix, the tubes are placed in a shaking incubator at 37°C for 30 minutes.

The incubation reaction is terminated by placing the tubes in an ice water bath and adding 200 μl of a 1 M borate solution containing 25 mg/ml EDTA (pH 10). At this point, 50 μl of a solution containing 1 mg/ml each of 3MT, M, and NM is added to each tube as cold carriers. If other catechol drugs are to be assayed, the appropriate cold carriers are added.

The labeled metabolites are extracted by addition of 5 ml of a 3:2 toluene to isoamylalcohol ratio to each tube. The tubes are capped and shaken for 10 minutes on a reciprocating shaker. The organic and aqueous phases are separated by centrifugation at 6,000 rpm for 5 minutes. The organic phase is pipetted or the tubes are placed in an ethanol-dry ice bath containing sufficient ethanol to just cover the aqueous layer in the tubes. When the aqueous phase freezes, the organic phase is decanted into a second 100 × 13 mm polypropylene tube which contains 250 μl of 0.1 N acetic acid.

The tubes containing the organic phase and acetic acid are capped, shaken for 10 minutes and centrifuged, and the organic phase is aspirated and discarded. The remaining acid phase is washed one time with water-saturated 3:2 toluene to isoamyl alcohol, shaken for 5 minutes, and then separated by centrifugation. After removal of the organic phase, the aqueous layer is lyophilized to dryness in an evaporating centrifuge (Savant). The samples may be frozen at this point.

The freeze-dried residue is taken up into 210 μl of a mixture of 90% ethanol-5% 0.002 N HCl by vortexing for 10 seconds. This solution is then drawn up into 500-μl glass hypodermic syringe (Hamilton) fitted with 1½ inch 25-gauge needle. The tubes are washed a second time with 200 μl ethanol, and this solution is drawn into the same syringes. They are placed on a TLC multispotter (Analytical Instrument Co., Libertyville, IL) and slowly spotted onto the preabsorbant area of a silica gel TLC plate (Whatman LD6DF, 250-μl thickness) containing a fluorescent indicator. A stream of warm air is directed at the preabsorbant area of the plate by two hairdryers during the procedure. With 16 samples spotted on each plate, the spotting procedure requires about 45 minutes per plate.

The plates are developed in closed TLC jars lined with chromatography paper to equalize solvent vapor. Development at 2 to 4°C increases development time by 30% but helps concentrate the spots and decreases crossover. After air-drying, the spots are visualized with ultraviolet light and marked with a pencil. From the origin the bands are in order: normetanephrine (NM), metanephrine (M), then 3-methoxytyramine (3MT; Fig 2.1).

The 3MT spots are scraped and placed directly into 7-ml counting vials. To each vial is then added: 500 μl of 1 M sodium acetate buffer (pH 5.5), 200 μl fluor (Concentrol), and 5 ml of 1% diethylhexylphosphoric acid (DEHP) in toluene. The vials are capped and shaken for 5 minutes before counting.

Compounds containing a β-hydroxyl group are converted to vanillin by periodation. The M and NM are, therefore, scraped and placed into 10 × 75 mm culture tubes. The compounds are eluted by addition of 0.5 ml of 2 N NH_4OH and vortexed for 10 seconds. After centrifugation at 6,000 rpm for 4 minutes, the NH_4OH is decanted into 7-ml counting vials. The elution procedure is repeated one time and the NH_4OH combined with that from the first elution. The silica pellet is then discarded. For periodation 50 μl of freshly prepared 4% $NaIO_4$ is added to each vial. After exactly 5 minutes the periodation reaction is terminated by addition of 50 μl 10% glycerol. To each tube is then added 200 μl of 10 N acetic acid, 200 μl fluor, and 4 ml toluene. The vials are capped and shaken for 5 minutes before counting. The MN and M spots may be scraped directly into counting vials and the periodation reaction carried out there. This gives a few chemoluminescent counts during liquid scintillation spectrometry which diminish over 1 to 2 days.

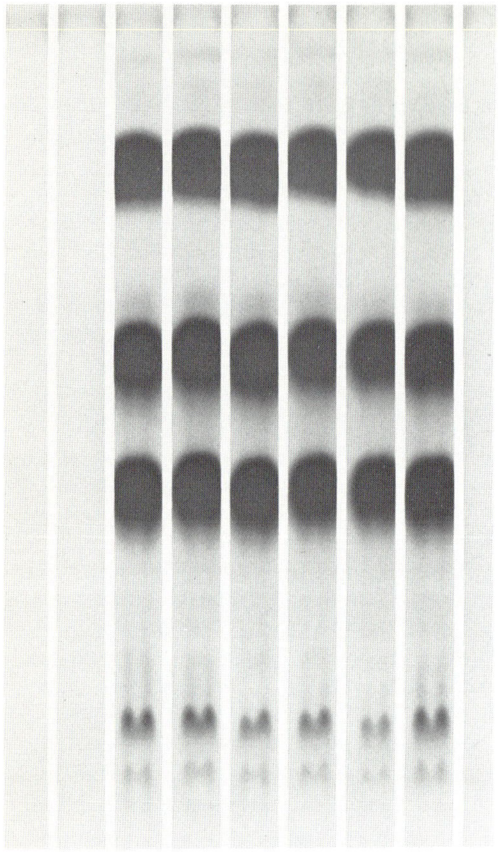

Figure 2.1. A thin layer chromatography plate with the origin on the bottom and solvent front at the top. The *dark spots* are, from top to bottom: 3-methoxytyramine, metanephrine, and normetanephrine. Samples were spotted on Whatman LK6DF plates and developed in ethylamine (70%) to chloroform to ethanol ratio of 2:16:3 at 20°C.

Calculations

The following formula is used for calculation of values:

$$\frac{[CPM \text{ of sample}] - [CPM \text{ of blank}]}{[CPM \text{ of (sample + standard)}] - [CPM \text{ of sample}]} \times \frac{\text{pg standard added}}{\text{Sample volume}} = \text{pg/ml}$$

Separate standards must be used for each type of tissue assayed, since different tissues cause different amounts of inhibition of COMT enzymatic activity.

A great number of variations on the COMT assay for catecholamines have been devised. The assay outlined in this chapter incorporates those techniques best suited to measurement of plasma levels of norepinephrine and epinephrine in man. While this is the most frequent use of the assay, there are other situations which require different levels of sensitivity, selectivity, economy and speed. The many reports of different assay techniques make it possible to choose between various methods so that the assay is most suited to the individual requirements of a particular laboratory.

Sensitivity

In some specific circumstances, for example when only very small sample sizes are available, or when one wishes to measure plasma levels of circulating dopamine, increased sensitivity of the assay is helpful. Sensitivity may be increased somewhat by using [^{3}H]SAM of higher specific activity whch has recently become available through New England Nuclear and Amersham. However, this increases the expense of the assay, and the higher [^{3}H]SAM degrades more rapidly on storage. Sensitivity of the assay may also be enhanced by using a more efficient solvent extraction procedure for isolation of the [^{3}H]metabolites. The technique of DaPrada and Zürcher (13) provides a substantial improvement of extraction efficiency over the method detailed here. However, it involves the use of ether and requires transfer of the metabolites to a strong acid phase which can interact with the TLC plates and lead to streaking and poor separation of the metabolites. If there is adequate sample volume, then concentration of the sample can increase sensitivity of the assay. Gauchy *et al.* (19) describe a technique for concentration of catechols on alumina with an extraction procedure that avoids sample contamination with aluminum ion. Weise and Kopin (39) similarly use an alumina concentration step, but their technique allows for some inhibition of COMT enzyme activity by aluminum, so that the potential for an increase in sensitivity is diminished. Thus, it is possible to increase the sensitivity of this assay somewhat, but these techniques may lead to increased effort or decreased selectivity.

Selectivity

Crossover between catecholamines can be diminished by several maneuvers. Very careful spotting on chromatography plates with no preadsorbant layer can improve chromatographic separation and increase selectivity. However, any variation in spotting technique will lead to variable crossover of one catecholamine metabolite into the other, and this inconsistency may make it impossible to routinely subtract calculated crossover of one catecholamine into the other. Developing chromatography plates at lower temperatures will generally lead to a longer development time but will decrease the size of the metabolite spots and diminish crossover between metabolites (Figs. 2.1 and 2.2). It is possible to select solvent systems to obtain the maximal distance between metabolites of choice on TLC and thus diminish crossover between these metabolites (Table 2.2). We have found that radiolabeled metabolites that cannot be adequately separated by TLC can usually be separated by high pressure liquid chromatography (HPLC).

Economy and Speed

If bought in kit form, reagent assay of an individual sample may cost as much as $15.00, but it is possible to perform this assay much more cheaply. Several authors have advised running a standard for each sample assayed in order to correct for the sample's inhibition of COMT activity. Obviously, this will double the cost of the assay. It is essential to run individual standards for each type of tissue assayed since, for example, cerebrospinal fluid may inhibit COMT activity only half as much as plasma. However, the variability between individual plasma samples is usually smaller than the variability introduced by interposing another assay in the calculations for each sample. If the catecholamine levels are fairly high and it is not necessary to differentiate between them, then the TLC step can be eliminated with a considerable savings in time and expense. Since periodate will oxidize NE and E to vanillin, but oxidizes DA much less well, it is possible to separate DA from NE and E without performing thin layer chromatography (9). If tissue samples are being assayed and great sensitivity is not necessary, but separation of catechols is required, then the method of Saller and Zigmund (34) is adequate and eliminates solvent extraction steps for the assay.

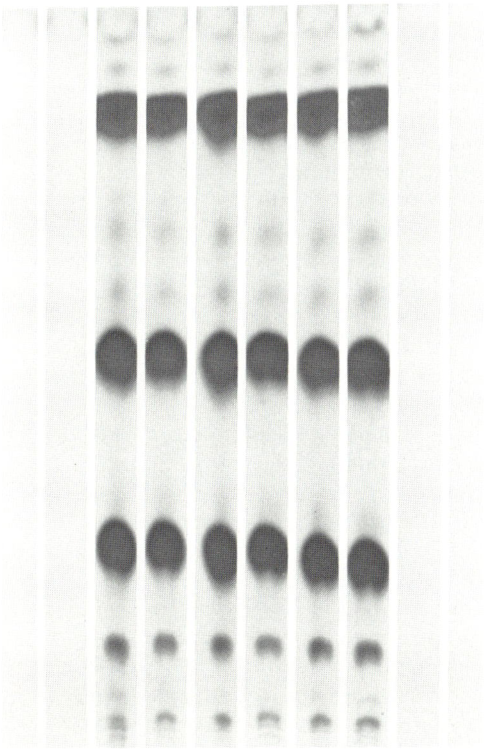

Figure 2.2. A thin layer chromatography plate processed as in Figure 2.1, but developed at 2°C. Several minor bands are visible due to cold carrier impurities (bands near origin) or solvent and TLC plates (bands near solvent front).

Some investigators prefer to remove proteinaceous material from plasma samples by acid precipitation before incubation. The addition of strong acid at this point, however, necessitates subsequent addition of higher buffer and salt concentrations which reduce COMT activity. We obtain better COMT activity in our incubation phase by not deproteinizing the samples and using more dilute buffer. If it is necessary to deproteinize samples, this should be done by bringing them to 0.3 N $HClO_4$, homogenizing, and centrifuging. The supernatant (100 μl) is then assayed with a buffer solution containing 2 M Tris HCl at pH 8.8 in the incubation mix (rather than 0.2 M Tris HCl).

COMMONLY USED CATECHOL DRUGS

COMT is not a highly specific enzyme (1, 10, 14) and will methylate a variety of compounds which contain a catechol moiety. While this has presented a problem of interference to investigators in the past, the nonspecificity of COMT allows simultaneous measurement of endogenous catecholamines and catechol drugs. Figure 2.3 shows the structure of a variety of commonly used catechol drugs and their apparent K_m values in this

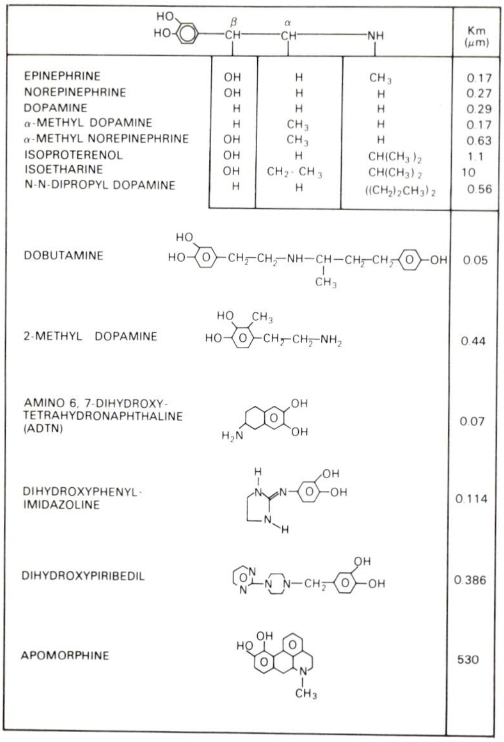

Figure 2.3. The chemical structure of several catechol drugs *o*-methylated by rat liver COMT. The apparent *Km* values obtained by the COMT assay are listed in the *right-hand column.* Note that all the compounds except apomorphine are avid substrates for COMT.

Table 2.1
Upper Limits of Linearity and Sensitivity (Estimated as Twice Blank Value) for Various Catechol Drugs Assayed by the COMT Technique

Drug	Upper Limits of Linearity (ng/ml)	Sensitivity (ng/ml)
Epinephrine	40	0.003
Norepinephrine	30	0.002
Dopamine	100	0.003
α-Methyldopamine	150	0.004
α-Methylnorepinephrine	175	0.004
Isoproterenol	150	0.006
Isoetharine	200	0.010
N-N-dipropyldopamine	160	0.010
Dobutamine	50	0.006
2-Methyldopamine	100	0.005
ADTN	150	0.004
Dihydroxyphenylimidazoline	50	0.004
Dihydroxypiribedil	100	0.005

assay. Michaelis constants for our incubation conditions were obtained by Lineweaver-Burk analysis of five to seven duplicate data points for each drug. As shown in the table, substitution on the side chain or about the catechol ring had little effect on affinity for the enzyme. Dobutamine, which contains a bulky side chain, was methylated as well as E. The presence of a methyl group on the α-carbon also had no apparent effect on methylation. Apomorphine was the only catechol-containing compound tested of sufficient size and/or inflexibility to result in a low affinity for the enzyme. The lower limits of sensitivity for some of these drugs are shown in Table 2.1.

Applications and Limitations

In spite of prior reports showing methylation of catechol drugs by COMT *in vitro* (10, 26, 2), very few investigators have made use of this assay in pharmacokinetic studies (26). Pharmacokinetic analysis of catechol drugs has been difficult due to low plasma levels of the drugs, interference by the endogenous catecholamines, and rapid metabolism of these compounds *in vivo*. Such studies have been considered of sufficient importance, however, to warrant administration of radiolabeled drugs to human subjects (4, 22). The present assay allows determination of catechol levels while circumventing such problems.

The greatest difficulty in simultaneous assay of these drugs and the endogenous catecholamines is the separation process. Separation of the drugs listed in Table 2.2 by TLC can be accomplished with high specificity by manipulation of the solvent systems and development at 2°C. As shown in Figure 2.2, development in the cold results in greater separation of M, NM, and 3MT spots and concentrates the spots into smaller areas, thus aiding in separation of metabolites. The three major bands seen in Figure 2.2 are always visible to the naked eye. The presence of the minor bands was detected only by high contrast film after exposure of TLC plates to light for several days. The appearance of visible minor bands is dependent upon the purity of the cold carriers (bands near origin) and solvents (bands near solvent front).

Several different solvent systems have been reported which give good separation of M, NM, and 3-MT. Solvents 1 and 2 (Table 2.2 footnote) provide excellent separation of most of the commonly used catechol drugs from the en-

Table 2.2
Rf Values and Crossover into Endogenous Catecholamines of Various Catechol Drugs Assayed by the COMT Technique[a]

	Ethylamine Solvent				Methylamine Solvent			
		Crossover				Crossover		
Drug	R_f	E	NE	DA	R_f	E	NE	DA
Epinephrine	0.35		0.014	0.064	0.32		0.014	0.03
Norepinephrine	0.24	0.006		0.007	0.18	0.02		0.06
Dopamine	0.53	0.005	0.002		0.43	0.025	0.018	
Isoproterenol	0.53	0.05	0.01	1.0	0.39	0.045	0.002	0.045
Isoetharine	0.60	0.013	0.014	0.013	0.48	0.15	0.005	0.56
Dobutamine	0.55	0.07	0.01	1.0	0.46	0.03	0.03	1.0
α-Methyldopamine	0.77	0.06	0.02	0.07	0.46	0.23	0.03	1.0
α-Methylnorepinephrine	0.34	1.0	0.005	0.005	0.37	1.0	0.005	0.02
2-Methyldopamine	0.44	0.02	0.05	0.01	0.39	0.57	0.03	0.05
ADTN	0.77	0.01	0.01	0.77	0.46	0.04	0.04	1.0
DPI	0.68	0.09	0	0.9	0.24	0.02	1.0	0.11
Dihydroxypiribedil	0.87	0.05	0.005	0.19	0.51	0.19	0.33	0.19
N-N-dipropyldopamine	0.91	0.005	0.005	0.01	0.60	0.01	0.02	0.01
Apomorphine	0.91	0.005	0.005	0.01	0.67	0.005	0.005	0.005

[a] Drug metabolites were separated at room temperature on Whatman LK6DF TLC plates using a 2:16:3 ratio of ethylamine (70%) to chloroform to ethanol or using a 3:6:2 ratio of methylamine to *t*-amyl alcohol to benzene.

dogenous catecholamines. Consider isoproterenol as an example. In the ethylamine solvent system there is 100% overlap with DA as shown by crossover of 1.0 in Table 2.2. However, the methylamine solvent system provides good separation of 3-methoxytyramine and 3-0 methyl isoproterenol with a crossover of less than 5%. The data shown in Table 2.2 was obtained at room temperature; crossover values may be lowered by development at 2°C. The use of chromatography plates with preadsorbent area results in consistent interference among those spots with significant crossover because separation is not dependent upon the size of the spot applied. This permits a reliable calculated correction for small levels of interference.

Another aspect of the TLC procedure to be considered is the time involved in development. Other systems or techniques such as two-dimensional TLC (15) may provide better separation if certain compounds cause crossover problems, but may also decrease efficiency by prolonging development time and/or decreasing the number of samples per plate. In those few cases where we have not been able to get adequate separation between metabolites by TLC, high pressure liquid chromatography has provided good separation. More details of this technique are given in Chapter 3.

SUMMARY

Radioenzymatic assays based on the use of COMT and [^{3}H]SAM have evolved into a highly sensitive and efficient method for measuring catecholamines in body tissues and fluids. In addition, the technique described here measures several commonly used catechol drugs and/or metabolites simultaneously with the endogenous catecholamines. This method will measure most of these compounds at levels below 10 pg with a high specificity and low interassay variability. This assay has been studied in sufficient detail so that it can be tailored to measure a variety of catechols in different tissues. We have outlined a basic assay technique and modifications useful in some tissues and for specific catechol drugs.

References

1. Axelrod J, Tomchick R: Enzymatic O-methylation of epinephrine and other catechols. *J Biol Chem* 233:702–705, 1958.
2. Bauce L, Thornhill J, Cooper K, Veale W: A radioenzymatic assay for femtomole determination of catecholamines using α-methyldopamine as an internal standard. *Life Sci* 27:1921–1928, 1980.

3. Ben-Jonathon N, Porter J: A sensitive radioenzymatic assay for dopamine, norepinephrine and epinephrine in plasma and tissue. *Endocrinology* 98:1497–1507, 1976.
4. Blackwell EW, Briant RH, Conolly ME, Davies DS, Dollery CT: Metabolism of isoprenaline after aerosol and direct intrabronchial administration in man and dog. *Br J Pharmacol* 50:587–591, 1974.
5. Brecht H, Schoeppe W: Relation of plasma noradrenaline to blood pressure, age, sex and sodium balance in patients with stable essential hypertension and in normotensive subjects. *Clin Sci Mol Med* 55:81s–83s, 1978.
6. Caruthers M, Taggart P, Conway N, Bates D, Somerville W: Validity of plasma catecholamine estimations. *Lancet* 2:62, 1970.
7. Champlain J, Farley L, Cousineau D, Van-Amerigan M: Circulating catecholamine levels in human and experimental hypertension. *Circ Res* 38:109–114, 1976.
8. Christensen NJ: Plasma catecholamines in long-term diabetics with and without neuropathy and in hypophysectomized subjects. *J Clin Invest* 51:779–787, 1972.
9. Coyle JT, Henry D: Catecholamines in fetal and newborn rat brain. *J Neurochem* 21:61–67, 1973.
10. Creveling CR, Moris N, Shimizu H, Ong H, Daly J: Catechol-O-methyltransferase. IV. Factors affecting m- and p-methylation of substituted catechol. *Mol Pharmacol* 8:398–409, 1972.
11. Cryer PE, Santiago J, Shah S: Measurement of NE and epinephrine in small volumes of human plasma by a single isotope derivative method: Response to upright posture. *J Clin Endocrinol Metab* 39:1025–1029, 1974.
12. Cuello A, Hiley R, Iverson L: Use of catechol-O-methyltransferase for the enzyme radiochemical assay of dopamine. *J Neurochem* 21:1337–1340, 1973.
13. DaPrada M, Zürcher G: Simultaneous radioenzymatic determination of plasma and tissue adrenaline, noradrenaline and dopamine within the femtomole range. *Life Sci* 19:1161–1174, 1976.
14. Durrett LR, Ziegler MG: A sensitive radioenzymatic assay for catechol drugs. *J Neurosci Res* 5:587–598, 1981.
15. Engelman K, Portnoy B: A sensitive double-isotope derivative assay for norepinephrine and epinephrine. *Circ Res* 26:53–57, 1970.
16. Engelman K, Portnoy B, Lovenbert R: A sensitive and specific double-isotope derivative method for the determination of catecholamines in biological specimens. *Am J Med Sci* 255:259–268, 1968.
17. Evans M, Halter J, Porte D: Comparison of double- and single-isotope enzymatic derivative methods for measuring catecholamines in human plasma. *Clin Chem* 24:567–579, 1978.
18. Fokete M, Kanyicska, Herman J: Simultaneous radioenzymatic assay of catecolamines and DOPAC: Comparison of the effects of drugs on the tuberoinfundibular and striatal dopamine metabolism and on plasma prolactin level. *Life Sci* 23:1549–1556, 1978.
19. Gauchy C, Tassin J, Glowinski J, Cheramy A: Isolation and radioenzymatic estimation of picogram quantities of dopamine and norepinephrine in biological samples. *J Neurochem* 26:471–480, 1976.
20. Hortnagel H, Benedict C, Grahame-Smith P: A sensitive radioenzymatic assay for adrenaline and noradrenaline in plasma. *Br J Clin Pharmacol* 4:553–558, 1977.
21. Jones D, Hamilton C, Reid J: Plasma noradrenaline, age and blood pressure: A population study. *Clin Sci Mol Med* 55:73s–75s, 1978.
22. Kadar I, Tang HY, Conn AW: Isoproterenol metabolism in children after intravenous administration. *Clin Pharmacol Ther* 18:789–795, 1974.
23. La Brosse EH, Axelrod J, Kety SS: O-methylation, the principal route of metabolism of epinephrine in man. *Science* 128:593–594, 1958.
24. Lake CR, Ziegler MG, Kopin IJ: Use of plasma norepinephrine for evaluation of sympathetic neuronal function in man. *Life Sci* 18:1315–1326, 1976.
25. Moerman E, Bogaert M, De Schaepdryuer A: Estimation of plasma catecholamines in man. *Clin Chim Acta* 72:89–96, 1976.
26. Murphy P, Williams T, Dan K: Disposition of dobutamine in the dog. *J Pharmacol Exp Ther* 199:423–431, 1976.
27. Nechay B, Jackson R, Ziegler MG, Neldon S, Thompson J: Effects of chronic digitalization on cardiac and renal $Na^+ + K^+$-dependent adenosine triphosphatase activity and circulating catecholamines in the dog. *Circ Res* 49:655–660, 1981.
28. Nikodejevic B, Daly J, Creveling CR: Catechol-O-methyltransferase. I. An enzymatic assay for cardiac norepinephrine. *Biochem Pharmacol* 18:1577–1584, 1969.
29. Nikodejevic B, Senoh S, Daly J, Creveling CR: Catechol-O-methyltransferase. II. A new class of inhibitors of COMT; 3,5-dihydroxy-4-methoxybenzoic acid and related compounds. *J Pharmacol Exp Ther* 174:83, 1970.
30. Palkovits M, Brownstein M, Saavedra J, Axelrod J: Norepinephrine and dopamine content of hypothalamic nuclei of the rat. *Brain Res* 77:137–149, 1974.
31. Passon P, Peuler J: Radioenzymatic assay for plasma norepinephrine and epinephrine. *Analyt Biochem* 51:618–631, 1973.
32. Peuler J, Johnson G: Simultaneous single isotope radioenzymatic assay of plasma norepinephrine, epinephrine and dopamine. *Life Sci* 21:625–636, 1977.
33. Pisano J: Assay of catecholamine metabolites based on vanillin formation. In Rall JE, Kopin IJ: *The Thyroid and Biogenic Amines*. New York, American Elsevier, 1972.
34. Saller C, Zigmond M: A radioenzymatic assay for catecholamines and dihydroxyphenylacetic acid. *Life Sci* 23:1117–1130, 1978.
35. Siggers D, Salter C, Toseland P: A double isotope dilution method for differential determination of adrenaline and noradrenaline in plasma. *Clin Chim Acta* 30:373–376, 1970.
36. Sole M, Hussain N: A simple, specific radioen-

zymatic assay for the simultaneous measurement of picogram quantities of norepinephrine, epinephrine and dopamine in plasma and tissues. *Biochem Med* 18:30–37, 1977.

37. Vlachakis N, Alexander N, Velasquez M, Maronde R: A radioenzymatic assay for simultaneous measurement of catecholamines and their deaminated metabolites. *Biochem Med* 22:323–331, 1979.
38. Weeks J: Cardiovascular Techniques Using Anesthetized and Freely Roaming Rat. Kalamazoo, Mich, Upjohn 1011, 1973.
39. Weise V, Kopin IJ: Assay of catecholamines in human plasma: Studies of a single isotope radioenzymatic procedure. *Life Sci* 19:1673–1686, 1976.
40. Wyatt R, Portnoy B, Kupfer D, Snyder F, Engelman K: Resting plasma catecholamine concentrations in patients with depression and anxiety. Arch Gen Psychiatry 24:65–70, 1971.
41. Ziegler MG, Lake CR, Foppen F, Shoulson I, Kopin IJ: Norepinephrine in cerebrospinal fluid. *Brain Res* 108:436–440, 1976.
42. Ziegler MG: Postural hypotension. *Ann Rev Med* 31:239–245, 1980.
43. Zürcher G, DaPrada M: Radioenzymatic assay of femtomole concentrations of DOPA in tissues and body fluids. *J Neurochem* 33:631–639, 1979.

CHAPTER THREE

Rapid Liquid Chromatographic Methods for Assay of Norepinephrine, Epinephrine, and Dopamine in Biological Fluids and Tissues

RONALD E. SHOUP, Ph.D., PETER T. KISSINGER, Ph.D., and DAVID S. GOLDSTEIN, M.D., Ph.D.

INTRODUCTION

As the most recent approach to the measurement of catecholamines (CA) in biological samples, liquid chromatography with electrochemical detection (LCEC) provides unique capabilities to the neuroscientist. The method couples modern liquid chromatography, a technique of high resolving power, with sensitive thin-layer electrochemistry. The advantages of this union include subpicomole detection limits, excellent selectivity, and uncomplicated sample clean-up procedures. The instrumentation and reagents are inexpensive, and the number of sample manipulations necessary can often be reduced when compared to gas phase, fluorescence, or radioenzymatic methods. In addition, LCEC is capable of measuring any nonconjugated tyrosine or tryptophan metabolite directly without the use of fluorophors or radiolabeled derivatives. This wide ranging applicability is unique to LCEC.

More than 850 papers are now published in the scientific literature using LCEC, and the majority of these deal with diverse neurochemical applications of the technique (17). In this chapter we will limit the discussion to the determination of CA in body fluids and tissues.

ANALYTICAL CONCEPTS

A primary route to the measurement of CA has been their conversion of fluorescent trihydroxyindole derivatives, the product of their oxidation in base, to a quinone and subsequent cyclization. Generally, the first step of this same reaction is used in LCEC. CA may be oxidized at a carbon electrode to form orthoquinones (18). The electrochemical analog used for their detection can be seen in Figure 3.1.

The advantage of this approach is that the reaction may be followed *directly* by monitoring the passage of electrons onto the electrode surface. The rate of electron flow per unit time (current) is easily measured using modern electronic amplifiers, even in the low picoampere range.

The oxidation reaction must be initiated in a manner controllable by the scientist. In homogeneous solution, oxidizing agents such as iodine or ferricyanide may be used. In electrochemistry, the electrode acts as the oxidizing or reducing agent. A potential set by the operator is applied between the detector and reference electrodes. As the working electrode's potential becomes more positive relative to the solution, the electrode surface becomes a better oxidizing agent. Conversely, an electrode is a stronger reducing agent as the applied potential becomes more negative. Most reports cite applied potentials of approximately +500 to +700 mV (*vs.* Ag/AgCl). These potentials will vary due to differing pH values of the reaction medium (mobile phase).

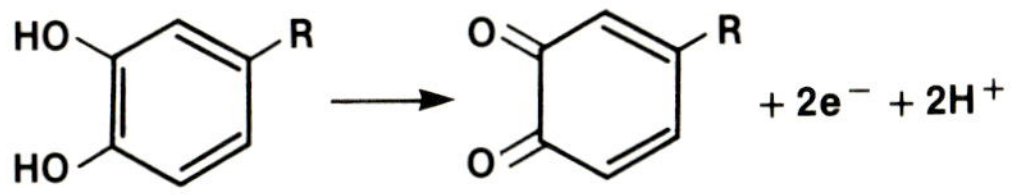

Figure 3.1 Electrochemical oxidation of CA.

Unfortunately, the identity of the R substituent in Figure 3.1 provides little analytical differentiation among norepinephrine (NE), epinephrine (E), and dopamine (DA). All oxidize at approximately the same potential. A modern reverse phase column capable of resolving the individual CA into discrete elution zones provides a suitable solution. A complete LCEC analyzer for CA is shown in Figure 3.2. A high pressure, dual piston pump provides a stable, reproducible flow of the mobile phase, whose composition provides conditions suitable for both the separation and detection processes. The separation is carried out isocratically (without gradients) to avoid baseline drift at high sensitivity. Sample or sample extract is introduced using a syringe into a conventional six-port injection valve.

An optimally configured electrochemical flow cell for this prior liquid chromatographic separation employs very little dead volume. This is conveniently arranged by using a thin-layer sandwich, as shown in Figure 3.3, where the detection ("working") electrode is situated along one side of a channel defined by a 50-μm thick Teflon gasket separating the two plastic blocks (usually Kel-F). The working electrode is either carbon paste or glassy carbon. Carbon paste forms from finely divided graphite powder mixed with an organic binding agent such as paraffin oil; the mixture is packed into the electrode cavity and then smoothed to form the surface. Glassy carbon is a hard, solvent impervious material which is permanently embedded in the electrode block. The surface may be cleaned by simply rinsing and wiping with alcohol. The choice of electrode surface seems to be mostly a matter of personal preference. Two other electrodes are required by the electronic controller, the reference and the auxiliary electrodes. As mentioned earlier, the controller provides a stable source of oxidizing power from the potential selected by the operator. It acts as a potentiostat, and the potential selected is relative to the reference electrode potential (e.g., +0.55 V *vs*. Ag/AgCl). The auxiliary electrode completes the current path through the detector cell; for every oxidation occurring at the working electrode, a forced reduction to conserve charge balance consumes an equal number of electron equivalents at the auxiliary electrode.

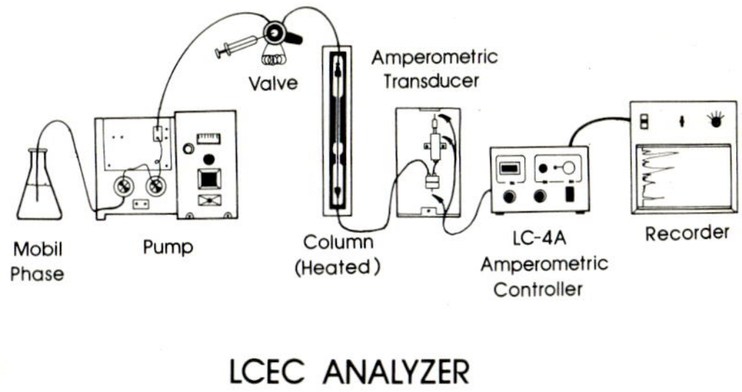

Figure 3.2 A complete LCEC analyzer for CA provides a pump capable of pulseless flow over flow rates from 0.5 to 3.0 ml/min, a syringe-loaded high pressure injection valve, analytical reverse phase column for the separation, and a thin-layer electrochemical flow cell with associated electronics. The separation is accomplished with a fixed mobile phase composition for maximum sensitivity.

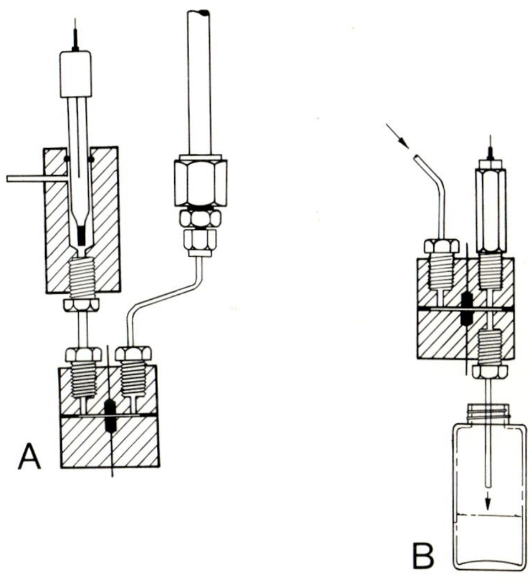

Figure 3.3 *(A)* Conventional thin-layer electrochemical flowcell with working, auxiliary, and reference electrodes. The dead volume in the thin-layer zone is ca. 3.5 μL. *(B)* Special configuration available for fraction collection, as in turnover studies. The oxidation efficiency is only 3 to 5%, so most of the analyte is available intact.

For very low current measurements in buffers of moderate ionic strength, the placement of either the reference or auxiliary electrodes is not critical. After detection in the thin-layer region, the eluent may pass into a larger volume compartment without effect.

For the scientist who wishes to collect fractions for subsequent analysis, the cell is better configured as shown in Figure 3.3*B*.

The liquid chromatographic separation of CA is performed on either reverse phase or cation exchange stationary phases. Early separations employed pellicular, 30- to 50-μm diameter cation exchange materials such as VYDAC SCX or ZIPAX SCX. These columns were relatively inexpensive but offered separation efficiencies of no more than 1000 plates/m. The advantages of modern 5- to 10-μm reverse phase and cation exchange packing materials have largely outdated the former approach. Efficiencies of 25,000 to 50,000 plates/m are realistically achieved with greater security against interference and shorter analysis times per sample.

Since ordinary buffers do not permit adequate separation of CA on reverse phase materials, surface modifiers such as sodium octylsulfate or sodium heptylsulfonate are added to the mobile phase (7). These hydrophobic anionic agents are strongly absorbed to the stationary phase and in effect transform it to a cation exchange column. Although this modification appears roundabout, a column transformed in this manner is superior to commercially available 10-μm cation exchange columns when comparing efficiency and peak symmetry. The former are also more versatile, as they are suitable for separations of neutral and anionic substances as well (e.g., 3-methoxy-4-hydroxyphenylglycol [MHPG], homovanillic acid [HVA], 5-hydroxyindole-3-acetic acid [5-HIAA], etc.).

A significant advantage of the LCEC technique is its ability to quantitate any nonconjugated tyrosine or tryptophan metabolite by simply changing the applied potential and/or stationary or mobile phases (Fig. 3.4). Nearly all separations can now be effected using a single reverse phase column with applied potentials between +0.45 and +1.1 V (vs. Ag/AgCl).

SAMPLE COLLECTION AND STORAGE

CA suffer from chemical instability from a variety of sources. The catechol function is prone to oxidation from dissolved oxygen; keeping the sample in mildly acidic solution (see Fig. 3.1) represses this tendency. Degassing of sample reagent solutions helps but is seldom carried out. Chelating agents such as EDTA bind metal ions which catalyze the oxidation or act as essential cofactors. Samples should be kept frozen to retard decomposition prior to sample preparation and analysis. All of the recommendations outlined below ensure chemical stability for at least 2 months.

Tissue Samples

Brain or other tissue should be removed as rapidly as possible, transferred to a 1.5 mL conical centrifuge tube, frozen in liquid nitrogen, and stored on solid CO_2. The weight is obtained by difference before freezing. To assay CA, a small volume (0.5 to 1.0 mL) of 0.1 M $HClO_4$ and a reproducible amount of internal standard (optional, usually 1 to 2 nmoles of 3,4-dihydroxybenzylamine) are added, and the tissue is then disrupted by sonication (11).

Plasma Catecholamines

Blood is drawn via an indwelling intravenous (i.v.) needle or catheter into chilled heparinized, evacuated tubes and spun in a refrigerated centrifuge at 4°C. The plasma is removed, frozen in plastic tubes without further additives or protein precipitation, and stored in liquid nitrogen or at −70°C until the time of assay (3).

Urinary Catecholamines

Timed or 24-hour urine samples are collected and the volume noted. For 24-hour samples, a large jug is provided containing 20 mL of 5 M HCl. Homogeneous 20-ml aliquots are withdrawn, 150 μL each of 10% disodium ethylenediaminetetraacetate (Na_2EDTA) and 4% thioglycolic acid are added, and the samples frozen at −35°C.

SAMPLE EXTRACTION

The goal of any bioanalytical method is to extract information selectively from a complex sample matrix containing a great deal of uninteresting "noise." At some point, the method will most likely have to differentiate between compounds of very similar structure. This is particularly true if one is to examine the metabolites of tyrosine. These can be conveniently classified as bases, neutrals, or acids. A simple pH adjustment followed by solvent extraction or liquid-solid adsorption will permit isolation of these classes for individual determinations.

It is important to recognize that the proce-

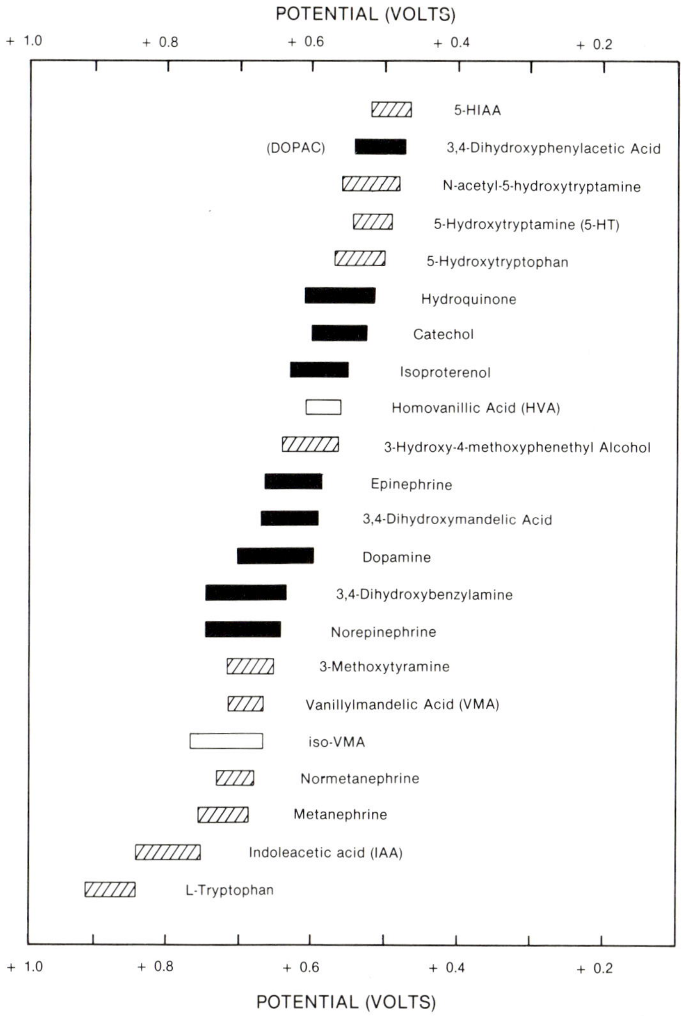

Figure 3.4 Oxidative cyclic voltammetric data for common tyrosine and tryptophan metabolites and related compounds. The peak potentials, obtained in 90% 0.1 M citrate buffer (pH 4)/10% ethanol on a carbon paste electrode, provide estimates of detection potentials for LCEC.

dure used will depend greatly on the particular sample and the ultimate analytical objective. For example, screening patient body fluids for highly elevated concentrations of biogenic amines or metabolites associated with tumors or severe stress can be a very simple matter, whereas determining the same substance in a healthy individual at rest may require more elaborate procedures.

Tissue Catecholamines

In spite of its complex chemical composition, tissue is a remarkably clean matrix for LCEC determinations, and publications using the technique in brain tissue are now commonplace. Recently, such diverse samples as eye, heart, lung, venous, and uterine samples have been assayed (17).

A common means of sample preparation for tissue CA is to employ a simple liquid-solid extraction onto alumina followed by back extraction into a small volume of acid (2, 6, 11, 19). The latter serves as the injectable extract.

Samples are generally homogenized or sonicated in 0.1 M $HClO_4$. The diluted acid inhibits the oxidation of CA and also denatures sample proteins. Antioxidants such as sodium metabisulfite or ascorbic acid may be added to the 0.1 M $HClO_4$ as a precautionary measure prior to storage, but these are not apparently necessary in many cases. Metabisulfite must be kept cold and be made fresh every 1 to 2 weeks because decomposition to SO_2 inevitably occurs.

In order to assay CA, samples are spiked with an appropriate amount of internal standard (usually the synthetic CA, 3,4-dihydroxybenzylamine). The internal standard elutes in the middle of the chromatogram between E and DA and provides a convenient reference peak for both sample extraction and injection precision. A reproducible amount of acid-washed aluminum oxide, usually 10 to 50 mg, is added to each sample vial. The pH is elevated by the rapid addition of concentrated Tris buffer, usually 1 to 3 M, pH 8.6, which also contains 1% Na_2EDTA. The vial is immediately capped and shaken for 5 minutes to facilitate adsorption. After shaking, the alumina is allowed to settle and the residual supernatant aspirated to waste. The complexed CA are washed twice with water and aspirated after each washing. At this stage, the CA content of the homogenate is quantitatively adsorbed onto a small mass of alumina; the sample's volume has been virtually eliminated. To provide an injectable extract, the CA are desorbed from the alumina using a small volume of dilute acid, usually 0.1 to 1.0 M $HClO_4$ (Fig. 3.5). Other acids such as phosphoric, sulfuric, hydrochloric, acetic, etc. can also be employed.

The procedure outlined above provides two advantages. First, it is fairly specific for catechols. Metabolites such as HVA or 3-methoxytyramine are not extracted. The chromatogram is highly simplified, and samples can be pushed through the system rapidly. Also, the procedure provides reduction of the original sample volume to the smallest volume consistent with good recovery, precision, and maximum amount injected. Most assays of this nature permit absolute recoveries of 60 to 80%, depending on the specific eluting acid, amount of alumina used, etc. Since the assay involves few manipulations and/or transfers, the precision in even the sub-ng/mL range is still very acceptable, usually 3 to 8% relative standard deviation. Detection limits on the order of 10 to 20 pg injected on-column are achievable. Such performance has permitted LCEC to be used for the mapping of individual brain parts and nuclei as small as 0.5 mg. One study demonstrated significant lateral concentration differences in the NE content of human thalamus by measuring concentrations in punchouts obtained from thin cross-sectional slices (14). A three-dimensional CA density plot could be reconstructed. Many research laboratories use the assay, with minor modifications, as an inexpensive or more reliable alternative to radioenzymatic or fluorescence procedures.

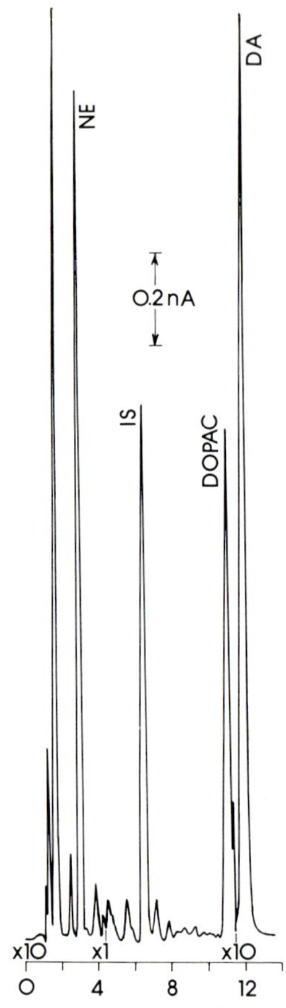

Figure 3.5. CA in rat brain after a simple alumina extraction. Concentrations in this tissue sample were: NE, 157 ng/gm; E, 2.4 ng/gm; DA 602 ng/gm. DOPAC was separated but not quantitated. Conditions: stationary phase, Biophase ODS 5μ (Bioanalytical Systems); mobile phase, 0.15 M monochloroacetate buffer (pH 3.0) with 2 mM Na_2EDTA and 28 mg/L sodium octylsulfate; flow rate, 2.2 mL/min; instrument: LC-304 with TL-3/CP-O electrode (Bioanalytical Systems); potential: +650 mV *vs.* Ag/AgCl.

The high selectivity of the LCEC analyzer permits even easier sample preparation under normal circumstances. One of the problems with alumina assays is their failure to retain important neurochemicals such as HVA, 5-HIAA, or serotonin (5-HT) in the final extract. If the initial tissue homogenate is sufficiently concentrated, 50 to 100 μL may be injected directly into the LCEC analyzer after filtration to remove particulate matter (8). Filtration is conveniently carried out using centrifugal microfilters (Bioanalytical Systems). An intermediate approach, which still permits measurement of 5-HT as well as CA, employs extraction of the tissue homogenate into butanol followed by back extraction into dilute HCl. Sasa and Blank (16) devised this technique to assay NE, DA, and 5-HT in whole brain and brain parts using pellicular cation exchange chromatography. Shea and co-workers (10) described an extraction scheme whereby tissue parts were homogenized in a formic acid/acetone mixture and centrifuged. The supernatant was then washed with heptane/chloroform and the aqueous portion dried down in a vacuum centrifuge. The residue was then reconstituted in 100 μL of mobile phase for injection into the LCEC analyzer. The separation took place on a microparticulate reverse phase column (μBondapak C_{18}, Waters Associates) using dilute pH 3.5 citric acid/phosphate buffers containing sodium octylsulfate and methanol. A glassy carbon electrode maintained at +800 mV (*vs.* Ag/AgCl) allowed the detection of NE, DA, 5-hydroxytryptophan (5-HTP), 5-HT, and 5-HIAA in brain tissue. The assay demonstrated that 5-HT and 5-HIAA in specific nuclei within the hypothalamus of food-deprived rats are differentially affected.

A comprehensive bibliography details many applications of LCEC in neurochemistry (17). Space limitations and the scope of this chapter do not permit more detailed discussion here.

Cerebrospinal Fluid

Most applications of LCEC in the analysis of cerebrospinal fluid (CSF) have dealt with monoamine metabolites such as 5-HIAA, HVA, MHPG, or 3,4-dihydroxyphenylacetic acid (DOPAC). 5-HT and tryptophan have also been determined in CSF (9). Since CSF is a relatively uncomplicated sample compared to brain tissue or urine, it is common practice to inject as much sample as possible directly into the LCEC analyzer. Wightman *et al.* (20) showed the practicality of this simple technique following the *in vivo* release of DA and 5-HT metabolites after stimulation by drug or electricity. Using repetitive sampling from an indwelling cannula, the time variation was easily followed in HVA release after discrete injections of its precursor, L-dopa. The method was extremely reliable owing to its simplicity.

Plasma Catecholamines

Like brain and CSF, plasma or serum is reasonably "clean" chemically, and sample preparation is often similar to that for brain tissue. The primary problem encountered with plasma CA is their very low endogenous concentrations in healthy individuals: NE usually runs between 100 and 300 pg/mL in resting humans, while E and DA fall below 50 pg/mL in most cases. Since sample size is usually limited to 2 mL or less in humans and much less in laboratory animals such as rats, the total CA content is often only a few picomoles or less. The goal of sample preparation is to obtain a simple extract of the smallest possible volume with highest absolute recovery, then inject as much of this extract as possible to maximize signal. In plasma, the concentrations of all three CA are sufficiently minute so that sources of chromatographic interference, such as improper reagents, collection tubes, sample vials, etc., must be carefully avoided.

The development of a successful LCEC plasma CA assay was first described by Hallman *et al.* (4) and later improved by Hjemdahl *et al.* (5). The sample preparation scheme resembled an earlier one for brain parts and punchouts. Deproteinization is apparently not necessary. Frozen samples thaw and remain on ice during preparation. Sequentially, the following reagents are added: (1) an internal standard, usually α-methyldopamine or 3,4-dihydroxybenzylamine, (2) antioxidants such as metabisulfite or dithiothreitol, (3) acid-washed aluminum oxide, and (4) concentrated Tris buffer, pH 8.6. Upon addition of Tris, the samples are shaken immediately and treated as before with two to three wash/aspirate cycles. For improved recovery, the CA may be centrifugally eluted by placing the alumina in disposable pipette tips or microfilters and then desorbing with a small volume of acid. Amounts as small as 50 to 100 μL of 0.1 M $HClO_4$ have provided useable results. Both laboratories (4, 5) employed cation exchange columns for liquid chromatography and carbon paste working electrodes for detection. In their study, Hjemdahl *et al.* (5) demonstrated excellent correlation between the radioenzymatic and LCEC methods in baboon plasma. Further correlative data will be described in a later section.

More assays for plasma CA have surfaced since these initial reports. It is now popular to employ microparticle reverse phase chromatography and ion pairing agents such as sodium octylsulfate to effect the separation. In most cases, the sample preparation procedure is the same as that described earlier for the cation exchange packing materials. Davis, Shoup, and Kissinger (1) outlined several strategies for separation on Biophase ODS 5-μm columns (250 mm × 4 mm, Bioanalytical Systems), with chromatographic efficiencies of approximately 40,000 plates/m in real samples. Mefford *et al.* (12) used a 0.1 M sodium acetate/0.01 M citric acid buffer containing 10% methanol, 100 mg/L sodium sulfate, and 50 mg/L disodium EDTA as mobile phase on a spherical 5-μm ODS stationary phase (Ultrasphere ODS, Altex Scientific). Goldstein *et al.* (3) (Figs. 3.6 and 3.7) described the use of irregularly shaped 10-μm ODS stationary phase (μBondapak C_{18}, Waters Associates) with a 0.05 M sodium acetate buffer containing 100 mg/L disodium EDTA, 1.0 gm/L heptanesulfonic acid, and 70 mL/L acetonitrile. As discussed earlier, the effect of the alkyl sulfate or sulfonates is to increase retention through a dynamic ion exchange process. The more concentrated the reagent in the mobile phase, the greater the capacity factors for CA, since the stationary phase contains a greater surface concentration of anionic sites. Conversely, solvents such as acetonitrile or methanol reduce capacity factors. These reductions are most prominent for strongly retained solutes such as DA. Thus, although these mobile phases may appear vastly different in terms of the solvent and alkyl sulfate concentrations, the balance between these two ingredients provides roughly equivalent separations on the most popular reverse phase columns for CA.

Not every reverse phase column will yield satisfactory results. Inadequate chromatographic efficiencies and insufficient retention of NE are two of the problems. Most separations of plasma CA are now provided by spherical 5-μm reverse phase particles. These are regarded as more mechanically stable and offer resolution superior to the irregularly shaped 10-μm bonded phases first introduced to the marketplace. Reverse phase columns

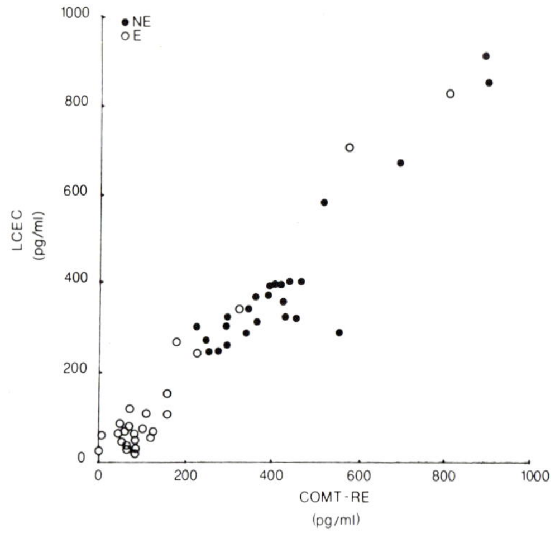

Figure 3.7. Correlation between the LCEC and radioenzymatic COMT procedure for catecholamines in human plasma (3). The *solid circles* are NE; *open circles* represent epinephrine.

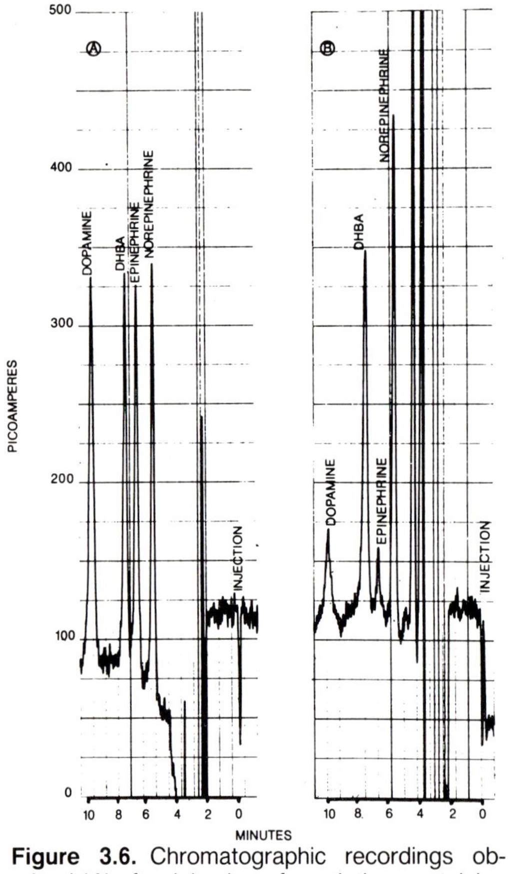

Figure 3.6. Chromatographic recordings obtained (*A*) after injection of a solution containing a mixture of 250 pg each of NE, E, 3,4-dihydroxybenzylamine, and DA; and (*B*) after injection of 50 μL of a plasma extract. Conditions: mobile phase, 93% 0.05 M sodium acetate buffer (pH 4.8) with 100 mg/L Na_2EDTA and 1 gm/L heptanesulfonic acid/7% acetonitrile; stationary phase, μBondapak C_{18} (Waters Associates); detector, LC-4/LC-17 detector (Bioanalytical Systems) with TL-5 glassy carbon electrode; potential, +0.72 V *vs.* Ag/AgCl; flow rate, 1.0 mL/min.

also differ strongly by the carbon content of the bonded packing material. Generally, the higher the amount of carbon provided (as alkyl C_{18} chains), the greater the retention of a particular CA under otherwise identical conditions. It is wise to select a stationary phase which provides the greatest possible retention. Then, as the column degrades with time, the mobile phase composition has sufficient latitude to compensate for reduced retention. High coverage, spherical stationary phases are recommended.

One problem with plasma CA samples is uric acid, present at concentrations approximately 10,000-fold higher than the biogenic amines. Although the alumina support is fairly specific for CA, a small portion of the uric acid is extracted, and it elutes as a major component in the chromatogram between the void volume response and NE. With some stationary phases the separation of uric acid and NE may be marginally acceptable, particularly as the column ages.

Davis *et al.* (1) described a two-stage extraction procedure for plasma CA based on cation exchange mini-columns for initial cleanup followed by alumina adsorption as before. The serum or plasma is deproteinized first and the pH then raised to 6.5 before sample application on the cation exchange resin. The columns are rinsed with water and then eluted with dilute acid, which is collected for the alumina step as before. The advantages of a preliminary ion exchange cleanup include a simplified chromatogram, reduced interference problems with uric acid and minor plasma components, and longer useable column lifetimes due to less stringent efficiency and retention requirements. The additional extraction stage lengthens the overall procedure approximately 30 minutes.

Urinary Catecholamines

Urinary CA concentrations are generally 1000-fold greater than plasma values. Sample size is also not limited. Therefore, the assay for urinary CA present no problems in maintaining adequate sensitivity. The complexity of the urinary sample, however, prevents simplistic preparation schemes (such as the one-step alumina extraction) from being practical. Although the detector and chromatograph have enough selectivity, the problem is chiefly a matter of sample throughput. Simple extracts are still complex. When they are chromatographed, the elution time for the last peak may approach a few hours. Since the CA elute in only 10 to 15 minutes, such time is wasted between sample injections. A step-flush of the system with a 50% CH_3CN/50% buffer would rinse the column efficiently, but the long time required for reequilibration would be counterproductive. The necessary additional sample clean-up may be accomplished in about the same time it would take one crude extract to elute. These measures may be done with up to 20 samples in parallel.

Two schemes have proved popular for the measurement of CA in urine. Riggin and Kissinger (15) described a two-step ion-exchange/aluminum oxide isolation procedure similar to the one described earlier for plasma (Fig. 3.8). The pH of thawed or freshly acidified samples was raised to pH 6.5 and then applied to weakly acidic cation-exchange mini-columns. The beds were rinsed with water to flush nonretained components, and the CA were then eluted with dilute H_2SO_4 and $(NH_4)_2SO_4$ solutions. This eluent was collected for the usual batch-type alumina adsorption maneuver described for plasma. Samples could be processed at the rate of 50 per day. The method was linear over the 1 to 1000 μg/L range and offered detection limits of 1 μg/mL for quantitation, well below the concentrations encountered in human urine specimens.

A more complicated two-stage method devised by Moyer *et al.* (13) relies on small alumina columns followed by a batch extraction onto a specially prepared celluloseamidophenylborate resin (13). Twenty-four-hour urine specimens are collected, the volumes noted, and a 4-mL aliquot assayed. The sample is washed with ethylacetate, the pH raised to 8.5, and the urine applied to a 7 × 80 mm column

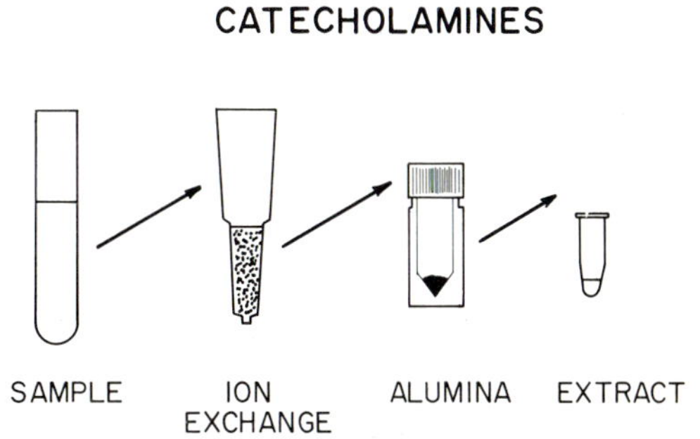

Figure 3.8. Schematic diagram for assay of urinary CA. A two-stage cleanup procedure involving miniature cation exchange followed by a batch of alumina purification of the eluant provides a satisfactory extract for LCEC analysis.

of alumina. After drainage, the column is rinsed twice with water and then eluted with 0.05 M acetic acid. The pH of the collected eluant is again raised to 8.7, and 1.0 mL of the celluloseamidophenylborate suspension is added. The CA adsorb onto the resin as borate complexes. The gel is washed by repeated wash/centrifuge/aspirate cycles. Finally, CA are eluted into 800 μL of boric acid. The latter solution is injected into the LCEC analyzer.

Both procedures provide useful results. Samples may be put through the system every 8 to 10 minutes and an internal standard conveniently incorporated into every sample. NE, E, and DA may be quantitated simultaneously. Riggin's method is particularly attractive due to its straightforward simplicity and use of inexpensive, readily available reagents. Both schemes have been used to profile CA in patients presenting with pheochromocytoma, neuroblastoma, hypertension, etc. (see Chapters 22 and 32).

CONCLUSION

In summary, LCEC can detect all the endogenous CA and several other neurotransmitters and their metabolites in samples of tissue, CSF urine, and blood. It is very versatile and can quantitate numerous substances in a single sample. LCEC is in wide use for assay of tissue samples, but under optimal conditions it can reliably measure plasma CA.

References

1. Davis GC, Kissinger PT, Shoup RE: Strategies for determination of serum or plasma norepinephrine by reverse-phase liquid chromatography. *Anal Chem* 53:156–159, 1981.
2. Felice LJ, Felice JD, Kissinger PT: Determination of catecholamines in rat brain parts by reverse-phase ion-pair liquid chromatography. *J Neurochem* 31:1461–1465, 1978.
3. Goldstein DS, Feuerstein G, Izzo JL Jr, Kopin IJ, Keiser HR: Validity and reliability of liquid chromatography with electrochemical detection for measuring plasma levels of norepinephrine and epinephrine in man. *Life Sci* 28:467–476, 1981.
4. Hallman H, Farnebo LO, Hamberger B, Jonsson G: A selective method for the determination of plasma catecholamines using liquid chromatography with electrochemical detection. *Life Sci* 23:1049–1052, 1978.
5. Hjemdahl P, Daleskog M, Kahan T: Determination of plasma catecholamines by high performance liquid chromatography with electrochemical detection: Comparison with a radioenzymatic method. *Life Sci* 25:131–138, 1979.
6. Keller R, Oke A, Mefford I, Adams RN: Liquid chromatographic analysis of catecholamines. Routine assay for regional brain mapping. *Life Sci* 19:995–1004, 1976.
7. Kissinger PT: Reverse-phase ion-pair partition chromatography. *Anal Chem* 49:883, 1977.
8. Koch DD: Analytical approaches to the determination of tryptophan and its metabolites in physiological media. Ph.D. Thesis, Purdue University, W. Lafayette, Ind, 1979.
9. Langlais PJ, McEntee WJ, Bird ED: Rapid liquid-chromatographic measurement of 3-methoxy-4-hydroxyphenylglycol and other monoamine metabolites in human cerebrospinal fluid. *Clin Chem* 26:786–788, 1980.
10. Loullis CC, Felten DL, Shea PA: HPLC determination of biogenic amines in discrete brain areas in food deprived rats. *Pharmacol Biochem Behav* 11:89–93, 1979.
11. Mefford IN, Gilberg M, Barchas JD: Simultaneous determination of catecholamines and unconjugated 3,4-dihydroxyphenylacetic acid in brain tissue by ion pairing reverse phase high performance liquid chromatography with electrochemical detection. *Anal Biochem* 104:469, 1980.
12. Mefford IN, Ward MM, Miles L, Taylor B, Chesney MA, Keegan DL, Barchas JD: Determination of plasma catecholamines and free 3,4-dihydroxyphenylacetic acid in continuously collected human plasma by high performance liquid chromatography with electrochemical detection. *Life Sci* 28:477–484, 1981.
13. Moyer TP, Jiang NS, Tyce GM, Sheps SG: Analysis for urinary catecholamines by liquid chromatography with amperometric detection: Methodology and clinical interpretation of results. *Clin Chem* 25:256–263, 1979.
14. Oke A, Keller R, Mefford I, Adams RN: Lateralization of norepinephrine in human thalamus. *Science* 200:1411–1413, 1978.
15. Riggin RM, Kissinger PT: Determination of catecholamines in urine by ion-pair reverse-phase liquid chromatography with electrochemical detection. *Anal Chem* 49:2109–2111, 1977.
16. Sasa S, Blank CL: Simultaneous determination of norepinephrine, dopamine, and serotonin in brain tissue by high-pressure liquid chromatography with electrochemical detection. *Anal Chem Acta* 104:29–45, 1979.
17. Shoup RE: *Bibliography of Recent Reports on Electrochemical Detection*. West Lafayette, Ind, Bioanalytical Systems Inc., 1981.
18. Sternson AW, McCreery R, Feinberg B, Adams RN: Electrochemical studies of adrenergic neurotransmitters and related compounds. *J Electroanalyt Chem Interfacial Electrochem* 46:313–321, 1973.
19. Wagner J, Palfreyman M, Zraika M: Determination of dopa, dopamine, dopac, epinephrine, norepinephrine, α-monofluoromethyldopa and α-difluoromethyldopa in various tissues of mice and rats using reversed-phase ion pair liquid chromatography with electrochemical detection. *J Chromatogr* 164:41–54, 1979.
20. Wightman RM, Plotsky PM, Strope E, Delcore RJ, Adams RN: Liquid chromotographic monitoring of CSF metabolites. *Brain Res* 131:345–349, 1977.

SECTION 2

Anatomy

CHAPTER FOUR

The Structure of the Peripheral Adrenergic Synapse and Its Functional Implications

WYATT R. HUME, D.D.S., Ph.D., and JOHN A. BEVAN, M.D.

INTRODUCTION

The sympathetic nervous system (SNS) contributes substantially to the control of the body's internal environment. In general, nerve activity in this system causes rapid change in effector cells through neuronal release of norepinephrine (NE). However, the density of innervation varies considerably, as does the structure of synapses and the nature and density of postsynaptic receptors, even within a single organ system of a particular species. This chapter describes the structure of the final synapse of the sympathetic motor pathway, the variations which occur in this structure, and their functional consequences.

Although details of the relative activities of different branches in the SNS are difficult to elucidate, motor control of the blood vascular tree is likely to form a major part of the system, functionally as well as anatomically. The synapse between sympathetic postganglionic nerves and the smooth muscle of the blood vessel wall has been intensively studied. Because of the proximity of this synapse to circulating blood, it is a potential source of plasma catecholamines (CA). For these reasons, this chapter will focus primarily on the adrenergic synapse in blood vessels and will then describe this synapse in other systems.

THE INNERVATION PATTERN OF BLOOD VESSELS

Nonmyelinated, postganglionic sympathetic nerves form the predominant innervation of the vascular system (9). Generally, the nerves form two plexuses in the tunica adventitia of larger vessels: an outer layer of fiber bundles, arranged longitudinally, which supply more distal parts of the vessel and its branches, and an inner network, the terminal plexus, usually confined to the adventitia near the junction with the medial layer. The distribution patterns of CA-containing nerves may be determined using the fluorescent histochemical technique of Falck *et al.* (16) which converts CA to compounds which fluoresce under ultraviolet light. Figure 4.1 shows the sympathetic nerve distribution in a typical medium-sized muscular artery, the central artery of the rabbit ear. Examination of different vessel types in the blood vascular system in several species shows that the terminal plexus is most commonly confined to the external surface of the medial layer. Intramedial innervation at

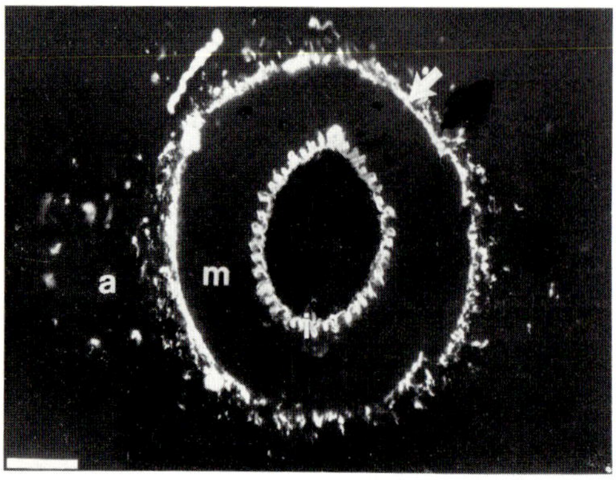

Figure 4.1. Transverse section of rabbit ear artery treated with the technique of Falck *et al.* (16) for the demonstration of catecholamine fluorescence. Illuminated with ultraviolet light, and examined through a 470 mμ barrier filter. Scale, 100 μm. *a*, adventitia; *m*, media; *i*, intima. *Arrow*, Medial-adventitial border. The fluorescent structures in the adventitia are catecholamine-containing nerves; the tunica intima contains autofluorescent elastin.

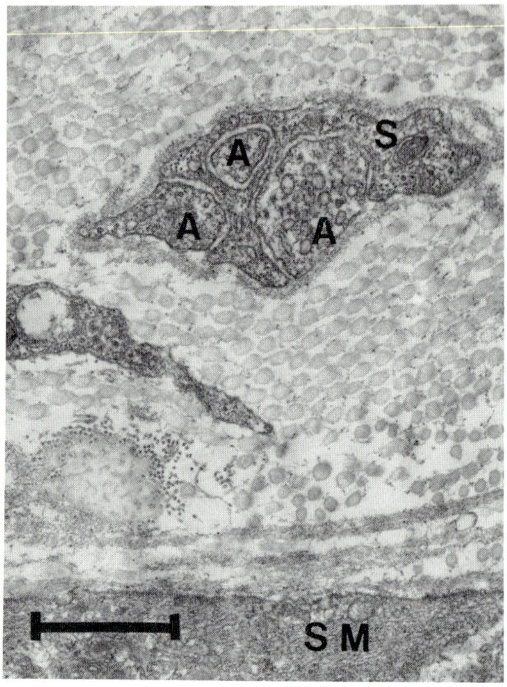

Figure 4.2. Electron micrograph of a thin section of rabbit ear artery in the region of the medial-adventitial border. Longitudinal section. *A*, axon; *S*, Schwann cell; *SM*, smooth muscle. Scale, 1,000 nm (approximate). (Reproduced with permission from W. R. Hume and J. G. Waterson.)

various depths, however, has been noted in several medium-sized and large arteries and veins (4).

COMPONENTS AND DIMENSIONS OF THE ADRENERGIC SYNAPSE

Electron microscopic examination of ultrathin sections reveals detail of the axons in the terminal region and of the synaptic cleft. Again, the rabbit ear artery has typical features (25) (Fig. 4.2); the ultrastructure of the terminal axons appears to be similar in both vascular and nonvascular tissues (35). The terminal plexus is composed of nonmyelinated axons, 150 to 200 nm in diameter, which are almost always accompanied by a Schwann cell sheath. The axons are varicose, with swellings to about 500 nm in diameter at intervals of about 10,000 nm. The varicosities, or nodes, contain storage vesicles and mitochondria. Most of the vesicles are small (35 to 55 nm in diameter), with a dense core (small granular), and are presumed to contain NE. Small agranular vesicles and large (up to 90 nm diameter) vesicles are also present.

The appearance of the vesicles in the axon terminal region varies with the method of preparation of the tissue, and in particular with the fixative used. Visualization of the dense core of small vesicles is good with permaganate fixation (22) and also with formaldehyde-glutaraldehyde-acrolein (25). Treatment of tissues with 5-hydroxydopamine (5-OHDA) before fixation enhances the granularity of the vesicles. Entry of this osmiophilic compound into the vesicles indicates that they are sites of CA storage (42). Vesicle granularity decreases following treatment with reserpine, and reappears following NE infusion (35).

Nerve axons in the terminal region may be totally surrounded by the Schwann cell, but commonly a bare area of the varicosity is seemingly directed towards the adjacent smooth muscle cell (4). No consistent evidence exists of postsynaptic structural specialization in blood vessels. The synaptic cleft is wide compared to synapses in general and tends to differ in various blood vessels. Nerve-to-muscle distances vary from 4,000 nm in the dog pulmonary artery (17) to 60 nm or less in some small vessels (38) (Table 4.1). Generally, the synaptic cleft is narrower in smaller vessels. The overall impression from the literature is

Table 4.1
Minimum Neuromuscular Cleft Dimensions in Representative Blood Vessels[a]

Vessel	Cleft Width (nm)	Source
Pulmonary artery, dog	4,000	Fillenz (17)
Aorta, rabbit	2,000	Bevan (3)
Pulmonary artery, rabbit	1,900	Bevan (3)
Ear artery, rabbit	500	Hume and Waterson (25)
Mesenteric artery, rat	500	Devine and Simpson (13)
Uterine artery, guinea pig	200	Bell and Vogt (2)
Portal vein, rat	100	Booz (8)
Arterioles, precapillary sphincters	60	Rhodin (38)

[a] The method of estimating minimum cleft width varied considerably between the above studies.

that in larger vessels, the adrenergic terminal plexus forms either unspecialized or loosely specialized synapses. Although the concept of a "synaptic cleft" is useful in the following discussions, the great distance between the nerve varicosity, and even the closer smooth muscle cells in such vessels, suggests a system of regional transmitter release, rather than a single nerve to single effector cell relationship. There appears to be no integration of pre- and postsynaptic structures into a single unit, as occurs in the skeletal neuromuscular junction. In vessels with smaller cleft dimensions, a single nerve-to-muscle cell relationship may be imagined, but morphological evidence of postsynaptic specialization is lacking.

NEUROTRANSMITTER RELEASE AND RECEPTOR ACTIVATION

Investigators generally agree that the probable mechanism for NE release from the synaptic nerve varicosity involves arrival of a nerve action potential, influx of Ca^{2+}, and then exocytosis of some of the vesicles' contents (40). Strong evidence exists for calcium dependence in the release process (24, 29). The simultaneous and directly proportional release of soluble, but not membrane-bound, dopamine-β-hydroxylase (DBH) with NE argues strongly in favor of expulsion of the vesicle contents, but not the vesicle membrane, during release (39). Some data seem to suggest that each varicosity does not release a vesicle on the arrival of every action potential, and that the ratio of action potential trains to vesicles released per varicosity may be of the order 30:1 (10). It may be that the necessary preconditions for vesicle release include close contact between a vesicle and the axon membrane, an event which occurs only a small fraction of the time.

Postsynaptic receptors, either α- or β-adrenergic, lead to mechanical responses in blood vessels, which, by analogy with the effects of exogenously applied NE, are presumed to be concentration-dependent within a certain range.

An inhibitory feedback modulation of NE release via presynaptic α_2-like receptors in sympathetic nerves has been well-documented (9, 18, 32). It is reasonable to postulate that, if anatomical situations exist where the degree or rate of diffusion of NE away from the nerve varies (see text under Consequences of Variation in Synaptic Cleft Width), then the degree of activation of the inhibitory receptors might also vary.

THE FATE OF RELEASED NEUROTRANSMITTER

Released NE leaves the synaptic cleft by at least four mechanisms or pathways (Fig. 4.3).

1. Reuptake into the sympathetic nerve terminal (uptake$_1$, neuronal uptake) appears to be a major mechanism of NE clearance from the cleft region (26, 34). This uptake depends on active cellular processes and is stereospecific, sodium ion-dependent, and temperature-

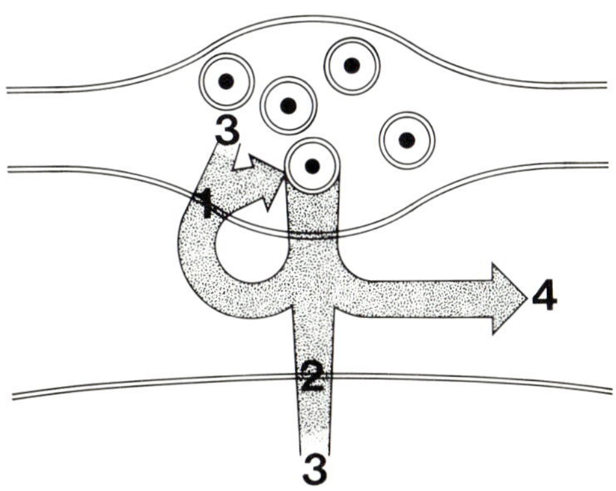

Figure. 4.3 Local disposition pathways of neurotransmitter in the adrenergic synaptic cleft. *1*, reuptake into the nerve terminal; *2*, uptake into smooth muscle; *3*, local metabolism; *4*, overflow.

sensitive. It is blocked by cocaine, imipramine, phenoxybenzamine, and other compounds. The degree of neuronal uptake of exogenous NE varies with NE concentration, time of exposure, and width of the synaptic cleft (3). Blockers of neuronal uptake potentiate the response of blood vessels to endogenous NE (11) and increase the nerve stimulus-induced overflow of [^{3}H]-labeled neuronal NE from isolated blood vessels. The degree of these increases is less than expected from data on the effects on exogenous NE, probably because when $uptake_1$ is blocked, more negative feedback inhibition of release occurs, due to an increase in NE concentration near the neuronal membrane.

2. Norepinephrine is actively taken up by vascular smooth muscle cells (1, 41) and also binds to extracellular components of the blood vessel wall (36). The former process, called $uptake_2$, is temperature-sensitive and partially sodium-dependent, though not stereospecific (4, 21). The latter is unaltered by factors which influence active cellular processes (36). $Uptake_2$ may be blocked with some steroids (12) and α-adrenergic blocking agents (21). Such blockade potentiates arterial responses to NE, indicating that $uptake_2$ is significant in the removal of NE from the region of the postsynaptic receptors (12, 14). There is evidence however, that $uptake_2$ has only a minor effect on the clearance rate of NE from the circulation (15).

3. Local inactivation of NE by degradation is catalyzed by the enzymes catechol-O-methyltransferase (COMT) and monoamine oxidase (MAO) (27, 28), although other inactivation pathways may exist (30). Both MAO and COMT are found within vascular smooth muscle, and a small but functionally important proportion of MAO is found within the nerve endings (11). O-Methylation, following $uptake_2$, is thought to be the more important route of metabolic deactivation of NE.

4. Diffusion of NE away from the synapse, or overflow, can contribute substantially to NE removal from the synapse. Up to 50% of NE released from the adrenergic plexus in some isolated blood vessels emerges intact into the surrounding medium (4). This extracellular NE has the potential to move into the blood circulation *in vivo*, either via the medial layer to the luminal blood, or via the adventitia to the periadventitial capillary plexus. The NE, however, is subject to uptake and degradation as it moves across the vascular wall (see below).

CONSEQUENCES OF VARIATION IN SYNAPTIC CLEFT WIDTH

Upon release, NE diffuses radially in all directions in the extracellular space until anatomically confined. In general, the concentration of NE developing in the nerve-muscle cleft varies with the cleft width. It is intuitively apparent that the degree of feedback inhibition of release, the extent of receptor activation, and the balance between the various disposition pathways may all be influenced by the width of the synaptic cleft. Having left the cleft, NE moves toward the vessel surfaces. The diffusion coefficient for NE in the adventitia (9×10^{-6} cm^2/sec) resembles that for free diffusion (3), while this value for the media of rabbit aorta is 7.3×10^{-7} cm^2/sec (6), and for the ear artery where there are no elastic laminae, it is 1.4×10^{-6} cm^2/sec (7). Thus, the cells of the medial layer are a barrier to diffusion.

A theoretical construct of these factors is presented below with some of the available evidence. It should be noted that predictions are based on mechanisms as they are presently understood, and that these may have to be modified as further information comes to hand. For example, the theoretical predictions of Bevan and Su (5) were made without the benefit of data on α-adrenergic-mediated inhibitory feedback mechanisms in release, and therefore required reevaluation. It is to be expected that this process of refinement will continue.

We assume that the diffusion coefficients for NE in the extracellular space, the avidity of each uptake process, and the sensitivity of the receptor-effector complex are constant, irrespective of cleft width. These assumptions are not necessarily valid, but there are presently insufficient data to state which variations exist, and if they do exist, their magnitude. The assumption is also made that nerve action is continuous, *i.e.*, that a steady state of neurotransmitter release and disposition occurs. For the individual synapse, it is very unlikely that this steady state assumption is valid, since release is episodic. However, it is useful to consider the integrated actions of the large family of synapses, *i.e.*, the terminal plexus, of the whole tissue as achieving an effective steady state, which results in continuous transmitter release and a maintained effector cell response.

For the narrower synaptic cleft, egress of NE by diffusion is thought to be limited by the smooth muscle cells on one side and by the neural complex on the other. Comparison of responses to nerve stimulation with responses to exogenous NE during partial α-adrenoreceptor blockade (to eliminate the possible response to lower concentrations of NE away from the synapse) allows estimates to be made of the intrasynaptic NE concentration during nerve action. In the narrow-cleft rat portal vein (100 nm), intrasynaptic NE was 1.4×10^{-5} M, while in the broader-cleft rabbit ear artery (500 nm) and pulmonary artery (1900 nm), the NE concentrations were 5.6×10^{-7} M and 1.5×10^{-7} M, respectively (5). The trend of these data is of fundamental importance to the concepts which follow.

In the narrower cleft, we might expect a more rapid development of endogenous NE concentration during release, and therefore a greater degree of negative feedback modulation. Bevan compared data from various vascular and nonvascular tissues where synaptic cleft width varied from 20 to 2,000 nm (3). Blockade of α-adrenoreceptors caused a much greater increase in transmitter overflow in the narrower cleft tissues. This indicated more active feedback in narrower clefts. With 20-nm cleft width, the increase in overflow was 35-fold (23) and with 20,000 nm, 3-fold (33). This modulation buffers the consequences of variation in synaptic cleft on NE concentration near the neuronal membrane; however, less total NE release is necessary to achieve that level.

As transmitter concentration varies inversely with distance from the NE source, the concentration of NE reaching the postsynaptic receptor will be higher in the narrower cleft. If saturation levels are not exceeded, we would expect a greater degree of uptake$_2$ and metabolic degradation by COMT in the narrower cleft, since these processes obey Michaelis-Menten kinetics. The influence of uptake$_1$ on transmitter disposition would also be proportionally greater in such a cleft. The local NE concentration need not be higher, but the physical confinement of NE would make a greater fraction of the released transmitter available for uptake. Tissues with a broad cleft take up much more exogenous [^{3}H]-NE per unit amount of nerve than those with a narrow junction (3), which suggests that local cell architecture limits the rate of entry into the cleft region. This is consistent with the suggestion that endogenously released NE would be subject to a greater degree of reuptake in a narrow cleft arrangement, because the same limits to diffusion apply in the opposite direction. The net result of greater efficiency in these uptake processes is that less intact NE would be lost from the narrow cleft by diffusion or overflow away from the synapse region. The narrower cleft might therefore be considered a more efficient mechanism for information transfer in terms of NE utilization, and one less likely to contribute to plasma NE levels.

Thus, in small vessels, little endogenous NE would be expected to reach the blood during normal function, due to the more efficient, tighter cleft. In larger vessels, more NE would escape from local neuronal disposal mechanisms and diffuse, on one hand, towards the vessel lumen. NE would be subject to extraneuronal degradation during this passage across the media. On the other hand, it could readily reach the circulation via adventitial and periadventitial capillaries (Fig. 4.4). Lake, Ziegler, and Kopin (31) noted higher levels of NE in the venous outflow than in the arterial supply of the arm of human subjects at rest, indicating that NE from sympathetic nerve sources does reach the circulation under such conditions.

During intense sympathetic activity, a different pattern of disposition would be expected to occur. It has been well established in *in vitro* studies of isolated blood vessels in which [^{3}H] NE has been incorporated in neuronal stores, that overflow of NE into the surrounding medium increases markedly with increased stimulus frequency (3, 4). The same is

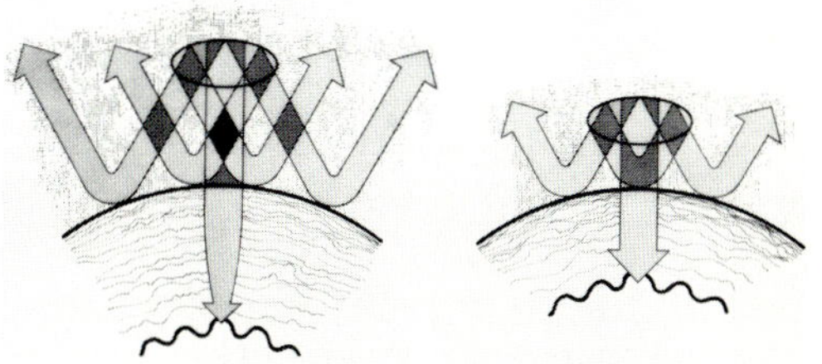

Figure 4.4. Diagrammatic comparison of the loss of NE to periadventitial capillaries and to the arterial lumen between the broad and narrow synaptic clefts of arteries of different size.

true of the whole animal. During sustained stimulation of the entire sympathetic outflow at 3 Hz in pithed rats, plasma NE rose about 20-fold (44). Surgical removal of the adrenal medulla had only a slight effect on the increase, which was abolished by bretylium, indicating that this rise was due to overflow from sympathetic nerve endings. A less dramatic increase in NE, about 5-fold, was seen in coronary sinus blood of anesthetized dogs during maximal stimulation of the cardio-accelerator nerve (45). In the heart, the sympathetic cleft width is relatively narrow—of the order of 10 to 30 nm—which may account for the quantitative difference between the heart and whole rat increase (9).

In local muscle function, the concentration of NE within the narrow cleft probably means that perisynaptic NE has little biological activity, and that excitation would be restricted to the cell subjacent to the nerve terminal. It seems likely that under these circumstances, subsequent rapid spread of excitation by direct cell-to-cell communication may occur. In contrast, in larger clefts more NE will diffuse to surrounding cells and result in a slow, but direct, transmitter control of a larger proportion of cells.

In a study comparing the rate of change in diameter of the femoral artery and resistance vessels of anesthetized dogs, Gero and Gerova (20) confirmed these suggestions. The smaller vessels changed diameter much more rapidly following electrical stimulation of the sympathetic trunk. The responses of the rabbit ear artery to sympathetic stimulation are consistent with a combination of the two mechanisms (7), as might be expected for a vessel of intermediate cleft width. When considering the rapidity of change in the narrow cleft, the more efficient removal of the transmitter from the region by the (neuronal) uptake$_1$ process will bring about a more rapid termination of its effect. Small vessels, therefore, appear to be better equipped for rapid adjustments in vascular resistance, while larger vessels seem more suited to slower and more persistent changes.

THE NORADRENERGIC SYNAPSE IN NONVASCULAR TISSUES

The adrenergic innervation of the vas deferens typifies the morphology of nonvascular tissue and is well documented (9, 19, 37, 43). The nature of the nerve axons, their varicosities, and the vesicular contents appear similar to those in blood vessels. The dimension of the neuromuscular cleft and the distribution of the terminal axons are the principal differences. The cleft is in the order of 10 to 30 nm in width, less than in even the smallest blood vessels. Nerve terminals commonly lie in grooves on the surface of the smooth muscle cells, and have been observed completely enfolded (19, 46). The terminal axons are dispersed relatively homogenously throughout the muscle layer.

With few exceptions, other nonvascular tissues show patterns of innervation similar to the vas deferens. In the pupil, nictitating membrane, and salivary glands, the neuromuscular gap is also in the order of 10 to 30 nm; in gut and respiratory muscle, broader clefts have been noted (9). One might expect, in light of the considerations detailed above, that close approximation between nerve endings and effector cells would be reflected in a generally close and rapid information transfer, with relatively little loss of neurotransmitter to the surrounding tissues or blood by diffusion.

SUMMARY

With the morphological arrangement of sympathetic nerve endings in blood vessels, nerve activity has the potential to contribute substantially to CA levels in the blood. The factors considered in this chapter demonstrate that the broader synaptic cleft in large vessels is the more likely source of NE in the circulation, and that the tight clefts of nonvascular sympathetic effector systems are less probable sites of origin. Direct evidence of NE movement into blood from nonadrenal, sympathetically innervated sources has been presented (44). The intriguing possibility that changes in vascular architecture which occur in hypertension (4) might alter the movement of NE into the circulation has yet to be explored.

References

1. Avakian V, Gillespie JS: Uptake of noradrenaline by adrenergic nerves, smooth muscle and connective tissue in isolated perfused arteries and its correlation with the vasoconstrictor response. *Br J Pharmacol* 32:168–184, 1968.
2. Bell C, Vogt M: Release of endogenous noradrenaline from an isolated muscular artery. *J Physiol* (London) 215:509–520, 1971.
3. Bevan JA: Some functional consequences of variation in adrenergic synaptic cleft width and in

nerve density and distribution. *Fed Proc* 36:2439–2443, 1977.

4. Bevan JA, Bevan RD, Duckles SP: Adrenergic regulation of vascular smooth muscle. In Bohn D, Somlyo A, Sparks HV: *The Cardiovascular System, Vascular Smooth Muscle. Handbook of Physiology*. Washington DC, American Physiological Society, 1980, sect 2, vol 2, pp 515–566.
5. Bevan JA, Su C: Variation of intra- and perisynaptic adrenergic transmitter concentrations with width of synaptic cleft in vascular tissue. *J Pharmacol Exp Ther* 190:30–38, 1974.
6. Bevan JA, Torok J: Movement of norepinephrine through the media of rabbit aorta. *Circ Res* 27:325–331, 1970.
7. Bevan JA, Waterson JG: Biphasic constrictor response of the rabbit ear artery. *Circ Res* 28:655–661, 1971.
8. Booz KH: Xur innervation der autonom pulsierendon Vena portae der weissen Ratte. Eine hisochishe und electronenmicroskopishe Untersuchung. *Z Zellforsch Mikroskol Anat* 117:294–418, 1971.
9. Burnstock G, Costa M: *Adrenergic Neurons: Their Organization, Function and Development in the Peripheral Nervous System*. New York, Wiley, 1975.
10. Dahlstrom A, Haggendal J, Hokfelt T: The noradrenaline content of the varicosities of sympathetic adrenergic nerve terminals in the rat. *Acta Physiol Scand* 67:289–294, 1966.
11. De la Lande IS, Hodge RL, Lazner MA, Jellett LB, Waterson JG: Pharmacological implications of the fate of noradrenaline in the artery wall. *Circ Res* 26/27(Suppl 2):41–48, 1970.
12. De la Land IS, Johnson SM: Action of deoxycorticosterone acetate (DOCA) on the rabbit ear artery. *Proc Aust Physiol Pharmacol Soc* 4:149–150, 1973.
13. Devine CE, Simpson FO: The fine structure of vascular sympathetic neuromuscular contacts in the rat. *Am J Anat* 121:153–174, 1967.
14. Eckert E, Henseling M: The effect of corticosterone on extraneuronal uptake in, and efflux of, $^{3}H(\pm)$-noradrenaline (NA) from rabbit aortic strips. *Naunyn Schmiedebergs Arch Pharmacol* 277(Supp):R14, 1973.
15. Esler M, Jackman G, Bobik A, Leonard P, Kelleher D, Skews H, Jennings G, Korner P: Norepinephrine kinetics in essential hypertension. *Hypertension* 3:149–156, 1981.
16. Falck B, Hillarp NA, Theime G, Torp A. Fluorescence of catecholamines and related compounds with formaldehyde. *J Histochem Cytochem* 10:348–354, 1962.
17. Fillenz M: Innervation of pulmonary and bronchial blood vessels of the dog. *J Anat* 106:449–461, 1970.
18. Fleisch JH: Further studies on the effect of aging on alpha-adrenoceptor activity of rat aorta. *Br J Pharmacol* 42:311–313, 1971.
19. Furness JA, Iwayama T: The arrangement and identification of axons innervating the vas deferens of the guinea pig. *J Anat* 113:179–196, 1972.
20. Gero J, Gerova M: *In vivo* studies of sympathetic control of vessels of different function. In Bevan JA, Furchgott RF, Maxwell RA, Somylo AP: *Physiology and Pharmacology of Vascular Neuroeffector Systems*. Basel, Karger, 1971, pp 86–94.
21. Gillespie JS: The uptake of noradrenaline by smooth muscle. *Br Med Bull* 29:136–141, 1973.
22. Hökfelt T, Jonsson G: Studies on reaction and binding of monoamines after fixation and processing for electron microscopy with potassium permanganate. *Histochemie* 16:45–67, 1968.
23. Hughes J: Evaluation of mechanisms controlling the release and inactivation of the adrenergic transmitter in the rabbit portal vein and vas deferens. *Br J Pharmacol* 44:472–491, 1972.
24. Hukovic S, Muscholl E: Die Noradrenalin-Abgabe aus dem isolierten Kaninchenherzen bei sympathischer Nervenreizung und ihre pharmakologische Beeinflussung. *Naunyn Schmiedebergs Arch Exp Pathol Pharmakol* 244:81–96, 1962.
25. Hume WR, Waterson JG: The innervation of the rabbit ear artery. An ultrastructural study. *Blood Vessels* 15:348–364, 1978.
26. Iversen LL: *The Uptake and Storage of Noradrenaline in Sympathetic Nerves*. Cambridge University Press, London, 1967, pp 169–171.
27. Jarrott B: Occurrence and properties of monamine oxidase in adrenergic neurons. *J Neurochem* 18:7–16, 1971.
28. Jarrott B: Occurrence and properties of catechol-O-methyltransferase in adrenergic neurons. *J Neurochem* 18:17–27, 1971.
29. Kirpekar SM, Misu Y: Release of noradrenaline by splenic nerve stimulation and its dependence on calcium. *J Physiol* (London) 188:219–234, 1967.
30. Kuchel O, Buu NT, Hamet P, Larochelle P, Genest J: Conjugation of catecholamines, a variable to be accounted for in human hypertension. *Bund Progn Biochem Pharmacol*, in press, 1983.
31. Lake CR, Ziegler MG, Kopin IJ: Use of plasma norepinephrine for evaluation of sympathetic neuronal function in man. *Life Sci* 18:1215–1226, 1976.
32. Langer SZ, Adler E, Enero MA, Stefano JE: The role of the alpha receptor in regulating noradrenaline overflow by nerve stimulation. *25th Proceedings of the International Congress of Physiological Science*. Munich, German Physiological Society, 1971, vol 9, p 335.
33. McCulloch MW, Bevan JA, Su C: Effects of phenoxybenzamine and norepinephrine on transmitter release in the pulmonary artery of the rabbit. *Blood Vessels* 12:122–133, 1975.
34. Paton DM: Characteristics of uptake of noradrenaline by adrenergic neurons. In Paton DM: *The Mechanism of Neuronal and Extraneuronal Transport of Catecholamines*. New York, Raven Press, 1976, pp 49–66.
35. Pellegrine de Iraldi A, Saburo AM: Two compartments in the granulated vesicles of the pineal nerves. In Wolstenholme GEW, Knight J: *The Pineal Gland*. London, Churchill, 1971, pp 177–195.
36. Powis G: Binding of catecholamines to connec-

tive tissue and the effect upon the responses of blood vessels to noradrenaline and to nerve stimulation. *J Physiol* (London) 234:145–162, 1973.

37. Richardson KC: The fine structure of autonomic nerve endings in smooth muscles of the rat vas deferens. *J Anat* 96:427–442, 1962.
38. Rhodin JAG: The ultrastructure of mammalian arterioles and precapillary sphincters. *J Ultastruct Res* 18:181–223, 1967.
39. Smith AD: Mechanisms involved in the release of noradrenaline from sympathetic nerves. *Br Med Bull* 29:123–129, 1973.
40. Smith AD, Winkler H: Fundamental mechanisms in the release of catecholamines. In Blaschko H, Muscholl E: *Catecholamines. Handbook of Experimental Pharmacology*. Berlin, Springer-Verlag, 1972, vol 33, pp 538–617.
41. Somlyo AP, Somlyo AV: Vascular smooth muscle. I. Normal structure, pathology, biochemistry and biophysics. *Pharmacol Rev* 20:197–272, 1968.
42. Tranzer JP, Thoenen H: Electronmicroscopic localization of 5-hydroxydopamine (3,4,5-trihydroxy-phenylethylamine), a new "false" sympathetic transmitter. *Experientia* 23:743–745, 1967.
43. Wanatabe H: Electron microscopic observations on the innervation of smooth muscle in the guinea pig vas deferens. *Am J Anat* 130:305–329, 1969.
44. Yamaguchi I, Kopin IJ: Plasma catecholamine and blood pressure responses to sympathetic stimulation in pithed rats. *Am J Physiol* 237:H-305-H310, 1979.
45. Yamaguchi N, de Champlain J, Nadeau RA: Regulation of norepinephrine release from cardiac sympathetic fibers in the dog by presynaptic alpha- and beta-receptors. *Circ Res* 41:108–117, 1977.
46. Yamauchi A, Burnstock G: Postnatal development of the innervation of the mouse vas deferens. A fine structural study. *J Anat* 104:17–32, 1969.

CHAPTER FIVE

Organization of Descending Noradrenergic Systems

KARIN N. WESTLUND, Ph.D., ROBERT M. BOWKER, V.M.D., Ph.D., MICHAEL G. ZIEGLER, M.D., and JOE DAN COULTER, Ph.D.*

INTRODUCTION

Functional and pharmacological studies have shown that noradrenergic pathways have an important role in mediating descending influences on spinal cord sensory, motor, and autonomic activity. However, until recently, little was known of the precise origins and the anatomical organization of the noradrenergic innervation of the spinal cord.

Early studies to localize central adrenergic neurons in the brainstem became possible with the development of histochemical fluorescence techniques for the demonstration of cellular catecholamines (4, 10, 16, 25, 28, 37). Dahlström and Fuxe (21) identified fluorescent catecholamine neurons as comprising a number of distinct cell groups. From caudal to rostral in the brainstem, cell groups were termed A_1-A_{10}. Later improvements in identifying noradrenergic and adrenergic neurons, as distinct from other catecholamine-containing cells, came with the use of labeled antibodies directed against endogenous dopamine-β-hydroxylase (DBH), the enzyme that converts dopamine to norepinephrine (33, 34, 35, 77), and phenylethanolamine-N-methyltransferase (PNMT), the enzyme converting norepinephrine to epinephrine (38, 65). Subsequently, using either histofluorescence techniques or immunocytochemical methods, fairly complete descriptions of the major noradrenergic cell groups and the distribution of noradrenergic terminals in the lower brainstem and spinal cord have been obtained for a number of species including the rat, cat, dog, monkey, and human (12, 19, 22, 26, 27, 29, 30, 43, 55, 57, 58, 77, 83).

More recent studies have begun to address the questions of which noradrenergic cell groups contribute descending spinal projections to the spinal cord and to what extent differences exist in the spinal terminations of these different systems.

SPINAL NORADRENERGIC PATHWAYS

Tracing Descending Noradrenergic Pathways

In defining the precise origins and terminations of descending noradrenergic pathways,

---

Abbreviations used in Chapter 5 are: bc, brachium conjunctivum; bp, brachium pontis; C, central canal; CG, central grey; IML, intermediolateral cell column; GC, nucleus gigantocellularis; KF, nucleus of Kölliker-Fuse; LPB, lateral parabrachial nucleus; ml, medial lemniscus; mlf, medial longitudinal fasciculus; MPB, medial parabrachial nucleus; MoV, motor nucleus of the trigeminal; py, pyramid; SC, subcoeruleus nucleus; SO, superior olivary nucleus; Vl, lateral vestibular nucleus; Vm, medial vestibular nucleus; V, trigeminal nucleus; VII, facial nucleus.

* We wish to thank Merry Sullivan for technical assistance and Debbie Pavlu, Susan Arena, and Phyllis Waldrop for help with manuscript preparation. Supported by NS12481 and NS11255.

considerable progress has been made employing axoplasmic transport tracing techniques and various histochemical and immunocytochemical staining methods.

Studies using the retrograde cell marker horseradish peroxidase (HRP) were the first to identify large numbers of cells giving rise to descending spinal projections within brainstem regions known to contain noradrenergic cell bodies (7, 17, 20, 32, 42, 47, 48, 52, 70, 83). In these studies, comparisons of the locations and cellular morphology of retrogradely labeled neurons with noradrenergic neurons suggest that virtually all of the major noradrenergic cell populations, from the medulla through the pons, contribute descending projections to the spinal cord. However, a crucial problem with these experiments concerns whether or not the retrogradely labeled neurons descending to the spinal cord are, in fact, noradrenergic. Neurons scattered within the various noradrenergic cell groups could contain a transmitter substance(s) other than norepinephrine.

Although the sites of termination of noradrenergic fibers in the spinal cord have been described in some detail using histofluorescence techniques or enzyme immunocytochemistry (12, 19, 22, 27, 30, 43, 57, 83), a more difficult problem arises in attempting to determine the contribution a particular brainstem noradrenergic cell group makes to terminals in a specific spinal cord region. Previous attempts to trace the pathways from different noradrenergic cell groups to their terminal endings in the spinal cord have used degeneration methods or combined lesions with various histochemical staining techniques (13, 22, 27, 41, 57, 64, 71). This approach suffers from the problem that lesions are difficult to control and may interrupt fibers of passage or fibers that are not noradrenergic, making the pattern of degeneration in the spinal cord difficult to interpret. Combining lesions with histofluorescence methods to demonstrate loss of noradrenergic terminals, while giving more useful information, also has drawbacks, since adequate controls for loss of the staining are difficult to devise for more caudally located cell groups. When fluorescent staining appears to be reduced but not eliminated by a particular lesion or treatment, quantitation of the loss of staining is also difficult, compromising interpretation of results.

More recently, anterograde autoradiographic tracing methods involving axonal transport of [^{3}H] amino acids (18) have been employed to examine the projections of descending noradrenergic pathways (for review see Westlund *et al.* (82)). While this approach has a number of advantages over the lesion and degeneration techniques, the question arises whether transport of the radioactive tracer has occurred in fibers other than those of noradrenergic neurons. This problem is obviated to some extent by comparing the pattern of fibers and terminals revealed by autoradiography with the distribution pattern of terminals stained by catecholamine histofluorescence or DBH immunocytochemical techniques. Where good correlations exist between the two sets of results and where the isotopically labeled amino acids have been injected into relatively confined brainstem sites known to contain descending noradrenergic neurons, the results of such studies can be interpreted as being a reasonably accurate description of the noradrenergic pathway being examined. However, it must be remembered that the data are only correlative and not definitive.

Because of the need to employ a variety of approaches to define and trace the projections of noradrenergic neurons in the central nervous system, detailed descriptions of the descending noradrenergic innervation of the spinal cord have recently begun to emerge. In a number of areas the data remain incomplete, and, in some instances, results obtained using different methods have produced conflicting or contradictory findings. Nevertheless, two important features of the anatomical organization of the descending noradrenergic systems have appeared: (1) the nucleus locus coeruleus (A_6 group) and the subcoeruleus complex (A_7 group), the two largest pontine noradrenergic cell groups, give rise to descending pathways which differ significantly in their patterns of termination in the brainstem and spinal cord (83); (2) medullary noradrenergic cell groups (groups A_1 and A_2), which were thought to have descending projections, do not contribute to the noradrenergic innervation of the spinal cord (81). The following discussion summarizes these and other recent findings.

Origins of Descending Noradrenergic Pathways

Using the retrograde HRP transport technique, cells descending to the spinal cord have been localized in the nucleus locus coeruleus

(A_6 cell group), the nucleus subcoeruleus, the medial and lateral parabrachial nucleus, and the nucleus of Kölliker-Fuse of the lateral pons (A_7 cell group), the ventrolateral brainstem dorsal and lateral to the superior olivary complex and the facial nucleus (A_5 group), the dorsomedial medulla around the nucleus of the solitary tract and dorsal motor nucleus of the vagus (A_2 group), and the ventrolateral caudal medulla in the region of the lateral reticular nucleus (A_1 group) (7, 17, 20, 32, 42, 47, 48, 52, 70, 83).

The nucleus locus coeruleus of the monkey, *Macaca fascicularis*, and the locations of spinally projecting neurons and neurons stained for the enzyme dopamine-β-hydroxylase within the nucleus are illustrated in Figure 5.1. Cells of the locus coeruleus labeled by retrograde transport of HRP following injections in the spinal cord are oval- to fusiform-shaped cells, characteristic of the ventral part of the nucleus (Fig. 5.1*B* and *C*). They are similar in size and appearance to those neurons of the ventral locus coeruleus which show positive staining with the DBH antibody (Fig. 5.1*C* and *D*).

Another major population of spinally projecting neurons is found in the nucleus subcoeruleus (A_7 cell group), immediately ventral to the locus coeruleus, extending in a broad band ventrolaterally in the pontine tegmentum. Both HRP- and DBH-labeled cells in the nucleus subcoeruleus are scattered, elongated cells with their axons oriented in the dorsomedial to ventrolateral direction (Fig. 5.2). Labeled neurons having a similar appearance are also located in the adjacent medial parabrachial nucleus.

Smaller populations of spinally projecting neurons are also located in other regions of the pons known to contain noradrenergic neurons (Fig. 5.3). These include neurons of the nucleus of Kölliker-Fuse, located in the lateral pontine tegmentum, and cells in the areas around the nucleus of the trapezoid body and the facial nucleus (A_5 cell group). In the lower brainstem, spinally projecting cells are also found in the region of the A_2 noradrenergic cell group in the vicinity of the solitary nuclear complex and the dorsal motor nucleus of the vagus, and in the A_1 group around the lateral reticular nucleus.

Based on findings with the HRP technique, it would appear that each of the major brainstem noradrenergic cell groups contribute descending projections to the spinal cord. However, this conclusion is based on simple morphological comparisons and appears to be in error in view of more recent data.

Recently, two methods have been developed for identifying the precise cells of origin of noradrenergic pathways. One method involves the retrograde axonal transport of the antibody to dopamine-β-hydroxylase (DBH) to label noradrenergic neurons (40). This approach has significant advantages over other methods in that the labeling is specific for cells which contain the DBH enzyme. DBH is known to be axoplasmically transported in central noradrenergic and peripheral adrenergic neurons (73, 85). Presumably, the transport of the antibody to DBH is via binding to endogenous DBH, since transport of the antibody occurs only in noradrenergic (and possibly adrenergic) neurons known to contain DBH. In experiments using retrograde transport of the DBH antibody, the antiserum (whole serum) is injected directly into the spinal cord using methods similar to those for retrograde transport of HRP. Following an appropriate survival interval (1 to 3 days) to allow for transport, the brainstem sections can be processed by modifications of conventional immunocytochemical staining procedures (81).

A second method for identifying the origins of descending noradrenergic pathways involves combining the retrograde HRP technique with conventional immunocytochemical staining for DBH on the same tissue sections (11). This double labeling method can indicate whether neurons projecting to the spinal cord contain DBH immunoreactivity. An additional advantage of this technique is that relationships between HRP-labeled spinally projecting cells, DBH-containing neurons, and neurons containing both markers can be analyzed quantitatively.

Using these two newer methods, the cells of origin of descending noradrenergic systems have been mapped in the rat (81, 82) and monkey as reported here. The results complement and extend information obtained previously by combining various other histochemical staining methods or following lesions or sectioning of the descending fibers to determine the sources of the noradrenergic innervation of the spinal cord (1, 13, 22, 41, 57, 64, 70, 71, 74).

Methods employing the retrograde axonal transport of the antibody to dopamine-β-hy-

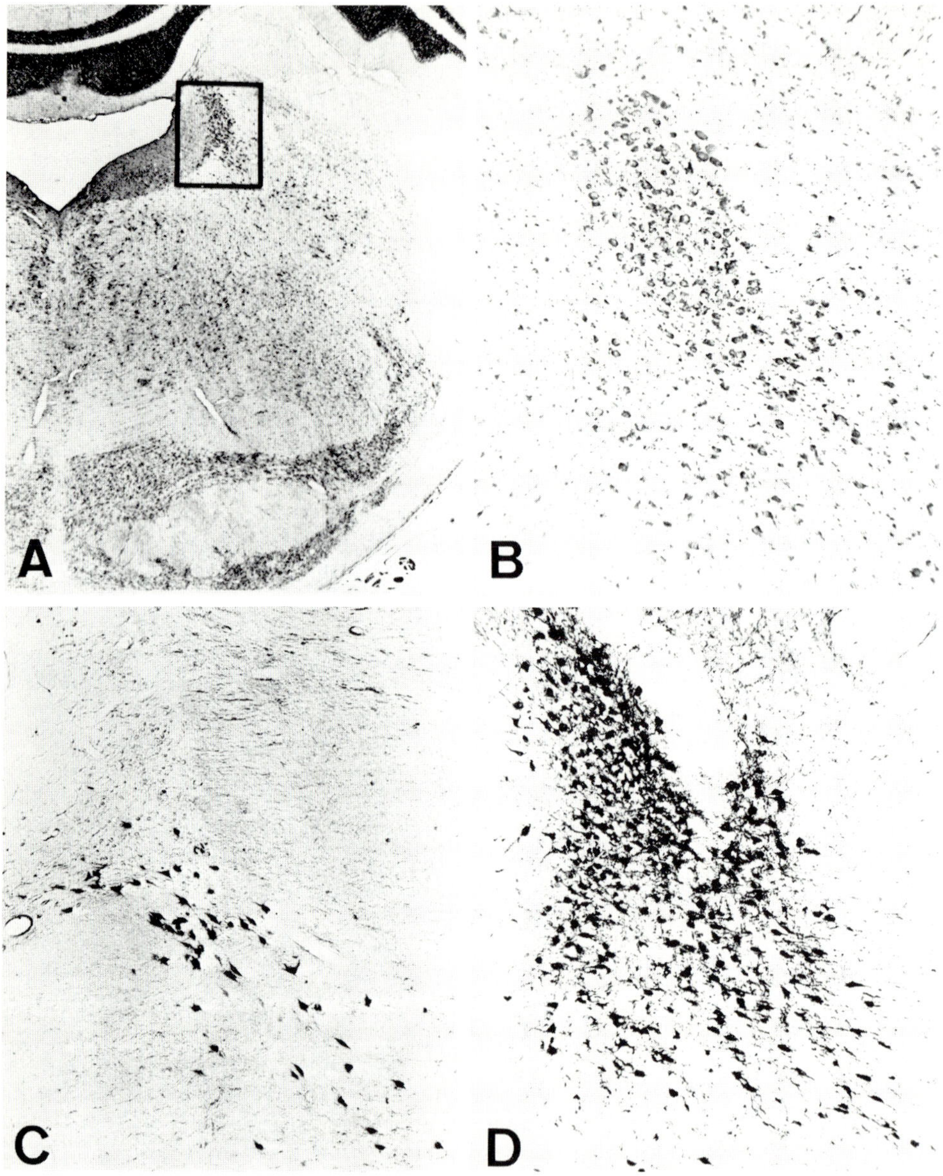

Figure 5.1. Locus coeruleus of monkey: cytoarchitecture, spinally projecting neurons, and DBH-immunoreactive neurons. (*A*) Pontine hemisection through the caudal portion of the nucleus locus coeruleus (*boxed area*) shows darkly stained neurons of the nucleus (15 μm paraffin section, cresyl violet acetate stain). (*B*) Nucleus locus coeruleus, at higher magnification, showing densely aggregated, darkly staining neurons of the dorsal division of the nucleus and the less densely packed cells of the ventral division. The division between the ventral portion of the nucleus locus coeruleus and the nucleus subcoeruleus is in the extreme lower right (15 μm paraffin section, cresyl violet acetate stain). (*C*) Neurons retrogradely labeled with HRP from the sacral spinal cord in the ventral locus coeruleus and extending into the subjacent nucleus subcoeruleus (*lower right*) (50 μm frozen section, tetramethylbenzidine used as chromagen, no counterstain). (*D*) DBH-immunoreactive neurons in the nucleus locus coeruleus. A few cells of the nucleus subcoeruleus appear in the *lower right*. DBH antiserum dilution: 1:2,000, 50 μm frozen sections, no counterstain. Calibration for *B*, *C* and *D*: 0.3 mm. (Reproduced with permission from K. N. Westlund and J. D. Coulter (83).)

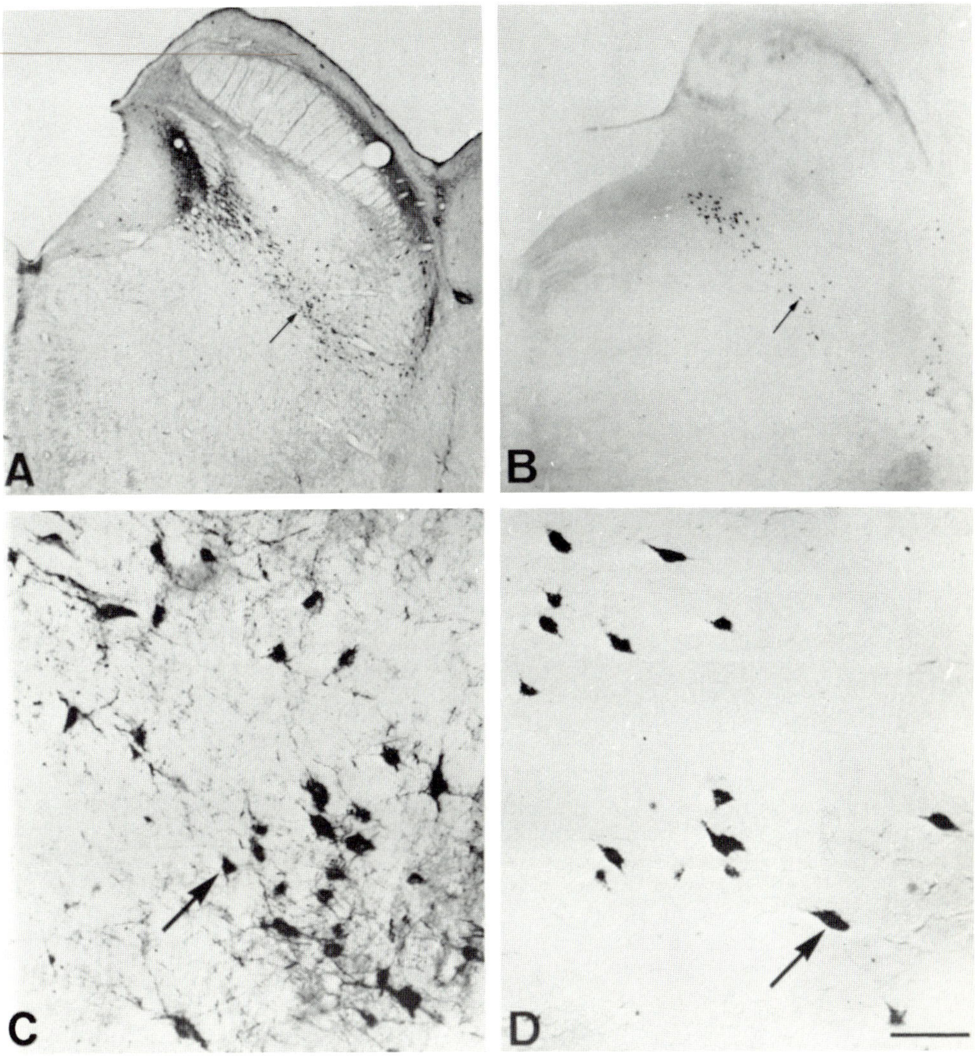

Figure 5.2. Subcoeruleus/medial parabrachial nuclei of monkey: DBH-immunoreactive neurons and spinally projecting neurons. (*A*) DBH-immunoreactive neurons of the dorsolateral pons showing stained neurons in the nucleus locus coeruleus, and subcoeruleus/medial parabrachial nuclei. *Arrow* indicates stained neurons in the nucleus subcoeruleus. DBH antiserum dilution:1:2000, 50 μm frozen section, no counterstain. (*B*) Neurons of the nucleus locus coeruleus and subcoeruleus/medial parabrachial nuclei retrogradely labeled with HRP from a large injection in the lumbosacral spinal cord. *Arrow* indicates labeled neurons in the nucleus subcoeruleus. Note the similarity in distribution patterns of the HRP-labeled cells in the ventral locus coeruleus and subcoeruleus/medial parabrachial nuclei with the DBH-stained neurons shown in *A* (50 μm frozen section, tetramethylbenzidine used as the chromagen, no counterstain) (*C*) DBH-immunoreactive neurons of the nucleus subcoeruleus, at higher magnification. *Arrow* indicates the same cell group shown in *A*. (*D*) Retrogradely labeled neurons of the nucleus subcoeruleus, at higher magnification. *Arrow* indicates the same cell as in *B*. Calibration for *C* and *D*: 100 μm. (Reproduced with permission from K. N. Westlund and J. D. Coulter (83).)

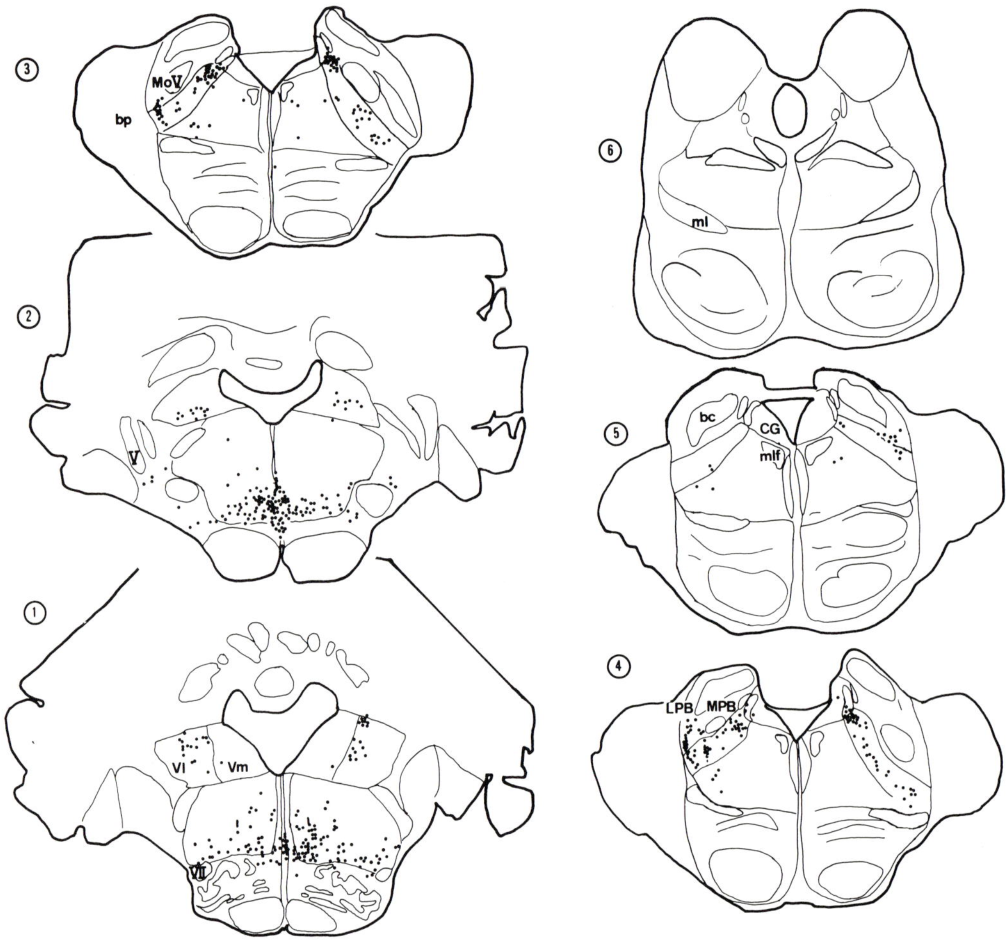

Figure 5.3. Distribution of labeled cells in six coronal planes of section in the pons of the monkey following a bilateral injection of HRP in the cervical spinal cord. Each dot represents a single HRP-labeled cell observed on an individual 50 μm section. Labeled neurons are located in the locus coeruleus, subcoeruleus, parabrachial nuclei, and the pontine tegmentum.

droxylase provide a more definitive way to identify the origins of noradrenergic projections than conventional HRP tracing (73, 85). Using this technique, noradrenergic neurons have been identified which project to the cerebellum (59) and to the spinal cord (81). In neurons retrogradely labeled from the spinal cord with the transported DBH antibody, the immunocytochemical staining with the modified peroxidase-antiperoxidase procedure yields a granular reaction product in labeled neurons of the brainstem (Fig. 5.4). The small, punctate reaction granules are distributed in the cytoplasm and in the proximal dendrites of labeled neurons. The appearance of staining resembles the reaction product produced by retrograde transport of HRP rather than the homogeneous pattern of cytoplasmic staining seen with conventional DBH immunocytochemical procedures.

Retrograde transport of antibody to dopamine-β-hydroxylase from the spinal cord is seen within cells in only those pontine nuclei known to contain noradrenergic neurons. In the pons, the cell groups containing labeled neurons include the nucleus locus coeruleus, the nucleus subcoeruleus, the medial and lateral parabrachial nuclei, the Kölliker-Fuse nucleus, and the region dorsal and lateral to the superior olive and the facial nucleus. As illustrated in Fig. 5.5, the caudal-most group of retrogradely labeled cells are located in the ventrolateral reticular formation at the caudal pole of the superior olivary complex in the A_5 cell group (Fig. 5.1). Cells labeled in this region are medium-sized and fusiform-shaped with some multipolar cells present. By the level of the facial nucleus, retrogradely labeled cells in the ventral, caudal part of the subcoeruleus nucleus (A_7 cell group) are seen scattered among the exiting fibers of the facial nerve (Fig. 5.2). Labeled cells become more numerous rostrally in the subcoeruleus nucleus (Fig. 5.3 to 5.5) and lie within a diagonal band approximately 1 mm wide through the pontine tegmentum in a position ventral to the motor nucleus of the trigeminal nerve. It is at this level that labeled cells extend dorsally to join other labeled cells in the caudal and ventral part of the nucleus locus coeruleus (A_6 cell group). The nucleus locus coeruleus contains large numbers of labeled cells which are concentrated in the ventral and caudal parts of the nucleus. At the rostral pole of the locus coeruleus, only an occasional retrogradely labeled cell is observed in the central grey of the dorsolateral pontine tegmentum (Fig. 5.5).

In the medulla, however, our studies have shown that none of the cells of the A_1 and A_2 noradrenergic cell groups are retrogradely la-

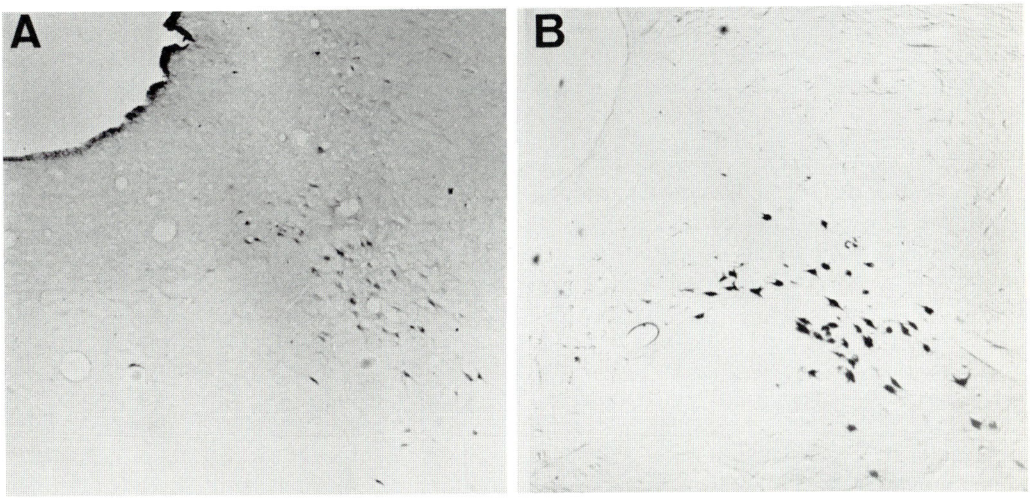

Figure 5.4. (*A*) Neurons of the nucleus locus coeruleus and subjacent nucleus subcoeruleus retrogradely labeled with transported antibody to DBH following injection of the antibody into the spinal cord of monkey. (*B*) A comparable section taken from a monkey injected with HRP in the spinal cord. Note the similarity in the location and numbers of retrogradely labeled cells within the locus coeruleus and the subcoeruleus nuclei compared to those labeled by retrograde transport of DBH antibody.

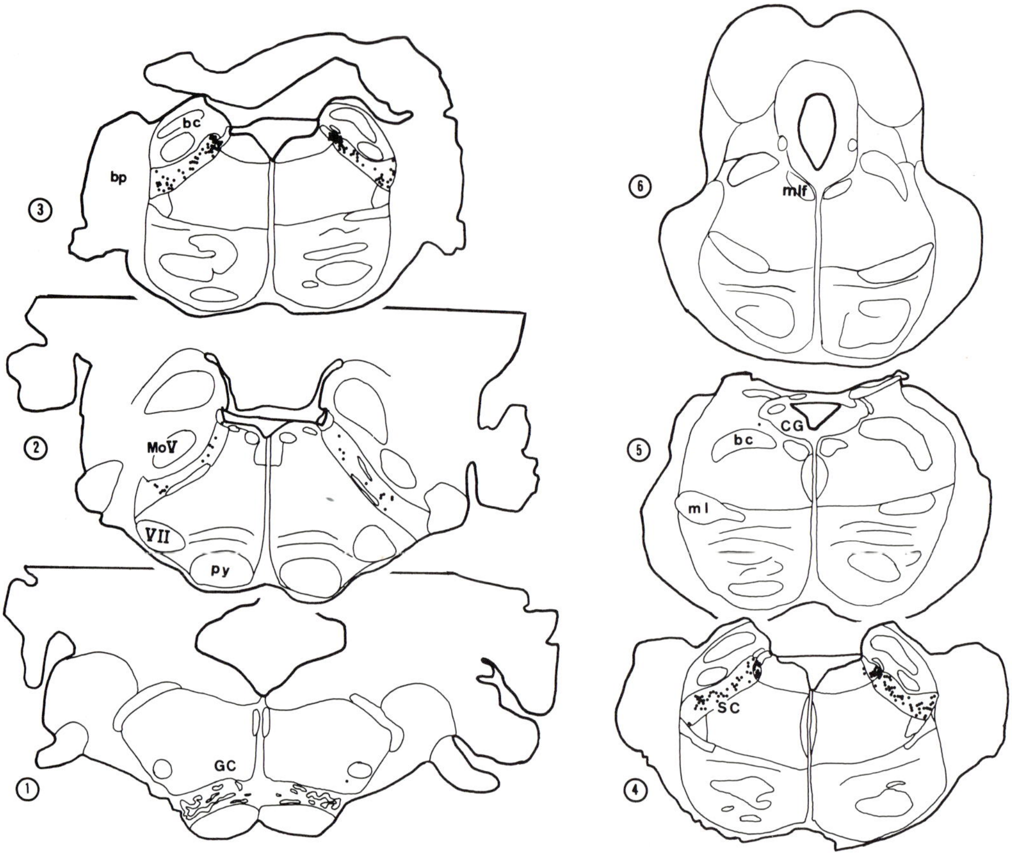

Figure 5.5 Summary plot illustrating retrogradely labeled cells in the monkey pons (six coronal planes of section) following injection of antibody to DBH into the cervical spinal cord. Each dot represents a single retrogradely labeled neuron. Many cells are found in the ventral portion of the locus coeruleus, the nucleus subcoeruleus, nucleus of Kölliker-Fuse, and the parabrachial nuclei. A few cells are also found caudally around the superior olive and facial nucleus corresponding to the region of the A_5 cell group.

beled from the spinal cord by transport of the DBH antibody (81, 82). This observation is somewhat unexpected, since it is widely believed that these medullary noradrenergic cell groups have descending projections. The failure to retrogradely label cells in the medulla by injections of DBH antibody in the spinal cord is not due to an inability of these neurons to transport the DBH antibody, since injections of the antibody into the hypothalamus result in heavy labeling of neurons in the A_1 and A_2 cell groups (K. N. Westlund, R. M. Bowker, and J. D. Coulter, unpublished observations, 1981).

The nucleus locus coeruleus and nucleus subcoeruleus account for about 90% of the neurons labeled from the spinal cord divided almost equally between them. The region around the superior olive and facial nucleus (A_5 cell group) contains roughly 5 to 10% of labeled spinal noradrenergic neurons, with the remainder distributed in the nucleus of Kölliker-Fuse and the parabrachial nuclei.

Injections of the antibody at different spinal cord levels results in only small variations in the numbers of retrogradely labeled cells in the different pontine noradrenergic cell groups. Furthermore, the basic distribution pattern of labeled cells in the pons remains the same, regardless of the spinal cord segmental level that is injected with the DBH antibody. This indicates that the majority of noradrenergic neurons descend the entire length of the spinal cord and probably give off terminals in the spinal grey matter along their entire course.

The findings obtained using the retrograde

axonal transport of the DBH antibody have been confirmed and supplemented with a second technique (82) that employs retrograde tracing with HRP combined with immunocytochemical staining for DBH on the same tissue sections (11). The double labeling method allows determination of the relative numbers of spinally projecting cells within a given cell group along with the number of neurons which also contain DBH immunoreactivity. From these studies it appears that virtually all of the spinally projecting neurons located in the three major noradrenergic cell groups of the pons (A_7, A_6, and A_5) contain the DBH enzyme and hence are noradrenergic. Furthermore, confirming the results obtained by the retrograde DBH antibody transport technique, the double labeling method shows that none of the A_1 or A_2 noradrenergic cells project to the spinal cord, although the HRP-labeled spinally projecting neurons are heavily intermingled with the DBH-positive neurons.

The cells in the A_1 and A_2 cell groups which project to the spinal cord may be epinephrine-containing neurons. A recent report has identified cells projecting to the spinal cord in the A_1 cell group which stain positively for phenylethanolamine-N-methyltransferase, the enzyme that converts norepinephrine to epinephrine (65). Previous results obtained using monoamine oxidase and histofluorescent techniques also suggest that these cells may be adrenergic (70, 74). It is unclear, however, why the retrograde DBH transport method or the double labeling method using conventional DBH immunocytochemistry fails to localize these descending epinephrine-containing neurons, since DBH is presumably present in these neurons as well as in noradrenergic cells. Perhaps the enzyme is not present in detectable quantities in epinephrine-containing cells or is present in a different form not immunoreactive with the DBH antibody.

In summary, the most recent and definitive data show that the origins of descending noradrenergic pathways to the spinal cord arise exclusively from noradrenergic cells of the pons and that no projections to the spinal cord arise from medullary noradrenergic cells. The main source of noradrenergic projections to the cord are the nucleus locus coeruleus and nucleus subcoeruleus which correspond, respectively, to the A_6 and A_7 catecholamine cell groups (21). Smaller numbers of noradrenergic neurons located in the parabrachial nuclei and the nucleus of Kölliker-Fuse, both of which belong to the A_7 cell group, plus cells scattered in the vicinity of the superior olive in the caudal pons (A_5 cell group) also descend to the spinal cord. In view of their proximity and morphological similarity to cells of the subcoeruleus nucleus, the descending noradrenergic cells scattered in these latter nuclear groups appear to be an extension of the descending noradrenergic system that arises from the subcoeruleus nucleus. This observation is consistent with recent findings showing strong similarities in the spinal terminations of the descending parabrachial (A_7) and A_5 cell group pathways with the projections of the nucleus subcoeruleus (49, 67). Since there are no spinal projections from medullary noradrenergic cell groups, descending projections from the locus coeruleus and subcoeruleus and adjacent areas comprise essentially two major systems descending to the spinal cord. Consideration of the spinal terminations of the descending pathways from the nucleus locus coeruleus and from the nucleus subcoeruleus gives further support to the division of the descending noradrenergic pathways into two distinctive systems.

Noradrenergic Innervation of the Spinal Cord

In the spinal cord, norepinephrine localized by catecholamine histofluorescence, DBH immunocytochemistry, and by various regional radiometric assays is known to be concentrated in a number of specific sites (1, 12, 19, 20, 27, 30, 41, 43, 57, 83, 86). These include spinal cord regions of the thoracic and sacral spinal segments which give rise to preganglionic autonomic fibers. In addition, dense concentrations of noradrenergic fibers are present in the superficial part of the spinal dorsal horn, the ventral horn, and the region around the central canal over the entire length of the cord.

The pattern of fibers in the spinal cord of monkey revealed by DBH immunocytochemistry using the peroxidase-antiperoxidase (PAP) staining method (75) is illustrated in Figures 5.6 and 5.7. The laminar borders of the spinal grey matter are based on Rexed's descriptions (63) for the cat. DBH-containing fibers possessing large, darkly staining terminal varicosities course diffusely over all of the spinal grey matter in the monkey, but are especially dense in the region of the interme-

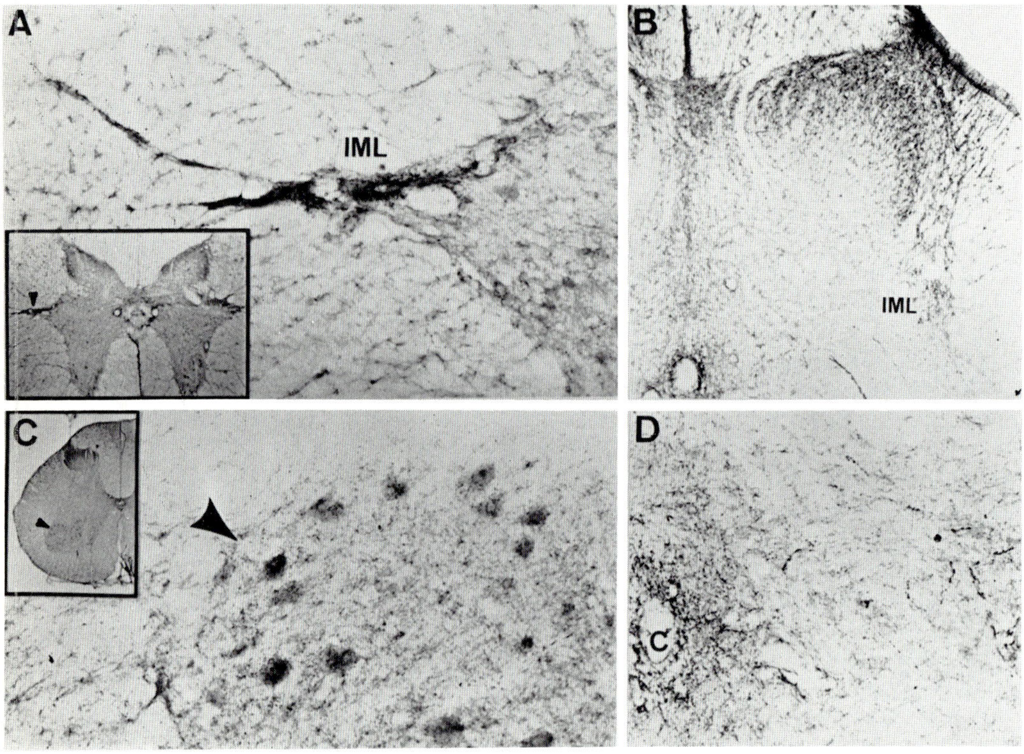

Figure 5.6. Immunocytochemical localization of DBH terminals in the primate spinal cord. (*A.*) Dense DBH-immunoreactive terminals are observed in the intermediolateral cell column (*IML*) at thoracic spinal cord levels. Photomicrograph is higher power of the portion of the thoracic cord indicated by the *arrowhead* in the *inset*. (*B*) DBH-positive terminals in the dorsal horn and in the intermediolateral cell column (*IML*) of the sacral spinal cord. (*C*) DBH terminal varicosities among motoneurons of the ventral horn of the cervical spinal cord. The *arrow* indicates the edge of the spinal grey matter. (*D*) DBH-immunoreactive terminals surrounding the central canal (*C*) of the sacral cord.

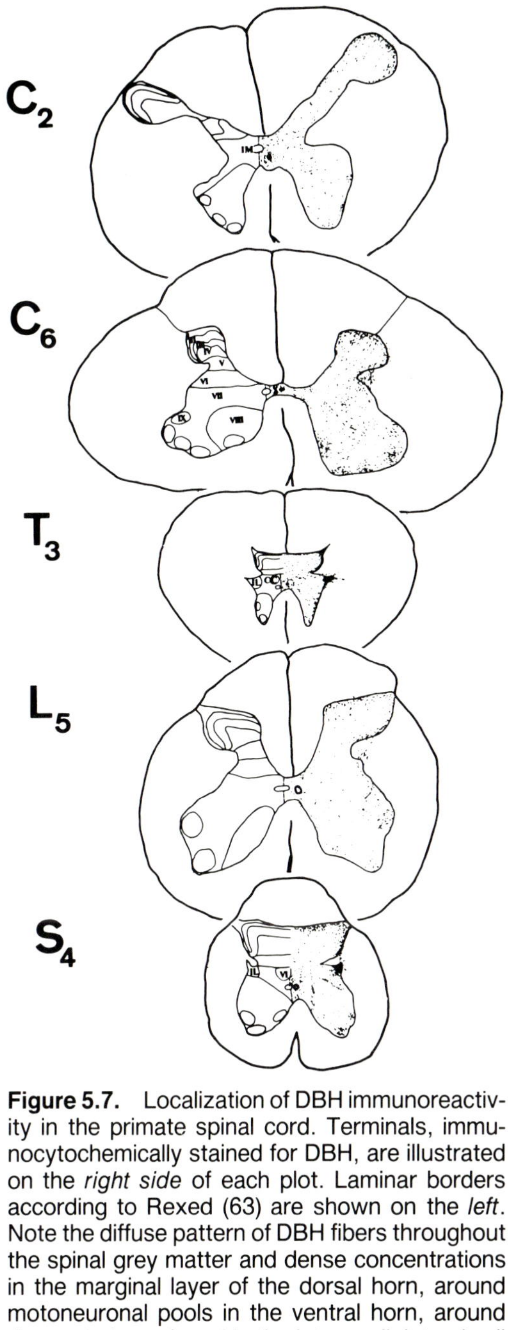

Figure 5.7. Localization of DBH immunoreactivity in the primate spinal cord. Terminals, immunocytochemically stained for DBH, are illustrated on the *right side* of each plot. Laminar borders according to Rexed (63) are shown on the *left*. Note the diffuse pattern of DBH fibers throughout the spinal grey matter and dense concentrations in the marginal layer of the dorsal horn, around motoneuronal pools in the ventral horn, around the central canal, and in the intermediolateral cell column of both the thoracic and sacral cord.

diolateral cell column (IML) of the thoracic cord (Fig. 5.6*A*). In the thoracic spinal segments in the rat, DBH-immunoreactive terminals often extend medially from the intermediolateral cell column into the region of the central autonomic and intercalated nuclei which also contains cells giving rise to preganglionic sympathetic fibers (30, 82). Similar to the pattern at thoracic levels, the regions of the sacral spinal cord containing the cells of the preganglionic parasympathetic fibers possess dense concentrations of DBH-immunoreactive terminals (Figs. 5.6*B* and 5.7). In the upper cervical cord as well as in the spinal enlargements and the thoracic and sacral cord, concentrations of DBH-containing fibers are found in the superficial dorsal horn (lamina I and II) (Fig. 5.6*B* and 5.7), and in the ventral horn, especially among groups of large motoneurons (Fig. 5.6*C*). Other fibers extend into the region around the central canal in lamina X (Fig. 5.6*D*). Thus, the extensive distribution of DBH terminals throughout various parts of the spinal grey matter suggests that descending noradrenergic systems have potentially important and widespread connections with spinal neurons involved in regulating autonomic, sensory, and somatic motor activities.

THE LOCUS COERULEUS

Descending Locus Coeruleus System

The two major sources of noradrenergic fibers in the spinal cord are the nucleus locus coeruleus and the nucleus subcoeruleus, plus adjacent areas, termed here the "subcoeruleus complex." Previous studies have suggested that the neurons of the locus coeruleus provide descending noradrenergic projections to the spinal dorsal horn, the ventral horn, and the region around the central canal, but that it is not a major source of fibers to the intermediolateral cell column of the thoracic cord (13, 22, 57, 64). This observation is of particular interest in light of more recent studies employing the anterograde autoradiographic tracing technique in which the locus coeruleus and the subcoeruleus complex have been shown to give rise to differential projections to certain spinal autonomic cell groups (83).

Following injection of [^{3}H]amino acids into the locus coeruleus, descending terminations are observed to be concentrated in regions of the brainstem and spinal cord controlling parasympathetic autonomic outflows. The site of an injection of [^{3}H]amino acids into the region of the locus coeruleus in the pons of a monkey and the resulting pattern of labeling seen in the spinal cord is illustrated in Figures 5.8*A* and 5.9*A*, respectively. Descending fibers from the locus coeruleus occupy a ventrolateral position in the ipsilateral spinal cord. This pattern is maintained through cervical, thoracic, lumbar, and sacral cord levels. Bundles of labeled fibers in the spinal white matter fill much of the lateral funiculus, concentrate in the ventral part, and extend around medially through most of the ventral funiculus. Heavier labeling appears in the lateral funiculus at lower lumbar and sacral cord levels.

As shown in Figure 5.9*A*, heavy terminal labeling of projections from the locus coeruleus appears bilaterally at all spinal segmental levels over the ventral horn in the region equivalent to Rexed's laminae VII-IX. The labeled terminals are especially dense over motoneurons. Terminal labeling also appears in lamina X around the central canal. Lighter label is seen in the dorsal horn in the marginal layer (lamina I) and extends into the superficial part of the substantia gelatinosa (equivalent to Rexed's lamina II). Labeling is also present deeper in the spinal intermediate zone corresponding to Rexed's laminae IV–VI.

However, a major site of termination of the descending projections from the locus coeruleus is in the sacral spinal cord at levels S_2–S_4. The terminal labeling is observed bilaterally in the lateral spinal grey in the location corresponding to the lateral parasympathetic preganglionic cell column (23, 60). At these same cord levels, labeling present over the dorsal horn and around the central canal is generally heavier than at other cord levels.

There is a remarkably close correlation between the pattern of spinal terminations of the locus coeruleus fibers shown by autoradiography (Fig. 5.9*A*) and the pattern of DBH-stained terminals revealed by immunocytochemistry (Fig. 5.7). Thus, the projections of the locus coeruleus traced by the autoradiographic technique are likely an accurate description of the contribution the descending locus coeruleus pathway makes to the noradrenergic innervation of the spinal cord. It is of interest, however, that the region of the intermediolateral cell column of the thoracic cord contains a dense concentration of DBH-positive fibers, but receives only a sparse projection from the locus coeruleus. Noradrenergic fibers terminating in the thoracic intermediolateral cell column arise from noradrenergic neurons of the subcoeruleus complex, rather than from the nucleus locus coeruleus.

Descending Subcoeruleus System

Previous studies have indicated that the thoracic spinal cord, specifically the intermediolateral cell column and the central autonomic and intercalated nuclei, receive major descending noradrenergic projections (22, 30). The A_5 cell group located in the caudal pons is, in part, the origin of these descending noradrenergic projections (49). Another major

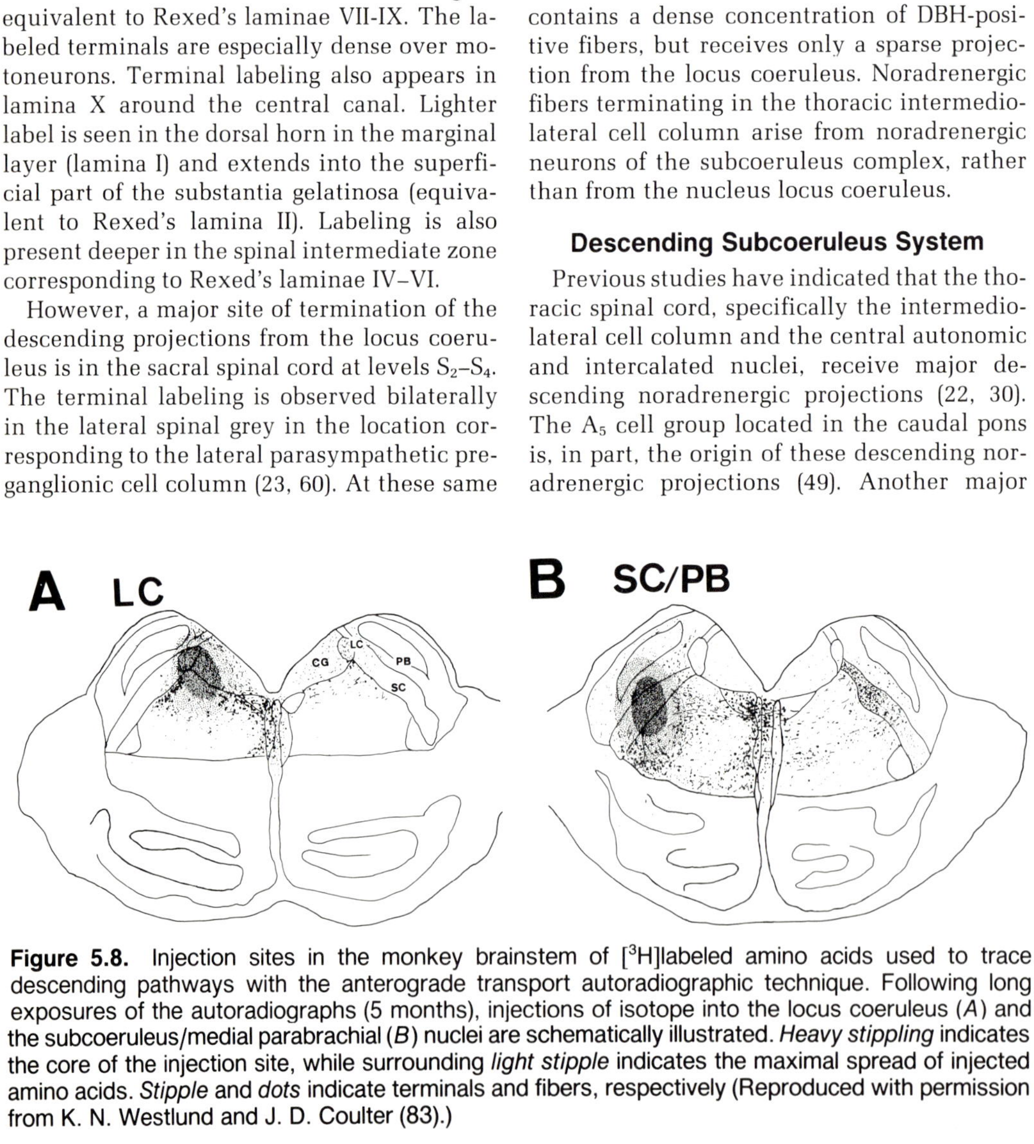

Figure 5.8. Injection sites in the monkey brainstem of [^{3}H]labeled amino acids used to trace descending pathways with the anterograde transport autoradiographic technique. Following long exposures of the autoradiographs (5 months), injections of isotope into the locus coeruleus (*A*) and the subcoeruleus/medial parabrachial (*B*) nuclei are schematically illustrated. *Heavy stippling* indicates the core of the injection site, while surrounding *light stipple* indicates the maximal spread of injected amino acids. *Stipple* and *dots* indicate terminals and fibers, respectively (Reproduced with permission from K. N. Westlund and J. D. Coulter (83).)

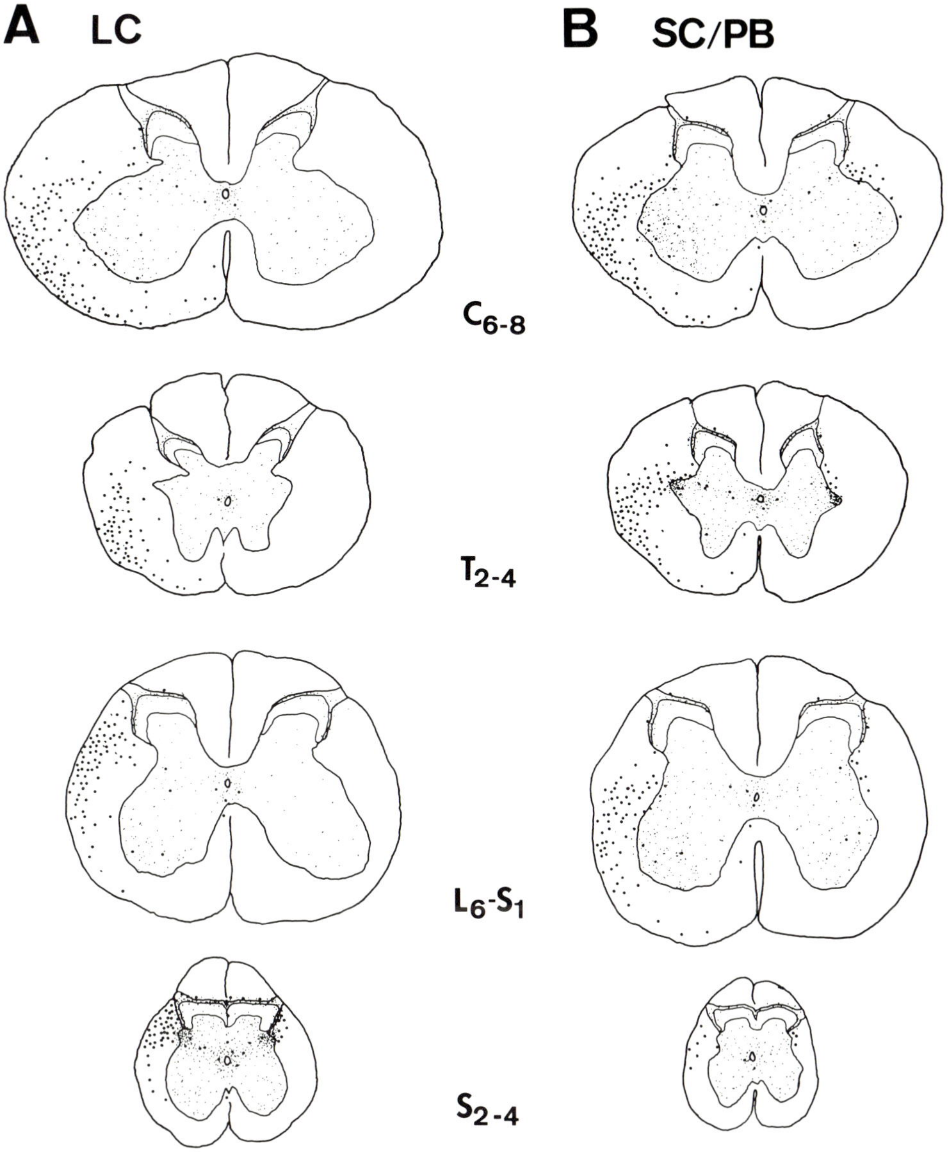

Figure 5.9. Spinal cord terminations of descending projections from the locus coeruleus and subcoeruleus/medial parabrachial nuclei of monkey. Distribution of anterograde axonal (*large dots*) and terminal labeling (*fine stipple*) are plotted from autoradiographs on representative sections of the spinal cord taken from within the indicated spinal segmental levels. (*A*) Labeling pattern from injection of isotope into the left locus coeruleus (*LC*) shows heaviest concentrations of terminals bilaterally over the sacral cord segments (S_{2-4}) in the intermediolateral cell column and in the dorsal horn. Labeling is also seen at all spinal cord levels in the superficial dorsal horn, spinal intermediate zone, the ventral horn, and around the central canal. (*B*) Labeling pattern from injections of isotope into the left subcoeruleus/medial parabrachial nuclei (*SC/PB*) shows heaviest concentrations of terminals bilaterally in the thoracic cord segments (T_{2-4}), as well as in the superficial dorsal horn, the ventral horn, and around the central canal. Note the presence of labeled fibers in the dorsal part of the lateral funiculus contralateral to injections of the subcoeruleus/medial parabrachial nuclei. In the drawings, the borders of the substantia gelatinosa (laminae II and III) of the dorsal horn are indicated (Reproduced with permission from K. N. Westlund and J. D. Coulter (83).)

source of noradrenergic fibers to the thoracic intermediolateral cell column arises in the nucleus subcoeruleus and medial parabrachial nuclei of the pons (67, 83). In Figure 5.8*B* the injection site of [^{3}H]amino acids into the subcoeruleus medial parabrachial region in the monkey is illustrated, and in Figure 5.9*B* the resulting pattern of labeling in the spinal cord, demonstrated by autoradiography, is shown.

Descending fiber bundles from the subcoeruleus/medial parabrachial nuclei are labeled mainly ipsilaterally in the lateral and ventrolateral spinal white matter. A small projection, which apparently crosses in the brainstem, descends contralaterally in the dorsal part of the lateral funiculus. Labeled fiber bundles can be traced the entire length of the spinal cord. The pattern of terminal labeling in the spinal cord grey matter is remarkably similar to that of the descending locus coeruleus projections in that terminal label is seen in the dorsal horn (laminae I and II), in the ventral horn and in the region around the central canal (lamina X) at all spinal cord levels. However, in contrast to the locus coeruleus pathway, the subcoeruleus/medial parabrachial nuclei make heavy terminations in the region of the sympathatic preganglionic cell column of the thoracic cord. Only a few fibers extend to the sacral cord where terminations are sparse.

ORGANIZATION OF THE NORADRENERGIC SYSTEMS

Previous studies of ascending noradrenergic pathways have indicated a segregation of ascending projections originating in the locus coeruleus and ascending pathways arising from the other noradrenergic cell groups of the brainstem. The locus coeruleus system is considered to form the dorsal noradrenergic bundle which gives off terminations in the midbrain central grey, the Edinger-Westphal nucleus, and certain hypothalamic sites before ascending to provide a widespread thalamic and cortical innervation. Other noradrenergic cell groups, considered to form the "subcoeruleus system" (lateral tegmental system) (for review see Moore and Bloom (53)) comprised of the A_7, A_5, and medullary noradrenergic cell groups (A_1 to A_4), innervate mainly the periventricular areas of the hypothalamus and the preoptic area. The noradrenergic innervation of the spinal cord originates from two distinct collections of cells in the locus coeruleus and subcoeruleus complex, consistent with the organization of the ascending noradrenergic projecting systems.

The organization of the noradrenergic system into ascending and descending components is depicted in Figure 5.10. The results of both the retrograde DBH antibody study presented here and the combined HRP-DBH immunocytochemical study (82) concur in showing that pontine noradrenergic cells are the primary source of spinal noradrenergic terminals. As shown schematically, the pontine noradrenergic cell groups of the nucleus locus coeruleus (A_6), the A_7 cell group of the subcoeruleus, parabrachial, and Kölliker-Fuse nuclei, and the A_5 cell group lateral to the superior olive give rise to descending noradrenergic projections to all spinal cord levels, including the sacral cord. The descending fibers terminate in the spinal grey matter, particularly in the marginal layer, among motoneurons in the ventral horn and the region around the central canal at all spinal levels. There is a differential distribution of descending locus coeruleus projections to regions of parasympathetic outflow of the brainstem and sacral cord, while the subcoeruleus complex, including the A_5 and the parabrachial cell groups, project to the sympathetic intermediolateral cell column.

Descending Modulation of Autonomic Function

With regard to possible involvement of the descending noradrenergic systems in central autonomic regulation, the differences in their organization appear to be especially significant. In particular, the projection patterns to brainstem and spinal cord autonomic cell groups distinguish the locus coeruleus system from the subcoeruleus pathway.

Previous physiological studies have implicated various regions of the dorsolateral pons, including the locus coeruleus and subcoeruleus/medial parabrachial nuclei, in the control of central autonomic function (78, 79, 80). The distribution of descending noradrenergic nerve terminals suggests a major role in central vasomotor function. Dense noradrenergic terminations are seen in the nucleus of the tractus solitarius, in the dorsal motor nucleus of the vagus, and in the sympathetic lateral column of the spinal cord (2, 12, 27, 82, 83, 86). Several investigators have suggested that sympathetic preganglionic neurons receive a di-

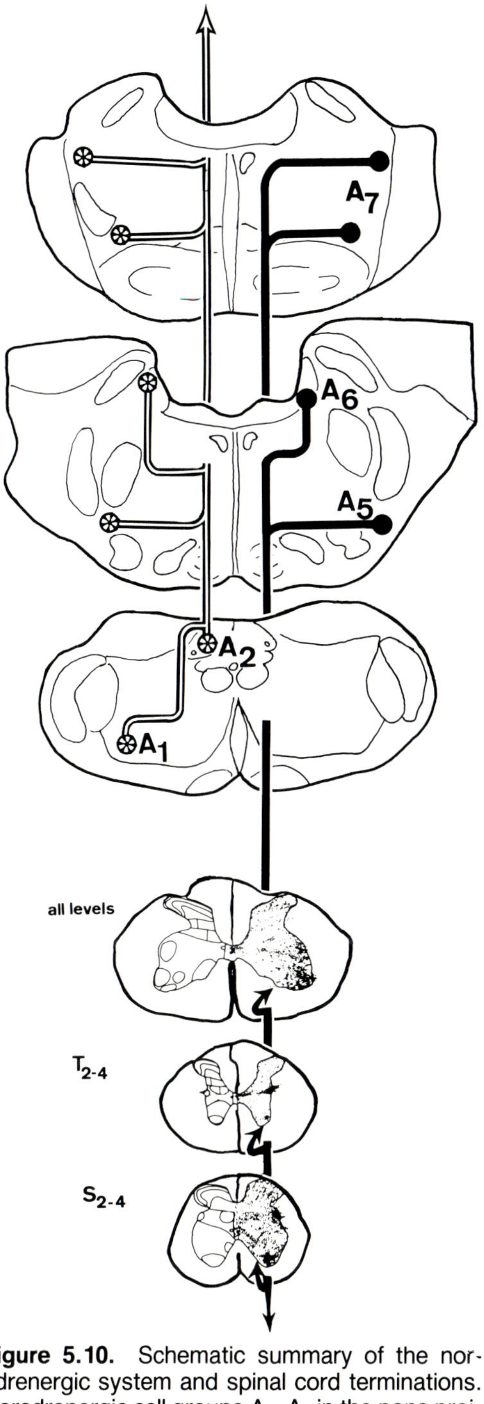

Figure 5.10. Schematic summary of the noradrenergic system and spinal cord terminations. Noradrenergic cell groups A_5–A_7 in the pons project axons to the spinal cord dorsal and ventral horns, as well as to regions of autonomic preganglionic outflows in the thoracic and sacral cord. Ascending noradrenergic projections arise from these same cell groups (A_5–A_7) along with ascending projections from medullary noradrenergic cell groups, A_1 and A_2.

rect inhibitory input from descending noradrenergic pathways (14, 15, 36). Another report, however, indicates that noradrenergic pathways may have excitatory influences on sympathetic outflows (54). Still another series of studies have concluded that the locus coeruleus system is capable of differentially modulating depressor and pressor responses (79, 80). Considering the stimulation sites illustrated in these studies, one could suggest that "pressor" responses originate in the subcoeruleus region and "depressor" responses originate in the locus coeruleus proper. Whether the anatomical separation of the descending noradrenergic pathways into two functional systems can be unraveled by physiological studies remains to be investigated, however.

Particular attention has been given to the spinal projections from the region of the locus coerulus involved in micturition (5, 6, 45, 46). Electrical stimulation in the dorsolateral pontine tegmentum excites preganglionic neurons of the sacral cord resulting in micturition, whereas bilateral lesions result in urinary retention (6, 23, 71). However, it appears unlikely that the locus coeruleus corresponds to the so called "micturition center," since norepinephrine administered iontophoretically to sacral preganglionic neurons inhibits their firing (66). This is not in accord with the view that the pontine micturition center exerts an excitatory action on the sacral parasympathetic neurons innervating the bladder. A more likely candidate is the nucleus tegmentalis laterodorsalis which lies adjacent to the nucleus locus coeruleus and provides a parallel nonnoradrenergic projection to the sacral cord (50, 71).

Descending Modulation of Sensory Functions and Pain

The locus coeruleus and the subcoeruleus/medial parabrachial nuclei both give rise to descending projections with extensive bilateral terminations in the spinal cord dorsal horn, especially to the superficial marginal layer and substantia gelatinosa. Sensory information processing in this part of the dorsal horn has been widely linked to pain sensations (see Willis and Coggeshall (84)). It seems likely that descending noradrenergic pathways are involved in modulating the flow of sensory impulses through the spinal dorsal horn, including activity in ascending pathways conveying information perceived as pain. Iontophoresis of norepinephrine onto dorsal horn neurons, in general, produces inhibition of neuronal firing (8, 24, 61). Electrical stimulation in the dorsolateral pons in the region of the locus coeruleus and the subcoeruleus/medial parabrachial nuclei gives a similar effect (68). These findings may be correlated with the loss of responsiveness to noxious stimuli which can be produced in conscious animals by electrical stimulation of the locus coeruleus and the inhibition of the analgesic effect by administration of noradrenergic blocking agents directly to the spinal cord (62).

Excitation of locus coeruleus neurons by noxious stimuli has been reported by several investigators (9, 44). This effect can be inhibited by morphine applied parenterally or iontophoretically and made to reoccur by naloxone treatment. Lesions of the locus coeruleus reduce morphine analgesia (69), whereas stimulation of the nucleus can produce analgesia (72). Thus, it is tempting to speculate that the descending noradrenergic neurons of the locus coeruleus form part of a complex feedback circuit involved in modulating spinal sensory pain transmission, which includes neurons containing opioid-like peptides.

Descending Modulation of Motor Function

Descending noradrenergic projections have also been implicated in spinal cord motor control. Brainstem stimulation in the vicinity of the locus coeruleus increases lumbar monosynaptic reflexes in the decerebrate cat (76). This is consistent with the finding that direct stimulation of noradrenergic receptor sites by clonidine increases the flexor hindlimb reflex in spinalized animals, an effect which can be abolished by noradrenergic receptor blocking agents, such as phenyoxybenzamine (2). Nygren and Olson (56) found increased sensitivity of flexor reflexes following 6-hydroxydopamine (6-OHDA) destruction of catecholamine innervation. However, this effect was not mediated by direct connections with motoneurons or primary afferents, and at least one interneuron appears to be interposed. The precise localization of these interneurons remains uncertain, as does the overall relationship of descending noradrenergic effects to neurons directly involved in motor control.

CONCLUSION

Unlike the relatively clear-cut functions ascribed to other descending pathways, conflicting and opposite functions have often been attributed to descending noradrenergic pathways. Our present lack of understanding may be partially explained by the as yet unknown microanatomical relationships of noradrenergic neurons to other classes of cells. Based on anatomical considerations alone, conflicting physiological and pharmacological data can be attributed to the influence of nonnoradrenergic systems interacting with noradrenergic systems or to the coactivation or inhibition of two noradrenergic systems which influence areas of contradictory function. Future studies of the descending noradrenergic systems can make increasing use of the emerging anatomical data on the organization of these systems to clarify the functional properties of these important pathways.

References

1. Adèr J-P, Postema F, Korf J: Contribution of the locus coeruleus to the adrenergic innervation of the rat spinal cord: A biochemical study. *J Neural Transm* 44:159–173, 1979.
2. Andén N-E: Distribution of monoamines and dihydroxyphenylalanine decarboxylase activity in the spinal cord. *Acta Physiol Scand* 64:197–203, 1965.
3. Andén N-E, Corrodi H, Fuxe K, Hökfelt B, Hökfelt T, Rydin C, Svensson T: Evidence for a central noradrenaline receptor stimulation by clonidine. *Life Sci* 9:513–523, 1970.
4. Axelsson S, Björklund A, Falck B, Lindvall O, Svensson L-Å: Glyoxylic acid condensation: A new fluorescence method for the histochemical demonstration of biogenic monoamines. *Acta Physiol Scand* 87:57–62, 1973.
5. Barrington FJF: The nervous mechanism of micturition. *Q J Exp Physiol* 8:33–71, 1914.
6. Barrington FJF: The effect of lesions of the hind- and midbrain on micturition in the cat. *Q J Exp Physiol* 15:81–102, 1925.
7. Basbaum AI, Fields HL: The origin of descending pathways in dorsolateral funiculus of the spinal cord of the cat and rat: Further studies on the anatomy of pain modulation. *J Comp Neurol* 187:513–532, 1979.
8. Belcher G, Ryall RW, Schaffner R: The differential effects of 5-hydroxytryptamine, noradrenaline and raphe stimulation on nociceptive and non-nociceptive dorsal horn interneurons in the cat. *Brain Res* 151:307–331, 1978.
9. Bird SJ, Kuhar MJ: Iontophoretic application of opiates to the locus coeruleus. *Brain Res* 122:523–533, 1977.
10. Björklund A, Lindvall O, Svensson L-Å: Mechanisms of fluorophore formation in the histochemical glyoxylic acid method for monoamines. *Histochemie* 32:113–131, 1972.
11. Bowker RM, Steinbusch HWM, Coulter JD: Serotonergic and peptidergic projections to the spinal cord demonstrated by a combined retrograde HRP histochemical and immunocyto-

chemical staining method. *Brain Res* 211:412–417, 1981.

12. Carlsson A, Falck A, Fuxe K, Hillarp N-Å: Cellular localization of monoamines in the spinal cord. *Acta Physiol Scand* 60:112–119, 1964.
13. Commissiong JW, Hellstrom SO, Neff NH: A new projection from locus coeruleus to the spinal ventral columns: Histochemical and biochemical evidence. *Brain Res* 148:207–213, 1978.
14. Coote JH, MacLeod VH: The influence of bulbospinal monoaminergic pathways on sympathetic nerve activity. *J Physiol* 241:453–475, 1974.
15. Coote JH, MacLeod VH: The spinal route of sympatho-inhibitory pathways descending from the medulla oblongata. *Pflugers Arch* 359:335–347, 1975.
16. Corrodi H, Jonsson G: The formaldehyde fluorescence method for the histochemical demonstration of biogenic monoamines. A review on the methodology. *J Histochem Cytochem* 15:65–78, 1967.
17. Coulter JD, Bowker RM, Wise SP, Murray EA, Castiglioni AJ, Westlund KN: Cortical, tectal, and medullary descending pathways in the cervical spinal cord. Reflex control of posture and movement. *Prog Brain Res* 50:263–279, 1979.
18. Cowan WM, Gottlieb DI, Hendrickson AE, Price JL, Woolsey TA: The autoradiographic demonstration of axonal connections in the central nervous system. *Brain Res* 37:21–51, 1972.
19. Crutcher KA, Bingham WG: Descending monoaminergic pathways in the primate spinal cord. *Am J Anat* 153:159–164, 1978.
20. Crutcher KA, Humbertson AO, Martin GF: The origins of brainstem-spinal pathways in the North American opossum (*Didelphis virginiana*). Studies using the horseradish peroxidase method. *J Comp Neurol* 179:169–194, 1978.
21. Dahlström A, Fuxe K: Evidence for the existence of monoamine-containing neurons in the central nervous system. I. Demonstration of monoamines in the cell bodies of brain stem neurons. *Acta Physiol Scand* 62(Suppl 232):1–55, 1964.
22. Dahlström A, Fuxe K: Evidence for the existence of monoamine neurons in the central nervous system. II. Experimentally induced changes in the interneuronal amine levels of the bulbospinal neuron systems. *Acta Physiol Scand* 64(Suppl 247):1–36, 1965.
23. deGroat WC: Nervous control of the urinary bladder of the cat. *Brain Res* 87:201–211, 1975.
24. Engberg I, Ryall RW: The inhibitory actions of noradrenaline and other monoamines on spinal neurons. *J Physiol (Lond)* 185:298–322, 1966.
25. Falck B, Hillarp N-Å, Thieme G, Torp A: Fluorescence of catecholamines and related compounds condensed with formaldehyde. *J Histochem Cytochem* 10:348–354, 1962.
26. Felten DL, Laties AM, Carpenter MB: Monoamine-containing cell bodies in the squirrel monkey brain. *Am J Anat* 139:153–166, 1974.
27. Fuxe K: Evidence for the existence of monoamine-containing neurons in the central nervous system. IV. Distribution of monoamine nerve terminals in the central nervous system. *Acta Physiol Scand* 64(Suppl 247):37–85, 1965.
28. Fuxe K, Hökfelt T, Jonsson G, Ungerstedt U: Fluorescence microscopy in neuroanatomy. In Ebbesen S, Nauta W: *Contemporary Research in Neuroanatomy*, New York, Springer, 1970, pp 275–314.
29. Garver DL, Sladek JR Jr: Monoamine distribution in primate brain. I. Catecholamine-containing perikarya in the brain stem of *Macaca speciosa*. *J Comp Neurol* 159:289–304, 1975.
30. Glazer EJ, Ross LL: Localization of noradrenergic terminals in sympathetic preganglionic nuclei of the rat: Demonstration by immunocytochemical localization of dopamine-β-hydroxylase. *Brain Res* 185:39–49, 1980.
31. Goode GE, Humbertson AO, Martin GF: Projections from the brain stem reticular formation to laminae I and II of the spinal cord. Studies using light and electron microscopic techniques in the North American opossum. *Brain Res* 189:327–342, 1980.
32. Hancock MB, Fougerousse C: Spinal projections from nucleus locus coeruleus and nucleus subcoeruleus in the cat and monkey as demonstrated by the retrograde transport of horseradish peroxidase. *Brain Res Bull* 1:229–234, 1976.
33. Hartman BK: Immunofluorescence of dopamine-β-hydroxylase. Application of improved methodology to the localization of the peripheral and central noradrenergic nervous system. *J Histochem Cytochem* 21:312–332, 1973.
34. Hartman BK: Localization of the noradrenergic nervous system in human brain. *J Psychiatr Res* 11:283–288, 1974.
35. Hartman BK, Udenfriend S: Immunofluorescent localization of dopamine-β-hydroxylase in tissues. *Mol Pharmacol* 6:85–94, 1970.
36. Henry JL, Calaresu FR: Origin and course of crossed medullary pathways to spinal sympathetic neurons in the cat. *Exp Brain Res* 20:515–526, 1974.
37. Hillarp N-Å, Fuxe K, Dahlström A: Demonstration and mapping of central neurons containing dopamine, noradrenaline and 5-hydroxytryptamine and their reactions to psychopharmaca. *Pharmacol Rev* 18:727–741, 1966.
38. Hökfelt T, Fuxe K, Goldstein M, Johansson O: Immunohistochemical evidence for the existence of adrenaline neurons in the rat brain. *Brain Res* 66:235–251, 1974.
39. Holstege F, Kuypers HGJM, Boer RC: Anatomical evidence for direct brainstem projections to the somatic motoneuronal cell groups and autonomic preganglionic cell groups in cat spinal cord. *Brain Res* 171:329–333, 1979.
40. Jacobowitz DM, Ziegler MG, Thomas JA: *In vivo* uptake of antibody to dopamine-β-hydroxylase into sympathetic elements. *Brain Res* 91:165–170, 1975.
41. Karoum F, Commissiong JW, Neff NH, Wyatt RJ: Biochemical evidence for uncrossed and crossed locus coeruleus projections to the spinal cord. *Brain Res* 196:237–241, 1980.
42. Kniesley LW, Biber MP, LaVail JH: A study of the origin of brain stem projections to monkey spinal cord using the retrograde transport method. *Exp Neurol* 60:116–139, 1978.
43. Konishi M: Fluorescence microscopy of the spinal cord of the dog, with special reference to the autonomic lateral horn cells. *Arch Histol Jpn* 30:33–44, 1968.

44. Korf J, Bunney BS, Aghajanian GK: Noradrenergic neurons: Morphine inhibition of spontaneous activity. *Eur J Pharmacol* 25:165–169, 1974.
45. Kuru M: Nervous control of micturition. *Physiol Rev* 45(3):425–494, 1965.
46. Kuru M, Yamamoto H: Fiber connections of the pontine detrusor nucleus (Barrington). *J Comp Neurol* 123:161–185, 1964.
47. Kuypers HGJM, Maisky VA: Retrograde axonal transport of horseradish peroxidase from spinal cord to brainstem cell groups in the cat. *Neurosci Lett* 1:9–14, 1975.
48. Loewy AD, McKellar S: The neuroanatomical basis of central cardiovascular control. *Fed Proc* 39:2495–2503, 1980.
49. Loewy AD, McKellar S, Saper CB: Direct projections from the A_5 catecholamine cell group to the intermediolateral cell column. *Brain Res* 174:309–314, 1979.
50. Loewy AD, Saper CB, Baker RP: Descending projections from the pontine micturition center. *Brain Res* 172:533–538, 1979.
51. Martin GF, Humbertson AO Jr, Laxson LC, Panneton WM: Dorsolateral pontospinal systems. Possible routes for catecholamine modulation of nociception. *Brain Res* 163:333–338, 1979.
52. Martin GF, Humbertson AO Jr, Laxson LC, Panneton WM, Tschismadia I: Spinal projections from the mesencephalon and pontine reticular formation in the North American opossum: A study using axonal transport techniques. *J Comp Neurol* 187:373–400, 1979.
53. Moore RY, Bloom FE: Central catecholamine neuron systems: Anatomy and physiology of the norepinephrine and epinephrine systems. *Ann Rev Neurosci* 2:113–168, 1979.
54. Neumayr RJ, Hare BD, Franz DN: Evidence for bulbospinal control of sympathetic preganglionic neurons by monoaminergic pathways. *Life Sci* 14:793–806, 1974.
55. Nobin A, Björklund A: Topography of the monoamine neuron systems in the human brain as revealed in fetuses. *Acta Physiol Scand* Suppl 388:1–40, 1973.
56. Nygren L-G, Olson L: On spinal noradrenaline receptor supersensitivity: Correlation between nerve terminal densities and flexor reflexes various times after intracisternal 6-hydroxydopamine. *Brain Res* 116:455–470, 1976.
57. Nygren L-G, Olson L: A new major projection from locus coeruleus: The main source of noradrenergic nerve terminals in the ventral and dorsal columns of the spinal cord. *Brain Res* 132:85–93, 1977.
58. Palkovits M, Jacobowitz DM: Topographic atlas of catecholamine- and acetylcholinesterase-containing neurons in the rat brain. II. Hindbrain (mesencephalon, rhombencephalon). *J Comp Neurol* 157:29–42, 1974.
59. Pasquier D, Gold MA, Jacobowitz DM: Noradrenergic perikarya (A_5-A_7 subcoeruleus) projections to the rat cerebellum. *Brain Res* 196:270–275, 1980.
60. Petras JM, Cummings JF: Autonomic neurons in the spinal cord of the Rhesus monkey. A correlation of the findings of cytoarchitectonics and sympathectomy with fiber degeneration following distal rhizotomy. *J Comp Neurol* 146:189–218, 1972.
61. Phillis JW, Tebēcis AK, York DH: Depression of spinal motoneurones by noradrenaline, 5-hydroxytryptamine and histamine. *Eur J Pharmacol* 4:471–475, 1968.
62. Reddy SVR, Yaksh TL: Spinal noradrenergic terminal system mediates antinociception. *Brain Res* 189:391–401, 1980.
63. Rexed B: A cytoarchitectonic atlas of the spinal cord in the cat. *J Comp Neurol* 100:297–379, 1954.
64. Ross RA, Reis DJ: Effects of lesions of locus coeruleus on regional distribution of dopamine-β-hydroxylase activity in rat brain. *Brain Res* 73:161–166, 1974.
65. Ross CA, Armstrong DM, Ruggiero DA, Pickel VM, Joh TH, Reis DJ: Adrenaline neurons in the rostral ventrolateral medulla innervate thoracic spinal cord: A combined immunocytochemical and retrograde transport demonstration. *Neurosci Lett* 25:257–262, 1981.
66. Ryall RW, deGroat WC: The microelectrophoretic administration of noradrenaline, 5-hydroxytryptamine, acetylcholine, and glycine to sacral parasympathetic preganglionic neurones. *Brain Res* 37:345–347, 1972.
67. Saper CB, Loewy AD: Efferent connections of the parabrachial nucleus in the rat. *Brain Res* 197:291–317, 1980.
68. Sasa M, Munekiyo K, Ikeda H, Takaori S: Noradrenaline-mediated inhibition by locus coeruleus of spinal trigeminal neurons. *Brain Res* 80:443–460, 1974.
69. Sasa M, Munekiyo K, Osumi Y, Takaori S: Attenuation of morphine and analgesia in rats with lesions of the locus coeruleus and dorsal raphe nucleus. *Eur J Pharmacol* 42:53–62, 1977.
70. Satoh K, Tohyama M, Yamomoto K, Sakumoto T, Shimizu N: Noradrenaline innervation of the spinal cord studied by the horseradish peroxidase method combined with monoamine oxidase staining. *Exp Brain Res* 30:175–186, 1977.
71. Satoh K, Tohyama M, Sakumoto T, Yamamoto K, Shimizu N: Descending projections of the nucleus tegmentalis laterodorsalis to the spinal cord; studied by the horseradish peroxidase method following 6-hydroxy-dopa administration. *Neurosci Lett* 8:9–15, 1978.
72. Segal M, Sandberg D: Analgesia produced by electrical stimulation of catecholamine nuclei in the rat brain. *Brain Res* 123:369–372, 1977.
73. Silver MA, Jacobowitz DM: Specific uptake and retrograde flow of antibody to dopamine-β-hydroxylase by central nervous system noradrenergic neurons *in vivo*. *Brain Res* 167:65–75, 1979.
74. Smolen AJ, Glazer EJ, Ross LL: Horseradish peroxidase histochemistry combined with glyoxylic acid-induced fluorescence used to identify brainstem catecholaminergic neurons which project to the chick thoracic spinal cord. *Brain Res* 160:353–357, 1979.
75. Sternberger LA: The unlabeled antibody peroxidase-antiperoxidase (PAP) method. In Sternberger L: *Immunocytochemistry*. New York, John Wiley and Sons, 1979, pp 104–169.
76. Strahlendorf JC, Strahlendorf HK, Kingsley RE,

Gintautas J, Barnes CD: Facilitation of the lumbar monosynaptic reflexes by locus coeruleus stimulation. *Neuropharmacology* 19:225–230, 1980.

77. Swanson LW, Hartman BK: The central adrenergic system: An immunofluorescence study of the location of cell bodies and their efferent connections in the rat utilizing dopamine-β-hydroxylase as a marker. *J Comp Neurol* 163:467–505, 1975.
78. Wang SC, Ranson SW: Autonomic responses to electrical stimulation of the lower brainstem. *J Comp Neurol* 71:437–455, 1939.
79. Ward DG, Gunn CG: Locus coeruleus complex: Elicitation of a pressor response and a brain stem region necessary for its occurrence. *Brain Res* 107:401–406, 1976.
80. Ward DG, Gunn CG. Locus coeruleus complex: Differential modulation of depressor mechanisms. *Brain Res* 107:407–411, 1976.
81. Westlund KN, Bowker RM, Ziegler MG, Coulter JD: Origins of spinal noradrenergic pathways demonstrated by retrograde transport of antibody to dopamine-β-hydroxylase. *Neurosci Lett* 25:243–249, 1981.
82. Westlund KN, Bowker RM, Ziegler MG, Coulter JD: Descending noradrenergic projections and their spinal terminations. *Prog Brain Res*, 57:219–238, 1983.
83. Westlund KN, Coulter JD: Descending projections of the locus coeruleus and subcoeruleus/medial parabrachial nuclei in monkey: Axonal transport studies and dopamine-β-hydroxylase immunocytochemistry. *Brain Res Rev* 2:235–264, 1980.
84. Willis WD, Coggeshall RE: *Sensory Mechanisms of the Spinal Cord.* New York, Plenum Press, 1978.
85. Ziegler MG, Thomas JA, Jacobowitz DM: Retrograde axonal transport of antibody to dopamine-β-hydroxylase. *Brain Res* 104:390–395, 1976.
86. Zivin JA, Reid JL, Saavedra JM, Kopin IJ: Quantitative localization of biogenic amines in the spinal cord. *Brain Res* 99:293–301, 1975.

CHAPTER SIX

Some Anatomical and Physiological Interactions among Noradrenergic Systems and Opioid Peptides

J. MICHAEL WALKER, Ph.D., HENRY KHACHATURIAN, Ph.D., and STANLEY J. WATSON, M.D., Ph.D.

INTRODUCTION

Prior to the discovery of endogenous *opiate peptides* (OP), there was considerable interest in the interactions between various neurotransmitters and morphine. At that time, the effects of *morphine* were explained on the basis of neurotransmitter modulation. In many cases these findings have been reinterpreted, together with immunocytochemical data, to delineate the functions of endogenous opiate systems and their interactions with other transmitters. This chapter presents an overview of the anatomical and physiological relationships between the OP and norepinephrine (NE). Space limitations have forced us to overlook a large amount of data in this area; for example, opiate-NE interactions in neuronal development (89), recovery from lesions, and many endocrine or pharmacological model systems cannot be covered. Comprehensive reviews of the subject have been presented elsewhere (61). We will focus on *anatomical* interactions among endogenous opiates and NE and where possible, the pharmacological and physiological implications of the anatomy.

BIOSYNTHESIS AND DISTRIBUTION OF OPIOID PEPTIDES (OP)

The discovery of *opiate receptors*, together with the logical link between analgesia produced by electrical stimulation of the brain and an endogenous morphine-like factor, led to the eventual identification of the enkephalins. Since that time, several other OP and a releaser of opiate-like peptides have been characterized (123). Consequently, there are at least three major systems of endogenous opiates: *β-endorphin*, met- and *leu-enkephalin*, and *dynorphin*. Each of these resemble morphine when evaluated by pharmacological and behavioral criteria and despite their identical N-terminal sequences, each has a unique biosynthetic route and distribution.

The biosynthesis of virtually all peptide hormones involves the production of a precursor protein which is then cleaved to form smaller active hormones and intermediate products.

Abbreviations used in Chapter 6 are: NE, norepinephrine; OP, opioid peptides; ACTH, adrenocorticotropic hormone; LPH, β-lipotropic hormone; LC, locus coeruleus; NTS, nucleus tractus solitarius; MSH, melanocyte stimulating hormone; HPLC, high pressure liquid chromatography; LHRH, luteinizing hormone-releasing hormone; cAMP, cyclic adenosine monophosphate; AMPT, α-methyl-p-tyrosine; CA, catecholamines; DA, dopamine; CRF, corticotropin-releasing factor; CNS, central nervous system; CSF, cerebrospinal fluid; DBH, dopamine-β-hydroxylase.

Each of the three major OP systems has a different precursor.

Met- and Leu-Enkephalin

The enkephalins were the first two endogenous OP discovered in the mammalian brain (58). *Met-enkephalin* constitutes the first five amino acids at the N-terminus of β-endorphin (58). However, both met- and leu-enkephalin form a separate neuronal and biosynthetic system distinct from β-endorphin and dynorphin (18, 149).

The biosynthesis of met-enkephalin in the adrenal has been studied extensively. While there is some evidence that biosynthesis in the brain may be different (59), the identification of the enkephalin precursor in the periphery is nearly complete. The precursor is approximately 30,000 daltons and has six copies of met-enkephalin and one copy of leu-enkephalin (69). In addition, several new peptide sequences have been identified within this precursor protein. These compounds may also have biologic activity as neurotransmitters/neuroregulators.

Enkephalin-containing cell bodies and fibers are widely distributed throughout several *limbic structures, basal ganglia* (*striatum* and *pallidum*), *hypothalamus, periventricular-periaqueductal gray areas,* and *spinal cord* (for a schematic representation of enkephalin cell bodies in the brain, see 35, 53, 54, 62, 108, 116, 130, 131, 145, 147, 149). The anatomical distribution of these two peptides is somewhat similar. Nonetheless, there is evidence to suggest that they might form separate neuronal systems (71).

β-Endorphin

β-Endorphin was discovered soon after the enkephalins (21, 45, 78) and was shown to be present in the same cells of the pituitary that store *adrenocorticortropic hormone* (ACTH) and *β-lipotropic hormone* (*LPH*) (17, 98). Furthermore, immunocytochemical studies suggest that hypothalamic neurons also produce these peptides. β-Endorphin, β-LPH, and ACTH-containing cell bodies are located primarily within the hypothalamic arcuate region, with fiber projections to the hypothalamus and other limbic areas, periventricular *thalamus*, periaqueductal central gray of mesencephalon, the *locus coeruleus* (LC), and *medullary reticular formation* (Fig. 6.1) (5, 6, 16, 18, 19, 148–150, 161). More recently, cell bodies containing β-endorphin or ACTH have also been demonstrated in the region of the *nucleus tractus solitarious* (*NTS*) of *colchicine* pretreated rats (68, 113).

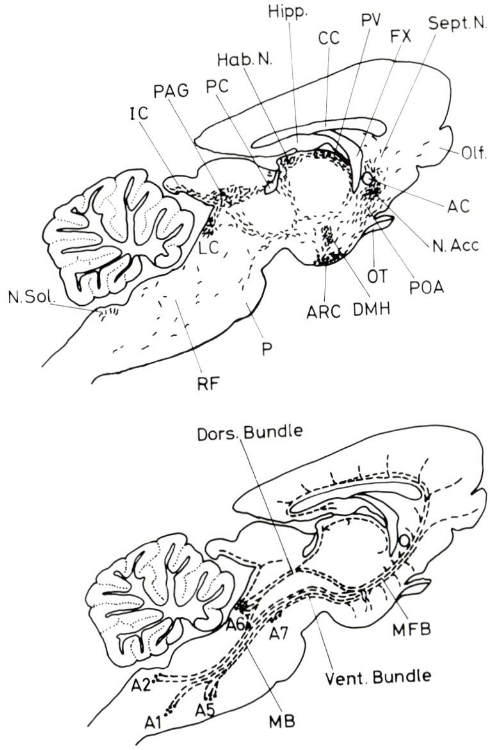

Figure 6.1. These schematic drawings represent β-endorphin (*top panel*) and NE (*bottom panel*) cell bodies (*solid circles*) and fiber projections (*dashed lines*) in parasagittal planes through the rat brain. The NE projections are modified from Ungerstedt (1971). The β-endorphin projections were drawn from our recent systematic studies and represent a composite of several levels through the brain. Note the many areas of possible interaction between these two neurotransmitter/neuromodulator systems. *A1* to *A7*, noradrenergic cell groups; *AC*, anterior commissure; *ARC*, arcuate nucleus; *CC*, corpus callosum; *DMH*, dorsomedial nucleus of hypothalamus; *FX*, fornix; *Hab. N.*, habenular nucleus; *Hipp.*, hippocampus; *IC*, inferior colliculus; *MB*, medullary catecholamine bundle; *MFB*, medial forebrain bundle; *N. Acc*, nucleus accumbens; *N. Sol.*, nucleus tractus solitarius; *Olf.*, olfactory bulb; *OT*, optic tract; *P*, pons; *PAG*, periaqueductal gray; *PC*, posterior commissure; *POA*, preoptic area; *PV*, periventricular nucleus of thalamus; *RF*, reticular formation; *Sept. N.*, septal nucleus.

β-Endorphin derives from a precursor which also contains the full sequence of $ACTH_{1-39}$

and β-LPH. cDNA sequencing of the mRNA for the entire precursor identified a new N-terminal sequence, containing a peptide known as γ-melanocyte-stimulating hormone γ-MSH (85, 95).

Posttranslational processing of this 31K dalton glycoprotein, now termed *pro-opiomelanocortin*, is complex and still under investigation. It is clear, however, that the brain, anterior pituitary, and intermediate lobe of the pituitary each process pro-opiomelanocortin differently. Thus, the final products take several forms. In the anterior pituitary, $ACTH_{1-39}$, β-LPH_{1-91} and β-endorphin are formed. About 30% of the β-endorphin-immunoreactive material is β-endorphin size, runs with β-endorphin on high pressure liquid chromatography (HPLC), and is biologically active (1).

The intermediate lobe processes pro-opiomelanocortin in a quite different manner. The ACTH portion is cleaved, acetylated, and amidated to yield α-MSH which is identical to N-acetyl-$ACTH_{1-13}$-NH_2 and corticotropin-like-intermediate lobe-peptide (CLIP or $ACTH_{18-39}$). Similarly the β-LPH portion is cleaved to yield mostly β-endorphin size material. However, little authentic β-endorphin is formed. The processing is carried further to yield primarily N-acetyl-β-$endorphin_{1-31}$, β-$endorphin_{1-27}$, and N-acetyl-β-$endorphin_{1-27}$ (called N-acetyl-C, C′ and N-acetyl C′, respectively) (118). At least one other product is created, but complete structural analysis of these forms has not yet occurred.

In the brain, the minute quantity of β-endorphin-like material has made it difficult to determine how pro-opiomelanocortin is processed there. Nevertheless, combined radioimmunoassay and HPLC analysis suggest that the brain processes pro-opiomelanocortin primarily to form α-MSH and β-endorphin. Thus, the brain seems to handle pro-opiomelanocortin molecules in part like the *anterior lobe* of the pituitary and in part like the *intermediate lobe*.

The combined storage and release of these peptides raise many questions about synaptic transmission and pituitary secretion, such as why several different peptides are packaged together and how they might interact with their target cells.

In general it appears that the pituitary does *not* alter the intracellular or released proportions of its costored hormones, though the studies on this aspect of the brain are not complete. Functionally, nonopiate pro-opiomelanocortin products sometimes antagonize the effects of opiates (160), but when injected into the terminal fields of β-endorphin fibers, the nonopioid products appear to either synergize with, or functionally mimic, the effects of β-endorphin (139, 142).

Dynorphin

This OP, discovered by Goldstein *et al.* (43), shares a common amino acid sequence with leu-enkephalin (as its N-terminus), much in the same way that β-endorphin contains the sequence of met-enkephalin. The C-terminus of dynorphin, on the other hand, is unique to that peptide.

While data are still accumulating on the $dynorphin_{1-13}$ sequence, the full 1-17 peptide has now been identified (44). Caution is indicated when interpreting data on the shortened fragment, because of drastic changes in the activity of β-endorphin fragments that are shortened at the C-terminal (29). While dynorphin does have naloxone-reversible effects *in vitro*, the behavioral effects often have components that are either not reversed by naloxone, or are incompletely reversed (141). This suggests that rapid breakdown of $dynorphin_{1-13}$ may produce active nonopiate metabolites or that another active sequence in dynorphin causes binding to some unique nonopiate receptor. Such a possibility seems especially likely in view of the similarity of the effects of dynorphin and the opiate-inactive fragment des-tyrosine-dynorphin (143, 144). In addition, the high degree of potency of $dynorphin_{1-13}$ *in vitro* stands in marked contrast to its ephemeral behavioral effects that follow a quite different pattern from those of classic opiates. For example, low doses of $dynorphin_{1-13}$ antagonize the analgesic effects of morphine and β-endorphin and produce *barrel-rolling behavior* which is *not* reversed by high doses of naloxone (73).

In contrast to the endorphins and enkephalins, dynorphin has been localized in the vasopressinergic magnocellular neurons of the hypothalamic supraoptic and *paraventricular nuclei* (Fig. 6.2), as well as medullary and spinal cord cells (153, 154).

DISTRIBUTION OF OPIATE RECEPTORS

The discovery of membrane receptors that reversibly bind opiates not only provided a major impetus for the discovery of endorphins,

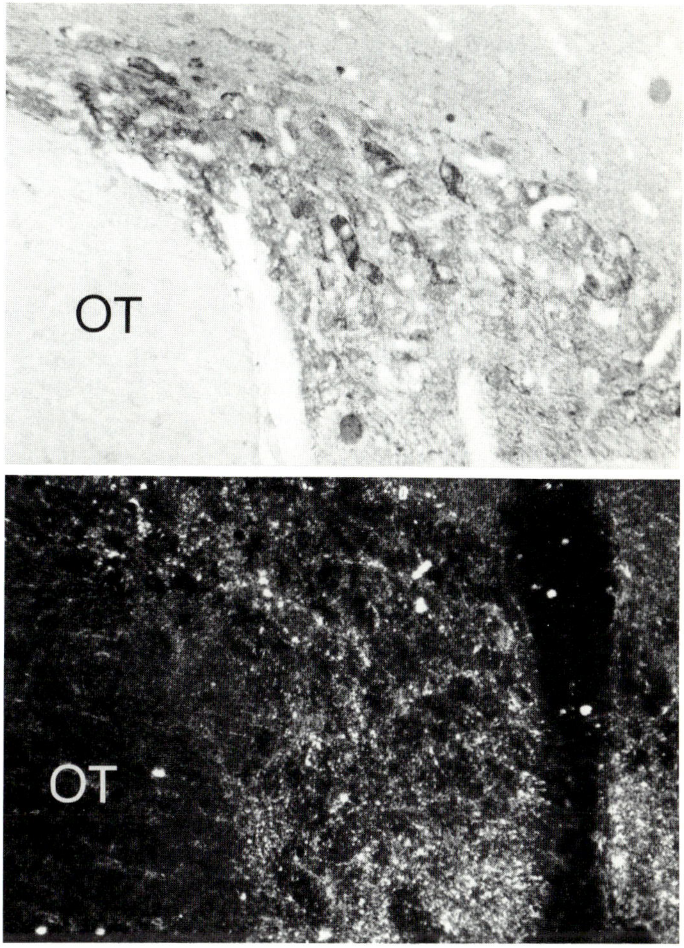

Figure 6.2. These nonconsecutive (frontal) sections show the distribution of dynorphin cell bodies (*top*) and NE-containing varicosities (*bottom, dark field*) in the supraoptic nucleus of the hypothalamus. Dynorphin is localized within vasopressin-producing neurons of this nucleus. *OT*, optic tract.

but remains useful in characterizing their effects. Current knowledge strongly suggests that several types of opiate receptors exist and have different distributions in the brain (84, 86). Several lines of evidence suggest that opiate receptors form clusters on the cell membrane, in a manner similar to *luteinizing hormone-releasing hormone* (LHRH). The clustering was first indicated by Coy *et al.* (25) who demonstrated unusual analgesic potency of a dimeric analog of enkephalin, which presumably binds to two receptor sites simultaneously. Later, clusters were visualized in an *in vitro* preparation using fluorescent techniques (49).

The correlation between opiate receptors and the enkephalins was noted early on, but as the number of endogenous opiates and opiate receptor types increased, it has not been possible to couple a particular receptor type to a particular endogenous opiate. Nevertheless, the mapping of opiate receptors using autoradiographic techniques has added considerably to our understanding (9, 50, 75, 101). Thus, in the same way that immunocytochemical techniques suggest interactions between NE and endogenous opiates, it is significant that opiate receptors have been visualized in several brain regions that involve NE (9, 75, 102). These include the LC, the hypothalamus, and the cerebral cortex. The cortical distribution of opiate receptors corresponds especially well with ascending noradrenergic fibers. Moreover, OP inhibit NE stimulation but not basal levels of cAMP in cortex (126). We have recently been able to visualize leu-enkephalin-containing neurons in the limbic and olfactory-associated cortical regions (68), but to date no other OP has been localized to the cortex. Nevertheless, biochemical techniques suggest the existence of several species of OP in this region.

PITUITARY ACTH, β-ENDORPHIN AND NOREPINEPHRINE

Anterior Pituitary

Much attention has been given to NE modulation of ACTH (41) which apparently is released with β-endorphin from the anterior pituitary, and α-MSH which is apparently released with β-endorphin from the intermediate lobe. More recently, this work has been extended to show parallel release of β-endorphin in many cases. Therefore, it is of considerable interest that, as shown in Figure 6.6, the hypothalamus is well-innervated by NE-secreting terminals arising primarily from the LC.

Intraventricular α-methyl-p-tyrosine (AMPT), at a dose that is not effective systemically, depletes hypothalamic catecholamines (CA) and raises corticosterone levels, presumably by increasing ACTH/β-endorphin output (111). The drug FLA-63 is a specific inhibitor of NE which produces a fall in hypothalamic NE without any change in dopamine (DA) levels. It also results in increased corticosterone levels (111). Furthermore, administration of the levo isomer of NE into the third ventricle reduces plasma corticosteroid content, whereas the dextro isomer of NE is ineffective. When taken together, these results strongly suggest an inhibitory noradrenergic control over ACTH (pro-opiomelanocortin) secretion. Since administration of phentolamine into the third ventricle increases plasma corticosteroids, it seems likely that α-adrenergic receptors are responsible for the effect. β-Adrenergic receptor antagonists are ineffective (112).

It is generally agreed that hypothalamic releasing factors control secretion of peptide hormones from the anterior pituitary. Corticotropin-releasing factor (CRF) is a potent releaser of ACTH (134), and NE decreases the activity of acetylcholine-stimulated CRF activity. Moreover, the α-adrenergic blocker phentolamine also inhibits CRF release (52). In summary, the anatomical convergence of NE terminals into the mediobasal hypothalamus, the potent pharmacological effects of NE and related agents, together with the *in vitro* findings concerning CRF, strongly implicate central NE systems in ACTH/β-endorphin control. The additional data showing common release of β-endorphin with ACTH leads to the conclusion that activation of noradrenergic cells from LC results in decreased release of CRF with a concomitant reduction in plasma β-endorphin and ACTH.

The direct effect of NE on cultured pituitary cells is opposite to the central nervous system (CNS) effect. Large, parallel changes in β-endorphin, β-LPH, and ACTH suggest that at this level NE *stimulates* the release of pro-opiomelanocortin products (105, 133). Once again, *phentolamine* blocks the effect, suggesting an α-adrenergic mechanism. It is apparent that factors other than CRF influence the secretion of β-endorphin-ACTH. In view of the evidence for NE in the median eminence (40), it seems possible that the response of the cultured cells mimics a physiologically relevant response. If so, two parallel influences of NE on β-endorphin secretion must be considered: an excitatory one mediated by CRF and a direct inhibitory effect on the corticotrophs.

Intermediate Lobe

While most of the β-endorphin immunoreactive material from the intermediate lobe is *not* authentic β-endorphin (it is NH_2 terminal-acetylated or COOH terminal-shortened), some genuine β-endorphin is costored with α-MSH (1). Assuming that these substances are coreleased, the NE influence on *α-MSH* apply to other β-endorphin products. The function of β-endorphin in the intermediate lobe is not well understood.

Many factors control the release of α-MSH/β-endorphin from the intermediate lobe. Sectioning of the infundibular stalk causes a lasting increase in α-MSH/pro-opiomelanocortin levels but only a temporary drop in ACTH secretion, suggesting a strong portal system influence on the anterior lobe and a strong neural influence on the intermediate lobe (36). Consistent with this suggestion is the finding that the pars intermedia is well-innervated by fibers originating in the hypothalamus. In fact, CA fluorescence histochemistry indicates that most cells of the pars intermedia are surrounded by fluorescent fibers that disappear after treatment with reserpine. Other transmitters, including peptide transmitters, may also be present.

Catecholaminergic effects upon the pars intermedia may affect MSH/β-endorphin release at several levels. Moreover, both excitatory and inhibitory influences are present. Thus, in the intermediate lobe, β-adrenergic receptors appear to be involved in an α-MSH/β-endorphin secretion, whereas α-adrenergic receptors appear to be involved in the inhibition of α-MSH/β-endorphin. In the hypothalamus, melanocyte-stimulating hormone inhibitory factor release is stimulated by NE, which results in a reduction of α-MSH/pro-opiomelanocortin output.

NOREPINEPHRINE AND STIMULATION-PRODUCED ANALGESIA

One impetus for the isolation of endogenous morphine-like factors was the finding that electrical stimulation of a number of periventricular structures results in a dramatic reduction in *pain sensitivity* (106). Early research noted that the electrical stimulation resembled the effects of morphine in several ways. Most important was the finding that *naloxone*, a specific opiate antagonist, reduces the effectiveness of the stimulation and that tolerance to morphine was also accompanied by a decreased effectiveness of the stimulation (4). Those areas sensitive to stimulation-produced analgesia are richly innervated by OP, and pro-opiomelanocortin products are present in these areas. Enkephalin analogs as well as β-endorphin and its analogs produce analgesia upon central administration (24, 25, 103, 137, 138). Other products of pro-opiomelanocortin also appear to play a role in pain sensitivity. α-MSH, $ACTH_{1-24}$, and various ACTH analogs produce analgesia (139, 142). Moreover, ACTH analogs combine with β-endorphin to enhance its analgesic effects upon administration in the *periaqueductal gray* (140). Cerebrospinal fluid (CSF) levels of these peptides (enkephalin, β-endorphin, α-MSH) are increased after electrical stimulation for pain relief (1). These findings suggest that enkephalins and pro-opiomelanocortin products may, in part, function normally to inhibit pain sensitivity.

The importance of NE in these events has received considerable attention. In particular, it has been suggested that brain NE inhibits and brain DA facilitates the production of stimulation-produced *analgesia*. Thus, DA receptor blockade with pimodize decreases stimulation-produced analgesia, while administration of L-Dopa increases stimulation-produced analgesia. On the other hand, selective depletion of NE with *disulfiram* results in an *increase* in stimulation-produced analgesia, and when NE levels are depressed and DA levels are increased (AMPT + L-DOPA), stimulation-produced analgesia is strongly increased (2). These findings are highly consistent with known NE-enkephalin and NE-pro-opiomelanocortin interactions, at the level of the periaqueductal gray and at the major source of NE—the LC.

INTERACTION BETWEEN BRAIN NOREPINEPHRINE AND OPIOID SYSTEMS

The nucleus LC, located just underneath the floor of the fourth ventricle, is a primary source of central noradrenergic innervation (Fig. 6.3). Axonal projections from this nucleus contribute to three major projections: the *dorsal tegmental bundle*, the *dorsal periventricular system*, and the *medial forebrain bundle*. These in turn project widely throughout the neuraxis (8, 79, 80).

Many of the brain areas that have been shown to contain OP are also rich in NE (Figs. 6.4 to 6.8). The evidence for an interaction between the endogenous opioid systems and the noradrenergic systems is extensive. Several investigators, while mapping the enkephalins and endorphins in the brain, have noticed an anatomical proximity between these peptides and ascending noradrenergic systems (68, 104, 148, 152). For example, a potential for extensive contact between the β-LPH system and the LC noradrenergic system was initially noted by Watson *et al.* (148).

β-Endorphin cells of the *arcuate nucleus* appear to be surrounded by terminals containing *dopamine-β-hydroxylase* (*DBH*), the enzyme responsible for the synthesis of NE (152). The latter study demonstrated an intimate interaction between the LC noradrenergic cells and processes of descending β-endorphin-containing fibers, as well as at several points along the pathway from the LC to the limbic structures (*i.e., periaqueductal gray area*, medial forebrain bundle, *periventricular thalamus*, *supraoptic nucleus*, *paraventricular nucleus*, *arcuate nucleus of hypothalamus*, and *lateral septum*). The periaqueductal gray (Fig. 6.4), periventricular region of the thalamus (Fig. 6.5), dorsomedial nucleus of the hypothalamus (Fig. 6.6), and nucleus accumbens septi (Fig. 6.7) represent areas of particularly heavy interaction between these two neuronal systems (68, 150). In addition to β-endorphin, an enkephalinergic input to LC has also been demonstrated immunocytochemically (Fig. 6.8) (64, 104, 108, 131).

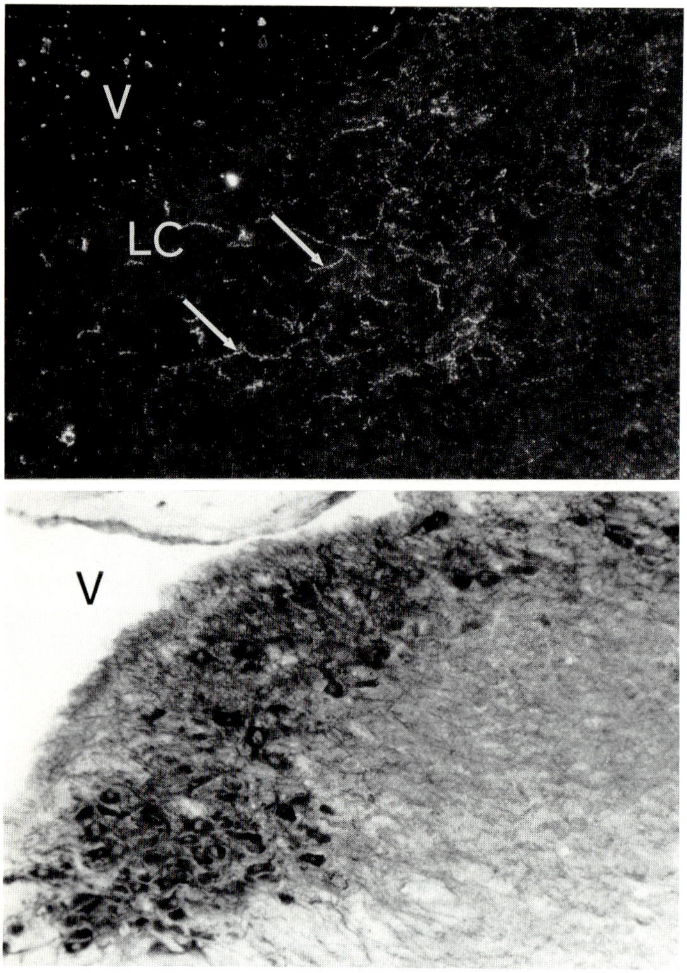

Figure 6.3. This plate shows the comparative (serial parasagittal sections) distribution of β-endorphin fibers (*top*) and DBH-containing cell bodies and fibers (*bottom*) in the nucleus locus coeruleus (*LC*). Note that the bulk of β-endorphin fibers (*arrows*) are somewhat rostral to the bulk of LC, in a region primarily occupied by LC axonal projections. *V*, ventricle.

It has been demonstrated that LC neurons possess opiate receptors (13, 14, 101) and that their spontaneous firing rate is decreased with iontophoretically applied morphine (13). Naloxone-reversible decreases in the rate of firing of NE-containing cell bodies of the LC have also been demonstrated by parenteral administrations of morphine (70). Also, morphine can reduce NE-mediated LC inhibitory effects upon cells of the *spinal trigeminal nucleus* (109, 110). Likewise, morphine has also been shown to block any increases in the rate of firing of midbrain reticular cells caused by nociceptive stimulation (46). These effects appear to be due to membrane hypopolarization from increased conductance (99).

The OP are also capable of altering cellular activity in the brain. Microiontophoretic applications of met- and leu-enkephalin and β-endorphin were all shown to cause a depression of the firing rates of responsive cells in such NE-containing regions as LC, periaqueductal gray, thalamus, caudate nucleus, and frontal cortex (38, 96, 158, 159).

The mesencephalic periaqueductal gray appears to play a primary role in the mediation of opiate-induced analgesia (51, 63, 76, 100, 129, 156), as well as stimulation-produced analgesia (76, 87, 88). Stimulation-produced analgesia is partially antagonized by naloxone (4). Segal and Sandberg (115) elicited stimulation-produced analgesia from both the LC and *substantia nigra*. Interestingly, stimulation of the *arcuate nucleus* resulted in the inhibition of LC firing, an effect that is naloxone-reversible (119). Thus, the β-endorphin system appears to be inhibitory to the noradrenergic system of the LC; however, it is not known whether or not NE is also inhibitory to arcuate β-endorphin neurons. Watson *et al.* (144) im-

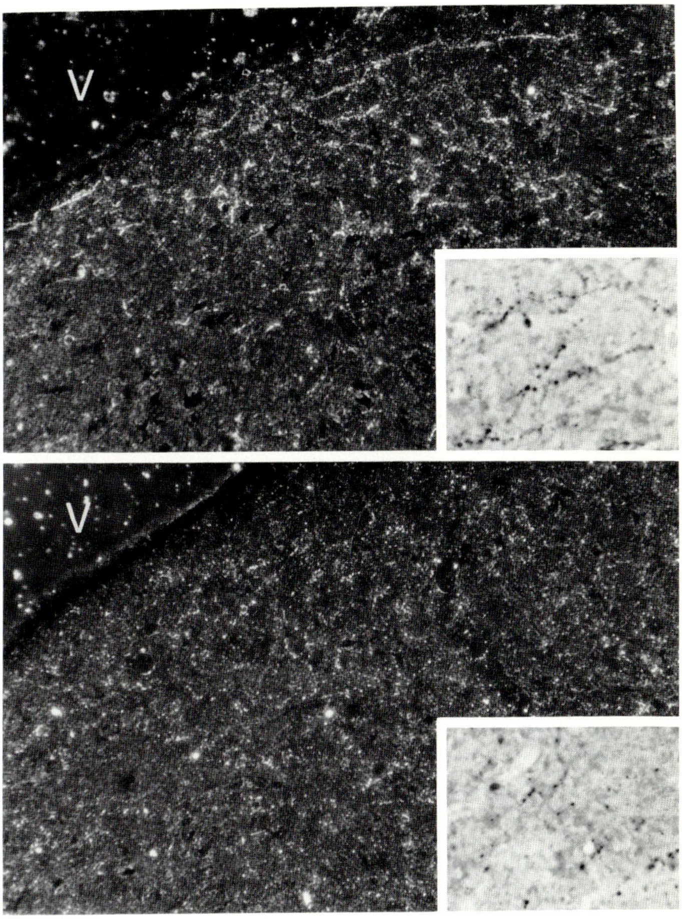

Figure 6.4. Comparative (serial parasagittal sections) distribution of β-endorphin axonal projections (*top*) and DBH-fibers (*bottom*) in the periaqueductal central gray of mesencephalon. *Insets* in both panels show higher magnification of fibers. *V*, ventricle.

plicate the dorsal periventricular CA-containing bundle as a possible major neuroanatomic substrate for stimulation-produced analgesia. The latter investigators further note that the production of analgesia correlates well with the proximity of the stimulating electrode to this CA-containing bundle. Thus, it seems likely that the noradrenergic LC system is involved in the production of analgesia, whether it is stimulation-produced or morphine-induced. Furthermore, a strong probability exists for an interaction between the noradrenergic and β-endorphin systems in the periaqueductal gray.

INTERACTION OF NOREPINEPHRINE AND OPIATES IN THE HYPOTHALAMIC MAGNOCELLULAR SYSTEM

Another area of possible interaction between an OP and ascending noradrenergic systems is the hypothalamic supraoptic nucleus and paraventricular nucleus. These nuclei are composed of *magnocellular neurons* that synthesize the neurohormones *vasopressin* and *oxytocin* and which project to the neural lobe of the hypophysis (30, 33). Several investigators have demonstrated the existence of enkephalins within magnocellular neurons of both rat and cat hypothalamus (35, 53, 91, 107, 108). Enkephalin-containing neurons appear to form a separate yet parallel system to those containing vasopressin and oxytocin (91). Recently, however, the OP dynorphin has not only been localized to the supraoptic and paraventricular nuclei, but it appears to be uniquely located within the vasopressin-producing neurons (Fig. 6.2) (153, 154).

Extensive evidence points to a central noradrenergic control of vasopressin release. The nucleus LC (to a minor extent) and other brain stem noradrenergic cell groups contribute to the innervation of the supraoptic nucleus and paraventricular nucleus. The existence of a

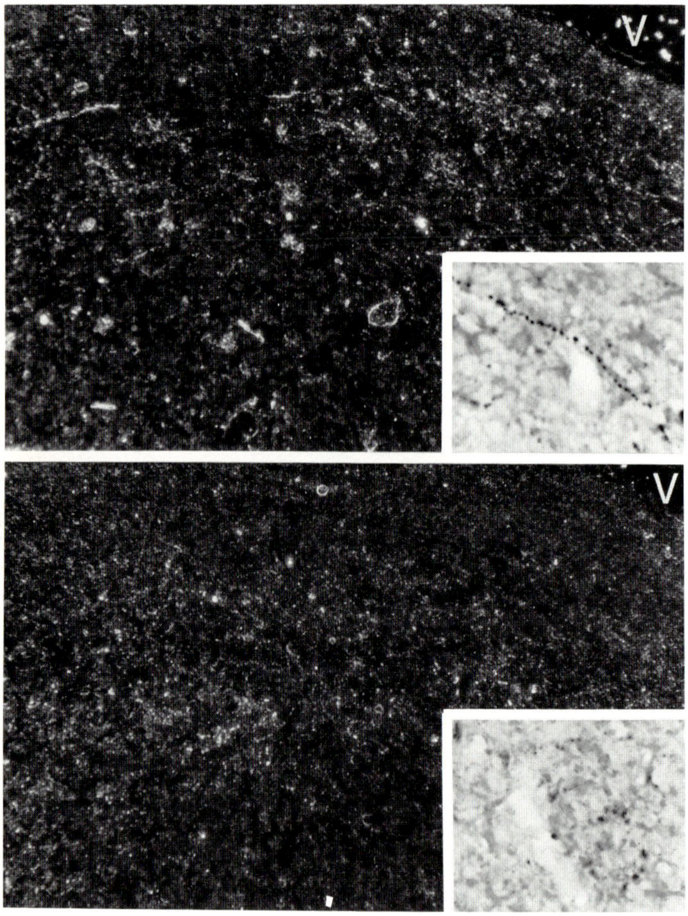

Figure 6.5. These panels show β-endorphin fibers (*top*) and DBH-containing projections (*bottom*) in the periventricular nucleus of the thalamus, in serial parasagittal sections. Insets in both panels are higher power photomicrographs depicting representative fibers. *V*, ventricle.

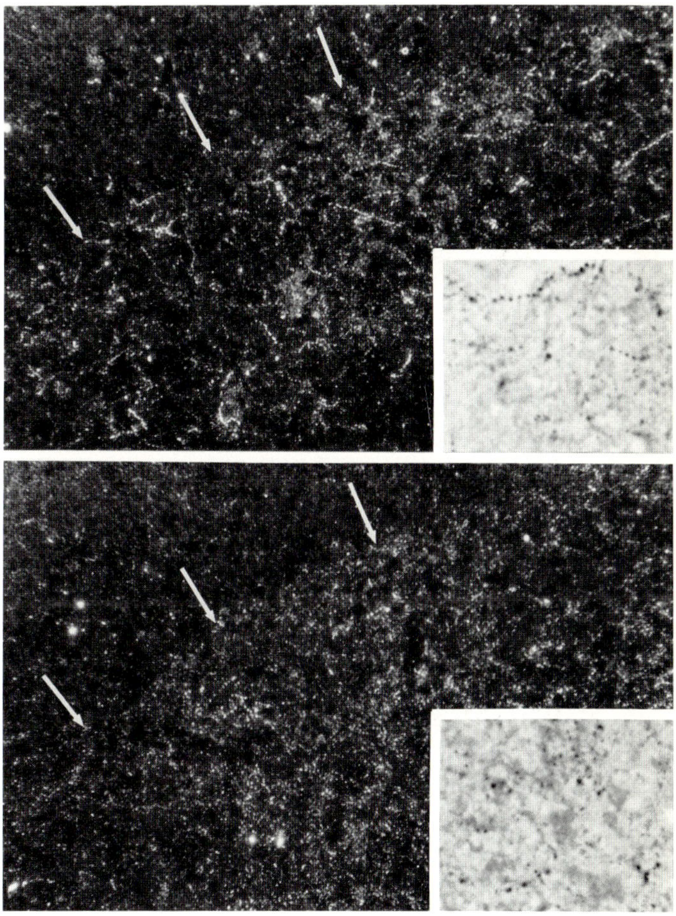

Figure 6.6. Comparative (serial parasagittal sections) photomicrographs of β-endorphin (*top*) and DBH (*bottom*) fibers in the dorsomedial nucleus of the hypothalamus. *Arrows* show the boundaries of the nucleus. *Insets* show higher power photographs of the respective fibers.

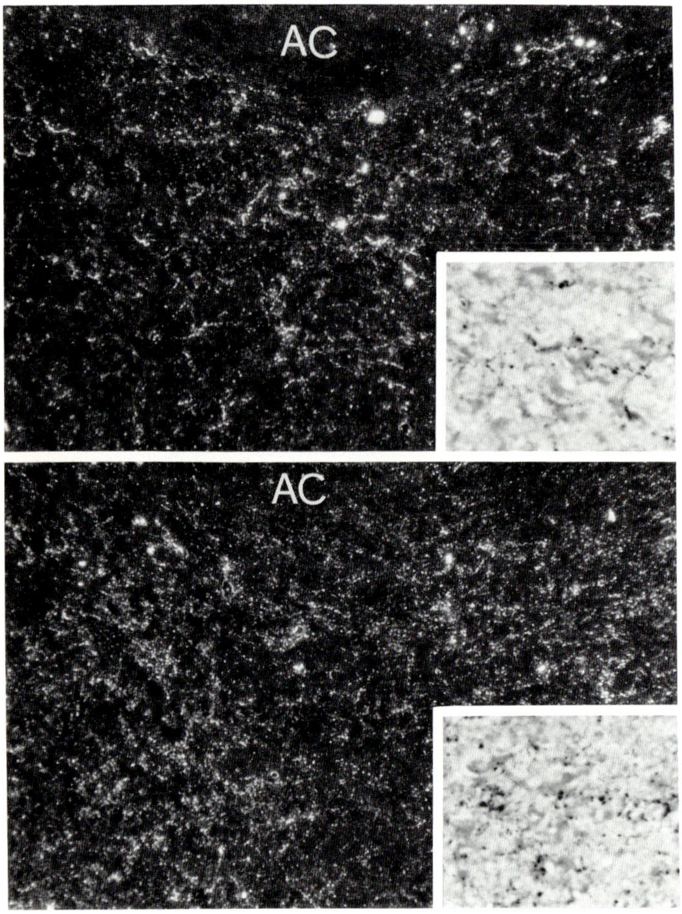

Figure 6.7. These panels show comparative (serial parasagittal sections) distribution of β-endorphin (*top*) and DBH-containing (*bottom*) projections to the nucleus accumbens septi. *Insets* are higher power photomicrographs in each panel showing finer detail of fibers. *AC*, anterior commissure.

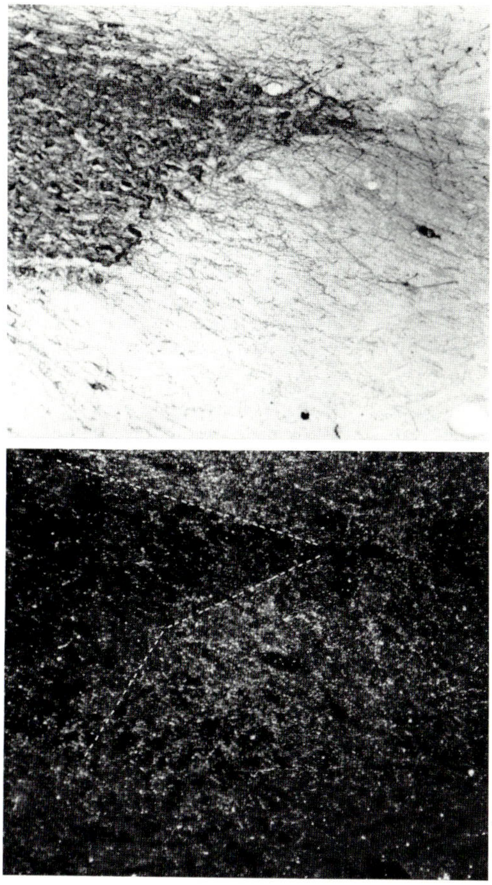

Figure 6.8. Comparative distribution (in serial parasagittal sections) of tyrosine hydroxylase (representing NE) cell bodies (*top*) and leu-enkephalin fibers/varicosities (*bottom*, *dark field*) in the locus coeruleus. *Dotted lines* outline the nucleus. Note that the bulk of enkephalin projections are dorsal and rostral to the nucleus as is the case with β-endorphin in locus coeruleus (Fig. 6.3).

dense NE input to these nuclei has been demonstrated by fluorescence-histochemical studies (39, 55, 79, 90, 132). Furthermore, the possibility of synaptic associations between the noradrenergic terminals and magnocellular neurons has been suggested by ultrastructural (74, 120, 159, 160), microiontophoretic (10, 11, 26, 93, 94), and organ culture (117) studies.

Recently, Khachaturian and Sladek (67) have performed quantitative analyses of noradrenergic input to the supraoptic magnocellular neurons. Their results suggest a differing degree of control of magnocellular secretory processes by NE neurons; the vasopressinergic perikarya appeared to be more heavily innervated than their oxytocinergic counterparts. The results of these findings, when coupled with the observation of dynorphin immunoreactivity within vasopressin-containing neurons, strongly suggest yet another site of possible interaction between noradrenergic and opioid peptide systems.

SUMMARY

We have discussed several brain regions in which there appear to be significant interactions among NE systems and opiate peptidergic systems. In general, there is a surprisingly good overlap in the fiber pathways of these systems, though the nuclei are generally removed from each other. Significant physiological interactions have been observed and are consistent with the early observation that NE systems have extensive fiber projections that affect many brain systems.

NOTE ADDED IN PROOF

Since the original submission of this manuscript, several new results have appeared that are highly germane to this paper. Notably, the precursor for dynorphin has been sequenced and found to contain the α-neo/β-neo-endorphin sequence and an additional dynorphin-like sequence, previously called rimorphin (65a). Moreover, some CNS dynorphin appears to be cleaved to dynorphin 1-8 and the fate of dynorphin 9-17 is thus unknown (154a). In that C-terminal cleavages of ACTH and β-endorphin appear to be releasable (as CLIP or a dipeptide), further investigation into these C-terminal peptides might be fruitful. All the prodynorphin peptides appear to share the property of being selective for κ-opioid receptors (60a), but structure-activity studies suggest nonopioid effects as well.

Anatomically, we now know a great deal more about the distribution of opioid peptides in relation to the catecholamines (see refs. 68a to 68d). There is a very substantial projection of dynorphin-containing fibers from the striatum to the dopamine cells of the substantia nigra (136a). Interestingly, the main effect of prototypic κ-receptor agonists is sedation (41a). Since dynorphin is very potent at κ-receptors, it will be interesting to see if the sedation is attributable to interactions with the dopamine cells of the extrapyramidal motor system.

References

1. Akil H, Shiomi H, Walker JM, Watson SJ: Characterization of multiple forms of β-E in pituitary and brain: effects of stress. *Adv Biochem Psychopharmacol* 33:61–67, 1982.
2. Akil H, Liebeskind JC: Monoaminergic mechanism of stimulation-produced analgesia. *Brain Res* 94:279–296, 1975.
3. Akil H, Mayer DJ: Antagonism of stimulation-produced analgesia by p-CPA, a serotonin synthesis inhibitor. *Brain Res* 44:692–697, 1972.
4. Akil H, Mayer DJ, Liebeskind JC: Antagonism of stimulation-produced analgesia by naloxone, a narcotic antagonist. *Science* 191:961–962, 1976.
5. Akil H, Watson SJ, Berger PA, Barchas JD: Endorphins, beta-LPH, and ACTH: Biochemical, pharmacological and anatomical studies. In Costa E, Trabucchi EM: *The Endorphins: Advances in Biochemical Psychopharmacology.* New York, Raven Press, 1978a, pp 125–139.
6. Akil H, Watson SJ, Levy R, Barchas JD: Beta-endorphin and other 31K fragments: Pituitary and brain systems. In Van Ree JM, Terenius L: *Characteristics and Function of Opioids.* Amsterdam, Elsevier/North-Holland, 1978b, pp 123–134.
7. Akil H, Watson SJ: Neuromodulatory functions of the brain pro-opiocortin system. In Costa E, Trabucchi M: *Neural Peptides and Neuronal Communications.* New York, Raven Press, 1980 pp 435–445.
8. Amaral DG, Sinnamon HM: The locus coeruleus, neurobiology of a central noradrenergic nucleus. *Prog Neurobiol* 9:147–196, 1977.
9. Atweh SF, Kuhar MJ: Autoradiographic localization of opiate receptors in rat brain. II. The brain stem. *Brain Res* 129:1–12, 1977.
10. Barker JL, Crayton JW, Nicoll RA: Supraoptic neurosecretory cells: Adrenergic and cholinergic sensitivity. *Science* 171:208–210, 1971a.
11. Barker JL, Crayton JW, Nicoll RA: Noradrenaline and acetylcholine responses of supraoptic neurosecretory cells. *J Physiol* 218:19–32, 1971b.
12. Beaudet A, Tremeau O, Menez A, Droz B: Visualization of opiate receptors in the locus coeruleus of rats: High resolution radioautography after administration of a tritiated met-enkephalin analog. *C R Acad Sci* [D] (Paris) 289(7):591–594, 1979.
13. Bird S, Atweh S, Kuhar MJ: Microiontophoretic study of the effects of opiates on autoradiographically localized opiate receptors. In Kosterlitz HW: *Opiates and Endogenous Opioid Peptides.* Amsterdam, Elsevier/North-Holland, 1976, pp 199–204.
14. Bird SJ, Kuhar MJ: Iontophoretic application of opiates to the locus coeruleus. *Brain Res* 122:523–533, 1977.
15. Bjorklund A, Moore RY, Nobin A, Stenevi U: The organization of tubero-hypophyseal and reticulo-infundibular catecholamine neuron systems in the rat brain. *Brain Res* 51:171–191, 1973.
16. Bloch B, Bugnon C, Fellman D, Lenys D: Immunocytochemical evidence that the same neurons in the human infundibular nucleus are stained with anti-endorphins and antisera of other related peptides. *Neurosci Lett* 10:147–152, 1978.
17. Bloom FE, Battenberg E, Rossier J, Ling N, Leppaluoto J, Vargo RM, Guillemin R: Endorphins are located in the intermediate and anterior lobes of the pituitary gland, not in neurohypophysis. *Life Sci* 20:43–48, 1977.
18. Bloom FE, Battenberg E, Rossier J, Ling N, Guillemin R: Neurons containing beta-endorphin in rat brain exist separately from those containing enkephalins: Immunocytochemical studies. *Proc Natl Acad Sci USA* 75:1591–1595, 1978a.
19. Bloom FE, Rossier J, Battenberg E, Bayon A, French E, Henricksen SJ, Siggins GR, Segal D, Browne R, Ling N, Guillemin R: Beta-endorphin: Cellular localization, electrophysiological and behavioral effects. In Costa E, Trabucchi M: *The Endorphins: Advances in Biochemical Psychopharmacology.* New York, Raven Press, 1978b, pp 89–109.
20. Bloom FE, Segal D, Ling N, Guillemin R: Endorphins, profound behavioral effects in rats suggest new etiological factors in mental illness. *Science* 194:630–632, 1976.
21. Bradbury AF, Feldberg WF, Smyth DG, Snell CR: Lipotropin C-fragment: An endogenous peptide with potent analgesic activity. In Kosterlitz HW: *Opiates and Endogenous Opioid Peptides.* Amsterdam, Elsevier/North-Holland Press, 1976, pp 9–17.
22. Chretien M, Benjannet S, Dragon N, Seidah NG, Lis M: Isolation of peptides with opiate activity from sheep and human pituitaries: Relationship to beta-lipotropin. *Biochem Biophys Res Commun* 72:472–478, 1976.
23. Cox B, Opheim KE, Teschemacher H, Goldstein A: A peptide-like substance from pituitary that acts like morphine. *Life Sci* 16:1777–1782, 1975.
24. Coy DH, Kastin AJ, Schally AV, Morin O, Caron NG, Labrie F, Walker JM, Fertel R, Berntson GG, Sandman CA: Synthesis and opioid activities of stereoisomers and other D-amino acid analogs of methionine enkephalin. *Biochem Biophys Res Commun* 73:632–638, 1976.
25. Coy DH, Kastin AJ, Walker JM, McGivern RF, Sandman CA: Increased analgesic activities of a fluorinated and a dimeric analogue of [D-Ala-2]-methionine-enkephalinamide. *Biochem Biophys Res Commun* 83:977, 1978.
26. Cross BA, Moss RL, Urban I: Effect of iontophoretic application of acetylcholine and noradrenaline to antidromically identified paraventricular units. *J Physiol* 214:28p–30p, 1971.
27. Cuello AC, Paxinos G: Evidence for a long leu-enkephalin striopallidal pathway in rat brain. *Nature* 271:178–180, 1978.
28. Dahlstrom A, Fuxe K: Evidence for the existence of monoamine-containing neurons in the central nervous system. I. Demonstration of monoamines in the cell bodies of brain stem neurons. *Acta Physiol Scand* 62(232):1–55, 1964.
29. Deakin JF, Dostrovsky JO, Smyth DG: Influence

of N-terminal acetylation and C-terminal proteolysis on the analgesic activity of beta-endorphin. *Biochem J* 89:501–506, 1980.
30. Defendini R, Zimmerman EA: The magnocellular neurosecretory system of the mammalian hypothalamus. In Reichlin S, Baldessarini RJ, Martin JB: *The Hypothalamus. New York, Raven Press, 1978, pp 137–152.*
31. De Wied D: Peptides and behavior. *Life Sci* 20:195, 1977.
32. DiChiara G, Vargiu L, Porceddu ML, Longoni R, Mulas A, Gessa GL: Indirect activation of the DA system as a possible mechanism for the stimulatory effects of narcotic analgesics. *Adv Biochem Psychopharmacol* 16:571–575, 1977.
33. Dierickx K: Immunocytochemical localization of the vertebrate cyclic nonapeptide neurohypophyseal hormones and neurophysins. *Int Rev Cytol* 68:197–249, 1980.
34. Dupont A, Cusan L, Garon M, Labrie F, Li CH: β-endorphin, stimulation of growth hormone release *in vivo*. *Proc Natl Acad Sci USA* 74:358–359, 1977.
35. Elde R, Hökfelt T, Johansson O, Terenius L: Immunohistochemical studies using antibodies to leucine-enkephalin: Initial observations on the nervous system of the rat. *Neuroscience* 1:349–351, 1976.
36. Etkin W: Relation of the pars intermedia to the hypothalamus. In Academic Press, Martin L, Ganong WF: vol 2, pp 261–282.
37. Ferland L, Fuxe K, Enroth P, Gustafsson JA, Skett P: Effects of methionine-enkephalin on prolactin release and catecholamine levels and turnover in the median eminence. *Eur J Pharmacol* 43:89–90, 1977.
38. Frederickson RCA, Norris FH: Induced depression of single neurons in brain areas with opiate receptors, antagonism by naloxone. *Science* 194:440–442, 1976.
39. Fuxe K: Evidence for the existence of monamine neurons in the central nervous system. IV. The distribution of monoamine terminals in the central nervous system. *Acta Physiol Scand* 64(Suppl 247):37–85, 1965.
40. Fuxe K, Hökfelt T, Jonsson G, Lofstrom M: Recent morphological and functional studies on hypothalamic, dopaminergic and noradrenergic mechanisms. In Snyder S, Usdin E: *Frontiers in Catecholamine Research*. New York, Pergamon Press, 1973, pp 787–794.
41. Ganong WF: Control of adrencorticotropin and melanocyte stimulating hormone secretion. In Martin L, Motta M, Frachini F: *The Hypothalamus*. New York, Academic Press, 1980.
41a. Gilbert PE, Martin WR: The effects of morphine-like drugs in the non-dependent, morphine-dependent and cyclazocine-dependent chronic spinal dog. *J Pharmacol Exp Ther* 198:66–82, 1976.
42. Gold MS, Redmond DC, Kleber HD: Noradrenergic hyperactivity in opiate withdrawal supported by clonidine reversal of opiate withdrawal. *Am J Psychiatry* 136:100–102, 1979.
43. Goldstein A, Tachibana S, Lowney LI, Hunkapiller M, Hood L: Dynorphin$_{1-13}$, an extraordinarily potent opioid peptide. *Proc Natl Acad Sci USA* 76:6666–6670, 1979.
44. Goldstein A, Fischli W, Lowney LI, Hunkapillar M, Hood L: Porcine pituitary dynorphin: Complete amino acid sequence of the biologically active heptadecapeptide. *Proc Natl Acad Sci USA* 78:7219–7223, 1981.
45. Guillemin R, Ling N, Burgus R: Endorphins, peptides d'origine hypothalamique et neurohypophysaire d'activite morphinomimetique. Isolement et structure moleculaire d'alpha-endorphine. *C R Hebd Seances Acad Sci Ser D* 282:783–785, 1976.
46. Haigler HJ: Morphine, ability to block neuronal activity by a nociceptive stimulus. *Life Sci* 19:841–858, 1976.
47. Harston CT, Morrow A, Kostrezewa RM: Enhancement of sprouting and putative regeneration of central noradrenergic fibers by morphine. *Brain Res Bull* 5:(4):421–424, 1980.
48. Hayes RR, Newton PG, Rosencrans JA, Mayer DJ: Reduction of stimulation produced analgesia by lysergic acid diethylamide, a depressor of serotonergic neural activity. *Brain Res* 122:367–372, 1977.
49. Hazum E, Chang J-K, Cuatrecasas P: Opiate (enkephalin) receptors of neuroblastoma cells: Occurrence in clusters on the cell surface. *Science* 206:1077–1079, 1979.
50. Herkenham M, Pert CB: *In vitro* autoradiography of opiate receptors in rat brain suggests loci of opiatargic pathways. *Proc Natl Acad Sci USA* 77:5532–5536, 1980.
51. Herz A, Albus K, Metys J, Schubert P, Teschemacher H: On the central sites for the antinociceptive action of morphine and fentanyl. *Neuropharmacology* 9:539–551, 1970.
52. Hillhouse EW, Burden J, Jones MT: The effect of various putative neurotransmitters on the release of corticotrophin releasing hormone of the rat *in vitro*. *Neuroendocrinology* 17:1–11, 1975.
53. Hökfelt T, Elde R, Johansson O, Terenius L, Stein L: The distribution of enkephalin-immunoreactive cell bodies in the rat central nervous system. *Neurosci Lett* 5:25–31, 1977a.
54. Hökfelt T, Ljungdahl A, Terenius L, Elde R, Nilsson G: Immunohistochemical analysis of peptide pathways possibly related to pain and analgesia: Enkephalin and substance P. *Proc Natl Acad Sci USA* 74:3081–3085, 1977b.
55. Hökfelt T, Elde R, Fuxe K, Johansson O, Ljungdahl A, Goldstein M, Luft R, Efendic S, Nilsson G, Terenius L, Gante D, Jeffcoate SL, Rehfeld J, Said S, Perez de la Mora M, Possani L, Tapia R, Teran L, Palacios R: Aminergic and peptidergic pathways in the nervous system with special reference to the hypothalamus. In Reichlin S, Baldessarini RJ, Martin JB: *The Hypothalamus*. New York, Raven Press, 1978, pp 69–135.
56. Hong JS, Yang H-YT, Costa E: On the location of methionine enkephalin neurons in rat striatum. *Neuropharmacology* 16:451–453, 1977.
57. Hong JS, Yang HYT, Fratta W, Costa E: Determination of methionine enkephalin in discrete regions of rat brain. *Brain Res* 134:383–386, 1977.
58. Hughes J: Isolation of an endogenous compound

from brain with pharmacological properties similar to morphine. *Brain Res* 88:295–308, 1975.

59. Hughes J: Biogenesis, release and inactivation of enkephalins and dynorphins. *Br Med Bull* 39:17–24, 1983.

60. Hughes J, Smith J, Kosterlitz H, Fothergill LA, Morgan BA, Morris HR: Identification of two related pentapeptides from the brain with potent opiate agonist activity. *Nature* 258:577–579, 1975.

60a. Huidobro-Toro JP, Yoshimura K, Lee NM, Loh HH, Way EL: Dynorphine interaction at the kappa opiate site. *Eur J Pharmacol* 72:265–266, 1981.

61. Iwamoto EI, Way E: Opiate actions and catecholamines. Neurochemical mechanisms of opiates and endorphins. *Adv Biochem Psychopharmacol* 20:357–407, 1979.

62. Jacobowitz DM, Silver MA, Soden WG: Mapping of leu-enkephalin containing axons and cell bodies of the rat forebrain. In Usdin E: *Endorphins in Mental Health Research.* New York, Oxford University Press, 1979, pp 62–74.

63. Jacquet YF, Lajtha A: Morphine action at central nervous system sites in rat: Analgesia or hyperanalgesia depending on site and dose. *Science* 182:490–492, 1973.

64. Johnson RP, Sar M, Stumpf WE: Relationships between the catecholaminergic and enkephalinergic systems demonstrated by a combined technique of formaldehyde-induced fluorescence and immunocytochemistry. *Neurosci Lett* 14:321–326, 1979.

65. Jones MT, Hillhouse EW, Burden J: Effect of various putative neurotransmitters on the secretion of corticotropin-releasing-hormone from the rat hypothalamus *in vitro*: A model of the neurotransmitters involved. *J Endocrinol* 69:1–10, 1976.

65a. Kakidani H, Furutani Y, Takahashi H, Noda M, Morimoto Y, Hirose T, Asai M, Inayama S, Nakanishi S, Numa S: Cloning and sequence analysis of c-DNA for porcine β-neo-endorphin/dynorphin precursor. Nature 298:245–249, 1982.

66. Kastin AJ, Olson RD, Schally AV, Coy DH: CNS effects of peripherally administered brain peptides. *Life Sci* 25:401, 1979.

67. Khachaturian H, Sladek JR Jr: Quantitation of noradrenergic input to magnocellular perikarya in the rat supraoptic nucleus. *Soc Neurosci Abstr* 6:199, 1980.

68. Khachaturian H, Watson SJ: Some perspectives on monoamine-opioid peptide interaction in rat central nervous system. *Brain Res Bull* 9:411–462, 1982.

68a. Khachaturian H, Lewis ME, Hollt V, Watson SJ: Telencephalic enkephalinergic systems in the rat brain. *J Neurosci* 3:844–855, 1983.

68b. Khachaturian H, Lewis ME, Tsou Kang, Watson SJ: Beta-endorphin, alpha-MSH, ACTH and related peptides. In Hökfelt T, Bjorklund A: *Neuropeptides in the CNS,* part 1. *Handbook of Chemical Neuroanatomy,* vol 3. New York, Elsevier Biomedical Publishers, in press 1984.

68c. Khachaturian H, Lewis ME, Watson SJ: Enkephalin systems in diencephalon and brain stem of the rat. *J Comp Neurol* 220:310–320, 1983.

68d. Khachaturian H, Watson SJ, Lewis ME, Akil H, Coy D, Goldstein A: Dynorphin immunocytochemistry in the rat central nervous system. *Peptides* 3:941–954, 1982.

69. Kimura S, Lewis RV, Stern AS, Rossier J, Stein S, Udenfriend S: Probable prescursors of Leu- and Met-enkephalin in adrenal medulla: Peptides of 3–5 kilodaltons. *Proc Natl Acad Sci USA* 77:1681–1685, 1980.

70. Korf J, Bunney BS, Aghajanian GK: Noradrenergic neurons, morphine inhibition of spontaneous activity. *Eur J Pharmacol* 25:165–169, 1974.

71. Larsson LI, Childers S, Snyder SH: Methionine and leucine-enkephalin occur in separate neurons. *Nature* 22:407–410, 1979.

72. Lazarus LH, Ling N, Guillemin R: β-Lipotropin as a prohormone for the morphinomimetic peptides, endorphins and enkephalins. *Proc Natl Acad Sci USA* 73:2156–2159, 1976.

73. Lee NM: Regulation of beta-endorphin activity. *Adv Biochem Psychopharmacol,* in press, 1983.

74. Leranth C, Zabeorszky L, Marton J, Palkovits M: Quantitative studies on the supraoptic nucleus in the rat. I. Synaptic organization. *Exp Brain Res* 22:509–523, 1975.

75. Lewis ME, Mishkin M, Bragin E, Brown R, Pert CB, Pert A: Opiate receptor gradients in monkey cerebral cortex: Correspondence with sensory processing hierarchies. *Science* 211:1166–1169, 1981.

76. Lewis VA, Gebhart GF: Evaluation of the periaqueductal central gray (PAG) as a morphine specific locus of action and examination of morphine-induced and stimulation-produced analgesia at coincident PAG loci. *Brain Res* 124:283–303, 1977.

77. Li CH, Chung D: Isolation and structure of an untriakontapeptide with opiate activity from camel pituitary glands. *Proc Natl Acad Sci USA* 73:1145–1148, 1976.

78. Li CH, Yamashiro D, Chung D, Doneen BA, Loh HH, Tseng LF: Isolation structure, synthesis and morphine-like activity of β-endorphin from human pituitary glands. *Ann NY Acad Sci* 297:158–166, 1977.

79. Lindvall O, Bjorklund A: The organization of the ascending catecholamine neuron systems in the rat brain as revealed by the glyoxylic acid fluorescence method. *Acta Physiol Scand* 92(Suppl 412):1–48, 1974.

80. Lindvall O, Bjorklund A, Steneri U: The adrenergic innervation of the rat thalamus as revealed by the glyoxylic and fluorescence method. *J Comp Neurol* 154:317–348, 1974.

81. Lindvall O, Bjorklund A: Organization of catecholamine neurons in the rat central neurons systems. In Iversen LL, Iversen SD, Snyder SH: *Chemical Pathways in the Brain. Handbook of Psychopharmacology.* New York, Plenum Press, 1978, vol 9, pp 139–231.

82. Loh HH, Li CH: Biologic activities of beta-en-

dorphin and its related peptides. *Ann NY Acad Sci* 297:115–130, 1977.

83. Loh HH, Tseng L-F, Wei E, Li CH: Beta-endorphin is a potent analgesic agent. *Proc Natl Acad Sci USA* 73:2895–2898, 1976.
84. Lord JAH, Waterfield AA, Hughes J, Kosterlitz HW: Multiple opiate receptors. In Kosterlitz HW: *Opiates and Endogenous Opioid Peptides*. Amsterdam, Elsevier/North Holland Press, 1976, pp 275–180.
85. Mains RE, Eipper BA, Ling N: Common precursor to corticotropins and endorphins. *Proc Natl Acad Sci USA* 74:3014, 1977.
86. Martin WR, Eades GG, Thompson JA, Huppler RE, Gilbert PE: The effects of morphine and nalorphine-like drugs in the nondependent and morphine-dependent chronic spinal dog. *J Pharmacol Exp Ther* 197:517–532, 1976.
87. Mayer DJ, Liebeskind JC: Pain reduction by focal electrical stimulation of the brain: An anatomical and behavioral analysis. *Brain Res* 68:73–93, 1974.
88. Mayer DJ, Wolfe TL, Akil H, Carder B, Liebeskind JC: Analgesia from electrical stimulation in the brainstem of the rat. *Science* 174:1351–1354, 1971.
89. McGinty JF, Ford DH: Effects of prenatal methadone on rat brain catecholamines. *Dev Neurosci* 3(4–6):224–234, 1980.
90. McNeill TH, Sladek JR Jr: Simultaneous monoamine histofluorescence and neuropeptide immunocytochemistry. II. Correlative distribution of catecholamine varicosities and magnocellular distribution of catecholamine varicosities and magnocellular neurosecretory neurons in the rat supraoptic and paraventricular nuclei. *J Comp Neurol* 193:1023–1033, 1980.
91. Micevych P, Elde R: Relationship between enkephalinergic neurons and the vasopressin-oxytocin neuroendocrine system of the cat: An immunohistochemical study. *J Comp Neurol* 190:135–146, 1980.
92. Miller RJ, Pickel VM: Immunohistochemical distribution of enkephalins: Interactions with catecholamine containing systems. *Adv Biochem Psychopharmacol* 25:349–359, 1980.
93. Moss RL, Dyball REJ, Cross BA: Responses of antidromically identified supraoptic and paraventricular units to acetylcholine, noradrenaline and glutamate applied iontophoretically. *Brain Res* 35:573–575, 1971.
94. Moss RL, Urban I, Cross BA: Microelectrophoresis of cholinergic and aminergic drugs on paraventricular neurons. *Am J Physiol* 223:310–318, 1972.
95. Nakanishi A, Inoue A, Kita T, Nakamura M, Chang ACY, Cohen SN, Numa S: Nucleotide sequence of cloned cDNA for bovine corticotropin-β-lipotropin precursor. *Nature* 278:423, 1979.
96. Nicoll RA, Siggins GR, Ling N, Bloom FE, Guillemin R: Neuronal actions of endorphins and enkephalins among brain regions: A comparative microiontophoretic study. *Proc Natl Acad Sci USA* 74:2584–2588, 1977.
97. Pasternak GW, Goodman R, Snyder SH: An endogenous morphine-like factor in mammalian brain. *Life Sci* 16:1765–1769, 1975.
98. Pelletier G, Leclerc R, Labrie F, Cote J, Chretien M, Les M: Immunohistochemical localizations of beta-lipotropin hormone in the pituitary gland. *Endocrinology* 100:770–776, 1977.
99. Pepper CM, Henderson G: Opiates and opioid peptides hyperpolarize locus coeruleus neurons *in vitro*. *Science* 209(4454):394–395, 1980.
100. Pert A, Yaksh T: Sites of morphine induced analgesia in the primate brain: Relation to pain pathways. *Brain* 80:135–140, 1974.
101. Pert CB, Kuhar MJ, Snyder SH: Autoradiographic localization of the opiate receptor in brain. In Goldstein A: *The Opiate Narcotics: Neurochemical Mechanisms in Analgesia and Dependence*. New York, Pergamon Press, 1975, pp 97–101.
102. Pert CB, Kuhar MJ, Snyder SH: Opiate receptor: Autoradiographic localization in rat brain. *Proc Natl Acad Sci USA* 73:3729–3733, 1976.
104. Pickel VM, Joh TH, Reis DJ, Leeman SE, Miller RJ: Electron microscopic localization of substance P and enkephalin in axon terminals related to dendrites of catecholaminergic neurons. *Brain Res* 160:387–400, 1979.
105. Raymond V, Lepine J, Giguere V, Lissitzky JC, Cote J, Labrie F: Parallel stimulation of ACTH, beta-LPH + beta-endorphin and alpha-MSH release by alpha-adrenergic agents in rat anterior pituitary cells in culture. *Mol Cell Endocrinol* 22:295–303, 1981.
106. Reynolds DV: Surgery in the rat during electrical analgesia induced by focal brain stimulation. *Science* 164:444–445, 1969.
107. Rossier J, Battenberg E, Pittman Q, Bayon A, Koda L, Miller R, Guillemin R, Bloom F: Hypothalamic enkephalin neurons may regulate the neurohypophysis. *Nature* 277:653–655, 1979.
108. Sar M, Stumpf WE, Miller RJ, Chang K-J, Cuatrecasas P: Immunohistochemical localization of enkephalin in rat brain and spinal cord. *J Comp Neurol* 182:17–37, 1978.
109. Sasa M, Munekiyo K, Osumi Y, Takaori S: Attenuation of morphine analgesia in rats with lesions of the locus coeruleus and dorsal raphe nucleus. *Eur J Pharmacol* 42:53–62, 1977.
110. Sasa M, Munkeiyo K, Takaori S: Morphine interference with noradrenaline-mediated inhibition from the locus coeruleus. *Life Sci* 17:1373–1380, 1976.
111. Scapagnini V, Preziosi P: Role of brain norepinephrine and serotonin in the tonic and phasic regulation of hypothalamic hypophyseal adrenal axis. *Arch Int Pharmacodyn* 196(Suppl): 205–220, 1972.
112. Scapagnini V, Van Loon GR, Moberg GP, Ganong WF: Effect of α-methyl-p-tyrosine on the circadian variation of plasma corticosterone in rats. *Eur J Pharmacol* 11:266–268, 1972.
113. Schwartzberg DG, Nakane PK: Pro ACTH/endorphin antigenicities in medullary neurons of the rat. *Soc Neurosci Abstr* 7:224, 1981.
114. Segal M: Morphine and enkephalin interactions with putative neurotransmitters in rat

hippocampus. *Neuropharmacology* 16:587–592, 1977.
115. Segal M, Sandberg D: Analgesia produced by electrical stimulation of catecholamine nuclei in the rat brain. *Brain Res* 123:369–372, 1977.
116. Simantov R, Kuhar MJ, Uhl GR, Snyder SH: Opioid peptide enkephalin: Immunohistochemical mapping in rat central nervous system. *Proc Natl Acad Sci USA* 74:2167–2171, 1977.
117. Sladek CD, Sladek JR Jr: Noradrenergic control of vasopressin release by the organ cultured rat hypothalamo-neurohypophyseal system. *Soc Neurosci Abstr* 5:459, 1979.
118. Smyth DG, Massey DE, Zakarian S: Endorphins are stored in biologically active and inactive forms: Isolation of α-*N*-acetyl peptides. *Nature* 279:252–254, 1979.
119. Strahlendorf HK, Strahlendorf JC, Barnes CD: Endorphin-mediated inhibition of locus coeruleus neurons. *Brain Res* 191:284–288, 1980.
120. Swanson LW, Connelly MA, Hartman BK: Further studies on the fine structure of the adrenergic innervation of the hypothalamus. *Brain Res* 151:165–174, 1978.
121. Taborsky GJ Jr, Halter JB, Porte D Jr: Morphine: Dual effects on plasma catecholamines. *Endocrinology* 109:319–321, 1981.
122. Takagi H, Shiomi H, Kuraishi Y, Fukui K, Ueda H: Pain and the bulbospinal noradrenergic system: Pain-induced increase in normetanephrine content in the spinal cord and its modification by morphine. *Eur J Pharmacol* 54:99–107, 1979.
123. Takagi H, Shiomi H, Ueda H, Amano H: A novel analgesic dipeptide from bovine brain is a possible met-enkephalin releaser. *Nature* 282:410–412, 1979.
124. Terenius L, Wahlstrom A: Search for an endogenous ligand for the opiate receptor. *Acta Physiol Scand* 94:74–81, 1975.
125. Thody AJ: *The MSH Peptides.* New York, Academic Press, 1980.
126. Tsang D, Tan AT, Henry JL, Lal S: Effect of opioid peptides on L-noradrenergic-stimulated cyclic AMP formation in homogenates of rat cerebrocortex and hypothalamus. *Brain Res* 152:521–527, 1978.
127. Tseng L-F, Loh HH, Li CH: Beta-endorphin as a potent analgesic by intravenous injection. *Nature* 263:239–240, 1976a.
128. Tseng L-F, Loh HH, Li CH: Beta-endorphin: Cross-tolerance and cross-physical dependence on morphine. *Proc Natl Acad Sci USA* 73:4187–4189, 1976b.
129. Tsou K, Jang CS: Studies on the site of analgesic action of morphine by intracerebral microinjection. *Sci Sin* 13:1099–1109, 1964.
130. Uhl GR, Kuhar MJ, Snyder SH: Enkephalin containing pathway amygdaloid efferents in the stria terminalis. *Brain Res* 149:223–228, 1978.
131. Uhl GR, Goodman RR, Kuhar MH, Childers SR, Snyder SH: Immunohistochemical mapping of enkephalin-containing cell bodies, fibers and nerve terminals in the brain stem of the rat. *Brain Res* 166:75–94, 1979.
132. Ungerstedt U: Stereotaxic mapping of the monoamine pathways in the rat brain. *Acta Physiol Scand* 82(Suppl 367):1–48, 1971.
133. Vale W, Rivier C, Yang L, Minick S, Guillemin R: Effect of purified hypothalamic corticotropin-releasing factor and other substances on the secretion of adrenocorticotropin and beta-endorphin-like immunoreactivities *in vitro*. *Endocrinology* 103:1910–1915, 1978.
134. Vale W, Spiess J, Rivier C, Rivier J: Characterization of a 41-residue ovine hypothalamic peptide that stimulates secretion of corticotropin and beta-endorphin. *Science* 213:1394–1397, 1981.
135. Van Loon GR, Appel NM, Ho D: Beta-endorphin-induced increases in plasma epinephrine, norepinephrine and dopamine in rats: Inhibition of adrenomedullary response by intracerebral somatostatin. *Brain Res* 212:207–214, 1981.
136. Van Loon GR, Appel NM, Ho D: Beta-endorphin-induced stimulation of central sympathetic outflow: Beta-endorphin increases plasma concentrations of epinephrine, norepinephrine, and dopamine in rats. *Endocrinology* 109:319–321, 1981.
136a. Vincent S, Hökfelt T, Christensson I, Terenius L: Immunohistochemical evidence for a dynorphin immunoreactive striato-nigral pathway. *Eur J Pharmacol* 85:251–252, 1982.
137. Walker JM, Berntson GG, Sandman CA, Coy DH, Schally AV, Kastin AJ: An analog of enkephalin having prolonged opiate-like effects *in vivo*. *Science* 196:85, 1977.
138. Walker JM, Sandman CA, McGivern R, Coy DA, Kastin AJ: Endorphin analogs with potent and long-lasting analgesic effects. *Pharmacol Biochem Behav* 7:543, 1977.
139. Walker JM, Berntson GG, Sandman CA, Kastin AJ, Akil H: Induction of analgesia by central administration of ORG 2766, an analog of ACTH(4–9). *Eur J Pharmacol* 69:71–79, 1981a.
140. Walker JM, Akil H, Watson SJ: Electrophysiological analysis interaction among products of the beta-endorphin precursor. *Neurosci Abstr* 7:128, 1981b.
141. Walker JM, Katz RJ, Akil H: Behavioral effects of dynorphin-(1–13) in the mouse and rat: Initial observations. *Peptides* 1:341–345, 1980.
142. Walker JM, Akil H, Watson SJ: Evidence for homologous actions of pro-opiocortin products. *Science* 210:1247–1249, 1980.
143. Walker JM, Moises HC, Coy DH, Baldrighi G, Akil H: Non-opiate effects of dynorphin and des-Tyr-dynorphin. *Science* 218:1136–1138, 1982.
144. Walker JM, Moises HC, Coy DH, Young EA, Watson SJ, Akil H: Dynorphin (1–17): Lack of analgesia but evidence for non-opiate electrophysiological and motor effects. *Life Sci* 31(16–17):1821–1824, 1982.
145. Wamsley JK, Young WS, Kuhar MJ: Immunocytochemical localization of enkephalin in rat

forebrain. *Brain Res* 190:153–174, 1980.

146. Watson SJ, Akil H, Barchas JD: A possible role for the dorsal periventricular catecholamine bundle in stimulation produced analgesia, a behavioral and histochemical study. *Brain Res* 130:335–342, 1977.
147. Watson SJ, Akil H, Sullivan S, Barchas JD: Immunocytochemical localization of methionine-enkephalin: Preliminary observations. *Life Sci* 25:733–738, 1977a.
148. Watson SJ, Barchas JD, Li CH: Beta-lipotropin: Localization of cells and axons in rat brain by immunocytochemistry. *Proc Natl Acad Sci USA* 74:5155–5158, 1977b.
149. Watson SJ, Akil H, Richard CW, Barchas JD: Evidence for two separate opiate peptide neuronal systems and the coexistence of beta-lipotropin, beta-endorphin, and ACTH immunoreactivities in the same hypothalamic neurons. *Nature* 275:226–228, 1978a.
151. Watson SJ, Akil H, Berger PA, Barchas JD: Some observations in the opiate peptides and schizophrenia. *Arch Gen Psychiatry* 36:35–41, 1979.
152. Watson SJ, Richard CW, Ciaranello RD, Barchas JD: Interaction of opiate peptide and noradrenalin systems: Light microscopic studies. *Peptides* 1:23–30, 1980.
153. Watson SJ, Akil H, Ghazarossian VE, Goldstein A: Dynorphin immunocytochemical localization in brain and peripheral nervous system: Preliminary studies. *Proc Natl Acad Sci USA* 78:1260–1263, 1981.
154. Watson SJ, Goldstein A, Ghazarossian VE, Zimmerman E, Nilaver G, Greidanus JW: Dynorphin and vasopressin are in the same magnocellular neurons. *Science* 216:85–87, 1982.
154a. Weber E, Evans CJ, Barchas JD: Predominance of the amino terminal octapeptide fragment of dynorphin in rat brain regions. *Nature* 299:77–79, 1982.
155. Wei ET, Tseng L-F, Loh HH, Li CH: Comparison of the behavioral effects of beta-endorphin and enkephalin analogs. *Life Sci* 21:321–328, 1977.
156. Yaksh TL, Yeung JC, Rudy TA: Systematic examination in the rat of brain sites sensitive to the direct application of morphine: Observations of differential effects within the periaqueductal gray. *Brain Res* 114:83–103, 1976.
157. Yaksh TL: Direct evidence that spinal serotonin and noradrenaline terminals mediate the spinal antinociceptive effects of morphine in the periaqueductal gray. *Brain Res* 160:180–185, 1979.
158. Young WS, Bird SJ, Kuhar MJ: Iontophoresis of methionine-enkephalin in the locus coeruleus area. *Brain Res* 129:366–370, 1977.
159. Zaborszky L, Leranth C, Makara GB, Palkovits M: Quantitative studies on the supraoptic nuclei in the rat. II. Afferent fiber connections. *Exp Brain Res* 22:525–540, 1975.
159. Zaborszky L, Leranth C, Makara GB, Palkovits M: Quantitative studies on the supraoptic nuclei in the rat. II. Afferent fiber connections. *Exp Brain Res* 22:525–540, 1975.
160. Zaborszky L, Leranth C, Marton J, Palkovits M: Afferent brain stem pathways to hypothalamus and to limbic system in the rat. In Lissak K: *Hormones and Brain Function.* New York, Plenum Press, 1973, pp 449–457.
161. Zieglgansberger W, Fry JP: Actions of enkephalin on cortical and striatal neurones of naive and morphine tolerant/dependent rats. In Kosterlitz HW: *Opiates and Endogenous Opioid Peptides.* Amsterdam, Elsevier, 1976, pp 231–238.
162. Zimmerman E, Krivoy WA: Antagonism between morphine and the polypeptides ACTH, ACTH(1–24) and α-MSH in the nervous system. Drug effects on neuroendocrine regulation. *Prog Brain Res* 39:383–394, 1973.
163. Zimmerman EA, Liotta A, Krieger DT: Beta-lipotropin in brain: Localization in hypothalamic neurons by immunoperoxidase technique. *Cell Tissue Res* 186:393–398, 1978.

SECTION 3

Physiology and Function

Central Noradrenergic Nervous System

CHAPTER SEVEN

Anatomy and Physiology of Locus Coeruleus Neurons: Functional Implications

GARY ASTON-JONES, Ph.D., STEVEN L. FOOTE, Ph.D., and FLOYD E. BLOOM, M.D.

INTRODUCTION

During the past several years, many approaches have been used to analyze the functions of the pontine nucleus locus coeruleus (LC). Anatomical studies have demonstrated a uniquely divergent efferent organization, in which fibers from this relatively small number of norepinephrine (NE)-containing neurons project widely throughout the CNS and provide the sole NE input to neo- and paleocortices (97). The potential physiological manifestations of such an ubiquitous brain system utilizing NE as its transmitter agent are implicit in the extensive clinical, behavioral, and pharmacological effects produced by agents which modify the action of NE (97). At the present time the NE-containing locus coeruleus (NE-LC) system is probably the most fruitful focus for functional speculation because more is known anatomically, physiologically, and clinically about this particular NE neuronal system than any other in the brain. In this chapter we review this body of information to critically evaluate functions proposed for brain NE.

As methodological advances have allowed

more diverse investigation of the NE-LC system, additional hypotheses have been advanced for its function. The purpose of this chapter is to critically examine previous proposals for NE-LC function in terms of the known anatomy and physiology of its constituent neurons. Clinically relevant theories of NE-LC function will be reviewed, as well as established aspects of NE-LC afferent, efferent, and intrinsic anatomy, the physiological effects of NE on LC target neurons, and the discharge characteristics of NE-LC neurons themselves. These anatomical and physiological attributes will then be used to compare and criticize proposed functions for the NE-LC. Finally, we will integrate these various data into an updated and coherent view of the role of the NE-LC system in overall brain and behavioral activities.

THEORIES OF NOREPINEPHRINE-LOCUS COERULEUS FUNCTION

Of the numerous theories concerned with possible NE-LC functions, we have chosen to describe seven of those which have generated, at one time or another over the past 15 years, the most intense experimentation aimed at empirical evaluation of their basic premises. In this section we review briefly those observations which motivated and substantiated some of these conceptualizations, as well as observations which are most at variance with a particular theory.

Arousal and Affective Disorders

The most popular version of this theory states that increased release or effectiveness of NE at brain synapses leads to elevated mood or behavioral activity. In early studies, the vast majority of antidepressant (e.g., monoamine oxidase inhibitors, tricyclic antidepressants), sedative and tranquilizing (e.g., reserpine, lithium, chlorpromazine), and stimulant (e.g., amphetamine) drugs appeared to affect brain NE efficacy in both animals and humans in a manner consistent with this hypothesis (reviewed in refs. 126 and 127). Also, corresponding abnormalities in NE utilization were found to accompany clinical depression and mania (also reviewed in ref. 127). The arousal theory obtained further support from infusion and lesion studies in animals. Segal and Mandell (129) found that intraventricular infusion of NE led to increased behavioral activity, as measured by general activity levels or locomotor activity in a free-field situation, and by performance on a continuous avoidance task. Jones *et al.* (69) reported that electrolytic lesions in or near LC in cats produced a decrease in percent waking time that was correlated with the percent depletion of forebrain NE. On the basis of this observation and related pharmacological studies, Jones (68) proposed that the NE-LC system is part of a larger neural network which maintains the waking state. Recently, Adams and Geyer (1) have found that bilateral LC lesions induced by the catecholamine-specific neurotoxin 6-hydroxydopamine (6-OHDA) in rats reduce startle responses to an air puff and that the lesioned animals habituate more rapidly in this testing situation.

Thus, there are numerous observations in animals and humans that are congruent with the hypothesis that whole brain NE, and perhaps the NE-LC system in particular, participates in controlling arousal levels and that abnormalities in this system may participate in the generation of related pathological states. However, many of these studies involve pharmacological or lesion manipulations whose effects are not specific to NE systems, and "arousal" remains a poorly defined term which has been measured by many different, perhaps conflicting, methods (see ref. 28 for a more detailed review of these points). For example, pharmacological manipulations often alter dopamine efficacy, as well as NE utilization. Antelman and Caggiula (8) have argued that there is a complex interaction between dopamine (DA) and NE systems which is evident in the literature just reviewed and that viewing NE as a modulator of DA systems may explain certain contradictions and discrepancies. Additionally, over the last decade the effects of depleting forebrain NE by lesioning ascending NE-LC projections in the dorsal NE bundle (DB) with 6-OHDA have been assessed for many different behavioral tasks (see, for example, the following sections on learning, memory, and reinforcement) without substantial arousal deficits being found. These observations with more specific lesions are not compatible with a major role for NE in maintaining the waking state. Finally, Robinson et al. (120) have reported that 6-OHDA DB lesions or blockade of NE synthesis by alpha-methylparatyrosine injections did not prevent the occurrence of neocortical low voltage fast activity and hippocampal rhythmical slow activity, two electrographic measures of arousal. They also found that LC stimulation was not as effective as reticular formation stimulation in producing these measures.

Learning or Memory

The hypothesis that the NE-LC system plays a crucial role in learning or memory has been more often acknowledged in reports of data incompatible with these notions than in the presentation of substantiating data. In an early report, Anlezark *et al.* (7) reported that rats with bilateral electrolytic LC lesions, which depleted cortical NE to about one-third of control levels, exhibited what the authors interpreted as a learning deficit. Sham-operated and cerebellum- or brainstem-lesioned groups ran faster each day in a runway to obtain a food reward, but LC lesioned animals, while improving each day, did not do so nearly as rapidly. It should be noted that this same result could be equally well interpreted by certain versions of "arousal" or "reinforcement" theories of LC function. Amaral and Foss (4) found that rats with bilateral LC lesions learn to make correct choices in an olfactory nerve-cued T-maze discrimination as rapidly as controls. As in the study of Anlezark *et al.* (7), LC-lesioned animals exhibited longer running times. Since these animals also exhibited decreased activity, Amaral and Foss (4) interpreted the longer running times to be the result of motor deficiencies.

There have been numerous reports of no acquisition deficits on learning tasks following various types of NE-LC lesions. Mason and Iversen (90) found that severe depletions (90% or more) of hippocampal and cortical NE by either 6-OHDA lesions in the DB or by systemic administration of 6-OHDA while the animals were neonates led to no impairment in the acquisition of a runway response for food in adult animals. Roberts *et al.* (119) evaluated the effects of electrolytic LC lesions or 6-OHDA DB lesions on maze learning. LC lesions did not disrupt acquisition of a running response for food reward in an L-maze or acquisition of a position habit in a T-maze. DB-lesioned animals exhibited normal activity levels and habituation to a novel environment. Similarly, Sessions *et al.* (136) performed bilateral electrolytic LC lesions which depleted cortical NE by 37%. They found no deficits in learning one-way active or passive shock avoidance, or in acquisition or extinction of a conditioned taste aversion, or in acquisition of a bar-pressing response for food. Once again, running for food was slower, although times decreased each day; this was interpreted to be the result of a motor deficit. There have been numerous other demonstrations of "normal" acquisition following NE-LC lesions, many of which are mentioned in this chapter in the context of other theories (e.g., see Attention, where the one other example of an acquisition deficit is also mentioned). Much of the available evidence is reviewed by Mason and Iversen (93) and by Clark (28). A recent review by Roberts (118) suggests that there may be certain aspects of the learning process that can be related to NE-LC function.

There have been suggestions that brain NE plays an essential role in memory consolidation. Gilbert (53) postulated that the LC releases increased NE onto cerebellar Purkinje cells following a "satisfactory" movement, altering synaptic strength and producing long-term facilitation of the movement. This proposal has yet to be adequately tested. Crow and Wendlandt (35) have reported that DB-lesioned or diethyldithiocarbamate (a dopamine-β-hydroxylase inhibitor)-treated rats showed normal acquisition of a passive avoidance response, but both groups exhibited a retention deficit. The authors propose that NE is required for formation of long-term memory. Zornetzer and colleagues (11, 156) have reported retention deficits and an alteration of the period during which electroconvulsive shock was effective in producing retrograde amnesia following NE-LC lesions. However, many of the experiments summarized in this and following sections demonstrate normal retention on many different tasks following similar lesions.

Reinforcement

The hypothesis that the NE-LC system is a crucial component in mediating the effects of reinforcement on behavior has typically utilized intracranial self-stimulation (ICSS) as a model behavior. For example, Wise and Stein (153) reported that ICSS supported by electrodes in the medial forebrain bundle was suppressed by drugs which blocked the synthesis of NE. This effect was reversed by intraventricular injections of L, but not D, NE. They reasoned that NE availability may be crucial for ICSS at this site. Crow *et al.* (34) reported that ICSS was supported by bipolar electrodes "closely related to the region of the locus coeruleus and the mesencephalic root of the trigeminal nerve," while other nearby electrodes did not support ICSS. Ritter and Stein (117) reported that 20 of 21 electrodes localized to the LC supported high and stable rates of ICSS. This behavior was increased by amphetamine, decreased by chlorpromazine and alpha-methylparatyrosine, and unaffected by pimozide. Crow and Arbuthnott (32) proposed that the LC receives gustatory information, that it may be activated by consumption of food, and that it may be part of the brain mechanisms controlling reinforcement. German and Bowden (51), in reviewing the literature then available, summarized it as indicating that ICSS induces catecholamine (CA) release, that lesions reduce ICSS in proportion to the severity of CA damage, and that pharmacological effects are compatible with the "CA hypothesis." They noted that the data suggested that both the NE-LC and DA systems supported ICSS.

Conflicting data and interpretations were soon to emerge, however. Amaral and Routtenberg (5), using a standardized protocol which produced ICSS from other brain sites, failed to obtain ICSS from within or adjacent to LC. Clavier and Routtenberg (30) reported that ICSS supported by electrodes in the midbrain tegmentum was not attenuated by ipsilateral or bilateral electrolytic LC lesions, which depleted cortical NE by 80%, or by lesions of the NE ventral bundle. Clavier *et al.* (29) and Corbett *et al.* (31) found that ICSS supported by LC electrodes was not disrupted by lesions of ascending LC projections, and Koob *et al.* (76) found that posterior hypothalamic ICSS was actually potentiated by electrolytic LC lesions. In reviewing the data to date, R. A. Wise (154) pointed out the difficulties posed for the CA theory of reinforcement. First, the observation of ICSS from electrodes in CA systems is correlational in that lesions of these systems do not necessarily disrupt this ICSS. Secondly, although the pharmacological data are consistent with this theory, it has not yet been demonstrated that the disruption of ICSS is a primary effect of any drug treatment. Thirdly, ICSS may not directly reflect the activation of the same reward system that mediates food and other types of reinforcment. To this final point might be added the observation that those learning tasks summarized in the preceding section certainly depend upon reinforcement, and the many cases noted there of normal acquisition are *prima facie* evidence that NE-LC lesions do not disrupt the type of reinforcement essential to the acquisition and performance of that particular task.

Sleep

One of the earliest attempts to identify the NE-LC system with a particular behavior was the hypothesis that this system is crucial for the generation of rapid eye movement (REM) sleep. Jouvet (72–74) postulated that the raphe-serotonin system actively produces slow-wave sleep and that the pontine NE system actively produces, along with other systems, REM sleep. The initial evidence for this hypothesis was the suppression of REM sleep by MAO inhibitors and by LC lesions (see also refs. 68 and 69). One variation on this hypothesis is the proposal that the function of REM sleep is to maintain homeostasis of transmitter levels in NE systems (62). More recent lesion experiments have produced data incompatible with the hypothesis that the NE-LC system actively generates REM sleep. Lidbrink (82) injected 6-OHDA into the mesencephalon in rats and then assessed sleep-waking behavior and EEG. He observed that EEG waking (as characterized by low voltage fast activity) was markedly suppressed following these lesions of the ascending dorsal CA bundle, being replaced by periods of high voltage slow waves and more synchronized activity. Behavioral waking, however, seemed unaffected. There was no change in slow wave sleep time and only a small decrease in REM sleep time. Jones *et al.* (70) made very large radiofrequency lesions of the LC region in cats, causing profound losses of NE in neo- and paleocortices (85% depletion) and thalamus and midbrain (60% depletion). EEG signs of waking occupied the same percentage of time on postlesion day 3 and thereafter. REM sleep time returned to normal by the second week postlesion. Muscle atonia was permanently absent in 7 of 8 animals, and phasic events characteristic of REM sleep were distributed across all stages. Jones *et al.* (70) concluded that "these results indicate that the noradrenaline locus coeruleus neurons are not necessary for the tonic maintenance of EEG activation that occurs in normal wakefulness. . . (or) for the occurrence of paradoxical sleep." Many of the recent results incompatible with an active role for the NE-LC system in generating REM sleep are reviewed by Ramm (113).

A proposal by Hobson and McCarley (63, 94) is based on their observation that many LC neurons slow their discharge rates prior to, and cease discharge during, REM sleep. They postulate that the NE-LC system plays a crucial but permissive role in REM sleep generation, with its reduced activity permitting other neuronal ensembles to elicit the events of REM sleep. Certain aspects of this model concerned with the selective discharge of reticular formation neurons during REM sleep have been challenged (137, 151), and it has been found that LC neurons are not unique in exhibiting slower rates during slow wave sleep and being nearly silent during REM sleep (McGinty and Harper) (95).

Anxiety and Nociception

The most elaborate form of this hypothesis has been proposed by Redmond and collaborators (reviewed in refs. 114 and 115), who suggest that high levels of activity in NE-LC neurons generate anxiety. Their proposal is based primarily on the following behavioral studies in stump-tailed macaque monkeys. Redmond *et al.* (116) reported that unilateral electrical stimulation of the LC in chair-restrained animals initially produced alerting, followed by behaviors such as scratching and yawning. These behaviors were also elicited in anxiety-producing situations and by threat behaviors directed toward the animal. Bilateral LC lesions were found to produce opposite effects, such as the loss of certain fear behaviors (114). Redmond (114) reports that a number of drugs which are known to alter LC or NE function produce appropriate behavioral effects: drugs which enhance such function produce effects similar to LC stimulation and drugs which decrease such function mimic the effects of LC lesions. For example, the benzodiazepines, which have been shown to have anxiolytic effects in humans, monkeys, and rats, have also been shown to decrease brain NE turnover in rats. Another type of data which has been used to buttress this theory is derived from LC neuron recordings in anesthetized rats. These data are reviewed elsewhere in this chapter.

Other investigators have described behavioral effects in rats following 6-OHDA lesions which seem to conflict with the anxiety hypothesis. For example, Crow *et al.* (33) tested several behaviors following bilateral 6-OHDA injections into the DB and the lateral LC. In a social interaction test, the lesioned animals showed a decrease in social contacts in an unfamiliar situation (interpreted as a response to anxiety) of similar magnitude to that seen in the control group. The lesioned animals did engage in more "aggressive" behavior (boxing and wrestling) than did controls. File *et al.* (42) tested rats with similar lesions on social interaction, anxiosoif, and home-cage intruder paradigms. No differences between controls and lesioned animals were seen on the first two tests, and only limited increases in mild agonistic behaviors were seen in the third. Mason and Fibiger (86) have reviewed the data available concerning the anxiety hypothesis, including their finding (87) of no deficits on Sidman avoidance or conditioned emotional response following bilateral DB lesions. These authors suggest that LC lesions may cause a cognitive deficit in monkeys that leads to the observed lack of response to threats. Redmond and Huang (115) have suggested ways in which certain of these new findings could be incorporated into an expanded theory of LC function. In this conceptualization, the LC would function at low levels as a novelty detector or attention focuser, while at higher levels of activity it would perform a combination of cautionary and related cognitive functions.

Cerebral Blood Flow and Metabolism

The hypothesis that the NE-LC system participates in the control of cerebral blood flow has generated much controversy. The observation that cerebral blood vessels are indeed the recipients of neural input immediately generated questions about the source(s) of these fibers. Most input to cerebral vessels is clearly from the superior cervical ganglia (SCG), but the possibility remains that in addition there is innervation from one or more brain sites (these observations and questions are reviewed in ref. 107). Edvinsson *et al.* (39) investigated the distribution of CA fibers in relation to central blood vessels in rats with SCGs removed. In these animals some NE fibers were observed in close approximation to vessels at the light microscopic level. However, the authors caution that "this mere topographical relation has not been considered by us to satisfy criteria for a true innervation of brain vessels by way of these central adrenergic nerves." Swanson *et al.* (148) presented ultrastructural evidence of central monoaminergic innervation of blood vessels in the paraventricular nucleus of the hypothalamus. Rats with SCGs removed were injected intraventricularly with 5-OHDA to enhance visualization of dense-core vesicles. Of 600 complete blood vessel profiles examined, five cases were found of varicosities with dense-core vesicles making contact with the basal lamina surrounding capillaries. No membrane specializations were apparent adjacent to the basal lamina. The varicosities were thought to resemble NE or E varicosities. Raichle *et al.* (111) tested the hypothesis that the NE-LC system is analogous to the peripheral sympathetic system in controlling blood flow and vascular permeability. These parameters were measured in monkeys with and without SCGs removed. Carbachol stimulation of LC reduced hemispheric cerebral blood flow and increased vascular permeability to water. Intraventricular administration of phentolamine (an alpha-adrenergic blocker) had the opposite effects. The authors interpreted these data as supporting their hypothesis.

Another hypothesis, forwarded by Harik and collaborators (60), is that the NE-LC system is a crucial component in controlling cerebral oxidative metabolism. LaManna *et al.* (79) reported that unilateral LC lesions did not produce a change in cortical "resting" metabolic balance, as determined by *in situ* reflection spectrophotometry of the redox state of cytochrome oxidase. However, when cortex was electrically stimulated, the transient metabolic response was slowed, in terms of both cytochrome oxidase and oxidized nicotinamide adenine dinucleotide (NAD^+) states. The authors interpret this result as indicating that NE has a role in modulating the response to increased metabolic demand in the cerebral cortex. However, Harik *et al.* (59) have also reported that although this deficit was evident 2 weeks after the NE-LC lesion, by 4 weeks after the lesion all of these parameters had returned to normal.

Much of the vigor of hypotheses that the NE-LC system plays an essentially "autonomic" function in the brain has been drawn from a claim that the axons of this nucleus usually do not form "classical synapses" (36, 37). Recent data (75, 106) suggest that a high percentage of NE varicosities do indeed form synapses in many brain regions.

Attention

Much interest has been focused on this hypothesis in the last few years. The most current form in which it appears is that the NE-LC system acts to filter out distracting or irrelevant stimuli while the animal is engaged in a given behavior requiring attention to some particular set of stimuli. Much of the theorizing about the role of NE in attention is centered around the "dorsal bundle extinction effect" (DBEE, see ref. 93 for review). Mason and Iversen (90, 91) found that rats with DB lesions exhibited prolonged extinction of a runway response and on a bar-press continuous reinforcement schedule. They also found that these animals exhibited perseveration in a successive light-dark discrimination task. A similar deficit was reported by Mason and Fibiger (85) for a spatial reversal task. Tremmel *et al.* (150) reported a similar prolongation of extinction following DB lesions. Mason and Iversen (92, 93) suggested an attentional basis for the DBEE, specifically that the function of the DB is to focus attention to relevant stimuli. The DB lesion is hypothesized to prevent this, and the animal attends to many irrelevant stimuli during acquisition. When nonreward conditions are instituted, extinction occurs more slowly because the animal must sample many possible cues as having been associated with reward. This develops an earlier suggestion by Roberts *et al.* (119), who had reported that DB lesioned animals were more distractable during performance of a previously trained response and suggested that the NE-LC system might be involved in attention. Oke and Adams (105) reported that animals treated neonatally with 6-OHDA were more distracted by stimuli irrelevant to a learned task than were controls. In two studies, DB lesioned animals have actually performed with greater accuracy than controls on tasks where a lack of focused attention would be predicted to enhance performance (84, 89).

In one of the few pieces of contrary data to date, Crow *et al.* (33) found that DB-lesioned rats exhibited normal habituation and distractability in exploration and water-licking paradigms.

Given this brief description of previously hypothesized functions for the NE-LC, it is clear that these data alone lead to no single unifying model of NE-LC function. We feel that a cellular physiological approach to this question promises more lucrative results. In particular, certain recent results on the anatomy, postsynaptic physiology, and normal discharge properties of NE-LC neurons lead us to a new functional viewpoint for understanding this system. Our hypothesis is briefly outlined following a review of the most pertinent recent data.

ANATOMY OF THE NOREPINEPHRINE-LOCUS COERULEUS SYSTEM

The anatomy of the locus coeruleus, including its efferent projections, has been reviewed extensively several times in the last few years and will be described in only a cursory fashion in this chapter (for reviews see refs. 6, 28, and 97). Following this very brief overview, more detailed attention will be given to recent developments in NE-LC neuroanatomy which have functional implications.

Overview

The locus coeruleus of the rat, monkey, or human is a well-delineated cluster of neurons adjacent to the fourth ventricle in the pontine brainstem (Figs. 7.1 and 7.2). These neurons have been shown by chemical and cytochemical methods to contain NE and the enzymes necessary for its synthesis. The NE-LC consists of some 1600 neurons in the rat (147), some 8000 neurons in *Macaca mulatta* (67), and some 15,000 neurons in man (19, 152). Swanson (147) delineated dorsal and ventral subdivisions of the rat LC, while Grzanna and Molliver (57) have described four subdivisions. This relatively small nucleus has been shown to project to nearly every major CNS region, from the spinal cord to the neocortex. Although the pathways and terminal arborization patterns of this massive efferent system have been studied extensively at the light-microscopic level, certain aspects of NE-LC efferent anatomy are poorly understood. For

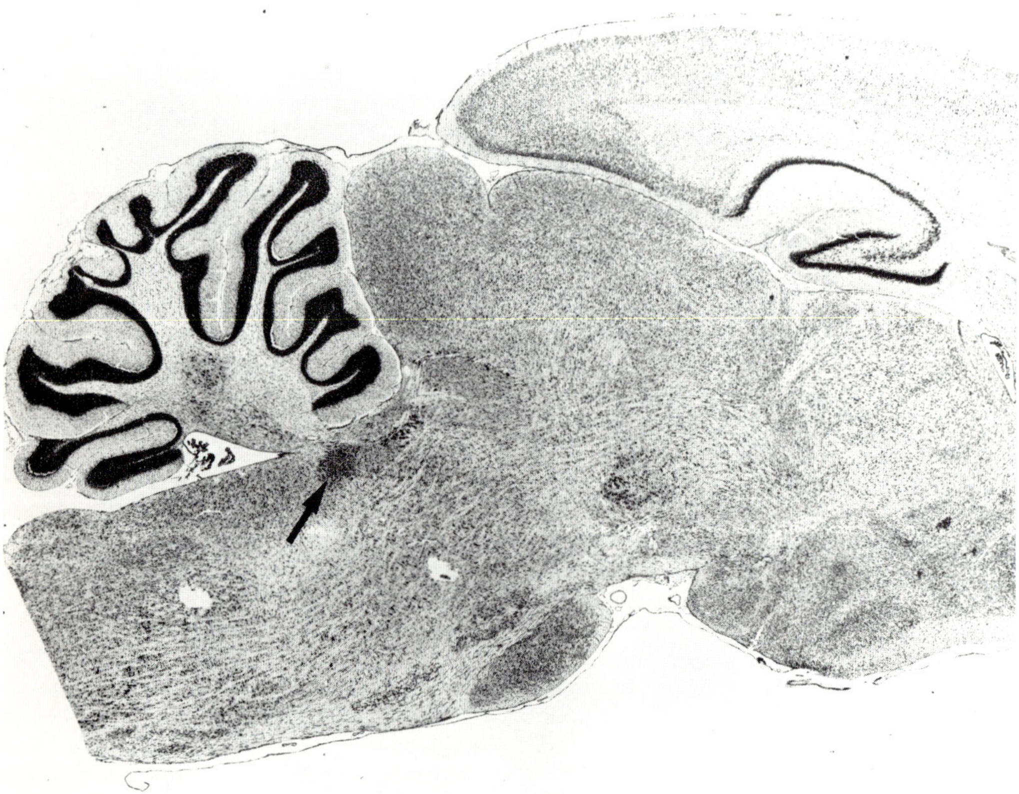

Figure 7.1. A Nissl-stained parasagittal, 12-μm section through the locus coeruleus of a rat brain. The compact nucleus of NE-LC neurons in the dorsorostral pons is indicated by the *arrow*.

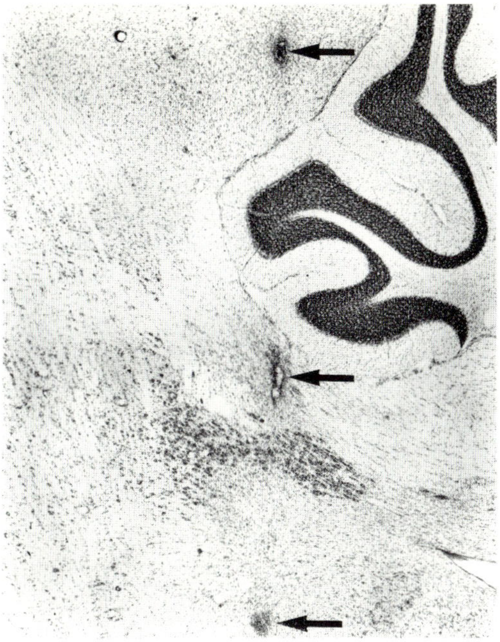

Figure 7.2. A Nissl-stained sagittal section through the locus coeruleus of a squirrel monkey brain with marking lesions made during a microelectrode penetration indicated by *arrows.* The compact nucleus of NE-LC neurons is located along the electrode tract between the middle and bottom marking lesions. (Reproduced with permission from S. L. Foote *et al.* (43).)

example, the organization of LC with respect to which neurons project to which terminal areas has been studied in only an exploratory way (46). At the ultrastructural level, it is unknown which particular neurons in a given terminal region receive NE-LC input. The afferent projections to LC are likewise poorly understood. It seems that numerous brain regions of unknown or poorly understood function project to LC (see 25 for the most extensive study of LC afferents to date). Since the functional properties of these afferent sources are unknown and presumably diverse, they yield few simple deductions about the possible function of the NE-LC system.

Recent Developments

Two recent advances in describing NE-LC anatomy, one at the light-microscopic level and one at the ultrastructural level, have provided a substantially new anatomical perspective on this system. At the light-microscopic level, Morrison and collaborators (98–100) have provided a detailed description of the NE-LC innervation of neocortex in the rat. These recent reports utilize DBH immunocytochemistry and describe a dense and highly ordered innervation of this region (Fig. 7.3). The NE innervation is both laminated and tangential, characteristics with specific functional implications in terms of the potential impact of the NE-LC system on neocortical information processing. Caviness and Korde (22) have discussed some possible implications of this laminar neocortical innervation by monoamine fibers in terms of which cortical cell types may be the primary recipients of monoamine input.

At the ultrastructural level, Olschowka *et al.* (106) have used DBH immunocytochemistry to describe the morphology of NE varicosities in several terminal regions. No DBH-pos-

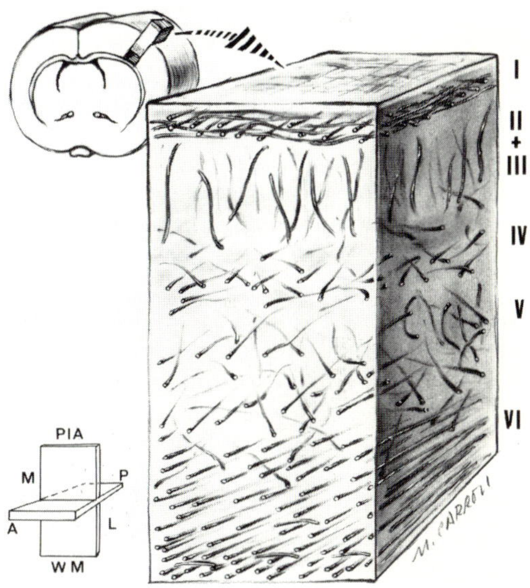

Figure 7.3. This schematic 3-dimensional illustration shows the orderly geometric pattern of noradrenergic innervation common to all areas of neocortex in the rat. The prevailing orientation of the NA axons within each cortical layer is emphasized. Layer I contains long, tangential fibers oriented at right angles to each other. Layers II and III are typically traversed by numerous straight radial fibers which rarely extend through layer I or IV. Layers IV and V contain many short oblique axon segments, while layer V also has occasional radial and tangential fibers. The NA fibers in layer VI are, with few exceptions, tangential and have an anteroposterior orientation. Cortical layers are indicated with roman numerals. *A*, anterior; *P*, posterior; *M*, medial; *L*, lateral; *WM*, white matter; *PIA*, pia mater. (Reproduced with permission from J. H. Morrison *et al.* (98).)

itive varicosities were observed in contact with intracerebral capillaries. 58% of all DBH-positive varicosities were found to form axo-dendritic synapses characterized by specialized junctional appositions. The authors conclude that "the present demonstration that NA axons typically form synapses suggests that the action of central NE is mediated by conventional neurotransmission and that central NE neuronal circuitry is characterized by a high degree of functional specificity that is determined by the spatial distribution of specialized junctions between presynaptic varicosities and specific neurotransmitter receptors."

These two lines of anatomical research stand in direct contradiction to the perspective that the NE-LC system is diffusely organized or that it must be supposed to serve some very general function with little anatomical or informational resolution.

POSTSYNAPTIC PHYSIOLOGY OF CENTRAL NORADRENERGIC CIRCUITS

The major qualitative effect of iontophoretically applied NE is to depress the spontaneous activity of test cells in nearly every region of the brain (17, 97). These effects are antagonized by beta-adrenergic blockers, prostaglandins, certain antipsychotics, and acute application of lithium. Recent evidence suggests that excitatory actions of iontophoretic NE on principal cells of the rat lateral geniculate are antagonized by alpha-adrenergic blockers (121), as are inhibitory responses of mitral cells in the olfactory bulb (18). Despite the overall congruity of iontophoretic results, the ability to study responses of identified NE-LC target cells during LC activation provides data which may be more reflective of the means by which LC regulates the responsiveness of these target cells to their other afferent synaptic connections.

Activation of the Coeruleocerebellar NE Pathway

Purkinje cells show remarkably uniform inhibitory responses to stimulation of the locus coeruleus at a latency of approximately 125 msec (66). Antipsychotic phenothiazines (47), cobalt, lanthanum, lead (48), lithium (139), as well as prostaglandins (66) antagonize the effects of LC stimulation. Intracellular recordings from Purkinje cells during single shock stimulation of LC reveal small, long-latency (about 150 msec) hyperpolarizations. With trains of pulses to LC, larger hyperpolarizations are observed, outlasting the stimulation period. Input resistance either increases or does not change during these LC-evoked hyperpolarizations. Identical effects, *i.e.*, hyperpolarization of the membrane potential with either minimal change or increased input impedance of the membrane, characterize the actions of NE during intracellular recordings from cerebellar Purkinje cells (66, 142, 143) spinal motor neurons (40) and hippocampal pyramidal neurons (80, 81, 135).

In physiological studies of hippocampus (132–134), septum (130), and spinal trigeminal nucleus (123, 124), the effects of LC activation are generally comparable to those described for the cerebellum with regard to overall rate changes; these are most simply summarized, at first glance, as depressing activity.

Molecular Mechanisms of Central Catecholamine Synapses

With the demonstration that rat cerebellar Purkinje and hippocampal pyramidal neurons are contacted by NE-LC axons and give uniform inhibitory responses to NE and to LC stimulation, subsequent studies investigated further the mechanism of this inhibitory response. In both test systems, cyclic 3′,5′-adenosine monophosphate (cyclic AMP) mimics the ability of NE to depress spontaneous activity (64, 66, 131, 132, 135, 142, 143).

To pursue the possibility that this action of cyclic AMP in the CNS might reflect a "second messenger" (145) for central noradrenaline synapses, many additional experiments have been conducted (15, 16, 64–66). Biochemical evidence had already suggested that NE could elevate cerebellar cyclic AMP levels through beta-adrenergic receptors (112; also see refs. in ref. 16). Recently, methods for direct microscopic localization of beta-receptors (108) show that Purkinje neurons in the cerebellum and pyramidal neurons in the hippocampal and cerebral cortex seem to be the major cell classes showing beta-receptors. Furthermore, parenteral administration of phosphodiesterase inhibitors such as aminophylline or theophylline potentiates noradrenergic depression of Purkinje cells, while iontophoretic administration of these methylxanthines, and of papaverine, converts weak excitant actions of iontophoretic cyclic AMP into pronounced depressions (64–66, 140). These observations led to the proposal that the action of NE (64) and of NE-LC synaptic projections to Purkinje cells (142, 143) could be mediated by cyclic AMP.

Subsequently, this proposal has been strengthened by observations that the actions of NE (142), the NE-LC pathway (66, 143), and cyclic nucleotides (142) all hyperpolarize Purkinje cells through similar membrane actions in which conductance to passive ion flow is decreased or unchanged. The cyclic AMP mediation of the NE actions also finds support from the observations that prostaglandins and nicotinate (140, 141) selectively block NE effects on Purkinje cells and hippocampal pyramidal cells (131, 132), as they do in the cyclic AMP-mediated adrenergic responses of adipocytes (145). In fact, all the substances shown to block the response of Purkinje neurons to iontophoretic NE also show very good antagonism toward the activation of cerebellar adenylate cyclase by NE (102, 103). Even more direct confirmation of the second messenger hypothesis stems from the observation that topically applied NE or stimulation of the NE-LC pathway increases the number of Purkinje cells reacting positively to an immunocytochemical method detecting bound intracellular cyclic AMP (16).

Finally, cyclic AMP is known to activate a class of enzymes termed protein kinases (56), which phosphorylate-specific protein substrates; this process may be the mode of expressing altered cyclic AMP levels (56). In cerebellum, chemical analogs of cyclic AMP mimic the inhibitory effects of NE in direct correlation to their ability to activate brain cyclic AMP-dependent protein kinase (138). The hyperpolarizing effect of NE-LC input on cerebellar Purkinje cells to increase membrane resistance contrasts markedly with changes seen with classical inhibitory transmitters such as GABA (142); in such cases, the hyperpolarization is associated with a decrease in membrane resistance. Intracellular recordings indicate that the hyperpolarization produced by NE, on the other hand, is accompanied by, and perhaps generated by, a decrease in passive ionic conductance across the membrane (that is, an increase in membrane resistance) (143). In this respect NE activity differs from conventional postsynaptic responses which are mediated by increased ionic conductance. The action of NE may be due to a decrease in conductance to some ion such as sodium or calcium, or to activation of an electrogenic pump (81, 109). However, the depressant actions of NE do not require extracellular calcium (50), nor do inhibitory substances, which also "antagonize" calcium actions, interfere selectively with NE inhibitory effects (48).

The NE-LC System and Cellular Integration

To recapitulate the preceding details, NE-LC activity inhibits simple spike activity in Purkinje cells and other identified target cells. The mechanism of synaptic action of LC on its target cells may have some rather unconventional aspects, such as coupling to cyclic AMP formation and possible selective effects on pacemaker activity. These properties suggest that the NE-LC may have a more complex impact on the postsynaptic integration of synaptic signals than "conventional" synaptic inhibition. One proposal to explain the overall array of cellular controls exerted by the LC system is termed an"enabling" hypothesis (17, 97), by which coactivity in LC terminals enables other systems converging on the same target neurons to transmit more effectively during the period of simultaneous activity. This same end point has been stated differently as an enhancement of signal-to-noise ratios for the effects of specific components within the afferent system simultaneously active with the NE-LC (45, also see refs. 134 and 155).

Available evidence suggests that the NE-LC system may be capable of enhancing, or "enabling" (17) rather than inhibiting, the effects of stimulation via other synaptic pathways. First, as discussed above, the effect of LC activity on Purkinje cells does involve an apparent decrease in membrane ionic conductance. It has been shown in Aplysia sensory neurons (20), and in neurons of the bullfrog sympathetic ganglion (128), that the generation of conductance decreases by one synaptic pathway results in increased transmission through other synaptic pathways functioning by means of classical conductance-increase mechanisms. Facilitory effects of NE or LC stimulation on the response of Purkinje cells to synaptic stimulation have been demonstrated (49), as have the enhancing effects of similar manipulations for both excitations and inhibitions of hippocampal pyramidal neurons (133, 134).

However, such effects may not be fully explained by the observed conductance decrease. It has been shown that facilitory effects of NE iontophoresis or LC stimulation on synaptic responsiveness outlast effects on tonic firing rate and can be demonstrated even when LC stimulation does not detectably alter the spontaneous rate of cerebellar Purkinje cells. Moreover, in cerebellum, such facilitory effects of NE occur for some transmitters (e.g. gamma-aminobutyric acid (GABA), endogenous to cerebellum) but not others (e.g., glycine, not found in cerebellum) (155). The mechanisms mediating these heterosynaptic interactions are not known.

The fact that NE-LC input reduces endogenous or "background" simple spike activity, while facilitating evoked activity, has led to the proposal that LC may act to increase the "signal-to-noise" (S/N) ratio of transmission via the Purkinje cells (49). The same proposal was made earlier for NE effects on auditory cortical neurons in the monkey (45). Such an interpretation seems consistent with our proposed role for the NE-LC in signalling or regulating levels of vigilance (9–11, 43; also, see text below). Mossy fiber input to cerebellar cortex is relayed by parallel fibers and excites Purkinje cells which lie on the parallel fiber "beam." Parallel fibers also excite basket and stellate cells, which exert an inhibitory effect on Purkinje cells located off the parallel fiber "beam." This system has been suggested to provide an "inhibitory surround" for incoming excitatory signals (38). This effect is considered (38) to be analogous to the lateral inhibition found in limulus eye, for example, where it serves a contrast-enhancing function (61). Since NE appears to potentiate both the on-beam excitatory effects and off-beam GABA-mediated inhibition (see refs. 96, 155), LC activity may also serve to regulate the strength of lateral inhibition in the cerebellar cortex.

Lateral inhibitory systems appear to provide sharp separation of input activity (61). Activation of LC might restrict the area of cerebellar cortex activated by a given mossy fiber input and enhance the spatial precision with which mossy fiber systems act on the cerebellar cortex. Likewise, a decrease in LC activity may decrease the precision of, and increase the interaction between, neighboring mossy fiber projections. Under this hypothesis, LC could act to change the character of the signal and the way it is processed, rather than simply favoring the transmission of evoked activity over putative "noise." The cell-cell functional abstractions of "contrast-enhance," signal/noise bias, or "enable" all bespeak of integrative properties congruent with the widely distributed anatomy of NE-LC efferents. Although this integrative conceptualization of LC function cannot yet be conclusively documented, further support for such views can be derived from consideration of the conditions under which this LC system is active behaviorally.

DISCHARGE CHARACTERISTICS OF NE-LC NEURONS

Anesthetized and Paralyzed Preparations

In anesthetized preparations, NE-LC cells have been reported to discharge spontaneously in a slow, regular, tonic fashion (2, 3, 14, 21, 23, 41, 54, 101, 104, 146, 149). The only physiological sensory stimuli reported to affect this pattern even slightly are strongly noxious. Painful stimuli, such as pressure applied to the tail and paws (23) or electrical stimulation to the sciatic nerve (12, 149), generate a transient increase followed by a more prolonged decrease in activity (Fig. 7.4). Light strokes on the body, bright flashes into the eyes, or loud auditory stimuli produce little, if any, response in NE-LC discharge (24,54; unpublished observations). Although most studies have used chloral hydrate anesthesia, there have been a few reports using urethane anesthesia (41, 101, 149) or gallamine-induced paralysis (21, 77), and we have made unpublished observations using halothane anesthesia. The basal firing rate of NE-LC neurons varies with these different treatments: mean rates of 1.1 Hz, 2.6 Hz, 6 to 30 Hz, and 2.4 Hz are seen with chloral hydrate, urethane, gallamine, and halothane, respectively. Responses are observed to noxious peripheral stimuli but not to visual, auditory, or nonnoxious somatosensory stimuli (where tested), regardless of the anesthetic agent used. Increased sensitivity to somatosensory stimuli is reported in unanesthetized paralyzed preparations (77). Electrical stimulation of peripheral (vagal, splanchnic, and sciatic nerves) and central (bed nucleus of the stria terminalis, olfactory bulb, preoptic region, ventromedial nucleus of the hypothalamus, ventral tegmentum, central grey) sites has been demonstrated to produce converging or-

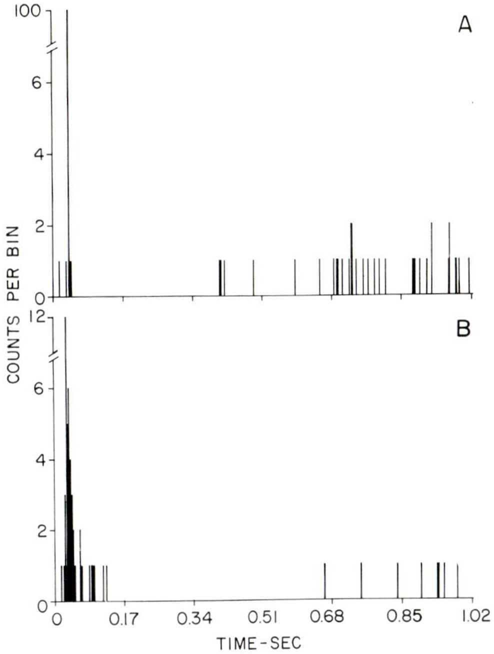

Figure 7.4. *Panel A* is a poststimulus time histogram for a rat NE-LC neuron, accumulated over 100 sweeps of antidromic stimulation from the cingulate cortex. The large sharp peak in the histogram shows antidromically driven spikes at a latency of about 40 msec. Note the pronounced decrease in activity shortly following driven spikes. *Panel B* is a poststimulus time histogram accumulating 100 sweeps for the same rat NE-LC neuron following electrical stimulation to the contralateral foot. Orthodromically driven spikes are apparent in the peak of activity 25 to 50 msec following stimulation. Note the pronounced suppression of activity following orthodromic driving, similar to that seen following antidromic activation in *panel A*.

thodromic excitatory and inhibitory influences onto NE-LC neurons (2, 149).

NE-LC neurons have also been antidromically activated by electrical stimulation of many of their target areas (2, 13, 41, 52, 101, 149). This property not only identifies these cells, it also permits analysis of various aspects of their physiology. LC neurons in the rat conduct slowly with a velocity in the range of 0.4 to 1.3 m/sec. Stimulation of either the cingulate cortex or midbrain NE-LC fiber areas reliably produces a pronounced reduction in activity at slightly longer latencies than antidromically driven spikes, even in cases where the recorded cell is not one that is antidromically activated. This postexcitatory pause in discharge, similar to that seen in the response to noxious peripheral stimulation (Fig. 7.4), is altered by various pharmacological manipulations in a manner consistent with a mechanism of recurrent collateral inhibition among NE-LC neurons (2).

We have studied (13) the impulse conduction properties of NE-LC axons utilizing patterns of activity characteristically found in our recordings from behaving animals (see Unanesthetized Animals). Our results demonstrated that NE-LC axons vary impulse conduction velocity as a function of previous activity along the fiber. The manner of conduction velocity fluctuation allowed us to conclude that short-lasting, high frequency bursts of impulses should be the maximally effective pattern of NE-LC activity for affecting target cells (Fig. 7.5). These findings are useful in describing optimal parameters for direct stimulation of NE-LC pathways, as well as for describing the patterns of activity seen in behaving animals that may reflect maximal information flow in the NE-LC system (see Unanesthetized Animals).

Numerous studies describe the effects of various injected or iontophoresed pharmacological agents on the electrical activity of NE-LC neurons in anesthetized rats. Systemically administered amphetamine (21, 54), morphine (77), clonidine (3, 146), and tricyclic antidepressants (104) slow discharge rates, while systemic piperoxane, an alpha-adrenergic antagonist, increases neural activity (2). In more direct tests of the pharmacological sensitivity of NE-LC neurons themselves, iontophoretic NE, epinephrine, GABA or isoproterenol (23), clonidine (23, 146), and opiates (14) typically reduce firing rates of NE-LC neurons, while iontophoretically delivered piperoxane (23), substance P (58), and acetylcholine (14) produce increased activity.

In conclusion, from the data available on the physiology of NE-LC neurons in anesthetized animals, it might be surmised that in the awake intact animal NE-LC neurons would be behaviorally unresponsive, except for some involvement with pain responses and mechanisms (however, data from behaving rats and monkeys in our laboratory have shown that this is not the case; see below). In the pharmacological data, the fact that NE-LC neurons are sensitive to opiates is consistent with such

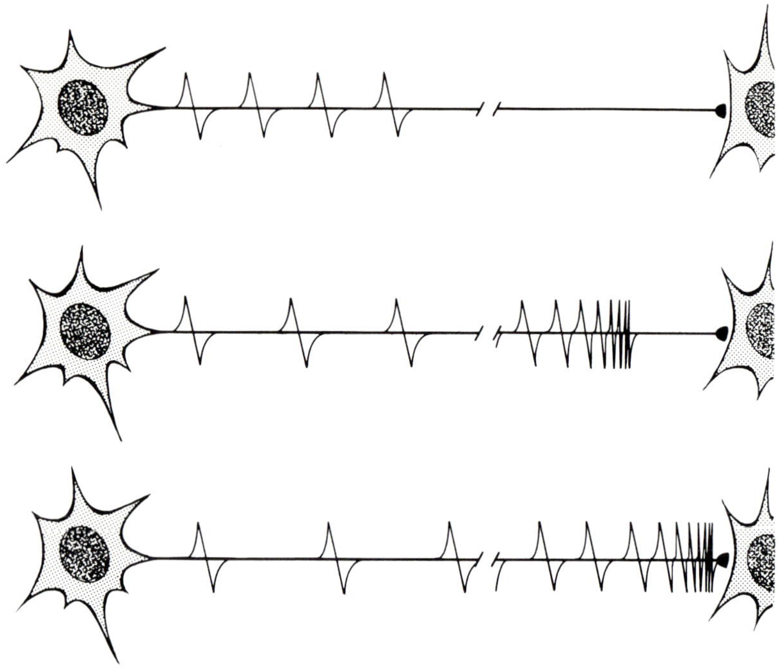

Figure 7.5. A schematic diagram to illustrate the postsynaptic effects of the variability of impulse conduction along NE-LC axons. As a high-frequency train of activity leaves the soma (*top panel*) and traverses the fiber of an NE-LC neuron, interspike intervals change as a result of variation in impulse conduction velocities. As shown in the *middle* and *bottom* panels, early spikes in the train will show increased velocity and bunch together, while later spikes in the same train will exhibit decreased velocities and therefore tend to spread apart in time. Such a result predicts that high frequency, short duration trains of activity would be most effective of NE-LC target cells (see Aston-Jones et al, 1980).

a hypothesis, while the other drug treatments lead to no clear physiological or behavioral hypotheses for NE-LC functions.

Unanesthetized Animals

Hobson and McCarley and collaborators (94, 144) have described 21 cells recorded in the cat LC region which possess the distinctive, and nearly unique, property of exhibiting lower discharge rates in REM sleep than in waking or slow wave sleep (SWS). These cells exhibit slow regular activity during waking, which is slightly reduced during SWS, while only occasional action potentials are observed during REM sleep. They report, however, that other cells within the LC region do not exhibit this "REM-Off" property and that some REM-Off cells are located outside LC. Recording site localizations did not include cytochemical characterizations. This physiological heterogeneity may reflect the anatomical heterogeneity of cat LC where NE and non-NE neurons are loosely interdigitated. In our own recordings from rat LC, which is composed entirely of NE neurons, we found that NE-LC neurons homogeneously exhibit such REM-Off characteristics (see below). Hobson and McCarley have not reported whether these REM-Off (or other) LC neurons alter discharge rate within the waking state as a function of arousal or sensory stimulation.

Sakai (122) reports recordings from 108 REM-Off neurons in the cat LC region. Such cells were recorded only from regions shown in separate animals to contain NE neurons, and cytoarchitectonic areas containing many NE neurons exhibited a higher percentage of REM-Off cells than areas containing few NE neurons. However, many REM-On cells were recorded in the same areas as REM-Off cells, and cytochemistry of recording sites was not done. No description is given of the activity of REM-Off neurons in waking, except their mean discharge rate of 0.5 to 2.5 Hz.

The only other descriptions of neuronal activity within cat LC are those of Chu and

Bloom (26, 27). These investigators attempted to verify recordings from NE neurons histochemically by creating microlesions or Prussian blue spots at recording sites and processing the tissue for catecholamine fluorescence. If the marking spot or lesion was within a cluster of fluorescent neurons, that particular recording was considered to have originated from an NE-containing neuron. Chu and Bloom (27) report that only one-fourth of their NE neurons showed their slowest rates during REM sleep, and they also observed such cells in non-NE recording sites. Of the three-fourths of the NE cells that exhibited substantial activity during REM sleep, some cells were phasically active while others were tonically active. For a subset of NE neurons, generally enhanced activity was observed during active waking (*i.e.*, vigilant surveillance of the environment) as opposed to quiet (*i.e.*, nonattentive) waking.

A study in rhesus monkey primarily reported that NE-LC neurons are antidromically activated from forebrain sites which support self-stimulation; here, German and Fetz (52) recorded the activities of 64 locus-coeruleus and subcoeruleus region neurons. They reported that cells within LC had slow (5 ± 3 Hz) discharge rates during quiet waking, while subcoeruleus cells exhibited higher (15 ± 2 Hz) rates. These rates did not change when the animals engaged in operant responding for an apple sauce reward. Activity during sleep was not studied. The accuracy of histological verification of recording sites is difficult to evaluate for this study since, in the two monkeys used, 46 and 40 penetrations were made over 6 and 2 months, respectively. Both the large number of penetrations and the long interval between recording and sacrifice would seem to render accurate reconstruction nearly impossible.

In summary, these studies demonstrate that in the cat there are REM-Off cells in the LC region whose distribution is roughly the same as that of NE-containing neurons. However, the cytochemical heterogeneity of neurons in the cat LC prevents assignment of the REM-Off property to NE or to non-NE neurons. In addition, in none of these studies were there systematic measurements of changes in LC activity during waking as a function of arousal changes, sensory stimulation, or behavior.

We have succeeded in recording unitary action potentials from NE-LC neurons in behaving squirrel monkeys (43, 44) and rats (9–11, 43, 77) over periods sufficiently long to permit detailed analysis. During recording, animals were presented with an extensive array of sensory stimuli, were allowed to remain undisturbed for periods of up to 2 hours while spontaneous sleep and waking were observed, were presented food (monkey), and their motor behaviors were closely observed. In both species, LC neurons exhibited slow tonic discharge rates (0 to 10 Hz) which tended to be higher during periods of cortical EEG desynchronization and behavioral alertness than during periods of drowsiness and EEG synchronization (see Figs. 7.6 to 7.8). During drowsiness and slow-wave sleep, long periods (many seconds) without a single action potential were common. During paradoxical sleep (PS), a state we have studied only in the rat, NE-LC neurons were virtually silent (Fig. 7.7). In both species, NE-LC neurons responded with slow bursts of action potentials to both simple (tone, light flash, mild touch; Fig. 7.8) and complex (food, novel objects; Fig. 7.9) stimuli. Poststimulus time histogram (PSTH) displays computed from some of the rat data indicate response latencies of 15 to 25 msec to tone pips and about 60 msec to light flashes (Fig. 7.8). Also, these responses decreased in intensity with decreasing levels of vigilance (11). Although NE-LC neurons frequently burst during orienting movements, we have not observed a consistent relationship between discharge activity and the performance of any specific motor act. Finally, we have found that the activity patterns of NE-LC neurons are largely homogeneous throughout the nucleus and that neurons in rat multiple-unit populations show markedly synchronous discharge during periods of intense activity (10, 11). Thus, NE-LC neurons appear to function in concert as an homogeneous ensemble, varying their discharge as a function of vigilance changes and nonnoxious (as well as noxious) sensory stimuli.

CRITICAL ANALYSIS OF PREVIOUS THEORIES

The above data are useful in reevaluating many of the prominent hypotheses of NE-LC function. In particular, these data are most relevant to proposals implicating this system in arousal and affective disorders, sleep, anxiety and nociception, metabolism and cerebral blood flow, and attention.

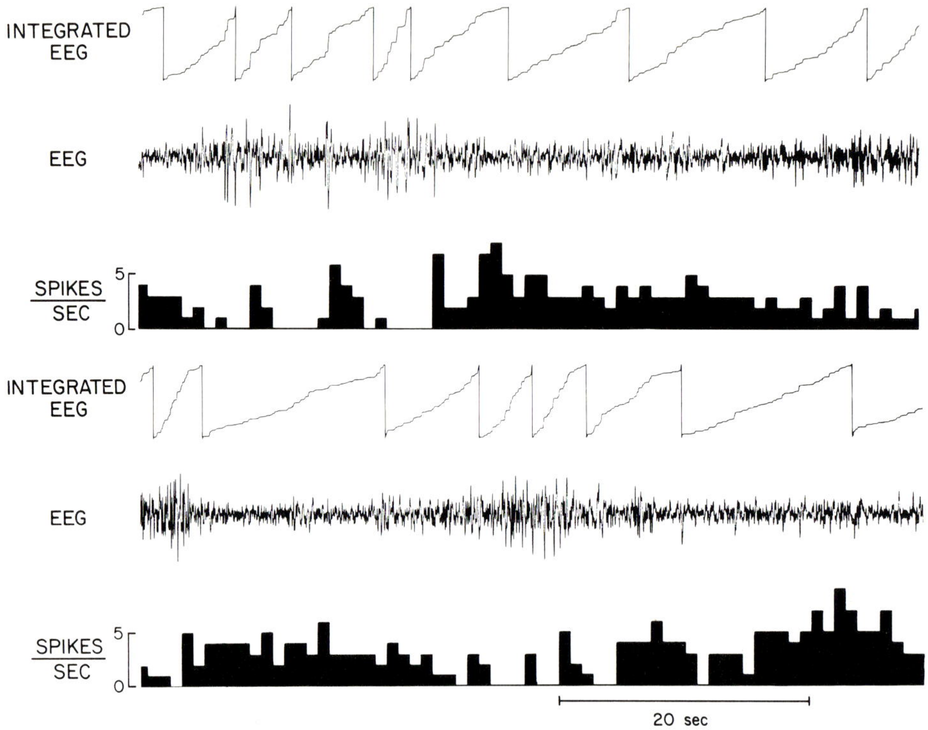

Figure 7.6. Typical relationship between monkey LC neuronal activity (third and sixth traces) and cortical EEG during an episode of behavioral wakefulness. Unit activity is higher during periods of low amplitude EEG. This observation was quantified by correlating the number of action potentials in each 1-sec epoch with the integrated EEG value (*i.e.*, mean amplitude) for that epoch. Changes in the activity of rat and monkey LC neurons often anticipated spontaneous EEG changes.

Arousal and Affective Disorders

The hypothesis that the NE-LC system is involved in the etiology of affective disorders is closely related to and dependent upon an integral role for these cells in general arousal processes. In the recent data reviewed above, it is clear that both the postsynaptic physiology of NE and the discharge characteristics of NE-LC neurons serve to strengthen the view that this system may be critically involved in the control of arousal. The increased S/N ratios in target neurons subsequent to a release of NE could serve to enhance arousal by globally increasing neural transmission in specific channels throughout the CNS. The fact that NE-LC discharge is closely related to changes in vigilance (and, in fact, usually anticipates such changes) (10) forms a tight link between arousal and the release of NE, in addition to the relationship between arousal and the postsynaptic effects of released NE. These recent data provide tantalizing but inconclusive evidence for a causal relationship between disease processes and NE-LC function.

Sleep

One of the oldest and best substantiated hypotheses for NE-LC function is its role in the control of sleep-waking processes. Our data on the discharge of NE-LC neurons in behaving animals are consistent with the popular view that the NE-LC system participates in controlling cortical and behavioral arousal. However, our results may help to discriminate among several specific hypotheses within this general arousal framework. From lesion studies in cat, Jouvet (74) concluded that, in addition to maintaining tonic behavioral and cortical arousal, NE-LC neurons may be executive elements for PS, such that this stage would be critically dependent upon their robust discharge. However, NE-LC neurons markedly decrease their discharge prior to and throughout PS, directly opposite to the pattern re-

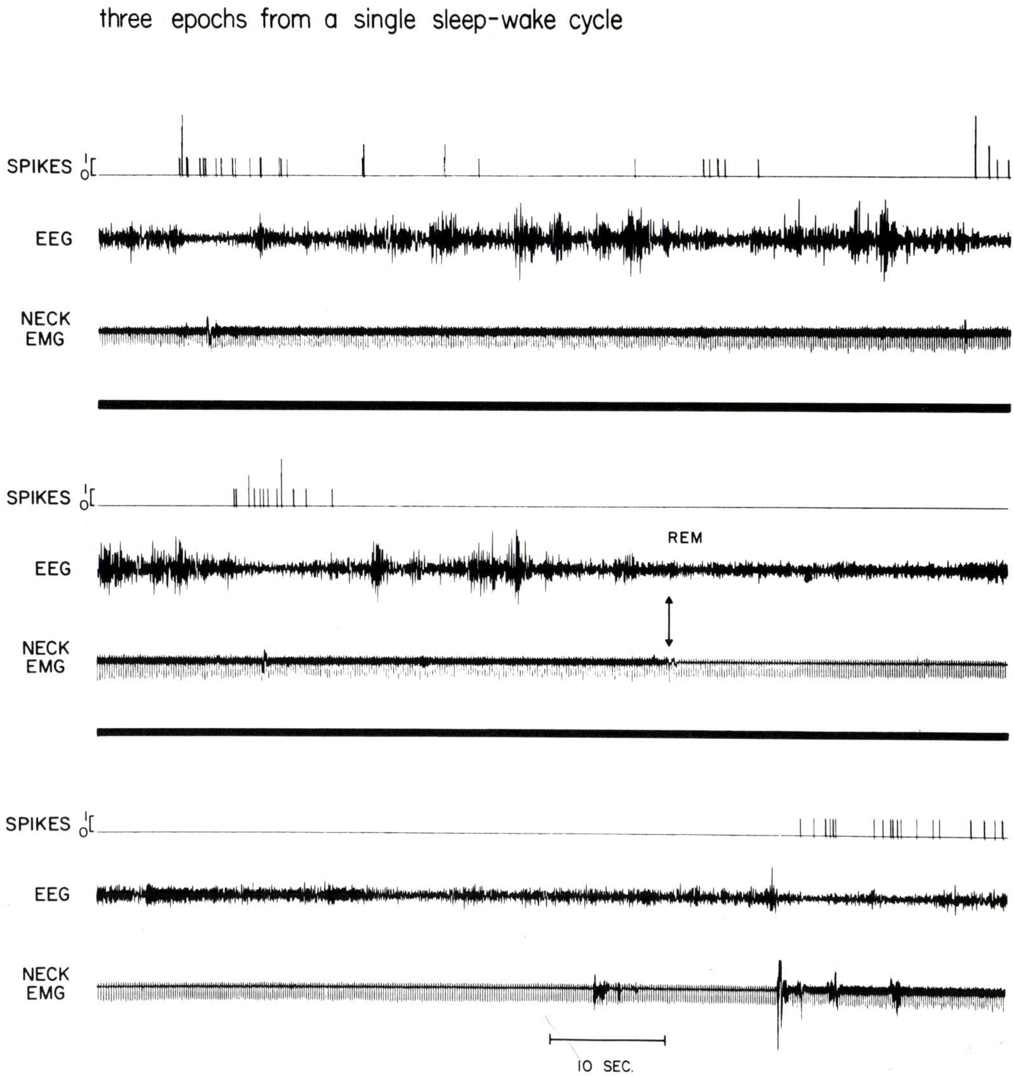

Figure 7.7. A single sleep cycle record demonstrating typical NE-LC single unit during sleep and waking. During paradoxical sleep (PS) periods, animals lay immobile and exhibited fast irregular breathing and facial twitches. Note the period of predominant theta in the EEG (at *arrow*, marked as REM) coincident with twitches and reduced muscle tonus in the EMG trace, during which there is no activity in the spike trace. Slow wave sleep (SWS) epochs contained high amplitude synchronized EEG while the animal lay immobile with eyes closed. While the rat was alert, LC neurons discharged at higher rates than during SWS, which in turn contained more LC activity than did REM periods. Anecdotally, LC neurons usually exhibited their highest rates surrounding a transition from sleep to waking, with bursts of activity slightly preceding the SWS to waking transition but coincident with PS-to-waking transitions. (Reproduced with permission from G. Jones *et al.* (71).)

quired for hypothetical neurons which execute PS phenomena.

As well as demonstrating many properties previously reported for REM-Off neurons in cat LC, our studies revealed an interesting characteristic of NE-LC discharge during PS-to-waking transitions that is not described in studies of cat LC. Although discharge of corresponding cells in both species anticipated the return of tonic EMG activity following PS, additional observations in rat disclosed that increased discharge did not precede the onset

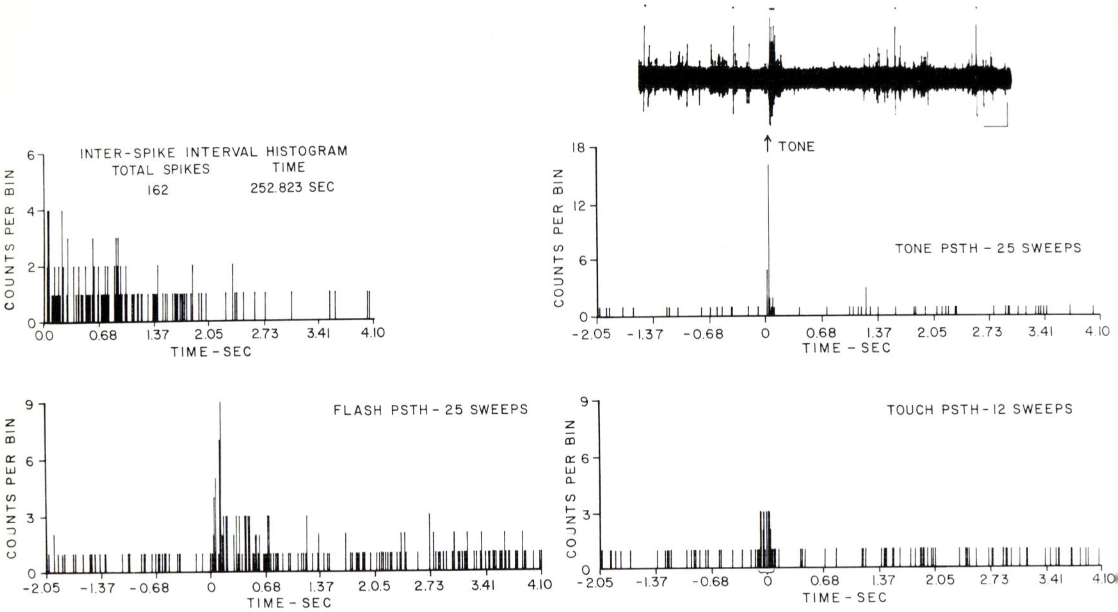

Figure 7.8. Spontaneous and sensory-evoked activity of a typical individual rat LC neuron. In the interspike interval histogram, there were 162 total counts from a sampling period of 253 sec. In the three poststimulus-time histogram (PSTH) displays, the stimuli were delivered at time 0. This neuron exhibited biphasic (excitatory-inhibitory) responses to stimuli in each sensory modality tested. The postexcitatory rate decrease was sometimes obscured in cumulative histograms when responses after the first 5 to 10 sweeps were attenuated in association with decreased vigilance (see ref. 11). (Inset, *upper right*) Analog record of this neuron's response to the first stimulus (*arrow*) of the tone PSTH display. *Dots* above action potentials are waveform discriminator outputs, which serve as the input to the computer. *Calibration marks* equal 50 μV and 400 msec. (Reproduced with permission from S. L. Foote et al. (43).)

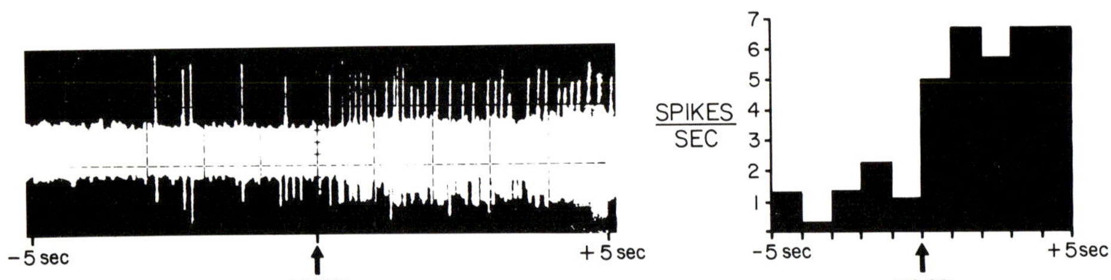

Figure 7.9. Response of a monkey LC neuron to the presentation of a preferred food. At the *arrow*, fruit juice in a syringe was presented visually to the monkey who had received juice in this fashion previously. The *left panel* is an oscilloscope trace from a single trial showing activity for 5 seconds before and after the food presentation. The panel on the right is a histogram displaying the mean response for this cell to 3 food presentations. The mean discharge rate was 1.3 spikes/second before food presentation and 6.1 after. (Reproduced with permission from S. L. Foote and F. E. Bloom (44).)

of characteristic waking signals in the EEG. This result contradicts the previous proposal (94) that NE-LC discharge serves to terminate PS episodes; in rat at least, other brain activity precedes or coincides with the return of NE-LC discharge at the end of PS. Therefore, although the NE-LC may have a role in terminating peripheral atonia, we conclude that discharge in these neurons is regulated by systems more directly responsible for the control of PS phenomena. Our results do support the proposal of McCarley and Hobson (94) that NE-

LC neurons play a permissive role in the generation of PS, enabling that state by virtually ceasing to discharge.

Anxiety and Nociception

The ineffectiveness of most sensory stimuli in anesthetized rats has led some investigators to propose that the NE-LC system is primarily involved in nociception, fear, or anxiety (55, 78, 115). Although some sensory response characteristics of NE-LC neurons in unanesthetized behaving rats resemble results reported for anesthetized rats, many fundamental properties differ markedly. Previous studies (2, 3, 14, 21, 23, 54), as well as our own observations, indicate that these cells in anesthetized animals respond only to strong noxious environmental stimuli. In contrast, we found that mild nonnoxious sensory stimuli of many modalities elicited pronounced NE-LC responses in unanesthetized behaving preparations. In our unanesthetized preparations, pronounced lability was observed in spontaneous NE-LC discharge, apparently resulting from responses to background environmental stimuli. In fact, increased sensory responsivity is the most prominent difference between behaving and anesthetized rats' NE-LC discharge. Reduced responsivity under anesthesia may correspond to the relationship between response amplitude and vigilance found in our studies. By maintaining a low vigilance level, anesthesia may decrease NE-LC sensory responsiveness much like sleep does in unanesthetized rats (11). Our results on NE-LC discharge in behaving rats indicate a much broader range of environmental influences on NE-LC discharge and a more general role for this system in brain and behavioral activity than involvement solely in nociception, fear, or anxiety.

Metabolism and Cerebral Blood Flow

As outlined above, one prevalent hypothesis of NE-LC function proposes that these cells primarily serve to control brain metabolism, either through or in addition to control over local cerebral blood flow. Previous data indicate that NE released from NE-LC neurons may have effects on target neuron metabolic processes (see Postsynaptic Physiology of Central Noradrenergic Circuits) but largely as part of the mechanism by which NE alters target neuron discharge. The more crucial question of NE-LC control over metabolism concerns whether these cells function primarily to control local blood flow in brain and thereby affect local cellular metabolism and only indirectly affect target neuron discharge. Recent data reviewed above that bear most directly on such proposals include those of Morrison and collaborators and of Olschowka and collaborators. Morrison *et al.* (98–100) demonstrated a highly ordered laminar distribution of NE-LC fibers in rat neocortex (Fig. 7.3). As cerebral blood vessels traverse the cortex in a radial fashion, exhibiting a pattern markedly distinct from that of NE-LC fibers, substantial innervation of blood vessels by these fibers appears unlikely. The findings of Olschowka *et al.* (106) that NE-LC neurons do form apparently functional synapses in cerebral cortex and that these synapses occur on neurons but not on blood vessels, argue strongly against the view that NE is either nonselectively "leaking" from presynaptic structures over a wide area of target tissue or that NE-LC neurons synapse directly onto blood vessels. In view of these data, it is difficult to see how the NE-LC could primarily be dedicated to controlling metabolism or blood flow locally in the CNS.

Attention

Previous data on the postsynaptic effects of NE on target neurons are largely consistent with proposals of a role for the NE-LC system in controlling or directing attentional processes. By increasing S/N ratios in target neurons, NE-LC transmitter release could provide a mechanism for selectively augmenting information flow through CNS channels most relevant to high priority sensory events, thereby affecting the focus of attention. In addition, our recent data demonstrating that NE-LC neurons are most responsive to sensory stimuli which are associated with a change in vigilance (11) indicate that these neurons normally release their transmitter in anticipation of attentional changes and that this system may well be an integral part of brain processes underlying such changes. However, these data also indicate that there is not a simple or linear relationship between NE-LC discharge and the level of arousal or attention; some focused, highly attentive behaviors are associated with a decrease in discharge, while others of a similar but different nature are anticipated by enhanced impulse activity (11). These rather subtle but important differences have led us to a different sort of attentional hypothesis of NE-LC function, as outlined below.

CONCLUSIONS

Recent anatomical and physiological data are useful in evaluating many prevalent theories of NE-LC function. These data strengthen the overall view that this system is involved in arousal processes and, therefore, possibly also in affective disorders. The discharge characteristics of NE-LC neurons in behaving animals indicate that these cells may play a role in instigating changes in arousal or vigilance. Such recordings during the sleep-waking cycle do not support Jouvet's view of these neurons as PS generators but are consistent with the proposal of Hobson and McCarley that the NE-LC may play a critically permissive role in PS processes, allowing PS to occur by virtually ceasing discharge. However, these same data make it equally clear that NE-LC neurons do not terminate PS epochs as these latter authors propose, but rather that the NE-LC is controlled by other brain processes during this stage transition. Previous proposals that the NE-LC may underlie fear, anxiety, or nociception are only partially supported by recent data: the wider range of sensory response properties seen for NE-LC neurons in awake *vs.* anesthetized animals indicates that these cells may participate in such processes but also that the NE-LC system is not limited exclusively to such a role. Recent data on the ultrastructure of NE-LC synapses in cortex reveal that these neurons are unlikely to have a direct influence over local blood flow but that their discharge in behaving animals is consistent with parallel blood pressure changes, cyclic AMP metabolic changes, and NE-LC target cell alterations. Finally, while recent data on the NE-LC strengthen the idea that this system may play an important role in attentional processes, they further indicate a more complex involvement in attention than has been previously proposed.

A New Global Hypothesis of NE-LC Function

It is apparent that, overall, the recent data on NE-LC neurons reviewed in this chapter are consistent with many previous hypotheses of NE-LC function but are not supportive of any one of them *in toto*. We conclude that each of these previous viewpoints represents a part of a more general function of the NE-LC system, but that the NE-LC does not primarily function in any one of these roles alone. Any overall hypothesis of NE-LC function must take into account several important, general previous results. (1) The ubiquitous NE-LC efferent fiber system indicates that these neurons serve a very global function, one that is generalizable over widely distant and disparate CNS processes. (2) The pronounced clinical and behavioral effects produced by pharmacological manipulations of NE activity in brain indicate a substantial and pervasive role for the NE-LC system. (3) Our recent results demonstrating physiological homogeneity, robust field potential activity, and synchronous discharge in multiple-unit populations in LC (10, 11) indicate that NE-LC neurons may function as a group, emitting phasically concerted fluctuations in overall output. This suggests that the NE-LC system provides a singular, discrete function in brain and behavioral processes. (4) The postsynaptic effects of NE-LC activity on target cells, by increasing S/N ratios and information flow in specific CNS channels, provides a means by which the extensive NE-LC efferent system may serve to coordinate activities in many different brain areas.

Although previous studies suggest general aspects of NE-LC function, they reveal little about specific factors which influence NE-LC discharge in physiologically normal brain activity. The data on NE-LC discharge in behaving animals extend the range of such information and indicate that these cells may be primarily influenced by two distinct input systems: excitatory afferents reflecting salient external stimuli and inhibitory afferents which reflect internally generated signals concerning tonic vegetative requirements (Fig. 7.10). The level of NE-LC discharge may predominantly be a function of the relative activities in these two proposed sets of inputs. In the absence of strong afferent inhibition, spontaneous as well as sensory-evoked NE-LC discharge would be enhanced. Certain behavioral states, however, may be incompatible with high levels of NE-LC discharge and may therefore require tonic inhibition of these neurons; our data indicate that the most intense inhibition may occur during PS. Thus, varying levels of afferent inhibition may underlie the observed state-dependent fluctuations in spontaneous and sensory-evoked NE-LC activity. In addition, the observation that robust discharge typically

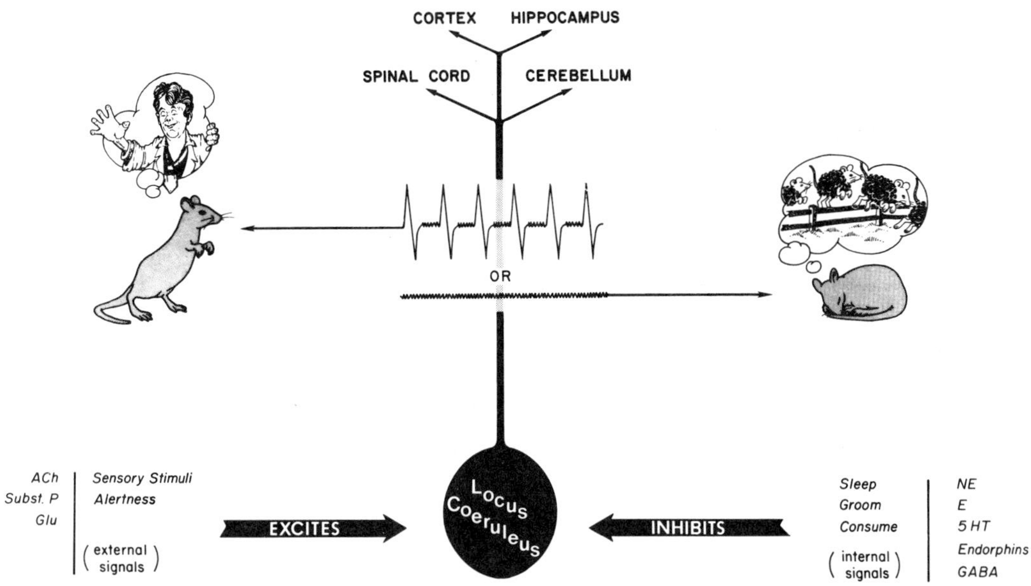

Figure 7.10. A schematic drawing illustrating our current view of NE-LC function. As indicated on the *lower left*, NE-LC neurons are excited by sensory stimuli or spontaneously occurring alertness, perhaps through systems with the transmitters indicated. Conversely, the *lower right* section of this drawing illustrates behavioral states and possible transmitters which are associated with suppressed NE-LC activity. Our hypothesis states that enhanced activity traversing the elaborate efferent projections of the NE-LC system (as a result of enhanced external signals) biases the orientation of brain and behavioral activities so as to deal preferentially with external events of phasically high priority. Conversely, diminished NE-LC activity would allow internally oriented, more instinctual brain and behavioral programs to be expressed.

anticipates interruptions of tonic vegetative behavior indicates that intense afferent excitation may terminate strong tonic inhibition and suggests that robust NE-LC activity has a role in generating the associated change in behavioral states.

Thus, NE-LC discharge may be controlled by phasic exogenously determined inputs as well as by tonic endogenously controlled afferents. Such powerful gating suggests specific behavioral consequences of NE-LC activity. Our working hypothesis states that robust NE-LC output may suppress CNS processes which have minimal value in phasically coping with unexpected systems primarily concerned with such immediate behavioral responses. Conversely, tonically reduced NE-LC discharge may enable the development of endogenously generated brain programs which mediate tonic vegetative behaviors. Figure 7.10 serves to illustrate this model for NE-LC function. In this way, the NE-LC system may bias the global orientation of behavior between the external and internal environments. Recent results of Livingstone and Hubel (83) on response properties of visual cortical neurons as a function of sleep and waking are consistent with this possibility.

The previously proposed theories of NE-LC function reviewed in this chapter are largely consistent with our hypothesis, as aspects of that more general viewpoint. Thus, the NE-LC may play a role in arousal and affective disorders as a consequence of its involvement in determining the global orientation of brain processing and behavioral activity. Similarly, the NE-LC system may have pronounced effects on sleep, anxiety and nociception, local brain metabolism and blood flow, and selective attention, all as aspects of its more general function of influencing the overall mode of brain information processing and, therefore, the global orientation of behavior.

References

1. Adams LM, Geyer MA: Effects of 6-hydroxydopamine lesions of locus coeruleus on startle in rats. *Psychopharmacology* 73:394–398, 1981.
2. Aghajanian GK, Cedarbaum JM, Wang RY: Evidence for norepinephrine-mediated collateral inhibition of locus coeruleus neurons. *Brain*

Res 136:570–577, 1977.
3. Aghajanian GK: Tolerance of locus coeruleus neurones to morphine and suppression of withdrawal response by clonidine. *Nature* 276:186–188, 1978.
4. Amaral DG, Foss JA: Locus coeruleus lesions and learning. *Science* 188:377–378, 1975.
5. Amaral DG, Routtenberg A: Locus coeruleus and intracranial self-stimulation: A cautionary note. *Behav Biol* 13:331–338, 1975.
6. Amaral DG, Sinnamon HM: The locus coeruleus: Neurobiology of a central noradrenergic nucleus. *Prog Neurobiol* 9:147–196, 1977.
7. Anlezark GM, Crow TJ, Greenway AP: Impaired learning and decreased cortical norepinephrine after bilateral locus coeruleus lesions. *Science* 181:682–684, 1973.
8. Antelman SM, Caggiula AR: Norepinephrine-dopamine interactions and behavior. *Science* 195:646–653, 1977.
9. Aston-Jones G: The Behavioral Physiology of Locus Coeruleus Neurons. Thesis, California Institute of Technology, Pasadena, California, 1981.
10. Aston-Jones G, Bloom FE: Activity of norepinephrine-containing locus coeruleus neurons in behaving rats anticipates fluctuations in the sleep-waking cycle. *J Neurosci* 1:876–886, 1981.
11. Aston-Jones G, Bloom FE: Norepinephrine-containing locus coeruleus neurons in behaving rats exhibit pronounced responses to non-noxious environmental stimuli. *J Neurosci* 1:887–900, 1981.
12. Aston-Jones G, Foote SL, Bloom FE: Low doses of ethanol disrupt sensory responses of brain noradrenergic neurones. *Nature* 296:857–860, 1982.
13. Aston-Jones G, Segal M, Bloom FE: Brain aminergic axons exhibit marked variability in conduction velocity. *Brain Res* 195:215–222, 1980.
14. Bird SJ, Kuhar MJ: Iontophoretic application of opiates to the locus coeruleus. *Brain Res* 122:523–533, 1977.
15. Bloom FE: Amine receptors in CNS. I. Norepinephrine. In Iversen LL, Iversen SD, Snyder SH: *Handbook of Psychopharmacology*. New York, Raven Press, 1975, pp 1–22.
16. Bloom FE: Role of cyclic nucleotides in central synaptic function. *Rev Physiol Biochem Pharmacol* 74:1–103, 1975.
17. Bloom FE: Is there a neurotransmitter code in the brain? *Adv Pharmacol Ther* 2:205–213, 1979.
18. Bloom FE, Costa E, Salmoiraghi GC: Analysis of individual rabbit olfactory bulb neuron response to microelectrophoresis of acetylcholine, norepinephrine and serotonin synergists and antagonists. *J Pharmacol Exp Ther* 146:16–23, 1964.
19. Bondareff W, Mountjoy CQ, Roth M: Selective loss of neurones of origin of adrenergic projection to cerebral cortex (nucleus locus coeruleus) in senile dementia. *Lancet* 1: 783–784, 1981.
20. Brunelli M, Castellucci V, Kandel ER: Synaptic facilitation and behavioral sensitivity in Aplasia: Possible role of serotonin and cyclic AMP. *Science* 194:1178–1181, 1976.
21. Bunney B, Walters J, Kuhar M, Roth R, Aghajanian G: D-and L-amphetamine stereoisomers: Comparative potencies in affecting the firing of central dopaminergic and noradrenergic neurons. *Psychopharmacol Commun* 1:177, 1975.
22. Caviness VS Jr, Korde MG: Monoaminergic afferents to the neocortex: A developmental histofluorescence study in normal and reeler mouse embryos. *Brain Res* 209:1–9, 1981.
23. Cedarbaum JM, Aghajanian GK: Noradrenergic neurons of the locus coeruleus: Inhibition by epinephrine and activation by the alpha-antagonist piperoxane. *Brain Res* 112:413–419, 1976.
24. Cedarbaum JM, Aghajanian GK: Activation of locus coeruleus neurons by peripheral stimuli: Modulation by a collateral inhibitory mechanism. *Life Sci* 23:1383–1392, 1978.
25. Cedarbaum J, Aghajanian G: Afferent projections to the rat locus coeruleus as determined by a retrograde tracing technique. *J Comp Neurol* 178:1–16, 1978.
26. Chu N-s, Bloom FE: Norepinephrine-containing neurons: Changes in spontaneous discharge patterns during sleeping and waking. *Science* 179:908–910, 1973.
27. Chu N-s, Bloom FE: Activity patterns of catecholamine-containing pontine neurons in the dorso-lateral tegmentum of unrestrained cats. *J Neurobiol* 5:527–544, 1974.
28. Clark TK: The locus coeruleus in behavior regulation: Evidence for behavior-specific *versus* general involvement. *Behav Neurol Biol* 25:271–300, 1979.
29. Clavier RM, Fibiger HC, Phillips AG: Evidence that self-stimulation of the region of the locus coeruleus in rats does not depend upon noradrenergic projections to telencephalon. *Brain Res* 113:71–81, 1976.
30. Clavier RM, Routtenberg A: Brain stem self-stimulation attenuated by lesions of medial forebrain bundle but not by lesions of locus coeruleus or the caudal ventral norepinephrine bundle. *Brain Res* 101:251–271, 1976.
31. Corbett D, Skelton RW, Wise RA: Dorsal noradrenergic bundle lesions fail to disrupt self-stimulation from the region of locus coeruleus. *Brain Res* 133:37–44, 1977.
32. Crow TJ, Arbuthnott GW: Function of catecholamine-containing neurones in mammalian central nervous system. *Nature New Biol* 238:245–246, 1972.
33. Crow TJ, Deakin JFW, File SE, Longden A, Wendlandt S: The locus coeruleus noradrenergic system—evidence against a role in attention, habituation, anxiety and motor activity. *Brain Res* 155:249–261, 1978.
34. Crow TJ, Spear PJ, Arbuthnott GW: Intracranial self-stimulation with electrodes in the region of the locus coeruleus. *Brain Res* 36:275–287, 1972.
35. Crow TJ, Wendlandt S: Impaired acquisition of a passive avoidance response after lesions induced in the locus coeruleus by 6-OH-dopa-

mine. *Nature* 259:42–44, 1976.
36. Descarries L, Lapierre Y: Noradrenergic axon terminals in the cerebral cortex of rat. I. Radioautographic visualization after topical application of DL-(3H)-norepinephrine. *Brain Res* 51:141–160, 1973.
37. Descarries L, Watkins KC, Lapierre Y: Noradrenergic axon terminals in the cerebral cortex of rat. III. Topometric ultrastructural analysis. *Brain Res* 133:197–222, 1977.
38. Eccles J, Ito M, Szentagothai J: *The Cerebellum as a Neuronal Machine.* Berlin, Springer-Verlag, 1967.
39. Edvinsson L, Lindvall M, Nielsen KC, Owman CH: Are brain vessels innervated also by central (non-sympathetic) adrenergic neurones? *Brain Res* 63:496–499, 1973.
40. Engberg I, Marshall K: Mechanism of noradrenaline hyperpolarization in spinal cord motoneurons of the cat. *Acta Physiol Scand* 83:142–144, 1971.
41. Faiers AA, Mogenson GJ: Electrophysiological identification of neurons in locus coeruleus. *Exp Neurol* 53:254–266, 1976.
42. File SE, Deakin JFW, Longden A, Crow TJ: An investigation of the role of the locus coeruleus in anxiety and agonistic behavior. *Brain Res* 169:411–420, 1979.
43. Foote SL, Aston-Jones G, Bloom FE: Impulse activity of locus coeruleus neurons in awake rats and monkeys is a function of sensory stimulation and arousal. *Proc Natl Acad Sci USA* 77:3033–3037, 1980.
44. Foote SL, Bloom FE: Activity of norepinephrine-containing locus coeruleus neurons in the unanesthetized squirrel monkey. In Usdin E, Kopin IJ, Barchas J: *Catecholamines: Basic and Clinical Frontiers.* New York, Pergamon Press, pp 625–627, 1979.
45. Foote SL, Freedman R, Oliver AP: Effects of putative neurotransmitters on neuronal activity in monkey auditory cortex. *Brain Res* 86:229–242, 1975.
46. Foote SL, Loughlin SE, Cohen PS, Bloom FE, Livingston RB: Accurate three-dimensional reconstruction of neuronal distributions in brain: Reconstruction of the rat nucleus locus coeruleus. *J Neurosci Meth* 3:159–173, 1980.
47. Freedman R, Hoffer B: Phenothiazine antagonism of the noradrenergic inhibition of cerebellar Purkinje neurons. *J Neurobiol* 6:277–288, 1975.
48. Freedman R, Hoffer B, Woodward D: A quantitative microiontophoretic analysis of the responses of central neurones to noradrenaline interactions with cobalt, manganese, verapamil and dichloroisoprenaline. *J Pharmacol* 54:529–539, 1975.
49. Freedman R, Hoffer BJ, Woodward DJ, Puro D: Interaction of norepinephrine with cerebellar activity evoked by mossy and climbing fibers. *Exp Neurol* 55:269–288, 1977.
50. Geller H, Hoffer B: Effect of calcium removal on monoamine-elicited depressions of cultured tuberal neurons. *J Neurobiol* 8:61–67, 1977.
51. German DC, Bowden DM: Catecholamine systems as the neural substrate for intracranial self-stimulation: A hypothesis. *Brain Res* 73:381–419, 1974.
52. German DC, Fetz EE: Responses of primate locus coeruleus and subcoeruleus neurons to stimulation at reinforcing brain sites and to natural reinforcers. *Brain Res* 109:497–514, 1976.
53. Gilbert P: How the cerebellum could memorise movements. *Nature* 254:688–689, 1975.
54. Graham AW, Aghajanian GK: Effects of amphetamine on single cell activity in a catecholamine nucleus, the locus coeruleus. *Nature* 234:100–102, 1971.
55. Gray J, McNaughton N, James D, Kelly P: Effect of minor tranquilizers on hippocampal theta rhythm mimicked by depletion of forebrain noradrenaline. *Nature* (Lond) 258:424–425, 1975.
56. Greengard P: Possible role for cyclic nucleotides and phosphorylated membrane proteins in postsynaptic actions of neurotransmitters. *Nature* 260:101–108, 1976.
57. Grzanna R, Molliver ME: The locus coeruleus in the rat: An immunohistochemical delineation. *Neuroscience* 5:21–40, 1980.
58. Guyenet PG, Aghajanian GK: Excitation of neurons in the nucleus locus coeruleus by substance P and related peptides. *Brain Res* 136:178–184, 1977.
59. Harik SI, Duckrow RB, LaManna JC, Rosenthal M, Sharma VK, Banerjee SP: Cerebral compensation for chronic noradrenergic denervation induced by locus coeruleus lesion: Recovery of receptor binding, isoproterenol-induced adenylate cyclase activity, and oxidative metabolism. *J Neurosci* 1:641–649, 1981.
60. Harik SI, LaManna JC, Light AI, Rosenthal M: Cerebral norepinephrine: Influence on cortical oxidative metabolism *in situ. Science* 206:69–71, 1979.
61. Hartline H, Ratliff F: Inhibitory interaction of receptor units in the eye of limulus. *J Gen Physiol* 40:357–376, 1957.
62. Hartmann E, Bridwell TJ, Schildkraut JJ: AMPT and sleep in the rat. *Psychopharmacologia* 21:157–164, 1971.
63. Hobson JA, McCarley RW, Wyzinski PW: Sleep cycle oscillation: Reciprocal discharge by two brainstem neuronal groups. *Science* 189:55–58, 1975.
64. Hoffer B, Siggins G, Bloom FE: Prostaglandins E1 and E2 antagonize norepinephrine effects on cerebellar Purkinje cells: microiontophoretic study. *Science* 166:1418–1420, 1969.
65. Hoffer BJ, Siggins GR, Bloom FE: Studies on norepinephrine-containing afferents to Purkinje cells of rat cerebellum. II. Sensitivity of Purkinje cells to norepinephrine and related substances administered by microiontophoresis. *Brain Res* 25:523–534, 1971.
66. Hoffer BJ, Siggins GR, Oliver AP, Bloom FE: Activation of the pathway from locus coeruleus to rat cerebellar Purkinje neurons: Pharmacological evidence of noradrenergic central inhibition. *J Pharmacol Exp Ther* 181:553–569,

1973.
67. Hwang B-H, Chiba T, Black AC Jr, Williams TH: Quantification of catecholamine-containing cell groups in the brainstem of the rhesus monkey. *Neuroscience* 1:563, 1975.
68. Jones BE: The respective involvement of noradrenaline and its deaminated metabolites in waking and paradoxical sleep: A neuropharmacological model. *Brain Res* 39:121–136, 1972.
69. Jones BE, Bobillier P, Pin C, Jouvet M: The effect of lesions of catecholamine-containing neurons upon monoamine content of the brain and EEG and behavioral waking in the cat. *Brain Res* 58:157–177, 1973.
70. Jones BE, Harper ST, Halaris AE: Effects of locus coeruleus lesions upon cerebral monoamine content, sleep-wakefulness states and the response to amphetamine in the cat. *Brain Res* 124:473–496, 1977.
71. Jones G, Segal M, Foote SL, Bloom FE: Locus coeruleus neurons in freely moving rats exhibit pronounced alterations of discharge rate during sensory stimulation and stages of the sleep-wake cycle. In: Usdin E, Kopin IJ, Barchas J: *Catecholamines: Basic and Clinical Frontiers.* New York, Pergamon Press, 1979, pp 643–645.
72. Jouvet M: Neurophysiology of the states of sleep. *Physiol Rev* 47:117–177, 1967.
73. Jouvet M: Biogenic amines and the states of sleep. *Science* 163:32–41, 1969.
74. Jouvet M: The role of monoamines and acetylcholine-containing neurons in the regulation of the sleep-waking cycle. *Ergeb Physiol Biol Chem Exp Pharmakol* 64:166–307, 1972.
75. Koda LY, Bloom FE: A light and electron microscopic study of noradrenergic terminals in the rat dentate gyrus. *Brain Res* 120:327–335, 1977.
76. Koob GF, Balcolm GJ, Meyerhoff JL: Increases in intracranial self stimulation in the posterior hypothalamus following unilateral lesion in the locus coeruleus. *Brain Res* 101:554–560, 1976.
77. Korf J, Bunney BS, Aghajanian GK: Noradrenergic neurons: Morphine inhibition of spontaneous activity. *Eur J Pharmacol* 25:165–169, 1974.
78. Lader M: The peripheral and central role of the catecholamines in the mechanisms of anxiety. *Int Pharmacopsychiatry* 9:125–137, 1974.
79. LaManna JC, Harik SI, Light AI, Rosenthal M: Norepinephrine depletion alters cerebral oxidative metabolism in the "active" state. *Brain Res* 204:87–101, 1981.
80. Langmoen IA, Segal M: The effect of norepinephrine on hippocampal pyramidal cells *in vitro. Acta Physiol Scand* 105:33–34a, 1979.
81. Langmoen IA, Segal M, Andersen P: Mechanisms of norepinephrine actions on hippocampal pyramidal cells *in vitro. Brain Res* 208:349–362, 1981.
82. Lidbrink P: The effect of lesions of ascending noradrenaline pathways on sleep and waking in the rat. *Brain Res* 74:19–40, 1974.
83. Livingstone MS, Hubel DH: Effects of sleep and arousal on the processing of visual information in the cat. *Nature* 291:554–561, 1981.
84. Lorden JF, Rickert EJ, Dawson R Jr, Pelleymounter MA: Forebrain norepinephrine and the selective processing of information. *Brain Res* 190:569–573, 1980.
85. Mason ST, Fibiger HC: Noradrenaline and spatial memory. *Brain Res* 156:382–386, 1978.
86. Mason ST, Fibiger HC: Current concepts. I. Anxiety: The locus coeruleus disconnection. *Life Sci* 25:2141–2147, 1979.
87. Mason ST, Fibiger HC: Noradrenaline, fear and extinction. *Brain Res* 165:47–56, 1979.
88. Mason ST, Fibiger HC: Regional topography within noradrenergic locus coeruleus as revealed by retrograde transport of horseradish peroxidase. *J Comp Neurol* 187:703–724, 1979.
89. Mason ST, Fibiger HC: Noradrenaline and selective attention. *Life Sci* 25:1949–1956, 1979.
90. Mason ST, Iversen SD: Learning in the absence of forebrain noradrenaline. *Nature* 258:422–424, 1975.
91. Mason ST, Iversen SD: Effects of selective forebrain noradrenaline loss on behavioral inhibition in the rat. *J Comp Physiol Psychol* 91:165–173, 1977.
92. Mason ST, Iversen SD: Reward, attention and the dorsal noradrenergic bundle. *Brain Res* 150:135–148, 1978.
93. Mason ST, Iversen SD: Theories of the dorsal bundle extinction effect. *Brain Res* 1:107–137, 1979.
94. McCarley RW, Hobson JA; Neuronal excitability modulation over the sleep cycle: A structural and mathematical model. *Science* 189:58–60, 1975.
95. McGinty DJ, Harper RM: Dorsal raphe neurons: Depression of firing during sleep in cats. *Brain Res* 101:569–575, 1976.
96. Moises H, Woodward D: Potentiation of GABA inhibitory action in cerebellum by locus coeruleus stimulation. *Brain Res* 182:327–344, 1980.
97. Moore RY, Bloom FE: Central catecholamine neuron systems: Anatomy and physiology of the norepinephrine and epinephrine systems. *Ann Rev Neurosci* 2:113–168, 1979.
98. Morrison JH, Grzanna R, Molliver ME, Coyle JT: The distribution and orientation of noradrenergic fibers in neocortex of the rat: An immunofluorescence study. *J Comp Neurol* 181:17–40, 1978.
99. Morrison JH, Molliver ME, Grzanna R: Noradrenergic innervation of cerebral cortex: Widespread effects of local cortical lesions. *Science* 205:313–316, 1979.
100. Morrison JH, Molliver ME, Grzanna R, Coyle JT: The intra-cortical trajectory of the coeruleo-cortical projection in the rat: A tangentially organized cortical afferent. *Neuroscience* 6:139–158, 1981.
101. Nakamura S, Iwama K: Antidromic activation of the rat locus coeruleus neurons from hippocampus, cerebral and cerebellar cortices. *Brain Res* 99:372–376, 1975.
102. Nathanson JA; Cyclic nucleotides and nervous

system function. *Physiol Rev* 57:157–256, 1977.

103. Nathanson J, Freedman R, Hoffer B: Lithium inhibits brain adenylate cyclase and blocks noradrenergic depression of Purkinje cell discharge independent of calcium. *Nature* 261:330–331, 1976.
104. Nyback HV, Walters JR, Aghajanian GK, Roth RH: Tricyclic antidepressants: Effects on the firing rate of brain noradrenergic neurons. *Eur J Pharmacol* 32:302–312, 1975.
105. Oke AF, Adams RN: Selective attention dysfunctions in adult rats neonatally treated with 6-hydroxydopamine. *Pharmacol Biochem Behav* 9:429–432, 1978.
106. Olschowka JA, Molliver ME, Grzanna R, Rice FL, Coyle JT: Ultrastructural demonstration of noradrenergic synapses in the rat central nervous system by dopamine-β-hydroxylase immunocytochemistry. *J Histochem Cytochem* 29:271–280, 1981.
107. Owman C, Edvinsson L (Eds): *Neurogenic Control of the Brain Circulation.* New York, Pergamon, 1977.
108. Palacios J, Kuhar MJ: Beta-adrenergic-receptor localization by light microscopic autoradiography. *Science* 208:1378–1380, 1980.
109. Phillis J, Lake N, Yarborough G: Calcium mediation of the inhibitory effects of biogenic amines on cerebral cortical neurons. *Brain Res* 53:465–469, 1973.
110. Prado de Carvalho L, Zornetzer SF: The involvement of the locus coeruleus in memory. *Behav Neurol Biol* 31:173–186, 1981.
111. Raichle ME, Hartman BK, Eichling JO, Sharpe LG: Central noradrenergic regulation of cerebral blood flow and vascular permeability. *Proc Natl Acad Sci USA* 72:3726–3730, 1975.
112. Rall T: Role of adenosine 3′,5′-monophosphate (cyclic AMP) in actions of catecholamines. *Pharmacol Rev* 24:399–409, 1972.
113. Ramm P: The locus coeruleus, catecholamines, and REM sleep: A critical review. *Behav Neurol Biol* 25:415–448, 1979.
114. Redmond DE Jr: New and old evidence for the involvement of a brain norepinephrine system in anxiety. In *Phenomenology and Treatment of Anxiety.* New York, Spectrum Publications, 1979, pp 153–203.
115. Redmond DE Jr, Huang YH: Current concepts. II. New evidence for a locus coeruleus-norepinephrine connection with anxiety. *Life Sci* 25:2149–2162, 1979.
116. Redmond DE Jr, Huang YH, Snyder DR, Maas JW: Behavioral effects of stimulation of the nucleus locus coeruleus in the stump-tailed monkey *Macaca arctoides. Brain Res* 116:502–510, 1976.
117. Ritter S, Stein L: Self-stimulation of noradrenergic cell group (A6) in locus coeruleus of rats. *J Comp Physiol Psychol* 85:443–452, 1973.
118. Roberts DCS: An evaluation of the role of noradrenaline in learning. In Cooper SJ: *Theory in Psychopharmacology,* vol 1. New York, Academic Press, 1983.
119. Roberts DCS, Price MTC, Fibiger HC: The dorsal tegmental noradrenergic projection: An analysis of its role in maze learning. *J Comp Physiol Psychol* 90:363–372, 1976.
120. Robinson TE, Vanderwolf CH, Pappas BA: Are the dorsal noradrenergic bundle projections from the locus coeruleus important for neocortical or hippocampal activation? *Brain Res* 138:75–98, 1977.
121. Rogawski M, Aghajanian G: Activation of lateral geniculate neurons by norepinephrine: Mediation by an alpha-adrenergic receptor. *Brain Res* 182:345–359, 1980.
122. Sakai K: Some anatomical and physiological properties of pontomesencephalic tegmental neurons with special reference to the PGO waves and postural atonia during paradoxical sleep in the cat. In Hobson J, Brazier M: *The Reticular System Revisited.* New York, Raven Press, 1980, pp 427–448.
123. Sasa M, Muneyiko K, Ikeda H, Takaori S: Noradrenaline-mediated inhibition by locus coeruleus of spinal trigeminal neurons. *Brain Res* 80:443–460, 1974.
124. Sasa M, Muneyiko K, Takaori S: Morphine interference with noradrenaline-mediated inhibition from the locus coeruleus. *Life Sci* 17:1373–1380, 1975.
125. Satoh K, Tohyama M, Yamamoto K, Sakumoto T, Shimizu N: Noradrenaline innervation of the spinal cord studied by the horseradish peroxidase method combined with monoamine oxidase staining. *Exp Brain Res* 30:175–186, 1977.
126. Schildkraut JJ: The catecholamine hypothesis of affective disorders: A review of supportive evidence. *Am J Psychiatry* 122:509–522, 1965.
127. Schildkraut JJ, Kety SS: Biogenic amines and emotion. *Science* 156:21–30, 1967.
128. Schulman J, Weight F: Synaptic transmission: Long-lasting potentiation by a postsynaptic mechanism. *Science* 194:1437–1439, 1976.
129. Segal DS, Mandell AJ: Behavioral activation of rats during intraventricular infusion of norepinephrine. *Proc Natl Acad Sci USA* 66: 289–293, 1970.
130. Segal M: Brain stem afferents to the rat medial septum. *J Physiol* 261:617–631, 1976.
131. Segal M, Bloom FE: The action of norepinephrine in the rat hippocampus. I. Iontophoretic studies. *Brain Res* 72:79–97, 1974.
132. Segal M, Bloom FE: The action of norepinephrine in the rat hippocampus. II. Activation of the input pathway. *Brain Res* 72:99–114, 1974.
133. Segal M, Bloom FE: The action of norepinephrine in the rat hippocampus. III. Hippocampal cellular responses to locus coeruleus stimulation in the awake rat. *Brain Res* 107:499–511, 1976.
134. Segal M, Bloom FE: The action of norepinephrine in the rat hippocampus. IV. The effects of locus coeruleus stimulation on evoked hippocampal unit activity. *Brain Res* 107:513–525, 1976.
135. Segal M, Oliver A: Transmembrane changes in hippocampal neurons: hyperpolarizing actions of norepinephrine, cyclic AMP and locus coeruleus. *Soc Neurosci* Abstr., 361, 1975.

136. Sessions GR, Kant GJ, Koob GF: Locus coeruleus lesions and learning in the rat. *Physiol Behav* 17:853–859, 1976.
137. Siegel JM, McGinty DJ, Breedlove SM: Sleep and waking activity of pontine gigantocellular field neurons. *Exp Neurol* 56:553–573, 1977.
138. Siggins GR, Henriksen S: Inhibition of rat Purkinje neurons by analogues of cyclic adenosine monophosphate: Correlation with protein kinase activation. *Science* 179:585–588, 1975.
139. Siggins GR, Henriksen S, Bloom FE: Iontophoresis of lithium antagonizes noradrenergic synaptic inhibition of rat cerebellar Purkinje cells. *Proc Natl Acad Sci USA* 76:3015–3018, 1979.
140. Siggins GR, Hoffer BJ, Bloom FE: Prostaglandin-norepinephrine interactions in brain: Microelectrophoretic and histochemical correlates. *Ann NY Acad Sci* 180:302–323, 1971.
141. Siggins GR, Hoffer BJ, Bloom FE: Studies on norepinephrine-containing afferents to Purkinje cells of rat cerebellum. III. Evidence for mediation of norepinephrine effects by cyclic 3′,5′-adenosine monophosphate. *Brain Res* 25:535–553, 1971.
142. Siggins GR, Hoffer BJ, Oliver AP, Bloom FE: Activation of a central noradrenergic projection to cerebellum. *Nature* 233:481–483, 1971.
143. Siggins GR, Oliver AP, Hoffer BJ, Bloom FE: Cyclic adenosine monophosphate and norepinephrine: Effects on transmembrane properties of cerebellar Purkinje cells. *Science* 171:192–194, 1971.
144. Steriade M, Hobson JA: Neuronal activity during the sleep-waking cycle. *Prog Neurobiol* 6:155–376, 1976.
145. Sutherland E, Robinson G, Butcher R: Some aspects of the biological role of adenosine 3′,5′-monophosphate (cyclic AMP). *Circulation* 3:279–306, 1968.
146. Svensson TH, Bunney BS, Aghajanian GK: Inhibition of both noradrenergic and serotonergic neurons in brain by the alpha-adrenergic agonist clonidine. *Brain Res* 92:291–306, 1975.
147. Swanson LW: The locus coeruleus: A cytoarchitectonic, Golgi and immunohistochemical study in the albino rat. *Brain Res* 110:39–56, 1976.
148. Swanson LW, Connelly MA, Hartman BK: Ultrastructural evidence for central monoaminergic innervation of blood vessels in the paraventricular nucleus of the hypothalamus. *Brain Res* 136:166–173, 1977.
149. Takigawa M, Mogenson GJ: A study of inputs to antidromically identified neurons of the locus coeruleus. *Brain Res* 135:217–230, 1977.
150. Tremmel F, Morris MD, Gebhart GF: The effect of forebrain norepinephrine depletion on two measures of response suppression. *Brain Res* 126:185–188, 1977.
151. Vertes RP: Selective firing of rat pontine gigantocellular neurons during movement and REM sleep. *Brain Res* 128:146–152, 1977.
152. Vijayashankar N, Brody H: A quantitative study of the pigmented neurons in the nuclei locus coeruleus and subcoeruleus in man as related to aging. *J Neuropathol Exp Neurol* 38:490–497, 1979.
153. Wise CD, Stein L: Facilitation of brain self-stimulation by central administration of norepinephrine. *Science* 163:299–301, 1969.
154. Wise RA: Catecholamine theories of reward: A critical review. *Brain Res* 152:215–247, 1978.
155. Woodward D, Moises H, Waterhouse B, Hoffer B, Freedman R: Modulatory actions of norepinephrine in the central nervous system. *Fed Proc* 38:2109–2116, 1979.
156. Zornetzer SF, Gold MS: The locus coeruleus: Its possible role in memory consolidation. *Physiol Behav* 16:331–336, 1976.

CHAPTER EIGHT

Norepinephrine and Dopamine-β-Hydroxylase in Cerebrospinal Fluid: Indicators of Central Noradrenergic Activity

LEE F. MAJOR, M.D., PAULINE LERNER, Ph.D., and
MICHAEL G. ZIEGLER, M.D.

INTRODUCTION

In the sympathetic nervous system the majority of norepinephrine (NE) is inactivated by reuptake and metabolized by catechol-O-methyltransferase. Some escapes to the circulation where blood levels reflect sympathetic nervous activity that occurred a few minutes previously and where the NE has a half-life of less than 3 minutes. NE is also an important neurotransmitter of the central nervous system (CNS) where it is also inactivated by reuptake and metabolism and where some escapes to the CSF.

Dopamine-β-hydroxylase (DBH), the biosynthetic enzyme for norepinephrine, is unique among enzymes—it is localized with NE in storage vesicles and is released with the transmitter during nerve firing. Although the exocytotic release of DBH has no readily apparent value to the cell, it is potentially of great value for neuroscientists. DBH released from noradrenergic tissue can be measured in plasma and CSF, and DBH levels are related to the level of neuronal exocytosis. In this article, we review and evaluate studies on NE and DBH in CSF, with special reference to the theory that NE and DBH activity in CSF may reflect central noradrenergic activity.

Basic Biology and Chemistry of NE and DBH

CELLULAR DISTRIBUTION

Noradrenergic pathways in the brain have been mapped histochemically with the use of antibodies to both norepinephrine (93), and DBH (38, 51), and the results are very similar.

The densest concentration of noradrenergic cell bodies in the brain, located in the caudal pontine central grey matter, is called the locus coeruleus. Fibers leaving the locus coeruleus form five major tracts. Three of these tracts (central tegmental tract, central dorsal longitudinal fasciculus tract, and ventral tegmental-medial forebrain bundle tract) ascend to innervate olfactory bulb, specific thalamic and hypothalamic nuclei, and all cortices. A fourth major tract ascends to cerebellar cortex, and the fifth tract descends into mesencephalon and spinal cord. A more detailed anatomical description of the coeruleo-cortical projection can be found in a recent paper by Morrison (93) and Chapters 6 and 7.

SUBCELLULAR LOCALIZATION AND FUNCTION

DBH, like other neuronal proteins, is synthesized in the cell body. It is then incorporated into vesicles and transported to nerve endings by axoplasmic flow. Within these vesicles, some of the DBH is membrane-bound, and some is soluble. The substrate for the DBH reaction, dopamine, is synthesized in the cytoplasm, and must be taken up into vesicles to be converted to norepinephrine (Fig. 8.1).

Molecular oxygen and ascorbic acid, as well as dopamine, are substrates for the DBH-catalyzed reaction (Fig. 8.2). Ascorbic acid is present in high concentration in adrenal medulla and neuronal tissue (1), where it is available for the reaction. DBH acts as a mixed function oxidase, *i.e.*, it inserts one oxygen atom from O_2 into the substrate and uses the other oxygen atom for the formation of H_2O. Ascorbic acid serves as the reductant for the reaction (72).

Several noradrenergic tissues contain compounds which can inhibit DBH (22, 63, 96, 101). It is not clear whether these compounds actually inhibit DBH *in vivo*, as they do *in vitro*. For a compound to be physiologically active as a DBH inhibitor, it must be localized in the noradrenergic vesicles with DBH, but the endogenous inhibitors are rather nonspecifically distributed (22, 63, 96, 101). However, exposure to drugs or chemicals may induce the formation of physiologically active DBH inhibitors. For example, the pollutant carbon disulfide reacts with endogenous catecholamines to form compounds which are effective DBH inhibitors and are also taken up by noradrenergic vesicles (89). A further discussion of subcellular localization of norepinephrine can be found in Chapter 6.

MOLECULAR PROPERTIES

DBH purified to homogeneity from bovine adrenals (4, 34, 35, 139), human adrenals (114), human plasma (36), or human pheochromocytoma (31, 125) has been characterized in

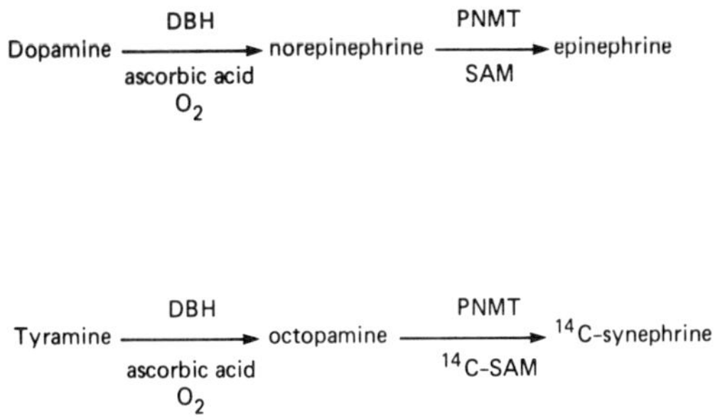

Figure 8.2. The reaction used in the radioenzymatic assay for dopamine-β-hydroxylase (DBH) (*bottom*) is analogous to the physiological reaction (*top*).

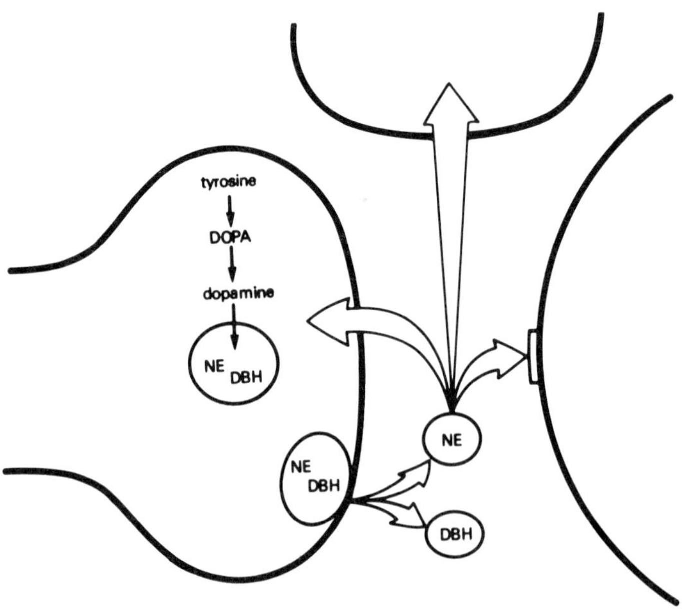

Figure. 8.1 Dopamine synthesized in cytoplasm is taken up into vesicles, where it is converted to norepinephrine by dopamine - β - hydroxylase (DBH). When noradrenergic cells fire, DBH and norepinephrine are released together by exocytosis. After release, norepinephrine can bind to postsynaptic receptors. Its synaptic action is terminated by cellular uptake and metabolism. There is no known physiological function or uptake mechanism for DBH.

detail. The native enzyme is a tetramer of molecular weight 290,000 (35). Enzymatically active dimers, as well as tetramers, exist in human plasma (36, 111). The tetramers can be dissociated into dimers without breaking covalent bonds, but dissociation of dimers into monomers requires the reduction of disulfide bonds (139). Each tetramer contains four copper atoms which form part of the active site and bind to molecular oxygen during the enzyme-catalyzed reaction (23, 138). The copper atoms are accessible to exogenous chelating agents, which are effective DBH inhibitors (45, 48). DBH is a glycoprotein, with about 4% of its weight composed of carbohydrates—mannose, glucosamine, galactose, glucose, fucose, and sialic acid (5, 139, 140).

It has not been possible to isolate DBH from CSF because of the very low concentration of the enzyme there. However, DBH in rabbit CSF has been partially characterized by differential gradient centrifugation (an indicator of molecular size and shape) and polyacrylamide gel electrophoresis (an indicator of molecular size and charge) (29). In both cases, the molecular properties of DBH in CSF were similar to those of DBH purified from bovine adrenal glands.

CSF NE ASSAY AND STABILITY

Norepinephrine can be accurately measured in CSF. When a phenylethanolamine-*N*-methyl transferase (PNMT)-based radioenzymatic assay for NE in CSF was compared with a gas chromatographic mass spectroscopic technique, the mean difference between the two methods was 12%, and the correlation between the two methods was $r = 0.95$ (104). NE in CSF can also be measured by the catechol-O-methyltransferase (COMT) technique. Although most experience with this COMT technique has been gained using plasma assays, CSF inhibits *O*-methylation by COMT less than does plasma, so that the technique is more sensitive when used to measure NE in CSF than in plasma.

Accurate assay techniques now exist for measuring NE in CSF, but there are still serious technical problems involved in the measurement. If NE is added to pH 7.4 water and left at room temperature, the solution will turn brown in less than an hour with NE degradation products. Degradation of NE in plasma does occur, but not quite so rapidly (32) as in CSF. We found that CSF from normal volunteers which was immediately mixed with ascorbic acid and frozen at −70°C had an NE level of 408 pg/ml, but that aliquots of the same CSF kept at 2°C for 30 minutes and then mixed with ascorbic acid and frozen had 242 pg/ml NE. Twenty-eight normal subjects had CSF collected into tubes containing ascorbic acid and frozen immediately, and NE in their CSF was 355 pg/ml. In another group of subjects CSF was kept at 2°C for 30 minutes, and their CSF had NE levels of 229 pg/ml, and in a third group, where the CSF was kept at 2°C for 90 minutes, the NE level was down to 91 pg/ml. There is thus a progressive and rapid decline of NE in CSF samples kept on ice. Unless assay, collection, and preservation techniques are rigidly standardized, it is impossible to compare values from one study to the next.

CSF DBH ASSAY AND STABILITY

A radioenzymatic method for the DBH assay (91) has been used to measure DBH in CSF from humans (43, 67, 70, 82, 83, 85), rabbits (28–30), cats (66, 68), and monkeys (71, 81). This assay method is based on reactions analogous to those occurring in adrenergic cells (Figs. 8.2 and 8.3). Tyramine is converted by DBH to octopamine, which is then methylated by phenylethanolamine-*N*-methyltransferase (PNMT) (purified from bovine adrenal glands) and radioactive *s*-adenosylmethionine (SAM) to give radioactive synephrine which can be quantitated by liquid scintillation counting. A spectrophotometric method (100) and a fluorometric method (37) have also been used to measure DBH in human CSF (Table 8.1). Two groups of investigators using the radioenzymatic method (43, 67) have found similar activities of DBH in CSF. The small group of samples measured spectrophotometrically had somewhat higher DBH (100). When the fluorometric assay was used, higher DBH values and a wider range of DBH values were found (37). Since the different assay methods have not been compared directly on the same samples, it is not certain whether the observed differences are caused by biological or chemical factors. In preliminary studies, DBH protein levels measured by radioimmunoassay in CSF correlate with DBH enzymatic activity measured by radioenzymatic assay. Several tissues and plasma (22, 31, 63, 96, 97, 101), but not CSF (66, 67), contain endogenous inhibitors of DBH. (It is possible to test for the presence of

Table 8.1
Comparison of Studies on DBH in Human CSF[a]

Range of DBH Values	Number of Patients	Assay Method	Reference
0.20-0.60	8	radioenzymatic	Goldstein and Cubeddu (1976)
0.07-1.69	48	radioenzymatic	Lerner etal. (1978)
0.6-5.34	4	spectrophotometric	Okada etal. (1976)
0.52-56.8	108	fluorometric	Fujita etal. (1977)

[a] Dopamine-β-hydroxylase (DBH) activity is expressed in nmole/ml/hour.

reversible inhibitors by adding a known amount of purified DBH to the sample and determining its recovery in the assay). CSF poses special problems because of the low concentration of DBH, its tendency to stick to glass containers, and its instability. The radioenzymatic DBH assay (91) has been modified for CSF studies (43, 67) by collection of CSF in polypropylene tubes, addition of bovine serum albumin to samples, and use of highly radioactive SAM. DBH in human CSF is remarkably unstable (67), losing a significant amount of activity after 1 day at 4°C or 4 days in liquid nitrogen (−196°C). We have recently found that the storage time for human CSF DBH samples can be lengthened to at least 2 weeks by the addition of a small amount of concanavalin A (Con A). Con A is a plant lectin which binds to glucose and mannose residues on DBH without affecting the catalytic activity of the enzyme (5, 140).

Exocytotic Release and Subsequent Fate of DBH and Norepinephrine

When NE is released from neuronal vesicles by exocytosis, soluble DBH is also released (Fig. 8.1). In some *in vitro* preparations NE and DBH are released proportionally in response to depolarizing agents (24, 42, 54, 136, 142). Although DBH and NE are released together, their subsequent fates differ. NE is inactivated by monoamine oxidase in presynaptic cells and by catechol-O-transferase in nonneuronal cells. There is no known uptake mechanism for DBH (Fig. 8.1). Thus the extracellular concentration of NE is determined by rate of release and rates of uptake and metabolism; the extracellular concentration of DBH is determined by rate of release.

Human plasma NE changes dramatically in response to acute changes in sympathetic activity, but plasma DBH does not (57). The large pool of DBH in human plasma and the slow turnover rate of the enzyme prevent substantial changes in plasma DBH from occurring when sympathetic activity changes (57). By analogy, it can be suggested that CSF serves as a "sink" for DBH and NE released in the central nervous system.

Origin of CSF NE

There are three potential sources for norepinephrine in the cerebrospinal fluid (CSF). It can be derived from the blood, from peripheral sympathetic nerves penetrating to the CNS, or from noradrenergic cells of central origin. It appears that the majority of CSF NE is derived from central noradrenergic cells. These cells originate in the brainstem and innervate most areas of the brain with a particularly rich supply to the hypothalamus and other periventricular areas (see Chapters 5 and 7). Spinal cord noradrenergic innervation is derived from the locus coeruleus, subcoeruleus, parabrachial nuclei, and Kolliker-Fuse nucleus. These cells of origin for spinal NE are found only in the pons and rostal medulla (144, 167) (see Chapter 5), and the noradrenergic terminals are concentrated in the cord gray matter (167). Most human CSF samples are obtained by lumbar puncture, and lumbar CSF NE likely represents release from these noradrenergic terminals derived from cells which originate in the brainstem nuclei. 3-Methoxy-4-hydroxyphenylglycol (MHPG) is the major brain metabolite of NE in man, and CSF levels are about 40 times as high as those of NE. MHPG in lumbar CSF is derived from spinal cord metabolism (55) in patients with spinal fluid block. Levels of MHPG in lumbar spinal fluid are unchanged unless there is spinal cord transection, in which case MHPG levels are considerably lower (104). This indicates that noradrenergic metabolites in lumbar CSF are derived from nerve terminals in the spinal cord, which have their cell bodies of origin above the spinal cord. NE is more labile than MHPG, so CSF NE levels should be even more greatly affected by local NE release.

NE does not ordinarily penetrate from blood into CSF (75, 65, 141). Patients with a pheochromocytoma may have enormously elevated plasma catecholamines and yet have normal or low CSF NE (32, 106). When radioactive NE was infused intravenously into a monkey, the peak level of radioactive NE in CSF reached only 2% of peak plasma levels (162). It is possible to break down the blood-brain barrier by infusing enormous levels of NE and thus causing severe hypertension (50). However, under most conditions (even with the level of hypertension produced by a pheochromocytoma), there is an effective blood-CSF barrier for NE.

A third possible influence on NE in CSF is peripheral sympathetic nervous activity. Any stimulus which alters the rate of formation of CSF can change the concentration of solutes such as NE, even though their rate of release might not change. A decrease in the rate of formation of CSF would have maximal influence on substances that are cleared slowly, such as DBH and MHPG, and should have a less marked influence on labile molecules such as NE, which undergo rapid reuptake into the brain from the CSF, irrespective of the rate of CSF formation. The choroid plexus is present in all four ventricles of the brain and constitutes the major site for the bulk formation of CSF. Approximately 0.5% of total CSF volume is replaced by newly formed fluid every minute (26). The choroid plexus is thoroughly innervated by adrenergic nerves which originate in the superior cervical ganglia (76), and stimulation of these adrenergic nerves diminishes the rate of CSF formation. This is mediated by a β-adrenergic receptor (77) stimulation of adenylate cyclase in the secretory cells of the choroid plexus (98). CSF formation is enhanced by cholinergic or β-adrenergic stimulation (52). The observation that adrenergic stimulation of the choroid plexus slows CSF formation might help explain findings by Post *et al.* (105) that exercise, an effective stimulus to peripheral sympathetic nervous activity, increases the level of CSF 5-hydroxyindolacetic acid (5HIAA), homovanillic acid (HVA), and MHPG, since the associated adrenergic stimulation might decrease the rate of CSF formation and thus diminish the rate at which these metabolites are flushed from CSF. These considerations are particularly important in measurements of adrenergic markers such as NE, MHPG, and DBH. Adrenergic stimulation of the choroid plexus could lead to an enhanced rate of release of these three substances from adrenergic nerve endings into the CSF and could also diminish the rate of CSF formation, thereby decreasing the rate of clearance. The parallel rate of increase in NE, DBH, and MHPG seen would then reflect peripheral, not central, adrenergic activities. When NE is injected into the CSF, it is predominantly taken up by noradrenergic neurons of central, not peripheral, origin (115). NE metabolites decrease in spinal CSF following cord transection. These studies and studies showing the enhanced release of cerebral NE by perfusion techniques (135) demonstrate that NE and its metabolites in CSF are released predominantly from the brain into CSF, but peripheral sympathetic nerves may have a similar influence on CSF NE and DBH.

Factors Affecting CSF DBH Activity

In vitro studies have shown that DBH and NE are released proportionally and that the quantities released are determined by neuronal activity (24, 42, 54, 137, 142). But is this also true in a more physiological situation, *i.e.*, in the CSF of a living animal? In addition to neuronal activity, several other factors which may affect the activity of DBH in CSF (Table 8.2) deserve consideration.

First, the tissue level of DBH may be a major determinant of CSF DBH. The effects of drug treatment on DBH activity in the brain have not been well studied because of the difficul-

Table 8.2
Factors Affecting CSF DBH Activity

1. Tissue levels of DBH
   - Amount of enzyme
   - Activity of enzyme (inhibitors)
2. Entry of DBH into CSF
   - Noradrenergic firing
   - Leakage from neurons
   - Blood – CSF barrier
3. Loss of DBH activity from CSF
   - Inactivation (inhibitors)
   - Clearance from CSF

ties inherent in DBH measurement. The high concentration of DBH inhibitors in brain makes meaningful interpretation of the radioenzymatic assay data difficult. Immunochemical techniques, which measure the number of DBH molecules rather than their catalytic activity, are unaffected by enzyme inhibitors. The immunochemical approach has been used in only a few studies on brain DBH to date (112, 113). These studies show that reserpine treatment causes an increase in the number of DBH molecules in the locus coeruleus by increasing the rate of synthesis of the enzyme (112). By contrast, axonal injury causes a decrease in the number of DBH molecules in this nucleus by decreasing the enzyme's synthesis rate (113). In both cases, changes in DBH are accompanied by similar changes in tyrosine hydroxylase, the rate-limiting enzyme for NE synthesis (112, 113). It will be interesting to see whether brain DBH levels are regulated in parallel with tyrosine hydroxylase levels after other drug treatments.

DBH may enter CSF from cells in the CNS or from blood. Tissue DBH is found only in cells which synthesize NE or epinephrine. In the CNS, most of the DBH is found in noradrenergic neurons, which greatly outnumber adrenergic neurons (51). Moreover, there is abundant evidence that it is released exocytotically (24, 42, 54, 137, 142). It is unlikely that a significant proportion of the DBH in CSF has "leaked" out of neurons, since tyrosine hydroxylase, an enzyme in noradrenergic neurons which is not released by exocytosis, is not detectable in CSF (29). Plasma is a potential source of DBH in CSF, since most of the protein in CSF comes from plasma through the blood-CSF barrier. The facility with which proteins cross the blood-CSF barrier is inversely related to their molecular size (79). DBH, with a molecular weight of 290,000 (139), is larger than most plasma proteins and should penetrate into CSF poorly. However, the specific activity of DBH (enzyme activity divided by total protein) is approximately 100 times higher in CSF than in plasma for both cats (Table 8.3) and rabbits (29). Therefore it is unlikely that plasma is a major donor to the CSF pool in these animals. In humans, the specific activity of DBH in CSF is about 25 times higher than in plasma, and there is a significant correlation between plasma DBH and CSF DBH ($r = 0.60$, $p < 0.05$, $n = 16$) (67). This suggests that in humans some of the DBH in CSF is derived from plasma and/or that DBH levels in both fluids are partially determined by the same genetic factors. Pharmacological studies in humans and in experimental animals show that DBH in CSF and in plasma are often affected differently by drug treatment (see Basic Pharmacological Studies), results which again suggest different sources for CSF DBH and plasma DBH.

Finally, the mechanism and rate of loss of DBH activity from CSF are unknown. DBH within CSF may lose activity by inhibition, denaturation, or degradation. Although the mode of clearance is not known, we can speculate that the carbohydrate moieties on the DBH molecule are recognized by a specific cellular uptake system; such recognition and uptake systems have been described for other glycoproteins in the periphery. In addition, DBH, like other proteins, may be removed from the CSF by bulk absorption (26).

Table 8.3
DBH and Protein Concentrations in Cat CSF and Plasma[a]

	CSF	Plasma
DBH (nmole/ml/hr)	0.16 ± 0.008(8)	0.41 ± 0.06(8)
Protein (mg/ml)	0.30 ± 0.03(8)	79 ± 3(8)
DBH specific activity (nmole/mg. protein/hr)	0.57 ± 0.05(8)	0.0052 ± 0.0008(8)

[a] Dopamine-β-hydroxylase (DBH) and protein were measured in CSF and plasma of unmedicated cats. Values shown are mean ± SEM for eight animals.

BASIC PHARMACOLOGICAL STUDIES

In order to be utilized as a quantitative index of central noradrenergic activity, CSF NE and DBH must be shown to be of central origin and to reflect activity of central noradrenergic neurons. The question of central origin has already been discussed under the headings Origin of CSF NE and Factors Affecting CSF DBH Activity. The remainder of this section will deal with the results of studies of drugs on levels of CSF NE and DBH.

Responsiveness of CSF NE to Pharmacological Manipulations

Previously available fluorescence assays were not sensitive enough to measure NE in CSF, so most early studies were performed with radioactive compounds. Radiolabeled NE, tyrosine, or L-dopa was perfused into the cerebral ventricles shortly before drug administration and CSF sampling, and radiolabeled norepinephrine in CSF was then measured (92, 20). Radioactive NE could be taken up and released by other than noradrenergic neurons (92, 20). When radioactive catecholamine precursors were perfused, only newly synthesized NE could be detected, and the newly synthesized NE was probably pharmacologically different from long-term stores (92, 20). The development of the sensitive radioenzymatic assays for norepinephrine (157, 19) made possible the measurement of the endogenous pool of norepinephrine in CSF (Fig. 8.3).

Three drugs that enhance the release and block the uptake of catecholamines—amphetamine, cocaine, and phenylethylamine—increase CSF NE levels. In a study on cats (19), the cerebroventricular system was perfused with *d*-amphetamine, and NE and dopamine were analyzed by a radioenzymatic assay using catechol-*O*-methyltransferase (COMT), similar to that described in Chapter 2.

The *d*-amphetamine treatment caused an increase in NE and a larger increase in dopamine in CSF. In rhesus monkeys given a single intramuscular injection of 0.75 mg/kg of amphetamine (102), NE increased 77% and dopamine increased 26%. Concentrations of both catecholamines remained elevated for at least 12 hours. Another study compared the effects of *d*- and *l*-amphetamine given intravenously to rhesus monkeys at 0.1 to 1.0 mg/kg (156). Both isomers caused similar increases in CSF NE, except for minor differences at the highest dose. NE remained elevated for 9 to 33 hours, depending on the dose. Both studies with rhesus monkeys showed that amphetamine, at doses in the therapeutic or abusive range for humans, caused marked and prolonged increases in CSF NE (102, 156). Cocaine induced a modest increase in CSF norepinephrine and a larger increase in CSF dopamine after cerebroventricular perfusion (19). Phenylethylamine (75 or 150 mg/kg, i.m.) caused long-lasting (at least 9 hours) increases in CSF NE in rhesus monkeys (102). The higher dose also caused long-lasting (9 hours) increases in CSF dopamine.

UPTAKE BLOCKERS

Imipramine and other tricyclic antidepressants block neuronal uptake of NE and serotonin after a single dose, but their mood-elevating effects require 2 to 3 weeks of treatment. The effects of long- and short-term treat-

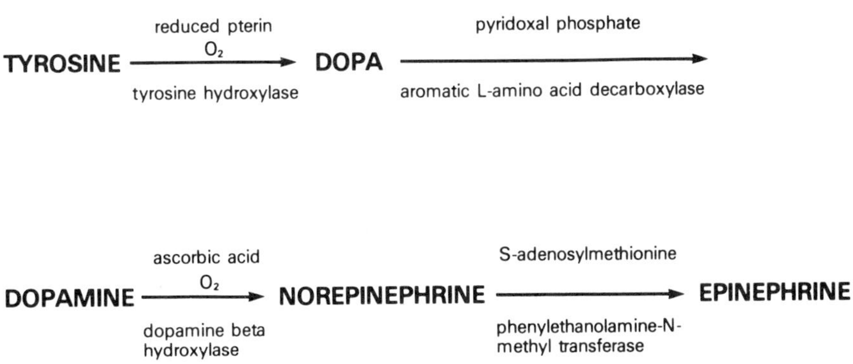

Figure 8.3. Catecholamine biosynthesis. Dopamine-β-hydroxylase (DBH) catalyzes synthesis of norepinephrine from dopamine.

ment with imipramine and NE in CSF have been studied in rhesus monkeys (71).

After a single injection of imipramine, there was a large increase in NE. After 7 to 14 days of drug treatment, NE levels returned to normal. On withdrawal, NE increased and later returned to control values. The increase in CSF NE after a single drug dose was probably due to uptake blockade (118). After 7 and 14 days of treatment, the tendency for NE increase (uptake blockade) was balanced by the tendency for NE decrease (decreased synthesis and firing), and there was no net change (87, 118, 129). The rebound of NE on drug cessation suggested that active, compensatory changes occurred during chronic drug treatment.

NOREPINEPHRINE PRECURSORS

The combination of L-dopa and carbidopa (a peripheral decarboxylase inhibitor) is commonly used to treat Parkinson's disease. This treatment increases brain dopamine, but central NE should also be increased. In both CSF and plasma, NE is increased by L-dopa/carbidopa.

MONOAMINE OXIDASE INHIBITORS

According to the catecholamine theory of affective disorders, monoamine oxidase inhibitors (MAOIs) and tricyclic antidepressants are effective in treating depression because they increase concentrations of NE and dopamine at central synapses. When depressed patients were treated with clorgyline or pargyline, there were no consistent changes in CSF NE (82). The MAOIs could increase CSF NE by inhibiting its breakdown or decrease NE by depressing firing of noradrenergic neurons (70). These two opposing processes could have caused a net increase in CSF NE in some patients and a net decrease in others. These changes were correlated with antidepressant response (82). A strong negative correlation was observed between drug-induced changes in CSF NE and the change in depression ratings and anxiety ratings. The association between CSF NE and mood is consistent with the catecholamine hypothesis of affective illness.

Deprenyl, like clorgyline and pargyline, had no consistent effect on CSF or plasma NE. However, it increased CSF dopamine, consistent with the claim that deprenyl enhances the therapeutic effect of exogenous L-dopa in the treatment of Parkinson's disease (33).

NORADRENERGIC ANTAGONISTS

Propranolol blocks both β_1- and β_2-receptors and lowers blood pressure by central and peripheral effects (27, 74, 131). Cats given a single subcutaneous injection of propranolol increase CSF NE and decrease plasma NE (69). Propranolol has opposite effects on the firing of central and peripheral noradrenergic neurons *in vivo*, and the difference in firing is reflected in CSF and plasma NE levels.

Propranolol also increased CSF NE in dogs (130). Propranolol was administered to the dogs by ventriculocisternal perfusion, and arterial pressure, heart rate, and CSF NE were measured. Both *dl*- and *d*-propranolol produced dose-dependent increases in CSF NE and concomitant decreases in mean arterial blood pressure which were highly intercorrelated. Pretreatment with the α-adrenergic antagonist phentolamine prevented the propranolol-induced hypotension and bradycardia. These data support the theory that the central hypotensive effect of propranolol was mediated by increased release or decreased uptake of central NE. The released NE would stimulate α-adrenergic receptors and cause a decrease in sympathetic outflow (130).

BROMOCRIPTINE

Bromocriptine is useful in the treatment of parkinsonism, but it lowers blood pressure. It is a dopaminergic agonist and an α-adrenergic antagonist. It stimulates presynaptic dopamine receptors that inhibit NE release. Patients treated with bromocriptine decrease their plasma and CSF NE by 50% (160). Likewise, parkinsonian patients treated with lergotrile decrease their CSF NE (132). The decrease in CSF NE is probably due to an inhibition of NE release, mediated by presynaptic dopamine receptors (160, 132). The decrease in plasma NE may be due to central or peripheral effects (160).

OTHER TREATMENTS

Probenecid is a weak organic acid that competitively interferes with the active transport of acidic metabolites by the kidney and the choroid plexus. It has been used to block the efflux of acidic monoamine metabolites from CSF when central neurotransmitter turnover in humans is studied. However, probenecid also increases NE in both CSF and plasma. The magnitude of the increase is less than that observed for the acid metabolites of dopamine and serotonin.

Angiotensin II is a vasopressor peptide which facilitates NE release and synthesis (7, 17). After ventriculocisternal perfusion of rabbits with angiotensin II CSF NE, arterial blood pressure and heart rate all increase (18). Arterial pressure and CSF NE correlate well after angiotension II, and this is opposite to the effect of propranolol (130).

Although a low monoamine diet causes significant decreases in 3-methoxy-4-hydroxyphenylethylglycol (an NE metabolite) in the urine of depressed patients (but not normal volunteers), NE in CSF is not significantly altered by diet (62).

Electrical stimulation of specific brain regions of neurological patients also affects CSF NE levels. Chronic cerebellar stimulation has been used in an effort to suppress cerebral epileptic activity in man. The effectiveness of this treatment is controversial (148). CSF NE is increased in the CSF of epileptic patients during chronic unilateral or bilateral cerebellar stimulation (148, 152), but NE decreases following cerebral cortical stimulation (148). Caudate nucleus stimulation decreases CSF NE (149–151). Since the caudate nucleus contains very little NE, this decrease is probably a secondary effect of the electrical stimulation.

Responsiveness of CSF DBH to Pharmacological Manipulations

NORADRENERGIC AGONISTS AND ANTAGONISTS

CSF DBH increases after acute administration of phenoxybenzamine (30, 66). Phenoxybenzamine blocks α-receptors (64), inhibits NE uptake (24), and increases firing rate. DBH and NE also increase in blood after phenoxybenzamine (24, 54). When it was given to rabbits (5 mg/kg, i.v.), CSF DBH increased gradually, reaching 500% of control value 3 hours after drug treatment (30) and remained elevated 2 to 15 hours following drug administration. These findings suggest a positive relationship between central noradrenergic firing rate and CSF DBH levels. Similar results were obtained using cats (66), where a single subcutaneous dose of phenoxybenzamine (6 mg/kg) produced a significant increase in CSF DBH 8 hours after treatment.

Clonidine, an α_2-adrenergic agonist, decreased CSF DBH, but only after 24 hours of drug treatment (66). The time lag is probably due to a slow clearance of DBH from CSF, not a slow onset of decreased central firing rate. Clonidine is a potent α_2 agonist which inhibits NE release from noradrenergic neurons and is known to lower plasma and CSF NE. When cats were injected with clonidine four times over a 19-hour period, CSF taken 5 hours after the last injection had consistently lower DBH than in a control period ($p < 0.005$). Clonidine inhibits the spontaneous firing of the locus coeruleus (128). The opposite effects of clonidine and phenoxybenzamine on CSF DBH support the proposed relationship between CSF DBH and central noradrenergic firing rate.

MONOAMINE OXIDASE INHIBITORS

Chronic administration of monoamine oxidase inhibitors (MAOIs) decreases CSF DBH in both humans (70) and animals (68). In a study performed on cats (68), two MAOIs were given chronically to individual groups of cats to study the time course of changes in the CNS. In this study, clorgyline (an MAO type-A inhibitor), pargyline (an MAO type-B inhibitor), or saline was administered to cats (1 mg/kg/day) for 24 days. CSF was removed prior to drug treatment and several times during drug treatment. Figure 8.4 illustrates group data for three groups of cats, indicating a chronic, but not acute, decrease in CSF DBH for both clorgyline and pargyline. The authors suggest that this decrease in CSF DBH results from a decrease in central noradrenergic firing rate—an effect which has been demonstrated in rats (13). There were no intergroup differences in brain tissue levels of DBH from autopsy on day 24, so the decreased CSF DBH levels were not due to enzyme depletion. There was a marked decrease in plasma DBH after 8 days of MAOI treatment—2 weeks before any significant changes in CSF DBH. This time lag argues against plasma being a major contributor to the pool of DBH in CSF. It is unclear why the decrease in CSF DBH was observed only after 22 days of MAOI treatment. This finding is of interest, however, in light of the delayed therapeutic response to these drugs, *i.e.*, antidepressant response occurs only after 2 to 3 weeks of drug treatment.

Clorgyline and pargyline also decrease human CSF DBH after chronic administration for 3 weeks (70). Results of this study parallel those from the animal studies.

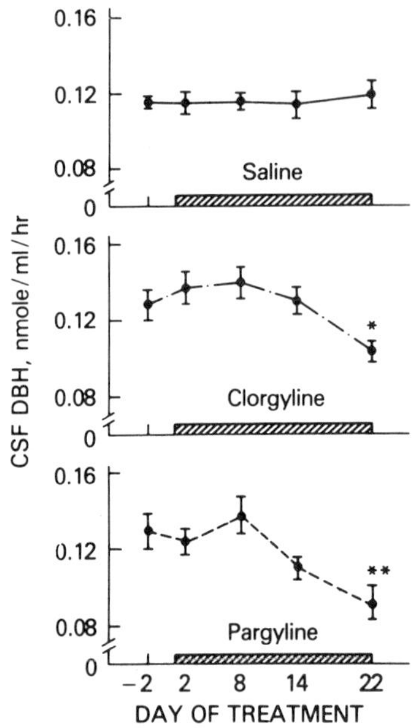

Figure 8.4. Effect of monoamine oxidase inhibitors on cat CSF dopamine-β-hydroxylase (DBH). DBH was measured in CSF taken from cats before treatment with saline (N = 5), clorgyline (N = 5), and pargyline (N = 5) as well as after 2, 8, 14, and 22 days of treatment with these monoamine oxidase inhibitors. Values are shown as mean ± SEM for five cats at each time. For each treatment group, DBH levels on treatment days were compared to pretreatment DBH levels using Student's-*t*-test for unpaired data. *, $p < 0.05$; **, $p < 0.02$.

IMIPRAMINE

Imipramine, like the MAOIs, shows therapeutic latency when given to humans. Imipramine blocks neuronal uptake of NE (116) and should thus increase NE levels in the synaptic cleft. If NE levels rise in the synaptic cleft, a compensatory decrease in presynaptic firing rate should follow.

In a study performed on rhesus monkeys (71), imipramine was given chronically at a dose of 4.5 mg/kg/day. CSF was removed prior to, during, and after the imipramine treatment period, with NE and DBH measured concurrently (Fig. 8.5). The large increase in NE 24 hours after the first imipramine injection is probably due to blockade of NE uptake. The concurrent decrease in DBH suggests a decrease in noradrenergic firing rate secondary to the abundance of NE in the synaptic cleft. Shifts in NE and DBH toward predrug levels later in the treatment period parallel the delayed onset of therapeutic effects of this drug, and shifts following cessation of treatment show withdrawal effects.

OTHER PHARMACOLOGICAL AND PHYSIOLOGICAL MANIPULATIONS

Central 6-hydroxydopamine injections destroy central noradrenergic neurons. Five days after injection there was a 50% decrease in CSF DBH and no change in plasma DBH. These data strongly suggest that the major source of the DBH in CSF is the central noradrenergic system and not plasma.

Two drugs which indirectly increase central noradrenergic activity, physostigmine and pentylenetetrazol, also increase DBH activity in rabbit CSF (30). Sciatic nerve stimulation and cold stress also increase rabbit CSF DBH (29). Although stimulation of a peripheral nerve increases DBH in CSF, it has not been determined whether direct electrical stimulation of central noradrenergic nerves causes an increase in CSF DBH.

Relationship between Central and Peripheral NE and DBH

We infused radioactive NE intravenously into a monkey and collected CSF samples for 2 days. Peak NE levels in CSF reached only 2% of peak plasma levels (162). This is not surprising because there is an effective blood-brain barrier for catecholamines (65, 75, 141). This includes an effective blood-CSF barrier for NE as demonstrated by one patient we measured (162) and several patients reported by Eide *et al.* (32) who had a pheochromocytoma with very high blood levels of NE and normal or low CSF levels of NE. It was surprising that 159 neurological patients had a strong relationship between NE in blood and CSF ($r = 0.78$, $p < 0.001$) (162). This relationship clearly cannot be explained by penetration of blood NE into CSF. Peripheral sympathetic innervation of the choroid plexus as discussed earlier might explain part of this relationship. However, the choroid plexus is present in the cerebral ventricles, and samples were taken by lumbar puncture from these patients. When NE is injected into the CSF, it is largely taken up by catecholaminergic neurons of central, not peripheral, origin (39). It may be that genetic and environmental influences that lead to a high rate of firing of central noradrenergic neurons also increase the firing

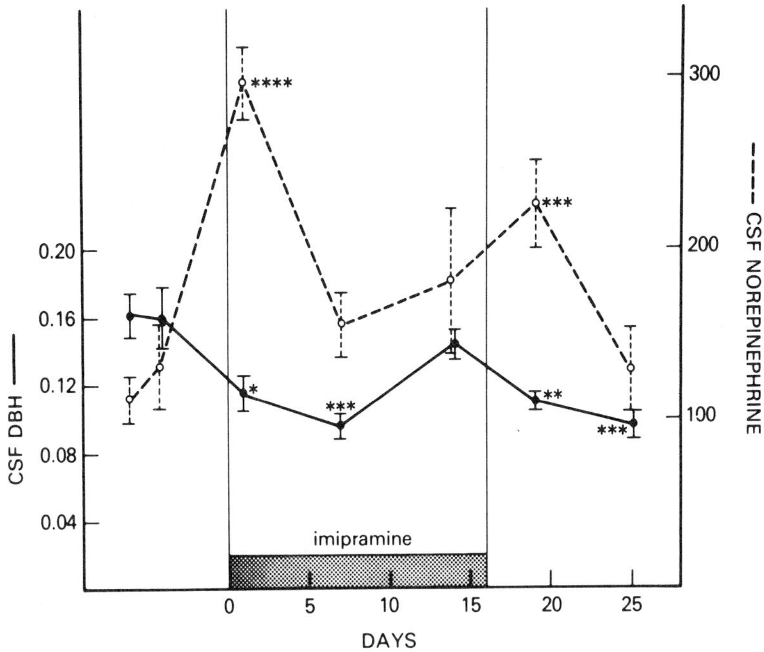

Figure 8.5. Effect of imipramine on dopamine-β-hydroxylase (DBH) and NE in CSF> DBH is expressed in nmole/ml/hr and NE in pg/ml. Data are shown as mean ± SEM. Changes in DBH and NE over time were evaluated by comparing mean values on each day to mean value from first baseline tap (Student's *t*-test for unpaired data). *, $p < 0.05$; **, $p < 0.01$; ***, $p < 0.005$; ****, $p < 0.001$.

rate of peripheral noradrenergic neurons. An individual who has very active mechanisms for clearance of NE from CSF should have similarly active mechanisms for clearance of NE from plasma. Thus, a high degree of correlation between CSF NE and plasma NE might exist without any direct relationship between these two compartments. At present, we do not know which, if any, of these explanations is the predominant influence in the strong correlation between CSF and plasma NE, but over 50% of the variance in CSF NE levels can be predicted by plasma NE levels. Whatever the explanation, it is the predominant influence on CSF NE levels.

Indirect evidence for the hypothesis that CSF DBH and plasma DBH come from different sources can be obtained from pharmacological studies. In several cases, drug treatment has different effects on DBH in CSF and in plasma (Table 8.4). The changes in CSF DBH are consistent with the known effects of the drugs on central noradrenergic activity. Plasma DBH changes, however, are not always consistent with changes in sympathetic activity. Plasma DBH, unlike CSF DBH, shows considerable species variation in its drug responsiveness.

There are two ways in which plasma DBH and CSF DBH can differ in their drug response (Table 8.4).

1. Direction of change. Increases or decreases in CSF DBH may not be accompanied by corresponding changes in plasma DBH. This is the case for phenoxybenzamine in cats, clonidine in cats, monoamine oxidase inhibitors in humans, disulfiram in humans and monkeys, and 6-hydroxydopamine (intraventricular) in rabbits.
2. Time of onset of change. When cats are treated chronically with monoamine oxidase inhibitors, plasma DBH is sharply decreased after 8 days, while CSF DBH decreases only after 22 days of treatment.

CLINICAL STUDIES

Lumbar Gradient for CSF NE and DBH

The concentration and rate of release of NE in brain tissue adjacent to CSF can have a

Table 8.4
Effects of Drugs on DBH in Plasma and CSF

Drug	Time	Species	Plasma	CSF
phenoxybenzamine	8 hours	cat	no consistent changes	increase
phenoxybenzamine	4 hours	rabbit	increase	increase
clonidine	24 hours	cat	low ones increase high ones decrease	decrease
monoamine oxidase inhibitors (clorgyline and pargyline)	21 days	man	no consistent changes	decrease
monoamine oxidase inhibitors (clorgyline and pargyline)	1-22 days	cat	decrease after 8 days	decrease after 22 days
disulfiram	4 weeks	man	increase	no change
disulfiram	6 weeks	monkey	increase after 3 weeks; long-lasting change	no change
pentylenetetrazole	3 hours	rabbit	increase	increase
6-hydroxydopamine (intraventricular)	5 days	rabbit	no change	decrease
physostigmine	3 hours	rabbit	increase	increase

marked effect on local levels of CSF NE. The concentration of NE in some brain areas is 1000 times as great as in CSF (158), so that NE release should be an important determinant of CSF NE. There is also an avid reuptake mechanism for NE, and after NE is injected into the cerebral ventricles, it is quickly taken up by neurons (39, 40). Since NE is a labile molecule subject to oxidation, it can be destroyed while in CSF. 3-methoxy-4-hydroxyphenylglycol (MHPG) is the major brain metabolite of NE, and its levels in spinal fluid are not altered if the flow of spinal fluid is blocked (104). However, if the cord is transected, thereby diminishing spinal levels of NE, then CSF MHPG levels are markedly diminished (104). This demonstrates that MHPG in lumbar CSF is primarily of spinal origin. The spinal subarachnoid space contains 30 ml of CSF (145), so that CSF obtained by routine lumbar puncture comes from the spinal subarachnoid space. We measured a gradient for CSF NE in seven patients with Huntington's chorea, in whom a total of 20 ml of CSF was taken. There was a gradient with increasing levels of NE until the 12th ml when there was no further increase in NE levels (158). In a further experiment with patients undergoing spinal drainage prior to neurosurgical procedures, a total of 38 ml CSF was removed, and there was an increase in CSF NE between the 1st and 16th ml, but not between the 16th and 35th ml. The last sample presumably contained a mixture of CSF from the cisterna magna, basal cisterns, and the cerebral ventricles, but NE levels did not increase in this sample. The NE gradient for lumbar CSF is of spinal origin and not dependent on brain areas containing NE. Spinal NE is localized mainly in gray matter (167), so the lower lumbar area in the region of the cauda equina should contribute little NE to the CSF, accounting for the lower levels of NE in the first 21 ml of CSF removed.

There is about 40 times as much MHPG in CSF as NE, and there is a parallel gradient for MPHG as with NE (165). Several earlier investigations did not detect this MHPG gradient (16, 47, 124), probably because of its small size. MHPG derived from NE is present in much higher concentrations than NE. This indicates that MHPG is cleared more slowly than NE, which explains why the gradient for MHPG is less steep than that for NE. The CSF gradient for NE is not as large as that reported for other substances, such as homovanillic acid (HVA), but is of sufficient size to be important in studies of CSF NE. The CSF gradient for MHPG is so small as to be of little importance in comparison with MHPG variation between individuals. In the case of CSF DBH, there are no reports in the literature looking at lumbar gradients for CSF DBH. We have, however, measured CSF DBH on the 3rd and 25th ml of CSF in the same individual and found them to be identical, suggesting no significant gradient for CSF DBH.

Circadian Rhythm for CSF NE and DBH

Brain areas heavily innervated with NE terminals have the highest levels of NE when animals are most active. Cervical spinal cord (110), midbrain (109), and hypothalamus (88) all have significant circadian variations in their NE levels. Atlhough these areas rich in NE terminals have their highest NE levels when animals are most active, the overall rate of synthesis of NE (166) and brain content of NE (21) are highest during the time of day when animals are least active. When we collected CSF continuously from the lateral cerebral ventricle of *Macaca mulatta* monkeys, which are most active in the daytime, NE content in the CSF was highest in mid-afternoon and lowest in early morning (163). This was a persistent and reproducible rhythm that endured through a variety of circumstances. On weekdays, the animals were exposed to a great deal of activity, but on weekends saw visitors only on two occasions for feedings. However, the rhythm for NE in CSF was the same on weekdays and weekends. One monkey confined to a box with a reverse light cycle did not detectably vary from his counterparts in a normal lighting cycle. This same diurnal rhythm is apparently present in man. When lumbar puncture was performed on patients with diverse neurologic diseases at 0300, 0900, and 1500, the NE levels were higher in the daytime and highest at 1500, as in the monkeys. When NE is infused into the cerebral ventricles, it increases motor activity, and the daily rhythm of NE in CSF parallels the functional effects of NE on activity. Monkeys who showed a distinct diurnal rhythm in their CSF NE failed to show any rhythm in the NE metabolites, MHPG, and vanillylmandelic acid (VMA) (103). These NE metabolites are not as chemically labile or metabolically active as NE, so alterations in their levels might be damped by their greater persistence in the CSF. There is no information available on circadian alterations of CSF DBH.

CSF NE in Hypertension

Animal models of hypertension show that central NE is involved in the development and maintenance of hypertension. The destruction of noradrenergic nerves by intracisternal injections of 6-hydroxydopamine prevents hypertension of renal (15), volume-dependent (108), or baroreceptor-mediated causes (14). In man, pharmacological data implicate noradrenergic hyperactivity (25, 119), as does one of two studies measuring MHPG (95, 115). Several diseases that affect central noradrenergic nerves also affect blood pressure in man. Parkinsonism affects dopaminergic nerves in the brain but also depletes NE. Parkinsonian patients have lower than normal cerebrospinal fluid NE and low blood pressure (132, 133). Huntington's chorea causes low levels of CSF NE and is associated with lower than normal blood pressure (164). On the other hand, amyotrophic lateral sclerosis is associated with elevated levels of CSF NE, elevated plasma NE, and increased blood pressure (155).

A decrease in blood pressure in experimental animals elicited by nitroprusside or bleeding leads to an increased rate of release of NE from the hypothalamus (123). Chronic sodium depletion in dogs increases CSF NE by over 40% (9). We have observed that administration of clonidine to monkeys lowers their CSF NE by 50%. Thus, varied treatments for hypertension could alter CSF NE, so therapy should be controlled in studies of hypertensive patients. While drug therapy can be controlled, the observation of Brosnihan *et al.* (9) that sodium depletion elevates CSF NE, is much harder to control for because hypertensives tend to have altered renal handling of sodium and patients who know they are hypertensive may voluntarily restrict their own dietary intake of sodium.

Eide *et al.* (32) measured NE in the cerebrospinal fluid of patients who underwent lumbar puncture for evaluation of neurologic symptoms. Seven of these patients had essential hypertension and had their antihypertensive medications discontinued 1 week prior to lumbar puncture. All the patients were hospitalized and had similar diets. Those with essential hypertension had CSF NE levels of 298 ± 60 pg/ml, while normotensives had NE levels of 167 ± 71 pg/ml. Both the CSF and plasma NE levels of the hypertensives were significantly higher than those of normotensives. Patients with hypertension due to secondary causes, including two subjects with pheochromocytoma and very high plasma NE levels, had low CSF NE levels.

We studied 128 patients (161) who underwent lumbar puncture for evaluation of neurologic disease. These patients were all hospitalized on the same ward, received similar diets, and underwent lumbar puncture about the same time of morning. Although many of the patients were hypertensive by usual criteria, only six of them had received treatment for hypertension at the time of the study. NE levels were measured on the 12th to the 16th ml of CSF removed. CSF NE correlated with diastolic blood pressure ($r = 0.41$, $p < 0.001$). In the 87 patients who were drug-free, the correlation was similar ($r = 0.36$, $p < 0.001$). In these patients, plasma NE did not correlate with blood pressure ($r = 0.04$). When these patients were divided according to neurologic diagnoses, some subgroups had significantly different CSF NE levels. For example, in the 35 patients with amyotrophic lateral sclerosis, mean CSF NE level was 375 ± 42, and mean blood pressure was 132/87. The 14 patients with no central nervous system disease, but who had peripheral myopathies, had CSF NE of 219 ± 55 and a mean blood pressure of 117/74. However, within each of these groups, there was a positive correlation of CSF NE with diastolic blood pressure. Thus, the overall, highly significant correlation between CSF NE and blood pressure in these patients was due to two causes: patients with certain neurologic diseases, such as amyotrophic lateral sclerosis, had significantly higher CSF NE levels and higher blood pressures than other patients; and within disease groups, CSF NE levels correlated with blood pressure.

The two studies just discussed used neurologic patients, and it appears that part of the relationship between CSF NE and blood pressure in these patients was due to their neurologic diseases. We felt that the results warranted performing lumbar puncture in patients who did not have neurologic diseases, but who had essential hypertension or who were normal control subjects. We measured CSF NE in 11 patients with essential hypertension (blood pressure 155/95) and in nine, healthy normotensive volunteers (blood pressure 116/73) who were of similar age (58). All of the hypertensives had essential hypertension, no subject took medication for 2 weeks prior to admission to a metabolic ward, and they all received the same diet for 7 days prior to lumbar spinal puncture. The lumbar puncture was performed at exactly the same time of day under the same conditions, and the 16th to 20th ml of CSF was collected into a tube containing ascorbic acid and frozen immediately on dry ice. Among these subjects, CSF NE correlated significantly with blood pressure ($r = 0.45$, $p < 0.05$). CSF NE levels were significantly higher in the hypertensives than in the normotensives. These same subjects had plasma NE levels measured on another day while recumbent, standing, and performing isometric exercise, and there was no significant difference between plasma NE levels under any of these conditions between normotensive and hypertensive subjects. None of these subjects had any neurologic disease or any illness other than hypertension, and they were evaluated under identical circumstances.

There were some revealing relationships between CSF NE and age in the patients in these three studies. Eide *et al.* (32) found that CSF NE correlated inversely with age in hypertensives, but not in normotensive patients. In the study of 128 neurologic patients who underwent lumbar puncture, there was a weak correlation of CSF NE and age ($r = 0.27$, $p < 0.01$). However, the correlation between CSF NE and blood pressure was much stronger in that half of the patient group less than 40 years of age ($r = 0.52$, $p < 0.001$). Interestingly, most of Eide's patients were under age 40. The 11 patients with essential hypertension and no other illness in the last study discussed were older (48 ± 4 years), and in them, CSF NE correlated positively but not significantly with age ($r = 0.43$). Among the whole group of subjects, CSF NE correlated positively with age ($r = 0.53$, $p < 0.02$). Thus, among normotensive subjects or in older hypertensives, there appears to be a positive correlation of CSF NE with age, as there is a positive correlation of plasma NE with age (159). However, the correlation between CSF NE levels and blood pressure is strongest in youngest subjects, and the young hypertensives studied by Eide (32) had by far the highest CSF NE levels. This suggests that CSF NE is highest in hypertensives early in the development of their disease.

NE in CSF is labile, difficult to measure, and subject to a wide number of variables. Several of these variables, such as the spinal gradient for CSF NE levels and the circadian rhythm of NE, can be easily controlled. Other variables, such as the effect of treatment on blood pressure, effects of plasma volume on CSF NE, and the effects of emotional status on central noradrenergic activity, are much more difficult to control. All of these phenomena seem to have been adequately controlled in at least one of these studies, however. Diet, medications, time of sampling, and aliquot of CSF sampled are accounted for in all of the studies. Although the knowledge that a person is being evaluated for hypertension may affect his diet, attitude, and level of anxiety, in one of the studies, most of the patients did not even know that they had hypertension, and none of them were hospitalized for their blood pressure.

It is unlikely that excess peripheral sympathetic nervous activity influenced either the rate of CSF formation or release of sympathetic nervous NE into the CSF, since in two of these studies, plasma NE did not differ between controls and hypertensives. There is thus a consistent finding that NE is elevated in the CSF of hypertensives, and, with our present knowledge of the dynamics of NE and CSF, this finding cannot be explained away as an artifact. The NE elevation appears to be greatest in young patients with essential hypertension and is not present in patients with secondary hypertension. The NE elevation is not simply a response to increased blood pressure. A decrease in blood pressure by vasodilators or hemorrhage increases NE release from the hypothalamus (123), and sodium depletion in dogs also increases CSF NE (9). Secondary hypertension, even severe hypertension caused by a pheochromocytoma, was associated with normal levels of CSF NE (32, 159). The elevated CSF NE is a phenomenon of essential hypertension that occurs early in the course of the disease and not simply as a reaction to the elevated blood pressure. The CSF NE elevation is temporally related to development of essential hypertension and appears to provide information about the pathogenesis of this disease.

CSF NE and DBH in Neuropsychiatric Disorders

Compared to the relatively few studies on CSF DBH in neuropsychiatric disorders, there are now a number of studies looking at CSF NE in these disorders. Since the results of these studies are discussed in various other chapters in this book, we will limit our discussion to CSF DBH, except in those cases where NE findings help to clarify the meaning of changes in CSF DBH.

CSF DBH IN AFFECTIVE DISORDERS

Pharmacological studies implicate the central noradrenergic system in the regulation of moods: drugs which deplete NE, such as reserpine, tend to cause depression, and drugs which release NE, such as amphetamine, cause stimulation. Further support for this hypothesis comes from the observation that clinically effective antidepressants such as the tricyclic antidepressants and the monoamine oxidase inhibitors increase brain NE in animals. These initial clinical observations, although not coupled with amine or amine metabolite data, were the basis for the catecholamine hypothesis, which suggested that an underlying abnormality in the central noradrenergic and dopaminergic systems were responsible for manic-depressive illness (46, 94, 116). Subsequent human studies examined the main metabolite of NE (MHPG) in both urine and CSF. Most urinary studies demonstrated significantly lower MHPG in bipolar depressed patients *vs.* controls (80, 117). Studies of MHPG in CSF were conflicting, with two reports showing a decrease (104, 127) and three reporting no change in depression but an elevation in mania (121, 122, 146).

Recent methodological advances have made possible a more direct assessment of central noradrenergic activity by the accurate measurement of the small amounts of NE and DBH present in cerebrospinal fluid. Post *et al.* (106) have reported 2-fold increases in NE in manic patients as compared to neurological controls or depressed patients. When these manic patients were treated and improved, their CSF NE levels tended to decrease toward normal. If these increased levels of NE in mania were the result of increased release, one would expect a concomitant increase in DBH, since both NE and DBH are released together upon neuronal firing. A report (67) that DBH among a group of manics is lower than among depressives is intriguing. However, a definitive statement about CSF DBH and mood state awaits the measurement of the enzyme in individuals switching from depression into mania. Mania may not be a result of increased neuronal firing but rather might represent an abnormality in the amount of NE incorporated into the vesicles, or an abnormality in NE reuptake. Although there is no direct evidence available to support either of these hypotheses, there is a pharmacological analog for each. Monoamine oxidase inhibitors which block the intraneuronal breakdown of NE have been shown by our group to increase the ratio of NE to DBH released in the CSF (70). This finding is of particular interest because MAO inhibitors have long been known to be capable of inducing mania in certain individuals. Imipramine, an NE uptake blocker which is also capable of inducing mania in certain individuals, has been shown to cause an increase in CSF NE while causing a decrease in CSF DBH in the same animals (71).

Further studies of CSF of affectively disturbed individuals will be necessary before conclusions about NE uptake and storage can be drawn.

CSF DBH IN SCHIZOPHRENIA

Although the most widely held biochemical hypothesis for schizophrenia is increased dopaminergic activity in certain areas of the brain, Wise and Stein (147) have postulated that schizophrenia may be a reflection of a progressive noradrenergic degeneration. In support of this hypothesis, they have reported decreased DBH levels in postmortem brain material from chronic schizophrenics. This finding, however, was not confirmed in a subsequent study by Wyatt *et al.* (154). Other investigators have also concluded that schizophrenics do not have lower CSF DBH levels than other psychiatric patients (67 and Table 8.5) or normal volunteers (125). Brain studies and CSF studies provide information on different aspects of central noradrenergic function. Analysis of brain gives total tissue content of DBH, while analysis of CSF gives an indication of the amount of DBH released from cells. Either the steady-state tissue levels of DBH or the rate of release of the enzyme could be affected in schizophrenia.

A potential difficulty in these studies is the heterogeneity of the patient populations. Patients had a diagnosis of schizophrenia but probably do not represent a single disease entity, and only certain subtypes of schizophrenia may result from a defect in the noradre-

Table 8.5
DBH in CSF of Psychiatric Patients[a]

Diagnostic Group	Number of Patients	DBH Activity in CSF	Range of DBH Values
Major depressive disorder	16	0.71±0.10	0.07-1.69
Unipolar	11	0.67±0.11	0.07-1.48
Bipolar	5	0.80±0.25	0.27-1.69
Schizophrenia and schizoaffective disorder	11	0.52±0.07	0.21-0.81
Alcoholism	16	0.57±0.09	0.13-1.39
Personality disorders	5	0.51±0.23	0.17-1.35

[a] Dopamine-β-hydroxylase (DBH) was measured in CSF from 48 drug-free psychiatric patients. For patients for whom more than one CSF sample was analyzed, average DBH was used. DBH activity (mean ± SEM) is expressed in nmole/ml/hr. There were no significant differences among mean DBH values of groups (Student's *t*-test for unpaired data).

nergic system. Support for such a hypothesis can be found in a recent study by Sternberg *et al.* (125), who reported a significant difference in CSF DBH between process and reactive schizophrenia.

Since CSF DBH levels might identify different subtypes of schizophrenia, we decided to look at nonschizophrenic individuals with both low and high DBH activity as a further means of investigating the possible mean of DBH activity in the CSF of man (85). Buchsbaum *et al.* (10) in their studies on monoamine oxidase activity, have described this research approach as a "high-risk paradigm," since group selection is based on the biological variable as opposed to diagnosis. The hypothesis underlying such an approach is that low enzyme activity may not be disease-specific, but rather may indicate a vulnerability either to psychopathology in general or to specific symptom constellations. In applying this "biological high-risk paradigm" to CSF DBH in alcoholics, we found that individuals with low DBH activity had significantly more psychopathology (as measured by the Minnesota Multiphasic Personality Inventory (MMPI)) than age-matched alcoholic individuals with high DBH activity (Fig. 8.6). These data suggest that low CSF DBH activity is associated with increased psychopathology in general, as opposed to a specific illness type. Consistent with this hypothesis were the results of another study on individuals who developed psychotic reactions to disulfiram (a DBH inhibitor). Patients with low initial levels of DBH in the cerebrospinal fluid were significantly more likely to develop psychotic reactions to disulfiram than those individuals with a normal or high level of DBH in the CSF (83). These findings also tended to support the hypothesis that low CSF DBH activity, as a reflection of either low noradrenergic activity centrally or of fewer noradrenergic neurons, was associated with an increased vulnerability to psychopathology.

CSF DBH IN ALCOHOLICS

The mean CSF DBH value for alcoholics does not differ significantly from the mean values for other psychiatric groups (67). However, more detailed analysis of the data from alcoholics provides interesting information on biological (83) and psychological (85) factors and their relationships to CSF DBH.

Disulfiram (tetraethylthiuram disulfide, Antabuse), a drug used in the treatment of alcoholism, causes adverse reactions when ethanol is consumed. These reactions are caused by

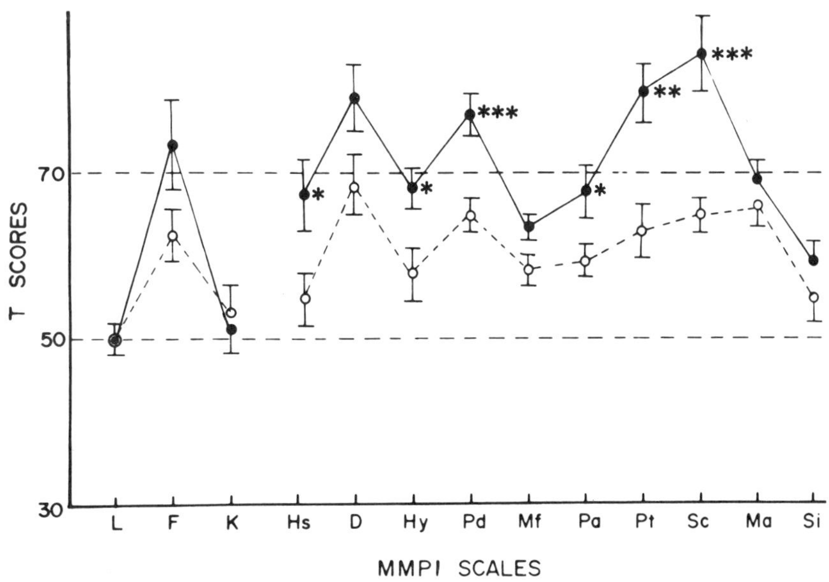

Figure 8.6. Relationship between Minnesota Multiphasic Personality Inventory (MMPI) and dopamine-β-hydroxylase (DBH) activity in the CSF. MMPI T scores in low (O---O, N = 16) and high (●---●, N = 16) CSF DBH supgroups. *, $p < 0.05$ low *vs.* high, two-tailed Student's *t-test;* **, $p < 0.01$ low *vs.* high, two-tailed Student's *t*-test; ***, $p < 0.001$ low *vs.* high, two-tailed Student's *t*-test.

inhibition of aldehyde dehydrogenase, an enzyme necessary for alcohol metabolism. Disulfiram is readily reduced, both *in vivo* and *in vitro*, to diethyldithiocarbamate (DDC), which inhibits DBH by chelating the copper associated with the enzyme. This inhibition is reversible, since after removal of DDC, DBH can reassociate with available copper ions and regain full activity (44, 45, 48). Another major metabolite of disulfiram, carbon disulfide, reacts with catecholamines to form compounds which are readily taken up by noradrenergic storage vesicles, where they act as DBH inhibitors. This inhibition is also reversible in the presence of copper ions (89). The disulfiram-related inhibition of DBH in brain may well be the cause of the psychotic reactions sometimes caused by disulfiram treatment.

In a study on alcoholics treated with disulfiram (83), there was no significant difference between CSF DBH in patients on drug and off drug. This is not surprising, since the disulfiram metabolites are reversible DBH inhibitors. The absence of such inhibitors in CSF from drug-treated patients was demonstrated when known quantities of purified DBH were added to the CSF and no enzyme inhibition occurred. However, disulfiram metabolites probably inhibited DBH effectively in brain tissue, as shown in animal studies (44).

In the same study (83), four of the patients receiving disulfiram (500 mg/day for more than 3 weeks) developed psychotic reactions. Another four patients on this drug regimen had minor, nonpsychotic reactions (fatigue, difficulty concentrating, general feelings of discomfort), while five additional patients had no adverse reactions. A comparison of the CSF DBH levels in these three groups (Fig. 8.7) showed low DBH in patients with psychotic reactions, intermediate DBH in those with nonpsychotic reactions, and higher DBH in those with no reactions. In this study, it is likely that CSF DBH values gave an indication of brain DBH values, and that individuals with low brain DBH were particularly susceptible to DBH inhibitors.

Just as low DBH in CSF was associated with vulnerability to disulfiram, low DBH in CSF was also associated with a psychological vulnerability in these alcoholics (85). Psychological traits were assessed on the MMPI.

The relationship of DBH in CSF to MMPI scores was examined by dividing the group of 32 patients into a high and a low DBH group (median split) and comparing their MMPI profiles. Compared to the high DBH group, the low DBH group showed statistically significant elevations on six of the MMPI subscales (Fig. 8.6). The low DBH group showed the "8-2-4 profile," which is associated with a general distrust of people, a suspicious questioning of their motives, and a fear of emotional involvement with others, in spite of an exaggerated need for attention (153). The data were also analyzed by comparing each individual's MMPI scores with his CSF DBH (Table 8.6). Significant negative correlations were found for seven of the MMPI subscales, again suggesting an association between low DBH in CSF and increased psychopathology. The results of this study suggest that CSF DBH values reflect some aspect of central noradrenergic function associated with psychological vulnerability (85). It would be of interest to see whether this relationship between DBH in CSF and MMPI scores exists in nonalcoholic populations.

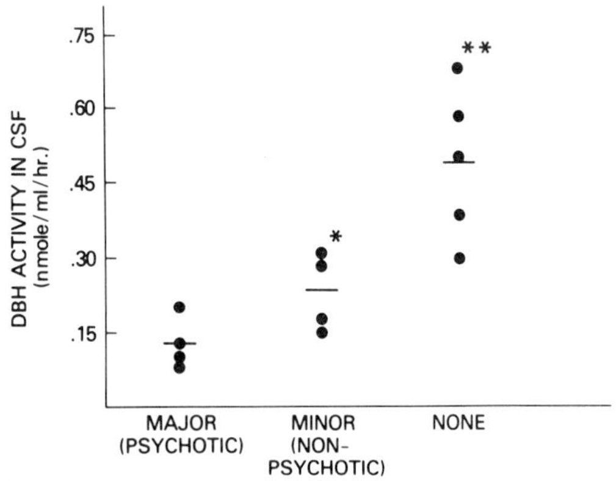

Figure 8.7. Relationship between CSF dopamine-β-hydroxylase (DBH) activity and adverse reactions to disulfiram. *, $p < 0.05$ compared to no reaction group; **, $p < 0.01$ compared to major (psychotic) group.

ANTIDEPRESSANT RESPONSE AND CHANGES IN CSF DBH AND NOREPINEPHRINE

Since DBH and norepinephrine are released together exocytotically, the extracellular concentrations of the two change coordinately in isolated, perfused organs or tissue slices (24, 42, 54, 137, 142). The *in vivo* situation is more

Table 8.6
Correlations between Personality Scales, Age, and Dopamine-β-Hydroxylase (DBH) Activity[a]

Personality Scales	CSF DBH	Age	Age-Corrected DBH	Age-Corrected, Log-Transformed DBH
Age	+.62 ($P < .001$)	1	.00	.00
Zung self-rating depression scale	−.29	−.18	−.24	−.48 ($P < .01$)
Zung self-rating anxiety scale	−.39 ($P < .02$)	−.34 ($P < .05$)	−.24	−.62 ($P < .01$)
MMPI T subscales				
L (Rigidity or deception)	−.17	.03	−.25	−.39 ($P < .02$)
F (Unusual answer)	−.30 ($P < .05$)	−.32 ($P < .05$)	−.14	−.39 ($P < .02$)
K (Test-taking attitude)	.24	.39 ($P < .02$)	−.01	−.22
Hs (Hypochondriasis)	−.19	.16	−.37 ($P < .02$)	−.52 ($P < .01$)
D (Depression)	−.41 ($P < .01$)	−.08	−.46 ($P < .01$)	−.62 ($P < .01$)
Hy (Hysteria)	−.18	.17	−.37 ($P < .02$)	−.60 ($P < .01$)
Pd (Psychopathic deviation)	−.35 ($P < .05$)	−.06	−.40 ($P < .02$)	−.57 ($P < .01$)
MF (Masculine-feminine)	−.04	−.07	−.11	−.54 ($P < .01$)
Pa (Paranoia)	−.42 ($P < .01$)	−.15	−.42 ($P < .01$)	−.51 ($P < .01$)
Pt (Psychastenia)	−.46 ($P < .01$)	−.19	−.44 ($P < .01$)	−.69 ($P < .01$)
Sc (Schizophrenia)	−.38 ($P < .02$)	−.31 ($P < .05$)	−.26	−.55 ($P < .01$)
Ma (Hypomania)	−.04	−.29	.19	−.34 ($P < .05$)
Si (Social introversion)	−.37 ($P < .02$)	−.20	−.32 ($P < .05$)	−.49 ($P < .01$)

[a] N = 32 patients. MMPI, Minnesota Multiphasic Personality Inventory. Pearson product-moment correlation coefficients were calculated between the personality scales, age, and CSF dopamine-β-hydroxylase (DBH) activity. The contribution of age differences to the variance of the DBH levels was removed by an analysis of covariance. Since CSF was not normally distributed, the values were log-transformed and the partial correlations recalculated using the log-transformed DBH values.

complex, with extracellular concentrations determined by rates of release, uptake, and metabolism. Measurements of DBH and NE in CSF provide different information. For instance, at certain times during chronic imipramine treatment, CSF DBH and NE change in different directions (see Fig. 8.5). Changes in DBH probably reflect alterations in central noradrenergic firing rate. NE changes are related to variations in release, uptake, and metabolism.

There is a strong correlation between DBH and NE ($r = 0.79$, $p < 0.005$, $n = 12$) in CSF of depressed patients on and off of MAO inhibitors (70). The NE to DBH ratio was significantly higher in MAO inhibitor-treated patients than in the same patients in the drug-free state. Since the MAO inhibitors block norepinephrine metabolism, the tissue content of NE would increase, allowing nonexocytotic leakage of NE from neurons without concomitant DBH release (70). MAO inhibitors cause a decrease in the central noradrenergic firing rate, so CSF DBH also decreases (70). These drugs can increase the amount of NE by inhibiting its breakdown and decrease NE by depressing firing of noradrenergic neurons. These two opposing processes cause variable NE changes (70, 82). Patients exhibiting the greatest increase of norepinephrine in CSF also have the best antidepressant response (82). These results support that aspect of the catecholamine theory of affective illness which associates depression with a deficit of catecholamines. It would be of interest to compare changes in CSF NE with antidepressant response in patients treated with tricyclic antidepressants.

SUGGESTIONS FOR FUTURE RESEARCH

The analysis of DBH in CSF is of great potential value for assessing central noradrenergic function in humans and in experimental animals. With CSF studies it is possible to monitor changes in individuals across time, thus allowing changes in pharmacological, physiological, and even psychological states to be investigated. Studies on DBH in CSF can be enhanced by concurrent measurements of other compounds from noradrenergic neurons (NE and its metabolites) and from other neurotransmitter systems. However, the data on CSF DBH are difficult to interpret because some basic questions are still unanswered. We would like to suggest some areas for further investigation.

Electrophysiological Studies

The demonstration that CSF DBH decreases dramatically 5 days after intraventricular 6-hydroxydopamine administration (30) supports the theory that CSF DBH comes largely from central noradrenergic neurons. In order to demonstrate more clearly a relationship between central noradrenergic firing and CSF DBH, one could stimulate the locus coeruleus, collect CSF, and measure DBH. This experimental paradigm could also be used to study the temporal relationship between noradrenergic firing and changes in CSF DBH.

DBH, Norepinephrine, and Antidepressant Response

When depressed patients are treated with MAO inhibitors, changes in CSF NE correlate with clinical improvement, while changes in CSF DBH probably reflect alterations in noradrenergic activity. Amphetamine increases CSF NE. It would be valuable to extend these observations to antidepressants such as desipramine, which block NE uptake and clomipramine, which has less effect on NE.

Mood State Changes in Affectively Ill Patients

DBH and NE levels correlate in CSF of depressed patients (70). It is not known if DBH and NE change in similar ways when depressed patients become manic or recover. A group of depressed patients had lower CSF NE than a group of manics (106); by contrast, another group of depressives had higher CSF DBH than a group of manics (67). It would be of interest to compare CSF DBH, NE, and their ratio as patients shift from one mood state to another. Such a study would provide valuable information on the catecholamine theory of affective disorders as well as on the metabolism of DBH and NE.

References

1. Adams RN: Probing brain chemistry with electroanalytical techniques. *Anal Chem* 48:1126A–1138A, 1976.
2. Adler-Graschinsky E, Langer SZ: Possible role of a beta adrenoceptor in the regulation of noradrenaline release by nerve stimulation through a positive feed-back mechanism. *Br J Pharmacol* 53:43–50, 1975.
3. Arnaiz JM, Garcia AG, Horga JF, Kirpekar SM: Tissue and plasma catecholamines and dopamine-β-hydroxylase activity of various animal species after neurogenic sympathetic stimulation. *J Physiol* 285:515–529, 1978.
4. Aunis D, Miras-Portugal MT, Mandel P: Bovine adrenal medullary dopamine-β-hydroxylase: Purification by affinity chromatography. *Biochim Biophys Acta* 327:313–327, 1973.
5. Aunis D, Miras-Portugal MT, Mandel P: Bovine adrenal medullary dopamine-β-hydroxylase: Studies on interaction with concanavalin A. *J Neurochem* 24:425–431, 1975.
6. Banerjee SP, Kung LS, Riggi SJ, Chandce SK: Development of beta adrenergic receptor subsensitivity by antidepressants. *Nature* (*Lond*) 268:455–456, 1977.
7. Boadle MC, Hughes J, Roth RH: Angiotensin accelerates catecholamine biosynthesis in sympathetically innervated tissues. *Nature* (*London*) 222:987–989, 1969.
8. Brimijoin S, Molinoff PB: Effects of 6-hydroxydopamine on the activity of tyrosine hydroxylase and dopamine-β-hydroxylase in sympathetic ganglia of the rat. *J Pharmacol Exp Ther* 178:417–425, 1971.
9. Brosnihan KB, Szilagyi JE, Ferrario CM: Effect of chronic sodium depletion on cerebrospinal fluid and plasma catecholamines. *Hypertension* 3:233–239, 1981.
10. Buchsbaum MS, Coursey RD, Murphy DL: The biochemical high-risk paradigm: Behavioral and familial correlates of low platelet monoamine oxidase activity. *Science* 194:339–341, 1976.
11. Bunney WE Jr, Davis JM: Norepinephrine in depressive reactions. *Arch Gen Psychiatry* 13:483–494, 1965.
12. Bunney WE Jr, Goodwin FK, Murphy DL: The "switch process" in manic-depressive illness. III. Theoretical implications. *Arch Gen Psychiatry* 27:312–317, 1972.
13. Campbell IC, Murphy DL, Gallagher DW, Tallman JF, Marshall EE: Neurotransmitter-related adaptation in the central nervous system following chronic monoamine oxidase inhibition. In Singer TP, Von Korff RW, Murphy DL: *Monoamine Oxidase: Structure, Function, and Altered Functions.* New York, Academic Press, 1980, pp 517–530.
14. Chalmers JP, Reid JL: Participation of central noradrenergic neurons in arterial baroreceptor reflexes in the rabbit. *Circ Res* 31:789, 1982.
15. Chalmers JP, Dollery CT, Lewis PJ, Reid JL: The importance of central adrenergic neurons in renal hypertension in rabbits. *J Physiol* 238:403, 1974.
16. Chase TN, Gordon EK, Ng LKY: Norepinephrine metabolism in the central nervous system of man: Studies using 3-methyoxy-4-hydroxyphenylethylene glycol levels in cerebrospinal fluid. *J Neurochem* 21:581–587, 1973.
17. Chevillard C, Duchene N, Alexandre JM: Selective release of newly synthesized cardiac norepinephrine induced by angiotension. II. *Eur J Pharmacol* 15:8–13, 1971.
18. Chevillard C, Duchene N, Pasquier R, Alexandre JM: Relationship of the centrally evoked pressor effect of angiotensin II to central noradrenaline in the rabbit. *Eur J Pharmacol* 58:203–206, 1979.

19. Chiueh CC, Kopin IJ: Radioenzymatic paper chromatographic assay for dopamine and norepinephrine in cerebroventricular cisternal perfusate of cat following administration of cocaine or d-amphetamine. *J Neurochem* 31:561–564, 1978.
20. Chiueh CC, Moore KE: *In vivo* release of endogenously synthesized catecholamines from the cat brain evoked by electrical stimulation and by d-amphetamine. *J Neurochem* 23:159–168, 1974.
21. Collu R, Jequir JC, Letarte J, Leboeuf G, Ducharne JR: Diurnal variations of plasma growth hormone and brain monoamines in adult male rats. *Can J Physiol Pharmacol* 51:890–892, 1973.
22. Coyle JT, Kuhar MJ: Subcellular localization of dopamine-β-hydroxylase and endogenous norepinephrine in the rat hypothalamus. *Brain Res* 65:475–487, 1974.
23. Craine JE, Daniels GH, Kaufman S: Dopamine-β-hydroxylase: The subunit structure and anion activation of the bovine adrenal enzyme. *J Biol Chem* 248:7838–7844, 1973.
24. Cubbedu LX, Barnes E, Langer SZ, Weiner N: Release of norepinephrine and dopamine-β-hydroxylase by nerve stimulation. I. Role of neuronal and extraneuronal uptake and alpha presynaptic receptors. *J Pharmacol Exp Ther* 190:431–450, 1974.
25. Davies D, Reid JL (ed): *Central Action of Drugs in Blood Pressure*. Baltimore, University Park Press, 1975.
26. Davson H: *Physiology of the Cerebrospinal Fluid*. Boston, Little, Brown, 1967.
27. Day MD, Roach AG: Centrally mediated cardiovascular effects of propranolol and other beta adrenoceptor antagonists in the conscious cat. In Davies DS, Reid JL, *Central Action of Drugs in Blood Pressure Regulation*. Baltimore, University Park Press, 1975, pp 215–224.
28. De Potter WP: The presence of dopamine-β-hydroxylase activity in rabbit cerebrospinal fluid and its increase after sciatic nerve stimulation. *J Physiol (London)* 258:26P–27P, 1976.
29. De Potter WP, Chanh CPH, De Smet F, De Schaepdryver AF: The presence of dopamine-beta-hydroxylase in the cerebrospinal fluid of rabbits and its increased concentration after stimulation of peripheral nerves and cold stress. *Neuroscience* 1:523–529, 1976.
30. De Potter WP, De Potter RW, De Smet FH, Fraeyman NN: Dopamine-β-hydroxylase in the cerebrospinal fluid and central noradrenergic activity. A pharmacological approach. *Arch Int Pharmacodyn Ther* 232:334–335, 1978.
31. Ebstein RP, Park DH, Freedman LS, Levitz SM, Chuchi T, Goldstein M: A radioimmunoassay of human circulatory dopamine-β-hydroxylase. *Life Sci* 10:769–774, 1973.
32. Eide I, Kolloch R, DeQuattro V, Miano L, Dugger R, Van der Meulen J: Raised cerebrospinal fluid norepinephrine in some patients with primary hypertension. *Hypertension* 1:255–259, 1979.
33. Eisler T, Calne DB, Ebert MH, Kopin IJ, Ziegler MG, Levine R, Murphy DL: Biochemical measurements during (−) deprenyl treatment of Parkinsonism. In Singer TP, von Korff RW, Murphy DL: *Monoamine Oxidase: Structure, Function and Altered Functions*. New York, Academic Press, 1979, pp 497–505.
34. Foldes A, Jeffrey PL, Preston BN, Austin L: Dopamine-β-hydroxylase of bovine adrenal medulla. *Biochem J* 126:1209–1217, 1972.
35. Friedman S, Kaufman S: 3,4-Dihydroxyphenylethylamine-β-hydroxylase: Physical properties, copper content and the role of copper in the catalytic activity. *J Biol Chem* 240:4763–4773, 1965.
36. Frigon RP, Stone RA: Human plasma dopamine-β-hydroxylase: Purification and properties. *J Biol Chem* 253:6780–6786, 1978.
37. Fujita K, Maruta J, Teradaira R, Beppu H, Shinpo K, Maeno Y, Ito T, Nagatsu T, Kato T: Dopamine-β-hydroxylase activity in human cerebrospinal fluid and serum. *J Neurochem* 29:1141–1142, 1977.
38. Fuxe K, Hokfelt T, Agnati LF, Johansson O, Goldstein M, Perez de la Mora M, Posanni L, Tapia T, Teran L, Palacios R: Mapping out central catecholamine neurons: Immunohistochemical studies on catecholamine-synthesizing enzymes. In Lipton MA, DiMascio A, Killiam KF: *Psychopharmacology: A Generation of Progress*. New York, Raven Press, 1978, pp 67–94.
39. Fuxe K, Ungerstedt Y: Localization of catecholamine uptake in rat brain after intraventricular injection. *Life Sci* 5:1817–1824, 1966.
40. Fuxe K, Ungerstedt Y: Histochemical studies on the effects of (+)-amphetamine, drugs of the imipramine groups, and tryptamine on central catecholamine and 5-hydroxytryptamine neurons after intraventricular injection of catecholamines and 5-hydroxytrypamine. *Eur J Pharmacol* 4:135–144, 1968.
41. Fuxe K, et al: *Central Action of Drugs in Blood Pressure Regulation*, Davies D, Reid J (eds). Baltimore, University Park Press, 1978.
42. Geffen LB, Livett BG, Rush RA: Immunological localization of chromogranins in sheep sympathetic neurons and their release by nerve impulses. *J Physiol (London)* 204:58P–59P, 1969.
43. Goldstein DJ, Cubeddu LX: Dopamine-beta-hydroxylase activity in human cerebrospinal fluid. *J Neurochem* 26:193–195, 1976.
44. Goldstein M, Anagnoste B, Lauber E, McKereghan MR: Inhibition of dopamine-beta-hydroxylase by disulfiram. *Life Sci* 3:763–767, 1964.
45. Goldstein M, Lauber E, McKereghan MR: Studies on the purification and characterization of 3,4-dihydroxyphenylethylamine-β-hydroxylase. *J Biol Chem* 240:2066–2072, 1965.
46. Goodwin FK, Bunney WF Jr: Depressions following reserpine: A re-evaluation. *Semin Psychiatry* 3:435–448, 1971.
47. Gordon EK, Oliver J: 3-Methoxy-4-hydroxyphenylethylene glycol in human cerebrospinal fluid. *Can Chim Acta* 35:145–150, 1971.
48. Green AL: The inhibition of dopamine-beta-

oxidase by chelating agents. *Biochim Biophys Acta* 81:394–397, 1964.
49. Greenacre JK, Teychenne PF, Petrie A, Calne DB, Leigh PN, Reid JL: The cardiovascular effects of bromocriptine in Parkinsonism. *Br J Clin Pharmacol* 3:571–574, 1976.
50. Haranth PS, Ramabhimaiah S, Venkatakrishna H: Passage of intravenously infused noradrenaline into the cerebrospinal space in anesthetized dogs. *Indian J Med Res* 64:218–223, 1976.
51. Hartman BK: Localization of the noradrenergic nervous system in human brain. *J Psychiatr Res* 11:283–288, 1974.
52. Haywood JR, Vogh BP: Some measurements of autonomic nervous system influence on production of cerebrospinal fluid in the cat. *J Pharmacol Exp Ther* 208:341–346, 1979.
53. Hortnagl H, Winkler H, Lochs H: Membrane proteins of chromaffin granules: Dopamine-β-hydroxylase, a major constituent. *Biochem J* 129:187–195, 1972.
54. Johnson DG, Thoa NB, Weinshilboum R, Axelrod J, Kopin IJ: Enhanced release of dopamine-β-hydroxylase from sympathetic nerves by calcium and phenoxybenzamine and its reversal by prostaglandins. *Proc Natl Acad Sci USA* 68:2227–2230, 1971.
55. Kessler JA, Fenstermacher JD, Patlak CS: 3-methoxy - 4 - hydroxy - phenylethyleneglycol (MHPG) transport from the spinal cord during spinal subarachnoid perfusion. *Brain Res* 102:131–141, 1976.
56. Kobinger W, Pichler L: Investigation into some imidazoline compounds, with respect to peripheral α-adrenoceptor stimulation and depression of cardiovascular centers. *Naunyn Schmiedebergs Arch Pharmacol* 291:175–191, 1975.
57. Kopin IJ, Kaufman S, Viveros H, Jacobowitz D, Lake CR, Ziegler MG, Lovenberg W, Goodwin FK: Dopamine-beta-hydroxylase: Basic and clinical studies. *Ann Intern Med* 85:211–223, 1976.
58. Lake CR, Gullner HG, Polinski RJ, Ebert MH, Ziegler MG, Bartter FC: Essential hypertension: Central and peripheral norepinephrine. *Science* 211:955–957, 1981.
59. Lake CR, Wood JH, Ziegler MG, Ebert MH, Kopin IJ: Probenecid-induced norepinephrine elevations in plasma and CSF. *Arch Gen Psychiatry* 35:237–240, 1978.
60. Lake CR, Ziegler MG, Coleman M, Kopin IJ: Lack of correlation of plasma norepinephrine and dopamine-beta-hydroxylase in hypertensive and normotensive subjects. *Circ Res* 41:865–869, 1977.
61. Lake CR, Ziegler MG, Kopin IJ: Use of plasma norepinephrine for evaluation of sympathetic neuronal function in man. *Life Sci* 18:1315–1326, 1976.
62. Lake CR, Ziegler MG, Shoulson I, Kopin IJ: No difference in cerebrospinal fluid levels of norepinephrine in patients when on high versus low monoamine diets. *Neuroscience* 3:252, 1977.
63. Lander J, Austin L: Subcellular distribution of dopamine-β-hydroxylase and inhibitors in the hippocampus and caudate nucleus in sheep brain. *J Neurochem* 26:661–673, 1976.
64. Langer SZ: The regulation of transmitter release elicited by nerve stimulation through a presynaptic feedback mechanism. In Usdin E, Snyder S: *Frontiers of Catecholamine Research.* New York, Pergamon Press, 1973, pp 543–549.
65. Leimdorfer A, Arana R, Hack MH: Hyperglycemia induced by the action of adrenaline on the central nervous system. *Am J Physiol* 150:589–595, 1947.
66. Lerner P, Dendel PS, Major LF: Dopamine-beta-hydroxylase in cerebrospinal fluid and plasma: Effects of α-adrenergic agents. *Brain Res* 189:183–191, 1980.
67. Lerner P, Goodwin FK, van Kammen DP, Post RM, Major LF, Ballenger JC, Lovenberg W: Dopamine-beta-hydroxylase in the cerebrospinal fluid of psychiatric patients. *Biol Psychiatry* 13:685–694, 1978.
68. Lerner P, Major LF, Dendel PS, Campbell IC, Murphy DL: Central and peripheral dopamine-beta-hydroxylase: Responses to long term treatment with monoamine oxidase inhibitors. *Neuropharmacology* 19:877–881, 1980.
69. Lerner P, Major LF, Dendel PS, Ziegler MG: Differential effects of *dl*-propranolol on norepinephrine in cerebrospinal fluid and plasma. *Brain Res* 223:444–447, 1981.
70. Lerner P, Major LF, Murphy DL, Lipper S, Lake CR, Lovenberg W: Dopamine-beta-hydroxylase and norepinephrine in human cerebrospinal fluid: Effects of monoamine oxidase inhibitors. *Neuropharmacology* 18:423–426, 1979.
71. Lerner P, Major LF, Ziegler MG, Dendel PS, Ebert MH: Central noradrenergic adaptation to long-term treatment with imipramine in rhesus monkeys. *Brain Res* 200:220–224, 1980.
72. Levin EY, Levenberg B, Kaufman S: The enzymatic conversion of 3,4-dihydroxy-phenylethylamine to norepinephrine. *J Biol Chem* 235:2080–2086, 1960.
73. Lew JY, Hata F, Ohashi T, Goldstein M: The interactions of bromocriptine and lergotrile with dopamine and alpha adrenergic receptors. *J Neural Transm* 41:109–121, 1977.
74. Lewis PJ: Propranolol—an antihypertensive drug with a central action. In Davies DS, Reid JL: *Central Action of Drugs in Blood Pressure Regulation.* Baltimore, University Park Press, 1975, pp 206–214.
75. Liljedahl SO, Von Euler US: Elimination of noradrenaline from the spinal fluid. *Acta Chir Scand* 108:163–169, 1954.
76. Lindvall M, Edvinsson L, Owman C: Histochemical, ultrastructural, and functional evidence for a neurogenic control of CSF production from the choroid plexus. *Adv Neurol* 20:111–120, 1978.
77. Lindvall M, Edvinsson L, Owman C: Effect of sympathomimetic drugs and corresponding receptor antagonists on the rate of cerebrospinal fluid production. *Exp Neurol* 64:135–145, 1979.
78. Lindvall M, Edvinsson L, Owman C: Sympathetic nervous control of cerebrospinal fluid

production from the choroid plexus. *Science* 201:176–178, 1978.

79. Link H, Zetterval O, Blennow G: Individual cerebrospinal fluid (CSF) proteins in the evaluation of increased CSF total proteins. *Z Neurol* 203:119–132, 1972.
80. Maas JW, Dekirmenjian H, Fawcett AJ: MHPG excretion by patients with affective disorders. *Int Pharmacopsychiatry* 9:14–26, 1974.
81. Major LF, Lerner P, Dendel PS, Glaser L, Post RM: The effects of prolonged disulfiram treatment on plasma dopamine-beta-hydroxylase in rhesus monkeys. *J Studies Alcohol* 43:146–152, 1982.
82. Major LF, Lake CR, Lipper S, Lerner P, Murphy DL: The central noradrenergic system and affective response to MAO inhibitors. *Prog Neuropsychopharmacol* 3:535–542, 1979.
83. Major LF, Lerner P, Ballenger JC, Brown GL, Goodwin FK, Lovenberg W: Dopamine-beta-hydroxylase in the cerebrospinal fluid: Relationship to disulfiram-induced psychosis. *Biol Psychiatry* 14:337–344, 1979.
84. Major LF, Lerner P, Dendel PS, Post RM: Dopamine-beta-hydroxylase in cerebrospinal fluid: Putative indicator of central noradrenergic activity. In Wood JH: *Neurobiology of Cerebrospinal Fluid* 2. New York, Plenum Press, 1983, ch 13, pp 179–196.
85. Major LF, Lerner P, Goodwin FK, Ballenger JC, Brown GL, Lovenberg W: Dopamine-beta-hydroxylase in cerebrospinal fluid: Relationships to personality measures. *Arch Gen Psychiatry* 37:308–310, 1980.
86. Major LF, Murphy DL, Lipper S, Gordon E: Effects of clorgyline and pargyline on deaminated metabolites of norepinephrine, dopamine, and serotonin in human cerebrospinal fluid. *J Neurochem* 32:229–231, 1979.
87. Mandell AJ: Neurobiological mechanisms of presynaptic metabolic adaptation and their organization: Implications for a pathophysiology of the affective disorders. In Mandell AJ: *Neurobiological Mechanisms of Adaptation and Behavior*. New York, Raven Press, 1975, pp 1–32.
88. Manshardt J, Wurtman RJ: Daily rhythm in the noradrenaline content of rat hypothalamus. *Nature (London)* 217:574–575, 1968.
89. McKenna MJ, DiStefano V: Carbon disulfide. II. A proposed mechanism for the action of carbon disulfide on dopamine-beta-hydroxylase. *J Pharmacol Exp Ther* 202:253–266, 1977.
90. McMillen BA, Warnak W, German DC, Shore PA: Effects of chronic desipramine treatment on rat brain noradrenergic responses to beta adrenergic drugs. *Eur J Pharmacol* 61:239–246, 1980.
91. Molinoff PB, Weinshilboum R, Axelrod J: A sensitive enzymatic assay for dopamine-beta-hydroxylase. *J Pharmacol Exp Ther* 178:425–431, 1971.
92. Moore KE, Chiueh CC, Zeldes G: Release of neurotransmitters from the brain *in vivo* by amphetamine, methylphenidate and cocaine. In Ellinwood EH Jr, Kilbey MM: *Cocaine and Other Stimulants*. New York, Plenum Publishing Corp, 1976, pp 143–160.
93. Morrison JH: The morphological organization of the coeruleo-cortical projection, Presented at the Annual Conference on Brain Research, 1980.
94. Murphy DL, Redmond DE: The catecholamines: Possible role in affect, mood and emotional behavior in man and animals. In Friedhoff A: *Catecholamines and Behavior*. New York, Plenum Press, 1975, pp 73–117.
95. Murray S, Jones DH, Davies DS, Dollery CT, Reid JL: Free and total 3-methoxy-4-hydroxyphenylethylene glycol in human cerebrospinal fluid: Relationship to blood pressure in hypertensives and patients with neuropsychiatric disease. *Clin Chim Acta* 79:63, 1977.
96. Nagatsu T, Kuzuya H, Hidaka H: Inhibition of dopamine-beta-hydroxylase by sulfhydryl compounds and the nature of the natural inhibitors. *Biochim Biophys Acta* 139:319–327, 1967.
97. Nagatsu T, Udenfriend S: Photometric assay of dopamine-beta-hydroxylase activity in human blood. *Clin Chem* 18:980–983, 1972.
98. Nathanson JA: Beta-adrenergic-sensitive adenylate cyclase in secretory cells of choroid plexus. *Science* 204:843–844, 1979.
99. Nelson DL, Harden TK, Molinoff PB: Adrenergic storage vesicles in rat heart: Quantitation of membrane-bound and membrane-enclosed dopamine-beta-hydroxylase. *J Pharmacol Exp Ther* 205:357–365, 1978.
100. Okada T, Ohta T, Shinoda T, Kato K, Ikuta K, Nagatsu T: Dopamine-beta-hydroxylase activity in serum and cerebrospinal fluid in neuropsychiatric diseases. *Neuropsychobiology* 2:139–144, 1976.
101. Orcutt JC, Molinoff PB: Endogenous inhibitors of dopamine-β-hydroxylase in rat organs. *Biochem Pharmacol* 25:1167–1174, 1976.
102. Perlow MJ, Chiueh CC, Lake CR, Wyatt RJ: Increased dopamine and norepinephrine concentrations in primate CSF following amphetamine and phenylethylamine administration. *Brain Res* 186:469–473, 1980.
103. Perlow M, Ebert MH, Bordon EK, Ziegler MG, Lake CR, Chase TN: The circadian variation of catecholamine metabolism in the subhuman primate. *Brain Res* 139:101–113, 1978.
104. Post RM, Gordon EK, Goodwin FK, Bunney WE Jr: Central norepinephrine metabolism in affective illness: MHPG in the cerebrospinal fluid. *Science* 179:1002–1003, 1973.
105. Post RM, Kotin J, Goodwin FK, Gordon EK: Psychomotor activity and cerebrospinal fluid amine metabolites in affective illness. *Am J Psychiatry* 130:67–72, 1973.
106. Post RM, Lake CR, Jimerson DC, Bunney WE, Wood JH, Ziegler MG, Goodwin FK: Cerebrospinal fluid norepinephrine in affective illness. *Am J Psychiatry* 135:907–912, 1978.
107. Reid JL, Jones DH, Fitzgerald G, Davies D, Boobis A: Catecholamine turnover in essential hypertension. *Clin Exp Hypertens* 2:395–408, 1980.
108. Reid JL, Zivin JA, Kopin IJ: Central and periph-

eral adrenergic mechanisms in the development of deoxycorticosterone-saline hypertension in rats. *Circ Res* 37:569, 1975.
109. Reis DJ, Weinbren M, Corvelli A: A circadian rhythm of norepinephrine regionally in cat brain: Its relationship to environmental lighting and to regional diurnal variations in brain serotonin. *J Pharmacol Exp Ther* 164:135–145, 1968.
110. Reis DJ, Wurtman RJ: Diurnal changes in brain noradrenaline. *Life Sci* 7:91–98, 1968.
111. Rosenberg RC, Lovenberg W: Active dimers of dopamine-beta-hydroxylase in human plasma. *Mol Pharmacol* 13:652–661, 1977.
112. Ross RA, Joh TH, Reis DJ: Increase in the relative rate of synthesis of dopamine-beta-hydroxylase in the nucleus locus coeruleus elicited by reserpine. *J Neurochem* 31:1491–1500, 1978.
113. Ross RA, John TH, Reis DJ: Reduced rate of biosynthesis of dopamine-beta-hydroxylase in the nucleus locus coeruleus during the retrograde reaction. *Brain Res* 160:174–179, 1979.
114. Rush RA, Kindler SH, Udenfriend S: Solid-phase radioimmunoassay on polystyrene beads and its application to dopamine-beta-hydroxylase. *Clin Chem* 21:148–150, 1975.
115. Saran RK, Sahuja RC, Gupta NN, Hasan M, Bhargava KP, Shanker K, Koshor K: 3-methoxy-4-hydroxyphenylglycol in cerebrospinal fluid and vanillylmandelic acid in urine of humans with hypertension. *Science* 200:317, 1978.
116. Schildkraut JJ: The catecholamine hypothesis of affective disorders: A review of supporting evidence. *Am J Psychiatry* 122:509–522, 1974.
117. Schildkraut JJ, Keeler BA, Papousek M, Hartmann E: MHPG excretion in depressive disorders: Relation to clinical subtypes and desynchronized sleep. *Science* 181:762–764, 1973.
118. Schildkraut JJ, Winokur A, Draskoczy PR, Hensle JH: Changes in norepinephrine turnover in rat brain during chronic administration of imipramine and protriptyline: A possible explanation for the delay in onset of clinical antidepressant effects. *Am J Psychiatry* 127:1032–1039, 1971.
119. Scrabine A, Clineschmidt B, Sweet C: Central noradrenergic control of blood pressure. *Ann Rev Pharmacol Toxicol* 16:113, 1976.
120. Severs WB, Daniels-Severs AE: Effects of angiotensin on the central nervous system. *Pharmacol Rev* 25:415–441, 1973.
121. Shaw DM, O'Keef R, Macsweeney DA, Brookbank BWL, Noguera R, Coppen A: MHPG in depression. *Psychol Med* 3:333–336, 1973.
122. Shopsin B, Wilk S, Shul M: An assessment of norepinephrine metabolism in affective disorders. *Arch Gen Psychiatry* 28:230–233, 1973.
123. Sinha JN, Deitl H, Philippu A: Effect of a fall of blood pressure on the release of catecholamines in the hypothalamus. *Life Sci* 26:1751–1760, 1980.
124. Sjostrom R, Ekstedt J, Anggard E: Concentration gradients of monoamine metabolites in human cerebrospinal fluid. *J Neurol Neurosurg Psychiatry* 38:666–668, 1975.
125. Sternberg DE: CSF dopamine-beta-hydroxylase and schizophrenia. Paper presented at American Psychiatric Association Annual Meeting, 1978.
126. Stone RA, Kirshner N, Reynolds J, Vanaman TC: Purification and properties of dopamine-beta-hydroxylase from human pheochromocytoma. *Mol Pharmacol* 10:1009–1015, 1974.
127. Subrahmanyam S: Role of biogenic amines in certain pathological conditions. *Brain Res* 87:355–362, 1975.
128. Svensson TH, Bunney BS, Aghajanian GK: Inhibition of both noradrenergic and serotonergic neurons in brain by the α-adrenergic agonist clonidine. *Brain Res* 92:291–306, 1975.
129. Svensson TH, Usdin E: Feedback inhibition of brain noradrenaline neurons by tricyclic antidepressants: Alpha receptor mediation. *Science* 202:1089–1091, 1978.
130. Tackett RL, Webb JG, Privitera PJ: Cerebroventricular propranolol elevates cerebrospinal fluid norepinephrine and lowers blood pressure. *Science* 213:911–913, 1981.
131. Tarazi RC, Dustan HP: Beta adrenergic blockade in hypertension. Practical and theoretical implications of long-term haemodynamic variations. *Am J Cardiol* 29:633–640, 1972.
132. Teychenne PF, Lake CR, Ziegler MG: Cerebrospinal fluid studies in Parkinson's disease: norepinephrine and gamma-aminobutyric acid concentration. In Wood JH: *Neurobiology of Cerebrospinal Fluid*. New York, Plenum Publishing Corp, 1980, pp 197–204.
133. Teychenne PF, Lake CR, Ziegler MG, Plotkin C, Wood JH, Calne DB: Central and peripheral deficiency of norepinephrine in Parkinson's disease and the effects of L-dopa therapy. *Neuroscience* 3:47, 1977.
134. Thoa NB, Wooten GF, Axelrod J, Kopin IJ: On the mechanism of release of norepinephrine from sympathetic nerves induced by depolarizing agents and sympathomimetic drugs. *Mol Pharmacol* 11:10–18, 1975.
135. Tilson HA, Rech RH, Sparber SB: Release of ^{14}C-norepinephrine into the lateral cerebroventricle of rats by exposure to a conditioned aversive stimulus. *Pharmacol Biochem Behav* 3:385–392, 1975.
136. Veersteeg DHG, Palkovits M, Van der Gugten J, Wijnin-Henk LJM, Smeets GWM, de Jong W: Catecholamine content of individual brain regions of spontaneously hypertensive rats. *Brain Res* 112:429, 1976.
137. Viveros OH, Argueros L, Kirshner N: Release of catecholamines and dopamine-beta-oxidase from the adrenal medulla. *Life Sci* 7:609–618, 1968.
138. Walker GA, Kon H, Lovenberg W: An investigation of the copper site(s) of dopamine-beta-hydroxylase by electron paramagnetic resonance. *Biochem Biophys Acta* 482:309–322, 1977.
139. Wallace EF, Krantz MJ, Lovenberg W: Dopamine-beta-hydroxylase: A tetrameric glycoprotein. *Proc Natl Acad Sci USA* 70:2253–2255, 1973.
140. Wallace EF, Lovenberg W: Studies on the carbohydrate moiety of dopamine-beta-hydroxy-

lase: Interaction of the enzyme with concanavalin A. *Proc Natl Acad Sci USA* 71:3217–3220, 1974.

141. Weil-Malherbe H, Axelrod J, Tomchik R: Blood-brain barrier for adrenaline. *Science* 129:1226–1227, 1959.
142. Weinshilboum RM, Thoa NB, Johnson DG, Kopin IJ, Axelrod J: Proportional release of norepinephrine and dopamine-beta-hydroxylase from sympathetic nerves. *Science* 174:1349–1351, 1971.
143. Werner U, Starke K, Schumann HJ: Actions of clonidine and 2-(2-methyl-6-ethylcyclohexylamino)-2-oxazoline on post ganglionic autonomic nerves. *Arch Int Pharmacodyn Ther* 195: 282–290, 1972.
144. Westlund KN, Bowker RM, Ziegler MG, Coulter JD: Origins of spinal noradrenergic pathways demonstrated by retrograde transport of antibody to dopamine-beta-hydroxylase. *Neurosci Lett* 25:243–249, 1981.
145. Weston PG: Sugar content of the blood and spinal fluid in insane subjects. *J Med Res* 35:199–208, 1916.
146. Wilk S, Shopsin B, Gershon S, Shul M: Cerebrospinal fluid levels of MHPG in affective disorders. *Nature* 235:440–441, 1972.
147. Wise CD, Stein L: Dopamine-beta-hydroxylase deficit in the brains of schizophrenic patients. *Science* 181:344–347, 1973.
148. Wood JH, Lake CR, Ziegler MG, Sode J, Brooks BR, Van Buren JM: Cerebrospinal fluid norepinephrine alterations during electrical stimulation of cerebellar and cerebral surfaces in epileptic patients. *Neurology* 27:716–724, 1977.
149. Wood JH, Lake CR, Ziegler MG, Van Buren JM: Neurophysiological and neurochemical alterations during electrical stimulation of human caudate nucleus. *J Neurosurg* 46:361–368, 1977.
150. Wood JH, Ziegler MG, Lake CR, Shoulson I, Brooks BR, Van Buren JM: Cerebrospinal fluid norepinephrine reductions in man after degeneration and electrical stimulation of the caudate nucleus. *Ann Neurol* 1:94–99, 1977.
151. Wood JH, Ziegler MG, Lake CR, Shoulson I, Brooks BR, Van Buren JM: Norepinephrine reductions in cerebrospinal fluid in man after electrical stimulation and degeneration of the caudate nucleus. *Trans Am Neurol Assoc* 101:1–5, 1976.
152. Wood JH, Ziegler MG, Lake CR, Sode J, Brooks BR, Van Buren JM: Elevations in cerebrospinal fluid norepinephrine during unilateral and bilateral cerebellar stimulation in man. *Neurosurgery* 1:260–265, 1977.
153. Works P, Seeman W, Haller D: *The Actuarial Use of the MMPI with Adolescents and Adults.* Baltimore, Williams & Wilkins, 1974.
154. Wyatt RJ, Schwartz MA, Erdelyi E, Barchas JD: Dopamine-beta-hydroxylase activity in brains of chronic schizophrenic patients. *Science* 187:368–370, 1975.
155. Ziegler MG, Brooks BR, Lake CR, Wood JH, Enna SJ: Norepinephrine and gamma-butyric acid in amyotrophic lateral sclerosis. *Neurology* 30:98–101, 1980.
156. Ziegler MG, Lake CR, Ebert MH: Norepinephrine elevations in cerebrospinal fluid after d- and l-amphetamine. *Eur J Pharmacol* 57:127–133, 1979.
157. Ziegler MG, Lake CR, Kopin IJ: Plasma noradrenaline increases with age. *Nature* 261:333–335, 1976.
158. Ziegler MG, Lake CR, Foppen RH, Shoulson I, Kopin IJ: Norepinephrine in cerebrospinal fluid. *Brain Res* 108:436–440, 1976.
159. Ziegler MG, Lake CR, Kopin IJ: Plasma noradrenaline increases with age. *Nature* 261:333–335, 1976.
160. Ziegler MG, Lake CR, Williams AC, Teychenne PF, Shoulson I, Steinsland O: Bromocriptine inhibits norepinephrine release. *Clin Pharmacol Ther* 25:137–142, 1979.
161. Ziegler MG, Lake CR, Wood JH, Brooks BR: Relationship between cerebrospinal fluid norepinephrine and blood pressure in neurologic patients. *Clin Exp Hypertens* 2:995–1008, 1980.
162. Ziegler MG, Lake CR, Wood JH, Brooks BR, Ebert MH: Relationship between norepinephrine in blood and cerebrospinal fluid in the presence of a blood-cerebrospinal fluid barrier for norepinephrine. *J Neurochem* 28:677–679, 1976.
163. Ziegler MG, Lake CR, Wood JH, Ebert MH: Circadian rhythm in cerebrospinal noradrenaline of man and monkey. *Nature (London)* 264:656–658, 1976.
164. Ziegler MG, Lake CR, Wood JH, Ebert MH: Norepinephrine in cerebrospinal fluid: Basic studies, effects of drugs and disease. In Wood JH: *Neurobiology of Cerebrospinal Fluid I.* New York, Plenum Publishing Corp, 1980, pp 141–149.
165. Ziegler MG, Wood JH, Lake CR, Kopin IJ: Norepinephrine and 3-methoxy-4-hydroxyphenylglycol gradients in human cerebrospinal fluid. *Am J Psychiatry* 134:565–568, 1977.
166. Zigmond MH, Wurtman RJ: Daily rhythm in the accumulation of brain catecholamines synthesized from circulating H-tyrosine. *J Pharmacol Exp Ther* 172:416–422, 1970.
167. Zivin JA, Reid JL, Saavedra JM, Kopin IJ: Quantitative localization of biogenic amines in the spinal cord. *Brain Res* 99:293–301, 1975.

CHAPTER NINE

Norepinephrine in Epilepsy

MATTHEW B. STERN, M.D., and JAMES H. WOOD, M.D.

INTRODUCTION

An important approach to the understanding of epilepsy has been the study of neurotransmitters and their metabolism. Electrophysiologic investigations suggest their role in synaptic mechanisms and neurotransmitters, being the mediators of synaptic communication, have been implicated in the initiation and spread of seizures. In addition, neurotransmitters are thought to modulate the neuronal response to epileptogenic stimuli.

Maynert *et al.* (113) and Morselli *et al.* (119a) have reviewed the role of neurotransmitters in the epilepsies. Metabolic abnormalities, hormone and vitamin metabolism, chemical agents, and electrical stimulation alter seizure thresholds, perhaps by influencing neuronal activity and altering the synthesis, storage, and release of neurotransmitters.

More specifically, the role of norepinephrine in epilepsy has received much recent attention because of the primarily inhibitory action of norepinephrine on spontaneously firing neurons (78, 95). The role of norepinephrine as a neuromodulator is discussed by Aston-Jones *et al.* (Chapter 7). The overwhelming majority of published studies suggest that brain norepinephrine reduces cerebral susceptibility to seizures. This chapter will review neurochemical studies that have attempted to enhance or inhibit cerebral noradrenergic activity and thereby alter seizure thresholds. A variety of approaches to the study of norepinephrine in epilepsy have been employed. These include studying physiologic variations in brain norepinephrine, altering cerebral norepinephrine by surgical, electrical, and pharmacologic manipulation, and employing a wide range of compounds with noradrenergic agonist and antagonist properties.

ELEVATION OF BRAIN NORADRENERGIC ACTIVITY

Systemic administration of norepinephrine is ineffective in raising brain levels of this neurotransmitter, because norepinephrine does not appear to penetrate the blood-CSF (189, 200, 202) or blood-brain (181) barrier. Systemic norepinephrine is also prone to cause severe metabolic disturbances, which may themselves alter seizure susceptibility.

Increased oral intake of tyrosine, the dietary precursor of norepinephrine, appears to augment neuronal catecholamine release (198) or synthesis (64) when amino acids competing for transport are reduced or central decarboxylation is inhibited (198). However, the morphological and enzymatic barrier mechanisms normally impede such perturbations (73).

L-3, 4-dihydroxyphenylalanine (L-dopa) is the direct precursor of both norepinephrine and dopamine, and can reverse the lowered thresholds induced by reserpine and tetrabenazine (11, 22, 106). The relative importance of norepinephrine, however, cannot be determined from these studies. Cox and Lomax (43) increased norepinephrine and reduced the severity of seizures in Mongolian gerbils treated with DL-threodihydroxyphenylserine. Dihydroxyphenylserine is decarboxylated to form norepinephrine (18), although it may not result in significant elevations in cerebral norepinephrine (13, 18).

As a result of these limitations, direct observations of the role of norepinephrine in epilepsy must be accomplished by biochemical studies of this transmitter system, measurements of norepinephrine levels in various brain regions following pharmacological manipulation, and neurophysiological recordings of the effects of intracerebral or intraventricularly administered norepinephrine.

Direct Effects of Centrally Administered Norepinephrine

Direct application of norepinephrine to cerebral cortical neurons inhibits their activity (95). Intracerebral administration of norepinephrine has protected several strains of mice from pentylenetetrazol-induced seizures (151), and microiontophoretic application of norepinephrine has been reported to inhibit epileptiform activity in hippocampal transplants to the anterior chamber of the eye (58). Although some investigators have found that monoamines injected intraventricularly are distributed locally (within 500 μm of the ventricular wall (62, 154), others have noted that intraventricularly administered norepinephrine is distributed throughout the catecholamine-rich regions of the brain (67, 139) and suppresses seizures induced by intraventricular injections of curare in cats (56, 67). Intraventricular norepinephrine also reduced or blocked the photomyoclonic syndrome in the baboon, *Papio papio* (2). Browning and Maynert (26) attributed their finding of seizure facilitation with intraventricular monoamine administration to changes in body temperature.

Pharmacologic Noradrenergic Neuronal Augmentation

Pharmacologic compounds with noradrenergic agonist properties have provided a useful tool for indirectly studying the effect of norepinephrine on seizure thresholds. Norepinephrine-releasing agents, inhibitors of reuptake and metabolism, as well as norepinephrine receptor agonists have been intensively investigated.

Tyramine promotes release of norepinephrine at the synapse, and its action is terminated primarily by presynaptic reuptake of norepinephrine. Krip and Vazquez (94) have shown that tyramine inhibits epileptiform afterdischarges in isolated strips of cat cerebral cortex. Amphetamine also promotes release of catecholamines (75) and has been shown to inhibit their reuptake (171). Blum and co-workers (21) suggest that *d*-amphetamine directly affects postsynaptic adrenergic receptors. These investigators demonstrated supersensitivity to norepinephrine and amphetamine following chemical sympathectomy of the brain by 6-hydroxydopamine. Ziegler (202) and others (12, 172) found increases in norepinephrine levels following administration of both *d* and l-amphetamine. The catecholaminergic properties of amphetamine support observations of its anticonvulsant properties (133, 144), though some investigators have noted seizure facilitation (174, 187). Riffee and Gerald (140, 141, 142) have carefully studied the relationship between amphetamine and norepinephrine in relation to seizure thresholds. This group initially observed that the acute administration of (+)-amphetamine in low doses reduces the susceptibility of mice to pentylenetetrazol-induced convulsions and increases endogenous norepinephrine and dopamine by 7 and 21%, respectively. The rates of biosynthesis of dopamine and norepinephrine were increased equally (141). These studies confirmed an important role for norepinephrine by demonstrating antagonism of the increased seizure threshold with low dose amphetamine after treatment with α and β-adrenergic receptor blockers. Haloperidol, a dopamine antagonist, only partially blocked this effect (140). In subsequent investigations (142), acute administration of high dose (+)-amphetamine lowered thresholds and decreased endogenous levels of dopamine and norepinephrine. Although the rates of synthesis of dopamine and norepinephrine were increased by acute and chronic amphetamine treatment, the subsequent alterations in seizure susceptibility could not be explained on the basis of the observed changes in monoamine metabolism. It is difficult to interpret studies employing amphetamine as a norepinephrine-releasing agent or uptake inhibitor because of its nonspecific effect on transmitter systems.

Tricyclic antidepressant drugs prevent reuptake of catecholamines into the presynaptic nerve terminal and increase free synaptic norepinephrine (7). The various tricyclic compounds have a differential effect upon norepinephrine and serotonin (5-hydroxytryptamine) reuptake and show some specificity in potentiating the physiologic action of norepinephrine. Desmethylimipramine and maprotiline have their greatest effect on norepinephrine (14, 62, 92), although imipramine has been shown to possess anticonvulsant properties (61, 98, 101, 164) and to increase brain norepinephrine coincident with its anticonvulsant effect (86). Although imipramine inhibits the reuptake of both norepinephrine and serotonin, Bhattacharya and co-workers (17) clearly showed that norepinephrine is the monoamine that mediates the anticonvulsant properties of imipramine during electroconvulsive stimulation of rats. Pretreatment with *p*-chlorophenylalanine, an inhibitor of serotonin synthesis, did not suppress the imipramine-induced anticonvulsant effect. Pretreatment with catecholamine synthesis inhibitors including α-methyl-*p*-tyrosine, α-methyl-dopa, and diethyldithiocarbamate (30), as well as the adrenergic receptor blockers, propranolol and phenoxybenzamine, significantly antagonized the anticonvulsant action of imipramine. Pretreatment with haloperidol had no effect. Pretreatment with desmethylimipramine protected animals treated with intracerebroventricular 6-hydroxydopamine from the marked facilitation of amygdaloid kindling (114) and audiogenic seizures (102) seen in animals treated with 6-hydroxydopamine alone (see below). Rats treated with desmethylimipramine also showed diminished afterdischarge evolution (114), giving further support to the protective effect of norepinephrine in comparison with dopamine and serotonin. Additional evidence was supplied by Koella and colleagues (92), who studied the epileptic phenomena in the rat induced by intravenous infusions of high doses of tricyclic compounds. Maprotiline resulted in high amplitude slow waves only with exceedingly high doses when compared with the epileptogenic doses of imipramine, clomipramine, and amitriptyline, which are moderately strong serotonin uptake inhibitors. Maprotiline in the rat is a specific inhibitor of norepinephrine reuptake in high doses (14, 92). The higher threshold for maprotiline-induced epileptiform activity is likely secondary to the inhibitory effect of increased synaptic norepinephrine.

Monoamine oxidase and catechol-O-methyltransferase (COMT) inactivate norepinephrine. While inhibition of monoamine oxidase probably results in increased levels of intraneural catecholamines and, indirectly, a greater concentration of monoamines available to their receptors (66), COMT appears to act extraneurally (8). Several studies (101, 148) have found that monoamine oxidase inhibition has a protective effect against epileptogenic stimuli, while others found no change in seizure thresholds in animals treated with monoamine oxidase inhibitors (103). Pfeifer and Galambos (132) noted that the lowered seizure thresholds seen in mice treated with reserpine were reversed by pretreatment with tranylcypromine, a monoamine oxidase inhibitor, and Chen and associates (32) noted increased thresholds to electrically induced seizures in animals treated with DOPA and monoamine oxidase inhibitors. The nonspecific effect of monoamine oxidase inhibition on catecholamines, however, precludes conclusions concerning the specific role of norepinephrine in these studies. Studies of COMT inhibitors have been similarly limited because of their nonspecific effect and inability to elevate cerebral catecholamines when administered systemically. Schlesinger and colleagues (150) did note lower levels of brain COMT in mice genetically susceptible to audiogenic seizures compared with other strains (see below) but found no difference in monoamine oxidase activity.

Noradrenergic Receptor Agonists

Two pharmacologic types of α-adrenergic receptors exist in the periphery. α_1-receptors are present on effector cells, while most α_2-

receptors are located on presynaptic neurons. Norepinephrine released from the prejunctional neurons acts as a feedback inhibitor on prejunctional α_2-receptors (99). Several investigators have studied the effects of adrenergic receptor agonists on seizure thresholds. Studies utilizing α_2-receptor agents have yielded inconsistent results. Clonidine and oxymetazoline are α_2-receptor agonists and inhibit the release of norepinephrine (161, 163). Clonidine has been shown to inhibit the spontaneous firing of norepinephrine-containing neurons in the rat locus coeruleus (169) and lower the metrazole convulsive threshold in mice (46). In other studies clonidine was either ineffective in altering seizure thresholds (29) or had a protective effect against convulsive stimuli (79, 88). In Horton's study of α_2-agents (77), clonidine, oxymetazoline, and UK14, 304 (a clonidine-like imidazole derivative) decreased the severity of audiogenic seizures in 19- to 26-day-old DBA/2 mice when administered intraperitoneally. The effect was reversed by yohimbine and piperoxan, α_2-receptor antagonists (45, 162), as well as by phentolamine. Although inhibition of noradrenergic neurons appears to lower seizure susceptibility, Horton and co-workers (77) aptly point out that at the doses used, these α_2-agonists may have been acting postsynaptically, as were the α_2-antagonists, yohimbine and piperoxan (69, 100).

Recent investigations have also demonstrated a role for β-adrenergic receptors in epilepsy. Isoproterenol (a β-receptor agonist) inhibits epileptiform activity in hippocampal transplants to the anterior chamber of the eye of host rats (58) and antagonizes the effect on electroshock seizures of the monoamine-depleting agent, R04-1284 (74). These pharmacologic actions support the finding of Segal and associates (156) that norepinephrine inhibited the spontaneous activity of hippocampal pyramidal cells by β-receptor stimulation.

Electrical Stimulation-Induced Augmentation

CEREBELLAR CORTEX

Cerebellar cortical stimulation has been reported to reduce experimentally induced cerebral epileptiform discharges (34, 80, 81, 121) and increase cerebrospinal fluid (CSF) norepinephrine (35, 77, 192). Wood and co-workers (192) found significant elevations in CSF norepinephrine after unilateral and unilateral plus bilateral cerebellar stimulation, although double-blind evaluation did not demonstrate enhanced clinical seizure control (176). Although low frequency (10/sec) cerebellar stimulation has been reported by others to be more effective in suppressing seizures than high-frequency stimulation (200/sec) (35, 77), higher frequency stimulation resulted in greater CSF increases in norepinephrine (192). High-frequency stimulation may increase neuronal firing rates and recruit additional, higher frequency dependent noradrenergic neurons. High-frequency cerebellar stimulation may induce the reported suppression of myoclonus and muscular tone (36, 37, 49) and reduction of somatosensory and dentate-evoked cortical potentials (77) via noradrenergic neurons.

The cerebellum indirectly mediates cerebral catecholamine metabolism (158), and noradrenergic neurons in the locus coeruleus send axons to the cerebellar cortex (129). Stimulation of the rat cerebellar cortex activates the locus coeruleus (124), while stimulation in the region of the locus coeruleus suppresses electrocortical epileptiform activity in rats (105). Increased activity of neurons in the locus coeruleus may therefore enhance noradrenergic inhibition on cortical neurons (95, 155) and may explain the observed inhibition of cerebral epileptiform discharges resulting from cerebellar stimulation (34, 80, 81, 121). The variance in clinical response to cerebellar stimulation may reflect a complex interrelationship between noradrenergic mechanisms and γ-aminobutyric acid (GABA, a putative neurotransmitter) and their effect on Purkinje cell output (19, 48, 78).

The fact that CSF norepinephrine levels are decreased after cerebral surface stimulation *in vivo* (191) (see below), while cerebellar surface stimulation results in increased CSF norepinephrine (35, 77, 192), suggests that there may be regional differences in noradrenergic response to electrical stimulation. It also provides evidence that the elevations in CSF norepinephrine observed following cerebellar stimulation are not due to nonspecific effects such as tissue destruction or breakdown of the blood-brain barrier.

LOCUS COERULEUS

Noradrenergic output from the locus coeruleus projects throughout the cortex and has the capacity to modulate neuronal activity simultaneously in many cortical regions (119). Stimulation of the locus coeruleus elevates CSF 3-methoxy-4-hydroxyphenylethylene glycol (MHPG) (1), the major metabolite of norepinephrine, and suppresses epileptiform discharges in the cortex of rats (105). Augmentation of noradrenergic activity in the locus coeruleus may therefore increase inhibition of cerebral neurons (155), suggesting that tonic noradrenergic output from the locus coeruleus may normally elevate seizure thresholds.

REDUCTIONS IN BRAIN NORADRENERGIC ACTIVITY

Tyrosine hydroxylase is the rate-limiting enzyme in norepinephrine synthesis. Farjo and Blackwood (55) noted decreased activity of this enzyme in the stimulated amygdala of kindled rats, but noted no differences in norepinephrine levels in the amygdala of kindled rats compared with controls. Callaghan and Schwark (29), however, found diminished levels of norepinephrine in the hippocampus, midbrain, limbic lobes, and frontal cortex of kindled rats, suggesting that kindling may be the result of a breakdown in the tonic inhibitory control exerted by norepinephrine (29, 38).

Decreased brain levels of norepinephrine were found during bicuculline-induced seizures in the rat (28). Diminished norepinephrine content was also noted in the cerebellum after administration of the convulsant, 3-mercaptopropionic acid (143), and in the focus of penicillin-induced spikes during the first and second stage of propagation (91).

Because decreased levels of norepinephrine in the brain may reflect increased functional activity of noradrenergic neurons, other investigators have studied the relationship of norepinephrine turnover to seizures. Although Wilkison and Halpern (185) noted increased turnover rate constants for dopamine but not norepinephrine in the forebrain of kindled rats, others have found increased norepinephrine turnover after electroshock seizures (89), and in strains of mice susceptible to audiogenic seizures (53, 157). Although increased rates of norepinephrine turnover with subsequent depletion of norepinephrine in epileptogenic foci may reflect an inherent attempt to limit the expression of enhanced neuronal excitability, these changes in norepinephrine turnover might also be an epiphenomenon of seizure activity.

Noradrenergic Blocking Agents

Drugs acting primarily as α-adrenergic blockers have been found to uniformly decrease seizure thresholds. Phentolamine and prazosin administered intracerebroventricularly induced spontaneous limb myoclonus in mice (79). Adrenergic blocking agents have also been noted to antagonize the elevated seizure thresholds induced by low dose amphetamine (140) and antagonize the anticonvulsant effect of imipramine (17) and phenobarbital (117). Phenoxybenzamine and phentolamine decreased seizure thresholds in mice (90, 145), as did the adrenergic blocking agent, dibenamine (125).

Although some investigators have noted anticonvulsant properties in β-adrenergic receptor blockade (90, 120, 199), recent studies have found seizure facilitation utilizing β-receptor antagonists. The anticonvulsant effect of imipramine (17) and amphetamine (140) is antagonized by β-receptor blockers. Propranolol (a β-receptor blocker) leads to facilitation of amygdaloid kindling (29). McNamara (115) further characterized the nature of β-receptors in seizures by demonstrating a transient decrease in the number of β-adrenergic receptor binding sites in the stimulated and contralateral amygdala of kindled rats. Bergstrom and coworkers (15) also noted decreased cerebral β-receptors following repeated electroshock seizures. This effect may be mediated by transient elevations in norepinephrine with subsequent diminution in β-receptors (116). The early increase in norepinephrine turnover following convulsive stimuli (89, 157) supports these observations.

Pharmacologic Inhibition

Drugs which inhibit norepinephrine synthesis, storage, and release have all been examined for their effect on seizures. Inhibition of dopamine-β-hydroxylase (68) (the enzyme responsible for the terminal step in norepinephrine synthesis) decreases cerebral norepinephrine content without diminishing dopamine levels (6, 18). Reduced thresholds for tonic seizures in mice were noted by Rudzik and

Johnson (144) after treatment with the dopamine-β-hydroxylase inhibitors, disulfiram and U14,624. Similarly, disulfiram (and the dopamine-β-hydroxylase inhibitor, FLA-63) decreased thresholds for electrically and chemically induced seizures and augmented evoked seizures in rats and mice (90). Furthermore, disulfiram blocks the return of electroshock seizure thresholds in mice treated with reserpine (184) and facilitates epileptogenesis in amygdaloid kindling (29). These studies lend strong support to the premise that norepinephrine is specifically involved in suppressing abnormal neuronal excitability.

Decarboxylase inhibitors such as α-methyl-p-tyrosine and α-methyl-dopa have not significantly contributed to our understanding of norepinephrine in convulsive disorders. Although these agents appear to deplete cerebral catecholamines and serotonin, several mechanisms of action may account for the inconsistent results of studies employing these compounds. Methyldopa may inhibit dopa decarboxylase, result in false neurotransmission, and act as a central α-adrenergic receptor agonist (reviewed by Frohlich (60)).

The inhibition of norepinephrine release is the mechanism of action of guanethidine; however, this agent probably does not cross the blood-brain barrier. Although no effect of guanethidine on cerebral norepinephrine was found by Cox and Maickel (44), Pfeifer and Galambos (132) found norepinephrine depletion and lowered thresholds to metrazol-induced convulsions in white mice treated with guanethidine. Pfeifer and Unyi (133) later confirmed these earlier results with guanethidine and demonstrated reversal of effect with monoamine oxidase inhibitors.

Reserpine and tetrabenazine, which block reuptake of serotonin as well as catecholamine neurotransmitters into storage vesicles, have been employed in numerous studies to investigate the role of catecholamine depletion on seizure disorders. In 1954, Chen and colleagues (31) first reported that reserpine lowers maximal electroshock and chemoshock seizure thresholds in mice. Subsequent studies have confirmed the seizure facilitative effect of reserpine in several species subjected to a variety of convulsive stimuli (2, 4, 11, 22, 32, 46, 86, 103, 132, 133, 148, 184). Similar findings resulted from treatment with the benzoquinolizine, RO4-1284 (a reserpine-like drug of short duration) (74, 86, 165, 166), and tetrabenazine (11, 32, 46, 148, 166). Jobe and associates (85) also suggested an inhibitory role for descending noradrenergic neurons terminating in the spinal cord by demonstrating an increased capacity of the spinal cord to produce self-sustaining discharges after treatment with RO4-1284. Although these investigators found that the significant decreases in spinal cord norepinephrine correlated with the facilitating effect of RO4-1284 on electrically induced spinal cord seizures, the relative importance of norepinephrine cannot be ascertained utilizing agents that also deplete dopamine and serotonin. Hatoum and colleagues (74), however, noted that the effect of RO4-1284 was antagonized by intracerebroventricular administration of norepinephrine. This antagonism was blocked by α- and β-adrenergic blocking agents. Other investigators claimed that low dose reserpine lowered seizure thresholds and decreased brain norepinephrine without significantly affecting dopamine (84, 132). Wenger and co-workers (184) showed that the reserpine effect was exacerbated by disulfiram, coincident with decreased brain norepinephrine and increased dopamine. These investigators suggested that the ability to synthesize and store neurotransmitters was crucial in the return of seizure thresholds after reserpine treatment. Other studies have found that seizure facilitation induced by reserpine is antagonized by pretreatment with the precursors L-dopa and 5-hydroxytryptophan (11, 22, 103), suggesting an important effect of serotonin and dopamine, as well as norepinephrine, in maintaining seizure thresholds.

Chemical Destruction of Noradrenergic Neurons

6-HYDROXYDOPAMINE

The ability to selectively damage catecholaminergic neurons with 6-hydroxydopamine (6-OHDA) has provided the most convincing evidence for the primary role of norepinephrine in maintaining seizure thresholds. The effect of intraventricularly administered 6-OHDA is depletion of brain catecholamines (27, 175) and facilitation of seizure activity from a variety of convulsive stimuli (24, 25, 27, 38, 39, 114, 118). Because the projections of dopamine and norepinephrine-containing neurons to the forebrain take different courses, investigators have subsequently been able to selectively destroy either noradrenergic or dopaminergic neurons. Studies have consistently shown that specific destruction of norepinephrine-containing neurons with subsequent depletion of brain norepinephrine results in facilitation of seizure activity.

Decreasing forebrain norepinephrine by selective destruction of ascending noradrenergic neurons, without altering levels of dopamine, serotonin, choline acetyltransferase, or glutamic acid accelerated the development of amygdaloid kindling (108) and increased the number and/or severity of both pentylenetetrazol and maximal electroshock seizures (110, 111). Nigrostriatal lesions with selective destruction of dopaminergic neurons had no effect on seizure thresholds (108, 110). Injection of 6-OHDA into fibers of the ascending noradrenergic bundles potentiates the intensity and duration of metrazol-induced seizures in rats (109) and facilitates both primary and secondary site kindling (40). These studies confirm the inhibitory role of noradrenergic innervation to forebrain areas (including amygdala and hippocampus) and suggest that forebrain norepinephrine depletion disinhibits the spread of seizure discharges.

Norepinephrine also exerts an inhibitory influence in areas other than the forebrain. Intravenous administration of 6-OHDA, which does not cross the blood-brain barrier, depletes peripheral norepinephrine and increases the number of lethal seizures in a tonic strain of mice (102). Oishi and co-workers (128) have shown that depletion of cortical, hypothalamic, and brainstem norepinephrine by intraventricular 6-OHDA increased susceptibility to pentylenetetrazol-induced seizures, while injections into the dorsal and ventral noradrenergic bundles, which spare brainstem norepinephrine, did not significantly alter seizure thresholds. This suggests that brainstem norepinephrine-containing neurons are primarily involved in inhibiting pentylenetetrazol seizures.

Spinal cord noradrenergic neurons also modulate neuronal excitability. Jobe and colleagues (83) pretreated rats with desipramine, which selectively prevents the uptake of 6-OHDA into noradrenergic neurons (170) and demonstrated antagonism of norepinephrine depletion and prevention of electrically induced spinal cord seizures after 6-OHDA treatment. Further evidence that spinal cord noradrenergic neurons influence seizure activity was supplied by other investigators, who showed that selective destruction of descending noradrenergic pathways with 6-OHDA suppresses decapitation convulsions in rats and depletes spinal cord norepinephrine (23, 112, 168). Decapitation seizures may be induced by sudden release from tonic, inhibitory noradrenergic tone, whereas release of the inhibitory system prior to decapitation abolishes the convulsive response (167).

KAINIC ACID

Injections of kainic acid can destroy nerve cell bodies without destroying passing axons. Nelson and co-workers (127) injected kainic acid into the hippocampus of rats, which resulted in a seizure disorder lasting several hours. Norepinephrine concentrations were significantly reduced in the hippocampus, lateral cortex, frontal cortex, and cerebellum, ipsilateral and contralateral to the side of injection, although brainstem levels were unaffected. Levels of GABA, serotonin, and dopamine were also unaffected. These investigators coincidentally noted 2-fold increases in the major norepinephrine metabolite, MHPG-sulfate, a finding which may again reflect increased norepinephrine turnover in the presence of seizures. Injections of kainic acid into the amygdala and thalamus caused seizures and decreased norepinephrine throughout the brain.

Surgical Lesions Decreasing Brain Norepinephrine Content

Surgical interruption of noradrenergic neurons is technically difficult, though several investigators have pursued this approach in attempting to alter cerebral norepinephrine. A lesion in the ventrolateral tegmentum significantly reduces telencephalic norepinephrine (76), while caudolateral tegmental lesions reduce forebrain norepinephrine (3), although it is unknown whether or not these lesions alter seizure susceptibility. Jerlicz and co-workers (82), however, demonstrated that bilateral lesions involving the locus coeruleus produced significant decreases in forebrain norepinephrine concentrations and increased susceptibility to audiogenic seizures. These authors concluded that norepinephrine arising from neurons in the locus coeruleus, which project diffusely to the cerebral cortex (119), inhibit seizures.

Seizures resulting from pinealectomy in gerbils (135) and seizures resulting from ischemia in animals subjected to carotid ligation (183) are associated with decreased cerebral norepinephrine, although it is unclear from these studies whether diminished norepinephrine is inducing seizures or is a secondary effect of epileptiform activity and/or destruction of brain tissue.

Electrical Stimulation-Induced Inhibition of Noradrenergic Activity

CEREBRAL SURFACE STIMULATION

Wood and co-workers (191) have installed subdural flap electrodes over the surface of the frontal lobe. These investigators demonstrated significantly decreased lumbar CSF norepinephrine levels by cerebral surface stimulation in seizure patients. Katz and Chase (87), however, found that electrical stimulation of isolated cerebral cortical slices caused release of norepinephrine into the incubation medium. It appears that the response to electrical stimulation of noradrenergic neurons in brain preparations *in vitro* does not correlate with *in vivo* observations. Perhaps these electrical stimulation-induced alterations in noradrenergic activity account for the precipitation of seizures during cortical stimulation occurring in some patients.

CAUDATE NUCLEUS STIMULATION

The organization of the caudate nucleus has been reviewed by Wood and colleagues (193). Although the striatum contains only small quantities of norepinephrine (16, 159) and no detectable dopamine-β-hydroxylase (159) (the enzyme necessary for norepinephrine synthesis), Wood *et al.* (194, 195, 196) found caudate stimulation-evoked reductions in CSF norepinephrine in man. In addition, Avakyan and Arushanyan (5) suggest that catecholaminergic drugs can modify the epileptogenic properties of the caudate nucleus. Caudate efferent projections may decrease the activity of noradrenergic neurons distant to the caudate nucleus, and may therefore play a role in modulating seizure thresholds (193).

Stach (160) demonstrated that electrical stimulation of the caudate-inhibited seizures in animals subjected to injection of picrotoxin into the amygdala, and exogenous norepinephrine injected into the caudate first potentiated and then inhibited seizures. Because only small amounts of norepinephrine exist in the caudate (16, 159) and because exogenous norepinephrine is taken up by both noradrenergic and dopaminergic neurons (159), it is more likely that these results were primarily mediated by dopaminergic neurons in the caudate.

PHYSIOLOGIC VARIATIONS IN BRAIN NOREPINEPHRINE CONTENT AND SEIZURE THRESHOLDS

Diurnal Variations

Circadian rhythms in norepinephrine and diurnal variation in seizure susceptibility have a parallel time course. Schreiber and Schlesinger (152) noted that several strains of mice were more susceptible to sound and electrically induced seizures at midnight. These investigators subsequently found that decreased levels of norepinephrine and serotonin in the brains of DBA/2J mice correlated with the times of greatest susceptibility to audiogenic seizures (153). Reversing the light cycle reversed the circadian rhythm in seizure susceptibility and levels of norepinephrine (and serotonin) in the cortex, diencephalon, and brainstem (153). Ziegler and associates (201, 202) have observed that CSF concentrations of norepinephrine in man are lower at night and in the early morning hours, when seizure thresholds are reduced.

Genetic Variations

Coleman (33) in 1960 proposed that differences in susceptibility to audiogenic seizures in certain strains of mice were due to differences in cerebral levels of monoamines. Altered phenylalanine metabolism in susceptible strains was proposed to affect monoamine metabolism. Schlesinger and associates (147) subsequently found low levels of norepinephrine in the brains of DBA/2 mice, a strain highly sensitive to audiogenic seizures. Norepinephrine levels were maximally depressed at 21 days of age, when genetic susceptibility to seizures is most pronounced (50, 147). Kellogg (88) also noted that forebrain and hindbrain norepinephrine levels were significantly lower in the genetically sensitive mice at 21 days compared to normal controls, suggesting that developmental differences in mechanisms of monoamine storage, synthesis, and/or release might result in deficient amounts of releaseable transmitter. Willott and colleagues (186), however, were unable to find significant differences in brain levels of norepinephrine between susceptible and nonsusceptible mice at the peak age of seizure susceptibility. Schaefer and Kuenzel (146) found low levels of cerebellar norepinephrine and high cerebral hemisphere norepinephrine in the seizure-prone chicks, *Gallus domesticus*. These investigators also observed a more rapid rise in brainstem norepinephrine with age in the susceptible animals.

Lovell (107) reported DBA/2 mice to be extremely sensitive to the monoamine-depleting effects of certain drugs. This finding was supported by Schlesinger *et al.* (148, 149), who noted that after treatment with tetrabenazine and reserpine, the rate of repletion of norepinephrine in 21-day-old mice was slower than in nonsusceptible controls. Schlesinger and colleagues (150) later found tha DBA/2J mice had significantly lower brain levels of catechol-O-methyltransferase than C57B1/6J nonsusceptible mice. No difference was noted in monoamine oxidase activity. Susceptible animals may have less norepinephrine available for synaptic transmission as well as lower levels of catechol-*O*-methyltransferase.

RELATIONSHIP OF CYCLIC AMP TO NOREPINEPHRINE WITH RESPECT TO EPILEPSY

Intracellular cyclic adenosine 3′,5′-monophosphate (cAMP) influences depolarization of neuronal membranes, prolongation of negative afterpotentials and neurotransmitter release and reuptake (47, 116, 126, 178). The metabolism of cyclic AMP has been studied for its effect on neuronal excitability and for its role in the mechanism of action of several anticonvulsants. In this chapter, cyclic AMP shall be briefly discussed with respect to norepinephrine and epilepsy, since recent studies have yielded conflicting results.

Dibutyryl cyclic AMP induces experimental seizures when administered intracerebroventricularly (63) or directly onto the cerebral surface (179). Elevated levels of cyclic AMP have also been found in epileptogenic cortical foci (180). Dibutyryl cyclic AMP, phosphodiesterase inhibitors and adenosine increase the uptake of norepinephrine (178), which may explain the observed epileptogenic properties of cyclic AMP.

Carbamazepine and phenytoin inhibit ouabain-induced elevations in cyclic AMP concentration (104), and phenytoin, carbamazepine, phenobarbital, and diazepam inhibit pentylenetetrazol-induced elevations in brain cyclic AMP levels (57, 131). Carbamazepine and phenobarbital have also been observed to suppress the accumulation of cyclic AMP evoked by norepinephrine (104, 130), while phenytoin and phenobarbital inhibit synaptic reuptake of norepinephrine (72, 182) and phenytoin blocks the cyclic-AMP mediated influx of calcium associated with neurotransmitter release (52). The inhibition of cyclic AMP accumulation and subsequent uptake of norepinephrine may be important mechanisms by which several of these compounds exert their anticonvulsant effect.

These observations are supported by Myllylä and colleagues (122, 123), who demonstrated elevations in CSF cyclic AMP within 3 days of a seizure (123) and diminished basal and postconvulsive CSF cyclic AMP content after treatment with phenytoin, phenobarbital, and carbamazepine (122). Wood and Brooks (190), however, found that baseline CSF cyclic AMP levels in intractable seizure patients treated with phenobarbital and phenytoin were similar to unmedicated control patients. With probenicid blockade, however, these investigators noted elevated CSF cyclic AMP content in medicated seizure patients, suggesting that these anticonvulsants augment cyclic AMP synthesis or affect transport into the CSF in seizure patients. Unfortunately, lumbar CSF cyclic AMP levels do not necessarily reflect brain cyclic AMP content (173, 188, 189).

The observations of Wood and Brooks (190) are more consistent with other investigators, who suggest that norepinephrine may modulate cerebral cyclic AMP levels, and that cyclic AMP has antiepileptic properties resulting in seizure termination (70). Iontophoretic application of cyclic AMP causes hyperpolarization and/or inhibition of central neurons (20, 134). Gross and Ferrendelli (70) found that reserpine, aminophylline, and propranolol all cause a more rapid appearance of tonic seizures and reduced accumulation of cyclic AMP in the brains of animals treated with the convulsant pentylenetetrazol. Similarly, Korf and Sebens (93) noted that α- and β-adrenergic receptor blocking agents suppress the rise in cerebral cyclic AMP resulting from electrical stimulation of the locus coeruleus. Norepinephrine released during abnormal neuronal discharges may be acting at α- and/or β-adrenergic receptor sites in the brain to elevate cyclic AMP and inhibit the spread of abnormal electrical activity.

NORADRENERGIC MECHANISMS OF ANTICONVULSANT DRUGS

Anticonvulsant agents undoubtedly exert their influence in neuronal convulsive thresholds by various mechanisms, including alterations in membrane permeability, synapses, and neurotransmitter metabolism (reviewed by Glaser *et al.* (65)). Because norepinephrine appears to be important in maintaining seizure thresholds, investigators have sought to ascribe some of the anticonvulsant properties of these agents to their effect on norepineprhine storage, release, and metabolism. Although phenytoin has been the most extensively studied anticonvulsant, some evidence suggests that the mechanism of action of other anticonvulsants may also relate to their effect on norepinephrine.

Phenytoin is a noncompetitive inhibitor of presynaptic norepinephrine reuptake (71, 182) and would thus be expected to enhance noradrenergic activity. Hadfield (71, 72) found that both *in vitro* and *in vivo*, phenytoin stimulated the uptake of norepinephrine into hypoxic synaptosomes, while inhibiting uptake into normal synaptosomes. Meyer and Frey (117) showed that the anticonvulsant effect of phenytoin in maximal electroshock seizures is reduced after treatment with the dopamine-β-hydroxylase inhibitor, disulfiram, while Rudzik and Jonson (144) found that (+)-amphetamine enhanced the anticonvulsant properties of phenytoin. Azzaro and co-workers (10) studied the effects of phenytoin on the accumulation of [^{3}H]norepinephrine in brain slices of rat cerebral cortex. These investigators found that low concentrations of phenytoin (10^{-5} to 10^{-4} M) were associated with facilitation of accumulation and reduced catabolism of [^{3}H] norepinephrine. This was consistent with the later observation that phenytoin inhibits monoamine oxidase (9). In higher concentrations (10^{-3} M), phenytoin reduced [^{3}H]norepinephrine accumulation (10). Phenytoin also inhibited calcium-dependent release of norepinephrine from brain slices (136), and inhibited the cyclic AMP-mediated calcium influx that is associated with neurotransmitter release (52). Woodbury (197) subsequently proposed that phenytoin has an inhibitory effect both on norepinephrine secretion and on norepinephrine reuptake and metabolism. The concentration of synaptic norepinephrine, therefore, depends upon the concentration of phenytoin, time after administration, and pathologic condition of the synapse.

Several investigators have studied the effect of carbamazepine on norepinephrine. Although carbamazepine is structurally related to imipramine, Purdy and associates (137) found that the inhibition of norepinephrine reuptake was insufficient to account for its anticonvulsant properties. Quattrone and Samanin (138), however, found that intraventricular 6-OHDA significantly reduced the anticonvulsant effect of carbamazepine, suggesting that catecholamines are important in the anticonvulsant action of this agent. Carbamazepine, in higher than therapeutic doses, inhibits both uptake and release of norepinephrine from brain synaptosomes (137). These observations do not support a uniform relationship between the mechanism of action of carbamazepine and norepinephrine, although carbamazepine may exert influence on transmitter systems by its effect on cyclic AMP (see above).

The anticonvulsant effect of phenobarbital on electroshock seizures is antagonized by treatment with α-methyltyrosine and phentolamine (117), and pentobarbital inhibits the presynaptic reuptake of norepinephrine (71, 182). Rudzik and Johnson (144) demonstrated that (+)-amphetamine enhanced the anticonvulsant effect of phenobarbital (as well as phenytoin and acetazolamide). Phenobarbital was also found to prevent seizures and norepinephrine depletion in the cortex and hippocampus of rats which had undergone kainic acid injection into the hippocampus (127). These studies indirectly support a relationship between norepinephrine and phenobarbital, and suggest that the anticonvulsant properties of the barbiturates may be moderated by norepinephrine metabolism (59).

Benzodiazepines have also been extensively studied in convulsive disorders. Benzodiazepines decreased turnover rates of catecholamines and were able to antagonize the increased norepinephrine turnover rate in animals subjected to various stresses (42). Although diminished catecholamine turnover has not been correlated with their anticonvulsant properties, the benzodiazepines have been found to diminish impulse flow in catecholaminergic neurons (41, 42). While benzodiazepines antagonized pentylenetetrazol-induced convulsions at doses which modify catecholamine turnover in the cerebellum and hypothalamus, no correlation was found between the anticonvulsant effect of these agents and hypothalamic norepinephrine metabolism (51).

The anticonvulsant sodium valproate (sodium *n*-dipropylacetic acid), a competitive inhibitor of both GABA-transaminase and succinic semialdehyde dehydrogenase (54, 97), also alters brain neurotransmitters, although the effect on dopamine, 5-hydroxyindoleacetic acid, and GABA appears to be more pronounced than the effect on norepinephrine (96).

CONCLUSIONS

There is convincing evidence that norepinephrine functions to reduce susceptibility to seizures induced by most epileptogenic stimuli. Elevation of noradrenergic activity by increasing brain norepinephrine, augmenting norepinephrine release, inhibiting norepinephrine reuptake, and stimulating noradrenergic receptors with norepinephrine agonists experimentally suppresses epileptiform activity. Similarly, reduction of noradrenergic activity by pharmacologic inhibition, surgical and chemical destruction of noradrenergic neurons, and electrical suppression promotes abnormal neuronal excitability. Susceptibility to convulsive stimuli may also be affected by diurnal and genetic variations in norepinephrine metabolism. Thus the inhibitory effect of norepinephrine on cortical neurons appears to be important in maintaining seizure thresholds, and may be relevant to the neuropathology of human epilepsy. Several major groups of anticonvulsant medication may suppress convulsions by augmenting noradrenergic activity.

Future investigations must approach the complexity of noradrenergic systems not only by studying the components of norepinephrine transmission and metabolism, but also by elucidating the relationship between norepinephrine and the other putative neurotransmitters, and their mutual effect on neuronal excitability. While sophisticated techniques such as radioimmunoassay, immunohistochemistry, and autoradiography have been invaluable in such studies using animal models, we are now approaching a technological stage that may provide the more accurate definition of neurotransmitter systems in man. Recent advances such as positron-emission tomography may allow the noninvasive visualization of specific synaptic activity, by incorporating radionuclides into neurotransmitters and their precursors or labeling their receptors. The possibility of directing pharmacologic treatments at specific, demonstrable defects in noradrenergic or other neurotransmitter systems is indeed intriguing. Valuable insight will undoubtedly be gained concerning the pathophysiology of seizure disorders, as we achieve a better understanding of the specific biochemical alterations in epilepsy.

References

1. Adér J-P, Aizenstein ML, Postema F, Korf J: Origin of free 3-methoxy-4-hydroxyphenylethylene glycol in the rat cerebrospinal fluid. *J Neural Transm* 46:279–290, 1979.
2. Altschuler HL, Killam EK, Killam KF: Biogenic amines and the photomyoclonic syndrome in the baboon, *Papio papio*. *J Pharmacol Exp Ther* 196:156–166, 1976.
3. Anden NE, Dahlstrom A, Foxe K, Larsson K, Olson L, Ungerstedt V: Ascending monamine neurons to the telencephalon and diencephalon. *Acta Physiol Scand* 67:313–326, 1966.
4. Arnold PS, Racine RJ, Wise RA: Effects of atropine, reserpine, 6-hydroxydopamine, and handling on seizure development in the rat. *Exp Neurol* 40:454–470, 1973.
5. Avakayan RM, Arushanyan EB: Effect of catecholaminergic drugs on epileptogenic properties of the caudate nucleus. *Neurosci Behav Physiol* 7:13–16, 1976.
6. Axelrod J: Dopamine-β-hydroxylase: Regulation of its synthesis and release from nerve terminals. *Pharmacol Rev* 24:233–243, 1972.
7. Axelrod J, Weinshilboun R: Catecholamines. *N Engl J Med* 287:237–242, 1972.
8. Axelrod J: Methylation reactions in the formation and metabolism of catecholamines and

other biogenic amines. *Pharmacol Rev* 18:95–113, 1966.

9. Azzaro A, Gutrecht J: Characteristics of the inhibitory action of diphenylhydantoin on brain monoamine oxidase. *Neurology (Minneap)* 25:378, 1975.
10. Azzaro AJ, Gutrecht JA, Smith DJ: Effect of diphenylhydantoin on the uptake and catabolism of L-^{3}H norepinephrine *in vitro* in rat cerebral cortex tissue. *Biochem Pharmacol* 22:2719–2729, 1973.
11. Azzaro AJ, Wenger GR, Craig CR, Stitzel RE: Reserpine-induced alterations in brain amines and their relationship to charges in the incidence of minimal electroshock seizures in mice. *J Pharmacol Exp Ther* 180:558–568, 1972.
12. Baldessarini RJ, Harris JE: Effects of amphetamines on the metabolism of catecholamines in the rat brain. *J Psychiatr Res* 11:41–43, 1974.
13. Bartholini G, Constantinidis J, Tissot R, Pletscher A: Formation of monoamines from various amino acids in the brain after inhibition of extracerebral decarboxylase. *Biochem Pharmacol* 20:1243–1247, 1971.
14. Baumann PA, Maitre L: Neurobiochemical aspects of maprotiline (Ludiomil) action. *J Int Med Res* 7:391–400, 1979.
15. Bergstrom DA, Iadarula P, Kellar KJ: Effect of electroconvulsive shock treatment on monoaminergic receptor binding sites in rat brain (abstr). *Neuroscience* 4:509, 1979.
16. Berther A, Rosengren E: Occurrence and distribution of dopamine in brain and other tissues. *Experimentia* 15:10–11, 1959.
17. Bhattacharya SK, Mukhopadhyay SN, Reddy PKSP, Das PK: Role of brain monoamines in the anticonvulsant effect of imipramine in albino rats. *Pharmacology* 14:428–434, 1976.
18. Blaschko H, Burn JH, Langemann H: The formation of noradrenaline from dihydroxyphenylserine. *Br J Pharmacol* 5:431–437, 1950.
19. Bloom FE, Hoffer BJ, Siggins GR: Studies on norepinephrine-containing afferents to Purkinje cells of rat cerebellum. I. Localization of the fibers and their synapses. *Brain Res* 25:501–521, 1971.
20. Bloom FE: The role of cyclic nucleotides in central synaptic function. *Rev Physiol Biochem Pharmacol* 74:1–103, 1975.
21. Blum K, Wallace JE, Schwertner HA, Meyer E, Morgan WW: Central supersensitivity to norepinephrine and amphetamine following chemical sympathectomy by 6-hydroxydopamine. *Life Sci* 20:1705–1714, 1977.
22. Boggan WD, Seiden LS: Dopa reversal of reserpine enhancement of audiogenic seizure susceptibility in mice. *Physiol Behav* 6:215–217, 1971.
23. Bosland MC, Versteeg DH, de Jong W: Selective depletion of spinal noradrenaline inhibits postdecapitation convulsions in rats. *Experientia* 36:224, 1980.
24. Bourn WM, Chin L, Picchioni AL: Effect of neonatal 6-hydroxydopamine treatment on audiogenic seizures. *Life Sci* 21:701–706, 1977.
25. Browning RA, Maynert EW: Increased seizure susceptibility in 6-hydroxydopamine-treated rats. *Fed Proc* 29:966, 1970.
26. Browning RA, Maynert EW: Effects of intraventricularly administered monoamines on seizure susceptibility and body temperature in rats. *Neuropharmacology* 17:649–653, 1978.
27. Browning RA, Maynert EW: Effect of intracisternal 6-hydroxydopamine on seizure susceptibility in rats. *Eur J Pharmacol* 50:97–101, 1978.
28. Calderini G, Carlsson A, Nordstrom C-H: Monoamine metabolism during bicuculline-induced epileptic seizures in the rat. *Brain Res* 157:295–302, 1978.
29. Callaghan DA, Schwark WS: Involvement of catecholamines in kindled amygdaloid convulsions in the rat. *Neuropharmacology* 18:541–545, 1979.
30. Carlsson A, Lindquist M, Foxe K, Hokfelt T: Histochemical and biochemical effects of diethyldithiocarbamate on tissue catecholamines. *J Pharm Pharmacol* 18:60–62, 1966.
31. Chen G, Ensor CR, Bohner BB: A facilitative action of reserpine on the central nervous system. *Proc Soc Exp Biol Med* 86:507–510, 1954.
32. Chen G, Ensor CR, Bohner B: Studies of drug effects on electrically-induced extensor seizures and clinical implications. *Arch Int Pharmacodyn Ther* 172:183–218, 1968.
33. Coleman DL: Phenylalanine hydroxylase activity in dilute and nondilute strains of mice. *Arch Biochem Biophys* 91:300–306, 1960.
34. Cooke PM, Snider RS: Some cerebellar influences on electrically-induced cerebral seizures. *Epilepsia* 4:19–28, 1955.
35. Cooper IS, Riklan M, Amin I, Cullinan T: A long-term follow-up study of cerebellar stimulation for the control of epilepsy. In Cooper IS: *Cerebellar Stimulation in Man*. New York, Raven Press, 1978, pp 19–38.
36. Cooper IS, Riklan M, Tabaddor K, Cullinan T, Amin I, Watkins ES: A long-term follow-up study of chronic cerebellar stimulation for cerebral palsy. In Cooper IS: *Cerebellar Stimulation in Man*. New York, Raven Press, 1978, pp 59–99.
37. Cooper IS, Upton GRM: Use of chronic cerebellar stimulation for disorders of disinhibition. *Lancet* 1:595–600, 1978.
38. Corcoran ME, Fibiger HC, McCaughran JA, Wada JA: Potentiation of amygdaloid kindling and metrazol-induced seizures by 6-hydroxydopamine in rats. *Exp Neurol* 45:118–133, 1974.
39. Corcoran ME, Fibiger HC, McGeer EG, Wada JA: Potentiation of Leptazol seizures by 6-hydroxydopamine. *J Pharm Pharmacol* 25:497–499, 1973.
40. Corcoran ME, Mason ST: Role of forebrain catecholamines in amygdaloid kindling. *Brain Res* 190:473–484, 1980.
41. Corrodi H, Foxe K, Hokfelt T: The effect of some psychoactive drugs on central monoamine neurons. *Eur J Pharmacol* 1:363–368, 1967.

42. Corrodi H, Fuxe K, Lidbrink P, Olson L: Minor tranquilizers, stress, and central catecholamine neurons. *Brain Res* 29:1–16, 1971.
43. Cox B, Lomax P: Brain amines and spontaneous epileptic seizures in the Mongolian gerbil. *Pharmacol Biochem Behav* 413:263–267, 1976.
44. Cox RH, Masckel RP: Effects of guanethidine on rat brain serotonin and norepinephrine. *Life Sci* 8:1319–1324, 1969.
45. Cubeddu LX, Barnes EM, Larger SZ, Weiner N: Release of norepinephrine and dopamine-β-hydroxylase by nerve stimulation. I. Role of neuronal and extraneuronal uptake and of alpha presynaptic receptors. *J Pharmacol Exp Ther* 190:431–450, 1974.
46. Dadkar VN, Dahanukar SA, Sheth UK: Role of dopaminergic and noradrenergic mechanisms in metrazole convulsions in mice. *Ind J Med Res* 70:492–494, 1979.
47. Daly JW: The formation, degradation and function of cyclic nucleotides in the nervous system. *Int Rev Neurobiol* 20:105–168, 1977.
48. Dauth GW, Dell S, Gilman S: Alterations of Purkinje cell activity from transfolial stimulation of the cerebellum in the cat. *Neurology (Minneap)* 28:554–560, 1978.
49. Davis RM, Cullen RF, Flitter M, Dueras D, Engle H, Ennis B: Control of spasticity and involuntary movements. *Neurosurgery* 1:205–207, 1977.
50. Deckard BS, Lieff B, Schlesinger K, DeFries JC: Developmental patterns of seizure susceptibility in inbred strains of mice. *Dev Psychobiol* 9(1):17–24, 1976.
51. Doteuchi M, Costa E: Pentylenetetrazol convulsions and brain catecholamine turnover rate in rats and mice receiving diphenylhydantoin or benzodiazepine. *Neuropharmacology* 12:1059–1072, 1973.
52. Dretchen KL, Standaert FG, Raines A: Effects of phenytoin on the cyclic nucleotide system in the motor nerve terminal. *Epilepsia* 18:337–348, 1977.
53. Eleftheriou BE: Regional brain norepinephrine turnover rates in four strains of mice. *Neuroendocrinology* 7:329–336, 1971.
54. Emson PC: Effects of chronic treatment with amino-oxyacetic acid or sodium-*n*-dipropylacetate on brain GABA levels and the development and regression of cobalt epileptic foci in rats. *J Neurochem* 27:1489–1494, 1976.
55. Farjo IB, Blackwood DHA: Reduction in tyrosine hydroxylase activity in the rat amygdala induced by kindling stimulation. *Brain Res* 153:423–426, 1978.
56. Feldberg W, Sherwood SL: Injections of drugs into the lateral ventricle of the cat. *J Physiol (Lond)* 123:148–167, 1954.
57. Ferrendelli JA, Kinscherf DA: Inhibitory effects of anticonvulsant drugs on cyclic nucleotide accumulation in brain. *Ann Neurol* 5:533–538, 1979.
58. Freedman R, Taylor DA, Seiger A, Olson L, Hoffer BJ: Seizures and related epileptiform activity in hippocampus transplanted to the anterior chamber of the eye: Modulation by cholinergic and adrenergic input. *Ann Neurol* 6:281–295, 1979.
59. Frey HH, Magnussen MP: A hitherto undescribed feature in the anticonvulsant effect of phenobarbital. *Pharmacology* 5:1–8, 1971.
60. Frohlich ED: Methyldopa: Mechanisms and treatment 25 years later. *Arch Intern Med* 140:954–959, 1980.
61. Fromm GH, Amores CY, Thies W: Imipramine in epilepsy. *Arch Neurol* 27:198–204, 1972.
62. Fuxe K, Ungerstedt U: Histochemical studies on the distribution of catecholamine and 5-hydroxytryptamine after intraventricular injections. *Histochemie* 13:16–28, 1968.
63. Gessa GL, Krishna B, Forn J, Tagliamonte A, Brodie BB: Behavioral and vegetative effects produced by dibutyryl cyclic AMP injected into different areas of the brain. In Greengard P, Costa E: *Role of Cyclic AMP in Cell Function: Advances in Biochemical Pharmacology*. New York, Raven Press, 1974, vol 3, pp 371–381.
64. Gibson CJ, Wurtman RJ: Physiological control of brain norepinephrine synthesis by brain tyrosine concentrations. *Life Sci* 22:1399–1406, 1978.
65. Glaser GH, Penry JK, Woodbury DM: *Antiepileptic Drugs: Mechanisms of action*. New York, Raven Press, 1980.
66. Glowinski J, Baldessarini RJ: Metabolism of norepinephrine in the central nervous system. *Pharmacol Rev* 18:1201–1238, 1966.
67. Glowinski J, Iversen LL: Regional studies of catecholamines in the rat brain. I. The deposition of H^3-norepinephrine, H^3-dopamine and H^3-DOPA in various regions of the brain. *J Neurochem* 13:655–666, 1966.
68. Goldstein M: Inhibition of norepinephrine biosynthesis at the dopamine-B-hydroxylase stage. *Pharmacol Rev* 18:77–82, 1966.
69. Greenberg PA, Snyder SH: Pharmacologic properties of ^{3}H-dihydroergotryptine binding sites associated with alpha noradrenergic receptors in rat brain membranes. *Mol Pharmacol* 14:38–49, 1978.
70. Gross RA, Farrendelli JA: Effects of reserpine, propranolol, and aminophylline on seizure activity and CNS cyclic nucleotides. *Ann Neurol* 8:296–301, 1979.
71. Hadfield MG: Uptake and binding of catecholamines: Effect of diphenylhydantoin and a new mechanism of action. *Arch Neurol* 26:78–84, 1972.
72. Hadfield MG, Boykin ME: Effect of diphenylhydantoin administered *in vivo* on ^{3}H-1-norepinephrine uptake in rat synaptosomes. *Res Commun Chem Pathol Pharmacol* 7:209–212, 1974.
73. Hardebo JE, Owman C: Barrier mechanisms for neurotransmitter monoamines and their precursors at the blood-brain interface. *Ann Neurol* 8:1–31, 1980.
74. Hatoum NS, Geiger PF, Jobe PC: Effects of adrenergic receptor stimulation and blockade on electrically-induced tonic extensor convul-

sions in RO4-1284 pretreated rats. *Res Commun Psychol Psychiatr Behav* 314:309–377, 1978.
75. Heikkila RE, Orlansky H, Mytilineou C, Cohen G: Amphetamine: Evaluation of *d*- and *l*-isomers as releasing agents and uptake inhibitors for ^{3}H-dopamine and ^{3}H-norepinephrine in slices of rat neostriatum and cerebral cortex. *J Pharmacol Exp Ther* 194:47–56, 1975.
76. Heller A, Bhatnagar RK, Moore RY: Progress in neuro-genetics, vol. 1. Proceedings of the Second International Congress of Neuro-Genetics and Neuropathology. *International Congress Series* No. 175. Amsterdam, Excerpta Medica, 1969, pp 283–288.
77. Hemmy DC, Larson SJ, Sances A, Millar EA: The effect of cerebellar stimulation on focal seizure activity and spasticity in monkeys. *J Neurosurg* 46:648–653, 1977.
78. Hoffer BJ, Siggins GR, Oliver AP, Bloom FE: Activation of the pathway from locus ceruleus to rat cerebellar Purkinje neurons: Pharmacologic evidence of noradrenergic central inhibition. *J Pharmacol Exp Ther* 184:553–569, 1973.
79. Horton R, Anlezark G, Meldrum B: Noradrenergic influences on sound-induced seizures. *J Pharmacol Exp Ther* 214:737–742, 1980.
80. Hutton JT, Frost JD, Foster J: The influence of the cerebellum in cat penicillin epilepsy. *Epilepsia* 13:401–408, 1972.
81. Iwata K, Snider RS: Cerebello-hippocampal influences on the electroencephalogram. *Electroenceph Clin Neurophysiol* 11:439–446, 1959.
82. Jerlicz M, Kostowski W, Bidzi'nski A, Hauptman M, Dymecki J: Audiogenic seizures in rats: Relation to noradrenergic neurons of the locus coeruleus. *Acta Physiol Pol* 29:409–412, 1978.
83. Jobe PC, Geiger PF, Ray RB, Woods TW, Mims ME: The relative significance of spinal cord norepinephrine and 5-hydroxytryptamine in electrically-induced seizure in the rat. *Neuropharmacology* 17:185–190, 1978.
84. Jobe PC, Picchioni AL, Chin L: Role of brain norepinephrine in audiogenic seizures in the rat. *J Pharmacol Exp Ther* 184:1–9, 1973.
85. Jobe PC, Ray TB, Geiger PF, Bourn WM: Effects of RO4-1284 on electrically-induced spinal cord norepinephrine and 5-hydroxytryptamine levels. *Res Commun Chem Path Pharmacol* 18:601–612, 1977.
86. Jobe PC, Stull RE, Geiger PF: The relative significance of norepinephrine, dopamine, and 5-hydroxytryptamine in electroshock seizure in the rat. *Neuropharmacology* 13:961–968, 1974.
87. Katz RI, Chase TN: Neurohumoral mechanisms in the brain slice. *Adv Pharmacol* 8:1–30, 1970.
88. Kellogg C: Audiogenic seizures: Relation to age and mechanisms of monoamine neurotransmission. *Brain Res* 101:87–103, 1976.
89. Kety SS, Javoy F, Thierry AM, Julou L, Glowinski J: A sustained effect of electroconvulsive shock on the turnover of norepinephrine in the central nervous system of the rat. *Proc Natl Acad Sci USA* 58:1249–1254, 1967.
90. Kilian M, Frey HH: Central monoamines and convulsive thresholds in mice and rats. *Neuropharmacology* 12:681–692, 1973.
91. Koboyashi K, Shirakube T, Kishikawa H, Mori A: Catecholamine levels in penicillin-induced epileptic focus of the cat cerebral cortex. *Acta Neurochir (Wien)* 23(Suppl):93–100, 1976.
92. Koella WP, Glatt A, Klebs K, Durst T: Epileptic phenomena induced in the cat by the antidepressants maprotiline, imipramine, clomipramine, and amitriptyline. *Biol Psychiatry* 14: 485–497, 1979.
93. Korf J, Sebens JB: Cyclic AMP in the rat cerebral cortex after activation of noradrenaline neurons of the locus coeruleus. *Neurochemistry* 32:463–468, 1979.
94. Krip G, Vazquez J: Effects of some sympathomimetic drugs and their antagonists on after discharges elicited in chronically isolated slabs of cerebral cortex. *Br J Pharmacol* 43:696–705, 1971.
95. Krnjevic K, Phillis JW: Actions of certain amines on cerebral cortical neurons. *Br J Pharmacol* 20:471–490, 1963.
96. Kukino K, Deguchi T: Effects of sodium dipropylacetate on γ-aminobutyric acid and biogenic amines in rat brain. *Chem Pharm Bull (Tokyo)* 25:2257–2262, 1977.
97. Kupferberg HJ, Lust WD, Penry JK: Anticonvulsant activity of dipropylacetic acid (DPA) in relation to GABA and cGMP brain levels in mice. *Fed Proc Fed Am Soc Exp Biol* 34:283, 1975.
98. Lange SC, Julien RM, Fowler GW: Biphasic effects of imipramine in experimental models of epilepsy. *Epilepsia* 17:183–196, 1976.
99. Langer SZ: Presynaptic receptors and their role in the regulation of transmitter release. *Br J Pharmacol* 60:481–497, 1977.
100. Lassen JB: Piperoxane reduces the effects of clonidine on aggression in mice and on noradrenaline dependent hypermotility in rats. *Eur J Pharmacol* 47:45–49, 1978.
101. Lehman A: Audiogenic seizures data in mice supporting new theories of biogenic amines in the central nervous system. *Life Sci* 6:1423–1431, 1967.
102. Lehmann A: Mechanisms underlying modifications in the severity of audiogenic convulsions. *Life Sci* 20:2047–2060, 1977.
103. Lessin AQ, Parkes MW: The effects of reserpine and other agents on leptazol convulsions in mice. *Br J Pharmacol* 14:108–111, 1959.
104. Lewin E, Black V: Cyclic AMP accumulation in cerebral cortical slices: Effect of carbamazepine, phenobarbitol and phenytoin. *Epilepsia* 18:237–242, 1977.
105. Libet B, Gleason CA, Wright EW, Feinstein B: Suppression of an epileptiform type of electrocortical activity in the rat by stimulation in the vicinity of locus coeruleus. *Epilepsia* 18:451–462, 1977.
106. London ED, Buterbaugh GG: Modification of electroshock convulsive responses and thresholds in neonatal rats after brain monoamine

reduction. *J Pharmacol Exp Ther* 208:81–90, 1978.
107. Lovell RA: Some neurochemical aspects of convulsions. In Lajtha A: *Handbook of Neurochemistry*. New York, Plenum Press, 1971, vol 6, pp 63–100.
108. Mason ST, Corcoran ME: Forebrain noradrenaline and metrazol-induced seizures. *Proceedings of the Epilepsy International Symposium*. Vancouver, Canada, 1978, p 206.
109. Mason ST, Corcoran ME: Forebrain noradrenaline and metrazol-induced seizures. *Life Sci* 23:167–172, 1978.
110. Mason S, Corcoran ME: Catecholamines and convulsions. *Brain Res* 170:497–507, 1979.
111. Mason ST, Corcoran ME: Depletion of brain noradrenaline, but not dopamine, by intracerebral 6-hydroxydopamine potentiates convulsions induced by electroshock. *J Pharm Pharmacol* 31:209–211, 1979.
112. Mason ST, Corcoran ME: Seizure susceptibility after depletion of spinal or cerebellar noradrenaline with 6-OHDA. *Brain Res* 166:418–421, 1979.
113. Maynert EW, Marczynski TJ, Browning RA: The role of neurotransmitters in the epilepsies. *Adv Neurol* 13:79–147, 1975.
114. McIntyre DC, Saari M, Pappas BA: Potentiation of amygdala kindling in adult or infant rats by injections of 6-hydroxydopamine. *Exp Neurol* 63:527–544, 1979.
115. McNamara JD: Selective alterations of regional β-adrenergic receptor binding in the kindling model of epilepsy. *Exp Neurol* 61:582–591, 1978.
116. McNamara JD: Complex neuronal systems: Approach to development of new strategies in the treatment of epilepsy. *Adv Neurol* 27:185–197, 1980.
117. Meyer H, Frey H-H: Dependence of anticonvulsant drug action on central monoamines. *Neuropharmacology* 12:939–947, 1973.
118. Morgan WW: Effect of 6-hydroxydopamine pretreatment on spontaneous convulsions induced by barbital withdrawal. *Speciala* 15:489–491, 1976.
119. Morrison JH, Molliver ME, Grzanna R, Coyle JT: The intracortical trajectory of the coeruleo-cortical projection in the rat: A tangentially organized cortical afferent. *Neuroscience* 6:139–158, 1981.
119a. Morselli PL, Lloyd KG, Löscher W, Meldrum B, Reynolds EH: *Neurotransmitters, Seizures, and Epilepsy*. New York, Raven Press, 1981.
120. Murmann W, Almirunte L, Saccani-Guelfi M: Central nervous system effects on four β-adrenergic receptor blocking agents. *J Pharm Pharmacol* 18:317–318, 1966.
121. Mutani R, Bergamini L, Dorigussi T: Experimental evidence of the existence of an extra-rhinencephalic control of the activity of the cobalt rhinencephalic epileptogenic focus. Part 2. Effect of paleocerebellar stimulation. *Epilepsia* 10:351–362, 1969.
122. Myllylä VV: Effect of convulsions and anticonvulsive drugs on cerebrospinal fluid cyclic AMP in rabbits. *Eur Neurol* 14:79–107, 1976.
123. Myllylä VV, Heikkinen ER, Vapaatulo H, Hokkanen E: Cyclic AMP concentration and enzyme activities of cerebrospinal fluid in patients with epilepsy or central nervous system damage. *Eur Neurol* 13:123–130, 1975.
124. Nakamura S, Iwarma K: Antidromic activation of the rat locus coeruleus neurons from the hippocampus, cerebral and cerebellar cortices. *Brain Res* 99:372–376, 1975.
125. Nasello AG, Marichich ES: Effects of some cholinergic, adrenergic, and serotonergic compounds, glutamic acid, and GABA on hippocampal seizures. *Pharmacology* 9:233–239, 1973.
126. Nathanson JA: Cyclic nucleotides and nervous system function. *Physiol Rev* 57:157–256, 1977.
127. Nelson MF, Zaczek R, Coyle JT: Effects of sustained seizures produced by intrahippocampal injection of kainic acid on noradrenergic neurons: Evidence for local control of norepinephrine release. *J Pharmacol Exp Ther* 214:694–702, 1980.
128. Oishi R, Suenaga N, Fukuda T: Possible involvement of brainstem norepinephrine in pentylenetetrazol convulsions in rats. *Pharmacol Biochem Behav* 10:57–61, 1978.
129. Olson L, Foxe K: On the projection from the locus coeruleus noradrenaline neurons: The cerebellar innervation. *Brain Res* 28:165–171, 1971.
130. Palmer GC: Interactions of antiepileptic drugs on adenylate cyclase and phosphodiesterases in rat and mouse cerebrum. *Exp Neurol* 63:322–335, 1979.
131. Palmer GC, Jones DJ, Medina MA, Stavinoha WB: Anticonvulsant drug actions on *in vitro* and *in vivo* levels of cyclic AMP in the mouse brain. *Epilepsia* 20:95–104, 1979.
132. Pfeifer AK, Galambos E: The effect of reserpine, α-methyl-m-tyrosine, prenylamine, and guanethidine on metrazol-convulsions and the brain monoamine level in mice. *Arch Int Pharmacodyn* 165:201–211, 1967.
133. Pfeifer AK, Unyi G: The role of brain monoamines in the susceptibility to seizures brought about by chemical and electrical stimulations. In Hormones and Brain Function. New York, Plenum Press, 1973, pp 259–272.
134. Phillis JW: The role of cyclic nucleotides in the CNS. *Can J Neurol Sci* 4:151–195, 1977.
135. Philo R, Reiter RJ, McGill JR: Changes in brain dopamine, norepinephrine and serotonin associated with convulsions induced by pinealectomy in the gerbil. *J Neural Transm* 46:239–252, 1979.
136. Pincus JH, Lee SH: Diphenylhydantoin and calcium. Relation to norepinephrine release from brain slices. *Arch Neurol* 29:239–244, 1973.
137. Purdy RE, Julien RM, Fairhurst AS, Terry MD: Effect of carbamazapine on the *in vitro* uptake and release of norepinephrine in adrenergic nerves of rabbit aorta and in whole brain synaptosomes. *Epilepsia* 18:251–257, 1977.

138. Quattrone A, Samanin R: Decreased anticonvulsant activity of carbamazepine in 6-hydroxydopamine-treated rats. *Eur J Pharmacol* 41:333–336, 1977.
139. Reivich M, Glowinski J: An autoradiographic study of the distribution of C^{14} norepinephrine in the brain of the rat. *Brain* 90:633–646, 1967.
140. Riffee WH, Gerald MC: Effects of amphetamine isomers and CNS catecholaminergic blockers on seizures in mice. *Neuropharmacology* 15: 677–682, 1976.
141. Riffee WH, Gerald MC: The effects of chronic administration and withdrawal of (+)-amphetamine on seizure threshold and endogenous catecholamine concentrations and their rates of biosynthesis in mice. *Psychopharmacology* 51:175–179, 1977.
142. Riffee WH, Gerald MC: Acute and chronic effects of (−)-amphetamine on seizure threshold and brain catecholamines in mice. *Biochem Pharmacol* 26:799–801, 1977.
143. Rodriguez de Lores Arnaiz G, Cardoni R, Pellegrinode Iraldi A: Levels of norepinephrine in rat cerebellum after administration of the convulsant 3-mercapto-propionic acid. *Int J Neurosci* 6:269–271, 1976.
144. Rudzik AD, Johnson GA: Effect of amphetamine and amphetamine analogs on convulsive thresholds. In Costa E, Garattini S: *International Symposium on Amphetamines and Related Compounds.* New York, Raven Press, 1970, pp 715–728.
145. Rudzik AD, Mennear JH: Antagonism of anticonvulsants by adrenergic blocking agents. *Proc Soc Exp Biol Med* 122:278–280, 1966.
146. Schaefer MM, Kuenzel WJ: Brain noradrenaline concentration in seizure-prone chicks, *Gallus domesticus. Poult Sci* 57:1052–1055, 1978.
147. Schlesinger K, Boggan W, Freedman D: Genetics of audiogenic seizures. I. Relation to brain serotonin and norepinephrine in mice. *Life Sci* 4:2345–2351, 1965.
148. Schlesinger K, Boggan W, Freedman D: Genetics of audiogenic seizures. II. Effects of pharmacologic manipulation of brain serotonin, norepinephrine, and gamma-aminobutyric acid. *Life Sci* 7:437–447, 1968.
149. Schlesinger K, Boggan W, Freedman D: Genetics of audiogenic seizures. III. Time response relationships between drug administration and seizure susceptibility. *Life Sci* 9:721–730, 1970.
150. Schlesinger K, Harkins J, Deckard BS, Paden C: Catechol-O-methyltransferase and monoamine oxidase activities in brains of mice susceptible and resistant to audiogenic seizures. *J Neurobiol* 6:587–596, 1975.
151. Schlesinger K, Staunes KL, Boggan WD: Modification of audiogenic and pentylenetetrazol seizures with gamma-aminobutyric acid, norepinephrine and serotonin. *Psychopharmacologia* 15:226–231, 1969.
152. Schreiber RA, Schlesinger K: Circadian rhythms and seizure susceptibility: Relation to 5-hydroxytryptamine and norepinephrine in brain. *Physiol Behav* 6:635–640, 1971.
153. Schreiber RA, Schlesinger K: Circadian rhythms and seizure susceptibility: Effects of manipulations of light cycles on susceptibility to audiogenic seizures and on levels of 5-hydroxytryptamine and norepinephrine in brain. *Physiol Behav* 8:699–703, 1972.
154. Schubert P, Ladisich W: Chronic administration of electroconvulsive shock and norepinephrine metabolism in the rat brain. I. Autoradiographic study of the distribution pattern of intracisternally injected H^3-norepinephrine in untreated rats. *Psychopharmacologia* 15:289–295, 1969.
155. Segal M, Pickel V, Bloom F: The projections of the nucleus locus coeruleus: An autoradiographic study. *Life Sci* 13:817–821, 1973.
156. Segal M, Bloom FE: The action of norepinephrine in the rat hippocampus. II. Activation of the input pathway. *Brain Res* 72:99–114, 1974.
157. Shaywitz BA, Yager RD, Gordon JW: Ontogeny of brain catecholamine turnover and susceptibility to audiogenic seizures in DBA/2J mice. *Dev Psychobiol* 11:243–250, 1978.
158. Snider SR, Snider RS: Phenytoin and cerebellar lesions: Similar effect of cerebral catecholamine metabolism. *Arch Neurol* 34:162–167, 1977.
159. Snyder SH: Catecholamines and serotonin. In Albers RW, Siegel GJ, Katzman R, Agranoff BW: *Basic Neurochemistry.* Boston, Little, Brown, 1972, pp 89–104.
160. Stach R: The influence of neuromediators injected into nucleus caudatus on bioelectrical seizure activity of amygdala. *Pol J Pharmacol Pharm* 28:167–180, 1976.
161. Starke K, Altmann KP: Inhibition of adrenergic neurotransmission by clonidine: An action on prejunctional α-receptors. *Neuropharmacology* 12:339–347, 1973.
162. Starke K, Borowski E, Endo T: Preferential blockade of presynaptic alpha adrenoreceptors by yohimbine. *Eur J Pharmacol* 34:385–388, 1975.
163. Starke K, Montel H: Influence of drugs with affinity for α-adrenoreceptors on noradrenaline release by potassium, tyramine, and dimethylphenylpiperazinium. *Eur J Pharmacol* 27:273–280, 1974.
164. Stilk G, Sayers A: The effect of antidepressant drugs on the convulsive excitability of brain structures. *Neuropharmacology* 3:605–609, 1964.
165. Stull RE, Jobe PC, Geiger PF: Brain areas involved in the catecholamine mediated regulation of electroshock seizure intensity. *J Pharm Pharmacol* 29:8–11, 1977.
166. Stull RE, Jobe PC, Geiger PF, Ferguson BG: Effects of dopamine receptor stimulation and blockade on RO-4-1284-induced enhancement of electroshock seizures. *J Pharm Pharmacol* 25:842–844, 1973.
167. Svenaga N, Oishi R, Fukuda T: The role of locus coeruleus in decapitation convulsions in

rats. *Brain Res* 177:83–93, 1979.

168. Svenaga N, Yamada K, Fukuda T: Correlation between central catecholamine level and post-decapitation convulsion in rats treated with 6-hydroxydopamine. *Brain Res* 122:165–169, 1977.
169. Svensson TH, Bonney BS, Aghayaman GK: Inhibition of both noradrenergic and serotonergic neurons in brain by the α-adrenergic agonist clonidine. *Brain Res* 92:251–306, 1975.
170. Taylor KM, Laverty R: The effects of drugs on the behavioral and biochemical actions of intraventricular 6-hydroxydopamine. *Eur J Pharmacol* 17:16–24, 1972.
171. Taylor KM, Snyder SH: Amphetamine: Differentiation of d- and l-isomers on behavior involving brain norepinephrine or dopamine. *Science* 168:1487–1489, 1970.
172. Thornburg JE, Moore KE: Dopamine and norepinephrine uptake by rat brain synaptosomes: Relative inhibitory potencies of l- and d-amphetamine and amantadine. *Res Commun Chem Pathol Pharmacol* 5:81–89, 1973.
173. Tsang D, Lal S, Sourkes TL, Ford RM, Aronoff A: Studies on cyclic AMP in different compartments of cerebrospinal fluid. *J Neurol Neurosurg Psychiatry* 39:1186–1190, 1976.
174. Turner TAR, Spencer PSJ: Effect of pretreatment with monoamine oxidase inhibitors or amphetamine on leptazole convulsions in mice and rats. *J Pharm Pharmacol Suppl* 20:122S–125S, 1968.
175. Uretsky NJ, Iversen LL: Effects of 6-hydroxydopamine on catecholamine-containing neurons in the rat brain. *J Neurochem* 17:269–278, 1970.
176. Van Buren JM, Wood JH, Oakley J, Hambrecht F: Preliminary evaluation of cerebellar stimulation by double-blind stimulation and biological criteria in the treatment of epilepsy. *J Neurosurg* 48:407–416, 1978.
177. Von Voigtlander PR, Moore KE: *In vivo* electrically evoked release of H^3-noradrenaline from cat brain (letter to the editor). *J Pharm Pharmacol* 23:381–382, 1971.
178. Walker JE, Goodman P, Jacobs D, Levin E: Uptake and release of norepinephrine by slices of rat cerebral cortex: Effect of agents that increase cyclic AMP levels. *Neurology (Minneap)* 28:900–904, 1978.
179. Walker JE, Lewin E, Moffitt B: Production of epileptiform discharges by application of agents which increase cyclic AMP levels in rat cortex. In Harris P, Mawdsley C: *Epilepsy: Proceedings of the Hans Berger Centenary Symposium.* New York, Churchill Livingstone, 1975, pp 30–36.
180. Walker JE, Lewin E, Sheppard JR, Cromwell R: Enzymatic regulation of adenosine 3′,5′-monophosphate (cyclic AMP) in the freezing epileptogenic lesion of rat brain and the homologous contralateral cortex. *J Neurochem* 21:79–85, 1973.
181. Weil-Malherbe H, Axelrod J, Tomchick R: Blood-brain barrier for adrenaline. *Science* 129:1226–1227, 1959.
182. Weinberger J, Nicklas WJ, Berl S: Mechanism of action of anticonvulsants: Role of the differential effects on the active uptake of putative neurotransmitters. *Neurology (Minneap)* 26: 162–166, 1976.
183. Welch KMA, Wang T-PF, Chabi E: Ischemia-induced seizures and cortical monoamine levels. *Ann Neurol* 3:152–155, 1978.
184. Wenger GR, Stitzel RE, Craig CR: The role of biogenic amines in the reserpine-induced alteration of minimal electroshock seizure thresholds in the mouse. *Neuropharmacology* 12:693–703, 1973.
185. Wilkison DM, Halpern LM: Turnover kinetics of dopamine and norepinephrine in the forebrain of kindled rats. *Neuropharmacology* 18:219–222, 1979.
186. Willott JF, Lints CE, Sze PY, Nenja LH: Inverse relationship between whole brain monoamine levels and audiogenic seizure susceptibility in mice: Failure to replicate. *Pharmacol Biochem Behav* 12:385–388, 1980.
187. Wolf HH, Stock GA Jr: Utility of two convulsant techniques as indicators of CNS excitability. *J Pharm Sci* 55:1455–1457, 1966.
188. Wood JH: Neurochemical analysis of cerebrospinal fluid. *Neurology (NY)* 30:645–651, 1980.
189. Wood JH: Physiologic neurochemistry of cerebrospinal fluid. In Lajtha A: *Handbook of Neurochemistry,* ed 2. New York, Plenum Press, 1982, vol 1, pp 417–487.
190. Wood JH, Brooks BR: Neurotransmitter, metabolite, and cyclic nucleotide alterations in cerebrospinal fluid of seizure patients. In Wood JH: *Neurobiology of Cerebrospinal Fluid.* New York, Plenum Publishing Co, 1980, vol 1, pp 259–276.
191. Wood JH, Lake CR, Ziegler MG, Sode J, Brooks BR, Van Buren JM: Cerebrospinal fluid norepinephrine alterations during electrical stimulation of cerebellar and cerebral surfaces in epileptic patients. *Neurology (Minneap)* 27:718–724, 1977.
192. Wood JH, Ziegler MG, Lake CR, Sode J, Brooks BR, Van Buren JM: Elevations in cerebrospinal fluid norepinephrine during unilateral and bilateral cerebellar stimulation in man. *Neurosurgery* 1:260–265, 1977.
193. Wood JH, Ziegler MG, Lake CR, Shoulson I, Brooks BR: Cerebrospinal fluid noradrenergic and behavioral alterations associated with stimulation and atrophy of the caudate nucleus. In Wood JH: *Neurobiology of Cerebrospinal Fluid.* New York, Plenum Press, 1980, vol 1, pp 81–169.
194. Wood JH, Ziegler MG, Lake CR, Shoulson I, Brooks BR, Van Buren JM: Cerebrospinal fluid norepinephrine reductions in man after degeneration and electrical stimulation of the caudate nucleus. *Ann Neurol* 1:94–99, 1977.
195. Wood JH, Ziegler MG, Lake CR, Shoulson I, Brooks BR, Van Buren JM: Norepinephrine reductions in cerebrospinal fluid in man after electrical stimulation and degeneration of the

caudate nucleus. *Trans Am Neurol Assoc* 101:68–72, 1976.

196. Wood JH, Lake CR, Ziegler MG, Van Buren JM: Neurophysiological and neurochemical alterations during electrical stimulation of human caudate nucleus. *J Neurosurg* 46:716–724, 1977.
197. Woodbury DM: Phenytoin: Proposed mechanisms of anticonvulsant action. *Adv Neurol* 27:447–471, 1980.
198. Wurtman RJ, Scally MC, Gibson CJ, Hefti F: Relation between brain tyrosine and catecholamine synthesis. In Usdin E, Kopin IS, Barchas JD: *Catecholamines: Basic and Clinical Frontiers*. New York, Pergamon Press, 1979, pp 64–66.
199. Yeoh PN, Wolf HH: Pharmacological evaluation of seizures induced by electrical stimulation of the hippocampus. *J Pharm Sci* 39:950–954, 1970.
200. Ziegler MG, Lake CR, Wood JH, Brooks BR, Ebert MH: Relationship between norepinephrine in blood and cerebrospinal fluid in the presence of a blood-cerebrospinal fluid barrier for norepinephrine. *J Neurochem* 28:877–879, 1977.
201. Ziegler MG, Lake CR, Wood JH, Ebert MH: Circadian rhythm in cerebrospinal noradrenaline of man and monkey. *Nature (London)* 264:646–658, 1976.
202. Ziegler MG, Lake CR, Wood JH, Ebert MH: Norepinephrine in cerebrospinal fluid: Basic studies, effects of drugs and disease. In Wood JH: *Neurobiology of Cerebrospinal Fluid*. New York, Plenum Press, 1980, vol 1, pp 141–152.

CHAPTER TEN

Behavioral Pharmacology of Drugs Affecting Norepinephrine Transmission

JAMES E. BARRETT, Ph.D., and RICHARD E. TESSEL, Ph.D.*

INTRODUCTION

This chapter addresses the general topic of the relationship between norepinephrine (NE) and various behavioral processes. The first section is devoted to a description of the neuropharmacology of NE and to the manner in which drugs affect NE in the central nervous system (CNS). The second section discusses various procedures used in behavioral research that can aid in delineating the behavioral pharmacology of NE. This section also describes experiments focusing on drugs that affect NE and attempts to review, identify, and integrate the role of NE in various behavioral processes. Thirdly, research dealing with changes in NE induced both by behavior and the environment is discussed. The emphasis throughout this chapter is primarily on pharmacological studies that have utilized experimentally conditioned behavior (*i.e.*, learned or operant) rather than unconditioned behavior (*i.e.*, instinctive or reflexive behavior such as spontaneous motor activity). The chapter ends with an evaluation of the status of the behavioral pharmacology of NE, implications of past studies, and potential directions for the future which are likely to be important to individuals interested in both basic and clinical aspects of NE research.

BEHAVIORAL NEUROCHEMISTRY

Presently, an understanding of the effects of drugs on behavior seems most likely to emerge from a knowledge of the reciprocal influences existing between behavior, neurochemistry, and the environment which, in turn, interact with a drug's mechanism of action. This is shown in Figure 10.1 where behavior is viewed as the product of both environmental and neurochemical factors. Although behavior is influenced by both the environment and neurochemistry, this relationship is reciprocal and dynamic. Behavior also acts on the environment to produce consequences that change future behavior (operant behavior). Similarly, behavioral changes stemming from environmental consequences appear to alter neurochemistry. Although there are indications that environmental factors can also affect neuro-

* Preparation of this chapter was supported in part, by PHS grants DA-02873 and DA-01614. We thank Myra J. Zimmerman for help in the preparation of the manuscript.

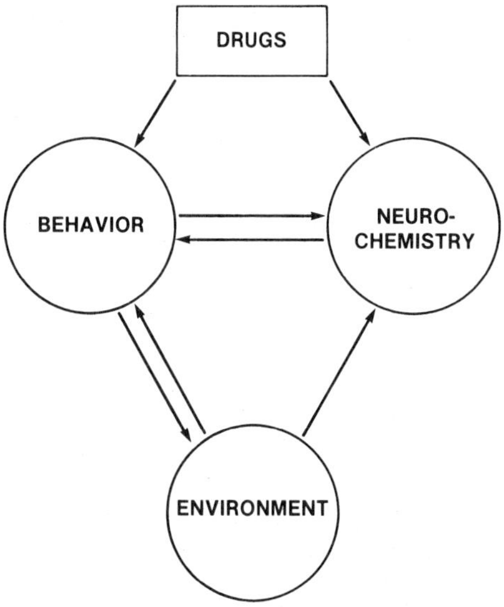

Figure 10.1. Representation of how the behavioral effects of drugs can depend on the reciprocal, dynamic, and interrelationships between behavior, the environment, and neurochemistry. According to this view behavior is influenced by environmental consesquences which, in turn, subsequently modify the environment. Similarly, behavior can both be modified by or can produce neurochemical changes. Although it seems likely that environmental factors can alter neurochemistry, it is unclear whether this is a direct effect or a byproduct of changes in behavior.

chemistry (e.g., nutritional factors), it is not presently clear whether the environment can affect neurochemistry directly without corresponding changes in behavior. Under such conditions behavior would mediate the influence of the environment on neurochemistry. The behavioral effects of drugs cannot be understood apart from the multiple interrelationships of these factors. Very little is actually known about the precise nature and extent of each of these complex interrelationships; however, there have been advances made in recent years in all these areas, some of which are discussed in this chapter. In spite of the fact that relevant variables are clearer, that more suitable techniques are available and that the outlook is optimistic, certain rather substantial difficulties exist which must be considered in order to understand neurochemical influences on both normal and pathological behavior. It is important at the outset to mention some of these problems briefly.

Although it would be desirable to demonstrate unequivocally that the behavioral effect of any drug is directly related to that drug's action on a *particular* neurotransmitter's disposition or receptors, in practice this is an exceptionally difficult task. Such a demonstration would require that the drug is: (1) *neurochemically specific* (i.e., it does not directly affect other neurotransmitter systems); (2) *pharmacologically specific* (i.e., similar actions occur with other drugs having the same neurochemical specificity but not with drugs having different neurochemical specificities); and (3) *behaviorally specific* (not every behavioral measure is affected in the same manner). In addition, at least when the drug is administered systemically, as is the case with most behavioral studies, one is forced to assume that either the drug affects only a small proportion of the total number of neurons containing the neurotransmitter or that all neurons of that population are affected equally and subserve similar if not identical behavioral functions. Unfortunately, few if any drugs currently available have been shown to have sufficient neurochemical specificity, or specificity of distribution within the brain, to allow unequivocal conclusions concerning its actions on a particular transmitter, let alone on a particular behavioral measure. Conversely, few if any behaviors are likely to be subserved by a single neurotransmitter-containing neuronal population or subpopulation, no matter how functionally specific the population's behavioral role may be.

One further complication is associated with what appears to be a commonly held assumption that drugs alter neurochemistry, with behavior being altered as a consequence. This assumes that the changed behavior *does not* itself result in neurochemical changes. As mentioned earlier, in all likelihood this is not true. There is growing evidence that behavioral and neurochemical changes are dynamic, a feature that poses certain experimental demands yet provides a rich source for the eventual understanding of the role of neurotransmitter systems in the regulation and control of various behavioral processes. Dynamic interactions between behavior and neurochemistry are also of obvious importance in clinical evaluations, suggesting that careful consideration be given to the individual's current state when evaluating the relative distribution of neurotransmitters in plasma or cerebrospinal fluid (CSF).

DRUGS AFFECTING NOREPINEPHRINE IN THE CENTRAL NERVOUS SYSTEM

Drugs that Act Primarily to Increase the Net Efflux of Norepinephrine from Nerve Endings

Drugs commonly included in this class are inhibitors of NE uptake such as the tricyclic antidepressants, indirectly and mixed-acting sympathomimetics, such as the psychomotor stimulants (e.g., amphetamine, ephedrine), and methylated xanthines, such as caffeine and theophylline. Uptake inhibition is believed to increase the net efflux of NE by blocking the removal of "spontaneously" released neurotransmitters from the synapse. Indirectly acting sympathomimetic amines, on the other hand, are believed to act by displacing intraneuronal NE from its binding sites. This increases the free pool of NE available for transport out of the cell by use of the same membrane-bound carrier molecule responsible for NE uptake. Inhibition of the reuptake of the displaced NE is considered secondary to the above effect. The effects of the xanthines on NE efflux are more complex and less well understood (see below).

UPTAKE INHIBITORS

Many drugs, including the psychomotor stimulants, have been shown to potently inhibit the accumulation of exogenous radiolabeled NE in various adrenergically innervated tissues, including brain tissue, both *in vivo* and *in vitro*. For purposes of discussion, however, the potent NE uptake inhibitor desmethylimipramine (DMI) will be used as a neurochemical prototype. Although the exact concentration depends on the tissue used, the *in vitro* concentration required to inhibit the uptake of NE by 50% (IC_{50}) ranges between 10^{9} and 10^{-7}/M (50). To determine the correspondence of the *in vitro* concentrations to doses given *in vivo*, Sugrue and Mireyless (89) administered the drug intraperitoneally to rats and subsequently isolated hypothalamic "pinched-off" nerve endings (*i.e.*, synaptosomes) 1 hour later. Having incubated the synaptosomes with a low concentration of ^{3}H-NE, they determined that 20.4 mg/kg of DMI inhibited uptake by 50%, an effect similar to that produced by 3.2×10^{-8} M DMI. This study is important from several standpoints. The drug was administered *in vivo*, and the brain tissue homogenized and centrifuged. Since the resulting supernatant was used as the synaptosomal fraction, some drug might have remained in the tissue pellet, causing an overestimation of the dose needed to inhibit uptake by 50%. Additionally, in this study as in most uptake studies, one must assume that only the NE neurons, and not those of other neuronal populations, accumulated the ^{3}H-amine. While the use of low concentrations of the ^{3}H-amine, and of tissues whose preponderate catecholamine (CA) is NE go far in satisfying the assumption, they cannot do so completely.

Based on binding studies, DMI also has an affinity, albeit a low one, for the presynaptic α_2-receptor (100) and chronically, but not acutely, DMI appears to markedly reduce the sensitivity of both α_2- (14, 15) and β-receptors (103). These compensatory reductions in sensitivity may occur as a consequence of the receptors' chronic exposure to uptake inhibition-mediated increases in synaptic NE concentrations and/or to direct drug-induced alterations in receptor sensitivity *per se*.

RELEASERS

Although some authors, particularly Raiteri and his colleagues (67), have obtained data suggesting that *d*-amphetamine is primarily an NE uptake inhibitor, the data of most others indicate that its releasing action is not dependent only on inhibition of the reuptake of "spontaneously" released NE, but requires entrance of the drug into the nerve ending. Amphetamine, at least at low concentrations ($<10^{-6}$ M), in spite of its high lipid solubility, apparently enters the neuron by being cotransported with sodium by the NE uptake carrier (2, 10, 99). It then displaces from and/or inhibits the access of dopamine (DA) and NE to intraneuronal storage vesicles and other binding sites (82, 96, 97). The carrier molecules now facing the intraneuronal environment (having been transported across the membrane by amphetamine) are thus available to transport the free neurotransmitter out of the neuron, decreasing the time available for the increased free NE to be metabolized by monoamine oxidase (MAO) (91). Amphetamine can only inhibit MAO directly at extremely high concentrations (72). No unequivocal evidence exists presently for a direct effect of amphetamine on a postsynaptic receptor.

Cocaine, on the other hand, although a local anaesthetic, generally enhances the efflux of NE only by inhibiting the reuptake of the "spontaneously" released transmitter. While this may be true in many if not most instances, it is important to note that a cocaine-induced increase in efflux may occur by other mechanisms (2). Tessel *et al.* (92) compared the effects of cocaine on spontaneously beating isolated guinea pig and rat atria with those of cocaine methiodide, a quaternary derivative of cocaine. Both cocaine and its derivative inhibited the uptake of ^{3}H-NE into the atria equally and shifted the positive chronotropic (*i.e.*, increased atrial beating rate) concentration-response curve of NE to the left. Yet, by itself, only cocaine had a positive chronotropic effect; the quaternary, cocaine methiodide, did not. Since the latter compound did not inhibit the rate-increasing effects of NE and, being a pH-independent and positively charged molecule does not readily pass through neuronal lipid membranes, the data suggest that the positive chronotropic effect and, presumably, NE-releasing effect of cocaine are not attributable to drug-induced uptake inhibition *per se*; influx of cocaine into the neuron is probably required. Thus, at least in these systems, cocaine has the characteristics of an amphetamine-like, indirectly acting sympathomimetic amine. Regardless, however, of the generality of the above mechanism, the pharmacological effects of DMI and cocaine certainly differ sufficiently to question whether the drugs produce their effects by the same mechanism.

It should also be noted that inhibitors of NE uptake can antagonize NE release induced by indirectly acting sympathomimetic drugs such as tyramine and amphetamine (2, 90, 98, 99). This antagonism may be attributable to an inhibition of the carrier-mediated influx of amphetamine (2, 10), of drug-induced carrier-mediated neurotransmitter efflux (66a), or both.

METHYLATED XANTHINES

It is most commonly believed that methylated xanthines, such as caffeine and theophylline, act to some extent by inhibiting the enzymes responsible for intracellular degradation of cyclic nucleotides (cyclic 3′, 5′-adenosine and 3′, 5′-guanosine monophosphatases; cAMP and cGMP, respectively) and phosphodiesterases (107). This results in increased cyclic nucleotide levels. Although the role of cGMP in exocytotic NE release is more controversial (110), cAMP and its analogs can indeed potentiate the exocytotic release of NE (16). However, because xanthines lack the potency to inhibit the above enzymes, and though other mechanisms may be involved, many investigators are now favoring the hypothesis that caffeine and its analogs act to modulate NE release and its postsynaptic actions by blocking presynaptic membrane adenosine receptors (17, 31, 42). If the hypothesis is correct, the pharmacology of the xanthines then depends upon the effect and amounts of adenosine within the synapse (88). Since these are multiphasic with both NE and adenosine concentrations, clear interpretations of the behavioral effects of caffeine, etc. using a noradrenergic model become markedly more difficult to obtain.

Adenosine, acting at surface receptors on synaptic membranes, can inhibit the exocytotic release of several neurotransmitters including NE (17). However, high concentrations are necessary for experiments because adenosine is rapidly removed from the synapse by uptake in nervous tissue and through degradation by adenosine deaminase (65). Adenosine can also inhibit adenylate cyclase at extremely low (nanomolar) concentrations yet stimulate it when present in the micromolar range. The former effects are inhibited by micro- and millimolar caffeine and theophylline concentrations, respectively (17). Finally, while stimulation by NE of the β-adrenoreceptor-mediated increase in adenylate cyclase activity (and thus cAMP accumulations) is apparently not modulated by adenosine, adenosine can potentiate α-adrenoreceptor-mediated increases in adenylate cyclase (41). Simplistically, this suggests that the ability of methylated xanthines to decrease the amount of NE in the synapse and to modulate the α-adrenoreceptor-mediated actions of this neurotransmitter depends on synaptic adenosine concentrations. More practically, interpretations of data using this hypothesis are probably invalid. Unfortunately, the variables influencing adenosine concentrations in the synapse are not well understood, and the location(s) in the brain (pre- or postsynaptic or even glial cells) of the increased cyclase activities are unclear. Additional difficulties for this hypothesis are presented by the recent data of Maggi *et al.* (53), which suggest that the number of α-adrenoreceptors, at least in the brain, may be increased reciprocally with decreases in the number of β-receptors. Whether this corresponds to an interconversion of these two receptor types is unclear (6).

Drugs that Act Primarily to Deplete the Neurotransmitter Content of Noradrenergic Neurons

Many drugs, including high doses of amphetamine, can reduce brain NE content, albeit by different mechanisms. In this section, only those drugs presumed to reduce synaptic concentrations of NE will be discussed. Included in this group are reserpine, fenfluramine, 6-hydroxydopamine (6-OHDA), and NE synthesis inhibitors.

Reserpine is a highly lipid-soluble molecule generally believed to diffuse into the neuronal ending by a noncarrier-mediated process (39, 104). It irreversibly binds to storage vesicles (some vesicles, however, may be reserpine-resistant) in a way that blocks the uptake of precursor DA and that inhibits access to vesicular DBH (72). The resultant immediate increase in free cytoplasmic DA (and NE "leaking" from storage vesicles) may further decrease content by inhibiting tyrosine hydroxylase (TH) (73). However, since cytoplasmic CA (derived either as above or from previously released and reaccumulated CA) have increased access to mitochondrial MAO, the resultant deamination is probably the major mechanism for reserpine-induced NE depletion after the drug's acute administration (104). The reduced NE content seen after chronic reserpine administration, however, is also associated with "disuse" supersensitivity of both pre- and postsynaptic adrenergic receptors (24, 66) and with a compensatory increase in the firing rate of noradrenergic neurons (38) until NE content is depleted below that required for neuronal function. Indeed, reserpine administration may result in the unmasking or even the generation of previously undetectable α_2-receptors (7).

The amphetamine congener fenfluramine (*meta*-trifluoro-methyl-N-ethyl-amphetamine) shows some of reserpine's pharmacological properties. Low concentrations of fenfluramine, like reserpine, apparently enter the noradrenergic nerve ending by a noncarrier-mediated diffusion and increase the intraneuronal deamination of NE (91, 111). At higher concentrations, however, fenfluramine can increase the net efflux of NE from the nerve ending and may also inhibit MAO (91).

6-OHDA is a CA whose effects on catecholaminergic neurons are consistent with its neurotoxicity (46, 94). Thus, endogenous NE uptake, storage, and synthetic capacity are decreased for days to weeks following its administration. Presumably, these changes are secondary to concomitant degenerative morphological changes observed in noradrenergic neurons (46). A concomitant increase in the sensitivity of adrenergic receptors supports the above view (37, 85, 101). Since pretreatment of animals with DMI antagonizes many of the effects of 6-OHDA, it is presumed that intraneuronal accumulation of the toxin by means of the NE uptake carrier is required for neurotoxicity to be manifested (46, 93).

Reductions in the amount of NE available for release into synapse are also presumed to result from drugs that inhibit the various steps in NE synthesis. α-Methylparatyrosine (AMPT) is believed to inhibit NE synthesis by competing with endogenous tyrosine for the active sites on TH (104). However, α-methylated products do not result even from high doses. In contrast, α-methyldopa is a substrate for the decarboxylase step and is metabolized to its corresponding α-methylated CA (12). Although reductions in NE content resulting from α-methyldopa administration may involve a small component of decarboxylase inhibition, the large amounts of both α-methyl DA and α-methyl NE that occur, and the insensitivity of these α-methylated CA to MAO, suggest that they are accumulated by storage vesicles replacing the natural transmitter NE and that they act as "false transmitters."

Drugs such as disulfiram can inhibit many copper-requiring enzymes, including DBH, and can presumably decrease NE content and increase the precursor DA pool by this mechanism (68). While the functional effect of this increased DA in NE neurons is unclear, it is conceivable that it could replace NE in the storage vesicles. This could result in the exocytotic release of the "false transmitter" DA instead of NE.

Drugs that Act Primarily on Noradrenergic Receptor Sites

The effects of drugs that act primarily on noradrenergic receptor sites, and the mechanisms by which they act, are better known in the sympathetic autonomic nervous system than in the brain. The lack of understanding of the effects of these drugs on the brain is due to a poor understanding of: (1) the location of the receptors with respect to the noradrenergic neuron (pre- vs. postsynaptic; glial); (2) the variables influencing noradrenergic receptor number, affinity, and receptor-stimulation efficacy; (3) the fact that most of these drugs are neither pure agonists nor antagonists; and (4) the heterogeneity of neuronal populations likely to be innervated by NE-containing neurons. Even these difficulties are not as profound as those that result from the absence of sufficient information to decide unequivocally what criteria should be used to *define* various receptor subtypes. This situation is perhaps clearest with the imidazoline, clonidine. Although recent literature emphasizes clonidine's ability to stimulate α_2-adrenergic receptors, the drug is more accurately described as a partial agonist at α_2-adrenergic receptors (18, 69–71, 87), at least at low doses, and can affect α_1 receptors at higher doses (87). Receptors of the α_2-type are thought to be located primarily on the noradrenergic nerve ending as described previously (59), but postsynaptic α_2-receptors may exist on certain peripheral tissues (42, 67–69, 93), as well as on noradrenergic cell bodies (47, 84). Receptors of the α_1-type, by contrast, appear to be located primarily postsynaptically. Thus, clonidine can conceivably possess full, partial, or no intrinsic activity at various α_2-receptors (and possible α_1-receptors) simultaneously.

Other drugs reputedly having marked selectivity at one or the other of the α-receptor subtypes include: yohimbine, an α-blocker with marked selectivity for the α_2-receptor subtype; prazosin, a selective α_1 blocker; and α-methylnorepinephrine (αMeNE) and phenylephrine, selective α_2 and α_1 agonists, respectively (48). Phentolamine and phenoxybenzamine are considered reversible and irreversible antagonists of α-receptors, respectively (49). They do not appear to have marked selectivity for either α-receptor subtype and can inhibit the uptake of NE into both noradrenergic neurons and extraneuronal tissues (40).

Propranolol blocks all β-adrenergic receptor subtypes nonselectively, the *levo* stereoisomer being the form predominantly, if not exclusively, responsible for the action. Effects of high doses of propranolol, however, may be attributable not to β blockade but to what has been variously called a "membrane-stabilizing" or "local anesthetic" action; the stereoisomers of propranolol are equipotent in this regard (107).

BEHAVIORAL TECHNIQUES

Schedule-Controlled Operant Behavior

The distinctive features of operant behavior, including its frequency of occurrence, idiosyncratic form, and further modification, depend principally on the consequences of past behavior acting with current environmental conditions. Operant behaviors interact directly with, and are maintained or modified by, the specific ways in which environmental consequences occur. Perhaps many of the important characteristics of operant behavior are seen best when behavioral consequences are arranged under schedules of reinforcement. Essentially, a schedule of reinforcement is the precise specification of the relationship between behavior and its consequences. Schedules have been used extensively in behavioral pharmacology and have contributed significantly to our understanding of both behavior and the behavioral effects of drugs. In addition, as will be described in a later portion of this chapter, reinforcement schedules not only change behavior but also appear to alter directly neurochemical events.

Under both laboratory and nonlaboratory conditions reinforcement is typically intermittent; that is, reinforcements will not follow each particular response. When reinforcements are intermittently scheduled, ongoing behavior typically assumes its most robust features in which the dynamic interactions between behavior and environmental consequences are conspicuously emphasized. This aspect of the dependence of operant behavior on environmentally arranged conditions is of both practical and theoretical importance for understanding drug effects on behavior and permits behavioral pharmacology to make a unique contribution to understanding drug action.

Different schedules of reinforcement generate and maintain different patterns and rates of behavior that are distinctive, stable, and reproducible across subjects, species, and types of reinforcing events. This can be of considerable importance when attempting to equate conditions under which drugs are studied, thereby minimizing potentially extraneous and confounding factors. Figure 10.2 shows how the similar scheduling of food to pigeons, monkeys, or rabbits can produce similar performances. In this case, response was maintained by a multiple fixed-ratio fixed-

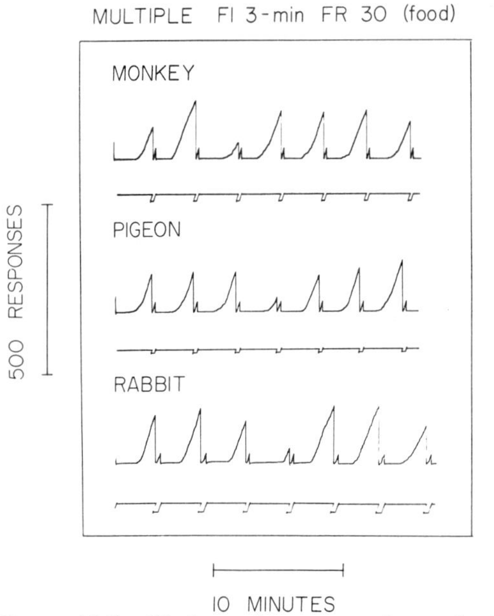

Figure 10.2. Similar performances of a squirrel monkey, pigeon, and rabbit under a multiple fixed interval 3-minute fixed ratio 30-response schedule of food presentation. The pen beneath each record was displaced during the fixed ratio component; the response pen was reset after each food presentation. The pigeon pecked a plexiglass key, whereas the monkey pressed and the rabbit lifted a lever. Note that patterns of responding were similar across the different species.

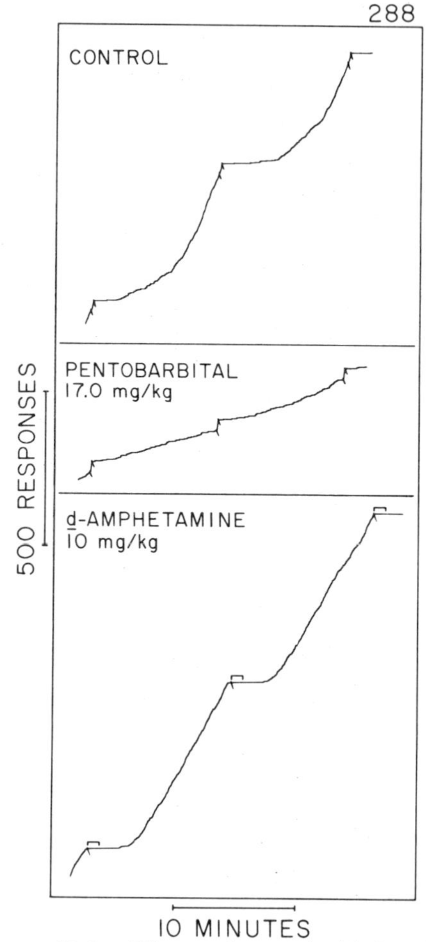

Figure 10.3. Effects of pentobarbital and *d*-amphetamine of responding under a multiple fixed interval 10-minute fixed ratio 31-response schedule of food presentation in the pigeon. The two schedule components alternated throughout the session and were each correlated with a different color keylight. If no response was made within 60 seconds of the end of the fixed interval, or if the 31 responses were not made within 60 seconds during the fixed ratio, the alternate schedule was placed in effect. Note that compared to control, pentobarbital markedly decreased responding under the fixed interval schedule, but had little effect on or increased fixed ratio responding. *d*-Amphetamine increased fixed interval performances but eliminated responding under the fixed ratio schedule (Reproduced with permission from J. W. McKearney and J. E. Barrett and Plenum Publishing Corp. (56a).)

interval schedule. In the presence of one stimulus (e.g., a colored keylight), the first response after 3 minutes produced food (fixed-interval schedule), whereas in the presence of a different stimulus, food was delivered after 30 responses were made. With all species, performance under the fixed-interval schedule was characterized by a low rate of response early in the interval followed by a gradual transition to a higher rate that was maintained until food was delivered. Under the fixed-ratio schedule, however, response rates followed a very brief pause and were initiated at an extremely high and sustained rate until food was delivered.

The characteristic rates and patterns shown under these schedules by the different species are significant not only because of the species-generality of schedule effects, but also because schedules can be an important factor in determining the behavioral effects of many drugs. Figure 10.3 illustrates how the effects of *d*-amphetamine or pentobarbital can differ depending on the schedule of reinforcement. Although pentobarbital markedly decreased responding under the fixed-interval schedule, response rates under the fixed-ratio schedule were unaffected or increased. With *d*-amphetamine, however, the opposite effect occurred in that responding under the fixed-interval schedule was increased substantially, and performance under the fixed-ratio component was eliminated.

There are two noteworthy aspects to this illustration. First, the effects of pentobarbital and *d*-amphetamine on the schedule-controlled performances were different. In order for schedule-controlled performances to yield useful information for understanding drug action, they must be sensitive to differences among drug classes. Second, the effects of both drugs in this example differed depending on the schedule. The finding that drugs such as *d*-amphetamine can either increase or decrease behavior depending on the characteristic features of the existing behavior prior to the drug's administration suggests that schedules can play an important role as modulators or determinants of drug action. This view places more than customary emphasis on behavior itself, since it is functionally related to its environmental consequences, as a direct determinant of drug effects.

Behavior Maintained by Different Events

The discussion of the relevance and importance of schedule-controlled behavior to the behavioral effects of drugs has emphasized thus far the uniformity and stability of schedule performances across different individuals, as well as across different species. It is also worth noting that radically different events can produce similar behavioral performances when scheduled similarly. Figure 10.4, for example, illustrates how events as different as food presentation, shock presentation, termination of stimuli associated with shock (escape performances), and intravenous cocaine can maintain essentially identical rates and patterns of behavior when scheduled to follow the first response after a 5-minute period has elapsed (5-minute fixed-interval schedule). Comparable behavioral performances maintained by dissimilar events allow questions about the influence different *types* of behavioral consequences have in determining drug effects. Though this question was raised several years ago, it is only recently that sufficient evidence has accumulated to indicate that the type of reinforcing event could play an important role in determining the behavioral effects of certain drugs (see review, ref. 5).

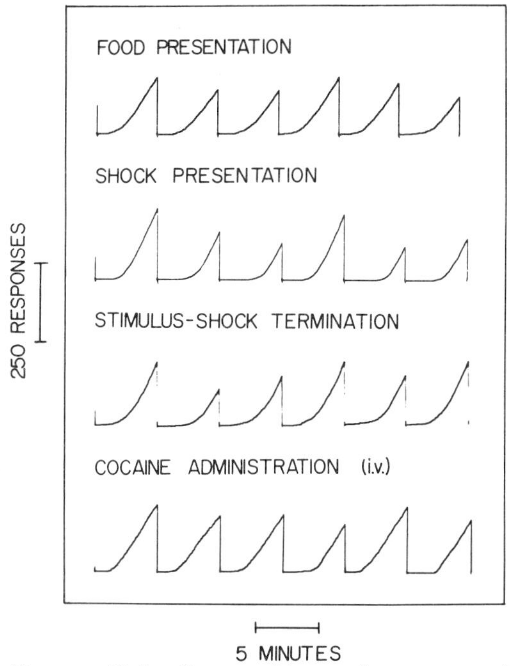

Figure 10.4. Comparable performances of squirrel monkeys responding under 5-minute fixed interval schedules with different maintaining events. The pens reset to baseline at reinforcement. Food presentation consisted of the delivery of a 300-mg Noyes banana-flavored pellet; shock presentation was a 200 msec 9-mA electric shock delivered to the shaven portion of the monkey's tail. Under the stimulus-shock termination schedule, 9-mA shocks were scheduled to occur beginning 1 second after the 5-minute interval had elapsed, terminating the prevailing stimuli and shock schedule and producing a 1-minute time out period. Cocaine hydrochloride was injected via an indwelling intravenous catheter connected to a motor-driven syringe; a response after the 5-minute fixed interval produced a 50 mg/kg infusion of cocaine. Note that performances maintained by the different events are quite similar despite their various characteristics and means of administration. (Reproduced with permission from J. E. Barrett and J. L. Katz and Academic Press (5).)

Punished Behavior

Research in behavioral pharmacology has also relied on the extensive use of behavior *suppressed* by consequent events, *i.e.*, *punishment*, to investigate a wide variety of drugs, including those affecting NE transmission. Punishment procedures typically involve first establishing a stable performance under a schedule of food presentation and then adding an additional schedule where responses occasionally produce a noxious stimulus such as electric shock. Under appropriate conditions, responding is suppressed and occurs at a lower than usual rate. Punished behavior is particularly sensitive to the effects of certain drugs, particularly the benzodiazepines and barbiturates which, at appropriate doses, produce marked increases in behavior. It is especially interesting that compounds such as the amphetamines, however, which normally increase low rates of response, do not usually increase low rates of punished responding.

Discriminative Stimulus Properties of Drugs

Drugs can serve as discriminative stimuli. For example, after the administration of a particular dose of amphetamine, responses on only one of two levers will produce food; responses on the alternate lever have no previously determined consequences. After saline, however, responses on the alternate lever produce food and responses on the "drug" lever have no consequences. Within a relatively short time, most animals rapidly respond appropriately and discriminate the two different conditions. When different drugs are substituted for the training drug, only those compounds sharing the training compound's pharmacological actions will produce the response characteristically performed under the training drug. For example, the discriminative effects of morphine generalize to other narcotic analgesics such as fentanyl and heroin, but not to drugs such as *d*-amphetamine, ketamine, and pentobarbital.

The ability to detect differences in receptor-mediated activity of pharmacological compounds based on their discriminative effects has important implications for receptor theories of drug action and for the development of effective, nonaddictive narcotic analgesics.

Dual Behavioral Effects of Drugs and Other Environmental Stimuli

Under certain conditions, environmental stimuli such as drugs, food, or electric shock can exert dual and, occasionally, opposite effects on behavior. Circumstances in which this occurs are particularly interesting because they suggest the importance of factors other than the events themselves.

The development of conditioned taste aversions is an experimental procedure that results in the suppression of behavior but is unlike that of punishment described above. Whereas with punishment the suppressing stimulus depends upon and follows a response, under conditioned taste aversions, suppression results when the stimulus (usually a drug injection) occurs independently of responses. In most studies of conditioned taste aversions, animals are presented with a novel-tasting substance such as saccharin-flavored water. After consuming the water, they are injected with a drug such as amphetamine. Drinking the available saccharin solution does not occur in subsequent tests, although control subjects, injected only with saline, will drink the solution avidly.

For present purposes, the noteworthy aspect of these studies is that under the procedures involved in conditioned taste aversion, a drug such as amphetamine suppresses responding. Under different procedures, however, this same drug will maintain high levels of responding when self-administered (108). Thus, amphetamine can have opposite effects on behavior, depending on the specific ways in which it is scheduled to occur.

Studies showing the maintenance and suppression of behavior by amphetamine are similar to those in which behavior has been maintained or suppressed by electric shock (63, 64). In these experiments, the same intensity of shock that maintains responding under one condition actually suppresses responding under a different condition. Figure 10.5 shows this effect in a single squirrel monkey. Responding was maintained initially by a food or shock presentation schedule in the presence of different stimuli (*top panel*). The addition of a fixed-ratio shock presentation schedule to the component in which food was delivered suppressed responding during that component only; responding was still maintained by shock in the alternate component (*lower panel*). Thus, as was true of amphetamine in the study described above, the same stimulus, shock, can serve as a punisher or reinforcer, depending on the conditions under which it is presented.

This section has been devoted to providing an overview of the use and importance of schedule-controlled behavior in the behavioral analysis of drug action. The conditions under which drugs are studied and the characteristic details of the particular behaviors under study can be of considerable importance in determining both qualitative and quantitative changes produced by drugs. Clearly, drugs do not have unitary effects on behavior. The array of factors involved in behavior and drug action is unquestionably complex. Environ-

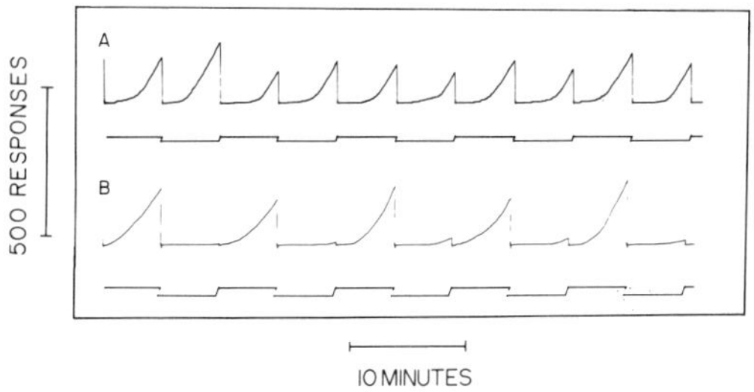

Figure 10.5. Maintenance and suppression of responding by response-produced presentation of electric shock (10 mA). In *A*, responding of squirrel monkeys was maintained under a multiple 5-minute fixed interval schedule of food or shock presentation, each correlated with a different stimulus. Comparable rates and patterns of responding were maintained by both events. In *B*, food-maintained responding was suppressed by the presentation of a 10-mA shock following every 13th response during that component; the fixed interval shock presentation schedule remained in effect during the alternate component. The lower event line in each record was displaced during the food presentation component, and the response pens were reset after each component. (Reproduced with permission from J. W. McKearney and J. E. Barrett and Plenum Publishing Corp. (56a).)

mental consequences act on existing behavior to accentuate some features and reduce others.

BEHAVIORAL EFFECTS OF DRUGS ALTERING NOREPINEPHRINE TRANSMISSION

The effects on schedule-controlled behavior of a number of drugs affecting NE transmission have been studied extensively. Unfortunately, as mentioned earlier, most of these compounds have effects on more than one neurotransmitter, thereby making definitive statements about neurotransmitter-behavior effects somewhat difficult. *d*-Amphetamine, for example, in addition to its effects on NE discussed earlier, also affects DA and serotonin (5-HT).

Principles of Behavioral Pharmacology

Schedules of reinforcement engender different rates and patterns of responding that are differentially sensitive to drug effects. The control rate with which a particular behavior occurs under a schedule can be a significant predictor of a drug's behavioral effects (20). Although demonstrated first with the amphetamines, these effects, referred to as *rate-dependency*, characterize the changes produced by other drugs as well. Generally, although not exclusively, the importance of response rate is seen under conditions where rates of responding vary from being relatively low to being relatively high, such as under a fixed-interval schedule. Under these conditions, the effects of amphetamine are inversely related to the control response rate: *i.e.*, relative to nondrug conditions, low rates of responding are increased and high rates decreased. This effect is shown in Figure 10.6. Different drugs, different doses of the same drug, and other environmental variables may maximize or minimize the influence of response rates.

Other behavioral determinants of drug effects include the individual's behavioral and pharmacological history, the environmental context in which the behavior occurs, and the type of event that maintains responding.

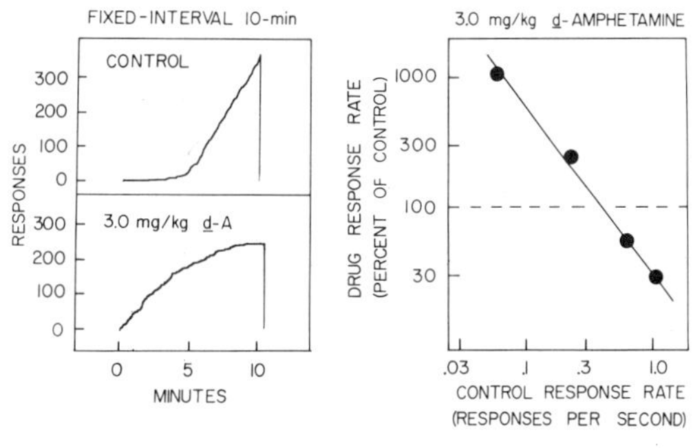

Figure 10.6. Dependence of the effects of *d*-amphetamine sulfate on response rate. The figure on the *left* is a cumulative record of responding under a 10-minute fixed interval (FI) food schedule under control conditions (*top*) and after administration of 3.0 mg/kg *d*-amphetamine (*bottom*). In the absence of amphetamine response rates are low at the beginning of the FI but increase as the interval elapses; towards the end of the interval responses occur at a fairly constant rate. The FI was divided into four successive 150-second segments, and response rates were calculated for each segment. Changes in these rates with 3.0 mg/kg *d*-amphetamine, expressed as percent of control are shown on the log-log function on the *right*. Note that amphetamine increased the lower control rates early in the FI but decreased higher rates toward the end. There is an inverse linear relationship between response rate and the effects of *d*-amphetamine. (Reproduced with permission from J. E. Barrett and Elsevier-North Holland Pub. Co. (3a).)

Antidepressants, Amphetamines, and Cocaine

In spite of the widespread clinical usage of the antidepressants, the effects of these compounds on schedule-controlled behavior have not been studied extensively. Further, when the results of the few experiments are compared, species differences appear that preclude many general statements. In an early study with pigeons, for example, Dews (21) reported that DMI and imipramine (IMI) increased responding by a factor of two to three under a four-component multiple fixed interval schedule in which the intervals were 3, 6, 9, and 10 minutes in duration. Increases obtained with IMI were slightly greater than those obtained with DMI. Further, the normally low response rates occurring early in the fixed intervals were increased to a greater extent with IMI than with DMI; IMI also increased the higher rates occurring at the end of the fixed interval, whereas DMI did not. The effects of methamphetamine in this study were similar to those of DMI. Increases in high response rates of pigeons after IMI have also been obtained under short (1 minute), fixed interval schedules as part of a multiple schedule with other fixed interval components of 3-, 5- and 10-minute durations (112). This experiment also examined the rate-dependent effects of *d*-amphetamine and cocaine. In contrast to IMI, neither of these drugs increased the high overall response rates under the 1 minute fixed interval schedule; all three drugs, however, did increase the relatively low response rates under the 3, 5, and 10-minute schedules. Generally, each drug increased the lowest rates under the 10-minute fixed interval schedule the most. When the effects of IMI were studied on high response rates maintained in pigeons under fixed ratio schedules, responding was decreased at doses that produce rate-dependent effects under fixed interval schedules (83, 102).

Although the antidepressants increase response rates of pigeons under fixed-interval schedules, typically in a rate-dependent manner, increases do not usually occur with these drugs in other species, such as the rat (56) or squirrel monkey (13). Increases in responding with the antidepressants may be idiosyncratic to the pigeon.

Except for the effects of IMI on high, fixed interval response rates, there seem to be minimal differences in pigeons between the effects of the antidepressant drugs studied thus far and the psychomotor stimulants. Marked differences do exist, however, between the effects of antidepressants and psychomotor stimulants in rats and squirrel monkeys. Whereas the antidepressants typically only decrease responding in rats and monkeys, amphetamine and cocaine produce marked increases under schedule conditions that maintain low-to-moderate response rates (see reviews in refs. 23, 60). The behavioral effects of certain narcotics (34) and many antipsychotic compounds (e.g., chlorpromazine and promazine) in the pigeon are also different from those found in the rat and squirrel monkey (22).

Other experiments have examined the effects of antidepressant drugs in rats under a schedule slightly different from that described thus far in that it specifically engenders low response rates by delivering reinforcement only if a minimal amount of time occurs between successive responses (differential reinforcement of low rate schedule). Both IMI (1, 45) and DMI (45) decrease overall response rates under this schedule. McGuire and Seiden (55) recently extended these earlier studies and also examined the effects of chronic IMI under the differential low rate reinforcement schedules with rats. As in the earlier studies, DMI, IMI, and chlorimipramine (CMI) decreased response rates. Chronic IMI administered over a 74-day period augmented the initial decreases in response rates as well as the increases in reinforcement rates. The unique effects of antidepressants on behavior of rats and monkeys under the differential reinforcement of low rate schedule may aid in delineating the neurochemical mechanisms and behavioral actions of the tricyclic antidepressants (55).

Although in mammals the behavioral effects of the antidepressants can be distinguished from cocaine and amphetamine, it is difficult to differentiate the behavioral effects of cocaine from amphetamine. Nevertheless, it appears that the neurochemical profile of cocaine differs from that of amphetamine in the action these drugs exert on NE (30). These neurochemical differences are not sufficient to produce different behavioral effects. For example, several experiments have examined the effects of cocaine and amphetamine on responding maintained by vastly different behavioral consequences such as food or electric shock presentation, drug self-administration, electrical stimulation of the brain, or the termination of stimuli correlated with periodic shock delivery (escape). Although schedule-dependent effects are occasionally found, under comparable schedules there appear to be few differences in the effects of cocaine or amphetamine on behavior maintained by different types of events (5).

Unfortunately, the antidepressant compounds have not been studied extensively under conditions where appropriate comparisons could be made across types of maintaining events. Antidepressants do not appear to differentially affect performances maintained by different events. IMI decreases responding under fixed-interval schedules of food or escape from continuous shock (13). Similarly, responding maintained under fixed interval schedules either by food or shock presentation is also decreased by IMI (unpublished observations).

Inhibitors of NE uptake antagonize the release of NE induced by amphetamine-like drugs. Ganousis and Tessel (27) observed that the same doses of amphetamine, which generally decreased the high response rates associated with a fixed ratio 30 schedule of food presentation, increased the low rates associated with fixed interval, 5-minute response. When animals were pretreated with doses of the potent and selective NE uptake inhibitor, nisoxetine (26, 44, 109) that had little effect on operant responding *per se*, the amphetamine-induced decreases in fixed ratio rates were potentiated, and the increases were inhibited or abolished. This latter effect was apparently unrelated to the former since the rate-increasing effect of amphetamine on fixed interval responding was not converted to a rate-decreasing one; it was simply blocked.

Other experiments have examined the effects of cocaine (28), amphetamine (29), and IMI (62) on behavior suppressed by punishment. Generally, all three drugs either did not affect or decreased punished behavior further. These studies extend the range of conditions under which the effects of these different compounds exert similar effects on schedule-controlled behavior. A recent experiment that examined combinations of amphetamine with noradrenergic blockers or with neuroleptics suggests that the effects of amphetamine on punished behavior may be due to its effects on the release of DA (51). In this study, the effects of *d*-amphetamine, which suppressed punished responding further, were not modified when given in combination with propranolol, phentolamine, and phenoxybenzamine. However, the rate suppressant effects of amphetamine on punished behavior were antagonized by spiroperidol, haloperidol, and clozapine.

Although amphetamine generally further decreases behavior reduced by punishment, there are conditions under which it does produce sizeable increases in punished behavior. In one experiment, for example, squirrel monkeys were trained first under a schedule in which food-maintained responding was punished by electric shock (3); *d*-amphetamine decreased this behavior (*left panel*, Fig. 10.7). After the dose-effect curves were determined, the monkeys were trained under a shock postponement (avoidance) schedule for approximately 2 weeks, and then returned to the original punishment schedule. When rates of punished responding were comparable to those under the original condition, *d*-amphetamine was again administered. In contrast to the initial effects, amphetamine produced marked increases in punished responding (*right panel*, Fig. 10.7). Thus, the effects of amphetamine can be modified by behavioral

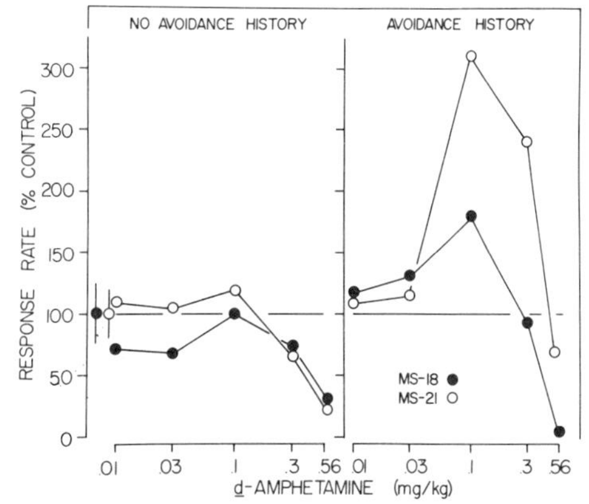

Figure 10.7. Effects of *d*-amphetamine sulfate on punished responding of two squirrel monkeys before and after exposure to an avoidance or shock postponement schedule. Control points ± SE are indicated on the *left*. Doses were determined at least twice. During drug testing responding was maintained under a 5-minute fixed interval food presentation schedule; every 13th response during the interval produced a 5 mA shock. *d*-Amphetamine did not increase punished responding before training under the avoidance schedule but produced sizeable increases in punished responding after exposure to the avoidance schedule, even though punished performances did not change from those under original condition. (Reproduced with permission from J. E. Barrett and the American Association for the Advancement of Science (3).)

experience under a particular set of environmental conditions.

Xanthines

Caffeine typically produces smaller rate-increasing effects on overall response rates than amphetamine (19, 57). Although amphetamine inevitably produces rate-dependent effects, caffeine does not typically do so in monkeys (19) or in rats (32). With pigeons, caffeine has been shown to increase responding under fixed interval schedules at doses that do not affect fixed ratio responding (58).

Recently, McKim (57) directly compared the effects of caffeine, theophylline, and amphetamine on fixed interval and fixed ratio responding of the mouse. Even though all three drugs increased responding under the fixed interval schedule and produced rate-dependent effects, there were quantitative differences between these drugs, suggesting that the effects of amphetamine differ from those of caffeine and theophylline. First, peak effects of amphetamine were greater under the fixed interval schedule than with the other two drugs, which were quite similar to one another. Secondly, intermediate doses of amphetamine that substantially increased responding under the fixed interval schedule markedly decreased responding under the fixed ratio schedule. With the methylxanthines, however, responding under both schedules was decreased at the same high dose. Thus, like the antidepressants and psychomotor stimulants, the methylxanthines can be differentiated behaviorally from other compounds affecting NE transmission. This difference also appears to be true with punished responding in that caffeine, unlike other drugs discussed thus far, appears to increase this behavior (61).

Caffeine and its analogs might modulate NE release by adenosine receptor blockade. Recent behavioral support for this possibility comes from a study in which the rate-decreasing effects of L-phenylisopropyl-adenosine (an adenosine analog) on the response of rats under a 30-response fixed ratio schedule were antagonized by caffeine (81). In addition to suggesting the role of caffeine as an adenosine receptor blocker, this study also demonstrates that adenosine analogs have very potent behavioral effects.

α- and β-Adrenergic Agonists and Antagonists

Clonidine is generally considered to function preferentially at α_2-receptors. Clonidine decreases avoidance behavior (33, 50) as well as behavior under a fixed ratio 20-response schedule of food presentation (54). Whereas the decreases in avoidance with clonidine were antagonized by the α-adrenergic blocking agents yohimbine and phentolamine, with yohimbine (an α_2 antagonist) being more effective than phentolamine, the β-adrenergic blocker propranolol produced greater decrements in combination with clonidine than did clonidine alone (33). Decreases in food-maintained responding produced by clonidine were also antagonized by yohimbine (54). Figure 10.8 shows the type of antagonism by yohimbine of clonidine-induced suppression of responding under a multiple three-minute fixed-interval 30-response fixed-ratio schedule. Although yohimbine effectively antagonized the suppression of responding produced by clonidine, prazosin, an α_1 antagonist, did not (4). Taken together, these studies suggest that the suppression of avoidance and food-maintained responding by clonidine are α_2 effects.

The effects of clonidine on performances of squirrel monkeys and pigeons have been studied under various schedules of food presentation, electric shock presentation, punishment, and stimulus-shock termination. Under all these conditions clonidine, as well as the α antagonists tolazoline and phentolamine, typically produced only dose-dependent decreases in responding (J. E. Barrett, unpublished data). However, in these same studies propranolol, a β antagonist, increased responding under fixed interval schedules of food presentation, shock presentation, and stimulus-shock termination. These findings showing rate-decreasing effects of the α agonists and antagonists and rate-increasing effects of the β antagonist propranolol indicate that there may be differential effects of α and β compounds on *nonpunished* schedule-controlled behavior. Differences between the α and β antagonists yohimbine and propranolol have also been found on *punished* behavior, although these are opposite in direction to those just discussed. Thus, yohimbine increases punished behavior (80), whereas propranolol does not (59, 79). On the basis of the limited data avail-

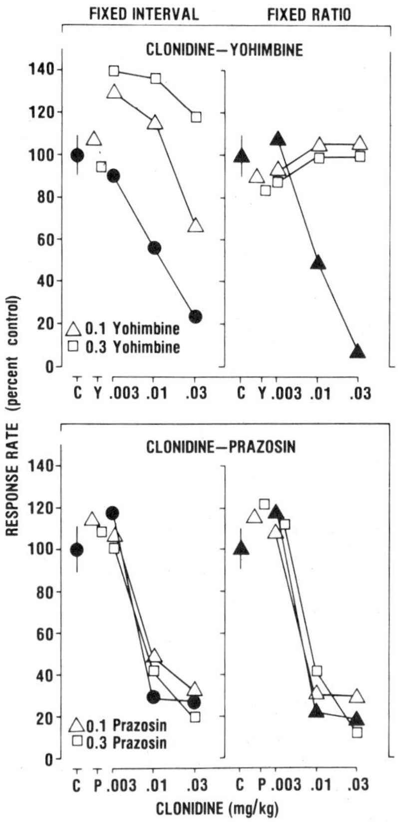

Figure 10.8. Antagonism of clonidine-induced suppression by yohimbine (*top*) but not prazosin (*bottom*). Performances of pigeons were maintained under a multiple 3-minute fixed interval 30-response fixed ratio schedule. *Solid symbols* represent effects of clonidine and saline injections on fixed interval (*left panels*) and fixed ratio performances (*right panels*). *Open symbols* denote effects of combinations of either yohimbine or prazosin with clonidine. *Y* represents doses of yohimbine alone; *P* represents doses of prazosin; *C* denotes control saline injections. Vertical bars are ± 1 SE.

able, it seems the effects of α- and β-noradrenergic compounds may differ consistently on punished and unpunished behavior.

A final comment on the relationship of α-adrenergic compounds and antidepressants seems appropriate. The behavioral studies described above indicate that the antidepressants and clonidine are similar in producing rate-decreasing effects. It has been reported (105) that the hypothermic response to clonidine can be antagonized by multiple doses of antidepressant compounds (e.g., IMI), as well as by α-adrenergic blocking agents (e.g., yohimbine). These effects correlate well with clinical efficacy, as well as with the delayed onset antidepressant drugs typically show, and suggest that effective antidepressant activity may be related to the ability to alter α-adrenergic sensitivity (15). Finally, this study also found that chronic IMI antagonized the slowing of NE turnover by clonidine.

The α- and β-noradrenergic compounds have clear effects on behavior. Behavior is selectively sensitive to α_1 and α_2 receptor subtypes. More extensive work should yield useful information for understanding the role of NE in various behavioral processes.

Discriminative Stimulus Properties of Drugs Affecting NE Transmission

Drugs have stimulus properties which, like other stimuli, are capable of controlling behavior across several dimensions. These properties include dose and pharmacological or behavioral mechanisms of action, as measured by the subject's tendency to respond in a similar manner to compounds substituted for the original training drug.

ANTIDEPRESSANTS

Howard *et al.* (36) were successful in establishing good discriminative control, without toxicity, with the relatively new phenylaminoketone antidepressant buproprion. Unlike typical tricyclic and MAO-inhibitor (MAO-I) antidepressants, buproprion does not inhibit type A and B forms of MAO. Further, it is a very weak inhibitor of biogenic amine uptake and is ineffective in altering the release of amines from synaptosomes.

When rats were trained to discriminate buproprion (20 mg/kg) from saline, the tricyclic compounds IMI, amitriptyline, nortriptyline, and desimipramine were poorly generalized. However, there was substantial transfer to higher doses of psychomotor stimulants cocaine (5 to 10 mg/kg), *d*-amphetamine (0.4 to 0.8 mg/kg), and caffeine (25–50 mg/kg). This finding is not surprising in view of the fact that buproprion is a structural analog of diethylproprion, a common psychomotor stimulant, and appears to be an atypical antidepressant. Despite the efficacy of using buproprion as a discriminative stimulus, this experiment concluded by pointing out that establishing antidepressant compounds as effective discriminative stimuli continues to be done only with some difficulty.

PSYCHOMOTOR STIMULANTS

There is considerable generalization of response across the psychomotor stimulant drugs. Animals trained to discriminate *d*-amphetamine from saline will also generalize this response to appropriate doses of cocaine, *d*, *l*-amphetamine, and methylphenidate (35). It also appears that training with any of these drugs, as well as with apomorphine, will produce generalization to each of the others. It is interesting that when other compounds have been substituted after discriminative control by stimulants has been established, only the DA receptor blockers produce significant antagonism (see review in ref. 35). Thus, it would appear that dopaminergic systems are predominantly responsible for discriminative effects of the psychomotor stimulants.

Colpaert *et al.* (11) have conducted extensive studies on the discriminative effects of cocaine and have suggested that some of these effects may be due to increasing levels of NE. This conclusion is based on the finding that some subjects trained to discriminate cocaine from saline showed partial response to the β antagonist propranolol. Similarly, a small percentage of subjects trained with propranolol also generalized to cocaine. However, the conclusion about the specific role of NE in cocaine's discriminative effects must remain suggestive at this point because there was no generalization to another β antagonist, practolol. Responding under the control of cocaine also did not generalize to the β agonists isoproterenol and salbutamol. At the present time, it appears that compounds which induce generalization to the psychomotor stimulants do so on the basis of their increasing levels of DA.

BEHAVIORALLY AND ENVIRONMENTALLY INDUCED CHANGES IN NOREPINEPHRINE

Neurochemical Changes Correlated with Schedule-Controlled Behavior

The material discussed thus far has been based on experiments in which exogenous compounds have been administered under carefully controlled conditions. Changes in behavior have served as the primary indicant of the role of NE. Meaningful inferences from experiments of this type depend on a knowledge of the particular drug's central mechanisms of action, an understanding of variables controlling behavior, the sensitivity of the behavior being studied, and the interaction of the drug with changes in neurotransmitter systems that may be dynamically responsive to changes in behavior and/or the environment. This final section discusses studies in which the emphasis has been on directly monitoring neurochemical changes correlated with behavior and the environment.

Rats were implanted with push-pull cannulae located in the lateral ventricle and were then trained to respond under either 5- or 15-response fixed ratio schedules of food reinforcement (84). One hour prior to the experimental session, tritiated NE (NE-^{3}H) was injected into the ventricle. Perfusate, collected while the animals were responding during the session, showed higher levels of radioactivity from NE under the higher (fixed ratio 15-response) reinforcement schedule. This schedule also engendered higher rates of response. When food delivery was discontinued (extinction), NE levels were transiently elevated under both schedules but were increased much more under the fixed ratio five schedule. In this same study, systemic administration of 2.5 mg/kg *d*-amphetamine and 15 mg/kg mescaline decreased performances maintained under a 30-response fixed ratio schedule. Only *d*-amphetamine, however, increased [^{3}H]NE.

The push-pull technique has been used to study homeostatic mechanisms involved in hunger and satiety. Yaksh and Myers (110) collected perfusate from hypothalamic sites of food-deprived rhesus monkeys. When this perfusate was cross-perfused into a homologous site in a satiated recipient monkey trained to obtain food under a fixed ratio 10-response schedule, responding was initiated. Perfusate from satiated donor monkeys had no effect. Assays of the hypothalamic perfusate suggests that the biologically active substance resembles NE and that the relevant sites are sensitive to microinjections of NE. Those loci responsible for these effects are also known to be rich in NE synapses.

Taken together, these experiments using push-pull cannulae indicate that NE plays a fundamental role in basic behavioral processes such as food-directed behavior, as well as in environmentally controlled behavior conditioned by its previous consequences.

Interactions between Behavior, the Environment, and Norepinephrine

It is quite clear that CA play a functional role in the maintenance of behavior, that brain CA metabolism can be modified by behavior, and that the interactions between behavior and the effects of drugs on CA may account for the dependence of drug effects on ongoing behavior and the environmental conditions under which the drug is given (75). For example, the functional role of DA and NE in behavior was demonstrated in an experiment (76, 77) in which reserpine-induced reduction of conditioned avoidance behavior was antagonized by L-dopa. Subsequent research indicated that this effect was mediated by DA and NE, thereby suggesting that these neurotransmitters play a significant role in the maintenance of conditioned avoidance behavior (75, 78).

Other experiments have amplified the importance of CA by demonstrating that the effects of α-methylparatyrosine (AMPT), which reduce CA levels, depend on the particular schedule of reinforcement (74). Decreases in response after AMPT treatment were directly related to the parameter value under fixed ratio schedules, whereas under fixed interval schedules, decreases were comparable across the range of parameter values studied. Brain DA and NE concentrations were reduced approximately 75 and 45%, respectively, with the dosage of AMPT studied. Other studies from the same laboratory have confirmed that engaging in operant behavior directly affects NE metabolism (52).

Water-deprived rats were maintained under a variable-interval, 20-second schedule of water reinforcement. Different groups of rats were studied with different magnitudes of reinforcement (0.01 to 0.1 ml). Despite comparable rates of response across groups, the effects of amphetamine depended on the reinforcer magnitude; those maintained with the larger magnitude showed little effect of amphetamine, whereas the response of other groups was decreased in a dose-dependent manner (75). NE depletion in pons-medulla by AMPT pretreatment typically was inversely proportional to the magnitude of reinforcement, suggesting an interaction between the magnitude of reinforcement and NE metabolism.

Changes in NE can be neuroanatomically specific and can depend on the experimental history of the subject (8, 9, 25). Rats performing under a variable-interval 30-second schedule showed increased NE metabolism in the pons-medulla (25). Systematic manipulation of several relevant variables demonstrated that alterations in NE metabolism did not depend on the lever press response or on increased motor activity, but on stimulus variables correlated with water presentation. NE metabolism was altered by periodic water presentation in subjects with a history of exposure to the water reinforcement schedule, even though at the time of testing subjects were nondeprived and did not drink.

The role of stimulus variables and conditioning history in altering neurochemical responses has also been emphasized in a study showing that previously neutral stimuli paired with footshock increases levels of 3-methoxy-4-hydroxyphenylglycol (MHPG), a major metabolite of NE in the brain (8, 9). The finding demonstrating that previously neutral stimuli can be effective in altering NE levels after appropriate conditioning experience is similar to the results just described with scheduled presentations of water. The importance of prior experience was also discussed earlier in this chapter when the effects of amphetamine on punished response were shown to depend on behavioral history.

SUMMARY

The studies summarized in this chapter have reviewed the behavioral pharmacology of drugs affecting NE transmission. Despite the recent increase in availability of selectively acting drugs and specific operant schedules, there are still areas in which very little is known about the behavioral effects of drugs that affect NE transmission. For instance, very little behavioral research has been done with the vast array of antidepressant compounds.

The task of differentiating behavioral effects of drugs on the basis of different neurochemical effects is also somewhat problematical. Drugs with different effects on NE often, but not always, produce different effects on behavior. It is presently unclear whether the behaviors studied are insensitive to these differences or whether the mixed and occasionally unknown actions of the drugs preclude a differentiation at the behavioral level. It is hardly surprising that not all behaviors are able to distinguish compounds which differentially affect NE transmission. When this is the case, perhaps, NE is not critical in that particular type of behavior. Neuropharmacological mechanisms of drug action are exceedingly complex and rarely unitary in their effects on neurotransmitter systems. Progress in understanding NE depends as much on a familiarity with and application of different types of behavior as it does on knowledge of the neuropharmacology of drugs affecting NE. The identification and isolation of behavioral variables is still evolving.

Although this chapter has dealt largely with basic research findings, there is little question that the results of these studies have direct clinical implications. The studies summarized in the last section of this chapter provide increasingly clearer evidence that the interrelationship between behavior, neurochemistry, and the environment is dynamic (see Introduction). The substrate upon which drugs act to change behavior is not static but subject to multiple influences. The possibility of a fluid or pliable population of neurons or receptor sites which modulate NE activity, as suggested by these findings, prompts further questions about the most effective means of altering NE activity for therapeutic benefits. For example, alterations in NE metabolism or synthesis, as a function of changes in behavior brought about by the environment, suggest alternative nonpharmacological methods of modifying NE activity. Drugs, behavior, the environment, and neurochemistry interact in ways not formerly considered and, as yet, still poorly understood. A search for answers which acknowledge the combined influences of pharmacological, neurochemical, behavioral, and environmental factors is likely to be profitable for the understanding and treatment of disorders in which norepinephrine plays a role.

References

1. Ando K: Profile of drug effects on temporally spaced responding in rats. *Pharmacol Biochem Behav* 3:833–841, 1975.
2. Azzaro AJ, Ziance RJ, Rutledge CO: The importance of neuronal uptake for amphetamine-induced release of ^{3}H-norepinephrine from isolated brain tissue. *J Pharmacol Exp Ther* 89:110–118, 1974.
3. Barrett JE: Behavioral history as a determinant of the effects of *d*-amphetamine on punished behavior. *Science* 198:67–69, 1977.

3a. Barrett JE: Behavioral pharmacology: Recent developments and new trends. *Trends Pharmacol Sci* 1:215–218, 1980.

4. Barrett JE: Antagonism of clonidine-induced suppression of operant behavior by yohimbine but not prazosin, manuscript in preparation, 1983.
5. Barrett JE, Katz JL: Drug effects on behaviors maintained by different events. In Thompson T, Dews PB, McKim WA: *Advances in Behavioral Pharmacology*. New York, Academic Press, 1981, vol 3, pp 119–168.
6. Benfey BG: The evidence against interconversion of α and β-adrenoceptors. *Trends Pharmacol Sci* 1:193–194, 1980.
7. Bylund DB, Martinez, JR: α_2-Adrenergic receptors appear in rat salivary glands after reserpine treatment. *Nature* 285:229–230, 1980.
8. Cassens G, Roffman M, Kuruc A, Orsulak PJ, Schildkraut JJ: Alterations in brain norepinephrine metabolism induced by environmental stimuli previously paired with inescapable shock. *Science* 209:1138–1140, 1980.
9. Cassens G, Kuruc A, Roffman M, Orsulak PJ, Schildkraut JJ: Alterations in brain norepinephrine metabolism and behavior induced by environmental stimuli previously paired with inescapable shock. *Behav Brain Res* 2:387–407, 1981.
10. Chiang CF, Rutledge CO: Evidence for sodium dependent uptake of amphetamine into rat brain synaptosomes. *Fed Proc* 40:244, 1981.
11. Colpaert FC, Niemegeers CJE, Janssen PAJ: Discriminative stimulus properties of cocaine: Neuropharmacological characteristics as derived from stimulus generalization experiments. *Pharmacol Biochem Behav* 10:535–546, 1979.
12. Conway, EL, Louis WJ, Jarrott B: Acute and chronic administration of α-methyldopa: Regional levels of endogenous and α-methylated catecholamines in rat brain. *Eur J Pharmacol* 52:271–280, 1978.
13. Cook L, Catania AC: Effects of drugs on avoidance and escape behavior. *Fed Proc* 23:818–835, 1964.
14. Crews FT, Smith CB: Presynaptic alpha-receptor subsensitivity after long-term antidepressant treatment. *Science* 202:322–324, 1978.
15. Crews FT, Smith CB: Potentiation of responses to adrenergic nerve stimulation in isolated rat atria during chronic tricyclic antiepressant administration. *J Pharmacol Exp Ther* 215:143–149, 1980.
16. Cubeddu LX, Barnes E, Weiner N: Release of norepinephrine and dopamine-beta-hydroxylase by nerve stimulation. IV. An evaluation of a role for cyclic adenosine monophosphate. *J Pharmacol Exp Ther* 193:105–177, 1975.
17. Daly JW, Bruns RF, Snyder SH: Adenosine receptors in the central nervous system: Relationship to the central actions of methylxanthines. *Life Sci* 28:2083–2097, 1981.
18. Davis JN, Maury W: Clonidine and related imidazolines are postsynaptic *alpha* adrenergic antagonists in dispersed rat parotid cells. *J Pharmacol Exp Ther* 207:425–430, 1978.
19. Davis TRA, Kensler CJ, Dews PB: Comparison of behavioral effects of nicotine, *d*-amphetamine, caffeine and dimethylheptyl tetrahydrocannibinol in squirrel monkeys. *Psychopharmacologia* 32:51–65, 1973.

20. Dews PB: Studies on behavior. IV. Stimulant actions of methamphetamine. *J Pharmacol Exp Ther* 122:137–147, 1958.
21. Dews PB: A behavioral output enhancing effect of imipramine in pigeons. *Int J Neuropharmacol* 1:265–272, 1962.
22. Dews PB: Interspecies differences in drug effects: Behavioral. In Usdin E, Forrest IS: *Psychotherapeutic Drugs*. New York, Marcel Dekker, 1976, pp 175–224.
23. Dews PB, Wenger GR: Rate-dependency of the behavioral effects of amphetamine. In Thompson T, Dews PB: *Advances in Behavioral Pharmacology*. New York, Academic Press, 1977, vol 1, pp 167–227.
24. Dixon WR, Mosiman WF, Weiner N: The role of presynaptic feedback mechanisms in regulation of norepinephrine release by nerve stimulation. *J Pharmacol Exp Ther* 209:196–204, 1979.
25. Emmett-Oglesby MW, Lewy AJ, Albert LH, Seiden LS: Role of lever responding and water presentation in altering rat brain catecholamine metabolism. *J Pharmacol Exp Ther* 204:406–415, 1978.
26. Fuller RW, Snoddy HD, Molloy BB: Blockade of amine depletion of nisoxetine in comparison to other uptake inhibitors. *Psychopharmacol Commun* 1:455–464, 1975.
27. Ganousis LG, Tessel RE: Interactions of catecholamine uptake inhibitors and receptor blockers with amphetamine-induced changes in locomotion, stereotypy and milk-maintained operant responding in mice. *Fed Proc* 40:276, 1981.
28. Geller I, Hartmann RJ, Blum K: The effects of low-dose combinations of *d*-amphetamine and cocaine on experimentally induced conflict in the rat. *Curr Ther Res* 14:220–224, 1972.
29. Geller I, Seifter J: The effects of meprobamate, barbiturates, *d*-amphetamine and promazine on experimentally-induced conflict in the rat. *Psychopharmacologia* 1:482–492, 1960.
30. Gropetti A, DiGiulio AM: Cocaine and its effects on biogenic amines. In Mule SJ: *Cocaine: Chemical, Biological, Clinical, Social and Treatment Aspects*. Cleveland, CRC Press, 1976, pp 93–102.
31. Harms HH, Wardeh G, Mulder AH: Adenosine modulates depolarization-induced release of ^{3}H-noradrenaline from slices of rat brain cortex. *Eur J Pharmacol* 49:305–308, 1978.
32. Harris RA, Snell D, Loh HH: Effects of stimulants, anorectics, and related drugs on schedule-controlled behavior. *Psychopharmacology* 56:49–55, 1978.
33. Hawkins M, Monti JM: Effects of pretreatment with 6-hydroxydopamine or noradrenergic receptor blockers on the clonidine-induced disruption of conditioned avoidance responding. *Eur J Pharmacol* 58:53–58, 1979.
34. Herling S, Woods JH: Discriminative stimulus effects of narcotics: Evidence for multiple receptor-mediated actions. *Life Sci* 28:1571–1584, 1981.
35. Ho BT, Silverman, PB: Stimulants as discriminative stimuli. In Colpaert FC, Rosecrans JA: *Stimulus Properties of Drugs: Ten Years of Progress*. New York, Elsevier/North-Holland, 1978, pp 53–68.
36. Howard JL, Jones CN, McBennett ST: Discriminative stimulus properties of antidepressants. In Colpaert FC, Rosecrans JA: *Stimulus Properties of Drugs: Ten Years of Progress*. New York, Elsevier/North-Holland, 1978, pp 157–166.
37. Huang M, Ho AKS, Daly JW: Accumulation of adenosine cyclic 3′, 5′-monophosphate in rat cerebral cortical slices. Stimulatory effects of alpha and beta adrenergic agents after treatment with 6-hydroxydopamine, 2, 3, 5-trihydroxyphenethylamine and dihydroxytryptamines. *Mol Pharmacol* 9:711–717, 1973.
38. Iggo A, Vogt M: Preganglionic sympathetic activity in normal and reserpine-treated cats. *J Physiol* 150:114–133, 1960.
39. Iversen LL: The inhibition of norepinephrine uptake by drugs. *Adv Drug Res* 2:5–23, 1965.
40. Iversen LL: Uptake processes for biogenic amines. *Handbook Pharmacol* 3:381–430, 1975.
41. Iversen LL: Catecholamine-sensitive adenylate cyclase in nervous tissues. *J Neurochem* 29:5–12, 1977.
42. Khan MT, Malik KU: Inhibitory effect of adenosine and adenine nucleotides on potassium-evoked efflux of [^{3}H]-noradrenaline from the rat isolated heart: Lack of relationship to prostaglandins. *Br J Pharmacol* 68:551–561, 1980.
43. Kobinger W, Pichler L: Relationship between central sympathoinhibitory and peripheral pre- and postsynaptic α-adrenoceptors as evaluated by different clonidine-like substances in rats. *Naunyn Schmiedebergs Arch Pharmacol* 315:21–27, 1980.
44. Koe BK: Molecular geometry of inhibitors of the uptake of catecholamines and serotonin in synaptosomal preparations of rat brain. *J Pharmacol Exp Ther* 199:649–661, 1976.
45. Kornetsky C: A comparison of the effects of desimipramine and imipramine on two schedules of reinforcement. *Int J Neuropharmacol* 4:13–16, 1965.
46. Kostrzewa RM, Jacobowitz DM: Pharmacological actions of 6-hydroxydopamine. *Pharmacol Rev* 26:200–288, 1974.
47. Langer SZ: Presynaptic adrenoreceptors and regulation of release. In Paton DM: *The Release of Catecholamines from Adrenergic Neurons*. New York, Pergamon Press, 1979, pp 59–88.
48. Langer SZ: Presynaptic regulation of the release of catecholamines. *Pharmacol Rev* 32:337–358, 1981.
49. Langer SZ, Briley MS, Raisman R: Regulation of neurotransmission through presynaptic receptors and other mechanisms: Possible clinical relevance and therapeutic potential. In Pepeu G, Kuhar MJ, Enna SJ: *Receptors for Neurotransmitters and Peptide Hormones*. New York, Raven Press, 1980, pp 203–212.
50. Laverty R, Taylor KM: Behavioral and biochemical effects of 2-(2, 6-dichlorophenylamine-2-imidazoline hydrochloride (St 155) on the central nervous system. *Br J Pharmacol* 35:253–259, 1969.

51. Lazareno S: *d*-Amphetamine and punished responding: The role of catecholamines and anorexia. *Psychopharmacology* 66:133–142, 1979.
52. Lewy AJ, Seiden LS: Operant behavior changes in norepinephrine metabolism in rat brain. *Science* 175:454, 1972.
53. Maggi A, U'Prichard DC, Enna SJ: β-Adrenergic regulation of α_2-adrenergic receptors in the central nervous system. *Science* 207:645–647, 1980.
54. McCleary PE, Leander JD: Clonidine analgesia and suppression of operant responding: Dissociation of mechanism. *Eur J Pharmacol* 69:63–69, 1981.
55. McGuire PS, Seiden LS: The effects of tricyclic antidepressants on performance under a differential-reinforcement-of-low-rates schedule in rats. *J Pharmacol Exp Ther* 214:635–641, 1980.
56. McKearney JW: The relative effects of *d*-amphetamine, imipramine and harmaline on tetrabenazine suppression of schedule-controlled behavior in the rat. *J Pharmacol Exp Ther* 159:429–440, 1968.
56a. McKearney JW, Barrett JE: Schedule-controlled behavior and the effects of drugs. In Blackman DE, Sanger DJ: *Contemporary Research in Behavioral Pharmacology*. New York, Plenum, 1978, pp 1–68.
57. McKim WA: The effect of caffeine, theophylline and amphetamine on operant responding of the mouse. *Psychopharmacology* 68:135–138, 1980.
58. McMillan DE: The effects of sympathomimetic amines on schedule-controlled behavior in the pigeon. *J Pharmacol Exp Ther* 160:315–325, 1968.
59. McMillan DE: Drugs and punished responding. IV. Effects of propranolol, ethchlorvynol and chloral hydrate. *Res Commun Chem Pathol Pharmacol* 6:167–174, 1973.
60. McMillan DE, Leander JD: Effects of drugs on schedule-controlled behavior. In Glick SD, Goldfarb J: *Behavioral Pharmacology*. St. Louis, Mosby, 1976, pp 85–139.
61. Morrison CF: The effects of nicotine on punished behavior. *Psychopharmacologia* 14:221–232, 1969.
62. Morse WH: Effect of amobarbital and chlorpromazine on punished behavior in the pigeon. *Psychopharmacologia* 6:286–294, 1964.
63. Morse WH, Kelleher RT: Schedules as fundamental determinants of behavior. In Schoenfeld WN: *The Theory of Reinforcement Schedules*. New York, Appleton-Century-Crofts, 1970, pp 139–185.
64. Morse WH, Kelleher RT: Determinants of reinforcement and punishment. In Honig WK, Staddon JER: *Handbook of Operant Behavior*. Englewood Cliffs, NJ, Prentice-Hall, 1977, pp 174–200.
65. Muller MJ, Paton DM: Presynaptic inhibitory actions of 2-substituted adenosine derivatives on neurotransmission in rat vas deferens: Effects of inhibitors of adenosine uptake and deamination. *Naunyn Schmiedebergs Arch Pharmacol* 306:23–28, 1979.
66. Palmer GC, Wagner HR, Putnam RW: Neuronal localization of the enhanced adenylate cyclase responsiveness to catecholamines in the rat cerebral cortex following reserpine injections. *Neuropharmacology* 15:695–702, 1976.
66a. Paton DM: *The Release of Catecholamines from Adrenergic Neurons*. New York, Pergamon Press, 1979.
67. Raiteri M, Levi G: Release mechanism for catecholamines and serotonin in synaptosomes. *Rev Neurosci* 3:77–130, 1978.
68. Rosenberg RC, Lovenberg W: Dopamine β-hydroxylase. In Youdin MBH, Sharman DF, Lovenberg W, Lagnado JR: *Essays in Neurochemistry and Pharmacology*. New York, John Wiley & Sons, 1980, vol 3, pp 163–210.
69. Ruffolo, RR Jr., Rosing EL, Waddel JE: Receptor interactions of imidazolines. I. Affinity and efficacy for *alpha* adrenergic receptors in rat aorta. *J Pharmacol Exp Ther* 209:429–436, 1979a.
70. Ruffolo, RR Jr., Dillard RD, Yaden EL, Waddell JE: Receptor interactions of imidazolines. II. Affinities and efficacies of hydroxy-substituted tolazoline derivatives in rat aorta. *J Pharmacol Exp Ther* 211:74–79, 1979b.
71. Ruffolo, RR Jr., Dillard RD, Waddell JE, Yaden EL: Receptor interactions of imidazolines. III. Structure-activity relationships governing *alpha* adrenergic receptor occupation and receptor activation for mono- and dimethoxy-substituted tolazoline derivatives in rat aorta. *J Pharmacol Exp Ther* 211:733–738, 1979c.
72. Rutledge CO: The mechanisms by which amphetamine inhibits oxidative deamination of norepinephrine in brain. *J Pharmacol Exp Ther* 171:188–95, 1970.
73. Rutledge CO, Weiner N: The effect of reserpine upon the synthesis of norepinephrine in the isolated rabbit heart. *J Pharmacol Exp Ther* 157:290–302, 1967.
74. Schoenfeld R, Seiden LS: Effect of α-methyltyrosine on operant behavior and brain catecholamine levels. *J Pharmacol Exp Ther* 167:319–327, 1969.
75. Seiden LS, Campbell AB: Catecholamines, drugs and behavior: Mutual interactions. In Usdin E: *Neuropsychopharmacology of Monoamines and Their Regulatory Enzymes*. New York, Raven Press, 1974, pp 325–338.
76. Seiden LS, Carlsson A: Temporary and partial antagonism by L-DOPA of reserpine-induced suppression of a conditioned avoidance response. *Psychopharmacologia* 4:418–423, 1963.
77. Seiden LS, Carlsson A: Brain and heart catecholamine levels after L-DOPA administration in reserpine-treated mice: Correlations with a conditioned avoidance response. *Psychopharmacologia* 5:178–181, 1964.
78. Seiden LS, MacPhail RC, Oglesby MW: Catecholamines and drug-behavior interactions. *Fed Proc* 34:1823–1831, 1975.
79. Sepinwall J, Grodsky FS, Sullivan JW, Cook L: Effects of propranolol and chlordiazepoxide on conflict behavior in rats. *Psychopharmacologia* 31:375–382, 1973.
80. Sethy VH, Winter JC: Effects of yohimbine and mescaline on punished behavior in the rat.

Psychopharmacologia 23:160–166, 1972.

81. Sirochman V, Carney JM: Behavioral pharmacology of adenosine analogs. *Fed Proc* 40:294, 1981.
82. Slotkin TA, Salvaggio M. Seidler FJ, Whitmore WL: Structural characteristics for inhibition of [^{3}H]norepinephrine uptake into rat brain synaptic vesicles by beta-carboline, indolealkylamine, phenethylamine and *N*-alkylamine derivatives. *Mol Pharmacol* 15:607–619, 1979.
83. Smith CB: Effects of *d*-amphetamine upon operant behavior of pigeons: Enhancement by reserpine. *J Pharmacol Exp Ther* 146:167–174, 1964.
84. Sparber SB: Neurochemical changes associated with schedule-controlled behavior. *Fed Proc* 34:1802–1812, 1975.
85. Sporn JR, Harden RK, Wolfe BB, Molinoff PB: Supersensitivity in rat cerebral cortex: Pre- and postsynaptic effects of 6-hydroxydopamine at noradrenergic synapses. *Mol Pharmacol* 13:1170–1180, 1977.
86. Starke K: Presynaptic regulation of release in the central nervous system. In Paton DM: *The Release of Catecholamines from Adrenergic Neurons.* New York, Pergamon Press, 1979, pp 143–184.
87. Starke K, Montel H, Gayk W, Merker R: Comparison of effects of clonidine on pre- and postsynaptic adrenoceptors in rabbit pulmonary artery. *Naunyn Schmiedebergs Arch Pharmacol* 285:133–150, 1974.
88. Stone TW: Physiological roles for adenosine and adenosine 5′-triphosphate in the nervous system. *Neuroscience* 6:523–555, 1981.
89. Sugrue MF, Mireyless SE: Effects of mazindol in rat brain synaptosomal monoamine uptake. *Biochem Pharmacol* 27:1843–1847, 1978.
90. Tainter ML, Chang DK: The antagonism of the pressor action of tyramine by cocaine. *J Pharmacol Exp Ther* 30:193–207, 1927.
91. Tessel RE, Rutledge CO: Specificity of release of biogenic amines from isolated rat brain tissue as a function of the *meta*-substituent of *N*-ethylamphetamine derivatives. *J Pharmacol Exp Ther* 197:253–262, 1976.
92. Tessel RE, Smith CB, Russ DN, Hough LB: The effects of cocaine HCl and a quaternary derivative of cocaine methiodide, on isolated guinea pig and rat atria. *J Pharmacol Exp Ther* 205:569–576, 1978.
93. Tessel RE, Kennedy LE, Burgess SK, Borchardt RT: Epinephrine in rat hypothalamus: Antagonism by desipramine of 6-hydroxydopamine-induced depletion. *Brain Res* 153:615–617, 1978.
94. Thoenen, H: Biochemical alterations induced by 6-hydroxydopamine in peripheral adrenergic neurons. In Malmfors T, Thoenen H: *6-Hydroxydopamine and Catecholamine Neurons.* Amsterdam, North-Holland, 1971, pp 75–86.
95. Timmermans PBMWM, Kwa HY, van Zwieten PA: Possible subdivision of postsynaptic α-adrenoceptors mediating pressor responses in the pithed rat. *Naunyn Schmiedebergs Arch Pharmacol* 310:189–193, 1979.
96. Trendelenburg U: Classification of sympathomimetic amines. *Handbook Exp Pharmacol* 33:336–362, 1972.
97. Trendelenburg U: Release induced by phenethylamines. In Paton DM: *The Release of Catecholamines from Adrenergic Neurons.* New York, Pergamon Press, 1979, pp 333–354.
98. Tyler TD, Tessel RE: Amphetamine's locomotor stimulant and norepinephrine-releasing effects: Evidence for selective antagonism by nisoxetine. *Psychopharmacology* 64:291–296, 1979.
99. Tyler TD, Tessel RE: Inhibitors of norepinephrine uptake as biochemically and behaviorally selective antagonists of the locomotor stimulation induced by indirectly-acting sympathomimetic amines in mice. *Psychopharmacology* 69:27–34, 1980.
100. U'Prichard DC, Greenberg DA, Snyder SH: Binding characteristics of a radiolabeled agonist and antagonist at central nervous system *alpha* noradrenergic receptors. *Mol Pharmacol* 13:454–473, 1977.
101. U'Prichard, DC, Yamamura HI, Reisine TD: Characterization and differential *in vivo* regulation of brain adrenergic receptor subtypes. In Pepeu G, Kuhar MJ, Enna SJ: *Receptors for Neurotransmitters and Peptide Hormones.* New York, Raven Press, 1980, pp 213–221.
102. Vaillant GE: A comparison of chlorpromazine and imipramine on behavior of the pigeon. *J Pharmacol Exp Ther* 146:377–384, 1964.
103. Vetulani J, Sulser F: Action of various antidepressant treatments reduces reactivity of noradrenergic cyclic AMP-generating system in limbic forbrain. *Nature* 257:495–496, 1975.
104. von Euler US: Synthesis, uptake and storage of catecholamines in adrenergic nerves; the effects of drugs. *Handbook Exp Pharmacol* 33:186–230, 1972.
105. Von Voigtlander PF, Triezenberg HJ, Losey EG: Interactions between clonidine and antidepressant drugs: A method for identifying antidepressant-like agents. *Neuropharmacology* 17:375–381, 1978.
106. Weiner N: The role of cyclic nucleotides in the regulation of neurotransmitter release from adrenergic neurons by neuromodulators. In Youdim MBH, Sharman DF, Lovenberg W, Lagnado JR: *Essays in Neurochemistry and Pharmacology.* New York, John Wiley & Sons, 1980, vol 3, pp 69–124.
107. Weiner N: Drugs that inhibit adrenergic nerves and block adrenergic receptors. In Gilman, AG, Goodman LS, Gilman A: *The Pharmacological Basis of Therapeutics.* New York, MacMillan Publishing Company, 1980, pp 176–210.
108. Wise RA, Yokel RA, Dewit H: Both positive reinforcement and conditioned aversion from amphetamine and from apomorphine in rats. *Science* 191:1273–1275, 1976.
109. Wong DT, Horng JS, Bymaster FP: *d*-α-*N*-methyl-3-(*o*-methoxy-phenoxy) 3-phenylpropylamine hydrochloride, Lilly 94939, a potent inhibitor for uptake of norepinephrine into rat brain synaptosome. *Life Sci* 17:755–760, 1975.
110. Yaksh TL, Myers RD: Neurohumoral sub-

stances released from hypothalamus of the monkey during hunger and satiety. *Am J Physiol* 222:503–515, 1972.

111. Ziance RJ, Azzaro AJ, Rutledge CO: Characteristics of amphetamine-induced release of norepinephrine from rat cerebral cortex *in vitro*. *J Pharmacol Exp Ther* 182:284–294, 1972.

112. Zuccarelli RR, Barrett JE: A comparison of the effects of *d*-amphetamine, cocaine, imipramine and pentobarbital on local and overall rates of responding maintained under a four-component multiple fixed-interval schedule. *Pharmacol Biochem Behav* 12:899–907, 1980.

CHAPTER ELEVEN

Are Catecholamines Involved in the Slow Inhibitory Postsynaptic Potential in Sympathetic Ganglia?

PATRICIA SHINNICK-GALLAGHER, Ph.D., ALISON E. COLE, Ph.D., and JOEL P. GALLAGHER, Ph.D.*

INTRODUCTION

In 1939, Marrazzi (45) demonstrated that the catecholamine epinephrine inhibited transmission in sympathetic ganglia. Since that time investigators have been searching for the functional role of catecholamines in ganglionic transmission.

When preganglionic fibers are stimulated, a fast excitatory postsynaptic potential (F-EPSP) followed by a slow inhibitory postsynaptic potential (S-IPSP) and a slow excitatory postsynaptic potential (S-EPSP) can be recorded intracellularly in principal neurons of sympathetic ganglia.† It is generally believed that the F-EPSP and S-EPSP are generated through activation of nicotinic and muscarinic cholinoceptors, respectively, but a subject of controversy has been the proposed role of catecholamines as the neurotransmitter mediating the S-IPSP.

The S-IPSP was first recorded in 1943 and 1944 in curarized cat superior cervical ganglia by Eccles (18, 19) and in turtle superior cervical ganglia in 1950 by Laporte and Lorente de No (34). Subsequently, Eccles (20) found that the anticholinesterase drug physostigmine potentiated the S-IPSP and suggested that the neurotransmitter mediating the S-IPSP was acetylcholine.

In 1961, Eccles and Libet (21) reported that dibenamine, an α-adrenoceptor antagonist, reduced the amplitude of the P potential (see below) and proposed that the S-IPSP was mediated by a disynaptic pathway involving the release of a catecholamine. This hypothesis suggested that an interneuron, in response to muscarinic excitation, released a catecholamine which initiated a S-IPSP in the principal ganglion cell. In 1970, Libet (35) proposed dopamine rather than norepinephrine as the neurotransmitter involved in the S-IPSP in rabbit superior cervical ganglia. An important anatomical corollary to the catecholamine hypothesis of the S-IPSP proposed that small cells containing chromaffin granules or small intensely fluorescent (SIF) cells found in sympathetic ganglia are the adrenergic interneurons that mediate the S-IPSP (63).

S-IPSPs have been recorded in cat (19), rat (17), and rabbit (21) superior cervical ganglia and in frog paravertebral ganglia (37, 50). A S-IPSP does not occur in the guinea pig (14, 35), except in the presence of neostigmine (16). S-IPSPs are most easily observed after curarization and repetitive preganglionic stimulation; however, they can occasionally be recorded in the absence of blocking agents with a single stimulus pulse (42, 58).

---

* The authors thank Drs. Mark Mayer and Betty Williams for their careful review of the manuscript.
† The extracellular correlates (*c.f.* Fig. 11.4) of the above potentials are the L wave, the P wave, and the LN wave, respectively. For the purposes of this discussion we will use the intracellular terminology.

The hypothesis that a catecholamine mediates the S-IPSP is based on evidence from electrophysiological and pharmacological studies performed primarily in rabbit superior cervical ganglia. We will examine this data in detail. In this review we will discuss only experiments in ganglia treated with a nondepolarizing nicotinic cholinoceptor antagonist. Treatment with nicotine induces a depolarizing desensitizing type of antagonism and may cause a retention of intracellular Na^+ (33) which may be responsible for differences in data between nicotinized and "curarized" preparations (33, 39).

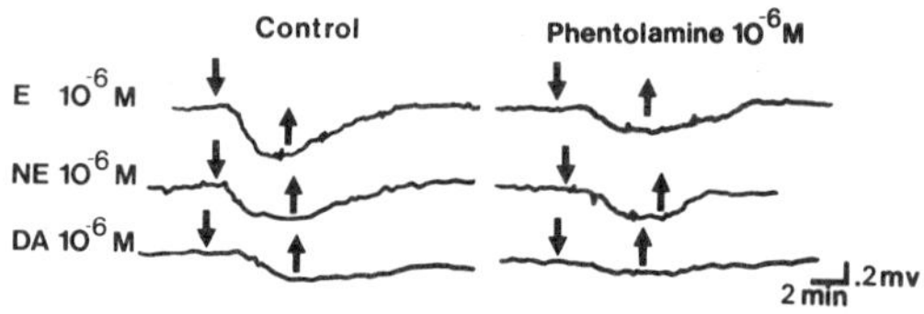

Figure 11.1. Membrane hyperpolarizations elicited by 10^{-6} M epinephrine (*E*), norepinephrine (*NE*), and dopamine (*DA*) in the rabbit superior cervical ganglion (*left*) and the effect of 10^{-6} M phentolamine on the hyperpolarizations (*right*). All traces are from sucrose-gap recordings; *downward* and *upward arrows* indicate drug application and withdrawal, respectively.

THE POSTGANGLIONIC CATECHOLAMINE RECEPTOR

It seems likely that if catecholamines (CAs) are involved in the S-IPSP, they must hyperpolarize the ganglion membrane. It was shown quite early that norepinephrine (NE) and epinephrine (E) hyperpolarized mammalian ganglia (10, 38, 44). The hyperpolarization was blocked by α-, but not β-adrenoceptor antagonists (10).

Libet (35) found that dopamine (DA) was "equally effective" as NE and E in hyperpolarizing rabbit SCG, and Dun *et al.* (13) further showed that haloperidol (0.1 to 1 μM) antagonized a dopamine-induced hyperpolarization. Brown and Caulfield (5) stated that the previous data did not support the involvement of a dopamine receptor because haloperidol is a strong α-adrenoceptor antagonist in the rabbit (28). We found that in rabbit SCG the concentration-response relationship for catecholamines had the following order of potency:E $\geq$ NE$\gg$DA; DA was about 10 times less potent than NE. We have also studied the effects of antagonists on the catecholamine hyperpolarizations (6, 8) and found that the antagonists phentolamine and yohimbine both depressed the NE-evoked hyperpolarization in a concentration-dependent manner. The IC_{50} for phentolamine (Fig. 11.1) and yohimbine were 6×10^{-6} M and 1.5×10^{-6} M, respectively. In contrast, the α-antagonist prazosin, the DA antagonists sulpiride and haloperidol, and the β-antagonist propranolol, all at 10^{-5} M, did not depress the NE-evoked (5×10^{-5} M) hyperpolarization. Similar results were previously reported in the rat (5). These results suggest that DA is not the best candidate for a neurotransmitter mediating the S-IPSP, since the receptors mediating the catecholamine responses are α-adrenoceptors, not dopamine receptors (5, 6, 8). According to Libet's (35) hypothesis, DA, the least potent catecholamine, would be released and activate α-adrenoceptors; however, this does not appear to be an efficient system. It should be noted that Libet and Kobayashi (39) found epinephrine was the most potent CA in amphibian and suggested it as the actual transmitter for the S-IPSP in frog ganglia.

THE EFFECT OF CATECHOLAMINE ANTAGONISTS

The α-adrenoceptor antagonist dibenamine (1 to 3 μg/ml) was shown to eliminate the S-IPSP response and depress the S-EPSP with little effect on the F-EPSP (21). Subsequently, Libet and Tosaka (43) stated "that phenoxybenzamine was more selective and much faster in blocking the S-IPSP than was dibenamine, while dihydroergotamine was much less effective and phentolamine had virtually no selective effects"; no data, however, were published. These same authors showed that exogenously applied dopamine enhanced the S-EPSP. They explained that the effects of the adrenergic antagonists could be due to inhibition of a facilitatory effect of synaptically released dopamine on the S-EPSP. That dibenamine and phenoxybenzamine were able to depress the S-IPSP may be due to their block of muscarinic receptors (2, 3, 4, 24).

In experiments employing extracellular recording methods, Libet and Kobayashi (39) reported that in curarized frog ganglia, both dihydroergotamine and phentolamine, at 4×10^{-5} M, depressed the amplitude of the S-IPSP by 25% and increased its duration. At a higher concentration (2×10^{-4}M) phentolamine nonspecifically depressed synaptic transmission. We have recently studied the effect of pharmacological antagonists on the S-IPSP in rabbit superior cervical ganglia (7, 8, 8a, 26) and observed that the α-adrenoceptor antagonists, phentolamine (10^{-6} M) and yohimbine (10^{-6} M), had no effect on the S-IPSP. After 30 minutes of exposure at higher concentrations (10^{-5} M to 10^{-3} M), these drugs depressed all ganglionic responses nonselectively. Similar results were observed in the rat and guinea pig by Dun and Karczmar (16). Weight and Smith (61) have also shown that dihydroergotamine (10^{-5} M) does not antagonize the S-IPSP in frog sympathetic ganglia.

The dopamine receptor antagonists sulpiride (10^{-8} –10^{-3} M) and haloperidol (10^{-8} to 10^{-3} M) had no effect on the S-IPSP (7, 8) and Nakamura (48) reported that the DA antagonists, chlorpromazine (10^{-5} M) and pimozide (10^{-5} M), did not affect the S-IPSP in rabbits. Thus, these data do not support the involvement of an α-adrenoceptor or a dopamine receptor in the S-IPSP.

A corollary to the interneuron hypothesis is that catecholamine antagonists should inhibit the hyperpolarizing response to exogenously applied acetylcholine (ACh). In curarized frog ganglia, dihydroergotamine (4×10^{-5} M) produced a 30% depression of a methacholine (MCh)-induced hyperpolarization, and a high concentration of phentolamine (2×10^{-4} M) completely abolished the MCh response recorded extracellularly. On the other hand, phentolamine produced only a small depression of the ACh-induced hyperpolarization; these authors did not address the apparent discrepancy in their results (38).

However, Dun and Karczmar (15) demonstrated that haloperidol (0.1 μM), a dopamine receptor antagonist, blocked an ACh-induced hyperpolarization. They concluded that postsynaptic dopaminergic sites mediated the ACh hyperpolarization and assumed a disynaptic mechanism was involved. We have studied the effect of sulpiride, a dopamine receptor antagonist, on the S-IPSP and ACh-induced hyperpolarization in the same cell (Fig. 11.2). In three of four cells sulpiride antagonized the large ACh hyperpolarization (5.6 ± 1.4 mV; mean ± SE) but did not affect the S-IPSP (1.4 ± 4.7 mV; mean ± SE). In seven additional cells sulpiride (10^{-8} to 10^{-3} m) had no effect on S-IPSPs.

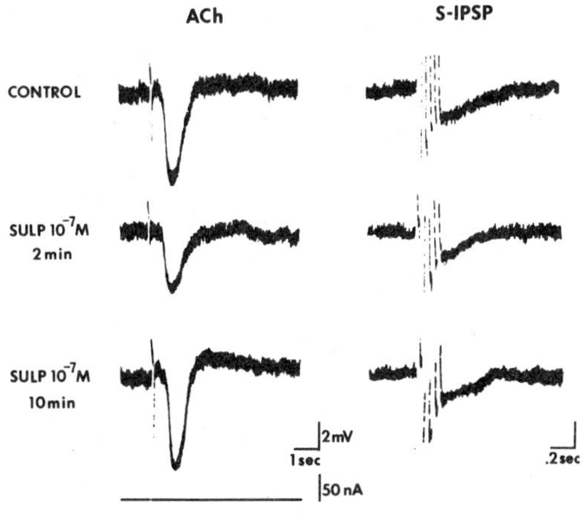

Figure 11.2. Effect of sulpiride (10^{-7} M) on the nerve-evoked S-IPSP and on the hyperpolarization induced by iontophoretically applied ACh. Control recordings are shown followed by the recording at 2 and 10 minutes after drug application. Both the ACh hyperpolarization and the S-IPSP were recorded in the same cell.

The fact that these agents depress large ACh responses and not the smaller S-IPSP responses suggests that they may act as noncompetitive blocking agents of ACh. Other agents (certain local anesthetics (56) and alcohols (25)) shorten the lifetime of the nicotinic cholinoceptor channel opened by the ACh without markedly affecting peak conductance. Under these conditions, miniature endplate potential amplitude is less affected, although the ACh depolarization is significantly depressed (11). It is thus possible that these agents, haloperidol and sulpiride, may be acting on channels. Most importantly, however, phentolamine and sulpiride (as well as haloperidol (48)) did not antagonize the endogenous response, the S-IPSP. Phentolamine (10^{-6} M) also did not antagonize the ACh-induced hyperpolarization recorded intracellularly (Fig. 11.3); only atropine is unique in blocking the S-IPSP at 10^{-7} M (7). These findings do not agree with the catecholamine hypothesis for the S-IPSP.

THE EFFECT OF BLOCKING TRANSMITTER RELEASE

According to the disynaptic hypothesis, an agent which inhibits transmitter release should block the hyperpolarizing response to exogenously applied acetylcholine or other muscarinic agonists.

Two methods that are available for inhibiting transmitter release are: (1) a low Ca^{2+}/high Mg^{2+} medium which impairs release of transmitter from virtually all chemical synapses (55); and (2) treatment with tetrodotoxin, an inhibitor of voltage-sensitive Na^{+} currents, which prevents action potential generation in the nerve terminal and subsequent release of transmitter.

Libet (35) showed with sucrose gap extracellular recording that in low Ca^{2+}/high Mg^{2+} medium there was no hyperpolarizing response to bethanechol, whereas one was recorded in other ganglia in the presence of normal Ca^{2+} concentrations. However, a bethanechol hyperpolarization was observed in the presence of tetrodotoxin (36), suggesting that bethanechol could release dopamine by acting directly on nerve terminals. Conversely, treatment with tetrodotoxin, as well as with low Ca^{2+}/high Mg^{2+}, blocked the ACh-induced hyperpolarization recorded with intracellular techniques (15).

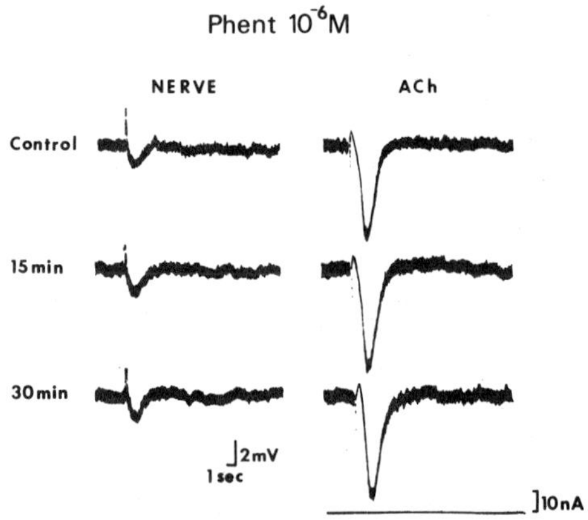

Figure 11.3. Effect of phentolamine (10^{-6} M) on the nerve-evoked S-IPSP and the hyperpolarization induced by iontophoretically applied ACh. Control recordings are shown followed by recordings at 15 and 30 minutes after drug application.

Using the sucrose gap technique, we compared the effect of a low Ca^{2+}/high Mg^{2+} solution and a solution containing TTX on the muscarine-induced hyperpolarization and with their effects on a S-IPSP. Under these conditions, the muscarine-evoked hyperpolarization was still present at the time synaptic transmission, *i.e.*, the S-IPSP, was blocked (Fig. 11.4). Furthermore, when ACh-induced hyperpolarizations and S-IPSPs were recorded from the same cell, treatment with low Ca^{2+}/high Mg^{2+} solution or TTX rapidly abolished the S-IPSP, whereas the ACh hyperpolarization persisted; however, the ACh-induced hyperpolarization was eventually blocked with longer treatment time (7).

The discrepancy in the time required to

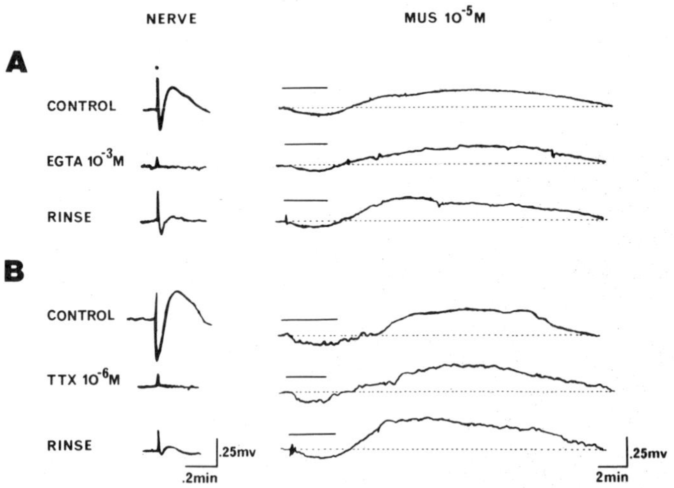

Figure 11.4. Chart recordings of the nerve-evoked S-IPSP and the hyperpolarization to muscarine (*MUS*) during normal transmission and during blockade of synaptic transmission. *Dot* indicates the onset of train stimulation, and the *bar* indicates the duration of MUS application. (*A*) The S-IPSP and MUS hyperpolarization in the absence and presence of zero Ca^{2+}, Mg^{2+}/EGTA solution. The zero Ca solution eliminated the S-IPSP, but MUS still hyperpolarized the membrane potential. (*B*) S-IPSP and MUS hyperpolarization in the absence and presence of TTX 10^{-6} M. TTX eliminates the S-IPSP, but the MUS hyperpolarization remains intact.

block synaptic transmission (*i.e.*, S-IPSP) *vs.* that required to block the ACh-induced hyperpolarization suggests that low Ca^{2+}/high Mg^{2+} medium or TTX may be affecting the ionic mechanism of the ACh-induced hyperpolarization at the postsynaptic site. It is important to note that the S-IPSP, proposed as a disynaptic event, always precedes the S-EPSP, regarded generally to be a monosynaptic process.

DEPENDENCE ON THE PRESYNAPTIC SUPPLY OF CATECHOLAMINE

Libet (35) suggested that the S-IPSP was dependent on the presynaptic supply of DA that could be decreased by prolonged treatment with a muscarinic agent, *i.e.*, by functionally stimulating the adrenergic interneurons to release DA (43). Indeed, bethanechol treatment for 30 minutes depressed the S-IPSP, and incubation with DA, harmine (a monoamine oxidase inhibitor), and ascorbic acid subsequently restored the S-IPSP (43). Libet (35) showed that NE has less effect in restoring the S-IPSP. He further reported that E had no effect, whereas dopa produced a significant but lesser restoration of the S-IPSP than DA.

In contrast, Dun (12) demonstrated that bethanechol (0.6 mM) treatment depressed the S-IPSP, and that simply washing out the agonist restored the S-IPSP. We observed that a muscarine-evoked hyperpolarizing response of the ganglion diminished with repeated exposure and found that application at 1-hour intervals avoided the phenomenon and elicited consistent reproducible responses (7). It therefore seems likely that the depression of the S-IPSP may have been due to desensitization of receptors, not functional release of catecholamine, and that restoration of the S-IPSP was due to removal of the high concentration of agonist, not replenishment of catecholamines.

SIF CELL INVOLVEMENT IN THE S-IPSP

The hypothesis that the anatomical basis for the adrenergic interneuron mediating the S-IPSP is the small granule-containing (SG) catecholamine-rich cells observed in electron micrographs (46, 63) or the small intensely fluorescent (SIF) cells observed with the Falck-Hillarp technique of formaldehyde-induced fluorescence for catecholamines (23, 53) has many discrepancies. For example, the rabbit has the smallest number of SIF cells (86/ganglion) (64) but the largest S-IPSP (1.91 ± 0.31 mV) (16), and the rat has a large number (388/ganglion) of SIF cells (64) and a small S-IPSP, 0.67 ± 0.21 mV (16). Libet and Owman (40) suggested that small numbers of SIF cells are able to reach every principal ganglion cell because of the extensive ramification of the SIF cell processes (46). However, Dail and Evan (9) observed that most of the adrenergic fibers surrounding the principal ganglion cells appeared to be axon collaterals and dendrites of adjacent principal ganglion cells. Another inconsistency is the observation that SIF cells appear to lack efferent synapses in the guinea pig (22) and frog (62) but S-IPSPs can be observed in both species (16, 37, 51). The cat superior cervical ganglion possesses type II SIF cells found in clusters around blood vessels and thought to have a neurosecretory function, but this ganglion is almost devoid of type I cells, thought to serve as the adrenergic interneuron (65) although the S-IPSP was first recorded in cats (18). There seems to be little correlation between SIF cells and the S-IPSP.

The SIF cells of rabbit SCG contain DA (40). The ability of preganglionic stimulation to release catecholamine from the SIF cells and the correlation between changes in DA content of SIF cells and the S-IPSP response has been studied with histofluorescent techniques (40).

A loss of the S-IPSP and DA fluorescence was observed after treatment with eserine, DL-α-methyl *p*-tyrosine and preganglionic stimulation, or after bethanechol treatment (40). However, when the ganglia were treated only with eserine and stimulated preganglionically, the S-IPSP decreased with no change in fluorescent intensity. Libet and Owman (40) offered no explanation as to why bethanechol treatment affected DA fluorescence while eserine and preganglionic stimulation did not. The concurrent changes in S-IPSP and DA fluorescence may be coincidental.

RELEASE OF CATECHOLAMINES

If CAs do mediate the S-IPSP they should be released with preganglionic stimulation. [^{3}H]-Norepinephrine and its metabolite [^{3}H]-normetanephrine were released from superfused rabbit superior cervical ganglia incubated in [^{3}H]tyrosine in response to preganglionic stimulation. Neither [^{3}H]dopamine nor its metabolite [^{3}H]3-methoxytyrosine could be detected (52) and most importantly, [^{3}H]NE release was not blocked by *d*-tubocurarine or by atropine. Apparently changes in NE release could not be correlated with cholinergic stimulation. Similarly there was no detectable release of [^{14}C]dopamine after incubation with [^{14}C]tyrosine following preganglionic stimulation (57). Altogether these data do not support a role for either DA or NE in the mediation of a S-IPSP.

However, Pearson and Sharman (54) detected significant increases in the concentration of dopamine metabolites, 3,4-dihydroxyphenylacetic acid (DOPAC) and homovanillic acid (HVA) in ganglia from pigs in response to preganglionic stimulation. In addition, Karoum *et al.* (29) found that stimulation of muscarinic receptors (by carbachol plus hexamethonium) increased DOPAC and DA concentrations, whereas nicotinic receptor stimulation (carbachol plus atropine) decreased DOPAC concentration only; carbachol alone increased DOPAC in decentralized but not sham-operated ganglia. These data suggest both a nicotinic and a muscarinic cholinoceptor modulation of dopamine synthesis and metabolism, but do not necessarily indicate a correlation with a physiological response—the S-IPSP.

IONIC MECHANISM OF ACTION

A neurotransmitter candidate should exhibit the same conductance changes and ionic mechanism of action as the transmitter mediating the endogenous response, the S-IPSP. In frog and rabbit sympathetic ganglia the S-IPSP can be generated without changes in membrane conductance in both nicotinized and curarized ganglia (30, 35, 51). There is a body of literature on the ionic mechanism of the S-IPSP in nicotinized ganglia (32, 33, 60), but treatment with nicotine probably increased intracellular Na^+ (33), so we will restrict our discussion to results obtained from ganglia treated with a nondepolarizing antagonist (*d*-tubocurarine or hexamethonium).

In most bullfrog sympathetic ganglia treated with *d*-tubocurarine (*d*-Tc), the S-IPSP was reduced, nullified, and then reversed with membrane hyperpolarization (1). When the external K^+ concentration was raised to 10 mM, the amplitude of the S-IPSP decreased and the reversal potential was closer to the resting membrane potential (1). Recently, Dodd and Horn (28a) clearly showed that the S-IPSP was due to direct muscarinic activation of a potassium conductance in bullfrogs. In curarized rabbit ganglia, the S-IPSP is increased by moderate hyperpolarization while further hyperpolarization decreases its amplitude (35). Furthermore, both in rabbit (35) and frog ganglia (38) depolarization depresses and eventually blocks the S-IPSP.

In amphibian ganglia, ouabain (2×10^{-6} M) has little effect on the S-IPSP (1), although a nonselective action on all postsynaptic potentials was observed with ouabain in rabbit and frog ganglia (30). Thus, synaptic activation of a Na^+-pump is not likely involved in the generation of the S-IPSP in curarized ganglia.

A norepinephrine-induced hyperpolarization was not accompanied by changes in membrane conductance in mammalian sympathetic ganglia and was depressed with membrane depolarization (31). In frog sympathetic ganglion cells 0.3 mM E produced no detectable change in the current-voltage relationship (and inversely conductance) (1, 47). The E-induced hyperpolarization was depressed and/or eliminated: (1) with strong hyperpolarization of the membrane; (2) when K^+ concentration was increased or decreased; (3) in Na^+-free Tris solution; or (4) with low temperature. Furthermore, the E hyperpolarization in rabbit (27) and amphibian ganglia (49) was not sensitive to ouabain at low concentration. A similar effect was observed with a DA-induced hyperpolarization in rabbit (41).

Although the E responses and the S-IPSP have common characteristics, the E-induced hyperpolarization appears to behave differently from the S-IPSP in a K^+-free solution, *i.e.*, the E hyperpolarization is depressed, whereas the S-IPSP is enhanced. These data do not support a role for catecholamines in the S-IPSP.

The ACh-induced hyperpolarizing potential was also not accompanied by a change in membrane resistance. Similarly, the amplitude of the ACh response was enhanced by hyperpolarizing up to membrane potentials of −80 mV. Further displacing the membrane potential to −90 mV reduced the amplitude, and depolarization of the membrane potential also reduced the ACh hyperpolarization (15).

We compared the conductance changes during a S-IPSP and ACh-induced hyperpolarization by constructing current-voltage plots prior to and at the peak of the responses. We found a small increase in conductance in both responses. Removal of K^+ from the superfusing solution enhanced both the S-IPSP and ACh hyperpolarization, but addition of a high K^+ solution depressed both responses (8a). These data are consistent with the hypothesis that the S-IPSP and ACh-induced hyperpolarization are due in part to an increase in conductance to K^+. However, since both TTX treatment and a zero Ca^{2+}/high Mg^{2+} (EGTA) solution block the ACh hyperpolarization (as well as the S-IPSP), Ca^{2+} and/or Na^+ ions may also be involved in the responses. The fact that the S-IPSP and an ACh hyperpolarization behave in a similar manner suggests that ACh may be acting directly on the principal ganglion cell.

SUMMARY

In conclusion, the following points do not support the catecholamine hypothesis for the S-IPSP.

1. Phentolamine (10^{-6} M) antagonized a NE hyperpolarization but did not depress a S-IPSP. These data suggest that the receptors involved in a NE hyperpolarization and a S-IPSP are different.
2. Only atropine selectively blocked a S-IPSP.
3. The ACh-induced hyperpolarization persisted when synaptic transmission (*i.e.*, a S-IPSP) was blocked in cells by a low Ca^{2+}/high Mg^{2+} or TTX-containing solution, suggesting that transmitter release from an interneuron is not an obligatory step in the generation of a S-IPSP.
4. After depression of the S-IPSP by bethanechol treatment, simply washing out bethanechol restores the S-IPSP; it is not necessary to incubate with a catecholamine.
5. There is no relationship between SIF cell number, type, or catecholamine content and the S-IPSP.
6. Changes in norepinephrine or dopamine release are not correlated with cholinergic stimulation; cholinergic modulation of dopamine synthesis and metabolism do not necessarily indicate a correlation with the physiological responses, the S-IPSP.
7. A NE hyperpolarization and S-IPSP behave differently in K^+-free solution, suggesting that their ionic mechanism may be different.

It appears that the evidence supporting the role of a catecholamine in the S-IPSP is quite weak. The alternative, that ACh acts directly on the principal ganglion cell, is the most likely mechanism.

References

1. Akasu T, Koketsu K: Ionic mechanism of the slow inhibitory postsynaptic potential of sympathetic ganglion cells in bullfrogs. *Jpn J Physiol* 33:279–300, 1983.
2. Beddoe F, Nicholls PJ, Smith HJ: Inhibition of muscarinic receptor by dibenamine. *Biochem Pharmacol* 20:3367–3776, 1971.
3. Benfey BG, Grillo SA: Antagonism of acetylcholine by adrenaline antagonists. *Br J Pharmacol* 20:528–533, 1963.
4. Boyd H, Burnstock G, Campbell G, Lowett A, O'Shea J, Wood M: The cholinergic blocking action of adrenergic blocking agents in the pharmacological analysis of autonomic innervation. *Br J Pharmacol* 20:418–435, 1963.
5. Brown DA, Caulfield MP: Hyperpolarization 'α-2'-adrenoceptors in rat sympathetic ganglia. *Br J Pharmacol* 65:435–445, 1979.
6. Cole AE, Shinnick-Gallagher P: Characterization of a postganglionic catecholamine receptor in the superior cervical ganglion (RSCG). *Soc Neurosci* 5:2485, 1979.
7. Cole AE, Shinnick-Gallagher P: Alpha-adrenoceptor and dopamine receptor antagonists do not block the slow inhibitory postsynaptic potential in sympathetic ganglia. *Brain Res* 187:226–230, 1980.
8. Cole AE, Shinnick-Gallagher P: Comparison of the receptors mediating the catecholamine hyperpolarization and slow inhibitory postsynaptic potential in sympathetic ganglia. *J Pharmacol Exp Ther* 217:440–444, 1981.

8a. Cole AE, Shinnick-Gallagher P: Muscarinic inhibitory transmission in mammalian sympathetic ganglia mediated by potassium conductance. *Nature*, in press, 1984.

9. Dail WG Jr, Evan AP: Ultrastructure of adrenergic terminals and SIF cells in the superior cervical ganglion of the rabbit. *Brain Res* 148:469–477, 1978.
10. deGroat WC, Volle RL: The actions of the catecholamines on the transmission in the superior cervical ganglion of the cat. *J Pharmacol Exp Ther* 154:1–13, 1966.
11. Dionne VE, Stevens CR: Voltage dependance of agonist effectiveness at the frog neuromuscular junction: Resolution of a parodox. *J Physiol* 251:245–270, 1975.
12. Dun, NJ: Ganglionic transmission: Electrophysiology and pharmacology. *Fed Proc* 39:2982–2989, 1980.
13. Dun NJ, Kaibara K, Karczmar AG: Dopamine and adenosine 3′,5′-monophosphate responses of single mammalian sympathetic neurons. *Science* 197:778–780, 1977.
14. Dun N, Karczmar AG: The presynaptic site of action of norepinephrine in the superior cervical ganglion of guinea pig. *J Pharmacol Exp Ther* 200:328–335, 1977.
15. Dun N, Karczmar AG: Involvement of an interneuron in the generation of the slow inhibitory postsynaptic potential in mammalian sympa-

thetic ganglia. *Proc Natl Acad Sci USA* 75:4029–4032, 1978.
16. Dun NJ, Karczmar AG: A comparative study of the pharmacological properties of the positive potential recorded from the superior cervical ganglia of several species. *J Pharmacol Exp Ther* 215:455–460, 1980.
17. Dunant Y, Dolivo M: Relations entre les potentials synaptique chez le rat. *J Physiol (Paris)* 59:281–294, 1967.
18. Eccles JC: Synaptic potentials and transmission in sympathetic ganglion. *J Physiol (Lond)* 101:465–483, 1943.
19. Eccles JC: The nature of synaptic transmission in a sympathetic ganglion. *J Physiol (Lond)* 103:27–54, 1944.
20. Eccles RM: Responses of isolated curarized sympathetic ganglia. *J Physiol (Lond)* 117:196–217, 1952.
21. Eccles RM, Libet B: Origin and blockade of the synaptic responses of curarized sympathetic ganglia. *J Physiol* (Lond) 157:484–503, 1961.
22. Elfin L-G, Hökfelt T, Goldstein M: Fluorescence microscopical, immunohistochemical and ultrastructural studies on sympathetic ganglia of the guinea pig, with special reference to the SIF cells and their catecholamine content. *J Ultrastruct Res* 51:377–396, 1975.
23. Eränkö O, Härkonen M: Monoamine-containing small cells in the superior cervical ganglion of the rat and an organ composed of them. *Acta Physiol Scand* 63:1511–1512, 1965.
24. Furchgott RF: The use of β-haloalkylamines in the differentiation of receptors and in the determination of dissociation constants of receptor-agonist complexes. In Harper NJ, Simmonds AB: *Advances in Drug Research*. New York, Academic Press, 1966, pp 21–55.
25. Gage PW, McBurney RN, Van Helden D: Octanol reduces end-plate channel lifetime. *J Physiol* 274:279–298, 1978.
26. Gallagher JP, Shinnick-Gallagher P, Cole AE, Griffith WH, Wiliams BJ: Current hypotheses for the slow inhibitory postsynaptic potential in sympathetic ganglia. *Fed Proc* 39:3009–3015, 1980.
27. Gebber GL, Volle RL: Mechanisms involved in ganglionic blockade induced by tetramethylammonium. *J Pharmacol Exp Ther* 152:18–28, 1966.
28. Gothert M, Lox HJ, Rieckesmann JM: Effect of butyrophenones on the sympathetic nerves of the isolated rabbit heart and on the postsynaptic α-adrenoceptors of the isolated rabbit aorta. *Naunyn Schmiedebergs Arch Pharmacol* 300:255–265, 1977.
28a. Dodd J, Horn JP: Muscarinic inhibition of sympathetic C neurones in the bullfrog. *J Physiol (Lond)* 334:271–291, 1983.
29. Karoum F, Garrison CK, Neff N, Wyatt RJ: Transsynaptic modulation of dopamine metabolism in the rat superior cervical ganglion. *J Pharmacol Exp Ther* 201:654–661, 1977.
30. Kobayashi H, Libet B: Generation of slow postsynaptic potentials without increases in ionic conductance. *Proc Natl Acad Sci USA* 60:1304–1311, 1968.
31. Kobayashi H, Libet B: Actions of noradrenaline and acetylcholine on sympathetic ganglion cells. *J Physiol (Lond)* 208:353–372, 1970.
32. Koketsu K, Nishi S: Characteristics of the slow inhibitory postsynaptic potential of bullfrog sympathetic ganglion cells. *Life Sci* 6:1827–1836, 1967.
33. Kuba K, Koketsu K: Synaptic events in sympathetic ganglia. *Prog Neurobiol* 11:77–169, 1978.
34. Laporte Y, Lorente de No R: Potential changes evoked in a curarized sympathetic ganglion by presynaptic volleys of impulses. *J Cell Comp Physiol* (Suppl) 35:61–106, 1950.
35. Libet B: Generation of slow inhibitory and excitatory postsynaptic potentials. *Fed Proc* 29:1945–1956, 1970.
36. Libet B: The SIF cells as a functional dopamine releasing interneuron in the rabbit superior cervical ganglion. In Eranko O: *SIF Cells-Structure and Function of Small Intensely Fluorescent Sympathetic Cells*. DHEW Publication No. 76-942, 1976, pp 163–177.
37. Libet B, Chichibu S, Tosaka T: Slow synaptic responses and excitability in sympathetic ganglia of the bullfrog. *J Neurophysiol* 31:383–395, 1968.
38. Libet B, Kobayashi H: Generation of adrenergic and cholinergic potentials in sympathetic ganglion cells. *Science* 164:1530–1532, 1969.
39. Libet B, Kobayashi H: Adrenergic mediation of slow inhibitory postsynaptic potential in sympathetic ganglia of the frog. *J Neurophysiol.* 37:805–814, 1974.
40. Libet B, Owman CH: Concomitant changes in formaldehyde-induced fluorescence of dopamine interneurons and in slow inhibitory postsynaptic potentials of rabbit superior cervical ganglia, induced by stimulation of preganglionic nerve or by a muscarinic agent. *J Physiol (Lond)* 237:635–662, 1974.
41. Libet B, Tanaka T, Tosaka T: Different sensitivities of acetylcholine-induced "after-hyperpolarization" compared to dopamine-induced hyperpolarization, to ouabain or to lithium replacement of sodium, in rabbit sympathetic ganglia. *Life Sci* 20:1863–1870, 1977.
42. Libet B, Tosaka T: Slow inhibitory and excitatory postsynaptic responses in single cells of mammalian sympathetic ganglia. *J Neurophysiol* 32:43–50, 1969.
43. Libet B, Tosaka T: Dopamine as a synaptic transmitter and modulator in sympathetic ganglia: A different mode of synaptic action. *Proc Natl Acad Sci USA* 67:667–673, 1970.
44. Lundberg A: Adrenaline and transmission in the sympathetic ganglion of the cat. *Acta Physiol Scand* 26:252–263, 1952.
45. Marrazzi AS: Electrical studies on the pharmacology of autonomic synapses. II. The action of a sympathomimetic drug (epinephrine) on sympathetic ganglia. *J Pharmacol Exp Ther* 65:395–404, 1939.
46. Matthews MR, Raisman G: The ultrastructure and somatic efferent synapses of small granule-containing cells in the superior cervical ganglion. *J Anat* 105:255–282, 1969.
47. Minota S, Koketsu K: Effects of adrenaline on

the action potential of sympathetic ganglion cells in bullfrogs. *Jpn J Physiol* 27:353–366, 1977.
48. Nakamura J: The effect of neuroleptics in the dopaminergic and cholinergic systems in sympathetic ganglia. *Kurume Med J* 25:241–253, 1978.
49. Nakamura M, Koketsu K: The effects of adrenaline on sympathetic ganglion cells of bullfrog. *Life Sci* 11:1165–1173, 1972.
50. Nishi S, Koketsu K: Early and late after discharges of amphibian sympathetic ganglion cells. *J Neurophysiol* 31:109–121, 1968a.
51. Nishi S, Koketsu K: Analysis of slow inhibitory postsynaptic potential of bullfrog sympathetic ganglion. *J Neurophysiol* 31:717–728, 1968b.
52. Noon JP, McAfee DA, Roth RH: Norepinephrine release from nerve terminals within the rabbit superior cervical ganglion. *Naunyn Schmiedebergs Arch Pharmacol* 291:139–162, 1975.
53. Norberg K-A, Sjoqvist F: New possibilities for adrenergic modulation of ganglionic transmission. *Pharmacol Rev* 18:743–749, 1966.
54. Pearson JDM, Sharmon DF: Increased concentrations of acidic metabolites of dopamine in the superior cervical ganglion following preganglionic stimulation *in vivo*. *J Neurochem* 22:547–550, 1974.
55. Rubin RP: The role of calcium in the release of neurotransmitter substances and hormones. *Pharmacol Rev* 22:389–428, 1970.
56. Steinback AB: Alteration of xylocaine (lidocaine) and its derivatives on the time course of the end plate potential. *J Gen Physiol* 52:144–161, 1968.
57. Steinberg ML, Keller CE: Enhanced catecholamine synthesis in isolated rat superior cervical ganglia caused by nerve stimulation: Dissociation between ganglionic transmission and catecholamine synthesis. *J Pharmacol Exp Ther* 204:384–399, 1978.
58. Tosaka T, Chichibu S, Libet B: Intracellular analysis of slow inhibitory and excitatory postsynaptic potentials in sympathetic ganglia of the frog. *J Neurophysiol* 31:396–409, 1968.
59. Tosaka T, Libet B: Additional (adrenergic) step in sympathetic ganglia. *Fed Proc* 29:716, 1970.
60. Weight FF, Padjen A: Acetylcholine and slow synaptic inhibition in frog sympathetic ganglion cells. *Brain Res* 55:225–228, 1973.
61. Weight FF, Smith PA: Histochemistry and cell biology and autonomic neurons, SIF cells and paraneurons. *Adv Biochem Psychopharmacol* 25:159–171, 1980
62. Weight FF, Weitsen NA: Identification of small intensely fluorescent (SIF) cells as chromaffin cells in bullfrog sympathetic ganglia. *Brain Res* 128:213–226, 1977.
63. Williams TH: Electron microscopic evidence for an autonomic interneuron. *Nature* 214:309–310, 1967.
64. Williams TH, Block AC, Chiba T, Jew JY: Species differences in mammalian SIF cells. In Costa E, Gessa GL: *Advances in Biochemical Psychopharmacology*. New York, Raven Press, 1977, pp 505–511.
65. Williams TH, Chiba T, Block AC Jr, Bhalla RC, Jew J: Species variation in SIF cells of superior cervical ganglia: Are there two functional types? In *SIF Cells, Structure and Function of the Small Intensely Fluorescent Sympathetic Cells*. DHEW Publication No. 76–942, 1976, pp 143–162.

CHAPTER TWELVE

Opioid Peptides in the Peripheral Sympathetic Nervous System and in Pheochromocytoma

O. HUMBERTO VIVEROS, Ph.D., and STEVEN P. WILSON, Ph.D.

INTRODUCTION

Sympathetic nervous system (SNS) content of endogenous opioid pentapeptides (enkephalins) was established by Hughes *et al.* (20) when they found opioid activity in extracts of the lumbar sympathetic chain and celiac ganglia. This finding was soon followed by the visualization of enkephalin-like immunoreactivity in small intensely fluorescent cells, in the perikaryon of a few noradrenergic ganglion cells, and in a terminal network within sympathetic ganglia (16, 57). At about the same time, several laboratories demonstrated the presence and secretion of enkephalins in the adrenal medulla (6, 17, 36–40, 55, 58, 67–69, 72–74, 79) and in human pheochromocytoma tumors (2, 13, 40, 63, 69, 74, 75).

The documentation of opioid peptides in the peripheral autonomic nervous system has led to numerous subsequent reports discussing the presence of opiate receptors and the actions of administered opioid peptides (agonists and antagonists) in peripheral tissues (see reviews in refs. 11, 19, 42, 43). Possible functions for the opioid peptides secreted from the adrenal medulla and sympathetic terminals are discussed.

OPIOID PEPTIDES IN THE ADRENAL MEDULLA

Chemical Characterization

The incubation of extracts containing opioid peptides from the adrenal medulla with general proteases such as pronase (67, 68, 79) produces a complete loss of enkephalin-like immunoreactivity and of the biological activity of the extracts (both in classical opiate assays and in the ability to compete with specific radioligands in binding to opiate receptors). This protease sensitivity indicates that all of the adrenal medullary opioid activity is due to peptides and proteins. Treatment of the extracts with trypsin followed by limited digestion with carboxypeptidase-B generates large amounts of Leu- and Met-enkephalin and greatly increases the amount of opioid peptides and immunoreactivity measured (36, 74). In this report, those peptides that have opiate activity revealed by trypsin and carboxypeptidase-B treatment but no opiate activity in their native form will be called peptides containing cryptic enkephalin sequences. Molecular sieve chromatography of bovine adrenal medulla extracts reveals six size classes of peptides containing cryptic enkephalin sequences and opioid peptides (37), ranging in approximate molecular mass from 500 daltons, *i.e.*, the pentapeptides Met- and Leu-enkephalin (62, 67, 68) to a 50 K-dalton protein that contains multiple copies of the Met- and Leu-enkephalin sequences in a ratio of 7:1 (36). The human (3a) and bovine adrenal medulla (14a, 45a) preproenkephalins have recently been cloned. The protein sequence is highly conserved, the deduced molecular weight for both is close to 30 K daltons, and the two precursors contain one leu-enkephalin sequence, four met-enkephalins, and one met-enkephalin-Arg^6-Gly^7-Leu^8 bracketed by pairs of basic amino acid residues (trypsin-like cleavage sites), and at the carboxyl terminus one met-enkephalin-Arg^6-Phe^7 preceded by a Lys-Arg. Several of the peptides containing cryptic enkephalin sequences and opioid peptide peaks from gel filtration are complex and can be resolved into multiple components; a few have been purified to homogeneity and sequenced (24, 28, 36–38, 44, 45, 60, 62). The reason for the heterogeneity in adrenomedullary peptides containing cryptic enkephalin sequences and opioid peptides is not clear. Many of these peptides may be intermediates in the synthesis of the enkephalins, but others have potent opiate activity and may be neurohormones (21, 28, 68). Peptide E (28) (25 aminoacids, with a Met-enkephalin at the amino terminus and a Leu-enkephalin at the carboxy terminus), the related peptides BAM-22, -20 and -12 (44, 45), and the heptapeptide Met-enkephalin [Arg^6, Phe^7] (21) are far more potent than the enkephalins on the guinea pig ileum or in nociceptive assays. No significant amounts of β-lipotropin or β-endorphin are found in adrenal medulla (37, 58, 67, 70, 79). Dynorphin-like peptides have been found in the bovine adrenal medulla (31a) and apparently are stored in the norepinephrine-containing cells (9a). There are approximately 10 times more enkephalin sequences obtained by digesting the peptides containing cryptic enkephalin sequences than there are native Met- and Leu-enkephalins in adrenal extracts (36, 62). Met- and Leu-enkephalin contribute 25% of the opioid peptides in adrenal extracts when the relative activities are measured with a radioreceptor assay. Met-enkephalin predominates over Leu-enkephalin by a ratio of 4:1; the next most abundant peptide in this group is Met-enkephalin [Arg^6, Phe^7] (50, 61, 62).

Costorage of Opioid Peptides and Catecholamines

Immunocytochemical studies show that enkephalin-like immunoreactivity is present in the same cells that contain catecholamines (CA); this immunoreactivity is also found in a sparse to moderate number of nerve fibers in the adrenal medulla and adrenal cortex (39, 40, 58). Radioimmuno- and radioreceptor assays indicate that opioid peptides are 15 to 35 times more concentrated in the medulla than in the cortex (17, 67). While the adrenal medullae of all vertebrates examined contain opioid peptides (Tables 12.1 and 12.2), the concentration between different species has more than a 100-fold range (17, 67). The number of enkephalin-like immunoreactive cells also varies with the species and is generally proportional to the content of opioid peptides. For example, rats have only small groups of enkephalin-like immunoreactive cells (58), while the dog medulla shows intense immunostaining of every chromaffin cell with both Met- and Leu-enkephalin antisera (R. I. Linnoila, E. J. Diliberto, Jr., S. P. Wilson, and O. H. Viveros, unpublished data). Met- and Leu-enkephalin

Table 12.1
Opioid Peptide Content in the Adrenal Medulla of Several Species[a]

Species	N	Opioid Peptides (nmoles/gm wet wt)	Catecholamines (μmoles/gm wet wt)	Molar Ratio (CA:OP)
Rat	(6)	0.29 ± 0.03		
Hamster	(5 pooled)	0.50		
Rabbit	(6)	0.85 ± 0.15		
Guinea pig	(12)	1.2 ± 0.15	7.7 ± 0.7	7,024 ± 638
Cat	(5)	3.2 ± 0.7	31.7 ± 2.9	9,906 ± 620
Ox	(8)	6.8 ± 0.7	58.6 ± 2.1	8,618 ± 570
Dog	(6)	21.2 ± 1.1	23.7 ± 2.7	1,118 ± 89

[a] Fresh adrenal medullae were rapidly dissected from the cortex in the cold and homogenized in 1 M acetic acid. Values represent the mean ± SEM of duplicate determination on medulla extracts. Opioid peptide values are expressed as Met-enkephalin equivalents. Data modified from Viveros *et al.* (67, 68).

Table 12.2
Opioid Peptide Content of Human Adrenal Medulla and Pheochromocytoma[a]

			Opioid Peptides (μmoles/gm wet wt)		Catecholamines (μmoles/gm wet wt)	
Tissue	N	Cause of Death	Mean ± SEM	Range	Mean ± SEM	Range
Adrenal medulla	7	Internal disease or sudden death	1.99 ± 0.32	1.15–3.61	12.1 ± 1.5	7.7–19.3
Adrenal medulla	8	Cranial trauma or internal hemorrhage	0.26 ± 0.06	0.08–0.59	6.4 ± 1.6	2.0–13.7
Pheochromocytoma	13		24.9 ± 10.5	0.6–116	47.9 ± 9.8	10.1–127.5

[a] Data for adrenal medulla from Saria *et al.* (55) and pheochromocytoma modified from Wilson *et al.* (75). CA values for pheochromocytoma obtained from samples from seven tumors.

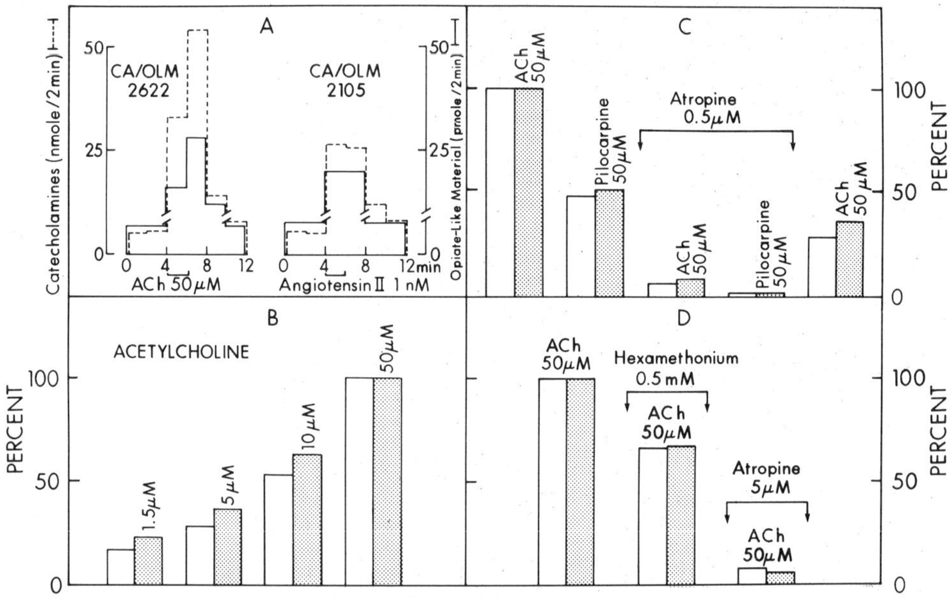

Figure 12.1. The effect of secretagogues and antagonists on the secretion of opioid peptides and CA from perfused dog adrenals. The effluent from the perfused glands was collected every 2 minutes and assayed for CA and opioid peptides. Each period of stimulation lasted 2 minutes. *Bars* in panels *B*, *C* and *D* correspond to the cumulative secretion of CA (*open bars*) and opioid peptides (*stippled bars*) during the 2 minutes of stimulation and the 6 minutes of washout; 100% is the effect of 50 μM acetylcholine. (Data from Viveros *et al.* (68).)

immunostaining is observed in about one-third of human adrenal medullary cells (39, 40), with a few cells positive exclusively to Leu-enkephalin antisera (39). Within the cells, the opioid peptides are concentrated almost exclusively in the chromaffin vesicles (67, 68) where they are stored with CA, soluble dopamine β-hydroxylase (DBH), chromogranins, and adenosine 5′-triphosphate (64). In addition to the enkephalins and other opioid peptides (67, 68), the adrenomedullary peptides containing cryptic enkephalin sequences are stored in the chromaffin vesicle (17, 37).

Cosecretion of Opioid Peptides and Catecholamines

Secretion from the adrenal medulla is a quantal, all-or-none, exocytotic release of the chromaffin vesicle content (64, 66). DBH, the largest known soluble component of the vesicle, is secreted with CA (64–66); thus it follows that all the other vesicular components will be secreted. As predicted, enkephalins are secreted in the same proportion as CA and in a molar ratio identical to the ratio within the gland (67–69, 74, 77) (see Fig. 12.1). The dog adrenal responds particularly to muscarinic agonists; the proportion of CA and opioid peptide secretion under the different experimental conditions remains constant. Secretion of opioid peptide is Ca^{2+}-dependent (67) as discussed below. In the perfused gland or in chromaffin cell cultures, opioid peptide secretion can be stimulated by the same secretagogues that induce secretion of CA (nicotine, K^+, Ba^{2+}, veratridine, ionomycin). The secretion of opioid peptides and CA is directly proportional to the intracellular concentration of the neurohormones. After reserpine treatment, which depletes CA and increases opioid peptide levels by inducing opioid peptide synthesis, a maximal dose of nicotine induces the secretion of a correspondingly larger amount of peptides and smaller amount of CA (74).

In vivo cat studies showed that increased splanchnic nerve discharge, induced reflexly by insulin hypoglycemia, is followed after a few hours by a proportional decrease in the amount of CA and opioid peptides remaining in the adrenal medulla (68). A similar situation occurs in individuals who die in shock as a consequence of traumatic injury (Table 12.2), indicating opioid peptide secretion *in vivo*.

Regulation of Opioid Peptide Synthesis

Adrenal medullary opioid peptide levels rapidly recover to normal or above normal following depletion from increased splanchnic stimulation (68). Increases in peptide levels also occur *in vivo* after reserpine administration and are blocked by inhibitors of protein synthesis (Fig. 12.2). The *de novo* synthesis of enkephalins and its induction by reserpine treatment have been confirmed in chromaffin

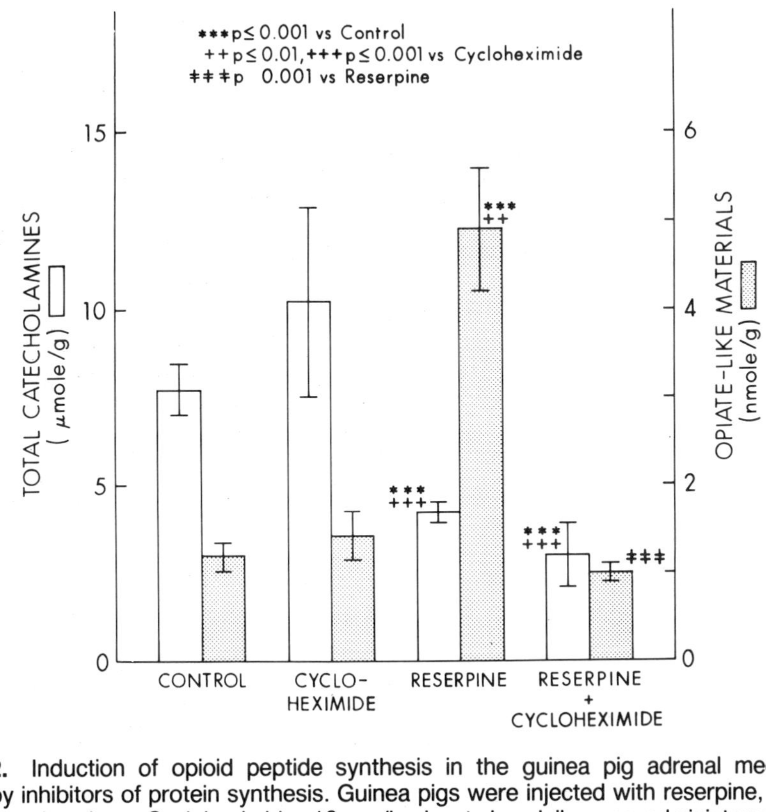

Figure 12.2. Induction of opioid peptide synthesis in the guinea pig adrenal medulla and its prevention by inhibitors of protein synthesis. Guinea pigs were injected with reserpine, 2 mg/kg i.p., on two successive days. Cycloheximide, 10 mg/kg i.p. twice daily, was administered from day 1 through day 4. Animals were sacrificed 12 hours after the last dose of cycloheximide or 3 days after the last reserpine injection. (Reproduced with permission from O. H. Viveros *et al.* (68).)

cells in culture (69, 72–74, 78). Reserpine also increases the content of the peptides containing cryptic enkephalin sequences (74). The synthesis of a specific 22,000-dalton peptide containing these sequences, preceding the labelling of the enkephalins, has also been demonstrated (51).

The signal to increase the synthesis of opioid peptides is not simply a decrease in total CA levels in the chromaffin cell (69, 74). The effect of a number of adrenergic drugs on CA and opioid peptide levels is shown in Table 12.3. Tetrabenazine, another CA-depleting agent, increases opioid peptide levels, but reserpine at low concentrations does so with no overall CA depletion. Similar effects can be obtained with inhibitors of monoamine oxidase (MAO) and synthetic dopamine (DA) receptor agonists (but not DA). Propranolol, phentolamine, and haloperidol do not block the effect of reserpine. More studies will be required to determine the mechanism of peptide regulation, but the evidence available suggests a critical role for an extravesicular pool of CA in this process.

OPIOID PEPTIDES IN HUMAN PHEOCHROMOCYTOMA

Several large peptides containing cryptic enkephalin sequences and opioid peptides, including Met-enkephalin, Leu-enkephalin, and Met-enkephalin [Arg^6, Phe^7], are identified in human pheochromocytoma tumors (2, 13, 63, 75). The opioid peptide content varies widely between and within tumors (Table 12.2). Since the CA content of different regions of a pheochromocytoma varies independently of the concentration of opioid peptides, the ratio of these two substances may differ as much as 4-fold between areas of the same tumor (68a). Some tumor cells may preferentially synthesize and store enkephalins with little or no CA and vice versa. This possibility is supported by the visualization of enkephalin-like immunoreactive cells with no DBH in some human pheochromocytomas, a situation never observed in human adrenal glands (39, 40).

As with the adrenal medulla, most of the opioid peptides cosediment with the CA-containing particles (75) and can be induced by reserpine or nicotine treatment in pheochromocytoma cells (74). As expected, a number of stimuli activate opioid peptide secretion from pheochromocytes in culture (68a, 77, 77a). Despite the similarities, the secretion of opioid peptides, CA, and preloaded [^{3}H]norepinephrine (NE) differs in several significant respects from the secretion of normal chromaffin cells (compare Figs. 12.3*A* and *C*). Adrenal medulla cells, under all conditions examined, secrete opioid peptides and CA in the same proportion, and the ratio of CA to opioid peptides secreted is the same as the ratio within the cell (67–69, 74) (see Figs. 12.3*C* and *D*). In pheochromocytoma, however, all secretagogues tested were more potent in inducing secretion of opioid

Table 12.3
Regulation of Chromaffin Cell Opioid Peptide Content

Experiment	Addition to Culture Medium[a]	Opioid Peptides (pmoles/culture)[b]	Catecholamines (nmoles/culture)[b]
I	None (n = 9)	17.8 ± 1.1	29.7 ± 0.8
	3 nM reserpine	23.4 ± 1.1[d]	31.7 ± 0.3[c]
	100 nM reserpine	39.4 ± 1.4[e]	9.6 ± 0.7[e]
II	None (n = 7)	31.7 ± 2.3	21.0 ± 0.7
	100 nM reserpine	56.7 ± 3.8[d]	6.2 ± 0.3[e]
	10 μM apomorphine	58.2 ± 7.2[c]	23.0 ± 1.6
	10 μM tranylcypromine	41.0 ± 2.4[c]	23.1 ± 0.2
	10 μM pargyline	43.9 ± 3.6[d]	23.2 ± 2.1
III	None	62.4 ± 1.9	206 ± 6
	1 mM 8-Br cyclic AMP	60.9 ± 1.7	176 ± 4[c]
	100 μM tetrabenazine	102.5 ± 6.8[d]	79 ± 2[e]

[a] Two days before harvest.
[b] Mean ± SEM; n = 3, except as noted.
[c] $p < 0.05$ compared to no addition.
[d] $p < 0.01$ compared to no addition.
[e] $p < 0.001$ compared to no addition.

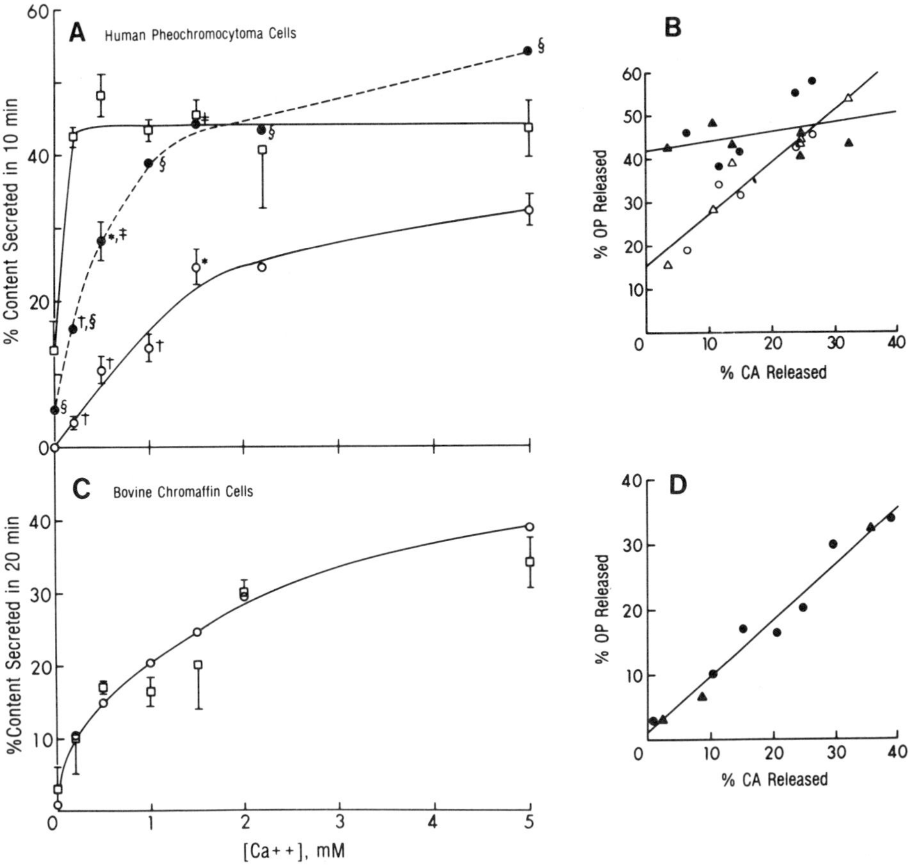

Figure 12.3. The dependence of opioid peptide secretion, endogenous CA, and [^{3}H]NE induced by nicotine on the external concentration of Ca^{2+}. (*A* and *B*) Human pheochromocytoma cells. (*C* and *D*) Bovine chromaffin cells in primary culture (78). The nicotine concentration was 10 μM for both cell types. In *A* and *C*, the *bars* representing the SEM are shown only when the value is larger than the symbol used for the mean. Endogenous CA (○), preloaded [^{3}H]NE (●), and opioid peptide (□) secretion are shown. (*, $p < 0.05$ compared with opioid peptide secretion; † $p < 0.01$ compared with opioid peptide secretion; ‡ $p < 0.05$ compared with CA secretion; § $p < 0.01$ compared with CA secretion). Panels *B* and *D* show the correlation between opioid peptide secretion and endogenous CA (○, △) and between secretion of [^{3}H]CA and endogenous CA (●, ▲). Regression coefficients are: chromaffin cells opioid peptides *vs.* CA, $r = 0.98$; pheochromocytoma cells ^{3}H *vs.* CA, $r = 0.94$; pheochromocytoma cells opioid peptides *vs.* CA, $r = 0.11$.

peptides than CA (77, 77a). In normal chromaffin cells, secretion of [^{3}H]CA is identical to the secretion of endogenous CA (25); in pheochromocytoma cells, the secretion of [^{3}H]NE is consistently greater than that of endogenous CA. Under maximal conditions of stimulation, it becomes the same as the secretion of opioid peptides.

The Ca^{2+} dependence of the secretory response to nicotine is identical for CA and opioid peptides in normal cells (Fig. 12.3), while in pheochromocytoma cells nicotine stimulates secretion of opioid peptides at much lower concentrations of extracellular Ca^{2+} than is required to elicit endogenous CA secretion. These results suggest heterogeneous storage pools of opioid peptides and CA in pheochromocytoma cells and even heterogeneity in the sites for recently stored [^{3}H]NE as compared to endogenous CA. This is more clearly seen in Figure 12.3*B*; a good correlation between [^{3}H]NE and CA secretion exists, although the high Y intercept indicates that a fraction of the stored [^{3}H] can be secreted in the absence of endogenous CA secretion. No significant correlation is found between the secretion of opioid peptides and CA. If these results are considered in light of the variable ratios of CA/opioid peptides in different regions of the same tumor and of the immunocytochemical observations on opioid peptides and DBH distribution (40) discussed above, it seems likely that there is a spectrum from highly responsive pheochromocytoma cells that contain only or mainly opioid peptides to less responsive cells that preferentially synthesize and store CA. The [^{3}H]NE may be taken up preferentially, but not exclusively, into the storage vesicles of the CA-containing cells.

In these comparative studies on secretion from normal and tumor cells, another important difference is that the sensitivity of pheochromocytoma cells to different secretagogues appears to be greater than that of medullary cells. Veratridine at 5 μM induces maximal secretion of opioid peptides, [^{3}H] and CA from pheochromocytoma (68a, 77, 77a), whereas half-maximal secretion of CA in normal cell is obtained at 50 μM veratridine and maximal stimulation at 250 μM (27). The sensitivity of opioid peptide secretion to stimulation by nicotine at low Ca^{2+} concentration is much greater in the tumor cells (see Fig. 12.3). This increased sensitivity to chemical stimuli may be important in the initiation of the paroxysmal secretion characteristic of these tumors.

The very high concentration of opioid peptides in some pheochromocytomas, the large size they can reach when compared to the adrenal medulla, and the low secretory threshold of the tumor cells may result in massive release of opioid peptides into the circulation which may be part of the clinical pleomorphism of this disease (5, 75). Pheochromocytoma patients may be under hypertensive therapy for prolonged periods of time. The treatment needed to control blood pressure (BP), however, may concurrently exacerbate the symptoms derived from the excessive production of opioid peptides, since CA-depleting agents as well as other autonomic-acting drugs induce the synthesis of opioid peptides.

COSTORAGE OF OPIOID PEPTIDES AND NOREPINEPHRINE IN POSTGANGLIONIC SYMPATHETIC NERVES

Recent advances in immunocytochemistry make it possible to study the simultaneous presence of more than one putative transmitter substance in the same neuron. Although only few examples of multiple neurotransmitter/neuromodulator coexistence have been demonstrated, this condition may be prevalent (18). Enkephalin-like immunoreactivity has been visualized in a small number of noradrenergic perikarya in rat sympathetic ganglia (57). However, it is difficult to assess the extent to which the initial biochemical characterization studies done in sympathetic ganglia (7, 20) reflect the biochemistry of the small intensely fluorescent cells, or the nerve terminal network which also contain opioid peptides.

The subcellular distribution of opioid peptides has been determined in sympathetic nerve fibers (69, 76). In the bovine splenic nerve, 98% of the fibers are sympathetic C fibers, and nerve vesicles can be obtained at high purity and yield (29). As in the adrenal medulla, 85% of the opioid peptides in a nerve homogenate sediment in a particulate fraction. When resolved by sucrose density centrifugation, 90% of this material is in the noradrenergic, large, dense-cored vesicles (Fig. 12.4). Since not all principal neurons in sympathetic ganglia show enkephalin-like immunoreactivity (57), the opioid peptides may be contained in only a fraction of the total large, dense-cored vesicle population. The exocytotic secre-

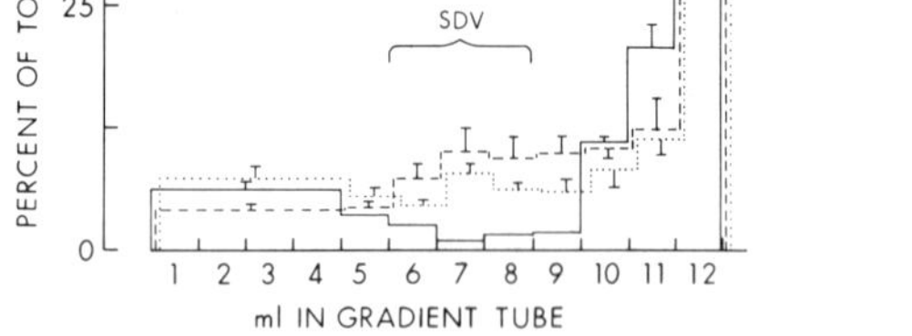

Figure 12.4. Distribution of opioid peptides, NE and DBH on density gradient centrifugation of bovine splenic nerve homogenate. After removal of the coarse debris from the nerve homogenate by differential centrifugation, the supernatant containing the noradrenergic vesicles and other particles is layered on top of the gradient (fractions 1 to 4) and centrifuged to equilibrium. SDV and LDV mark the area of the gradient where the small and the large dense-cored vesicles equilibrate. (Data from Viveros *et al.* (69) and Wilson *et al.* (76).)

tion of NE supports the prediction of cosecretion of NE and opioid peptides from those nerve terminals that contain both substances. Compared to the adrenomedullary chromaffin vesicles the large, dense-cored vesicles contain only a small amount of peptides containing cryptic enkephalin sequences relative to the receptor-active opioids. Also, more of the opioid peptides can be accounted for by Met- and Leu-enkephalin (50 to 60% of total opioid peptides) which are found in a ratio of approximately 2:1. Met-enkephalin [Arg^6, Phe^7] seems to be one of the minor opioid peptide components (76).

OPIOID PEPTIDES: SYMPATHETIC NEUROMODULATORS AND NEUROHORMONES

Adrenomedullary Opioids in the Circulation

Because peptides containing cryptic enkephalin sequences and opioid peptides are secreted from the adrenal medulla (26, 55, 67–69, 74) and probably from sympathetic nerves, they should be present in peripheral blood. There is no evidence for a cellular uptake mechanism for enkephalin (68, 69), the main route of opioid peptide disposition being enzymic inactivation. The half-life of a crude adrenal medullary opioid peptide extract and of the enkephalins in rat and dog plasma is 2 to 4 minutes (15, 68), slightly longer than the half-life of CA in the circulation (71). The adrenomedullary opioid peptides should then have adequate time to reach their peripheral targets. Met- and Leu-enkephalin are found in the peripheral plasma of humans, dogs, and rats (3, 31, 52, 53). Under basal conditions in the anesthetized dog, a substantial amount of the circulating enkephalins may originate from the gut (31); morphine administration, which increases splanchnic-adrenomedullary activity, also increases circulating enkephalin levels without raising levels in portal vein plasma. Insulin hypoglycemia markedly increases dog plasma enkephalin levels (52). Furthermore, a human patient had twice as much Met-enkephalin in a blood sample obtained by catheterization of the left adrenal vein as in a sample from the left internal jugular (3). The other adrenomedullary peptides with opioid activity (peptide E, BAMs-12P, -20, -22P, and Met-enkephalin [Arg^6, Phe^7]), which may be more resistant to enzymic degradation, have not yet been examined in the circulation.

Actions of Peripherally Administered Opioid Peptides

It is not possible in this chapter to review the many recently discovered actions of the natural and synthetic opioid peptides when administered outside the blood-brain barrier. Some of the most pronounced effects of opioid peptides secreted from adrenal medulla and sympathetic nerve terminals, however, will be discussed.

Intravenously injected enkephalins and their metabolically resistant synthetic analogs have marked endocrinological and metabolic effects (for reviews see refs. 42, 43); they not only modify release of anterior and posterior pituitary hormones, but also act directly on peripheral endocrine cells (22, 48, 59). In the gastrointestinal system, opioids affect intestinal motility, increase gastric acid secretion, raise gastric and intestinal blood flow and oxygen consumption, modify intestinal ion transport, and inhibit pancreatic bicarbonate and enzyme secretion (8, 19, 30, 47). In the peripheral autonomic nervous system, activation of opiate receptors inhibits the release of NE and acetylcholine (9, 12; for reviews see refs. 11, 19). Peripherally administered enkephalins change BP and heart rate (56), inhibit carotid body chemoreceptor discharge (41), inhibit the vagal baroreceptor reflex (56), and may inhibit the discharge of peripheral nociceptors (10) while inducing pulmonary J-receptor stimulation (54). Although their mechanism of action is not fully understood, peripherally administered enkephalins also produce a number of behavioral effects (23, 46, 49).

A complicating factor in understanding the role of opioid peptides in peripheral systems is that their effects are frequently of the modulatory rather than of the ON-OFF type as are the classical transmitters and neurohormones. For example, the enkephalins, depending on the dose, potentiate or inhibit the effect of glucose on the β-cell of the pancreas but do not directly stimulate insulin secretion (14, 22). There is considerable variability in how endogenous opioid peptides, and especially the synthetic peptides, affect individual peripheral tissues. This is probably the result of the heterogeneity of opiate receptors, the different relative affinities of the endogenous and synthetic opioid peptides for these receptors and the different resistances of these peptides to metabolic inactivation. For example, β-endorphin stimulates corticosterone synthesis and secretion in isolated adrenal cortical cells (59), while Met- and Leu-enkephalin inhibit these functions (48). The number of different adrenal medullary opioid peptides and the probable heterogeneity in their affinities for opiate receptors make extrapolation of the peripheral effects of Met- and Leu-enkephalin to a possible physiological role premature and of limited value.

Possible Functions for the Opioid Peptides That Are Secreted with Catecholamines

Although several roles are proposed for opioid peptides secreted with CA (4, 6, 12, 28, 35, 68, 69), only a few have been tested experimentally.

Membranes from the adrenal medulla and from chromaffin cells in culture (1, 4) contain binding sites for various natural and synthetic opioids. These sites have the characteristics of σ-receptors (4). Several opioid agonists partially inhibit nicotine-stimulated secretion of CA from the adrenal medulla, and morphine and β-endorphin inhibit the binding of α-bungarotoxin to membranes from cultured cells. The opiate-binding sites may, therefore, function as autoreceptors tightly coupled to nicotinic receptors and may regulate the release of CA from the adrenal medulla (4). This hypothesis is challenged because: (a) the very high concentrations of endogenous opioid peptides required to inhibit secretion (10^{-3} to 10^{-5}M for the enkephalins) far exceed the concentrations required for effective receptor occupancy; (b) inhibition of secretion is also obtained with opioid receptor antagonists (naloxone and naltrexone); and (c) dextrorphan (the inactive enantiomer) is as potent as levorphanol in inhibiting CA secretion (32).

The opioid peptides may be involved in the development of hypertension in the spontaneously hypertensive rat (SHR). These animals have a significant decrease in levels of Met- and Leu-enkephalin immunoreactivity in the adrenals and in the celiac and superior cervical ganglia but no change in hypothalamic opioid peptides. Since these differences precede the development of high BP, they do not appear to be secondary to the disease (6).

Adrenal medullary opioid peptides may play a role in stress-induced analgesia (35, 68, 68a). Lewis *et al.* (33, 34) recently characterized two forms of stress analgesia induced by inescapable foot shock in rats. A short exposure to continuous foot shock produces a nonopioid-mediated analgesia, whereas intermittent shock for 30 minutes causes profound analgesia that is antagonized by naloxone (33) and shows cross-tolerance to morphine (34). Adrenalectomy, adrenal demedullation, and adrenal medullary denervation antagonize the opioid but not the nonopioid stress analgesia. The naloxone-sensitive analgesia is significantly enhanced after administration of reserpine on a treatment schedule that depletes CA and increases opioid peptides (35, 68a). This strongly suggests that the opioid peptides secreted from the adrenal medulla are critically involved in certain forms of stress analgesia.

Establishing a role for opioid peptides in the adrenal medulla and the sympathetic nervous system (SNS) is difficult. Many possibilities exist for direct actions of opioid peptides on peripheral neuroeffector systems as well as possible interactions with cosecreted CA. Several more years of laboratory research are probable before establishing the physiological roles of this family of neuropeptides.

SUMMARY

Opioid peptides are present in the adrenal medulla in large amounts in man and dog. They are stored in chromaffin vesicles and secreted along with CA in a molar ratio identical to that of the adrenal medulla. Synthesis of opioid peptides is stimulated by reserpine, a drug that depletes CA.

Opioid peptides are present in pheochromocytoma tissue, but their distribution is variable, they are not released with CA in the same molar ratio as they occur in the tumor, and they are more easily released than CA by many pharmacologic stimuli.

Sympathetic nerves contain opioid peptides in NE storage vesicles. After release from nerves or adrenal, the opioid peptides persist in the circulation for several minutes. The opioid peptides have many potential functions in the peripheral circulation, but their physiologic role is not yet clearly defined.

References

1. Chavkin C, Cox BM, Goldstein A: Stereospecific opiate binding in bovine adrenal medulla. *Mol Pharmacol* 15:751–753, 1979.
2. Clement-Jones V, Corder R, Lowry PJ: Isolation of human met-enkephalin and two groups of putative precursors (2K-pro-met-enkephalin) from an adrenal medullary tumor. *Biochem Biophys Res Commun* 95:665–673, 1980.

3. Clement-Jones V, Lowry PJ, Rees LH, Besser GM: Met-enkephalin circulates in human plasma. *Nature* 283:295–297, 1980.

3a. Comb M, Seeburg PH, Adelman J, Eiden L, Herbert E: Primary structure of the human met- and leu-enkephalin precursor and its mRNA. *Nature* 295:663–666, 1982.

4. Costa E, Guidotti A, Hanbauer I, Hexum T, Saiani L, Stine S, Yang H-YT: Regulation of acetylcholine receptors by endogenous cotransmitters: Studies of adrenal medulla. *Fed Proc* 40:160–165, 1981.

5. DeCourcy JL, DeCourcy CB: *Pheochromocytoma and the General Practitioner*. Cincinnati, Barclay Newman, 1952.

6. DiGiulio M, Yang H-YT, Fratta W, Costa E: Decreased content of immunoreactive enkephalin-like peptide in peripheral tissues of spontaneously hypertensive rats. *Nature* 278:646–647, 1979.

7. DiGiulio AM, Yang H-YT, Lutold B, Fratta W, Hong J, Costa E: Characterization of enkephalin-like material extracted from sympathetic ganglia. *Neuropharmacology* 17:989–992, 1978.

8. Dobbins J, Racusen L, Binder HJ: Effect of D-alanine methionine enkephalin amide on ion transport in rabbit ileum. *J Clin Invest* 66:19–28, 1980.

9. Dubocovich ML, Langer SZ: Pharmacological differentiation of presynaptic inhibitory α-adrenoceptors and opiate receptors in the cat nictitating membrane. *Br J Pharmacol* 70:383–393, 1980.

9a. Dumont M, Day R, Lemaire S: Distinct distribution of immunoreactive dynorphin and leucine enkephalin in various populations of isolated adrenal chromaffin cells. *Life Sci* 32:287–294, 1983.

10. Ferreira S: Peripheral analgesia: Mechanism of the analgesic action of aspirin like drugs and opiate-antagonists. *Br J Clin Pharmacol* 10:237S–245S, 1980.

11. Frederickson RCA: Enkephalin pentapeptides—a review of current evidence for a physiological role in vertebrate neurotransmission. *Life Sci* 21:23–42, 1977.

12. Gaddis RR, Dixon WR: Modulation of norepinephrine release in the isolated perfused cat spleen by methionine-enkephalin. *Fed Proc* 40:286, 1981.

13. Giraud P, Eiden LE, Audigier Y, Gillioz P, Conte-Devolx B, Boudouresque F, Eskay R, Oliver C: Enkephalins, ACTH, α-MSH and β-endorphin in human pheochromocytomas. *Neuropeptides* 1:237–252, 1981.

14. Green IC, Perrin D, Pedley KC, Leslie RDG, Pyke DA: Effect of enkephalins and morphine on insulin secretion from isolated rat islets. *Diabetologia* 19:158–161, 1980.

14a. Gubler U, Seeburg P, Hoffman BJ, Gage LP, Udenfriend S: Molecular cloning establishes proenkephalin as precursor of enkephalin-containing peptides. *Nature* 295:206–208, 1982.

15. Hambrook JM, Morgan BA, Rance MJ, Smith CFC: Mode of deactivation of the enkephalins by rat and human plasma and rat brain homogenates. *Nature* 262:782–783, 1976.

16. Hervonen A, Pelto-Huikko M, Helén P, Alho H: Electron-microscopic localization of enkephalin-like immunoreactivity in axon terminals of human sympathetic ganglia. *Histochemistry* 70:1–6, 1980.

17. Hexum TD, Yang H-YT, Costa E: Biochemical characterization of enkephalin-like immunoreactive peptides of adrenal glands. *Life Sci* 27:1211–1216, 1980.

18. Hökfelt T, Lundberg JM, Schultzberg M, Johansson O, Ljungdahl Å, Rehfeld J: Coexistence of peptides and putative transmitters in neurons. In Costa E, Trabucchi M: *Neuronal Peptides and Neuronal Communication*. New York, Raven Press, 1980, pp 1–23.

19. Hughes J: Peripheral opiate receptor mechanisms. *Trends Biochem Sci* 2:21–24, 1981.

20. Hughes J, Kosterlitz HW, Smith TW: The distribution of methionine-enkephalin and leucine-enkephalin in the brain and peripheral tissues. *Br J Pharmacol* 61:639–647, 1977.

21. Inturrisi CE, Umans JG, Wolff D, Stern AS, Lewis RV, Stein S, Udenfriend S: Analgesic activity of the naturally occurring heptapeptide [Met]enkephalin-Arg^6-Phe^7. *Proc Natl Acad Sci USA* 77:5512–5514, 1980.

22. Ipp E, Dobbs R, Unger RH: Morphine and β-endorphin influence the secretion of the endocrine pancreas. *Nature* 276:190–191, 1978.

23. Izquierdo I, Paiva ACM, Elisabetsky E: Post-training intraperitoneal administration of leu-enkephalin and β-endorphin causes retrograde amnesia for two different tasks in rats. *Behav Neural Biol* 28:246–250, 1980.

24. Jones BN, Stern AS, Lewis RV, Kimura S, Stein S, Udenfriend S, Shively JE: Structure of two adrenal polypeptides containing multiple enkephalin sequences. *Arch Biochem Biophys* 204:392–395, 1980.

25. Kilpatrick DL, Ledbetter FH, Carson KA, Kirshner AG, Slepetis R, Kirshner N: Stability of bovine adrenal medulla cells in culture. *J Neurochem* 35:679–692, 1980.

26. Kilpatrick DL, Lewis RV, Stein S, Udenfriend S: Release of enkephalins and enkephalin-containing polypeptides from perfused beef adrenal glands. *Proc Natl Acad Sci USA* 77:7473–7475, 1980.

27. Kilpatrick DL, Slepetis R, Kirshner N: Ion channels and membrane potential in stimulus-secretion coupling in adrenal medulla cells. *J Neurochem* 36:1245–1255, 1981.

28. Kilpatrick DL, Taniguchi T, Jones BN, Stern AS, Shively JE, Hullihan J, Kimura S, Stein S, Udenfriend S: A highly potent 3200-dalton adrenal opioid peptide that contains both a [met]- and [leu]-enkephalin sequence. *Proc Natl Acad Sci USA* 78:3265–3268, 1981.

29. Klein RL, Thureson-Klein AK, Yen S-HC, Baggett JMcC, Gasparis MS, Kirksay DF: Dopamine β-hydroxylase distribution in density gradients: Physiological and artifactual implications. *J Neurobiol* 10:291–307, 1979.

30. Konturck SJ: Opiates and the gastrointestinal tract. *Am J Gastroenterol* 74:285–291, 1980.

31. Laasberg LH, Johnson EE, Hedley-Whyte J: Effect of morphine and naloxone on leu-enkephalin-

like immunoreactivity in dogs. *J Pharmacol Exp Ther* 212:496–502, 1980.

31a. Lemaire S, Chouinard L, Denis D: Mass spectrometric identification of various molecular forms of dynorphin in bovine adrenal medulla. *Biochem Biophys Res Commun* 16:51–58, 1982.

32. Lemaire S, Livett B, Tseng R, Mercier P, Lemaire I: Studies on the inhibitory action of opiate compounds in isolated bovine adrenal chromaffin cells: Noninvolvement of stereospecific opiate binding sites. *J Neurochem* 36:886–892, 1981.

33. Lewis JW, Cannon JT, Liebeskind JC: Opioid and nonopioid mechanisms of stress analgesia. *Science* 208:623–625, 1980.

34. Lewis JW, Sherman JE, Liebeskind JC: Opioid and nonopioid stress analgesia: Assessment of tolerance and cross-tolerance. *J Neurosci* 1:358–363, 1981.

35. Lewis JW, Tordoff MG, Sherman JE, Liebeskind JC: Role of the adrenal medulla in opioid stress analgesia. *Soc Neurosci Abstr* 7:735, 1981.

36. Lewis RV, Stern AS, Kimura S, Rossier J, Stein S, Udenfriend S: An about 50,000-dalton protein in adrenal medulla: A common precursor of [met-] and [leu]enkephalin. *Science* 208:1459–1461, 1980.

37. Lewis RV, Stern AS, Rossier J, Stein S, Udenfriend S: Putative enkephalin precursors in bovine adrenal medulla. *Biochem Biophys Res Commun* 89:822–829, 1979.

38. Lewis RV, Stern AS, Kimura S, Stein S, Udenfriend S: Enkephalin biosynthetic pathway: Proteins of 8000 and 14,000 daltons in bovine adrenal medulla. *Proc Natl Acad Sci USA* 77:5018–5020, 1980.

39. Linnoila RI, Diaugustine RP, Hervonen A, Miller RJ: Distribution of [met^5]- and [leu^5]-enkephalin, vasoactive intestinal polypeptide- and substance P-like immunoreactivities in human adrenal glands. *Neuroscience* 5:2247–2259, 1980.

40. Lundberg JM, Hamberger B, Schultzberg M, Hökfelt T, Granberg P, Efendić S, Terenius L, Goldstein M, Luft R: Enkephalin- and somatostatin-like immunoreactivities in human adrenal medulla and pheochromocytoma. *Proc Natl Acad Sci USA* 76:4079–4083, 1979.

41. McQueen DS, Ribeiro JA: Inhibitory actions of methionine-enkephalin and morphine on the cat carotid chemoreceptors. *Br J Pharmacol* 71:297–305, 1980.

42. Meites J, Bruni JF, Van Vugt DA, Smith AF: Relation of endogenous opioid peptides and morphine to neuroendocrine functions. *Life Sci* 24:1325–1336, 1979.

43. Morley JE: The endocrinology of the opiates and opioid peptides. *Metabolism* 30:195–209, 1981.

44. Mizuno K, Minamino N, Kangawa K, Matsuo H: A new endogenous opioid peptide from bovine adrenal medulla: Isolation and amino acid sequence of a dodecapeptide (BAM-12P). *Biochem Biophys Res Commun* 95:1482–1488, 1980.

45. Mizuno K, Minamino N, Kangawa K, Matsuo H: A new family of endogenous "big" met-enkephalins from bovine adrenal medulla: Purification and structure of docosa-(BAM-22P) and eicosapeptide (BAM-20P) with very potent opiate activity. *Biochem Biophys Res Commun* 97:1283–1290, 1980.

45a. Noda M, Furutani Y, Takahasi T, Toyosato M, Hirose T, Inayama S, Nakanishi S, Numa S: Cloning and sequence analysis of cDNA for bovine adrenal preproenkephalin. *Nature* 295:202–206, 1982.

46. Olson RD, Kastin AJ, Olson GA, Coy DH: Behavioral effects after systemic injection of opiate peptides. *Psychoneuroendocrinology* 5:47–52, 1980.

47. Pawlik WW, Walus KM, Fondacaro JD: Effects of methionine-enkephalin on intestinal circulation and oxygen consumption. *Proc Soc Exp Biol Med* 165:26–31, 1980.

48. Raćz K, Gláz E, Kiss R, Lada Gy, Varga I, Vida S, DiGleria K, Medzihradszky K, Lichtwald K, Vecsei P: Adrenal cortex—a newly recognized peripheral site of action of enkephalins. *Biochem Biophys Res Commun* 97:1346–1353, 1980.

49. Rigter H, Jensen RA, Martinez JL, Jr, Messing RB, Vasquez BJ, Liang KC, McGaugh JL: Enkephalin and fear-motivated behavior. *Proc Natl Acad Sci USA* 77:3729–3732, 1980.

50. Rossier J, Dean DM, Livett BG, Udenfriend S: Enkephalin congeners and precursors are synthesized and released by primary cultures of adrenal chromaffin cells. *Life Sci* 28:781–789, 1981.

51. Rossier J, Trifaró JM, Lewis RV, Lee RWH, Stern A, Kimura S, Stein S, Udenfriend S: Studies with [^{35}S] methionine indicate that the 22,000-dalton [met]enkephalin-containing protein in chromaffin cells is a precursor of [met]enkephalin. *Proc Natl Acad Sci USA* 77:6889–6891, 1980.

52. Ryder SW, Eng J: Radioimmunoassay of leucine-enkephalin-like substance in human and canine plasma. *J Clin Endocrinol Metab* 52:367–369, 1981.

53. Sachse G, Laube H: Leucine-enkephalin plasma levels after an oral or intravenous arginine stimulation in rats. *Horm Metab Res* 12:416–417, 1980.

54. Sapru HN, Willette RN, Krieger AJ: Stimulation of pulmonary J receptors by an enkephalin-analog. *J Pharmacol Exp Ther* 217:228–234, 1981.

55. Saria A, Wilson SP, Molnar A, Viveros OH, Lembeck F: Substance P and opiate-like peptides in human adrenal medulla. *Neurosci Lett* 20:195–200, 1980.

56. Schaz K, Stock G, Simon W, Schlör K-H, Unger T, Rockhold R, Ganten D: Enkephalin effects on blood pressure, heart rate, and baroreceptor reflex. *Hypertension* 2:395–407, 1980.

57. Schultzberg M, Hökfelt T, Terenius L, Elfvin L-G, Lundberg JM, Brandt J, Elde RP, Goldstein M: Enkephalin immunoreactive nerve fibers and cell bodies in sympathetic ganglia of the guinea pig and rat. *Neuroscience* 4:249–270, 1979.

58. Schultzberg M, Lundberg JM, Hökfelt T, Terenius L, Brandt J, Elde RP, Goldstein M: Enkephalin-like immunoreactivity in gland cells and nerve terminals of the adrenal medulla. *Neuroscience* 3:1169–1186, 1978.

59. Shanker G, Sharma RK: β-endorphin stimulates corticosterone synthesis in isolated rat adrenal

cells. *Biochem Biophys Res Commun* 86:1–5, 1979.

60. Stern AS, Jones BN, Shively JE, Stein S, Udenfriend S: Two adrenal opioid polypeptides: Proposed intermediates in the processing of proenkephalin. *Proc Natl Acad Sci USA* 78:1962–1966, 1981.
61. Stern AS, Lewis RV, Kimura S, Rossier J, Gerber LD, Brink L, Stein S, Udenfriend S: Isolation of the opioid heptapeptide met-enkephalin [Arg^6, Phe^7] from bovine adrenal medullary granules and striatum. *Proc Natl Acad Sci USA* 76:6680–6683, 1979.
62. Stern AS, Lewis RV, Kimura S, Rossier J, Stein S, Udenfriend S: Opioid hexapeptides and heptapeptides in adrenal medulla and brain; possible implications on the biosynthesis of enkephalins. *Arch Biochem Biophys* 205:606–613, 1980.
63. Sullivan SN, Bloom SR, Polak JM: Enkephalin in peripheral neuroendocrine tumors. *The Lancet* i:986–987, 1978.
64. Viveros OH: Mechanism of secretion of catecholamines from adrenal medulla. In Blaschko H, Sayers J, Smith AD: *Handbook of Physiology, Section of Endocrinology VI. Adrenal Gland.* Washington, DC, American Physiological Society, 1975, pp 389–426.
65. Viveros OH, Arqueros L, Kirshner N: Release of catecholamines and dopamine-β-oxidase from the adrenal medulla. *Life Sci* 7:609–618, 1968.
66. Viveros OH, Arqueros L, Kirshner N: Quantal secretion from adrenal medulla: All-or-none release of storage vesicle content. *Science* 165:911–913, 1969.
67. Viveros OH, Diliberto EJ Jr, Hazum E, Chang K-J: Opiate-like materials in the adrenal medulla: Evidence for storage and secretion with catecholamines. *Mol Pharmacol* 16:1101–1108, 1979.
68. Viveros OH, Diliberto EJ Jr, Hazum E, Chang K-J: Enkephalins as possible adrenomedullary hormones: Storage, secretion, and regulation of synthesis. In Costa E, Trabucchi M: *Neural Peptides and Neuronal Communication*, New York, Raven Press, 1980, pp 191–204.

68a. Viveros OH, Wilson SP, Cubeddu LX, Kirshner N: Opioid peptides in human adrenal medulla and pheochromocytoma. In Ehrenpreis S, Sicuteri F: *Degradation of Endogenous Opioids, Its Relevance in Human Pathology and Therapy.* New York, Raven Press, 1983, pp 11–24.

69. Viveros OH, Wilson SP, Diliberto EJ Jr, Hazum E, Chang K-J: Enkephalins in adrenomedullary chromaffin cells and sympathetic nerves. In Stark E, Makara EB, Halász B, Rappay Gy: *Endocrinology, Neuroendocrinology, Neuropeptides-II, Advances in Physiological Sciences.* New York, Pergamon Press, 1980, vol 14, pp 349–353.
70. Watson SJ, Akil H, Ghazarossian VE, Goldstein A: Dynorphin immunocytochemical localization in brain and peripheral nervous system: Preliminary studies. *Proc Natl Acad Sci USA* 78:1260–1263, 1981.
71. Whitby LG, Axelrod J, WeilMalherbe H: The fate of H^3-norepinephrine in animals. *J Pharmacol Exp Ther* 132:193–201, 1961.
72. Wilson SP, Abou-Donia MM, Chang K-J, Viveros OH: Reserpine increases opiate-like peptide content and tyrosine hydroxylase activity in adrenal medullary chromaffin cells in culture. *Neuroscience* 6:71–79, 1981.
73. Wilson SP, Chang K-J, Viveros OH: Synthesis of enkephalins by adrenal medullary chromaffin cells: Reserpine increases incorporation of radiolabeled amino acids. *Proc Natl Acad Sci USA* 77:4364–4368, 1980.
74. Wilson SP, Chang K-J, Viveros OH: Opioid peptide synthesis in bovine and human adrenal chromaffin cells. *Peptides* 2 (Suppl 1):83–88, 1981.
75. Wilson SP, Cubeddu LX, Chang K-J, Viveros OH: Met-enkephalin, leu-enkephalin and other opiate-like peptides in human pheochromocytoma tumors. *Neuropeptides* 1:273–281, 1981.
76. Wilson SP, Klein RL, Chang K-J, Gasparis MS, Viveros OH, Yang W-H: Are opioid peptides cotransmitters in noradrenergic vesicles of sympathetic nerve? *Nature* 288:707–709, 1980.
77. Wilson SP, Slepetis R, Chang K-J, Kirshner N, Viveros OH: Secretogogues that act by different mechanisms induce secretion of opioid peptides and catecholamines from chromaffin and pheochromocytoma cells. *Soc Neurosci Abstr* 7:103, 1981.

77a. Wilson SP, Slepetis R, Chang K-J, Kirshner N, Viveros OH: Differential secretion of opioid peptides and catecholamines from cultured cells of a human pheochromocytoma tumor. *Life Sci* 29:2257–2264, 1981.

78. Wilson SP, Viveros OH: Primary culture of adrenal medullary chromaffin cells in a chemically defined medium. *Exp Cell Res* 133:159–169, 1981.
79. Yang H-YT, Hexum TD, Majane E, Costa E: Opiate peptides in bovine adrenal gland. In Costa E, Trabucchi M: *Neural Peptides and Neuronal Communication.* New York, Raven Press, 1980, pp 181–190.

CHAPTER THIRTEEN

The Role of Prostaglandins in the Modulation of Adrenergic Nervous System Function

HANS-GEORG GÜLLNER, M.D.

INTRODUCTION

Most research on the interactions of prostaglandins with the adrenergic nervous system has originated from a relatively small number of laboratories worldwide. The most recent review of mechanisms of the actions of prostaglandins (PG) on adrenergic neurotransmission was in 1977 by Hedqvist (28).

PG are a family of fatty acids independently described in the early 1930s by von Euler (15) and Goldblatt (23). They were originally extracted from human seminal fluid and sheep vesicular glands, in which they are present in high concentrations, and were found to have potent hypotensive effects (14). Since then, PG have been found ubiquitously in mammalian organs. They are generated and act locally in response to specific stimuli and are tissue hormones rather than circulating hormones.

Within the last decade, studies of PG have shown that these compounds have a great variety of functions in many organs. It is unknown whether PG have direct actions or whether their primary action is to modulate the response of target organs to stimuli such as circulating hormones. The nature and the mechanism of their action are not understood. PG modulate sympathetic neuronal activity and are released in many sites including the vicinity of adrenergic neuroeffector junctions. Release may be spontaneous or in response to physical, chemical, or electrical stimuli. PG influence the release of adrenergic and cholinergic transmitters from nerve endings and the response of target organs to the transmitters. It is presently unknown if PG can also effect the release of other putative neurotransmitters like neuropeptides. The effects of PG (or of inhibitors of PG synthesis) on adrenergic neurotransmission *in vitro* are frequently different from those observed *in vivo*; the reasons for this discrepancy are unclear. Recent evidence proposes that PG may influence not only the release of catecholamines (CA), but also the number and possibly the function of adrenergic receptor sites.

EFFECTS OF PROSTAGLANDINS ON ADRENERGIC NEUROTRANSMISSION *IN VITRO*

In 1968, Hedqvist (35) and Brundin (6) independently described an inhibitory effect of PGE_1 on the pharmacological responses to nerve stimulation and to norepinephrine (NE). Hedqvist (35) showed in the *in vitro* perfused cat spleen that PGE_1 administration reduces the vasoconstriction caused by nerve stimulation or NE. Brundin (6) demonstrated that PGE_1 inhibits the luminal occlusion of the rabbit oviduct in response to stimulation of the hypogastric nerve. These dose-dependent and reversible effects could be produced with both PGE_1 and PGE_2 (36,38,51). Since then, it has been shown that PGE_1, PGE_2, PGD_2, and the endoperoxides PGG_2 and PGH_2 (32) inhibit adrenergic neurotransmission in various tissues *in vitro*. However, the PG of the E series are much more potent than the others. For example, PGG_2 and PGH_2 inhibit the release of NE induced by nerve stimulation in the guinea pig vas deferens, but they are less than half as potent as PGE_2. Moreover, PGD_2 inhibits the release of NE in this organ preparation only in a concentration 100 times higher than that of PGE_2. It has been suggested that the effects of PGD_2 may be entirely due to contamination by PGE_2 (28).

PG of the F series enhance rather than inhibit neurotransmission in vascular tissue (28,32). For example, in the dog spleen or in the tibial artery preparation (5,9), $PGF_{2\alpha}$ enhances the response to nerve stimulation without affecting the response to NE. $PGF_{2\alpha}$, PGA_2, and PGB_2 have been examined for their effects on NE release in various tissues. In most of these models, no effect could be observed. In contrast, $PGF_{2\alpha}$, PGA_2, and PGB_2 decrease NE release in the rabbit kidney or guinea pig vas deferens when given in concentrations that enhance the response to nerve stimulation (18,32). The effects of thromboxanes, metabolites of arachidonic acid which are potent vasoconstrictors, have not yet been studied.

The perfused spleen has been widely used to study PG-NE interactions because of its rich innervation with adrenergic fibers. PGE_2 inhibits the release of NE and the splenic capsular contraction produced by nerve stimulation of the cat spleen (29). PGE_2 effects are variable; low doses inhibit splenic vasoconstriction in response to nerve stimulation in a dose-dependent fashion, but this effect is diminished or reversed with higher doses (29). This biphasic response may be due to the prejunctional inhibitory actions of PGE_2 on transmitter release as well as postjunctional enhancement of vasoconstrictor responses with high doses. The vascular responses produced by epinephrine (E) in the dog spleen are enhanced after inhibition of endogenous PG synthesis by indomethacin, an inhibitor of the enzyme cyclo-oxygenase (16). However, PGE_1 and PGE_2 have little or no effect on dog spleen smooth muscle contraction (10). This finding may be evidence that the prejunctional actions of PGE_2 are more significant.

PG have little or no effect on cardiac function when administered directly to isolated heart preparations or in intact animals. However, in the isolated rabbit heart, PG of the E series can reversibly inhibit the chronotropic and inotropic effects of sympathetic nerve stimulation (40,62). Furthermore, the positive chronotropic effects of transmural stimulation of the sinoatrial node are blocked by PGE_2 (54). The inhibitory effects of PGE_2 in the heart appear to be due exclusively to a decrease in NE release from nerve endings (40,62). This view is supported by the finding that the PGEs inhibit the release of NE evoked by potassium or nicotine in the guinea pig heart (61).

Other major organ models in which the role of PG in adrenergic neurotransmission has been studied *in vitro* are the guinea pig vas deferens (1,3,13,37), a number of arterial vascular beds (8,19,31,33,41,46,47,49,50), the field-stimulated rat and rabbit iris (4,53), and the cat nictitating membrane (5,43). All have predominantly adrenergic innervation and are suitable for studying the effects of PG on adrenergic function. The results of experiments performed in these systems are generally compatible with those obtained in experiments with the isolated spleen or the isolated heart.

EFFECT OF INHIBITORS OF PROSTAGLANDIN SYNTHESIS

Certain nonsteroidal antiinflammatory drugs, such as indomethacin (17), inhibit PG synthesis by blocking the enzyme cyclo-oxygenase and thereby the conversion of arachidonic acid to PGG_2 and PGH_2. This makes it possible to examine the role of endogenously generated prostaglandins on adrenergic neurotransmission and on the release of NE from nerve endings. Substrate analogs compose a second class of prostaglandin synthetase inhibitors. Only one compound of this class, the acetylenic derivative of arachidonic acid 5,8,11,14-eicosatetrayonic acid (ETA), has thus far been examined for its effect on adrenergic transmission (11). In the rabbit heart, cat spleen and guinea pig vas deferens, ETA blocks the release of PG and simultaneously enhances the release of NE induced by nerve stimulation (34,39,55). ETA also enhances field-stimulated NE release in the rabbit iris (53).

In general, inhibitors of PG synthesis probably increase the release of NE per nerve impulse, although the mechanism of this action is poorly understood. Indomethacin and meclofenamate, two of the aspirin-like nonsteroidal antiinflammatory drugs, increase the release of NE induced by nerve stimulation in a large number of *in vitro* perfused organs

(7,20,21,30,59,60). No effect, however, was observed in the cat spleen (12,42).

Oral administration of indomethacin to rats increases NE turnover in spleen, heart, submandibular gland, and adipose tissue (22). These data have been interpreted as evidence that indomethacin increases NE release per nerve impulse, by blocking the PG-mediated feedback inhibition of transmitter release. In summary, inhibitors of PG synthesis generally appear to have effects on adrenergic neurotransmission opposite to those of PG. Contradictory results have been obtained, however, in the rat kidney and cat spleen. Indomethacin has been reported to enhance (52) or inhibit (50) the vascular response to nerve stimulation of rat kidney. It remains to be seen if experimental differences alone are responsible for these inconsistencies.

Interpreting the data obtained in studies with indomethacin can be difficult because of toxic effects and pharmacological effects of the drug on release or metabolism of CA independent of the effects of inhibition of PG synthesis. Effective doses of indomethacin have toxic effects, especially when given for more than 3 days. The drug causes gastritis, gastrointestinal bleeding, and perforation of the gastrointestinal tract. Indomethacin causes perforation of the gut when given to rats in a dose of 8 mg/kg for 3 days. Indomethacin (5 mg/kg/day) causes a reduction in urine volume, a decline in spontaneous motor activity, and even death in some rats (58). These animals had enlarged adrenal glands on autopsy. Obviously, such severe organ damage is likely to have a significant "stress" effect, which may be responsible for the increase in urinary NE excretion observed in rats treated with indomethacin. Therefore, the increase in urinary NE following indomethacin treatment may be caused by a blockade of PG synthesis or by stress.

Indomethacin inhibits several enzyme systems unrelated to PG synthesis. Consequently, it may affect enzymes involved in the regulation of CA synthesis and metabolism.

EFFECTS OF PROSTAGLANDINS ON ADRENERGIC NERVOUS SYSTEM FUNCTION *IN VIVO*

Very little work has been reported on the effects of PG or inhibitors of PG synthesis on the adrenergic nervous system in intact animals or in man. All published studies have used the pharmacological blockade of PG synthesis to examine the interrelationship between PG and adrenergic nervous system function; none have studied injection or infusion of PG. One major problem is assessment of adrenergic nervous system function in animals and man. Even the measurement of plasma NE and E allows the assessment of only one aspect of the interaction between PGE and the adrenergic nervous system.

Subcutaneous injection of indomethacin in conscious rats consistently increased urinary NE excretion (44), but no effect was observed with oral indomethacin given at 1 mg/kg/day (58). This suggests that the associated pain or stress of the injection may have contributed to the increase in urinary CA excretion. This is further supported by the finding that phosphate buffer given subcutaneously increased NE excretion in the urine as much as did indomethacin. Five mg/kg/day of oral indomethacin significantly increased urinary NE (58). However, those animals showed drug toxicity. Toxic effects of indomethacin may account for findings unrelated to PG synthesis. In the latter study, the NE content of spleen, heart, and adrenal glands in the surviving animals was significantly decreased; a satisfactory interpretation of this finding is not available.

Few studies have been published on the interaction between PG and the adrenergic nervous system in man. The increased plasma NE concentration in a patient with Bartter's syndrome could be normalized by treatment with indomethacin in a dose which usually lowers urinary PG excretion (56). Measurements of urinary PG excretion can be used to index the degree of inhibition of PG synthesis. In a second study, six patients with Bartter's syndrome (24) had normal plasma NE, but urinary excretion of both NE and E was increased significantly. Treatment with indomethacin (150 mg/day for 4 days) completely corrected the hypercatecholaminuria in these patients. The results suggest that NE release in patients with Bartter's syndrome is increased in an attempt to compensate for the PG resistance to NE.

In normotensive volunteers, indomethacin (2 mg/kg/day for 7 days) significantly decreased plasma NE concentration, both in the basal state and in the acute response to erect posture and exercise with a dynamometer (26).

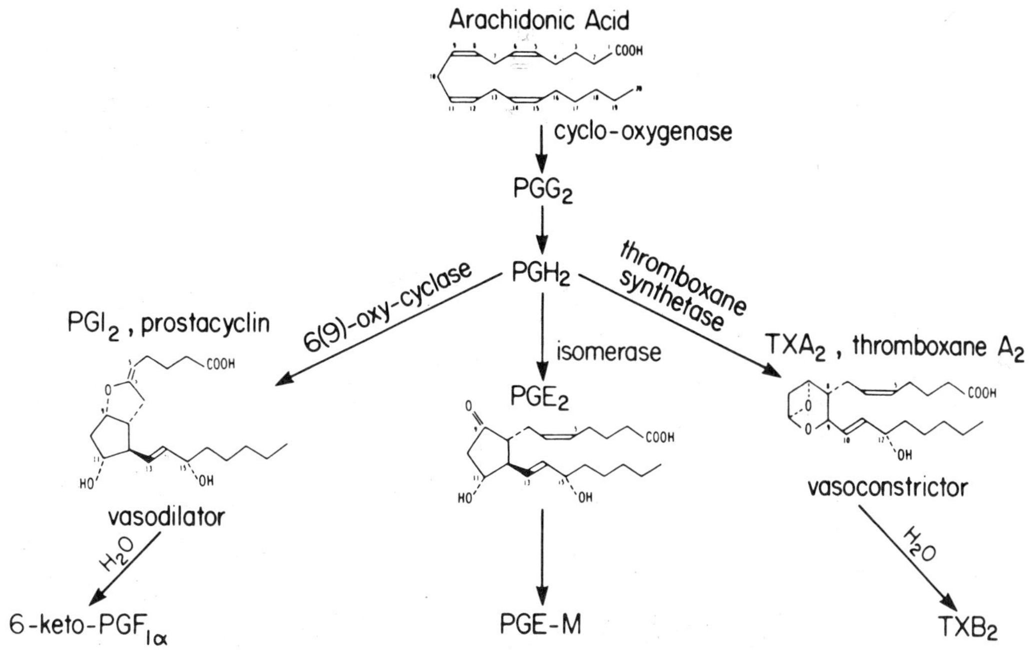

Figure 13.1. Pathways of arachidonic acid metabolism.

Urinary excretion of NE or E was not affected. Indomethacin decreased plasma NE in these two human studies, the opposite effect expected from *in vitro* studies in which PG decreased the release of NE from nerve endings. It has been postulated that plasma NE, a vasoconstrictor, was lowered via baroreceptor feedback in response to the decrease in PG vasodilators. The percentage increases of plasma NE after stimulation by posture or hand grip did not change, indicating that no change in baroreceptor sensitivity had occurred. This suggests an intact release mechanism for NE during inhibition of PGE synthesis.

The physiological effects of NE and E on their target organs are mediated by specific cell-surface receptors. Binding of the CA to these receptors is the first step mediating their biological actions. In the intact organism, α-adrenergic receptor number and function can be measured on platelet membranes, whereas β-adrenergic receptor number and function can be determined on leukocyte membranes. Inhibition of PG synthesis in normal volunteers by oral indomethacin (150 mg/day for 7 days) increases the number of β-adrenergic receptors by 92% (25). In contrast, the number of α-adrenergic binding sites did not change (25,26). Similarly, the function of either α- or β-adrenergic receptors, assessed by measuring cyclic AMP production in the isolated cell membranes in response to specific stimulating agents such as PGE or NE, did not change with indomethacin. The decrease in plasma NE with indomethacin may lead to an "up-regulation" of β-adrenergic receptor binding sites.

Obvious discrepancies exist between the results obtained in the intact organism and those obtained *in vitro*. Whereas PG appear to antagonize NE release and adrenergic neurotransmission *in vitro*, they seem to have agonistic effects on the release of NE *in vivo*; the reasons for this discrepancy are not understood. It should be pointed out that the PG concentrations used in most *in vitro* studies probably cannot be achieved in any organ under physiological conditions. Experimental evidence indicates that the *in vitro* inhibitory effects of PG on neurotransmission are predominantly mediated by PG of the E series, whereas other PG such as PGF and possibly PGB may have stimulating effects on adrenergic transmission (28). This explanation, however, can be argued. For instance, a dose of indomethacin in man, which decreased urinary excretion of

PGE-like material by 70 to 80% (26), had only variable effects on urinary PGF, but lowered plasma NE significantly. The physiological significance of PG-inhibiting actions *in vitro* may depend upon restriction of transmitter release only in states of increased sympathetic nervous activity. In most *in vitro* studies, NE release was stimulated above normal physiological levels.

MECHANISM OF PROSTAGLANDIN ACTION ON THE ADRENERGIC NERVOUS SYSTEM

Very little information is available concerning the mechanism by which prostaglandins affect release of NE from prejunctional nerve endings or adrenergic transmission in general. The available experimental evidence indicates that PG interfere somehow with stimulus-secretion coupling, *e.g.*, by impeding release or availability of calcium. Release of NE by nerve impulses is dependent on calcium. Depolarization of nerve fibers causes an inward movement of calcium bound to membranes which then causes release of NE into the synaptic cleft. In the cat spleen and in the guinea pig vas deferens, the inhibiting action of PGE_2 on stimulated NE release can be prevented by an increase in calcium ion concentration in the perfusion medium. NE release by tyramine, which is independent of calcium, is not affected by PGE_2. Finally, an increase in impulse frequency or a prolongation of the nerve action potential increases the actual concentration of calcium and diminishes the inhibitory effect of PGE_2 on NE release by nerve stimulation in the guinea pig vas deferens and in the rabbit heart and kidney (2,20,27,45,48,57).

SUMMARY

Prostaglandins are released in the vicinity of adrenergic receptors and alter NE release, receptor sensitivity to NE, and the end organs that have adrenergic receptors. Studies *in vitro* demonstrate both inhibition and stimulation of NE release, but the predominant effect of prostaglandin is to inhibit NE release and antagonize NE effects.

Studies *in vivo* have used indomethacin to block prostaglandin synthesis, but their interpretation is not clear because toxic effects of the drug can reflexly stimulate an adrenergic stress response. In man, inhibition of prostaglandin synthesis by indomethacin lowers plasma NE levels in control subjects and urinary NE in patients with Bartter's syndrome. This may be a reflex response to adrenoceptor antagonism and vasodilation produced by the prostaglandins.

References

1. Ambache N, Zar MA: An inhibitory effect of prostaglandin E_2 on neuromuscular transmission in the guinea pig vas deferens. *J Physiol (Lond)* 208:30P, 1970.
2. Baker PF, Hodgkin AL, Shaw TI: The effect of changes in internal ionic concentrations on the electrical properties of perfused giant axons. *J Physiol (Lond)* 164:335, 1962.
3. Baum T, Shropshire AT: Influence of prostaglandins on autonomic responses. *Am J Physiol* 221:1470, 1971.
4. Bergstrom S, Farnebo LO, Fuxe K: Effect of prostaglandin E_2 on central and peripheral catecholamine neurons. *Eur J Pharmacol* 21:362, 1973.
5. Brody MJ, Kadowitz PJ: Prostaglandins as modulators of the autonomic nervous system. *Fed Proc* 33:48, 1974.
6. Brundin J: The effect of prostaglandin E_1 on the response of the rabbit oviduct to hypogastric nerve stimulation. *Acta Physiol Scand* 73:54, 1968.
7. Chanh PH, Junstad M, Wennmalm Å: Augmented noradrenaline release following nerve stimulation after inhibition of prostaglandin synthesis with indomethacin. *Acta Physiol Scand* 86:563, 1972.
8. Chapnic BM, Paustian PW, Klainer E, Joiner PD, Hyman AL, Kadowitz PJ: Influence of prostaglandins E, A and F on vasoconstrictor responses to norepinephrine, renal nerve stimulation and angiotensin in the feline kidney. *J Pharmacol Exp Ther* 196:44, 1976.
9. Davies BN, Withrington PG: Actions of prostaglandins A_1, A_2, E_1, E_2, $F_{1\alpha}$ and $F_{2\alpha}$ on splenic vascular and capsular smooth muscle and their interactions with sympathetic nerve stimulation, catecholamines and angiotensin. In Mantegazza P, Horton EW: *Prostaglandins, Peptides and Amines*. London, Academic, 1969, p 53.
10. Davies BN, Withrington PG: The effects of prostaglandin E_1 and E_2 on the smooth muscle of the dog spleen and on its response to catecholamines, angiotensin and nerve stimulation. *Br J Pharmacol* 32:136, 1968.
11. Downing DT, Ahern DG, Bachta M: Enzyme inhibition by acetylenic compounds. *Biochem Biophys Res Commun* 40:218, 1970.
12. Dubocovich ML, Langer SZ: Evidence against a physiological role of prostaglandins in the regulation of noradrenaline release in the cat spleen. *J Physiol (Lond)* 251:737, 1975.
13. Euler USv, Hedqvist P: Inhibitory action of prostaglandins E_1 and E_2 on the neuromuscular transmission in the guinea pig vas deferens. *Acta Physiol Scand* 77:510, 1969.
14. Euler USv: Über die spezifische blutdrucksen-

kende Substanz des menschlichen Prostata-und Samenblasensekretes. *Klin Wochenschr* 14:1182, 1935.

15. Euler USv: Zur Kenntnis der pharmakologischen Wirkungen von Nativsekreten und Extrakten männlicher accessorischer Geschlechtsdrüsen. *Naunyn Schmiedelbergs Arch Exp Pathol Pharmakol* 175:78, 1934.
16. Ferreira SH, Moncada S, Vane JR: Indomethacin and aspirin abolish prostaglandin release from the spleen. *Nature [New Biol]* 231:237, 1971.
17. Flower RJ: Drugs which inhibit prostaglandin biosynthesis. *Pharmacol Rev* 26:33, 1974.
18. Frame MH: A comparison of the effects of prostaglandins A_2, E_2 and $F_{2\alpha}$ on the sympathetic neuroeffector system of the isolated rabbit kidney. In Samuelsson B, Paoletti R: *Advances in Prostaglandin and Thromboxane Research*. New York, Raven Press, 1976, p 369.
19. Frame MH, Hedqvist P, Aström A: Effect of prostaglandin E_2 on vascular responses of the rabbit kidney to nerve stimulation and noradrenaline, *in vitro* and *in situ*. *Life Sci* 15:239, 1974.
20. Frame MH, Hedqvist P: Evidence for prostaglandin mediated prejunctional control of renal vascular sympathetic tone. *Br J Pharmacol* 54:189, 1975.
21. Fredholm B, Hedqvist P: Increased release of noradrenaline from stimulated guinea pig vas deferens after indomethacin treatment. *Acta Physiol Scand* 87:570, 1973.
22. Fredholm BB, Hedqvist P: Indomethacin-induced increase in noradrenaline turnover in some rat organs. *Br J Pharmacol* 54:295, 1975.
23. Goldblatt MW: A depressor substance in seminal fluid. *J Soc Chem Ind London* 52:1056, 1933.
24. Güllner H-G, Gill JR, Bartter FC, Lake CR, Lakatua DJ: Correction of increased sympathoadrenal activity in Bartter's syndrome by inhibition of prostaglandin synthesis. *J Clin Endocrinol Metab* 50:857, 1980.
25. Güllner H-G, Kafka MS, Bartter FC: Indomethacin increases leukocyte β-adrenoreceptors in man. *Clin Sci* 59:397, 1980.
26. Güllner H-G, Lake CR, Bartter FC, Kafka MS: Effect of inhibition of prostaglandin synthesis on sympathetic nervous system function in man. *J Clin Endocrinol Metab* 49:552, 1979.
27. Hedqvist P: Aspects of prostaglandin and α-receptor mediated control of transmitter release from adrenergic nerves. In Usdin E, Synder S: *Frontiers in Catecholamine Research*. New York, Pergamon, 1973, p 1219.
28. Hedqvist P: Basic mechanisms of prostaglandin action on autonomic neurotransmission. *Ann Rev Pharmacol Toxicol* 17:259, 1977.
29. Hedqvist P: Control by prostaglandin E_2 of sympathetic neurotransmission in the spleen. *Life Sci* 9:269, 1970.
30. Hedqvist P: Effect of prostaglandins and prostaglandin synthesis inhibitors on norepinephrine release from vascular tissue. In Robinson HJ, Vane JR: *Prostaglandin Synthetase Inhibitors*. New York, Raven Press, 1974, p 395.
31. Hedqvist P: Inhibition of prostaglandin E_1 on vascular response to sympathetic nerve stimulation *in vivo*. *Acta Physiol Scand* 80:6A, 1970.
32. Hedqvist P: Prostaglandin action on transmitter release at adrenergic neuroeffector junctions. In Samuelsson B, Paoletti R: *Advances in Prostaglandin and Thromboxane Research*. New York, Raven Press, 1976, p 357.
33. Hedqvist P: Prostaglandin-induced inhibition of vascular tone and reactivity in the cat's hindleg *in vivo*. *Eur J Pharmacol* 17:157, 1972.
34. Hedqvist P: Prostaglandin mediated control of sympathetic neuroeffector transmission. *Adv Biosci* 9:461, 1973.
35. Hedqvist P: Reduced effector response to nerve stimulation in the cat spleen after administration of prostaglandin E_1. *Acta Physiol Scand* 74:7A, 1968.
36. Hedqvist P, Brundin J: Inhibition by prostaglandin E_1 of noradrenaline release and of effector response to nerve stimulation in the cat spleen. *Life Sci* 8:389, 1969.
37. Hedqvist P, Euler USv: Prostaglandin-induced neurotransmission failure in the field-stimulated, isolated vas deferens. *Neuropharmacology* 11:177, 1972.
38. Hedqvist P, Moawad A: Presynaptic α- and β-adrenoceptor mediated control of noradrenaline release in human oviduct. *Acta Physiol Scand* 95:494, 1975.
39. Hedqvist P, Stjärne L, Wennmalm Å: Facilitation of sympathetic neurotransmission in the cat spleen after inhibition of prostaglandin synthesis. *Acta Physiol Scand* 83:430, 1971.
40. Hedqvist P, Wennmalm Å: Comparison of the effects of prostaglandin E_1, E_2 and $F_{2\alpha}$ on the sympathetically stimulated rabbit heart. *Acta Physiol Scand* 83:156, 1971.
41. Hedwall PR, Abdel-Dayed WA, Schmid PG, Mark AL, Abboud FM: Vascular responses to prostaglandin E_1 in gracilis muscle and hindpaw of the dog. *Am J Physiol* 221:42, 1971.
42. Hoszowska A, Panczenko B: Effects of inhibition of prostaglandin biosynthesis on noradrenaline release from isolated perfused spleen of the cat. *Pol J Pharmacol Pharm* 26:137, 1974.
43. Illés P, Vizi ES, Knoll J: Adrenergic neuroeffector junctions sensitive and insensitive to the effect of PGE_1. *Pol J Pharmacol Pharm* 26:127, 1974.
44. Junstad M, Wennmalm Å: Increased renal excretion of noradrenaline in rats after treatment with prostaglandin synthesis inhibitor indomethacin. *Acta Physiol Scand* 85:573, 1972.
45. Junstad M, Wennmalm Å: Prostaglandin mediated inhibition of noradrenaline release at different impulse frequencies. *Acta Physiol Scand* 89:544, 1973.
46. Kadowitz PJ, Sweet CS, Brody MJ: Blockade of adrenergic vasoconstrictor responses in the dog by prostaglandins E_1 and A_1. *J Pharmacol Exp Ther* 179:563, 1971.
47. Kadowitz PJ: Effect of prostaglandins E_1, E_2 and A_2 on vascular resistance and responses to noradrenaline, nerve stimulation and angiotensin in the dog hindlimb. *Br J Pharmacol* 46:395, 1972.
48. Koketsu K: Action of tetraethylammonium chlo-

ride on neuromuscular transmission in frogs. *Am J Physiol* 193:213, 1958.

49. Lonigro AJ, Terragno NA, Malik KU, McGiff JC: Differential inhibition by prostaglandins of the renal actions of pressor stimuli. *Prostaglandins* 3:595, 1973.
50. Malik KU, McGiff JC: Modulation by prostaglandins of adrenergic transmission in the isolated perfused rabbit and rat kidney. *Circ Res* 36:599, 1975.
51. Moawad A, Hedqvist P, Bygdeman M: Noradrenaline release following nerve stimulation and its modification by prostaglandin E_2 in human and rabbit oviduct. *Acta Physiol Scand* 95:142, 1975.
52. Needleman P, Marshall GR, Johnson EM: Determinants and modification of adrenergic and vascular resistance in the kidney. *Am J Physiol* 227:665, 1974.
53. Neufeld AH, Page ED: Regulation of adrenergic neuromuscular transmission in the rabbit iris. *Exp Eye Res* 20:549, 1975.
54. Park MK, Dyer DC, Vincenzi FF: Prostaglandin E_2 and its antagonists: Effects on autonomic transmission in the isolated sino-atrial node. *Prostaglandins* 4:717, 1973.
55. Samuelsson B, Wennmalm Å: Increased nerve stimulation induced release of noradrenaline from the rabbit heart after inhibition of prostaglandin synthesis. *Acta Physiol Scand* 83:163, 1971.
56. Silverberg AB, Mennes PA, Cryer PE: Resistance to endogenous norepinephrine in Bartter's syndrome. *Am J Med* 64:231, 1978.
57. Stjärne L: Frequency dependence of dual negative feedback control of secretion of sympathetic neurotransmitter in guinea pig vas deferens. *Br J Pharmacol* 49:358, 1973.
58. Stjärne L: Hyperexcretion of catecholamines induced by indomethacin. *Acta Physiol Scand* 83:574, 1971.
59. Stjärne L, Gripe K: Prostaglandin-dependent and -independent feedback control of noradrenaline secretion in vasoconstrictor nerves of normotensive human subjects. *Naunyn Schmiedebergs Arch Pharmacol* 280:441, 1973.
60. Stjärne L: Prostaglandin-versus α-adrenoceptor-mediated control of sympathetic neurotransmitter secretion in guinea pig isolated vas deferens. *Eur J Pharmacol* 22:233, 1973.
61. Westfall TC, Brasted M: Specificity of blockade of the nicotine-induced release of ^{3}H-norepinephrine from adrenergic neurons of the guinea pig heart by various pharmacological agents. *J Pharmacol Exp Ther* 189:659, 1974.
62. Wennmalm Å: Studies on mechanisms controlling the secretion of neurotransmitters in the rabbit heart. *Acta Physiol Scand* 82:1, 1971.

CHAPTER FOURTEEN

Strategies for Dynamic Assessment of Sympathetic Activity from Plasma Catecholamine Measurements

PETER C. RUBIN, M.D., and JOHN L. REID, M.D.*

INTRODUCTION

Improved biochemical precision has made it possible to measure plasma norepinephrine and epinephrine concentrations with increasing sensitivity. Norepinephrine concentrations have been a source of considerable interest, particularly in blood pressure research, and it has often been assumed that plasma concentrations of norepinephrine can be used as an index of sympathetic nervous activity. However, the concentration of catecholamines in plasma is a static measurement which is determined by two dynamic processes: release into and removal from the circulation.

Norepinephrine in plasma is derived mainly from sympathetic nerves (32). Following release into the neuroeffector junction, most norepinephrine is either metabolized or removed by uptake$_1$ into the nerve terminal (23). A small proportion of the released norepinephrine escapes these processes and reaches the circulation. It is then removed either by metabolism in the liver or lung or by nonspecific uptake into nonneuronal tissue, referred to as uptake$_2$ (19) or uptake into another nerve. Changes in plasma norepinephrine concentration can therefore be the result of changes in the rate of release into the circulation or the rate of clearance by the various uptake and metabolic processes or both.

The importance of determining norepinephrine kinetic indices has been appreciated for some time. This chapter describes methods used in studying the kinetics of norepinephrine in plasma and discusses results of studies in which these approaches have been used. It compares these limited kinetic studies to the more extensive literature describing simple "static" measurements of plasma norepinephrine and discusses the interpretation of studies involving plasma norepinephrine, *i.e.*, choice of control or reference groups in studies of patient populations.

CHOICE OF "NORMAL" CONTROL GROUPS

In any experimental design the choice of appropriate controls is essential. Unfortunately, in many clinical studies there are difficulties in defining or obtaining an optimal control population with which to compare different pathophysiological groups or categories.

In early studies using sensitive radioenzymatic assays for plasma catecholamines, several groups reported that the mean catecholamine (usually norepinephrine) level in groups of patients with essential hypertension was significantly higher than "normal" controls (7, 8, 27). Although the "controls" were normotensive, very little additional information was published on the age range, sex distribution, and other characteristics of the reference groups in these studies.

It has been noted that while the mean value

* We would like to thank Mrs. Mary Wood for her assistance in preparing this manuscript and our numerous colleagues in Glasgow and London who contributed to the experiments described in this paper. We are grateful for a grant from Age Research.

of norepinephrine in hypertensives exceeds that of normals, in 50 to 60% of patients the levels in hypertensives are within the range described for normals. The variance of the "controls" is very much smaller than that of the patient groups in studies which observed increased norepinephrine levels (23).

In view of the complex release and clearance mechanisms determining plasma norepinephrine at steady state, it is not surprising that factors other than blood pressure can contribute to the level of amine in plasma. Ziegler *et al.* (34) reported that increasing age was associated with higher plasma norepinephrine. In a further study when age was taken into account, no difference was found between plasma norepinephrine in normotensive and hypertensive subjects (25). Several authors have subsequently confirmed the age-related increase in plasma norepinephrine (13, 30), but have failed to note any age-related increase in plasma epinephrine (1, 13).

Several more recent studies have reported significantly higher plasma norepinephrine levels, particularly in young hypertensives (5, 17, 22), but a large number of other studies have failed to find differences between hypertensive and normotensive groups (6, 13, 18, 20, 30, 31). Plasma epinephrine has been measured in a more limited number of studies and has been found to be elevated in hypertensive groups in the studies of Franco Morselli *et al.* (13) and Bertel *et al.* (1).

Full information is not available on the source of control subjects in all these studies, but it is likely that hypertensive patients, particularly in the early studies, were not carefully age-matched and that older patients were compared with younger controls. The most accessible group for use as healthy "controls" are often medical, scientific, and paramedical staff or students associated with the research project. These individuals are usually in their 20s or early 30s, in contrast to the patient group being studied. In addition to their relative youth, these "controls" are usually much more familiar with the investigational environment, personnel, instruments, and techniques used in research than a patient group being investigated.

Plasma norepinephrine was collected from supine and standing subjects by venipuncture from medical and paramedical staff familiar with clinical research projects. These samples were compared to those of two groups of outpatients (21). Both the normotensive group and the hypertensive group saw the same doctors the same number of times. Supine plasma norepinephrine after at least 10 minutes rest was significantly lower and the variance markedly smaller in the group of laboratory workers than in the patient groups. There were no clear differences in supine plasma norepinephrine between normotensive and hypertensive outpatients (Fig. 14.1). After standing erect for 5 minutes, plasma norepinephrine rose in all three groups to a similar level (Fig. 14.2). These results indicate that experienced laboratory personnel may reduce sympathetic activity to "basal levels" after 10 minutes rest while inexperienced subjects or patients may have a residual increase in sympathetic tone—a possible manifestation of anxiety or psychosocial stress. We have no evidence that longer periods of supine rest consistently reduce the raised levels of catecholamines in these subjects.

It may be relevant that norepinephrine levels in our laboratory controls are very similar to "normal controls" in studies where plasma norepinephrine was found to be elevated in hypertensives (15). Few could dispute

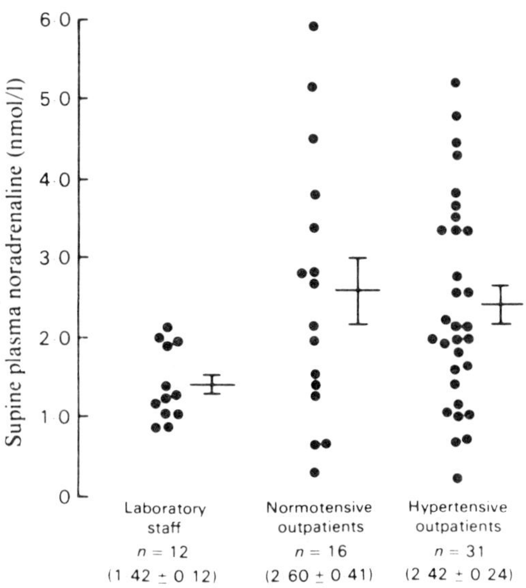

Figure 14.1. Plasma norepinephrine 10 mins after resting supine in normotensive laboratory staff and normotensive and hypertensive outpatients. (Reproduced with permission from D. H. Jones *et al.* (21).)

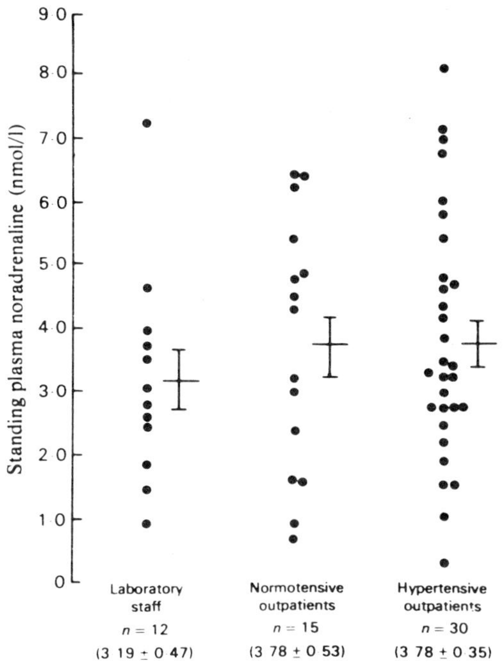

Figure 14.2. Plasma norepinephrine after 5 minutes standing in normotensive laboratory staff and normotensive and hypertensive outpatients. (Reproduced with permission from D. H. Jones *et al.* (21).)

that age-matched normotensive outpatients are a more appropriate control group for hypertensive patients than young laboratory workers, and we noted no differences between the two patient groups. Recent reviews have highlighted the deficiencies in design and execution of catecholamine studies in hypertension (15, 23) and will be discussed further in this volume.

Where possible, the optimal "control" group should therefore be selected from the same population as the group under study and if necessary, more than one "control" group included. Much more effort may be necessary to assemble the most appropriate "control" group than the test group, especially since the index being studied, plasma norepinephrine, is subject to many physical and psychological factors and its clearance (as discussed later) shows wide interindividual differences.

CATECHOLAMINE RELEASE AND CLEARANCE IN MAN

Some experimental strategies for assessing release and clearance of catecholamines include measurement of terminal metabolites in plasma and urine, use of radioactive or stable isotope precursors, and drug-induced perturbations of synthesis or clearance. Although many of these techniques are not readily applicable to man or to distinguishing between central and peripheral catecholamine turnover (29), two relatively simple methods applicable to clinical studies have been explored and are described in detail.

Infusion of Unlabeled *l*-Norepinephrine

This method is based on achieving steady-state concentrations by infusing norepinephrine at a constant rate. The rate of norepinephrine clearance can then be calculated, given a knowledge of the infusion rate, the preinfusion plasma norepinephrine concentration, and the steady-state norepinephrine concentration using the following relationship:

$$\text{Clearance} = \frac{\text{Infusion rate}}{\text{Steady-state concentration} - \text{basal concentration}}.$$

Once clearance rate is established, the endogenous rate of norepinephrine entry to the circulation can be calculated from a knowledge of basal norepinephrine concentrations according to the following relationship:

$$\text{Rate} = \text{Clearance} \times \text{basal concentration}.$$

This method is based on two critical assumptions. The first is that steady state has, indeed, been achieved at the time of blood sampling. Because of the extremely short half-life of norepinephrine, steady state is achieved in less than 10 minutes as shown by our studies and others (4, 12). The requirement for steady state implies that this method is suitable only for calculating kinetic indices under stable conditions and cannot be used to assess short-term alterations in release or clearance. The second assumption is that infusion of exogenous norepinephrine influences neither the release nor clearance of endogenous norepinephrine. Inevitably, the infusion of norepinephrine at a rate sufficient to elevate plasma concentrations will result in some hemodynamic changes. The question of whether acute hemodynamic alterations influence kinetic indices has been directly addressed (12), and modest changes in blood pressure (up to 20 mm Hg systolic and 14 mm Hg diastolic) do not appear to influence norepinephrine kinetics.

To carry out the infusion, norepinephrine is dissolved in normal saline containing 1% ascorbic acid and is then infused at the rate of 0.06 μg/kg/min using a constant rate infusion pump, syringe, and connecting tubing protected from light. The infusions performed by us lasted for 3 hours, and six blood samples were drawn during this time (12).

Infusion of Radiolabeled *l*-Norepinephrine

Recently a method has been described which uses high specific activity tritiated *l*-norepinephrine (9). The tritiated norepinephrine is infused until a plateau concentration is achieved, whereupon both plasma norepinephrine and specific activity are determined under steady-state conditions. This technique allows the calculation of norepinephrine kinetic indices based on the following relationships:

Apparent release rate

$$= \frac{[^3\text{H}]\text{norepinephrine infusion rate (dpm/min)}}{\text{Specific activity of plasma norepinephrine (dpm/pg)}}$$

Clearance

$$= \frac{[^3\text{H}]\text{norepinephrine infusion rate (dpm/min)}}{\text{Steady-state plasma } [^3\text{H}]\text{NE concentration (dpm/ml)}}$$

Again, the steady-state assumption is critical to this method, as it is for the technique described above. Another assumption is that recirculation of radiolabeled norepinephrine by release from nerve endings following uptake contributes insignificantly to the total norepinephrine release (26).

The advocates of this method indicate that low concentrations of norepinephrine are not likely to alter hemodynamic indices. Use of this technique assumes that all the analyzed tritium actually represents unmetabolized norepinephrine. The investigators tried to ensure that their extraction procedure did, indeed, exclude tritiated metabolites. It appears to take between 30 and 60 minutes to achieve a plateau of radioactivity in the plasma, which suggests the possibility that accumulation of tritiated metabolites contributes to the final observed steady state or that [^{3}H]norepinephrine is released back into plasma after uptake.

Studies on Catecholamine Kinetics in Normal Subjects

The first investigation of norepinephrine kinetics using an infusion of unlabeled substance was described in 1959 (4) when 16 normal subjects aged 21 to 29 received constant rate intravenous infusions of epinephrine or norepinephrine. The infusions lasted up to 65 minutes, and plasma analyses were performed using a fluorometric determination. Steady-state concentration of both epinephrine and norepinephrine increased linearly with infusion rate over an extremely wide range of infusion (up to 0.5 μg/kg/min). Both epinephrine and norepinephrine steady states were achieved within 10 minutes of starting any particular rate of infusion. The authors estimated that norepinephrine is released into the circulation at the rate of 0.04 μg/kg/min, while epinephrine is released at the rate of 0.01 μg/kg/min.

Another study 20 years later reported results of norepinephrine kinetics in normal subjects following the infusion of *l*-norepinephrine, but on this occasion the plasma analyses were performed using the radioenzymatic method (12). Initially, five normal young men received an infusion of *l*-norepinephrine, which was successively increased from 0.01 through 0.075 μg/kg/min, each infusion step lasting for 10 minutes. The second part of the study used an infusion rate that produced an increase in blood pressure of between 15 and 20 mm Hg. This infusion was carried out for 10 hours and blood drawn at 2.5, 7.5, and 10 minutes after start of infusion and at various times during steady state. The calculated rate of norepinephrine entry into the circulation did not differ significantly between the two study days and was found to average 0.83 μg/min. There was no difference in the calculated values for clearance between the earliest time of observation shortly after the infusion began and the later times. Since the blood pressure continued to rise during the first 10 minutes of the long-term infusion, the authors concluded that acute hemodynamic changes do not significantly affect the clearance of norepinephrine. Both the rate of norepinephrine entry to the circulation and its clearance showed considerable individual variation: the rate of norepinephrine entry ranged from 0.33 to 1.95 μg/min, and the norepinephrine clearance at steady state ranged from 2.97 to 7.25 min. In

Table 14.1
Summary of Studies Involving Norepinephrine Kinetics

Investigators	Reference	Method	Subjects	Conclusions
Cohen *et al.*	4	Unlabeled norepinephrine and epinephrine	Normal young	Norepinephrine release rate 0.04 μg/kg/min; epinephrine 0.01 μg/kg/min
Fitzgerald *et al.*	12	Unlabeled norepinephrine	Normal young	Entry to circulation range: 0.33–1.95 μg/min; clearance 2.97 to 7.25 L/min; small hemodynamic changes did not influence kinetics
Clutter *et al.*	3	Unlabeled epinephrine	Normal young	Clearance average 50 ml/min/kg; influenced by rate of infusion at lower rates
Rubin *et al.*	Submitted	Unlabeled norepinephrine	Normal young; normal elderly	Release rate greater in elderly; no difference in clearance (see text for details)
Esler *et al.*	9	[^{3}H]norepinephrine	Normal young	Release rate average 0.54 μg/m^2/min; clearance 3 L/min
Esler *et al.*	11	[^{3}H]norepinephrine	Normal: age range 20–69	Clearance decreased significantly with age

view of this considerable variance, using plasma norepinephrine concentrations as an index of sympathetic nerve activity within an individual at different times is probably valid. However, the considerable variations in release and clearance between individuals do not favor its use in comparisons between individuals. Studies in this laboratory have confirmed the wide variation in norepinephrine kinetic indices between healthy individuals within a narrow age range.

The method of infusing unlabeled catecholamines has also been employed to study the kinetics of epinephrine (3). This study involved six normal subjects who each received a 60-minute epinephrine infusion at 0.1, 0.5, 1.0, 2.5, and 5.0 μg/min. Like norepinephrine, steady state was achieved within 10 minutes of beginning the infusion. The basal clearance of epinephrine was around 50 ml/min/kg, but interestingly, the clearance rate increased significantly between infusion rates. The authors conclude from this that epinephrine influences its own clearance, in contrast to the findings described above for norepinephrine. However, some reservations that should be considered are: (a) The only difference in clearance is between the very lowest rates of infusion and all the other rates of infusion. Epinephrine clearance does not progressively rise with an increased infusion rate, even though systolic blood pressure progressively rose as the steady-state epinephrine concentration increased. (b) Clearance calculation at the lowest rates of infusion could have included some error since the authors indicate that epinephrine concentrations overlap the normal range. Whether the epinephrine concentration at steady state was actually significantly different from baseline is not entirely clear.

Norepinephrine kinetics in normal subjects have also been investigated using the technique of infusing tritiated norepinephrine (9). Ten normal male subjects of unspecified age received an infusion of tritiated norepinephrine at a constant rate of 0.35 μCi/m^2/min for 90 minutes. Blood samples were drawn for norepinephrine analysis and radioactive counting on two or three occasions following the attainment of steady state. Using this technique, norepinephrine release rate into the circulation was found to be 0.54 ± 0.2 μg/m^2/min, while clearance was found to average 3 L/min.

These investigations indicate that different methods of calculating norepinephrine kinetics reach very similar conclusions. Although it is certainly tempting to assume that each method to some extent validates the other, the exact meaning of the values obtained for clearance and rate of release must be carefully considered, particularly with regard to their relevance to events at the synaptic cleft. It must be clearly understood that rate of release refers to spillover of norepinephrine from the synaptic cleft into the circulation and does not necessarily equate with the rate of norepinephrine release from sympathetic nerve terminals. Similarly, most norepinephrine which is released into the synaptic junction is probably cleared locally, while at least a considerable proportion of infused norepinephrine will undergo clearance by the extraneuronal pathways. A further point worth noting is the substantial variance within apparently homogeneous groups, since this certainly has implications with regard to type 1 and type 2 statistical errors in comparing data from different groups.

Norepinephrine Kinetics in Idiopathic Autonomic Insufficiency

Norepinephrine kinetics were studied in four untreated patients with idiopathic peripheral autonomic insufficiency (9). These individuals were found to have a norepinephrine release rate about a third of that described by these workers for normal subjects. However, the clearance of norepinephrine was also found to be reduced from around 3 L/min in the normal subjects to less than 2 L/min in the patients. The decreased clearance achieved statistical significance and is of interest, since it implies that, other things being equal, neuronal uptake of norepinephrine plays a significant role in the clearance of norepinephrine from the circulation. Unfortunately, very little information is given about these patients. Their blood pressures and general medical conditions at the time of the study are not referred to. This interesting observation would seem to merit further investigation.

Norepinephrine Kinetics and Age

Plasma concentration of norepinephrine increases with age (34) and the kinetic background to this has been the subject of two studies with a different methodology being used in each. We have studied 16 male volunteers, 8 with ages of between 21 and 36 and 8 aged between 65 and 78 years. Unlabeled *l*-norepinephrine was infused at the rate of 0.06 μg/kg/min, and six blood samples were drawn from each subject during steady state. Figures 14.3 and 14.4 show that there was no significant difference in clearance between the two groups (instead, there was virtually complete overlap in clearance values), but the rate of norepinephrine entry into the circulation was significantly greater in the older group, both in the lying and standing positions. Since the normal elderly volunteers tended to have higher blood pressures than their younger counterparts, we analyzed the data for the whole group by a generalized linear interac-

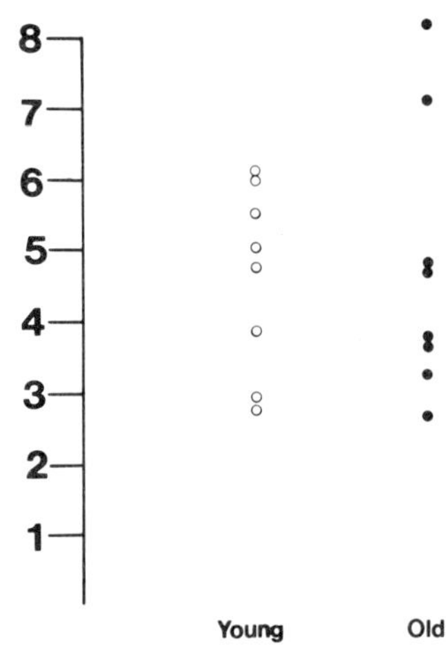

Figure 14.3. Clearance of norepinephrine (L/min) in young and old subjects. There is no significant difference between groups.

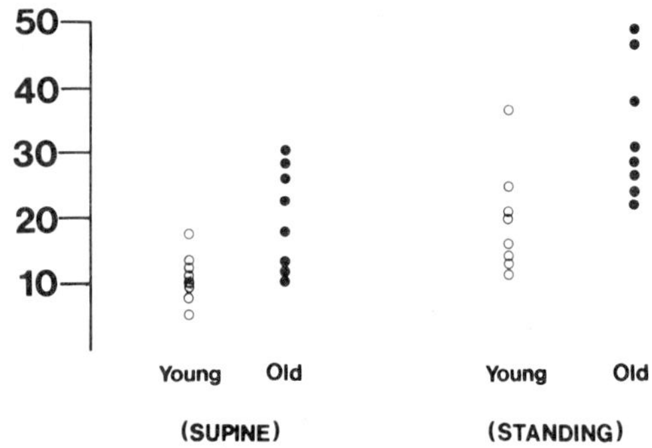

Figure 14.4. Rate of norepinephrine entry to the circulation (nmoles/min). The group of elderly subjects has a significantly greater rate of entry in both supine ($p < 0.05$) and standing position ($p < 0.01$). Data were compared by Wilcoxon rank sum test.

tive modelling program to assess the relative influence of blood pressure and age (entered as a discontinuous variable) on rate of norepinephrine release. We found that blood pressure had no significant effect on the observed relationship. Another group of investigators studied this problem using an infusion of tritiated norepinephrine but arrived at an opposite conclusion (11). This study involved 34 healthy subjects with ages ranging from 20 to 69 years. Each subject received a bolus injection of tritiated norepinephrine followed by a constant infusion for 90 minutes as described above. Although the rate of norepinephrine entry to the circulation was higher in the older age ranges, this did not reach statistical significance. In contrast, these investigators found a significant exponential relationship between age and norepinephrine clearance of the form

$$\text{Clearance} = 1.66\ e^{-0.0057\ \text{age}}.$$

The conflicting results of these two studies, which used different types of study design and data analysis, leave the question of norepinephrine kinetics in the elderly unresolved. Studies by B. G. Wallin (Chapter 28) which report that bursts of muscle sympathetic nerve activity, when adjusted for heart rate, increase with age (33) support the findings from our laboratory.

CONCLUSIONS

Sensitive specific assays that are now available for catecholamines in biological materials will not compensate for poor experimental design or inappropriate choice of control reference groups. Age, sex, smoking, and alcohol intake, physical and psychological aspects, and familiarity with the investigators and their laboratory must be taken into account. Relatively simple methods of assessing catecholamine release and clearance using intravenous infusions of *l*-norepinephrine radiolabeled or unlabeled, indicate that there are wide interindividual differences in clearance of catecholamines which will further confound interpretation of single measures of plasma catecholamines.

References

1. Bertel O, Buhler FR, Kiowski W, Lutold BE: Decreased beta adrenoceptor responsiveness as related to age, blood pressure and plasma catecholamines in patients with essential hypertension. *Hypertension* 2:130–138, 1980.
2. Bufano G, Vaona G, Storcick R: Approach to the pharmacokinetics of dl-7-H^3-noradrenaline. *Eur J Clin Pharmacol* 6:88–92, 1973.
3. Clutter WE, Bier DM, Shah SD, Cryer PE: Epinephrine plasma metabolic clearance rates and physiologic thresholds for metabolic and haemodynamic actions in man. *J Clin Invest* 66:94–101, 1980.
4. Cohen G, Holland B, Sha J, Goldenberg M: Plasma concentrations of epinephrine and norepinephrine during intravenous infusions in man. *J Clin Invest* 38:1935–1941, 1959.
5. Cousineau D, de Champlain J, Lapointe L: Circulatory catecholamines and systolic time intervals in labile and sustained hypertension. *Clin Sci Mol Med* 55:65s–68s, 1978.
6. De Leeuw PW, Wester A, Punt R, Falke HE, Birkenhager WH: Noradrenaline levels in essential hypertensives and normotenive controls. *Neth J Med* 22:145–148, 1979.
7. De Quattro V, Chan S: Raised plasma-catecholamines in some patients with primary hypertension. *Lancet* i:806–809, 1972.
8. Engleman K, Portnoy B, Sjoerdsma A: Plasma catecholamine concentration in patients with hypertension. *Circ Res* 26,27 (Suppl 1):141–145, 1970.
9. Esler M, Jackman G, Bobik A, Kelleher D, Jennings G, Leonard P, Skews H, Korner P: Determination of norepinephrine apparent release rate and clearance in humans. *Life Sci* 25:1461–1470, 1979.
10. Esler M, Leonard P, Kelleher D, Jackman G, Bobik A, Skews H, Jennings G, Korner P: Assessment of neuronal uptake of noradrenaline in humans: defective uptake in some patients with essential hypertension. *Clin Exp Pharmacol Physiol* 7:535–539, 1980.
11. Esler M, Skews H, Leonard P, Jackman G, Bobik A, Korner P: Age-dependence of noradrenaline kinetics in normal subjects. *Clin Sci* 60:217–219, 1981.
12. Fitzgerald GA, Hossman V, Hamilton CA, Reid JL, Davies DS, Dollery CT: Interindividual variation in kinetics of infused norepinephrine. *Clin Pharmacol Ther* 26:669–675, 1979.
13. Franco-Morselli R, Elghozi JL, Joly E, Di Giuilio S, Meyer P: Increased plasma adrenaline concentrations in benign essential hypertension. *Br Med J* 2:1251–1254, 1977.
14. Gitlow SE, Mendlowitz M, Bertani LM, Silk S, Wilk EK: Human norepinephrine metabolism—its evaluation by administration of tritiated norepinephrine. *J Clin Invest* 50:859–865, 1971.
15. Goldstein DS: Plasma norepinephrine in essential hypertension. A study of the studies. *Hypertension* 3:48–52, 1981.
16. Goodall MC, Kirshner N, Rosen L: Metabolism of noradrenaline in the human. *J Clin Invest* 38:707–714, 1959.
17. Hofman A, Boomsma F, Schalekamp MADH, Valkenburg HA: Raised blood pressure and plasma noradrenaline concentrations in teenagers and young adults selected from an open population. *Br Med J* 1:1536–1538, 1979.
18. Ibsen H, Christensen NJ, Hollnagel H, Leth A,

Kappelgaard AM, Giese J: Plasma noradrenaline concentration in hypertensive and normotensive forty year-old individuals: Relationship to plasma renin concentration. *Scand J Clin Lab Invest* 40:333–339, 1980.

19. Iversen LL, Jarrott B, Simmonds MA: Differences in uptake, storage and metabolism of (+) and (−) noradrenaline. *Br J Pharmacol* 43:845–855, 1971.
20. Jones DH, Hamilton CA, Reid JL: Plasma noradrenaline, age and blood pressure: A population study. *Clin Sci Mol Med* 55:73S–76S, 1978.
21. Jones DH, Hamilton CA, Reid JL: Choice of control groups in the appraisal of sympathetic activity in essential hypertension. *Clin Sci* 57:339–344, 1979.
22. Koboyashi K, Kolloch R, deQuattro V, Miano L: Increased plasma and urinary normetanephrine in young patients with primary hypertension. *Clin Sci* 57:173S–176S, 1979.
23. Kopin IJ, McCarty R, Yamaguchi I: Plasma catecholamines in human and experimental hypertension. *Clin Exp Hypertens* 2:379–394, 1980.
24. LaBrosse EH, Axelrod J, Kopin IJ, Kety SS: Metabolism of 7-H^3 epinephrine-d-bitartrate in normal young men. *J Lab Clin Invest* 40:253–260, 1961.
25. Lake CR, Ziegler MG, Coleman MD, Kopin IJ: Age-adjusted plasma norepinephrine levels are similar in normotensive and hypertensive subjects. *N Engl J Med* 296:208–209, 1977.
26. Landsberg L, Young JB: Fasting, feeding and regulation of the sympathetic nervous system. *N Engl J Med* 298:1295–1300, 1978.
27. Louis WJ, Doyle AE, Anavekar SN, Johnston CI, Geffen LB, Rush R: Plasma catecholamine, dopamine-beta-hydroxylase, and renin levels in essential hypertension. *Circ Res* 34 and 35 (Suppl 1):57–61, 1974.
28. Maas JW, Landis DH: The metabolism of circulating norepinephrine by human subjects. *J Pharmacol Exp Ther* 177:600–612, 1971.
29. Reid JL, Jones DH, Fitzgerald G, Davies D, Boobis A: Catecholamine turnover in essential hypertension. *Clin Exp Hypertens* 2:395–408, 1980.
30. Sever PS, Birch M, Osikowska B, Tunbridge RDG: Plasma-noradrenaline in essential hypertension. *Lancet* i:1078–1081, 1977.
31. Taylor AA, Pool JL, Lake CR, Ziegler MG, Rosen RA, Rollins DE, Mitchell JR: Plasma norepinephrine concentrations: No difference among normal volunteers and low, high or normal renin hypertensive patients. *Life Sci* 22:1499–1510, 1978.
32. von Euler V, von Euler US, Franksson C, Hellstrom J: Adrenaline and noradrenaline output in urine after unilateral and bilateral adrenalectomy in man. *Acta Physiol Scand* 51:19, 1954.
33. Wallin BG, Sundlof G: A quantitative study of muscle nerve sympathetic activity in resting normotensive and hypertensive subjects. *Hypertension* 1:67–77, 1979.
34. Ziegler MG, Lake CR, Kopin IJ: Plasma noradrenaline increases with age. *Nature* 261:333–335, 1976.

CHAPTER FIFTEEN

Caloric Intake and Sympathoadrenal Activity: Implications for Thermogenesis, Obesity, and Hypertension

LEWIS LANDSBERG, M.D. and JAMES B. YOUNG, M.D.

INTRODUCTION

The studies of Walter B. Cannon early this century clearly established the role of the sympathoadrenal system in the regulation of many diverse aspects of mammalian physiology. One area not investigated by Cannon, however, concerns the role of the sympathoadrenal system in the adaptation to changes in dietary intake. Although in retrospect a relationship between dietary intake and the sympathetic nervous system (SNS) seems obvious, the effect of fasting and overfeeding on sympathetic activity has only been investigated recently. Studies conducted in our laboratory during the last 5 years indicate that caloric restriction diminishes SNS activity, while overfeeding stimulates it. Glucose metabolism within the ventromedial portion of the hypothalamus, perhaps stimulated by insulin, appears to be one factor initiating changes in SNS activity in response to dietary changes. Diet-induced changes in SNS activity may play an important role in mediating some physiological alterations that accompany changes in caloric intake, and therefore may have implications for the pathogenesis and treatment of human obesity and hypertension.

NOREPINEPHRINE TURNOVER AS AN INDEX OF SNS ACTIVITY: THE EFFECT OF COLD EXPOSURE

Assessment of SNS Activity

Measuring SNS activity with precision and sensitivity is difficult. Our laboratory has utilized measurements of norepinephrine (NE) turnover in peripheral sympathetially innervated tissues to estimate SNS activity quantitatively in a variety of different physiological and pathophysiological states in the experimental animal. NE turnover in individual organs or tissues may be estimated from the rate of decline of NE specific activity after the neurotransmitter stores are labeled by a tracer dose of tritiated NE (11, 13) or from the disappearance rate of endogenous NE after inhibition of neurotransmitter biosynthesis with α-methyltyrosine (2, 11). The studies reviewed here were conducted predominantly with the tritiated NE technique; in many instances, however, the results were confirmed by similar experiments with α-methyltyrosine (24, 25). In addition, experiments with ganglionic blocking agents were utilized to relate measured changes in NE turnover to changes in central SNS activity.

The tritiated NE technique depends upon the axonal membrane transport system of the peripheral sympathetic nerve endings. After intravenous administration, a bolus of labeled tracer NE is rapidly cleared from the circulation, taken up in sympathetic nerve endings, and stored along with endogenous NE in subcellular storage granules within the nerve endings. Five minutes after administration, the levels of tracer in plasma are low compared to levels in sympathetically innervated tissues, so that subsequent relabeling of the tissue stores from the circulation is quantitatively insignificant; within 5 minutes the tracer has mixed with the NE pool, and the disappearance rate of tracer is a valid marker of NE release (13).

Effect of Cold Exposure on Cardiac NE Turnover in the Rat

The measurement of cardiac NE turnover during acute *cold exposure* in the rat is demonstrated in Figure 15.1. Tracer [^{3}H]NE was

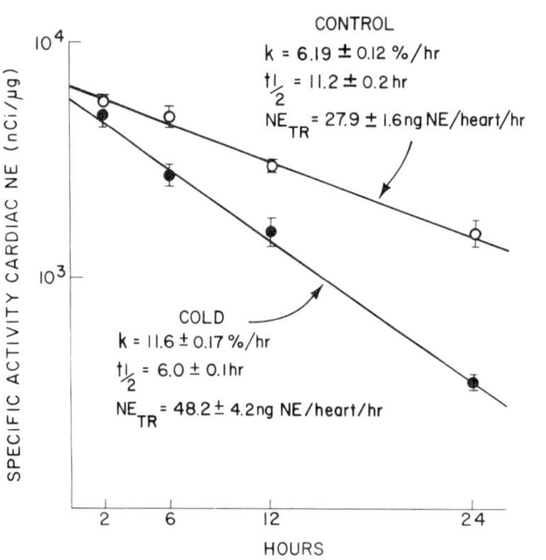

Figure 15.1. Increase in cardiac NE turnover during acute cold exposure in the rat. After intravenous administration of tracer NE at time zero, groups of animals were killed, and the hearts removed and analyzed for [^{3}H] and endogenous NE. The specific activity of cardiac NE is plotted as a semilogarithmic function of time according to the method of least squares. The rate of decline obeys first order kinetics. The slope (*k*) is equal to the fractional turnover rate; a NE turnover rate (NE_{TR}) can be calculated from the product of the slope and the endogenous NE concentration. Cold exposure increases cardiac NE turnover. (Reproduced with permission from *New England Journal of Medicine* 298:1295–1301, 1978.)

administered intravenously at time zero. Half the animals were placed in a 4°C environment, while the other half remained at ambient temperature. All the animals were allowed ad lib access to chow and water. At various preselected times, groups of animals were killed and the hearts rapidly removed and analyzed for tritiated and endogenous NE. The disappearance of the tracer as a function of pool size is represented by the rate of decline of specific activity over time; since the decline obeys first order kinetics, a slope can be calculated representing the fractional turnover rate (k) as well as half-time of disappearance ($t = ½$) of the transmitter (Fig. 15.1). NE turnover rate may be calculated from the product of the slope and the steady-state NE level (11, 13). The validity of this technique depends upon certain assumptions that are fulfilled generally in the experimental situations described here (26, 33). Since the major determinant of NE turnover is centrally mediated SNS activity, the turnover rate is related to impulse traffic in the peripheral sympathetic nerves innervating the organ.

In the experiment shown in Figure 15.1, the increase in SNS activity in cold-exposed animals is shown graphically by the line of steeper slope. The relationship between cardiac NE turnover and the SNS activity is established by experiments with ganglionic blocking agents. When SNS activity is increased, ganglionic blocking agents have a greater effect on NE turnover, because there are more impulses to block at the level of the ganglia. In cold-exposed animals the effect of ganglionic blockade on cardiac NE turnover is considerably greater than in ambient temperature controls, which is consistent with an increase in centrally mediated SNS activity during acute cold exposure (26).

DIETARY INTAKE AND SNS ACTIVITY

Fasting and Restricted Caloric Intake

The effect of fasting on cardiac NE turnover is shown in Figure 15.2. In this experiment, 180-gm Sprague-Dawley rats fasted for 2 days prior to and during the turnover study; they were allowed free access to a dilute electrolyte solution to prevent sodium depletion. Fasting was associated with a significant decrease in cardiac NE turnover (24). Studies involving ganglionic blockade, as described below, indicate that the decrease in NE turnover induced by fasting is secondary to a fall in centrally

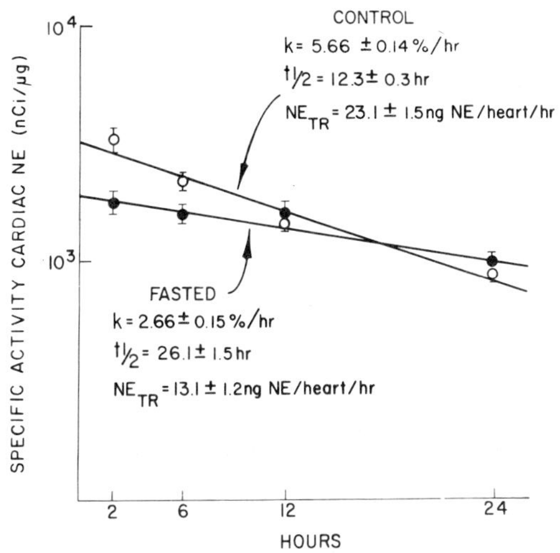

Figure 15.2. Suppression of cardiac NE turnover during fasting in the rat. The effect of 2 days of fasting on cardiac NE turnover is shown. As compared with ad lib-fed control animals, cardiac NE turnover was diminished. (Reproduced with permission from J. B. Young and L. Landsberg (24) and the American Association for the Advancement of Science.)

mediated SNS activity. Additional studies in the rat reveal that the suppressive effect of fasting on SNS activity: (a) is demonstrable in organs other than the heart (26, 33); (b) begins during the first day of fasting (15); (c) occurs throughout the life-span of the rat (6 weeks through 2 years) (14); and (d) is demonstrable when dietary intake is restricted to 30% of control group (16).

The finding that fasting suppressed SNS activity was surprising. Although the effect of caloric deprivation on SNS activity had not been adequately studied, the general presumption was that fasting was associated with enhanced SNS activity. This presumption was based on the fact that substrate mobilization occurs during fasting and that catecholamines (CA) are known to cause substrate mobilization from fuel depots (23), as well as on the vague notion that fasting is a stress and stress is associated with increased SNS activity. Although many pathophysiologic states are known to increase SNS activity, very few situations, if any, are known to be associated with SNS suppression.

The available evidence indicates that suppression of SNS activity occurs during fasting or caloric restriction in human subjects as well (5, 7, 9). Diminished plasma NE levels during fasting or hypocaloric feeding was demonstrated in obese subjects during therapeutic semistarvation (5, 9), as well as in starved patients with anorexia nervosa (7). Unpublished studies from our laboratory (Rosa, Young, & Landsberg) show a significant fall in urinary NE excretion during a 3-day fast in human subjects of normal weight.

Overfeeding Sucrose

A subsequent series of experiments has demonstrated that overfeeding sucrose is associated with increased SNS activity. When offered ad lib access to diluted sucrose solutions in addition to laboratory chow, rats increased their caloric intake approximately 30%. The effect of overfeeding sucrose on cardiac NE turnover is shown in Figure 15.3. In this experiment, 180-gm female Sprague-Dawley rats were offered 8% sucrose in addition to chow for 3 days prior to and during the turnover study periods. Sucrose feeding resulted in a significant increase in cardiac NE turnover (25). In another typical experiment shown in Figure 15.4, the effect of fasting and sucrose overfeeding is compared to feeding of a control group in a single study. The difference in cardiac NE turnover between sucrose overfeeding and fasting is particularly striking (Fig. 15.4).

Experiments with ganglionic blockade demonstrate the relationship between changes in

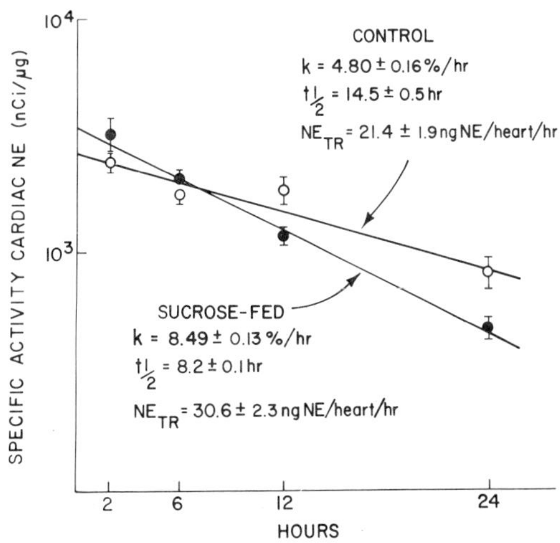

Figure 15.3. Increased cardiac NE turnover during sucrose overfeeding in the rat. The effect of 3 days of ad lib access to an 8% sucrose solution in addition to laboratory chow is shown. As compared with control-fed animals, sucrose feeding increased cardiac NE turnover. (Reproduced with permission from J. B. Young and L. Landsberg (25).)

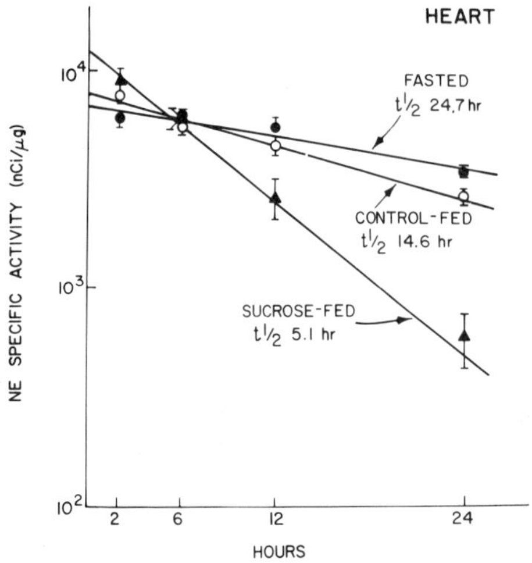

Figure 15.4. Diet-induced changes in cardiac NE turnover. The effects of fasting (2 days) and overfeeding sucrose (3 days) in the same experiment are shown. Both fasted and sucrose-fed animals differ significantly from the control fed group. (Reproduced with permission from J. B. Young and L. Landsberg (26).)

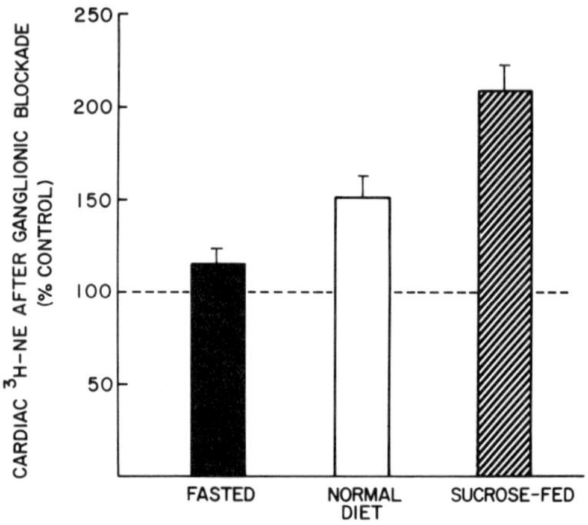

Figure 15.5. Effect of ganglionic blockade on diet-induced changes in cardiac NE turnover. Fasting and sucrose feeding protocols were as in Figure 15.4. The effect of ganglionic blockade on the retention of [^{3}H]NE is diminished in fasting and increased in sucrose-fed animals consistent with reduced and increased SNS activity, respectively. (Reproduced with permission from L. Landsberg *et al.* (12).)

cardiac NE turnover and centrally mediated SNS activity. As noted above, when SNS activity is increased, ganglionic blockade has a greater effect on NE turnover; conversely, when SNS activity is diminished, ganglionic blockade affects NE turnover less. In Figure 15.5, the effect of ganglionic blockade on NE turnover in 2-day fasted and 3-day sucrose-fed animals is compared to ad lib-fed controls; the fasting and overfeeding regimens were the same as those shown in Figure 15.4. In fasted animals, ganglionic blockade had a negligible effect on the retention of tracer NE in the heart. In contrast, in sucrose-fed animals, the effect of ganglionic blockade on the retention of labeled tracer was considerably increased. These experiments indicate that changes in NE turnover represent changes in central sympathetic activity.

Additional experiments with sucrose overfeeding indicate that increased SNS activity: (a) is demonstrable in a variety of organs in addition to the heart (26); (b) occurs during the first day of sucrose overfeeding and dissipates after 1 day of control diet (15); (c) persists through at least 8 days of sucrose overfeeding (15); (d) is demonstrable throughout the 2-year life-span of the rat (14).

Overfeeding a Mixed Diet

Preliminary evidence from our laboratory demonstrates that overfeeding a mixed, highly palatable "cafeteria" diet for periods as long as 10 days is associated with a significant increase in cardiac NE turnover (33). Ad lib access to this high fat and sugar diet usually results in a 50% increase in caloric intake. Additional experiments (I. L. Schwartz, J. B. Young, and L. Landsberg, unpublished observations) indicate that fat supplements also increase cardiac NE turnover, but the details of this effect and its relationship to overfeeding sucrose remain to be established. Thus, although sucrose has been studied most extensively, other nutrients, particularly fat, appear to affect SNS activity as well.

Data on the effect of feeding and overfeeding on SNS activity in human subjects are limited. Plasma NE levels rise after oral glucose ingestion (22, 32) in association with increased oxygen consumption and cardiovascular stimulation.

HYPOGLYCEMIA

How do the diet-induced changes in SNS activity described above relate to the effect of hypoglycemia on the sympathoadrenal system? It has been established for more than 50 years that hypoglycemia is associated with marked stimulation of the adrenal medulla. The effect of hypoglycemia on the SNS, however, has not been extensively studied. Recent studies from our laboratory indicate that several varieties of hypoglycemia (Table 15.1) are associated with suppressed SNS activity, despite marked stimulation of the adrenal medulla. In the normal nongravid rat, 4 days of fasting do not induce hypoglycemia; plasma glucose remains above 55 mg/dl. Moreover, SNS activity is suppressed (as shown above), and no demonstrable increase in adrenal medullary secretion occurs. When pregnant rats are fasted for 4 days, however, frank hypoglycemia develops (27), the adrenal medulla is markedly stimulated, but the SNS remains suppressed (Table 15.1). Under these circumstances, suppression of SNS activity cannot be attributed to the increase in adrenal medullary secretion, since it occurs even in the presence of bilateral splanchnic nerve section (27). A similar pattern of adrenal medullary stimulation and SNS suppression is seen in rats which have fasted and been treated with phlorizin (12) and in rats which have been fed ad lib and treated with 2-deoxyglucose (16). Thus,

Table 15.1
Effect of Hypoglycemia on Sympathoadrenal Activity in the Rat[a]

	Plasma Glucose (mg/dl)	Sympathetic Nervous System: Calculated Cardiac NE Turnover Rate (% Decrease)	Adrenal Medulla: Urinary E (% Increase)
Pregnancy			
Fed	87 ± 4	−66%	+420%
Fasted	36 ± 3		
Phlorizin			
Fed	116 ± 3	−17%	+1000%
Fasted	49 ± 3		
Control			
Fed	—		
2 DG-treated (fed ad lib)	—	−25%	+440%

[a] Mean ± SEM. Data are summarized from refs. 12, 16, and 27.

diet-induced changes in SNS activity do not involve the adrenal medulla. On the other hand, hypoglycemia is associated with SNS suppression and stimulation of the adrenal medulla.

THE SIGNAL: POSSIBLE ROLE OF GLUCOSE AND INSULIN

What is the signal that couples changes in dietary intake with changes in SNS activity? The prompt change in SNS activity when diet is altered is evidence that a change in body mass or body composition is not the signal. Furthermore, it is not established that body mass or composition is a variable that can be assessed by the central nervous system (CNS). The experiments described above are, however, consistent with a mechanism involving carbohydrate metabolism.

Role of Glucose

An important role for glucose utilization within the CNS may be inferred from the fasting, sucrose-feeding, and hypoglycemia experiments described above. Availability of glucose for cellular metabolism is reduced during hypoglycemia and fasting, and glucose utilization is disrupted by 2-deoxyglucose; glucose availability is increased in sucrose overfeeding. Is the plasma glucose level a sufficient determinant of intracellular glucose utilization so that the level itself is the signal? During hypoglycemia, the glucose level may be low enough to decrease glucose metabolism markedly. During fasting, when plasma glucose levels are kept above 50 to 55 mg/dl, and during sucrose feeding, when plasma glucose levels are not different from animals fed a normal diet, it seems unlikely that the plasma glucose level *per se* would be an adequate signal of carbohydrate intake.

Role of Insulin

There are several reasons for considering insulin, particularly insulin-mediated glucose metabolism, as a signal in the coupling of dietary intake with SNS activity. Unlike the plasma glucose level, which is maintained within narrow limits despite wide changes in dietary intake, the plasma insulin level varies widely and reflects, in a general sense, caloric intake and intake of carbohydrate. Furthermore, certain regions of the brain are believed to be sensitive to insulin (4). Since insulin is an important signal to tissues outside the CNS that calories have been assimilated, it is logical that insulin serves the same function for the brain as well.

In order to evaluate a potential role for insulin, studies were performed in normal human subjects utilizing insulin and glucose clamp techniques (20). These techniques, which depend upon priming and variable infusions of glucose or glucose and insulin, permit the attainment of either steady-state hyperglycemia (hyperglycemic clamp) or steady-state hyperinsulinemia with a normal glucose level (euglycemic clamp).

The results of a series of four clamp protocols in 12 normal volunteers aged 18 to 36 years are shown in Figure 15.6. SNS activity was assessed by measuring the plasma NE level. In the control clamp (Fig. 15.6, *open circles*), 0.45% saline was infused, and the phlebotomy was the same as in the experimental protocols. Over the course of the control study, plasma glucose and insulin concentrations were unchanged (81 ± 1.0 mg/dl (mean ± SEM) and 7.3 ± 1.1 μU/ml, respectively), plasma NE level did not change significantly (although there was a tendency to drift downward through the course of the study), and pulse and blood pressure (BP) were unchanged. In the hyperglycemic clamp (*closed circles*) the glucose was elevated to 125 mg/dl over basal (mean value 208 ± 1.6 mg/dl). At steady state, the mean plasma insulin concentration was 44 ± 0.7 μU/ml. Plasma NE levels were unchanged compared to the control infusion (Fig. 15.6). Pulse pressure increased significantly over the course of the infusion. In the 2 mU (2 mU/kg/min) euglycemic clamp, glucose and insulin were infused so that the mean steady-state plasma glucose level was 78.6 ± 0.5 mg/dl, with a mean insulin level of 154 ± 32 μU/ml; this was associated with a significant rise in plasma NE level (as shown in Fig. 15.6) and an increase in pulse rate, pulse pressure, and double product (systolic BP times pulse rate). In the 5 mU euglycemic hyperinsulinemic clamp, mean plasma glucose concentration was 80.2 ± 1.0 mg/dl; mean plasma insulin concentration was 601 ± 74 μU/ml. The plasma NE level increased significantly compared to the other studies and was associated with increased pulse pressure, double product, and mean arterial BP. It should be noted that the insulin infusions were not associated with hypoglycemia.

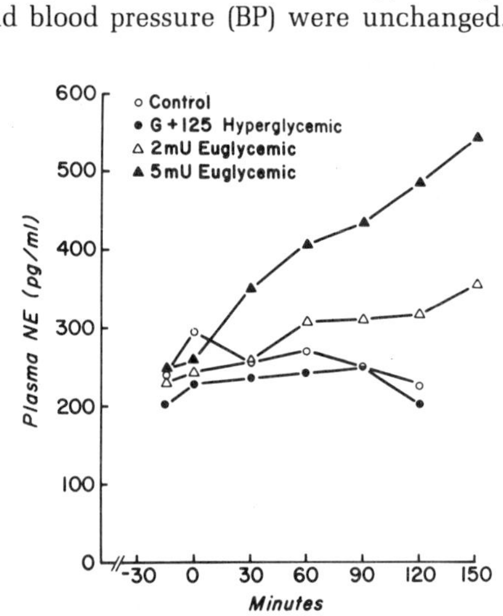

Figure 15.6. Plasma NE levels during hyperglycemic and euglycemic clamp studies. Hyperinsulinemia, without significant changes in plasma glucose level, increased plasma NE level. The rise in plasma NE concentration and the coexistent evidence of cardiovascular stimulation are consistent with stimulation of SNS activity by insulin. (Reproduced with permission from J. W. Rowe *et al.* (20).)

In separate studies, it was established that insulin infusion had no effect on either the metabolic clearance rate of NE or on the disappearance rate of circulating NE after NE infusion was terminated. These studies provide evidence for activation of the SNS by insulin. However, these data must be interpreted cautiously, since the insulin levels achieved during both euglycemic hyperinsulinemic clamps exceed those usually achieved during feeding. Nonetheless, the data indicate that hyperinsulinemia, without a change in blood glucose, has the capacity to stimulate SNS activity. Hyperglycemia, on the other hand, appears to have less effect.

CENTRAL INTEGRATION OF THE SIGNAL: THE GOLD THIOGLUCOSE-TREATED MOUSE

If insulin-mediated glucose metabolism within the CNS is important in the coupling of dietary intake and SNS activity, it seems likely that the ventromedial portion of the hypothalamus might be the brain region involved in integrating the signal and regulating SNS outflow. This portion of the hypothalamus is sensitive to insulin and glucose, is involved in the regulation of food intake (4), and is associated with the brainstem centers that regulate SNS activity. Since gold thioglucose is known to destroy this portion of the hypothalamus (1), the effect of gold thioglucose treatment on diet-induced changes in SNS activity in mice was evaluated (28).

Gold thioglucose-treated mice develop hyperphagia, hyperinsulinism, and marked obesity. The development of obesity in treated animals is therefore a reliable indication of hypothalamic destruction in the ventromedial nucleus. The effects of fasting and overfeeding sucrose on cardiac NE turnover in obese gold thioglucose-treated mice and untreated control animals are shown in Table 15.2. Gold thioglucose treatment completely blocked diet-induced changes in SNS activity. Both fasted and sucrose-fed, gold thioglucose-treated mice had cardiac NE turnover rates similar to those of sucrose-fed control animals. The suppression of SNS activity that usually occurs with fasting did not occur (Table 15.2). Additional studies indicated that ganglionic blockade had a significant effect on cardiac NE turnover in fasted, gold thioglucose-treated mice, whereas the effect was negligible in control mice (Fig. 15.7). The effect of ganglionic blockade was similar in both fasted and sucrose-fed gold thioglucose-treated animals. This experiment is consistent with impaired suppression of SNS activity during fasting in the treated animals. Fasts of longer duration (2 days) also failed to suppress SNS activity in the treated mice (28). Additional experiments indicated that failure to suppress SNS activity with fasting was not due to the nonspecific effect of gold, since gold thiomalate had no effect, nor did it reflect the nonspecific effects of obesity, since comparably obese (ob/ob)

Table 15.2
Effect of Gold-Thioglucose Pretreatment on Diet-Induced Changes in Sympathetic Activity[a]

	Fractional Turnover k (Heart) (%/hr)	Calculated NE Turnover Rate (ng NE/heart/hr)
Control		
Fast (24 hr)	1.7 ± 1.0	1.6 ± 1.0
Sucrose-fed (72 hr)	6.1 ± 1.0	4.6 ± 1.0
	($p < 0.001$)	
Gold-thioglucose		
Fast (24 hr)	5.9 ± 1.5	5.5 ± 1.6
Sucrose-fed (72 hr)	7.2 ± 0.7	5.4 ± 0.8
	(NS)	

[a] Means ± SEM. Data are from Ref 28.

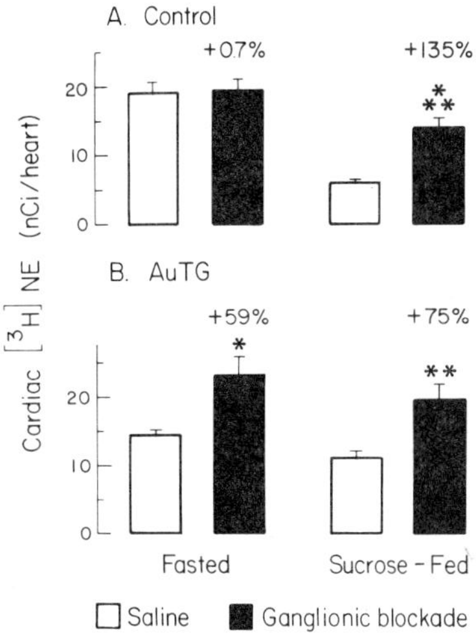

Figure 15.7. Effect of ganglionic blockade on cardiac NE turnover in gold thioglucose-treated and control mice. In control mice (*A*) the effect of ganglionic blockade on retention of [^{3}H]NE in fasting is quite different than in sucrose-fed animals. (*B*) In gold thioglucose-treated animals (*AuTG*) the effect of ganglionic blockade is similar in sucrose-fed and fasted animals. These data are consistent with failure of suppression of SNS activity in the gold thioglucose-treated animals. (Reproduced with permission from J. B. Young and L. Landsberg (28).)

mice suppress SNS activity normally in response to a fast (28). The specificity of the gold thioglucose effect, moreover, was emphasized by the SNS response to cold exposure which was preserved after gold thioglucose treatment, as was the adrenal medullary response to 2-deoxyglucose administration (28).

Gold thioglucose pretreatment, therefore, appears to disrupt the normal suppressive effect of fasting on SNS activity. This implies that under normal circumstances, hypothalamic neurons destroyed by gold thioglucose exert a suppressive effect on SNS outflow during fasting (Fig. 15.8) and that this suppression is an active process, not merely the passive withdrawal of SNS stimulation induced by feeding. These changes are consistent with the model outlined in Figure 15.8, which implies that descending inhibition of SNS activity during fasting is dependent upon stimulation of an inhibitory pathway or pathways between the hypothalamus and the brainstem. Conversely, stimulation of SNS activity by overfeeding may be associated with suppression of this inhibitory pathway, thus resulting in disinhi-

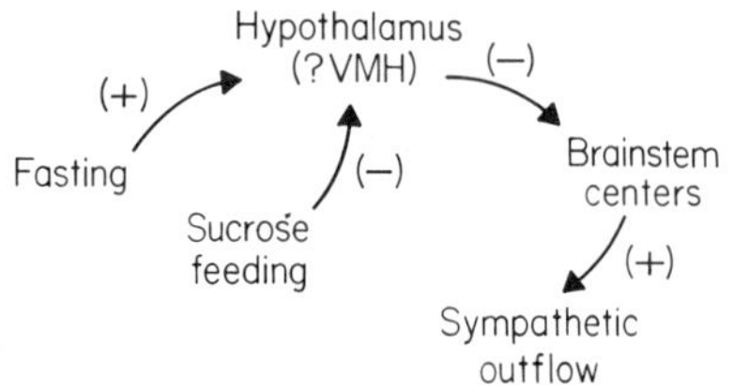

Figure 15.8. Model of hypothalamic regulation of sympathetic outflow in response to changes in diet. The experiments in the gold thioglucose-treated mice are consistent with a model based on descending inhibition. The model implies stimulation (+) of an inhibitory pathway (or pathways) between the hypothalamus and the brainstem during fasting. These pathways may involve the ventromedial portion of the hypothalamus (VMH). Overfeeding, according to this model, is associated with suppression of the inhibitory pathways and resultant disinhibiton of brainstem sympathetic centers. (Reproduced with permission from J. B. Young and L. Landsberg (28).)

bition of tonically active lower neurons. The results of these experiments, in conjunction with the insulin, 2-deoxyglucose, and hypoglycemia experiments described above, suggest that insulin-mediated glucose metabolism within gold thioglucose-sensitive neurons in the hypothalamus may be involved in initiating changes in SNS activity in response to changes in diet.

CALORIC INTAKE AND SNS ACTIVITY: IMPLICATIONS FOR THERMOGENESIS, OBESITY, AND HYPERTENSION

Dietary Thermogenesis

It is well established that changes in oxygen consumption coincide with changes in dietary intake. Debate continues as to whether these changes are adaptive and regulated or whether they merely reflect the chemical energy involved in the assimilation of nutrients or the alterations in body mass consequent to change in dietary intake. Recent evidence involving the SNS and brown adipose tissue, as described below, supports the concept that changes in thermogenesis accompanying changes in dietary intake are adaptive.

CA have long been recognized as playing a critical role in the mammalian thermogenic response to cold exposure; the potential role of CA in the regulation of dietary thermogenesis, however, has only recently been explored. Two lines of evidence support an important role for the SNS. Experiments by Rothwell and Stock (18, 19) have indicated an essential similarity between the processes of nonshivering thermogenesis (heat production unrelated to muscular activity which is increased in response to cold exposure) and dietary thermogenesis (heat production which is increased in response to augmented dietary intake). The studies of these investigators indicate that brown adipose tissue, the major site of NE-induced metabolic heat production (nonshivering thermogenesis) in cold-acclimated rats (6), is also an important source of heat production during overfeeding (18, 19). Since the SNS is known to have an important role in the initiation and regulation of nonshivering thermogenesis (8), the similarity between nonshivering thermogenesis and dietary thermogenesis suggests an important role for the SNS in the latter process as well.

The second line of evidence supporting an important role for the SNS in the regulation of dietary thermogenesis is the work reviewed here, which indicates that diet induces changes in SNS activity. Coupled with the known effects of CA on oxygen consumption, the evidence is consistent with a role of the SNS in the mediation of dietary thermogenesis. Additional evidence from our laboratory supports such a role. Fasting impairs the SNS response to cold exposure (29), thus suggesting that central integration of dietary and thermoregulatory responses affecting the SNS are interrelated. Furthermore, recent studies from our laboratory indicate that diet-induced changes in SNS activity occur in brown adipose tissue (33), the major thermogenic organ in the rat. A good case, therefore, can be made for the importance of the SNS in the regulation of diet-induced changes in thermogenesis.

A potential physiological role for diet-induced changes in SNS activity may be suggested. During periods of fasting or caloric restriction, suppression of SNS activity would result in a diminished metabolic rate and conservation of calories. The survival value of this adaptive change in oxygen consumption is obvious; conservation of calories at a time when caloric intake is limited is a metabolic adaptation certain to be favored by evolutionary pressures. Decreased caloric expenditure would prolong the time span that an organism could survive during famine. An increase in

SNS activity with overfeeding, on the other hand, would result in the dissipation of excess calories. The ability to expend calories consumed in excess of need might be of particular value to organisms surviving on a subsistence diet or marginal nutritional adequacy. Under these circumstances the need for *nitrogen* might be satisfied by increasing dietary intake of foodstuffs low in protein. The resultant increase in SNS activity would prevent excessive accumulation of fat while ensuring adequate nutrition. This interesting hypothesis warrants further study.

Implications for Obesity and Therapeutic Dieting

If the SNS does play an important role in diet-induced thermogenesis, a disorder in either SNS activation, or an impairment in the thermogenic effects of NE might predispose to the development of obesity. Either abnormality would be associated with diminished thermogenic responses to feeding; such individuals would have a "thrifty" metabolic trait resulting in efficient utilization of foodstuffs and food storage that would serve well during periods of caloric restriction or famine. In the face of an abundant food supply, however, these same individuals would be predisposed toward excessive fuel storage and obesity. Both SNS responses to feeding and the responsiveness of obese individuals to the thermogenic effects of CA (10) require further study.

Similarly, suppressed SNS activity during therapeutic dieting or fasting in obese subjects may limit weight loss by decreasing metabolic rate. Another major metabolic adaptation to starvation, the decrease in peripheral conversion of T_4 to T_3, would have the same net physiologic result. These metabolic adaptations to starvation may limit weight loss during therapeutic dieting and explain the beneficial effect of T_3 and amphetamines on weight loss during caloric restriction. CA and thyroid hormone, given exogenously during fasting, might overcome these adaptive responses and increase energy output. Questions about the safety of these drugs preclude their general use in the treatment of obesity at the present time.

Implications for Hypertension

The possibility also exists that diet-induced changes in SNS activity may have physiologic consequences unrelated to thermogenesis. Stimulation of the SNS during overfeeding may be associated with undesirable secondary effects on the cardiovascular system. The SNS may mediate, at least in part, changes in BP that occur in response to changes in caloric intake. Studies in the spontaneously hypertensive rat indicate that fasting or hypocaloric feeding, with maintenance of sodium intake, is associated with a prompt and significant fall in BP (31). Conversely, overfeeding sucrose on a constant sodium intake results in a significant increase in BP in this animal (30). The changes in BP in the normotensive control strain are in the same direction but much less marked.

The beneficial effect of caloric restriction on BP in obese hypertensive human subjects (3, 17, 21) may also relate to suppression of SNS activity. Similarly, the statistical association of elevated BP in individuals with recent weight gain may reflect the effect of increased caloric intake on SNS activity (3). The possibility also exists that excessive dietary intake and consequent SNS stimulation may contribute to hypertension even in nonobese patients.

References

1. Brecher G, Laquer GL, Cronkite EP, Edelman PM, Schwartz IL: The brain lesion of gold thioglucose obesity. *J Exp Med* 121:395–401, 1965.
2. Brodie BB, Costa E, Dlabac A, Neff NH, Smookler HH: Application of steady-state kinetics to the estimation of synthesis rate and turnover time of tissue catecholamines. *J Pharmacol Exp Ther* 154:493–498, 1966.
3. Chiang BN, Perlamn LV, Epstein FH: Overweight and hypertension: A review. *Circulation* 39:403–421, 1969.
4. Debons AF, Krimsky I, From A, Cloutier RJ: Rapid effects of insulin on the hypothalamic satiety center. *Am J Physiol* 217:1114–1118, 1969.
5. DeHaven J, Sherwin R, Hendler R, Felig P: Nitrogen and sodium balance and sympathetic-nervous-system activity in obese subjects treated with a low-calorie protein or mixed diet. *N Engl J Med* 302:477–482, 1980.
6. Foster DO, Frydman ML: Nonshivering thermogenesis in the rat. II. Measurements of blood flow with microspheres points to brown adipose tissue as the dominant site of the calorigenesis induced by noradrenaline. *Can J Physiol Pharmacol* 56:110–122, 1978.
7. Gross HA, Lake CR, Ebert MH, Ziegler MG, Kopin IJ: Catecholamine metabolism in primary anorexia nervosa. *J Clin Endocrinol Metab* 49:805–809, 1979.
8. Himms-Hagen J: Cellular thermogenesis. *Ann Rev Physiol* 38:315–351, 1976.
9. Jung RT, Shetty PS, Barrand M, Callingham BA,

James WPT: Role of catecholamines in hypotensive response to dieting. *Br Med J* 1:12–13, 1979.

10. Jung RT, Shetty PS, James WPT, Barrand MA, Callingham BA: Reduced thermogenesis in obesity. *Nature* 279:322–323, 1979.
11. Landsberg L, Axelrod J: Influence of pituitary, thyroid and adrenal hormones on norepinephrine turnover and metabolism in the rat heart. *Circ Res* 22:559–571, 1968.
12. Landsberg L, Greff L, Gunn S, Young JB: Adrenergic mechanisms in the metabolic adaptation to fasting and feeding: Effects of phlorizin on diet-induced changes in sympathoadrenal activity in the rat. *Metabolism* 29:1128–1137, 1980.
13. Neff NH, Tozer TN, Hammer W, Costa E, Brodie BB: Application of steady-state kinetics to the uptake and decline of H-3-NE in the rat heart. *J Pharmacol Exp Ther* 160:48–52, 1968.
14. Rappaport EB, Young JB, Landsberg L: Impact of age on basal and diet-induced changes in sympathetic nervous system activity of Fischer rats. *J Gerontol* 36:152–157, 1981.
15. Rappaport EB, Young JB, Landsberg L: Initiation, duration and dissipation of diet-induced changes in sympathetic nervous system activity in the rat. *Metabolism* 31:143–146, 1982.
16. Rappaport EB, Young JB, Landsberg L: Effects of 2-deoxy-D-glucose on the cardiac sympathetic nerves and the adrenal medulla in the rat: Further evidence for a dissociation of sympathetic nervous system and adrenal medullary responses. *Endocrinology* 110:650–656, 1982.
17. Reisen E, Abel R, Modan M, Silverberg DS, Eliahou HD, Modan B: Effect of weight loss without salt restriction on the reduction of blood pressure in overweight hypertensive patients. *N Engl J Med* 298:1–6, 1978.
18. Rothwell NJ, Stock MJ: A role for brown adipose tissue in diet-induced thermogenesis. *Nature* 281:31–35, 1979.
19. Rothwell NJ, Stock MJ: Similarities between cold- and diet-induced thermogenesis in the rat. *Can J Physiol Pharmacol* 58:985–991, 1980.
20. Rowe JW, Young JB, Minaker KL, Stevens AL, Pallotta J, Landsberg L: Effect of insulin and glucose infusions on sympathetic nervous system activity in normal man. *Diabetes* 30:219–225, 1981.
21. Tuck ML, Sowers J, Dornfield L, Kledzik G, Maxwell M: The effect of weight reduction on blood pressure, plasma renin activity and plasma aldosterone levels in obese patients. *N Engl J Med* 304:903–933, 1981.
22. Welle S, Lilavivanthana U, Campbell RG: Increased plasma norepinephrine concentrations and metabolic rates following glucose ingestion in man. *Metabolism* 29:806–809, 1980.
23. Young JB, Landsberg L: Catecholamines and intermediary metabolism. *Clin Endocrinol Metab* 6:599–631, 1977.
24. Young JB, Landsberg L: Suppression of sympathetic nervous system during fasting. *Science* 196:1473–1475, 1977.
25. Young JB, Landsberg L: Stimulation of the sympathetic nervous system during sucrose feeding. *Nature* 269:615–617, 1977.
26. Young JB, Landsberg L: Effect of diet and cold exposure on norepinephrine turnover in pancreas and liver. *Am J Physiol* 236:E524–E533, 1979.
27. Young JB, Landsberg L: Sympathoadrenal activity in fasting pregnant rats: Dissociation of adrenal medullary and sympathetic nervous system responses. *J Clin Invest* 64:109–116, 1979.
28. Young JB, Landsberg L: Impaired suppression of sympathetic activity during fasting in the gold thioglucose-treated mouse. *J Clin Invest* 65: 1086–1094, 1980.
29. Young JB, Landsberg L: Effect of concomitant fasting and cold exposure on sympathoadrenal activity in rats. *Am J Physiol* 240:E314–E319, 1981.
30. Young JB, Landsberg L: Effect of oral sucrose on blood pressure in the spontaneously hypertensive rat. *Metabolism* 30:421–424, 1981.
31. Young JB, Mullen D, Landsberg L: Caloric restriction lowers blood pressure in the spontaneously hypertensive rat. *Metabolism* 27:1711–1714, 1978.
32. Young JB, Rowe JW, Pallotta JA, Sparrow D, Landsberg L: Enhanced plasma norepinephrine response to upright posture and oral glucose administration in elderly human subjects. *Metabolism* 29:532–539, 1980.
33. Young JB, Saville E, Rothwell NJ, Stock MJ, Landsberg L: Effect of diet and cold exposure on norepinephrine turnover in brown adipose tissue of the rat. *J Clin Invest* 69:1061–1071, 1982.

CHAPTER SIXTEEN

The Effects of Anesthesia, Anesthetic Adjuvant Drugs, and Surgery on Plasma Norepinephrine*

MICHAEL F. ROIZEN, M.D., JONATHAN MOSS, M.D., Ph.D., and SHEILA M. MULDOON, M.D.

INTRODUCTION

Brewster *et al.* (17) and Price *et al.* (114) proposed a paradigm to account for differences in the cardiovascular effects of anesthetics. These effects were attributed to the influence of the anesthetic on sympathetic tone. Thus, cyclopropane supported the circulation by increasing sympathetic activity, while thiopental depressed the circulation by decreasing sympathetic activity. Subsequent investigations generally supported this view (5, 18, 30, 48, 115, 146, 150), with some exceptions (11). During the last 10 years, the paradigm has been expanded because cardiovascular effects of anesthetics are now attributed to: (a) central nervous system (CNS) control of vasoactive substance release; (b) local release or inactivation of the sympathetic neurotransmitters; (c) the effector organs (myocardium or vascular smooth muscle); and (d) reflex reactions. In addition, the ability of an anesthetic to depress stress- or pain-related increases in release of hormones or neurotransmitters appears to be an important determinant of the cardiovascular effects of surgery.

EFFECT OF ANESTHESIA AND ANESTHETIC ADJUVANT DRUGS ON PLASMA LEVELS OF NOREPINEPHRINE

Both old and recent data seem to agree that barbiturates depress adrenergic tone. Three early studies demonstrated that barbiturate infusion decreased or did not affect plasma levels of amines (37, 93, 121). More recent *in vivo* and *in vitro* studies support those findings. Skovsted *et al.* (137) demonstrated that thiopental and methohexital inhibited cervical preganglionic activity and weakened barostatic reflexes in cats. In studies of adrenal nerve stimulation and subsequent epinephrine (E) release, barbiturates inhibited neuronal release of catecholamines (CA) (67). Roizen *et al.* (125) showed that concentrations of norepinephrine (NE) and E in the blood were lower

* This work was supported in part by NIH Anesthesia Research Center grant GM 15571, Program Project 1PO1 AGO3104-0102, and NIH grant GM 25926.

when animals were anesthetized with pentobarbital than during handling or after decapitation. Sympathetic activity of the nerves in the skin decreases during induction with thiopental (161). Other investigators have found no change in plasma levels of NE or E during anesthesia with thiopental, administration of a muscle relaxant, or surgery (58, 114). In rats, pentobarbital anesthesia totally inhibits the catecholaminergic response to hemorrhagic shock (41). In a study on human subjects, i.v. thiopental decreased plasma levels of NE by about 50%, but did not change E (Fig. 16.1) (73).

We conclude that sympathoadrenal tone decreases during barbiturate anesthesia and that this decrease is at least partially responsible for the increased venous distensibility and arteriolar capacitance (and perhaps, decreased myocardial contractility) that occurs during barbiturate anesthesia. Thus, anesthesia with barbiturates affects plasma levels of CA in at least two ways: (a) it decreases resting levels; and (b) it decreases, in a dose-related fashion, the increase in plasma levels of NE and E that normally occurs with stress.

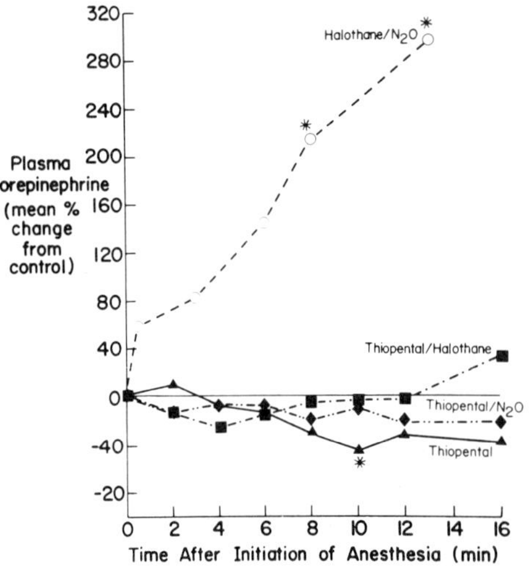

Figure 16.1. Mean percentage change in plasma levels of norepinephrine from control (mean) after initiation of anesthetic agents. Patients were randomly assigned to receive one of the following anesthetics: thiopental, 3 mg/kg iv, followed by inhalation of 100% oxygen and a continuous intravenous infusion of thiopental, 0.2 to 0.3 mg/kg/min, thiopental, 3 mg/kg, followed by inhalation of halothane (1.5% end-tidal concentration) in oxygen; or thiopental, 3 mg/kg, followed by inhalation of 70% nitrous oxide in oxygen. After thiopental injection, ventilation was controlled to maintain P_{CO_2} near control levels. (Reproduced with permission from J. T. Joyce and M. F. Roizen (73).)

VOLATILE AGENTS

Ether

Brewster *et al.* (17) based their paradigm on initial work that revealed increased plasma levels of CA during ether anesthesia. Comparing the cardiovascular response to ether in sympathectomized and intact animals, Brewster concluded that the sympathetic nervous system (SNS) has to be intact if cardiovascular function is to be stable during ether anesthesia. Similar conclusions had been reached earlier by Bhatia and Burn (12) and McAllister and Root (89). In essential agreement with these earlier observations, Anton *et al.* (5) and Hamelberg *et al.* (58) have reported increased levels of plasma CA with ether anesthesia during surgery. The cause of this increase is not clear.

Ether does not affect the uptake (19, 103) or metabolism (19) of NE, so it is probable that ether anesthesia causes sympathoadrenal stimulation that helps maintain cardiovascular function.

Cyclopropane

More than any other drug, cyclopropane was essential to the development of the Brewster-Price paradigm. However, results from current studies have rendered our understanding of the effects of cyclopropane on sympathoadrenal function less clear than for most anesthetics. Several investigators have reported increased plasma CA levels during anesthesia with cyclopropane (30, 59, 114, 144). Hamelberg and Stokke and their co-workers (58, 144) did not study the pharmacologic properties of an anesthetic *per se*, but rather, the effects of anesthesia during surgery. Painful stimuli that increased CA release may have contributed to their finding that CA levels were higher during cyclopropane anesthesia (and surgery) than during an awake control period (without surgery).

However, Millar and Morris (92) and Dobkin *et al.* (33) did not find significant changes in the concentration of E or NE upon cyclopropane administration to dogs. In another dog study, plasma CA levels were the same during cyclopropane anesthesia as after premedication with barbiturates (110). Because premedication has been shown to depress CA levels (see "Benzodiazepines" and "Barbiturates"), plasma CA levels were probably below baseline (control) values during cyclopropane anesthesia. Roizen *et al.* (125) found that during anesthesia with cyclopropane, total levels of CA and NE in plasma did not increase during either handling or decapitation. Furthermore, when the concentration of cyclopropane was increased, blood pressure (BP) decreased, and plasma levels of CA remained the same. However, after an additional 30 minutes of cyclopropane anesthesia, both CA and NE levels in plasma increased (Fig. 16.2). Thus, changes in sympathetic function cannot account for the changes in BP that occur during cyclopropane anesthesia.

Li *et al.* (81) measured NE in the myocardium of dogs and found that cyclopropane substantially increases myocardial NE. This could be the result of increased synthesis or uptake or decreased release or degradation of NE. However, cyclopropane did not affect uptake of tritiated NE in rat heart (107), cat iris (107), guinea pig (19), or cat atria (103). Other studies have demonstrated that cyclopropane does not potentiate or inhibit the effects of exogenous NE on atrial rate. Cyclopropane has not been shown to alter uptake of endogeneously released or exogenously administered NE.

Ngai and co-workers (107) examined the effects of cyclopropane on NE release in the intact dog heart. They labeled myocardial NE stores with intracoronary [^{3}H]NE and measured [^{3}H]NE release. Although cyclopropane (15%) did not significantly affect the release of [^{3}H]NE from myocardial stores when arterial pressure was maintained, when significant hypotension occurred, more NE was released, presumably because of compensatory increases in sympathetic outflow.

Using guinea pig vas deferens, Roizen and co-workers (127) also examined the effect of cyclopropane on release of endogenous NE. Release of NE was elicited using veratridine, a compound that depolarizes the nerve terminal by increasing the permeability of the cell membrane to a variety of ions, including sodium. Veratridine caused a release of NE that was proportional to release of dopamine-β-hydroxylase (DBH); this proportional release of NE and DBH was similar to that produced by electrical stimulation. (Phenoxybenzamine

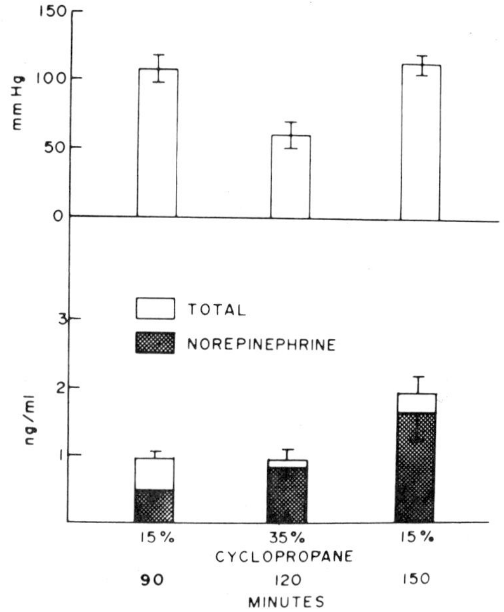

Figure 16.2. Effect of time and varying concentrations of cyclopropane on levels of catecholamines in plasma. Rats were anesthetized with cyclopropane, and ventilation was controlled. The first blood sample was obtained after 30 minutes of postsurgical anesthesia with 15% cyclopropane. The second blood sample was taken 30 minutes after the concentration of cyclopropane was increased to 35%. The third blood sample was taken after the concentration had been reduced to 15% for 30 minutes (150 minutes after induction of anesthesia). Results are mean (± SEM) for all animals. The plasma levels of both total catecholamines and norepinephrine in the final sample are significantly greater ($p < 0.02$) than in the two previous samples. The mean blood pressure during 35% cyclopropane anesthesia is significantly less ($p < 0.01$) than during 15% cyclopropane anesthesia. (Reproduced with permission from M. F. Roizen *et al.* (125).)

was added to the media to prevent reuptake of NE). Cyclopropane significantly decreased the release of NE evoked by veratridine, but did not affect the amount of DBH released. Sumikawa *et al.* (145) confirmed that cyclopropane inhibited exocytotic release of NE *in vitro.*

Gardier *et al.* (43) found that in a 100% cyclopropane atmosphere, the amount of metanephrine (M) formed from E is reduced significantly, but no changes in NE or E metabolism occur in a 30% cyclopropane atmosphere. The very high concentration of cyclopropane required to produce changes makes the clinical relevance of these results doubtful.

Since many investigators reported that cyclopropane does not influence release or uptake of NE at sympathetic nerve endings, and that at clinical concentrations it does not interfere with the metabolism of NE, Ngai *et al.* (108) examined the action of cyclopropane on the *in vitro* synthesis of NE in the heart and brain of rats. Cyclopropane anesthesia lasting as long as 2 hours does not alter the rates of NE synthesis in either the heart or brain.

Other results from *in vitro* studies indicate that cyclopropane (in the clinical dose range) decreases transmission across sympathetic ganglia and NE release approximately 50% (18, 28, 45, 80, 116).

Thus, it is not clear if cyclopropane has sympathoadrenal stimulating activity in the intact organism. Since it does not change NE synthesis (98, 108) or reuptake (19, 103, 107), and may decrease NE release, Ngai and co-workers (107, 108) inferred that the cardiovascular "stimulating" effect of cyclopropane is the result of its action on smooth muscle cells. Studies by Cristoforo and Brody (26) using an isolated limb preparation support this hypothesis; cyclopropane was found to have a direct vasoconstricting effect.

Halothane

Studies of halothane indicate a variety of effects on the SNS, including some actions that antagonize other actions. However, the predominant effect, recognized long ago, is that halothane inhibits the sympathetic response to surgery. Using relatively insensitive fluorometric assays, four groups (33, 58, 92, 114) found no change in plasma levels of CA during anesthesia with halothane. A more recent study of Stokke *et al.* (144) confirmed those results. However, Perry (110), Roizen (124), and their co-workers found that halothane decreases plasma NE in dogs and rats. In contrast, Joyce *et al.* (72) found that administration of halothane to humans initially increases plasma NE levels 100%. These concentrations return to baseline as administration of halothane continues. The increase was not due to placement of a face mask for induction of anesthesia (Fig. 16.3). The occurrence of signs of second stage anesthesia during induction of anesthesia with halothane and oxygen (*i.e.*, pupillary dilatation, hypertension, tachycardia, excitement, and involuntary activity) supports the possibility that halothane might initially cause sympathetic activation. However, there were increased NE levels into the third stage of anesthesia. After 1 hour of halothane administration, the plasma NE level decreased to control values.

Earlier, Rozien *et al.* (124) had demonstrated a dose-dependent decrease in plasma levels of CA during halothane administration in rats. However, blood samples were not obtained for at least 30 minutes after induction. Thus, these investigators probably missed the initial increase later observed by Joyce and colleagues. Halter *et al.* (57) found no change in plasma NE in humans but found a decrease in E concentrations 30 minutes after initiation of halothane anesthesia. However, they also administered pancuronium, a muscle relaxant that decreases the amount of NE and E in plasma (see below).

The idea that halothane could initially increase plasma NE was first proposed by Depocas and Behrens (29) in rat studies. Halothane initially increases spontaneous release of NE from *in vitro* cat spleen; subsequently, it blocks exocytotic release of NE in a dose-dependent fashion (126).

Other *in vitro* experiments have examined the range of possible effects of halothane on the sympathetic neuron, including synthesis, rapid axoplasmic transport, storage, metabolism, reuptake, and release of NE. The rate of synthesis of NE does not change in either the heart or brain of rats (108). Halothane also does not affect NE reuptake (19, 103, 107) but does increase MAO activity in guinea pig in a dose-dependent fashion (134). However, the concentration of halothane at which these effects occur is not given. We found that halothane has no effect on *in vitro* activity of liver-derived COMT (124).

Ngai and co-workers (107) examined the effects of halothane in the dog heart *in vivo* and in decentralized cat iris. In the former, myocardial stores of E were labeled by infusing intracoronary [^{3}H]NE. The sympathetic nerves were stimulated to evoke NE release. Halothane (0.75 to 1.0%) decreased arterial BP but did not affect release of [^{3}H] from the heart. Apparently, this concentration of halothane depresses circulatory compensation through barostatic mechanisms. This conclusion is supported by other studies in dogs (162).

Roizen *et al.* (124) found that halothane can interfere with the depolarization-induced release of endogenous NE from the vas deferens. Halothane (1 to 2%) depresses the electri-

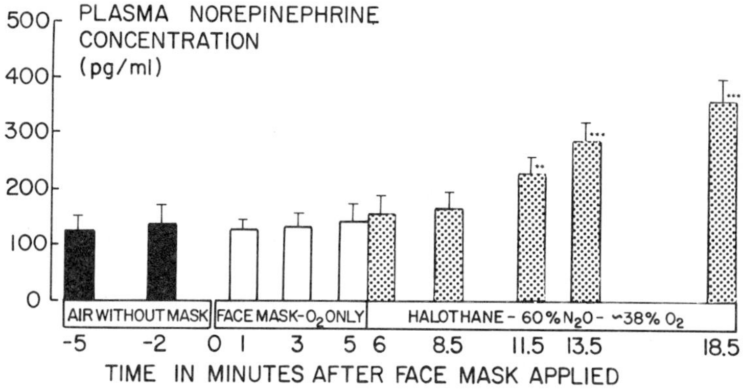

Figure 16.3. Plasma levels of norepinephrine during a control period (subjects breathing air without face mask), placement of a face mask, and administration of halothane-nitrous oxide. **, $p < 0.05$, increase in plasma levels of norepinephrine during halothane, compared with the control period and face-mask placement. ***, $p < 0.001$, increase in plasma levels of norepinephrine during halothane, compared with the control period and face mask placement. (Reproduced with permission from J. T. Joyce *et al.* (72).)

cally stimulated release of NE in a dose-dependent fashion, but does not decrease the release of DBH proportionately. Hexamethonium does not alter the effect on NE. In this preparation, halothane probably acted directly on the nerve terminals and not at nicotinic receptors. Halothane may enhance the affinity of NE to binding sites on the adrenergic vesicle, thereby diminishing the release of NE but not of DBH during exocytosis. Alternatively, exocytosis may not completely account for the release of DBH, and the enzyme may be released by a mechanism that does not involve release of NE (126).

Using the isolated rabbit heart perfused with physiological solution at a constant flow, Gothert (48, 51) studied the effects of halothane on NE release induced by depolarization of the nerve endings. NE release was evoked using several stimuli: submaximal electrical stimulation of the postganglionic nerves; KCl; the nicotinic receptor agonist 1,1 dimethyl-4-phenylpiperazine; and acetylcholine. At clinically useful concentrations, halothane was found to decrease NE output evoked by activation of nicotinic receptors; this decrease was concentration-dependent. Halothane similarly decreases the release of CA by the adrenal medulla. On the other hand, halothane does not affect NE release evoked by submaximal electrical stimulation or by raising the concentration of KCl in the perfusion medium. The release of NE evoked by all three of these stimuli is calcium-dependent. But, halothane blocks only NE output evoked by stimulation of nicotinic receptors. The reasons for this specificity are not clear; however, the three methods used by Gothert (48) to evoke NE release do not produce equivalent quantities of NE. Submaximal electrical stimulation releases modest amounts (45 ng/2 min of perfusion), while nicotinic agonists produce very large amounts (approximately 1000 ng/2 min of perfusion).

In another study, Muldoon *et al.* (101), using the isolated canine saphenous vein as a model of adrenergically innervated vascular smooth muscle, found that halothane causes a dose-related decrease in the response of veins to electrical stimulation. Halothane does not alter the response to exogenous NE; it increases the contractile response to tyramine. The effect of halothane on contractions of isolated veins caused by transmural electrical stimulation and by tyramine resembles the effect of acetylcholine (158). Acetylcholine inhibits the release of NE from nerve endings during electrical stimulation, but does not affect the release caused by the addition of tyramine. This suggests that halothane, like acetylcholine, acts on the nerve terminal to inhibit the release of NE that occurs with depolarization of the nerve cell. Although both electrical stimulation and tyramine liberate NE, the release resulting from stimulation of the nerve depends on the presence of extracellular calcium, while that produced by tyramine does not. The authors concluded that reduced influx of calcium may be the mechanism underlying the effect of halothane (79).

In intact animals, stress-induced increases in plasma CA are ameliorated by halothane (10, 42, 115, 122, 125, 144). In humans, halothane blocks (in a dose-dependent fashion) the increases in plasma NE levels and cardiovascular variables induced by surgical incision (122). Similar blockade by halothane of stress-related increases in CA occurred in pregnant ewes; the lower CA levels during "deeper" anesthesia correlate with increased uterine artery blood flow (136).

Thus, halothane anesthesia is associated with an initial increase in plasma NE followed by a decrease. Increasing the dose of halothane inhibits the cardiovascular, noradrenergic, and adrenergic responses to painful stimuli. These effects may be due in part to CNS effects and in part to direct action on sympathetic ganglia and nerve endings.

Enflurane

In vivo and *in vitro* studies demonstrate that enflurane decreases the release of E and NE. This has been shown after stimulation of splanchnic nerves in cats (50), on administration to rats (130) and humans (86), and from isolated adrenal glands after either spontaneous or depolarization-induced release (at 1 MAC enflurane, the reduction is approximately 50%) (50, 51). No effect of enflurane on synthesis, metabolism, or reuptake of amines has been demonstrated. Other investigators have found either no change in plasma CA levels or a change that was dependent on the initial value (that is, increases occurred when the initial value was low, and decreases occurred when the initial value was high) (122). Shnider *et al.* (136) found that adding enflurane (1.0%) to N_2O-O_2 anesthetized pregnant ewes blocks stress-related increases in plasma CA and increases uterine artery blood flow. In humans and cats, enflurane decreases, in a dose-dependent fashion, blood levels of NE produced by stimulation (50, 122). However, cardiovascular changes during enflurane anesthesia do not correlate with changes in NE or E levels (122). In fact, at times enflurane acts as a direct peripheral vasodilator. Nitroprusside, a potent peripheral vasodilator devoid of direct sympathetic effects, increases plasma CA levels, presumably as a reflex response to hypotension (119). We conclude that enflurane directly blocks, in a dose-dependent fashion, the release of CA that occurs in response to pain, and may decrease the amount of circulating CA in the resting individual. These effects may contribute to the minor effect of enflurane on myocardial contractility or to its major effects on peripheral vasodilatation.

Isoflurane

Isoflurane is a potent vasodilator with little effect on myocardial contractility. Only three studies have examined its effects on the SNS, and they were not consistent. Byles *et al.* (21) found that isoflurane has no effect on plasma NE in dogs, but in another study they found that it is associated with a 50% increase in man (32). Perry *et al.* (110) found that isoflurane decreases plasma CA by 50% in dogs. The first two studies used a fairly insensitive fluorometric assay, while the last study used a slightly more sensitive assay. Thus, the ability of isoflurane to decrease arterial pressure, increase heart rate, and maintain cardiac output appears unrelated to its ability to alter sympathetic function. Definitive studies, however, remain to be done.

NITROUS OXIDE

In one study, adding nitrous oxide to halothane or isoflurane increased heart rate and peripheral vascular resistance (138). These increases correlated with increases in levels of CA (as measured by fluorometric assay). In contrast, Joyce and Roizen (73) found that administering halothane plus nitrous oxide in oxygen does not produce changes in plasma NE levels different from those occurring with the same dose of halothane by itself. Furthermore, Joyce *et al.* (72) found that thiopental administration, followed by N_2O in oxygen, does not affect plasma CA levels (Fig. 16.1). We conclude that the cardiovascular effects of N_2O are not highly dependent upon their ability to alter sympathetic tone.

NARCOTICS

Morphine

The stability of cardiovascular variables after slow administration of morphine led to its use in open-heart surgery (8, 59, 84). Given in this fashion, morphine causes little change in rate-pressure product and no significant change in concentrations of CA in the plasma or urine (9, 87, 96, 123, 140). However, some of these studies report changes in plasma CA levels based on either the initial level of CA or on patient condition. Roizen *et al.* (122) found that changes in plasma CA levels after induction with morphine varies inversely with the initial level, higher initial levels falling and lower initial levels rising. Stanley *et al.* (140) found 35% increases in urinary CA in patients with atrial septal defects after morphine induction but decreases in patients with tetralogy of Fallot.

Morphine causes histamine-related hypotension when given rapidly and without prior sedative administration (65, 96). Perhaps as a response to hypotension or to the release of histamines, plasma levels of NE increase 2-fold and plasma levels of E 4-fold (65). Moss *et al.* (96) found no such changes in a study that was similar in design but for one important difference: their patients had received 2.5 to 5 mg of diazepam before being given morphine. In the study by Hoar *et al.* (65), concentrations of CA in the plasma returned to baseline 10 minutes after intravenous administration of diazepam (20 to 30 mg).

Anesthesia with morphine sulfate-N_2O-O_2 appears to decrease, in a dose-related fashion, increases in plasma levels of CA induced by surgical incision (Fig. 16.4) (122). These alterations in CA levels in plasma correlate with changes in cardiovascular variables (122).

Several studies indicate that morphine has a selective action on peripheral autonomic systems, that is, that morphine only affects certain neuroeffector junctions. In the nictitating membrane of the cat, morphine reduces the release of NE evoked by nerve stimulation (60, 68). In contrast, in the cardiovascular system, morphine fails to inhibit the release of NE (68, 95). Otto *et al.* (109) examined the effect of morphine on vascular reactivity and release of endogenous NE. Transmural electrical stimulation was used to evoke release of endogenous NE. Morphine (5×10^{-8} to 2×10^{-4} M) caused a dose-dependent depression of the contractile response to nerve stimulation, but did not cause a comparable change in the response to exogenous NE. Morphine (5×10^{-5} to 2×10^{-4} M) markedly depressed the release of NE. This inhibitory effect was not prevented by naloxone, which indicates that inhibition is not mediated by an opiate receptor sensitive to naloxone. The possible interaction between morphine and presynaptic α-receptors was also studied. The α-adrenergic receptor antagonists

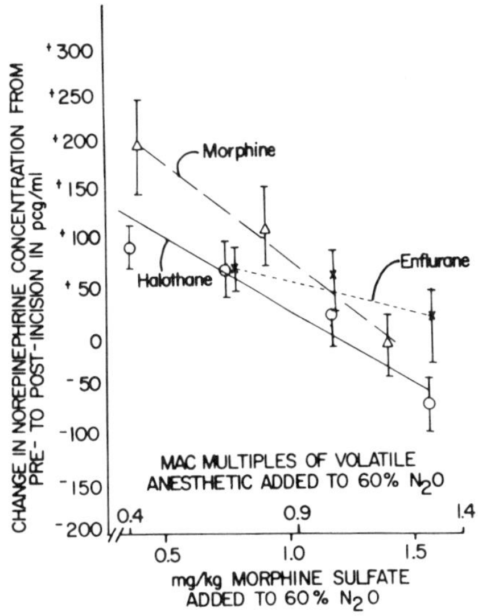

Figure 16.4. Absolute change in plasma levels of norepinephrine (pg/ml) after incision *vs.* the end-tidal dose of anesthesia the patient was receiving. The lines are described as follows: for morphine, the change in norepinephrine = (−207) (mg/kg dose of morphine) + 292 (r = 0.60); for halothane, the change in norepinephrine = (−199) (MAC dose of halothane including 60% N_2O) + 326 (r = 0.60); for enflurane, the change in norepinephrine = (−80) (MAC dose of enflurane including 60% N_2)) + 179 (r = 0.56). Note: the x-axis MAC values in the figure do not include the 0.57 MAC contribution of nitrous oxide. (Reproduced with permission from M. F. Roizen *et al.* (122).)

phenoxybenzamine and phentolamine increase the release of NE induced by stimulation, and this increase is inhibited by morphine (70). Thus, this report and similar studies (15, 159) indicate that morphine alters or inhibits neuronal release of CA in a dose-dependent fashion in isolated *in vitro* organ systems.

We conclude that morphine has three effects on plasma levels of CA:

1. It changes resting levels little when given in a dose and manner that do not affect hemodynamics.
2. It increases plasma levels of CA when given in a dose and manner that cause hypotension and/or release of histamines (the depression of arterial pressure is moderated by normal compensatory barostatic reflexes that increase sympathetic tone in response to hypotension).
3. It decreases, in a dose-related fashion, the increase in plasma CA levels that occur with

incision, an effect that correlates with cardiovascular changes with incision.

Meperidine and Fentanyl

Initial studies report an increase in sympathetic tone (as manifested by increased urinary CA) after administration of fentanyl (47, 82). Subsequent studies, however, do not bear this out.

Stanley *et al.* (141) measured the changes in CA and cortisol concentrations in response to administration of fentanyl-oxygen before and during surgery in 18 patients undergoing coronary artery revascularization. Values for hemodynamic variables (heart rate, cardiac output, mean arterial pressure, and peripheral vascular resistance) were determined. Administration of fentanyl (25 μg/kg) did not alter any cardiopulmonary variable except BP, which decreased. Plasma levels of E, NE, and cortisol were unchanged by endotracheal intubation. The administration of more fentanyl (75 μg/kg) significantly decreased plasma CA levels; however, after 15 minutes of cardiopulmonary bypass, plasma concentrations of E, NE, and cortisol increased. These values increased further after 60 minutes of bypass.

Balogh *et al.* (10) reported no effect of fentanyl anesthesia on CA prior to surgery. In that study, however, CA rose substantially during anesthesia and surgery with low-dose fentanyl. Thus, fentanyl probably inhibits surgery-related increases in plasma CA levels in a dose-related fashion.

Rorie *et al.* (129) examined the effects of fentanyl in superfused, isolated canine coronary arteries in which transmitter stores were prelabeled with [^{3}H]NE. Fentanyl (2×10^{-6} to 8×10^{-6} M) significantly increased the basal overflow of NE and metabolites into the superfusate. This increase was due to a small increase in [^{3}H]NE and a sizable increase in 3,4,dihydroxyphenylglycol (DHPG), the metabolite arising from the intraneuronal metabolism of NE. Because this metabolite is without postsynaptic α-agonist activity and the increases in NE were small, fentanyl had minimal effects on spontaneous adrenergic neurotransmission in this vascular bed. Similarly, these investigators found no effect on evoked release of NE. Thus, *in vitro* evidence confirmed that fentanyl does not affect the release of CA in unstimulated neurons. Studies of plasma CA levels after administration of meperidine (Demerol) report no effect (74, 151).

Mixed Agonists-Antagonists: Pentazocine

Few studies have examined the effects of the mixed "agonist-antagonist" narcotics on concentrations of CA in plasma. Tammisto *et al.* (151) found pentazocine transiently increased CA in plasma by 70%. In another study, total plasma CA increased 40% 5 minutes after administration of pentazocine. This increase was eliminated by epidural blockade to the T-6 dermatomal level (147). Therefore, the increase is probably due to central sympathetic stimulation.

BENZODIAZEPINES

Few studies have used benzodiazepines by themselves in anesthetic concentrations. However, when small doses are combined with narcotics or ketamine, diazepam decreases systemic arterial pressure, cardiac output and related indices, and concentrations of NE and E in the plasma and urine to baseline levels. This reduction in CA occurs regardless of the cause of the initial increase, *i.e.*, a reflex response to hypotension (caused by morphine) or an initial pharmacologic response (caused by ketamine) (65, 78, 82).

KETAMINE

Kumar and co-workers (78) examined the effect of ketamine (2 mg/kg) on arterial plasma NE and E during cardiac surgery for valvular replacement. Anesthesia was induced with diazepam, 0.3 mg/kg, intravenously. Before induction of anesthesia, plasma concentrations of NE and E were 0.39 ± 0.03 ng/ml and 0.10 ± 0.04 ng/ml, respectively. NE levels were significantly higher in these patients than in noncardiac patients. Neither mean systolic BP, heart rate, NE, nor E changes significantly from baseline values following diazepam or ketamine. However, mean diastolic pressure increased 12% following ketamine. This may have been due to a moderate increase in central sympathetic activity (154, 155) or to a direct effect of ketamine on postganglionic adrenergic neurons, where it has been shown to increase release (22, 160) and to decrease reuptake of NE (see below) (91, 95, 105).

These results (78) conflict with those obtained in an earlier study by the same investigators (165). In that study, ketamine (2 mg/kg) was administered to 12 patients who did not have cardiac disease. Five minutes after ketamine, NE concentrations increased from 0.4 ± 0.04 ng/ml to 0.8 ± 0.10 ng/ml. Mean systolic BP and pulse rate increased significantly 2, 5, and 10 minutes after ketamine. These increases did not occur when pancuronium was given concomitantly (88). However, pancuronium by itself causes a reflex decrease in plasma CA levels (see below) (123). In dogs, the amount of circulating CA did not change after administration of ketamine (61, 111).

Other studies reveal that plasma levels of E increase after ketamine and succinylcholine but not after ketamine alone (146). Using a radioenzymatic assay to determine concentrations of CA in plasma, Appel *et al.* (6) reported a 200 to 300% increase in NE and E after ketamine. Because concomitant increases in measured cardiovascular variables occurred, they concluded that ketamine is a sympathetic stimulant working through the CNS. Roizen *et al.* (125) found that ketamine inhibits the increase in plasma CA levels occurring after painful stimuli but less than did any other anesthetic tested.

To elucidate the mechanism by which these changes in CA occur after ketamine, Miletich *et al.* (91) examined its effects on uptake, release, and deaminative metabolism of NE in the isolated, perfused rat heart labeled with [^{14}C]NE. The effect of ketamine was dose-related; ketamine was 80% as effective as cocaine in blocking uptake of NE. These investigators also found that ketamine decreased NE metabolism. This process was lessened by a finding consistent with the inhibitory action of ketamine on neuronal uptake. However, blockade of neuronal uptake may not be enough by itself to increase plasma concentrations of CA (77).

In a more recent study, Salt and co-workers (132) examined the effect of ketamine on uptake-1 and uptake-2 (nonneuronal) in isolated rat hearts. Ketamine produced a concentration-dependent inhibition of both uptake-1 and uptake-2. The dose-response curves for inhibition of uptake-1 and uptake-2 overlapped to such an extent that both processes were seen to be affected almost identically at any given concentration of ketamine. However, the slopes of the dose-response curves differed markedly. The curve for inhibition of uptake-1 was much steeper than that for uptake-2. The results of this study, therefore, agree with those reported by Miletich *et al.* (91) and others (105), but not with those reported by Hill *et al.* (62). These last investigators demonstrated that ketamine did not potentiate the inotropic effect of isoprenaline (a substrate for uptake-2 sites in the heart).

Thus, ketamine seems to follow the Brewster-Price paradigm, *i.e.*, it appears to stimulate cardiovascular function through its autonomic effects. On the other hand, ketamine can inhibit further increases in cardiovascular variables and plasma levels of CA that occur with painful surgical stimuli.

NEUROLEPTIC AGENTS

Droperidol, a dopaminergic receptor antagonist, is frequently used with fentanyl to add "sedation" to the narcotic technique. The combination of fentanyl-droperidol-nitrous oxide produces no consistent change in basal plasma levels of CA (34). However, studies on the effect of droperidol alone on plasma NE have not been performed.

The effect of droperidol on release of CA from nerve endings was examined in isolated canine pulmonary artery and saphenous vein preparations previously incubated with [^{3}H] NE (99). In these unstimulated preparations, droperidol (3.9×10^{-7} M) increased the efflux of the neuronal deaminated metabolite, DHPG. At higher concentrations (3.9 ± 10^{-6} M), droperidol also increased the efflux of NE. These increases were caused by nonexocytotic release of NE from storage vesicles within the neuron. Thus, leakage of NE from storage vesicles may be the mechanism by which droperidol produces acute increases in BP in patients who have pheochromocytoma (14).

In other experiments, Muldoon *et al.* (99) examined the effect of droperidol on evoked tension and [^{3}H]NE release in the isolated canine pulmonary artery. Release of [^{3}H]NE was evoked by supramaximal electrical stimulation. Although the contractile response to 5 Hz was significantly depressed by droperidol, the evoked release of [^{3}H]NE was not. Depression of the contractile response to electrical stimulation is consistent with the α-receptor blocking action of droperidol *in vivo.* (164).

Hyatt *et al.* (70) examined whether droperidol can block presynaptic α-adrenoreceptors and increase the release of NE in other vascular tissue. Saphenous veins were incubated in [^{3}H]NE. When electrically stimulated veins were exposed to droperidol (3.9×10^{-6} M), tension decreased, but [^{3}H]NE efflux did not change. During electrical stimulation of veins, addition of unlabeled NE to the superfusing fluid inhibited the overflow of total radioactivity. Treatment of the veins with droperidol (3.9×10^{-6} M) did not alter this effect. These studies indicate that droperidol is an antagonist of post- but not presynaptic α-adrenoreceptors.

MUSCLE RELAXANTS AND ANTICHOLINERGICS

Muscle relaxants are often selected on the basis of their cardiovascular effects. The differences among muscle relaxants in cardiovascular effects are largely due to varying effects on the release of vasoactive substances. For example, administration of *d*-tubocurarine causes transient hypotension and tachycardia in animals and humans. Moss and colleagues (97) have shown that such tachycardia correlates with release of histamine. Ganglionic blockade occurs at a dose (1.35 mg/kg) that is two to three times greater than that used in clinical practice. Thus, after administration of tubocurarine, one expects plasma levels of CA not to change, or to increase only transiently, due to the effects of histamine or reflex-baroreflex mechanisms. However, Takki and Tammisto (148) reported a 120% "increase" (which was not statistically significant) in plasma CA concentrations after administration of *d*-tubocurarine (0.6 mg/kg); they had premedicated their subjects with atropine and had not established a control group. Thus, the response of CA to their anesthetic technique was not demonstrated. We cannot determine, therefore, if their "increase" would have been significant had their subjects been compared with subjects not given *d*-tubocurarine.

The ability of pancuronium to increase heart rate, cardiac output, and BP has been attributed to a vagolytic effect, or to a direct or indirect sympathomimetic effect. To investigate the role of the SNS on the cardiovascular effects produced by pancuronium, Roizen *et al.* (123) measured NE concentrations in plasma after three doses of pancuronium, one dose of atropine, or atropine followed by pancuronium in 45 unpremedicated patients anesthetized with halothane and 60% nitrous oxide. Concentrations of plasma NE and E and rate-pressure product (the product of heart rate and systolic BP) were determined 15 and 18 minutes after stable anesthesia was established. Either pancuronium (0, 8, 27, or 80 μg/kg) or atropine (5.7 μg/kg) was then given intravenously; 3 and 10 minutes later, NE concentrations and rate-pressure product were again determined. In patients who had not received pancuronium or atropine, NE concentrations did not change significantly with time. In those patients receiving pancuronium alone, NE levels decreased. This decrease correlated significantly with the dose of pancuronium and was inversely related to rate-pressure product. The unexpected decrease in plasma NE concentrations argues against a sympathomimetic mechanism for the circulatory response to pancuronium. The increase in rate-pressure product caused by atropine was also followed by a decrease in NE. In contrast to the results after pancuronium alone, when pancuronium was administered after premedication with atropine, rate-pressure product and NE levels did not change. Roizen *et al.* (123) postulated that the increase in rate-pressure product in response to pancuronium (or atropine) acts through baroreceptors to reduce sympathetic outflow.

Other studies (23, 75, 104, 166) reported increases or no change in plasma NE after pancuronium administration. The difference in the use of atropine between the protocol of Roizen and co-workers and others might account for differing results (23, 75, 94, 104). Atropine reduces or abolishes the increase in heart rate or BP that occurs after pancuronium and the resultant reflex changes that may affect sympathetic function.

Other studies report no change in plasma CA levels. Zsigmond *et al.* (166) premedicated patients with atropine and reported no change in pulse rate or BP after pancuronium was administered during thiamylal anesthesia. Thus, no reflex change in plasma NE levels was expected, and none was found. The same group (88) found no change in pulse or BP after administering a combination of pancuronium and ketamine. Ketamine by itself increased plasma NE level, while the combination of pancuronium (0.09 mg/kg) and ketamine (2 mg/kg) did not. Matsuki and co-workers (88) attributed this lack of effect to an anticholinergic blockade of NE release induced by pancuronium. Takki and Tammisto (148) premedicated their subjects with atropine and had no control group. They reported no change in NE after pancuronium, but the CA response to their anesthetic technique was not documented.

In contrast to the observations on its sympathetic effects, the vagolytic effects of pancuronium are well-substantiated and consistent. Pancuronium blocks the negative chronotropic and ionotropic actions of cholinergic agents in anesthetized dogs, but does not affect the cardiac response to potassium chloride. Therefore, pancuronium selectively blocks the muscarinic cholinergic receptors of the heart (133). It also blocks muscarinic receptors in the superior cervical ganglion (44).

Pancuronium does not affect the metabolism of NE by MAO or COMT *in vitro* (M. F. Roizen *et al.*, unpublished observations). On the other hand, it inhibits the uptake of NE in the myocardium, but only at doses 1,000 times greater than those used in clinical practice (71). Blood levels of pancuronium ranging from that occurring in clinical practice (0.1 nmol/L) (2, 3) to 100 times larger do not affect the uptake of NE (71).

Although succinylcholine administration is reported anecdotally to increase plasma CA, no data have substantiated that claim.

Of the newer neuromuscular blocking agents, ORG NC45 which has been tested in patients with pheochromocytoma, does not alter cardiovascular variables or plasma NE or E (46).

REGIONAL ANESTHESIA

Many studies (25, 86, 122, 139, 147) demonstrate that regional anesthesia to the T-7 dermatomal level decreases or does not affect plasma CA levels (there appears to be a reflex increase in sympathetic tone above the level of the block) (143). Regional anesthesia to a T-4 level, however, decreases cardiovascular function and sympathetic activity (a T-4 dermatomal level of anesthesia, as assessed by pin-prick, probably reflects blockade of sympathetic efferent fibers to the T-1 level, due to the relative susceptibility of these fibers to local anesthetics). Stanek *et al.* (139) report a decrease in plasma NE from about 250 pg/ml to 50 pg/ml and about 60 to 20 pg/ml in E 15 to 20 minutes after initiation of epidural blockade in dogs. No subsequent change in CA levels occurs in response to induced hypovolemia, indicating that regional anesthesia blocks reflex responses. Patients with adequate spinal anesthesia (*i.e.*, those who were pain-free on incision) had no NE or cardiovascular response to incision (113, 122). The decrease in NE concentration and height of spinal blockade correlate (113) (Fig. 16.5), as do the fall in NE and BP after spinal anesthetic (Fig. 16.6). Thus, regional anesthesia decreases resting NE levels and prevents the increase in responses to stress.

LOCAL ANESTHESIA

Physicians can frequently choose between a local or general anesthetic. For patients with compromised cardiovascular function, local anesthesia may be safe. A retrospective study of 10,278 patients at the Mayo Clinic demonstrated that even in a high-risk group of patients who recently had myocardial infarctions, ophthalmic surgery under local anesthesia caused no deaths or additional infarctions (7). In contrast, other retrospecive data from the same institution revealed a high perioperative mortality and the occurrence of subsequent infarctions in patients who had had prior myocardial infarctions and had received general anesthesia (142).

One concern, however, is that the psychological stress of being punctured with a needle during application of the local anesthetic, or of being aware during surgery, could cause a stress response as significant as that created by the surgical stimulus itself (31). Venipuncture significantly increases plasma NE concentrations, and these increases are related to changes in heart rate (77) (see Chapter 1).

If injection into sensitive areas is a significant stress, one would expect to find significantly higher levels of E and NE after administration of local anesthesia. Patients undergoing cataract operations did not change NE levels significantly following administration of local anesthetic (36). While E concentrations did increase in these patients, this increase was caused by diffusion of E from the local anesthetic solution (Tables 16.1 and 16.2). Some patients had marked increases in both NE and E after the retrobulbar block. These three patients had received technically inadequate local anesthesia, and their data were excluded from the calculations (Tables 16.1 and 16.2). Thus, successful local anesthesia does not alter plasma NE or E.

NE and its biosynthetic enzyme, DBH, are

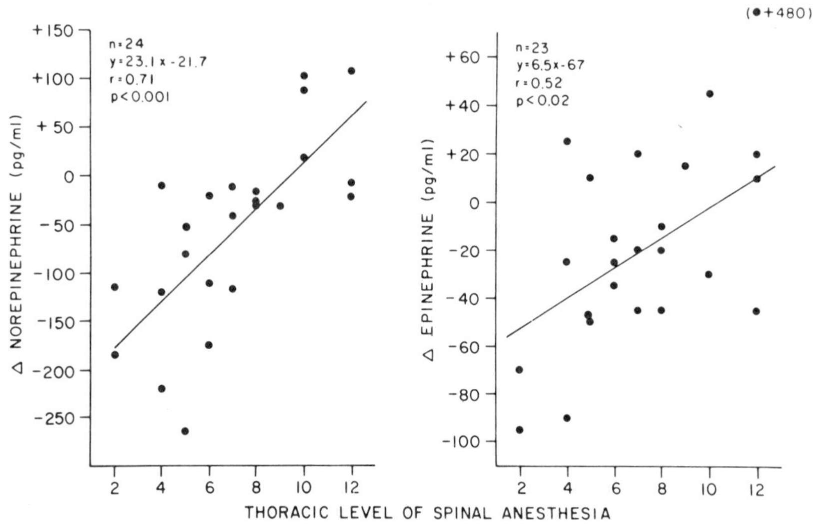

Figure 16.5. Regression analysis of the relationship between changes of plasma norepinephrine and epinephrine from preanesthesia baseline levels and the thoracic dermatone level of spinal anesthesia 30 minutes after subarachnoid injection of tetracaine. The data point in parentheses (high epinephrine level) was not included in the calculation of the regression line. The correlation between anesthesia level and Δ epinephrine is significant even when this data point is included (Spearman's rho = 0.56, $p < 0.01$). (Reproduced with permission from A. E. Pflug and J. B. Halter (113).)

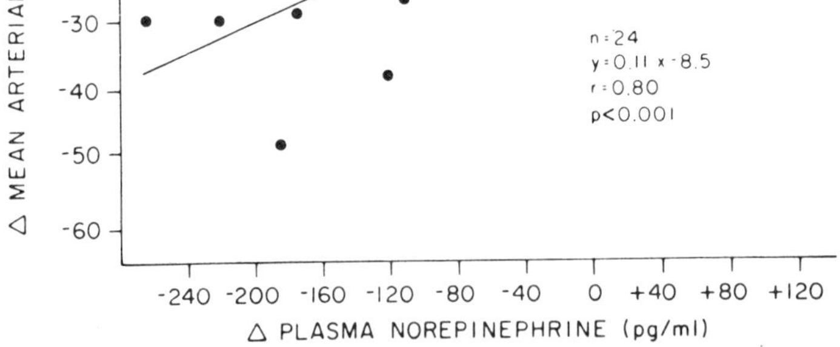

Figure 16.6. Regression analysis of the relationship between changes of plasma norepinephrine from preanesthesia baseline levels and changes of mean arterial blood pressure at 30 minutes after injection of tetracaine for spinal anesthesia. (Reproduced with permission from A. E. Pflug and J. B. Halter.)

transported from cell body to the synapse by axonal transport (24, 27) that uses microtubules. Considerable disagreement exists concerning the extent to which local anesthetics can interfere with microtubules and the rapid axonal transport. *In vitro* studies demonstrate that clinically useful concentrations of procaine and lidocaine interrupt rapid axonal transport of several amino acids *in vitro* (1, 13, 20) and *in vivo* (40). Local anesthetics interfere with microtubular integrity *in vitro* (4, 40, 64). One study, however, reports that anesthetics

Table 16.1
Changes in Plasma Epinephrine during Removal of Cataracts under Local Anesthesia[a]

Patient	Plasma Epinephrine (pg/ml) Control	2 min	7 min
1[b]	43	42	815
2[b]	48	2801	383
3	120	181	273
4	23	56	108
5	46	377	539
6	128	350	501
7	126	251	447
8	72	62	153
9	294	212	504
10	32	387	303
11	79	116	302
12	208	765	766
13[b]	114	780	944
14	66	94	124
15	109	106	152
16	279	360	538
17	180	263	383
Mean ± SE (n = 17)	116 ± 8	424 ± 65[c]	429 ± 25[c]
Mean ± SE (n = 14)	126 ± 9	256 ± 19[d]	368 ± 19[d]

[a] Adapted from Donlon and Moss (36).
[b] As per text, data for three patients with pharmacologic evidence of stress response were excluded from analysis.
[c] Significantly different, $p = 0.044$.
[d] Significantly different, $p = 0.002$.

Table 16.2
Changes in Plasma Norepinephrine during Removal of Cataracts under Local Anesthesia[a]

Patient	Plasma Norepinephrine (pg/ml) Control	2 min	7 min
1[b]	438	464	2675
2[b]	399	2300	440
3	218	226	212
4	251	288	325
5	451	570	692
6	376	388	325
7	923	919	607
8	752	803	1020
9	537	590	502
10	637	664	906
11	759	490	596
12	959	765	921
13[b]	168	311	540
14	341	421	411
15	401	485	227
16	773	834	680
17	806	594	390
Mean ± SE (n = 17)[c]	540 ± 25	653 ± 45	674 ± 57

[a] Adapted from Donlon and Moss (36).
[b] As per text, data for three patients with pharmacologic evidence of stress response were excluded from analysis.
[c] Not significantly different from control, $p = 0.647$.

fail to inhibit axonal transport of CA-synthesizing enzymes when applied *in vivo* to the sciatic nerve of guinea pigs (106). If local anesthetic agents do interfere with axoplasmic transport, one would expect a long-lasting depletion of biogenic amines to occur in presynaptic terminals, particularly when local anesthetics are infused continuously to a nerve trunk. This has not been demonstrated.

On the other hand, local anesthetic agents appear to increase the basal release of NE. Isolated canine saphenous veins incubated in [^{3}H]NE increased tension and moderately increased efflux of total radioactivity in response to etidocaine (100). Most of this increase resulted from increased formation of DHPG. This augmentation was dose-dependent; at the highest concentration tested (1.5×10^{-4} M), increases in DHPG were approximately 2-fold. Studies on lidocaine indicated a similar effect (102). However, when these investigators examined the effect of these local anesthetics on release of [^{3}H]NE in the canine saphenous vein, both etidocaine and lidocaine caused dose-dependent depression of the contractile response to nerve stimulation, exogenous NE, and K^+. Etidocaine decreased the release of [^{3}H]NE evoked by nerve stimulation and caused a parallel reduction in isometric tension responses. Such results help explain the vasodilator properties of these compounds in the intact organism. Release of [^{3}H]NE evoked by KCl was not altered by etidocaine, and tyramine-induced release was augmented.

Thus, etidocaine may augment basal leakage of NE from storage vesicles, making more CA available for intraneuronal deamination. On the other hand, the amide local anesthetics may block the release of NE induced by nerve stimulation.

Gothert (48) examined the effect of different concentrations of tetracaine on release of endogenous NE in the isolated rabbit heart preparation in response to KCl. NE liberation was not decreased at concentrations of tetracaine ranging from 2.6×10^{-8} to 2.6×10^{-5} gm/ml. This result conflicts with results reported by Haeusler *et al.* (55). Thus, *in vitro* it appears that local anesthetics increase the basal release of NE and inhibit, or do not affect, the stimulated release. Therefore, these *in vitro* results support similar changes that have been found to occur in plasma CA levels: (a) small or no changes in resting levels; and (b) inhibition of the normal increase in plasma CA levels that occurs with stress.

THE EFFECT OF SURGERY ON PLASMA LEVELS OF NOREPINEPHRINE

At least three mechanisms contribute to the stress response toward surgery: (a) anxiety; (b) pain from pressure or tissue destruction; (c) ischemia without tissue destruction.

The response to anxiety is discussed elsewhere; the ability of pain and ischemia from surgery to affect plasma CA is reviewed here.

THE EFFECTS OF PAIN AND ANESTHETIC DOSE

Pflug (113), Roizen (122), and their co-workers compared the response of plasma CA levels to surgery in two groups of patients. One group had spinal anesthesia, and one had general anesthesia. Those who received regional anesthesia and were pain-free upon incision had no change in plasma CA. Those who experienced pain on incision, or received low doses of general anesthesia, had increased levels of CA after incision. In fact, the ability of all three general anesthetics tested (morphine-N_2O, enflurane-N_2O, and halothane-N_2O) to prevent this increase was dose-related (Fig. 16.4) and correlated for halothane-N_2O and morphine-N_2O with the ability of those anesthetics to block the cardiovascular response to incision (122). Thus, neural afferents (from the site of tissue injury) that can be blocked by regional anesthesia mediate the neuroendocrine response to surgical stress.

With regional anesthesia, this blockade of the neuroendocrine response continued into the recovery room period; patients who had received general anesthesia had their highest level of NE in plasma in the postoperative period (56, 57, 113, 118). Thus, anesthesia may reduce the stress of surgery. In fact, one study demonstrated that when clinical doses of anesthesia are used, sternal incision is about as stressful in terms of NE response as venipuncture in the awake patient (128).

Only 7 years ago Selye (135) stated that the "stress of anesthesia is a well recognized phenomenon in surgery." Whereas some anesthetic maneuvers or techniques used to induce anesthesia (or lack of adequate anesthesia during intubation, for example) can elevate plasma CA levels, most data now suggest that anesthesia reduces stress in normal patients. In healthy rats, pregnant ewes, and normal human beings, the increase in NE levels associated with painful stimuli and/or surgical incision can be reduced or blocked completely by anesthesia. Other indices of stress in normal individuals—the release of antidiuretic hormone, growth hormone, or cortisol—decrease after initiation of anesthesia. However, no dose-response curves have been obtained for stress reactants other than NE and E. Even endotracheal intubation, if performed with the patient under deep anesthesia, does not elevate CA levels. Russell *et al.* (131) reported a 50% increase in plasma NE levels 3 minutes after endotracheal intubation in 16 patients given "sleep doses" of a barbiturate and paralyzing doses of a muscle relaxant. Takki *et al.* (149) found no increase in plasma CA 2 minutes after laryngoscopy in 30 patients given larger doses of either a barbiturate or a steroid anesthetic induction agent.

Other studies also indicate that the depth of anesthesia is an important determinant of the CA response to surgery (90, 141). After decapitation, all anesthetics tested blocked or ameliorated the response of plasma levels of CA to tissue destruction (125). Thus, "pain" traveling via neural afferents from the site of tissue injury or stimulation mediates the neuroendocrine response to surgical stress. Anesthetic agents decrease this response in a dose-related fashion.

The third mechanism that can contribute to the response of plasma levels of CA to surgery—ischemia, or local hypoperfusion—was brought to light during studies on surgery with cardiopulmonary bypass.

ISCHEMIA AND SURGERY INVOLVING CARDIOPULMONARY BYPASS

Cardiopulmonary bypass increases circulating CA in both animals and man (Table 16.3) (63, 66, 152, 153). The greatest increases occur during initiation of cardiopulmonary bypass

Table 16.3
Increases in Circulating Catecholamines during Cardiopulmonary Bypass (CPB)[a]

	Period				
	I: Awake (n = 14)	II: Anesthesia (n = 15)	III: Operation (n = 15)	IV: Conclusion of CPB (n = 15)	V: Chest Closed (n = 11)
Norepinephrine (pg/ml)	366 ± 38.0	470 ± 113.0	825 ± 123.0	1767 ± 290[c]	731 ± 112.0[b]
Epinephrine (pg/ml)	92 ± 27.0	17 ± 7.0[b]	81 ± 16.0[c]	519 ± 81.0[c]	88 ± 25.0[b]
Dopamine (pg/ml)	40 ± 17.0	97 ± 54.0	108 ± 23.0	151 ± 36.0[c]	106 ± 37.0
Plasma renin activity (ng/ml/hr)	4.0 ± 1.1	5.3 ± 1.5	5.3 ± 1.2	4.2 ± 1.0	4.6 ± 1.3

[a] Adapted from Hoar *et al.* (66). Values are means ± SE. Statistical analysis by Student's *t*-test for paired data comparing each study period to the one immediately preceding.
[b] $p < 0.05$.
[c] $p < 0.01$.

and upon rewarming. The extent to which these increases occur depends not only on the depth of anesthesia, but also on the mechanical properties of the bypass. Thus, by employing a deep level of anesthesia (152, 163) or pulsatile blood flow (76), one can attenuate the pharmacologic changes associated with cardiopulmonary bypass (112).

Although cardiopulmonary bypass increases plasma CA levels, this increase does not appear to be a reflex response to changes in hemodynamic function. Patients with high cardiac output whose total body perfusion with blood decreased on cardiopulmonary bypass are just as likely to have increased CA concentrations as are those with marginal hemodynamic performance whose perfusion with blood actually improves during cardiopulmonary bypass. Using computer modeling, Moss (unpublished data) could demonstrate no relationship between the absolute levels (or between changes in levels) of E and NE and perfusion. However, Reeves and co-workers (120) demonstrated an initial increase after myocardial ischemia. In an experimental model, Turley *et al.* (156, 157) reported that maintenance of high blood flow prevented any change in plasma levels of CA. In that study, deep hypothermia was employed to minimize metabolic demands of animals undergoing perfusion with low blood flow. Although Turley's study groups had similar pH values, changes in blood flow markedly altered plasma levels of CA. High blood flow slightly decreased CA levels, while total circulatory arrest massively increased such levels. Lack of blood flow to the head produced no change in NE or E levels, but lack of blood flow to the body significantly elevated both. These findings demonstrate a local response to hypoperfusion when metabolic demands are minimized.

One other aspect of cardiopulmonary bypass studies deserves mention. Several investigators have shown that very significant increases in CA can occur during rewarming. NE and E seem to increase when systemic vascular resistance increases (152, 163). Studies have excluded renin and other vasoactive hormones that increase systemic vascular resistance as contributing to this increase in vascular resistance during rewarming (66). Increases in plasma NE and E during cardiopulmonary bypass may cause a disjunction between plasma renin activity and sympathoadrenal activity (96). In several studies, plasma renin activity did not change significantly during initiation of cardiopulmonary bypass or during the vasoconstriction phase of rewarming (96). Furthermore, when patients with low cardiac output were studied, cardiopulmonary bypass was associated with a marked decrease in plasma renin activity (22.7 ± 4.3 to 10.4 ± 1.9), despite a modest increase in plasma CA (J. Moss, unpublished data). Prior studies have demonstrated that patients with low stroke volume and hemodynamic compromise appear to accommodate hemodynamic changes through the renin-angiotensin system rather than the sympathoadrenal system (96) (Fig. 16.7). Thus, cardiopulmonary bypass studies have been useful in defining the conditions during which plasma renin activity and sym-

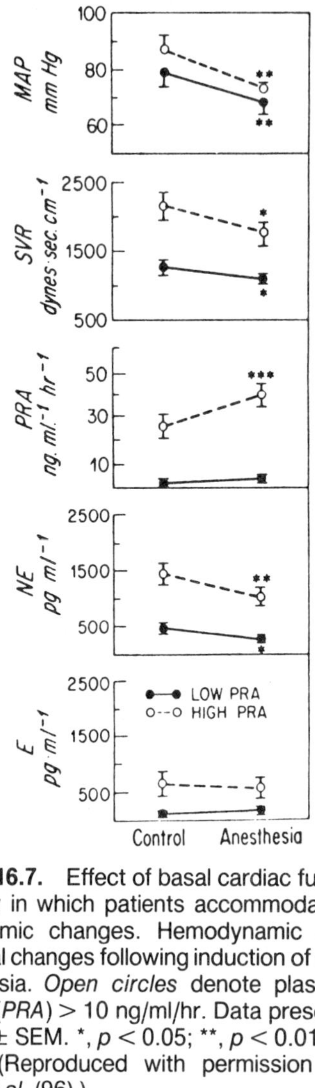

Figure 16.7. Effect of basal cardiac function on the way in which patients accommodate to hemodynamic changes. Hemodynamic and biochemical changes following induction of morphine anesthesia. *Open circles* denote plasma renin activity (*PRA*) > 10 ng/ml/hr. Data presented are means ± SEM. *, $p < 0.05$; **, $p < 0.01$; ***, $p < 0.001$. (Reproduced with permission from J. Moss *et al.* (96).)

pathoadrenal activity do not change similarly, and the role of local hypoperfusion in the response of plasma to CA levels to surgery.

SURGERY INVOLVING THE ADRENAL GLAND

Autopsies on patients with chronic hypertension reveal that 0.5% of the patients had pheochromocytomas (52, 54, 69). Excess NE secretion probably causes chronic hypertension. However, plasma concentrations of CA do not correlate with BP when considering all patients as one group (Fig. 16.8) (16). Indeed, normotension during periods when levels of circulating CA are very elevated is not uncommon. This lack of correlation has been attributed to lack of salt, production of a vasodilator compound such as dihydroxyphenylalanine (dopa) (85) or a prostaglandin (38), or "down-regulation" of vascular adrenoreceptors. This last process appears most probable and has been measured in lymphocytes (53). In fact, all three mechanisms may play a role in preventing a correlation between CA levels and BP across patient groups. In any one patient, however, there appears to be a correlation between plasma levels of NE and BP. Perhaps this is best demonstrated when NE levels change rapidly during resection of pheochromocytoma. Until the venous drainage is obliterated, manipulation of the tumor liberates vast amounts of CA in the plasma, e.g., 50,000 to 200,000 pg/ml of NE. Concomitantly, BP and peripheral vascular resistance increase dramatically, while cardiac output increases only slightly (M. F. Roizen, unpublished data).

Typical changes in BP and CA concentrations are shown in Figure 16.9. Note the maximum rise in plasma NE concentrations during tumor manipulation. We counteract the effects of this release of CA on BP with the peripheral vasodilator nitroprusside. It is clear, however, that circulating CA change cardiovascular variables. Obliteration of tumor vessels decreases plasma levels of CA, pulse pressure, systolic BP, and peripheral vascular resistance, and, surprisingly, changes cardiac output only slightly (M. F. Roizen, personal observation) (Fig. 16.9).

Although data are limited, surgery that involves the adrenal gland but not pheochromocytoma does not result in excessive or prolonged elevation of plasma CA, or in dramatic cardiovascular changes (16; M. F. Roizen, personal observation) (Fig. 16.10).

CONCLUSION: IMPLICATIONS FOR SURGERY AND THE PRACTICE OF ANESTHESIA

Anesthetics may either modify plasma CA levels directly or may modify these plasma CA responses to surgery. Although the direct effects of anesthetic and anesthetic adjuvant drugs on plasma levels of CA vary, anesthesia uniformly ameliorates or lessens the increase in plasma CA levels caused by surgery.

The conscious animal, including man, reacts to stress for survival. This reaction includes mobilization of energy stores, an increase in

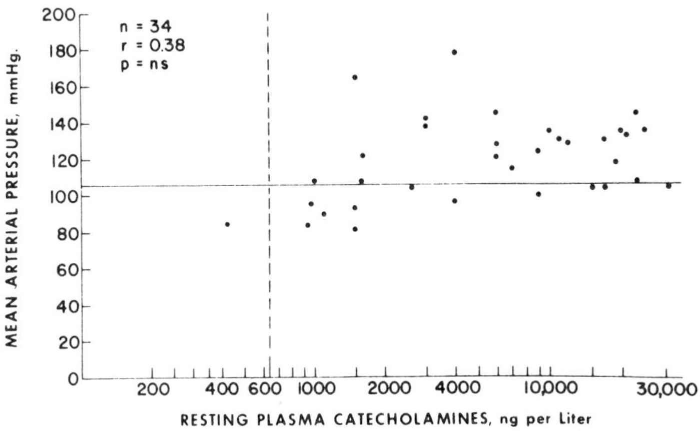

Figure 16.8. Lack of correlation between arterial blood pressure and plasma levels of catecholamines for 18 patients (34 observations) with pheochromocytomas. Plasma catecholamines are expressed on a log scale on the horizontal axis. The *vertical broken line* indicates plasma catecholamine values 3 SDs above the mean value in normal controls (620 ng/L, 3.66 nmoles/L). The *horizontal solid line* represents the upper limits of normal for mean arterial pressure corresponding to a blood pressure of 140/90 mm Hg. (Reproduced with permission from E. L. Bravo *et al.*)

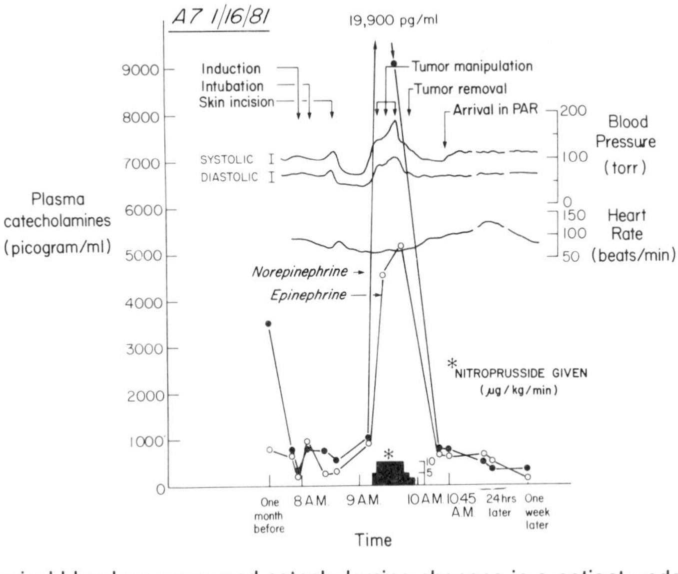

Figure 16.9. Typical blood pressure and catecholamine changes in a patient undergoing resection of a pheochromocytoma.

demand upon the heart and circulation, and a diversion of blood flow from tissues serving a less primary function (e.g., gut, liver, kidney). Although these responses may be vital to the conscious animal on a short-term basis, they may be detrimental or of no value to the anesthetized patient undergoing surgery. No functional gain may result, and the cost in myocardial work, the expenditure of energy, and the potential inadequacy of oxygen delivery relative to need may be significant. Studies in healthy animals and humans have not demonstrated that increasing or decreasing stress by altering anesthetic concentration is either harmful or beneficial. In pregnant ewes (136), the administration of less anesthesia, while associated with a decrease in uterine blood flow and an increase in plasma NE concentration, did not result in fetal compromise. The hypothesis that when one increases anesthesia

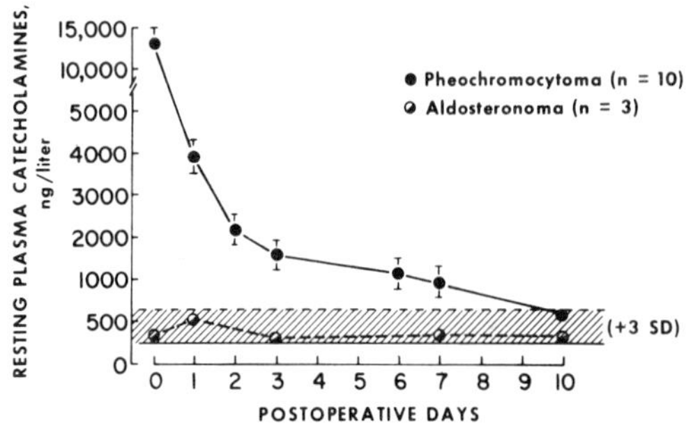

Figure 16.10. Effect of adrenal surgery with and without pheochromocytoma on plasma levels of catecholamines. In pheochromocytoma, high plasma concentrations of catecholamines linger for about a week after complete tumor resection. (Reproduced with permission from E. L. Bravo *et al.* (16).)

to decrease stress one also decreases morbidity has not been adequately tested.

This reasoning supposes that an "adequate" level of anesthesia should do more than merely immobilize the patient. If adverse effects (e.g., metabolic, hepatic, renal, or myocardial) result from a neuroendocrine response to stimulation, the hypothesis "the less anesthetic the better" may be wrong. The use of a dose of general anesthetic that blocks CA response to surgery, or high levels of spinal anesthesia, would markedly diminish or essentially abolish the intraoperative neuroendocrine stress response. However, greater depression of myocardial and respiratory function and an increased possibility of chemical (anesthetic) toxicity might result. Although the overall effect of using high anesthetic doses *vs.* a "the-less-anesthetic-the-better" approach on perioperative morbidity and mortality remains to be determined, the possibility of choosing a dose of anesthetic that results in better patient well-being has been facilitated by measuring levels of CA in plasma.

References

1. Aasheim G, Fink BR, Middaugh M: Inhibition of rapid axoplasmic transport by procaine hydrochloride. *Anesthesiology* 41:549–553, 1974.
2. Agoston S, Crul JF, Kersten UW, Scaf AH: Relationship of the serum concentration of pancuronium to its neuromuscular activity in man. *Anesthesiology* 47:509–512, 1977.
3. Agoston S, Vermeer GA, Kerstein UW, Meijer DKF: The fate of pancuronium bromide in man. *Acta Anaesthesiol Scand* 17:267–275, 1973.
4. Allson AC, Nunn JF: Effect of general anesthetics on microtubes: A possible mechanism of anesthesia. *Lancet* 2:1326–1329, 1968.
5. Anton AH, Gravenstein JS, What MW Jr: Extracorporeal circulation and endogenous epinephrine and norepinephrine in plasma, atrium, and urine in man. *Anesthesiology* 25:262–269, 1964.
6. Appel E, Dudziak R, Palm D, Wnuk A: Sympathoneuronal and sympathetic activation during ketamine anesthesia. *Eur J Clin Pharmacol* 16:91–95, 1979.
7. Backer CL, Tinker JH, Robinson DM: Myocardial reinfarction following local anesthesia. *Anesthesiology* 51:S61, 1979.
8. Bailey P, Berbode G, Garlington L: An anesthetic technique for cardiac surgery which utilizes 100% oxygen as the only inhalant. *Arch Surg* 76:437–440, 1958.
9. Balasaraswathi K, Glisson SN, El-Etr AA, Azad C: Effect of priming volume on serum catecholamines during cardiopulmonary bypass. *Can Anaesth Soc J* 27:135–139, 1980.
10. Balogh D, Hammerle AF, Hörtnagl H, Brucke T, Stadler-Wollfersgrün R: Plasma-Katecholamine bei halothan-N_2O-anaesthesie und neuroleptanalgesie. Intra- and post-operative vergleichsstudie. *Anaesthesist* 28:517–522, 1979.
11. Bartelstone HJ, Katz RL, Ngai SH: Effect of cyclopropane on reflexly induced circulatory responses in the dog. *Anesthesiology* 27:756–763, 1966.
12. Bhatia BB, Burn JH: The action of ether on the sympathetic system. *J Physiol* 78:257–270, 1933.
13. Bisby MA: Inhibition of axonal transport in nerves chronically treated with local anesthetics. *Exp Neurol* 47:481–489, 1975.
14. Bittar DA: Innovar-induced hypertensive crises in patients with pheochromocytoma. *Anesthesiology* 50:366–369, 1979.
15. Borrell J, Llorens I, Borrell S: Study of the effects of morphine on adrenal corticosteroids, ascorbic acid and catecholamines in unanaesthetized and anaesthetized cats. *Horm Res* 5:351–358, 1974.
16. Bravo EL, Tarazi RC, Gifford RW, Stewart BH: Circulating and urinary catecholamines in pheochromocytoma. Diagnostic and pathophysiologic implications. *N Engl J Med* 301:682–686, 1979.
17. Brewster WR, Isaacs JP, Wain-Andersen T: Depressant effect of ether on myocardium of the dog and its modification by reflex release of epinephrine and norpinephrine. *Am J Physiol* 175:399–414, 1953.
18. Briscoe TJ, Millar RA: The effect of cyclopropane, halothane and ether on sympathetic ganglionic transmission. *Br J Anaesth* 38:3–12, 1966.
19. Brown BR Jr, Tatum EN, Crout JR: The effects of inhalation anesthetics on the uptake and metabolism of L-^{3}H-norepinephrine in guinea pig atria. *Anesthesiology* 36:263–267, 1972.
20. Byers MR, Fink BR, Kennedy RD, Middaugh ME, Hendrickson AE: Effects of lidocaine on axonal morphology, microtubules, and rapid transport in rabbit vagus nerve *in vitro*. *J Neurobiol* 4:125–143, 1973.

21. Byles PH, Dobkin AB, Ferguson JH, Levy AA: Forane (Compound 469): Cross-over comparison with enflurane (Ethrane), halothane, and methoxy-flurane in dogs. *Can Anaesth Soc J* 18:376–386, 1971.
22. Chang P, Chan KE, Ganaenldran A: Cardiovascular effects of 2-(-O-chlorophyenyl)-2-methylamino cyclohexanone (C1-581) in rats. *Br J Anaesth* 41:263–267, 1969.
23. Coleman AJ, Downing JW, Leary WP, Moyes DG, Styles M: The immediate cardiovascular effects of pancuronium, alcuronium, and tubocurarine in man. *Anaesthesia* 27:415–422, 1972.
24. Coyle JT, Wooten GF: Rapid axoplasmic transport of tyrosine hydroxylase and dopamine-beta-hydroxylase. *Brain Res* 44:701–704, 1972.
25. Craft JB, Roizen MF, Dao SD, Edwards M, Gilman R: A comparison of T4 and T7 dermatomal levels of analgesia for caeserean section using the lumbar epidural technique. *Can Anaesth Soc J* 29:264–269, 1982.
26. Cristoforo MF, Brody MJ: Non-adrenergic vasoconstriction produced by halothane and cyclopropane anesthesia. *Anesthesiology* 29:44–56, 1969.
27. Dairman W, Geffin L, Marchelle M: Axoplasmic transport of aromatic amino acid-decarboxylase (EC4.1.1.26) and dopamine-hydroxylase (EC4.1.14.2.1) in rat sciatic nerve. *J Neurochem* 20:1617–1723, 1973.
28. de Jong RH, Freund FG, Robles R, Morikawa K-I: Anesthetic potency determined by depression of synaptic transmission. *Anesthesiology* 29:1139–1144, 1968.
29. Depocas F, Behrens WA: Effects of handling, decapitation, anesthesia, and surgery on plasma noradrenaline levels in the white rat. *Can J Physiol Pharmacol* 55:212–219, 1976.
30. Deutsch S, Linde HW, Price HL: Circulatory and sympathoadrenal responses to cyclopropane in the dog. *J Pharmacol Exp Ther* 135:354–357, 1962.
31. Dimsdale JE, Moss J: Plasma catecholamines in stress and exercise. *JAMA* 243:340–342, 1980.
32. Dobkin AB, Byles PH, Ghanooni S, Valbuena DA: Clinical and laboratory evaluation of a new inhalation anaesthetic: Forane (Compound 469) CHF_2-O-$CHClCF_3$ (1-chloro-2,2,2-trifluoroethyl difluoromethyl ether). *Can Anaesth Soc J* 18:264–271, 1971.
33. Dobkin AB, Byles PH, Neville JF Jr: Neuroendocrine and metabolic effects of general anesthesia during spontaneous breathing, controlled breathing, mild hypoxia, and mild hypercarbia. *Can Anaesth Soc J* 13:130–171, 1966.
34. Dobkin AB, Byles PH, Cho MH: Neuroleptanalgesics. 3. Effect of Innovar-nitrous oxide anaesthesia on blood levels of histamine serotonin, epinephrine, and norepinephrine, and on urine excretion. *Can Anaesth Soc J* 12:349–360, 1965.
35. Domenech JS, Garcia RC, Sasiain JMR, Loyola AQ, Oroz JS: Pancuronium bromide: An indirect sympathomimetic agent. *Br J Anaesth* 48:1143–1148, 1976.
36. Donlon JV, Moss J Jr: Plasma catecholamine levels during local anesthesia for cataract operations. *Anesthesiology* 51:471–473, 1979.
37. Duner H: The influence of blood glucose level on the secretion of adrenalin and noradrenaline from the supra-renal. *Acta Physiol Scand* 102(Suppl):422–432, 1953.
38. Dunn MJ, Hood VL: Prostaglandins and the kidney. *Am J Physiol* 233:F169–F184, 1977.
39. Farnebo LO, Hallman H, Hamberger B, Jonsson G: Catecholamines and hemorrhagic shock in awake and anesthetized rats. *Circ Shock* 6:109–118, 1979.
40. Fink BR, Kennedy RD, Hendrickson AE, Middaugh ME: Lidocaine inhibition of rapid axonal transport. *Anesthesiology* 36:422–432, 1972.
41. Fink BR, Kish SJ, Byers MR: Rapid axonal transport in trigeminal nerve of rat. *Brain Res* 90:85–95, 1975.
42. Freiberger KU, Hack G, Havers L: General anesthesia: The course of catecholamine excretion during combined anesthesia with halothane-thalamonal. *Anaesthesist* 24:465–470, 1975.
43. Gardier RW, Endahl GL, Hamelberg W: Cyclopropane: Effect on catecholamine biotransformation. *Anesthesiology* 28:677–679, 1967.
44. Gardier RW, Tsevdos EJ, Jackson DB: The effect of pancuronium and gallamine on muscarinic transmission in the superior cervical ganglion. *J Pharmacol Exp Ther* 204:46–53, 1978.
45. Garfield JM, Alper MH, Gillis RA, Flacke W: A pharmacologic analysis of ganglionic actions of some general anesthetics. *Anesthesiology* 29:79–92, 1968.
46. Gencarelli PJ, Roizen MF, Miller RD, Joyce J, Hunt TK, Tyrrell JB: ORG NC45 (Norcuron) and pheochromocytoma: A report of three cases. *Anesthesiology* 55:690–693, 1981.
47. Giesecke AH, Jenkins MT, Crout JR, Collett JM: Urinary epinephrine and norepinephrine during Innovar-nitrous oxide anesthesia in man. *Anesthesiology* 28:701–704, 1967.
48. Gothert M: Effects of halothane on the sympathetic nerve terminals of the rabbit heart: Differences in the membrane actions of halothane and tetracaine. *Naunyn Schmiedebergs Arch Pharmacol* 286:125–143, 1974.
49. Gothert M, Wendt J: Inhibition of adrenal medullary catecholamine secretion by enflurane. II. Investigation in isolated bovine adrenals—site and mechanism of action. *Anesthesiology* 46:404–410, 1977.
50. Gothert M, Wendt J: Inhibition of adrenal medullary catecholamine secretion by enflurane: I. Investigations *in vivo*. *Anesthesiology* 46:400–403, 1977.
51. Gothert M, Kennerknecht E, Thielecke G: Inhibition of receptor-mediated noradrenaline release from the sympathetic nerves of the isolated rabbit heart by anesthetics and alcohol in proportion to their hydrophobic property. *Naunyn Schmiedebergs Arch Pharmacol* 292:145–152, 1976.
52. Graham JB: Pheochromocytoma and hypertension. An analysis of 207 cases. *Int Abst Surg*

92:105–121, 1951.

53. Greenacre JK, Conolly ME: Desensitization of the beta-adrenoceptor of lymphocytes from normal subjects and patients with pheochromocytoma: Studies *in vivo*. *Br J Clin Pharmacol* 5:191–197, 1978.
54. Greer WER, Robertson CW, Wilkins RW: Pheochromocytoma: A discussion of symptoms, signs and procedures of diagnostic value. *N Engl J Med* 242:252–257, 1950.
55. Haeusler G, Thoenen H, Haefely W, Huerliman A: Electrical events in cardiac adrenergic nerves and noradrenaline release from the heart induced by acetylcholine and KC1. *Naunyn Schmiedebergs Arch Pharmacol Exp Pathol* 261:389–411, 1968.
56. Halter JB, Pflug AE: Relationship of impaired insulin secretion during surgical stress to anesthesia and catecholamine release. *J Clin Endocrinol Metab* 51:1093–1098, 1980.
57. Halter JB, Pflug AE, Porte D Jr: Mechanism of plasma catecholamine increases during surgical stress in man. *J Clin Endocrinol Metab* 45:936–944, 1977.
58. Hamelberg W, Sprouse JH, Mahaffey JE, Richardson JA: Catecholamine levels during light and deep anesthesia. *Anesthesiology* 21:297–301, 1960.
59. Hasbrouck JD: Morphine anesthesia for open heart surgery. *Ann Thorac Surg* 10:364–368, 1970.
60. Henderson G, Hughes J, Kosterlitz HW: The effects of morphine on the release of noradrenaline from the cat isolated nictitating membrane and the guinea pig ileum myenteric plexus-longitudinal muscle preparation. *Br J Pharmacol* 53:505–512, 1975.
61. Hensel I, Braun U, Kettler D, Knoll D, Martel J, Paschen K, Bretschneider HJ: Tierexperimentelle Untersuchungen, zur Frage der Katecholaminaktivitat unter Ketaminnarkose. In Gemperle M, Kreuscher H, Langrehr D: *Ketamin. Neue Ergbnisse in Forschung and Klinik. Anaesthesiology and Resuscitation 69.* Berlin, Springer-Verlag, 1973, pp 63–76.
62. Hill GE, Wong KC, Shaw CL, Sentker CR, Blatnick RA: Interactions of ketamine with vasoactive amines at normothermia and hypothermia in the isolated rabbit heart. *Anesthesiology* 48:315–319, 1978.
63. Hine IP, Wood WG, Mainwaring-Burton RW, Butler MJ, Irving MH, Booker B: The adrenergic response to surgery involving cardiopulmonary bypass, as measured by plasma and urinary catecholamine concentrations. *Br J Anaesth* 48:355–362, 1976.
64. Hinkler RE, Telser AG: The effects of halothane on microfilamentous systems in cultured neuroblastoma cells. In Fink BR: *Molecular Mechanisms of Anesthesia, Progress in Anesthesiology I.* New York, Raven Press, 1975, pp 103–118.
65. Hoar PF, Nelson NT, Mangano DT, Bainton CR, Hickey RF: Adrenergic response to morphine-diazepam anesthesia for myocardial revascularization. *Anesth Analg* 60:406–411, 1981.
66. Hoar PF, Stone JG, Faltas AN, Bendixen HH, Head RJ, Berkowitz BA: Hemodynamic and biochemical alterations produced by anesthesia and surgery in patients undergoing coronary artery bypass procedures. *J Thorac Cadiovasc Surg* 80:242–248, 1980.
67. Holmes JC, Schneider FH: Pentobarbitone inhibition of catecholamine secretion. *Br J Pharmacol* 49:205–213, 1973.
68. Hughes J, Kosterlitz HW, Leslie FM: Effect of morphine on adrenergic transmission in the mouse vas deferens: Assessment of agonist and antagonist potencies of narcotic analgesics. *Br J Pharmacol* 53:371–381, 1975.
69. Hutchinson GB, Evans JA, Davidson DC: Pitfalls in the diagnosis of pheochromocytoma. *Ann Intern Med* 48:300–309, 1958.
70. Hyatt M, Muldoon SM, Rorie DK: Droperidol, a selective antagonist of postsynaptic alpha-adrenoceptors in the canine saphenous vein. *Anesthesiology* 53:281–286, 1980.
71. Ivankovich AD, Miletich DJ, Albrecht RF, Zahed B: The effect of pancuronium on myocardial contraction and catecholamine metabolism. *J Pharm Pharmacol* 27:837–841, 1975.
72. Joyce JT, Roizen MF, Gerson JI, Grobecker H, Eger EI, Forbes AR: Induction of anesthesia with halothane increase plasma norepinephrine concentrations. *Anesthesiology* 56:280–290, 1982.
73. Joyce JT, Roizen MF: Application of face mask does not increase plasma norepinephrine levels. *Anesth Analg* 60:257–258, 1981.
74. Kehlet H, Nikki P, Jäättelä A, Takki S: Plasma catecholamine concentrations during surgery in unsupplemented glucocorticoid-treated patients. *Br J Anaesth* 46:73–77, 1974.
75. Kelman GR, Kennedy BR: Cardiovascular effects on pancuronium in man. *Br J Anaesth* 43:335–338, 1971.
76. Kono K, Philbin DM, Coggins CH, Moss J, Roscow CE, Schneider RC, Slater EE: Renal function and stress response during halothane and fentanyl anesthesia. *Anesth Analg*, 60:552–556, 1981.
77. Kopin IJ, Lake CR, Ziegler MG: Plasma levels of norepinephrine. *Ann Intern Med* 88:671–680, 1978.
78. Kumar SM, Kothary SP, Zsigmond EK: Effect of pancuronium on plasma free-norepinephrine and epinephrine in adult cardiac surgical patients. *Acta Anaesthesiol Scand* 22:423–429, 1978.
79. Lain RF, Hess ML, Gertz EW, Briggs FN: Calcium uptake activity of canine myocardial sarcoplasmic reticulum in the presence of anesthetic agents. *Circ Res* 23:597–604, 1968.
80. Larabee MG, Posternak JM: Selective action of anesthetics on synapses and axons in mammalian sympathetic ganglia. *J Neurophysiol* 15:91–114, 1952.
81. Li TH, Laasberg LH, Etsten BE: Effects of anesthetics on myocardial catecholamines. *Anesthesiology* 25:641, 1964.
82. Liu W-S, Bidwai AV, Lunn JK, Stanley TH: Urine catecholamine excretion after large doses of fentanyl, fentanyl and diazepam, and fentanyl, diazepam and pancuronium. *Can An-*

aesth Soc J 24:371–379, 1977.

83. Lorenz RR, Vanhoutte PM, Shepherd JT: Interaction between neuronal amine uptake and prejunctional alpha-adrenergic receptor activation in smooth muscle from canine blood vessels and spleen. *Blood Vessels* 16:113–125, 1979.
84. Lowenstein E, Hallowell P, Levine FH, et al: Cardiovascular response to large doses of intravenous morphine in man. *N Engl J Med* 281:1389–1393, 1969.
85. Louis WJ, Doyle AE, Heath WC, Robinson MJ: Secretion of dopa in phaeochromocytoma. *Br Med J* 4:325–327, 1972.
86. Madsen SN, Brandt MR, Engquist A, et al: Inhibition of plasma cyclic AMP, glucose and cortisol response to surgery by epidural analgesia. *Br J Surg* 64:669–671, 1977.
87. Martinez LR, von Euler C, Norlander OP: The sedative effect of premedication as measured by catecholamine excretion. *Br J Anaesth* 38:780–785, 1966.
88. Matsuki A, Zsigmond EK, Kelsch RC, Kothary SP: The effect of pancuronium bromide on plasma norepinephrine concentrations during ketamine induction. *Can Anaesth Soc J* 21:315–320, 1974.
89. McAllister FF, Root WS: The circulatory responses of normal and sympathectomized dogs to ether anesthesia. *J Physiol* 135:70–78, 1941.
90. Michaels J: Study of blood catecholamine levels before, during and after surgery, under different types of general anesthesia. *Acta Anaesthesiol Belg* 27(Suppl):93–114, 1976.
91. Miletich DJ, Ivankovich AD, Albrecht RF, Zahed B, Ilahi AA: The effect of ketamine on catecholamine metabolism in the isolated perfused rat heart. *Anesthesiology* 39:271–277, 1973.
92. Millar RA, Morris ME: Sympatho-adrenal responses during general anaesthesia in the dog and man. *Can Anaesth Soc J* 8:356–386, 1961.
93. Millar JW, Elliott HW: Plasma sympathin concentration of dogs. *Proc Soc Exp Biol Med* 87:487–488, 1954.
94. Miller RD, Eger EI II, Stevens WC, Gibbons R: Pancuronium-induced tachycardia in relation to alveolar halothane, dose of pancuronium, and prior atropine. *Anesthesiology* 42:352–355, 1975.
95. Montel H, Starke K: Effects of narcotic analgesics and their antagonists on the rabbit isolated heart and its adrenergic nerves. *Br J Pharmacol* 49:628–641, 1973.
96. Moss J, Lappas DG, Slater EE: Role of the renin catecholamine system in hemodynamic performance in man. In Usdin E, Kvetnansky R, Kopin IJ: *Catecholamines and Stress (Recent Advances)*. Amsterdam, Elsevier/North Holland, 1980, pp 513–519.
97. Moss J, Rosow CE, Savarese JJ, Philbin DM, Kniffen KJ: Role of histamine in the hypotensive action of *d*-tubocurarine in humans. *Anesthesiology* 55:19–25, 1981.
98. Mueller RA: Increases in tyrosine hydroxylase activity after exposure to cyclopropane and fluroxene. *Anesthesiology* 35:612–620, 1971.
99. Muldoon SM, Janssens WJ, Verbeuren TJ, Vanhoutte PM: Alpha-adrenergic blocking properties of droperidol on isolated blood vessels of the dog. *Br J Anaesth* 49:211–216, 1977.
100. Muldoon SM, Otto J, Thoa NB: Interaction between morphine and alpha adrenergic agents on evoked release of norepinephrine (NE) from isolated canine saphenous vein. *Fed Proc* 40:45, 1981.
101. Muldoon SM, Vanhoutte PM, Lorenz RR, Van Dyke RA: Venomotor changes caused by halothane acting on the sympathetic nerves. *Anesthesiology* 43:41–48, 1975.
102. Muldoon SM, Verbeuren TJ, Vanhoutte PM: Effects of amide-linked local anesthetics on adrenergic neuroeffector junction in cutaneous veins of dogs. *J Pharmacol Exp Ther* 196:723–736, 1976.
103. Naito H, Gillis CN: Anesthetics and response of atria to sympathetic nerve stimulation. *Anesthesiology* 29:259–266, 1968.
104. Nana A, Cardon E, Domokos M: Blood catecholamine changes after pancuronium. *Acta Anaesthesiol Scand* 17:83–87, 1973.
105. Nedergaard OV: Cocaine-like effect of ketamine on vascular adrenergic neurons. *Eur J Pharmacol* 23:153–161, 1973.
106. Ngai SH, Dairman W, Marchelle M: Effects of lidocaine and etidocaine on the axoplasmic transport of catecholamine-synthesizing enzymes. *Anesthesiology* 41:542–548, 1974.
107. Ngai SH, Diaz PM, Ozer S: The uptake and release of norepinephrine: Effects of cyclopropane and halothane. *Anesthesiology* 31:45–52, 1969.
108. Ngai SH, Neff NH, Costa E: The effects of cyclopropane and halothane on the biosynthesis of norepinephrine *in vivo*: Conversion of ^{14}C-tyrosine to catecholamines. *Anesthesiology* 31:53–60, 1969.
109. Otto J, Muldoon SM, Watson RL: Effect of morphine on vascular reactivity and endogenous norepinephrine release in innervated canine saphenous veins. *Pharmacologist* 202:196, 1980.
110. Perry LB, Van Dyke RA, Theye RA: Sympathoadrenal and hemodynamic effects of isoflurane, halothane, and cyclopropane in dogs. *Anesthesiology* 40:465–470, 1974.
111. Peter K, Altstaedt F, Hollmann G, Klose R, Mayr J: Verhalten der Katecholaminspiegel, im Blut wahrend intravenoser Narkose mit Ketamin beim Hund. In Gemperle M, Kreuscher H, Langrehr D: *Anaesthesilogie Resuscitation 69*. Berlin, Springer-Verlag, 1973, pp 59–76.
112. Philbin DM, Levine FH, Emerson CW, Kono K, Coggins CH, Moss J, Slater EE, Buckley MJ: Attenuation of the stress response to cardiopulmonary bypass by the addition of pulsatile flow. *Circulation* 64:808–812, 1981.
113. Pflug AE, Halter JB: Effect of spinal anesthesia on adrenergic tone and the neuroendocrine responses to surgical stress in humans. *Anesthesiology* 55:120–126, 1981.
114. Price HL, Linde HW, Jones RE, Black GW, Price ML: Sympatho-adrenal responses to general anesthesia in man and their relation to hemodynamics. *Anesthesiology* 20:563–574, 1959.

115. Price HL, Lurie AA, Black GW, Sechzer PH, Linde HW, Price ML: Modification by general anesthetics (cyclopropane and halothane) of circulatory and sympathoadrenal responses to respiratory acidosis. *Ann Surg* 152:1071–1077, 1960.
116. Price HL, Price ML: Relative ganglion-blocking potencies of cyclopropane, halothane and nitrous oxide and the interaction of nitrous oxide with halothane. *Anesthesiology* 28:349–353, 1967.
117. Price HL, Warden JC, Cooperman LJ, Millar RA: Central sympathetic excitation caused by cyclopropane. *Anesthesiology* 30:426–438, 1969.
118. Quasha A, Loeber H, Feeley TW, Ullyot DJ, Roizen MF: Postoperative respiratory care: A controlled trial of early and late extubation following coronary artery bypass graft surgery. *Anesthesiology* 52:135–141, 1980.
119. Rawlinson WAL, Loach AB, Benedict CR: Changes in plasma concentration of adrenaline and noradrenaline in anaesthetized patients during sodium nitroprusside-induced hypotension. *Br J Anaesth* 50:937–943, 1978.
120. Reves JG, Karp RB, Smith LR: Catecholamine changes during cardiopulmonary bypass: The stress response and the role of the lungs in modulating it. *Anesth Analg* 60:272, 1981.
121. Richardson JA, Woods EG, Richardson AK: Plasma concentrations of epinephrine and norepinephrine during anesthesia. *J Pharmacol Exp Ther* 119:378–384, 1957.
122. Roizen MF, Horrigan RW, Frazer B: Anesthetic doses blocking adrenergic (stress) and cardiovascular responses to incision—MAC BAR. *Anesthesiology* 54:390–398, 1981.
123. Roizen MF, Forbes AR, Miller RD, Lake CR: Similarity between effects of pancuronium and atropine on plasma norepinephrine levels in man. *J Pharmacol Exp Ther* 211:419–422, 1979.
124. Roizen MF, Moss J, Henry DP, Kopin IJ: Effects of halothane on plasma catecholamines. *Anesthesiology* 41:432–439, 1974.
125. Roizen MF, Moss J, Henry DP, Weise VW, Kopin IJ: Effects of general anesthetics on handling and decapitation-induced increases in sympathoadrenal discharge. *J Pharmacol Exp Ther* 204:11–18, 1978.
126. Roizen MF, Thoa NB, Moss J, Kopin IJ: Inhibition by halothane of release of norepinephrine, but not of dopamine-beta-hydroxylase from guinea pig vas deferens. *Eur J Pharmacol* 31:313–318, 1975.
127. Roizen MF, Thoa NB, Moss J, Kopin IJ: Inhibition by cyclopropane of release of norepinephrine, but not dopamine-beta-hydroxylase, from guinea pig vas deferens. *Anesthesiology* 44:54–56, 1976.
128. Roizen MF, Wilkinson PL, Chatterjee K, Hamilton WK, Moyers JR, Ports TA, Ullyot DJ, Frazer BM: Does anesthesia alter myocardial stress during coronary artery surgery? A-V myocardial norepinephrine difference, a new index of myocardial stress. In Kventnanski RJ, Usdin E, Kopin IJ: *Catecholamines and Stress (Recent Advances)*. New York, Elsevier, 1980, pp 811–216.
129. Rorie DK, Muldoon SM, Tyce GM: Effects of fentanyl on adrenergic function in canine coronary arteries. *Anesth Analg* 60:21–27, 1981.
130. Rosenberg PH, Klinge E: Some effects of enflurane anaesthesia on biogenic monoamines in the brain and plasma of rats. *Br J Anaesth* 46:708–713, 1974.
131. Russell WJ, Morris RG, Frewin DB: Changes in plasma catecholamine concentrations during endotracheal intubation. *Br J Anaesth* 53:837–840, 1981.
132. Salt PJ, Barnes PK, Beswick FJ: Inhibition of neuronal and extraneuronal uptake of noradrenaline by ketamine in the isolated perfused rat heart. *Br J Anaesth* 51:835–838, 1979.
133. Saxena PR, Bonta IL: Mechanism of selective cardiac vagolytic action of pancuronium bromide. Specific blockade of cardiac muscarinic receptors. *Eur J Pharmacol* 11:332–341, 1970.
134. Schneider D, O'Neill JJ: Monoamine oxidase activity in isolated guinea pig hearts: Halothane and triton X-100 (abstr). *Fed Proc* 30:441, 1971.
135. Selye H: *Stress without Distress*. Philadelphia, JB Lippincott, 1974, p 31.
136. Shnider SM, Wright RG, Levinson G, Roizen MF, Wallis KL, Rolbin SH, Craft JB: Uterine blood flow and plasma norepinephrine changes during maternal stress in the pregnant ewe. *Anesthesiology* 50:524–527, 1979.
137. Skovsted P, Price ML, Price HL: The effects of short-acting barbiturates on arterial pressure, preganglionic sympathetic activity and barostatic reflexes. *Anesthesiology* 33:10–18, 1970.
138. Smith NT, Eger EI II, Stoelting RK, Whayne TF, Cullen DJ, Kadis LB: The cardiovascular and sympathomimetic responses to the addition of nitrous oxide to halothane in man. *Anesthesiology* 32:410–421, 1970.
139. Stanek B, Schwarz M, Zimpfer M, Raberger G: Plasma concentrations of noradrenaline and adrenaline and plasma renin activity during extradural blockade in dogs. *Br J Anaesth* 52:305–311, 1980.
140. Stanley TH, Isern-Amaral J, Lathrop GD: The effects of morphine and halothane anesthesia on urine norepinephrine during surgery for congenital heart disease. *Can Anaesth Soc J* 23:58–67, 1976.
141. Stanley TH, Berman L, Green O, Robertson DH, Roizen MF: Fentanyl-oxygen anesthesia for coronary artery surgery: Plasma catecholamine and cortisol responses. *Anesthesiology* 51:S139, 1979.
142. Steen PS, Tinkers JH, Tarhan S: Myocardial reinfarction after anesthesia and surgery. *JAMA* 239:2566–2570, 1978.
143. Stevens WC, Caine WE, Hamilton WK: Circulatory studies during spinal anesthesia: Central and peripheral venous oxygen saturation before and after administration of vasopressors. *Anesth Analg* 47:725, 1968.
144. Stokke DB, Christense NJ, Hole P, Anderson PK, Juhl B: Plasma catecholamines during equipotent anaesthesia with cyclopropane and

halothane-N_2O in man. *Anaesthesist* 27:469–474, 1978.

145. Sumikawa K, Amakata Y, Kashimoto T, Izumi F: Effects of cyclopropane on catecholamine release from bovine adrenal medulla. *Anesthesiology* 53:49–52, 1980.
146. Takki S, Nikki A, Jäättelä A, Tammisto T: Ketamine and plasma catecholamines. *Br J Anaesth* 44:1318–1321, 1972.
147. Takki S, Nikki P, Tammisto T, Jäättelä A: Effect of epidural blockade on the pentazocine-induced increase in plasma catecholamines and blood pressure. *Br J Anaesth* 45:376–380, 1973.
148. Takki S, Tammisto T: Effect of pancuronium and tubocurarine on plasma catecholamines. *Ann Chir Gynaecol* 62:260–264, 1973.
149. Takki S, Tammisto T, Nikki P, Jäättelä A: Effect of laryngoscopy and intubation on plasma catecholamine levels during intravenous induction of anaesthesia. *Br J Anaesth* 44:1323–1328, 1972.
150. Tammisto T: Effect of neuroleptanesthesia on catecholamines. *Int Anesthesiol Clin* 2:185–195, 1973.
151. Tammisto T, Jäättelä AA, Nikki P, Takki S: Effect of pentazocine and pethidine on plasma catecholamine levels. *Ann Clin Res* 3:22–29, 1971.
152. Tan CK, Glisson SN, El-Etr AA, Ramakrishnaiah KB: Levels of circulating norepinephrine and epinephrine before, during, and after cardiopulmonary bypass in man. *J Thorac Cardiovasc Surg* 71:928–931, 1976.
153. Tan CK, Glisson SN, Younes SH: Adrenal responses to anesthetics during cardiopulmonary bypass. *Cardiovasc Med* 3:521–528, 1978.
154. Traber DL, Wilson RD, Priano LL: Blockade of the hypertensive response to ketamine. *Anesth Analg* 49:420–426, 1970.
155. Traber DL, Wilson RD: Involvement of the sympathetic nervous system in the pressor response to ketamine. *Anesth Analg* 48:248–252, 1969.
156. Turley K, Roizen MF, Vlahakes J, Graham B, Ebert PA: Catecholamine response to deep hypothermia and total circulatory arrest in the infant lamb. *Cardiovasc Surg* 62(Suppl):I175–I179, 1980.
157. Turley K, Roizen MF, Ebert PA: Deep hypothermia and total circulatory arrest: The effect of method of cooling on the catecholamine response to arrest. *J Surg Res* 30:379–383, 1981.
158. Vanhoutte PM, Lorenz RR, Tyce GH: Inhibition of norepinephrine ^{3}H release form sympathetic nerve endings in veins by acetylcholine. *J Pharmacol Exp Ther* 183:386–394, 1973.
159. Vasalle M: Role of catecholamine release in morphine hyperglycemia. *Am J Physiol* 200:530–534, 1961.
160. Virtue RW, Alanis JM, Mori M, Lafargue RT, Vogel JH, Metcalf DR: An anesthetic agent: 2-ortho chlorophenyl, 2-methylamino cyclohexanone HCl (CI-581). *Anesthesiology* 28:823–828, 1967.
161. Wallin BG, Konig U: Changes of skin nerve sympathetic activity during induction of general anaesthesia with thiopentone on man. *Brain Res* 103:157–160, 1976.
162. Wang HH, Epstein RA, Markee SJ, Bartelstone HJ: The effects of halothane on peripheral and central vasomotor control mechanisms of the dog. *Anesthesiology* 29:877–886, 1969.
163. Watkins WD, Moss J, Lappas DG, Levine L, Peterson MB: Vasoactive mediators and human cardiopulmonary bypass. *Anesthesiology* 51:S97, 1979.
164. Yelnosky J, Katz R, Dietrich EV: A study of some of the pharmacologic actions of droperidol. *Toxicol Appl Pharmacol* 6:37, 1964.
165. Zsigmond EK, Kelsch RC, Kothary SP, Vandnay L: Free norepinephrine concentrations during induction of anesthesia with ketamine. *Rev Brasil Anesth* 22:443–451, 1972.
166. Zsigmond EK, Matsuki A, Kothary SP, Kelsch RC: The effect of pancuronium bromide on plasma norepinephrine and cortisol concentrations during thiamylal induction. *Can Anaesth Soc J* 21:147–152, 1974.

CHAPTER SEVENTEEN

Conjugation of Norepinephrine and Other Catecholamines

OTTO KUCHEL, M.D., Sc.D., and NGUYEN THANH BUU, Ph.D.

CONJUGATION OF CATECHOLAMINES: GENERAL

Since Richter's observation in 1940 that a major fraction of orally ingested epinephrine (E) could be recovered as a conjugate in the urine, the *conjugation* of catecholamines (CA) was suspected to be an important pathway for the metabolism and urinary excretion of CA (112). The discovery of the predominant role of neuronal reuptake, oxidation, and O-methylation in the inactivation of CA (8) relegated this pathway to almost total oblivion. The consensus on conjugated CA was summed up about 10 years ago by this statement: "Conjugated E and norepinephrine (NE) are very minor diet-influenced dead-end metabolites that are not important enough to be included in metabolic schemes for E and NE as given in recent monographs and survey articles" (66). If mentioned at all in recent reviews, conjugated CA are referred to as metabolites of CA which probably facilitate their urinary excretion. The finding that high concentrations of conjugated CA themselves and some of their metabolites can be found in plasma (46, 52, 58) received little attention, probably because of methodological uncertainties surrounding these measurements and the possibility that they were artifacts originating in foodstuffs.

It appears that the main difficulty in the search for the role of conjugation in the physiology of CA is the heterogeneity of this process, which consists of both *glucurono-* and *sulfoconjugation*, with considerable species and organ differences in the distribution of the relevant enzymes and of the various species of conjugated CA.

Glucuronoconjugation has been demon-

Abbreviations used in Chapter 17 are: E, epinephrine; CA, catecholamine(s); NE, norepinephrine; PST, phenolsulfotransferase; PAPS, 3′-phosphoadenosine-5′-phosphosulfate; COMT, catechol-O-methyltransferase; M, metanephrine; NM, normetanephrine; DA, dopamine; ATP, adenosine-triphosphate; ADP, adenosine-diphosphate; MHPG, 3-methoxy-4-hydroxyphenylglycol; DCNP, 2,6-dichloro-4-nitrophenol; CNS, central nervous system; MAO, monoamine oxidase; TLC, thin-layer chromotography; RIA, radioimmunoassay; HPLC, high pressure liquid chromotography; [^{35}S]PAPS, 3′ phosphoadenosine-5′-phospho [^{35}S] sulfate; HVA, homovanillic acid; CSF, cerebrospinal fluid; DBH, dopamine-β-hydroxylase; PNMT, phenylethanolamine-N-methyltransferase; ACTH, adrenocorticotrophic hormone; VMA, vanillylmandelic acid; 5-HT, 5-hydroxytryptamine; BP, blood pressure; SNS, sympathetic nervous system.

strated only in the liver (130). *UDP-glucuronyltransferase* (EC 2·4·1·17) is a microsomal enzyme which catalyzes the formation of *glucuronides* from uridine diphosphate glucuronate and a phenolic substrate according to the following equation:

$$\text{R-OH} + \text{UDP-glucuronate} \xrightarrow{\text{enzyme}} \text{R-O-glucuronide} + \text{UDP}.$$

Sulfoconjugating activity, in contrast, is much more widely distributed. It is found in the liver, kidney, brain, adrenal, intestinal mucosa and placenta of several species. The *phenolsulfotransferase* (PST) (EC 2·8·2·1) (48) is a cytoplasmic enzyme which catalyzes the formation of sulfated derivatives by the transfer of a sulfate group from the main sulfate donor in the organism, 3′-phosphoadenosine-5′-phosphosulfate (PAPS), to a phenolic substrate:

$$\text{R-OH} + \text{PAPS} \xrightarrow{\text{enzyme}} \text{R-O-SO}_3 + \text{H}^+ + \text{PAP}.$$

The two modes of conjugation probably interact. Thus, 2,6-dichloro-4-nitrophenol inhibits the sulfoconjugation of harmol from 75 to 5% while increasing its glucuronoconjugation from 25 to 95% (99).

There are considerable differences in conjugation patterns between animal species (95) (see Table 17.1). Glucuronoconjugation predominates in the rat and sulfoconjugation in man and dog. Despite this predominance of one type of conjugation in each species, the wide distribution of PST in mammalian tissues suggests that most mammals studied (man, dog, rat, monkey, cat, guinea pig, and others) can sulfoconjugate phenolic substances (114). Sulfation occurs at the 3-*O*- and 4-*O*-hydroxyl groups of the CA molecule. Whenever the meta-hydroxyl group (3-O) is free, it is the preferred site of sulfoconjugation, but conjugation also occurs at the para-hydroxyl group (4-O) (61) (Fig. 17.1).

This well-documented (22) preferential sulfoconjugation at the meta-hydroxyl position suggests that PST competes with catechol-*O*-methyltransferase (COMT) and glucuronyltransferase for the released CA. The activities and localization of these enzymes and their different affinities for the CA probably determine the proportions of the resultant CA metabolites, namely metanephrines (M), CA sulfates, and CA glucuronides.

The specificity of the PST toward CA is still undetermined. However, the partially purified PST from blood platelets (53) and from central or peripheral neuronal tissue (93) has been able to sulfoconjugate CA and a large number of their metabolites; whether PST exists as one or many forms with different affinities for CA or their metabolites in different tissues remains to be investigated.

In several tissues such as rat brain (93), liver (130), and human platelets (53), dopamine (DA)

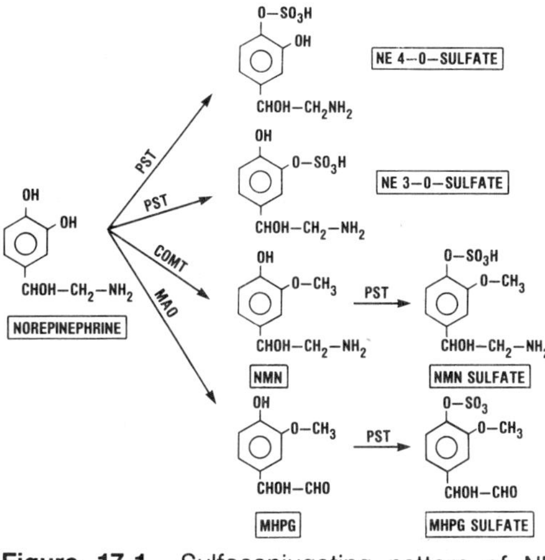

Figure 17.1. Sulfoconjugating pattern of NE and some of its metabolites. *PST*, phenolsulfotransferase.

Table 17.1
Metabolites Recovered in Urine after Administration of ^{14}C-Dopamine (15 mg/kg)[a]

	Percentage of ^{14}C Recovered in					
	Mouse	Guinea Pig	Rabbit	Monkey	Rat	Dog
DA glucuronide	24		2	2	34	3
DA sulfate	10	2	1	14	6	71
HVA free	2	13	16	18	17	16
conjugated	7			2		
DOPAC free	3	85	78	40	26	2
conjugated	53			18		

[a] Adapted from I. Merits (94).

showed the highest affinity to PST, followed by NE and E in that order. The tissue PST activity appears to increase with age: there is approximately a 7-fold increase in liver, brain, and kidney PST activity in rats between birth and 10 weeks of age (102).

The tissue distribution of free and sulfoconjugated DA, NE, and E in the dog, the species with a pattern of conjugation closest to man, is summarized in Table 17.2. NE is present only in the free form in the adrenal, liver, small intestine, and kidney. The same is true for E, except that one-third of its kidney content is in conjugated form (125). This suggests, in view of the high degree of conjugation (approximately 80%) of plasma NE and E, that most of the sulfoconjugation occurs by the action of platelet PST while NE and E are in transit through the circulation. This is borne out by catheterization studies in humans (80) (Fig. 17.2). *DA sulfate*, on the other hand, is present in the vesicular fraction of the *adrenal medullary granules* at a concentration six times that of free DA. An adrenal release of DA sulfate was demonstrated by measuring the arteriovenous differences across the adrenals (125) and from catheterization studies in man (73). No such adrenal release of NE or E sulfate has been found.

Platelets are considered to be the site where much NE sulfoconjugation occurs. Notably, the proportions of sulfates with NE, E, and DA in them are roughly equal to those found in plasma: 63% for NE, 57% for E, and 85% for DA (107). If we accept the claim that conjugated CA probably do not enter the platelets, it appears that these sulfates are synthesized within the platelets from CA taken up from the blood. PST activity in peripheral tissues appears to play a minimal role in the conjugation of plasma CA. Despite high PST activity in the dog liver, there is no arteriovenous difference across the liver in the proportion of conjugated DA (125). It is not clear what ratio of sulfoconjugation results in the generation of NE 3-sulfate and NE 4-sulfate. NE 4-*O*-sulfate is probably less stable than 3-*O*-sulfate if judged from DA 3-*O*-sulfate having a half-life of 3 to 4 hours compared to 1 to 3 minutes for DA 4-*O*-sulfate.

The release of physiologically active CA from CA sulfate has not been studied. In one case, Burns (26) found that arylsulfotransferase prepared from the microorganism *Aspergillus orizae* can function in the absence of phenolic acceptors, and also as arylsulfatase. This dual function of PST was, however, not found in

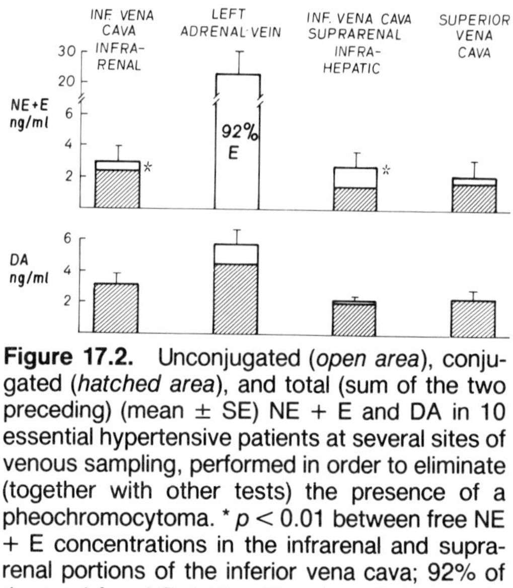

Figure 17.2. Unconjugated (*open area*), conjugated (*hatched area*), and total (sum of the two preceding) (mean ± SE) NE + E and DA in 10 essential hypertensive patients at several sites of venous sampling, performed in order to eliminate (together with other tests) the presence of a pheochromocytoma. * $p < 0.01$ between free NE + E concentrations in the infrarenal and suprarenal portions of the inferior vena cava; 92% of the total free NE + E outflow from the adrenals is E (for details see references 71 and 82).

Table 17.2
Free and Conjugated Catecholamines in Peripheral Organs in the Dog[a]

		Dopamine	Norepinephrine	Epinephrine
Adrenal, n = 7	Free	27 ± 8	439 ± 65.9	1530 ± 230
	Conjugated	149 ± 37	ND	ND
Liver, n = 5	Free	0.06 ± 0.01	0.77 ± 0.11	0.18 ± 0.09
	Conjugated	0.04 ± 0.02	ND	ND
Small intestine, n = 4	Free	0.03 ± 0.01	0.26 ± 0.03	0.10 ± 0.02
	Conjugated	0.01 ± 0.003	ND	ND
Kidney, n = 6	Free	0.06 ± 0.003	1.22 ± 0.10	0.009 ± 0.001
	Conjugated	0.11 ± 0.04	ND	0.03 ± 0.01

[a] Reprinted with permission from T. Unger *et al.* (124). Values are mean ± SEM in micrograms per gram of tissue. ND, not detectable.

mammals. The enzymes most responsible for the release of free CA in mammals are undoubtedly arylsulfatases (arylsulfate sulfohydrolase, EC 3.1.6.1). These ubiquitous enzymes are so designated on the basis of their activities with sulfate esters of phenol such as *p*-nitrocatechol sulfate (43). There are three types of *sulfatases* (A, B, C), two of which (A and B) are involved in the metabolism of CA. The latter are soluble and located in lysosomes. It is not clear which of these sulfatases is involved in the cleavage of sulfate from NE sulfates. The finding that both arylsulfatases act on the *p*-nitrocatechol sulfate substrate suggests that this is a nonspecific enzymatic system used in the desulfation not only of phenol sulfate but also of the sulfate esters of steroids, some lipids, chondroitin, and ascorbic acid, among others. Arylsulfatase A and B activities in rats have been found to be highest in the spleen, followed by the lung with a relative activity of 92% (compared to 100% in the spleen), the kidney with 78%, the liver with 65%, the heart with 60%, the brain and pancreas with 49% each, and intestinal mucosa with 41% (120). Syndromes with multiple sulfatase deficiency, the rare mucosulfatidose (45) could be an ideal model to study the effect of such an enzymatic defect on the stability of CA sulfates.

CA sulfates, like free CA, can originate from *exogenous sources*. Richter and MacIntosh (112) observed that a large dose of ingested E was excreted in the urine as the sulfo- or glucuronoconjugate of E, in contrast to small doses of parenterally administered E or NE, which were mostly metabolized by O-methylation and deamination (8). Common fruits, vegetables (122), and cereals containing *tyrosinase* (converting dietary tyrosine into dopa) (56) can induce an increased excretion of conjugated CA. Cereal-rich diet increases conjugates of DA and 3-O-methyldopamine, but decreases conjugated normetanephrine (NM) in rat urine (35). Thus, the possibility that some conjugated CA originates in exogenous dietary sources should not be dismissed. Crout and Sjoerdsma (37) observed that the ingestion of bananas increases the urinary excretion of conjugated NE. Eating bananas also increases plasma-conjugated NE and DA, and this increase persists 6 hours after consumption (41). In a control experiment to verify dietary effects on conjugated DA in plasma, blood samples from very obese subjects were taken at the same time every morning for 8 days while the patients starved (82). We observed relatively stable values of DA sulfates when they were repeatedly determined, while considerable short-lived variations were seen in patients on a free diet. This suggests that the known effect of meals, at least on plasma-conjugated DA, is of short duration and can be disregarded if the subjects fast for at least 12 hours before sampling. Free NE and DA in plasma and urine of normal subjects are not dependent on food intake (34). Milk diet increases free E, NE, DA, M, normetanephrine (NM), and *3-O-methyldopamine*, while cereal diet increases free NE, DA, and 3-O-methyldopamine excretions in rat urine (35).

Enzymatic systems, endogenous and exogenous *inhibitors and activators of PST and sulfatases* must all be considered. Substrate inhibition of these enzymes is probably their common characteristic. Some products of the sulfoconjugation reaction, such as PAPS (116) and *p*-nitrophenylsulfate, act as inhibitors (13). Nucleotides having structures similar to PAPS, such as ATP and ADP, also act as PST inhibitors (105). Of practical interest is the competitive inhibitory action on PST by some drugs, such as *salicylamide* (6) or *probenecid*, the latter being an activator or inhibitor of the sulfation of *3-methoxy-4-hydroxyphenylglycol* (MHPG), depending on the tissue studied (103). *2,6-Di-chloro-4-nitrophenol* (DCNP) specifically inhibits the PST *in vivo* and *in vitro* without affecting glucuronyltransferase (99). It could thus become a useful tool in determining the respective roles of glucurono- and sulfoconjugation. An inhibitor of sulfoconjugation may also potentiate the action of drugs which are normally extensively conjugated. Thus, in man, inhibition of sulfoconjugation is attributed to *vitamin C* (57), which is known to potentiate the effect of isoprenaline. Vitamin A-deficient rats show a low degree of sulfoconjugation of phenols, which can be restored by the addition of vitamin A (120).

THE PREDOMINANCE OF SULFOCONJUGATION IN MAN

The predominance of sulfoconjugation in man is demonstrated by the high proportions of CA sulfates (approximately 99% of DA, approximately 80% of NE and E) in plasma (29) and probably also in urine, in which two-thirds of CA conjugates were reportedly sulfates (51). Although NE as well as DA and E appear predominantly sulfoconjugated, some metabolites of NE such as MHPG are about equally sulfo- and glucuronoconjugated in plasma (67). In the CSF, MHPG is predominantly free (only 15% sulfated); however, it is 30% sulfated in the *hypothalamus* and 80% sulfated in the *substantia nigra*; thus, there are regional differences in the degree of sulfoconjugation (28). While there is some indication that the sulfate conjugate of MHPG originates almost exclusively in the central nervous system (CNS), measurements of MHPG or either of its conjugates in urine are a poor index of brain NE turnover in man (18). The recent demonstration that human ventricular fluid contains CA sulfate (121) opens new avenues of exploration for the role of sulfoconjugation of other CA in the brain. Even if conjugation were normally a minor CA inactivation pathway in the brain, it could assume an important role in clinical conditions where other pathways function abnormally or are pharmacologically blocked.

The sulfoconjugating enzyme PST is widely distributed in various healthy human tissues: adrenals, jejunal mucosa, kidneys, liver, ovaries, and spleen. In pathological tissues (cirrhotic liver, adrenal adenoma, hypernephroma, nephritic renal tissues) the levels of the enzyme activity equal those in normal tissues (19). A significant correlation has been found between the formation of *MHPG sulfate* by individual platelet preparations and that of sulfated products with one of the following substrates: *tyramine, DA, 5-OH tryptamine, acetaminophen*, and *α-methyldopa*. These results raise the possibility that human platelet PST activity measured with MHPG as substrate can reflect the degree of sulfate conjugation in a variety of substrates. PST activity in human platelets indicates that in addition to the normal pathway of inactivation of CA by monoamine oxidase (MAO) (113), an alternative route exists to terminate CA action: metabolism to the sulfate esters. The significant correlation between PST activity in fresh renal and jejunal tissues and that in platelets (4) raises the possibility that platelets may be used as diagnostic markers of the sulfotransferase activity in other organs (53). In contrast to the relatively high PST activity in platelets, the PST activity of erythrocytes is considerably lower, and its determination is plagued by the presence of endogenous inhibitors of PST (2). Platelet PST activity measured using DA as substrate (1) correlated well with that obtained with MHPG as substrate ($r = 0.73$, $p < 0.001$) (74). Preliminary results indicate that no correlation exists between the platelet PST activity and the plasma concentration of DA or E sulfates, but a positive correlation does exist between PST and the level of NE sulfate (74). It is improbable that the degree of sulfoconjugation is a simple reflection of the PST activity; probably many factors, such as the concentrations of CA at the conjugation sites and their respective affinities for the enzyme, are important in determining the levels of the sulfates of NE and other CA.

The degree of competition between inactivation pathways via COMT and PST remains unestablished. When CA methylation was inhibited *in vivo*, the synthesis of CA sulfates increases (96). The availability of inorganic sulfate does not appear to be a limiting factor in the sulfoconjugation (129), although it may become critical when the pool size of inorganic sulfate, very sensitive to dietary manipulation and stress, is low (98). Another factor affecting sulfoconjugation is the accessibility of the CA to sulfoconjugating sites. The respective roles of tissue and blood-borne PST activities in the inactivation of individual CA are not yet defined, but since sulfoconjugation seems to be predominant in the blood platelets (107), the accessibility of DA, NE, and E to blood platelets determines the relative proportions of their inactivation by sulfoconjugation. In this respect, the higher degree of plasma DA and E conjugation relative to NE (74) may reflect a more "hormonal" character of DA and E compared to the "neurotransmitter" character of NE, inactivated predominantly by local mechanisms and only incidentally leaking into the circulation (39). The presently limited knowledge of sulfoconjugation mostly concerns the metabolites of NE. One of the main NE metabolites, *NM sulfate*, has been found in higher concentrations in human urine (69) than NE sulfate itself.

METHODS FOR HYDROLYSIS OF CONJUGATED CA AND DETERMINATION OF PHENOLSULFOTRANSFERASE ACTIVITY

In general, CA sulfate concentrations have been determined by the difference between the CA contents of samples measured before hydrolysis (free CA) and those obtained after hydrolysis (total CA). The accuracy, specificity, and sensitivity of the conjugated CA assay thus depend on both the method of determining CA and the efficiency and specificity of hydrolyzing conjugated CA. The introduction of the radioenzymatic determination of CA, based on the O-methylation of CA by COMT in the presence of [^{3}H]SAMe (36), and the separation of the radiolabeled methylated CA by thin-layer chromatography (TLC) (40), have considerably increased the CA assay's sensitivity and specificity (see Chapter 2). More recent techniques using cation-exchange chromatography (68) or reverse-phase liquid chromatography (97) coupled with an electrochemical detection device have also been used for the measurement of free and conjugated CA (after hydrolysis), but their sensitivity remains less than that of the radioenzymatic assay (see Chapter 3). Most recently, radioimmunoassay (RIA) for NE and E in tissues and plasma (108) has been reported. Its sensitivity is comparable to that of the radioenzymatic assay.

Hydrolysis of Conjugated CA

The hydrolysis of conjugated CA to free CA remains difficult to standardize because of the lability of CA molecules in solutions, the difference in speed of hydrolysis of CA glucuronides and CA sulfates, and the different degrees of stability of CA in plasma and urine.

Plasma-conjugated CA are commonly extracted along with free CA in acid solution. However, they are not stable in an acid milieu and slowly hydrolyze even when stored at temperatures below 0°C. Hydrolysis can be accelerated by heating at 95° to 100°C for various periods of time. Recently, Nagel and Schumann (100) reported that the heating of 0.3 N $HClO_4$ extracts of plasma for 40 minutes can completely hydrolyze conjugated CA, with an excellent recovery (96 to 100%) of CA added before hydrolysis. Whether this method can also be used to hydrolyze urinary conjugated CA is unknown. It has been found that the dilution of urine prior to hydrolysis (66), the pH of the solution (55), the nature of the acid used (68), and the duration of the hydrolysis (88) are all important factors which can influence the hydrolysis. A complete hydrolysis of conjugated CA in human urine has been described (68) and consists of heating urine in 0.4 M hydrochloric acid at 85°C for only 30 minutes. It is, however, doubtful that such a mild hydrolysis could also hydrolyze CA glucuronides, which are more stable to acid hydrolysis than CA sulfates.

Conjugated CA can also be hydrolyzed in acid at low temperatures (29). With this method, plasma is deproteinized with 3 N $HClO_4$, while urine (human, dog, rat) is diluted 20-fold with 0.1 N $HClO_4$. After centrifugation, aliquots of the clear supernatant (to which dithiothreitol has been added) are freeze-dried. Hydrolysis occurs even at the lyophilization temperature when the conjugated CA are exposed to a medium or very high acid strength and when the solution is concentrated. The residue is reconstituted to its original volume with 0.1 N $HClO_4$ and the solution measured for its total CA content by the radioenzymatic assay, essentially as described by da Prada (40). Although this method of hydrolysis gives complete recovery for DA, the recovery for exogenous NE and E added to the sample is low. Values of conjugated CA obtained by this method of hydrolysis are, however, similar to those reported for methods of heating in dilute acid with reportedly higher recovery (Table 17.3).

The main disadvantage with an acid hydrolysis of conjugated CA lies in its nondifferentiation between CA sulfates and CA glucuronides. Such differentiation can be achieved by the use of specific hydrolyzing enzymes. In earlier methods, because of the difficulty in obtaining pure sulfatase, samples were usually incubated with a glusulase (sulfatase and glucuronidase) and with purified glucuronidase (for CA glucuronides). The difference between values represented the sulfated CA. This method, requiring incubations of the sample for long periods of time (12 to 24 hours), suffers from a large loss of CA and from the contaminating presence of large quantities of L-dopa and DA in the glusulase preparations (7). Recently, Johnson *et al.* (64) have found that adding a commercial sulfatase preparation (free of glucuronidase) to the incubation medium of a COMT radioenzymatic assay allows them to determine the sum of sulfated CA and free CA. Subtracting the values for free CA (determined without a sulfatase) from that sum yields the values of CA sulfates. These researchers' values of NE and E sulfate in human plasma resemble those obtained by acid lyophilization and acid heating (Table 17.4), indicating that most, if not all, of the conjugated plasma CA in man are CA sulfates.

Table 17.3
Comparison of Human Urinary Excretions of Unconjugated and Conjugated DA, NE, and E (ng/min) Obtained by Different Methods of Hydrolysis

Method of Hydrolysis and Determination	Dopamine			Norepinephrine			Epinephrine		
	Unconj.	Conj.	% Conj.	Unconj.	Conj.	% Conj.	Unconj.	Conj.	% Conj.
Acid hydrolysis + fluorometry (Mattok *et al.*, 1966; Kahane, *et al.*, 1967)	137	316	70	39 27	39 78	50 74	0.7 3.3	0.3 19	30 85
Acid hydrolysis + fluorometry (Hoeldtke and Sloan, 1970)	159	289	65	20	41	67	3.2	4.2	57
Acid lyophilization + radioenzymatic (Buu and Kuchel, 1977)	283	811	71	39	35	43	12	48	74
Acid hydrolysis[a] + radioenzymatic (Da Prada, 1980)	229	625	73	30	101	77	8	28	76
Acid hydrolysis[a] + HPLC (Kissinger *et al.*, 1975)	213[b]	588[b]	66	20[b]	42[b]	75	1.2[b]	11[b]	75
Acid hydrolysis + HPLC (Gilberg and Itzkovitz, 1983)	244	584	71	28	43	60	7	12	61
Sulfatase hydrolysis + radioenzymatic	149[b]	501[b]	77	28[b]	100[b]	78	10.5[b]	31[b]	74

[a] Values estimated from histograms.
[b] Nanograms per milligram creatinine.

Table 17.4
Comparison of Human Plasma Concentration of Free and Conjugated NE and E (ng/ml) Obtained by Different Methods of Hydrolysis (Mean ± SE)

Method of Hydrolysis and Determination	Norepinephrine			Epinephrine[a]		
	Unconj.	Conj.	% Conj.	Unconj.	Conj.	% Conj.
Acid hydrolysis + fluorometry (Haggendal, 1963)	0.3	0.7	70	ND	ND	ND
Acid hydrolysis + gas chromatography (Imai et al., 1973)	1.03	4.95	83	ND	ND	ND
Acid lyophilization + radioenzymatic (Buu and Kuchel, 1977)	0.23 ± 0.05	0.6 ± 0.2	71	0.04 ± 0.02	0.32 ± 0.1	88
Acid hydrolysis + radioenzymatic (Da Prada, 1980)	0.5 ± 0.1[b]	1.3 ± 0.4[b]	71	0.05 ± 0.01	0.30 ± 0.1	89
Acid hydrolysis + radioenzymatic (Nagel and Schümann, 1980)	0.33 ± 0.04	1.0 ± 0.2	68	0.08 ± 0.07	0.45 ± 0.1	80
Sulfatase hydrolysis + radioenzymatic (Johnson *et al.*, 1980)	0.67	1.29[c]	66	0.08	0.44[c]	83

[a] ND, nondetectable.
[b] Values estimated from histograms.
[c] Means of three human pools after subtracting the free from total NE and E values, respectively.

CA sulfates are, however, present as both 3-O and 4-O-sulfates. The predominance of the 3-O-sulfate over the 4-O-sulfate is documented in most mammalian systems for DA, but nothing is known of the NE or E sulfates. Since sulfoconjugation occurs predominantly at the 3-O-position (22, 105), it is likely that the 3-isomer predominates in both NE and E sulfates. At present, no easy, rapid methods exist for the routine differential measurement of CA sulfate isomers. Most recently, Elchisak and Carlson (44) have introduced a high pressure liquid chromatography (HPLC) method for analysis of DA sulfate isomers. Such a method is interesting, since it may indicate what fraction of DA is sulfoconjugated in competition with 3-O-methylation by COMT.

When free CA concentrations are much higher than those of conjugated CA, as in brain tissue, the determination of conjugated CA becomes difficult because of necessarily high dilution of the sample for radioenzymatic assay. Recently, however, we reported a new method for the measurement of brain CA sulfate (28) in which free CA are separated from conjugated CA by adsorption on alumina. The CA sulfates, partially purified by a passage through a Dowex resin, are measured by the radioenzymatic assay in the presence of both COMT and sulfatase. Such a method, as detailed in Figure 17.3, permits the detection not only of NE sulfate but also of DA and E sulfates in different areas of rat brain, and may be used for the determination of CA sulfates in areas rich in free CA.

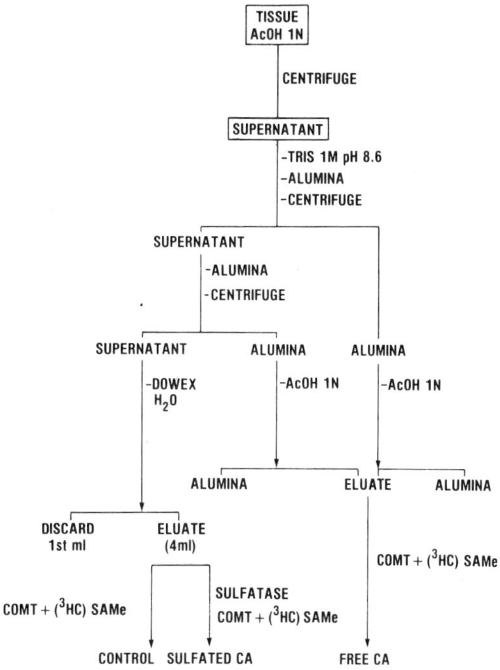

Figure 17.3. Flow sheet for the extraction of free and conjugated CA from brain tissue and their measurement by the radioenzymatic method using COMT and [^{3}H]SAMe. (Reproduced with permission from N. T. Buu *et al.* (28).)

Errors in CA Measurements Due to the Presence of CA Sulfate

The possibility that CA sulfates can be hydrolyzed by the presence of a sulfatase preparation and measured as free CA within the same incubation medium in the radioenzymatic assay underlines the need for caution in interpreting values of free CA. In particular, since DA is present in human plasma almost exclusively as DA sulfate (29), and the partially purified COMT preparation (11) used in the radioenzymatic assay may contain sulfatase activity, values of free DA obtained by this method (25) may be falsely high. Extraction and storage of free and conjugated CA in strong acid, even at low temperature, may also lead to hydrolysis of CA sulfate, particularly DA sulfate, and contribute to higher values for DA. On the other hand, since NE and E are easily oxidized by strong acids, storage in acid may decrease their concentrations.

Measurements of Phenolsulfotransferase Activity

Activity of PST has been detected in most mammalian tissues (93). The most sensitive method of measuring PST activity is described by Foldes and Meek (49) and is illustrated in Figure 17.4. Essentially, the high-speed supernatant of the tissue is incubated in the presence of a substrate with radioactive 3′-phosphoadenosine-5′-phospho[^{35}S]sulfate ([^{35}S] PAPS) which serves as a sulfate donor. At the end of the reaction, the excess of [^{35}S]PAPS is precipitated with $Ba(OH)_2$ and the high-speed supernatant measured for its radioactivity. This radioactivity corresponds well with the concentration of sulfated substrate, as shown by TLC for E (131) and by reverse-phase liquid chromatography for MHPG (3). To reduce the blank reading and increase the sensitivity of the assay, other workers have found it necessary to terminate the reaction by boiling at 95°C for 10 minutes (3) and/or to dilute the [^{35}S]PAPS with nonradioactive PAPS (32). High

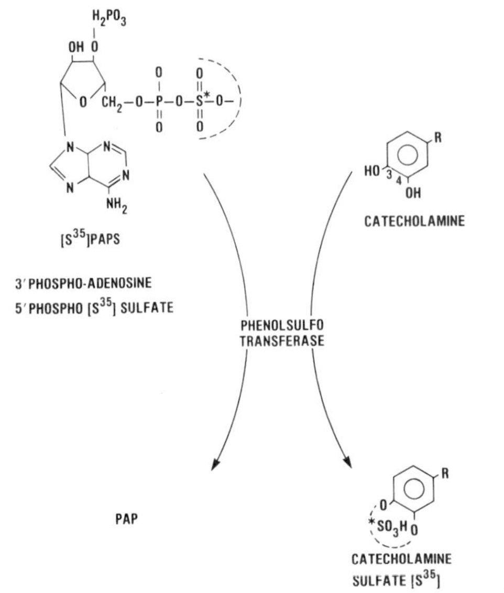

Figure 17.4. Pathway for the radioenzymatic determination of phenolsulfotransferase activity. For CA R = CHCH$_2$NH where x could be H OH, and y could be H or CH depending on the CA.

dilution of tissue sample can reduce the concentrations of possible endogenous inhibitors of the enzyme and also increase the sensitivity of the assay (2). Spectrophotometric determinations of PST based on the formation of paranitrophenol sulfate from *p*-nitrophenol (12) and spectrofluorometric methods using methylumbelliferin as substrate (126) are also used for the assay of PST, but their sensitivity and specificity are generally lower than that using radioactive PAPS. PST is not a specific enzyme, since it can accept a large variety of substrates ranging from monophenols to phenolic acids, phenolic amines, CA, and their O-methylated metabolites. The choice of substrate may be important in the determination of PST activity in different tissues because the enzyme may exist in different forms (107). Moreover, since measurement of the PST activity depends on the solubility of the sulfated product in aqueous media, the solubility of the sulfate ester of a compound in alkaline medium should be verified before it is used as substrate.

Platelet PST might have proven a useful marker of the degree of sulfoconjugation in man (5). However, because of the observed poor stability of platelet PST in storage even at low temperature, PST activity measurements must be standardized before a comparison of their activity becomes possible. Moreover, there seems to be no correlation between tissue PST activity and CA sulfate level in different areas of the rat brain. Thus, the activity of the enzyme *in vivo* may be different from that found *in vitro*.

THE BIOLOGICAL ROLE OF CONJUGATED CA

Because the sulfate esters of CA have always been considered inactive end products of CA metabolism destined for excretion, their own metabolism has seldom been explored: there are few studies on the fate of DA sulfate and none on that of DA glucuronide. It has been found, for instance, that E taken orally is excreted as E sulfate (112) and that the E sulfate loses all of its pressor effect on spinal cat nictitating membrane and intestine. DA sulfate administered to rats was also excreted quickly in their urine (61). However, CA sulfate esters are not necessarily inert; they can be metabolized, as shown recently by Merits (94). When radioactively labelled DA sulfate was injected into different animal species such as dogs, rats, and guinea pigs, radioactive metabolites of DA were excreted. Sulfate esters other than those of CA can be metabolized. Baulieu and coworkers (15) demonstrated that sulfate esters of androstenediol can serve as a substrate for 17-β-hydroxydehydrogenase and converted to dehydroisoandrosterone sulfate. Estrogen sulfate was recently found to act as an estrogen precursor (127).

Properties of CA Sulfates

The conflicting evidence regarding CA sulfates stems in part from their physical properties, which differ considerably from those of their parent compounds. These properties were elucidated from studies with DA sulfates, which are more readily synthesized (61) than the sulfate esters of NE or E.

Unlike free CA, MAO does not metabolize the sulfate esters of DA (111). Their stability is due possibly to the protective action, by steric hindrance or by bond formation, of the sulfate group on the amino group where the oxidation reaction takes place.

Sulfate esters of CA consist predominantly of the 3-O-isomers, as shown for DA sulfate by Bronaugh *et al.* (22) and for 3,4-dihydroxybenzoic sulfates by Pennings and van Kempen

(103), and are also resistant to COMT, since O-methylation occurs almost exclusively at the meta position (9). It is less clear why the 4-O-sulfate esters of DA and of 3,4-dihydroxybenzoic acid are resistant to COMT degradation despite their unoccupied metahydroxyl group (103). The presence of an adjacent and bulky sulfate group may prevent the molecule from fitting into the active site(s) of the COMT.

CA are largely excreted as sulfate esters of catechol acids (e.g., homovanillic acid (HVA) sulfate), catechol alcohols (e.g., MHPG sulfate) or 3-O-methylated CA (e.g., NM sulfate). Since the sulfate esters are stable toward both COMT and MAO, sulfoconjugation must have taken place after the oxidation and/or methylation of CA. It remains unknown why there is a need to conjugate these already physiologically inactive, readily excreted metabolites of CA and what is the significance of this process. Perhaps sulfoconjugation competes with MAO and COMT in the inactivation of CA. The extent of this inactivation pathway is not known, but quantitative analysis of NE clearance indicates that one quarter of the systemic clearance of NE is unaccounted for after inhibition of uptakes$_1$ and $_2$, COMT, and MAO (24) (Table 17.5). This unaccounted proportion of NE could be due to sulfoconjugation of NE.

Unlike their parent compounds, CA sulfate esters are highly polar. DA sulfates, for instance, can form an internal salt, which may explain their stability toward MAO. This property raises the possibility that the sulfate ester can penetrate cell membranes more easily than the free form (61). Whether it also allows the CA sulfate to pass freely across the blood-brain barrier remains to be studied. It has been shown that MHPG sulfate but not free MHPG can pass the blood-brain barrier in the rat (92). Recently, Tyce and co-workers (121) found DA sulfate in the cerebrospinal fluid (CSF) of patients with Parkinson's disease, whether or not they were on L-dopa therapy. They expected the presence of CA sulfates in the brain in view of the presence of large concentrations of free CA and of PST activity (32, 93) in different areas of the brain. The presence of sulfate esters of NE, E, and DA was recently demonstrated in the rat (28), but concentrations in the brain are small.

The presence of DA sulfate in the CSF and the small concentrations of CA sulfate in discrete areas of the rat brain raise the possibility that sulfoconjugation provides rapid transport of excess CA from the brain to the periphery. It is possible that the rapid removal of CA sulfates from the brain is necessary because of their effect on the CNS. Recently we found that injection of relatively large doses (200 μg/kg or more) of DA sulfate into the left lateral ventricle of Sprague-Dawley rats causes severe *convulsions* lasting several minutes (27). Such convulsive action is not reproducible by injections of smaller or larger doses of free DA, other CA, or hydrolyzed solutions of DA sulfate. Although the mechanism of such a convulsive effect awaits investigation, these findings imply that the CA sulfate in the CNS is far from biologically inactive.

DA Sulfate in CA Biosynthesis

The small concentration of CA sulfates in the brain contrasts markedly with the high concentrations in plasma and peripheral tissues, where they may serve as a transport and storage form of CA. The finding that CA sulfates can be metabolized suggests that sulfoconjugation is not an irreversible process. CA sulfates may be metabolized indirectly by being hydrolyzed to free CA which are then metabolized by the usual pathways, or they may be metabolized directly. The possible role of sulfatase in releasing free CA from parent compounds is uncertain. For instance, despite the widespread presence of high sulfatase activity (120) in kidney or adrenal medulla, there is nonetheless much less free DA than DA

Table 17.5
Quantitative Analysis of Norepinephrine Clearance[a]

Total Systemic Clearance of NE in Rabbits	
Baseline	69 ± 6 ml kg^{-1}min^{-1} (100%)
Inhibition of uptake I (desimipramine)	42 ± 5 ml kg^{-1}min^{-1} (61%)
Inhibition of uptake II (metanephrine)	59 ± 5 ml kg^{-1}min^{-1} (61%)
Inhibition of COMT (pyrogallol)	52 ± 6 ml kg^{-1}min^{-1} (75%)
1,2,3 + inhibition of MAO (pargyline)	16 ± 2 ml kg^{-1}min^{-1} (24%)

[a] Adapted from M. H. Brown *et al.* (24).

sulfate in these tissues. Moreover, concentrations of NE and E sulfates are higher than those of free NE and E in blood-irrigating tissues with sulfatase activity, while DA circulates almost exclusively as DA sulfate. Thus sulfatase, while playing an important role in hydrolyzing the phenolic sulfate or other sulfoconjugated products, seems to play only a limited and still undetermined role in the release of free CA from their sulfate precursors. This is also suggested by a finding that DA 3-*O*- and DA 4-*O*-sulfates are not readily desulfated in mammals *in vivo* (63).

On the other hand, DA sulfate may serve as a substrate for *dopamine-β-hydroxylase* (DBH) and become converted to free NE (30, 31). Recently, it has been shown that both DA 3-*O*-sulfate (30) and DA 4-*O*-sulfate (31) can be converted directly *in vitro* to free NE by the action of DBH. This conversion was completely blocked by the presence of *fusaric acid*, a DBH inhibitor, and exhibited different kinetic parameters with DA-3-*O*-sulfate or DA-4-*O*-sulfate as substrates (Table 17.6). Such a finding speaks against the possibility that NE was synthesized from free DA released from DA sulfates by a sulfatase, since sulfatase hydrolyzes the 3-*O*-sulfate ester more rapidly than the 4-*O*-sulfate (103). Whether such a conversion from DA sulfate to free NE can occur *in vivo* is not yet known. DA sulfate has been shown, however, to be present in the *vesicle fraction* of the dog adrenal medulla, and its concentration in the adrenal tissue is 10 times higher than that of free DA (125). Experiments with dogs in which the renal CA balance during surgery was calculated showed a large influx of conjugated DA (123) and strongly suggested that part of that influx was used for the synthesis of NE within the kidney. Injection of [^{3}H]DA sulfate in adrenalectomized rats results in the excretion of [^{3}H]NE in the urine (33). However, the difficulty of demonstrating direct formation of NE from DA sulfate *in vivo* lies in the role of sulfatase in the deconjugation of CA sulfate. Although the role of sulfatase seems to be minimal, the possibility that the injected DA sulfate is first hydrolyzed to free DA by a sulfatase cannot be eliminated. Administration of [^{3}H]DA sulfate into adrenalectomized rats was also accompanied by recovery of [^{3}H]E in their urine which was not detected when [^{3}H]DA was injected (33). The absence of adrenal phenylethanolamine-N-methyltransferase (PNMT) in these rats suggests that [^{3}H]E was synthesized elsewhere from DA sulfate, by a biosynthetic pathway which does not require PNMT. Although much work remains before demonstrating with certainty the role of DA sulfate as a precursor of NE and E, these findings suggest that such pathways may be active *in vivo*. The role of CA sulfates as a stable transport form for CA may represent an economical physiological use of the CA and justify the large amount of energy used in the synthesis of the PAPS required for the sulfoconjugation of the CA.

Table 17.6
Comparison of the Reactivity of Free Dopamine and Dopamine Sulfate towards Dopamine-β-Hydroxylase

Substrate (conc.)	NE formed[a] (μM ± SE)	K_m (mM)
Free dopamine (1.5 mM)	15.7 ± 0.5	0.1
Dopamine 4-*O*-sulfate (8 mM)	3.1 ± 0.2	2.6
Dopamine 3-*O*-sulfate (25 mM)	0.8 ± 0.1	17.0

[a] Expressed as concentration in the incubation medium.

THE SULFOCONJUGATING SYSTEM: A POSSIBLE HOMEOSTATIC ROLE

Limited evidence suggests that, in addition to the previously discussed roles of conjugation, the sulfoconjugating system has a role of its own as a site of some homeostatic interactions. These may regulate the proportion of the biologically active unconjugated CA. Although glucuronoconjugation is a more important mechanism in the detoxification of most drugs, tachyphylaxis in the action of some drugs with a phenolic structure (such as L-dopa) may also result from an induction of PST and increased sulfoconjugation (115). In this respect, sulfoconjugation may also have a "detoxifying" effect as part of the chemical "surveillance" system protecting the organism against exogenous substances. This chemical "surveillance" may apply also to endogenous phenols, as has been suggested for *pheochromocytoma* patients, in whom the degree of sulfoconjugation has an important biochemical and clinical effect (78, 79). Those patients with high conjugated CA do not have the "overflow" of free CA and sometimes have atypical crises which would have remained undiagnosed without measurements of conjugated CA (78). The CA excess can be "buffered" by enhanced sulfoconjugation, which acts as a defense system against the excess circulating CA, and is therefore a part of *homeostatic regulation*.

CA are also involved in regulating the release of some polypeptide and steroid hormones, the latter exemplified by an inhibitory action of DA (90) and a stimulatory action of NE (91) on *aldosterone* release from bovine adrenal slices *in vitro*. *Dexamethasone* administration suppresses plasma aldosterone levels in some patients with primary hyperaldosteronism, and plasma DA changes are indirectly correlated with those of plasma aldosterone (83). We found that after dexamethasone administration there was an increase in plasma DA and NE, while conjugated DA and NE decreased in most cases; this resulted in a decrease in the degree of plasma conjugation of DA (from 94 to 81%) and NE (from 89 to 61%) (74). It is not clear whether this is an effect of dexamethasone or of adrenocorticotrophic hormone (ACTH) suppression, or how it relates to the dexamethasone-induced increases in some tissue PST activities (89). In any case, it suggests that some factors, other than those previously described, can affect the process of sulfoconjugation and thereby indirectly affect the biological activity of the relevant CA.

In general, sulfoconjugation may participate not only in the biodegradation of phenolic drugs, but also in controlling the ratio between the biologically active NE and its conjugated fraction, which is either inactive or biologically important only in very high concentrations.

IMPLICATIONS OF CATECHOLAMINE CONJUGATION

Renal Excretory Mechanisms of Free and Conjugated CA

The polarity of conjugated CA (61) probably explains their increased excretion by the proximal tubular cells of the chicken kidney (110). Earlier investigations using radioactive substrates established that conjugation of CA is a consistent, major metabolic pathway in the kidney of various animal species (54), whereas enzymatic degradation by COMT and MAO in this organ appears to be variable and species-dependent (118). On the other hand, CA conjugates, particularly of DA, are not confined to the urine but represent a considerable portion of the hormones in the circulation. The reason for the high degree of conjugation in circulating CA (70, 85, and 99% for NE, E, and DA, respectively) and to a lesser degree in urinary NE (50 to 60%), E (70%) and DA (70%) is unknown. Species differences in degree of conjugation complicate the matter further, so that only animal studies limited to species closely similar to man, such as the dog, may help our understanding of the human physiology of CA excretion. Human (5) and dog kidneys (74) contain considerable tissue PST activity and are probably rich in the ubiquitous sulfatases. Even within these two species exhibiting sulfoconjugation, considerable quantitative differences in PST activity exist (approximately 10 times more PST activity in the dog kidney than in man (74)). The higher tissue PST activity in the dog probably results in a higher urinary proportion (87 to 92%) of conjugated DA (123) compared to approximately 60 to 70% in man (29). Perhaps as a consequence of this and the high affinity of DA for PST, unconjugated (free) urinary DA excretion in the dog equals or is less than that of free NE (123), while in man, free DA excretion is up to 10 times higher than that of free NE (29).

The renal excretory mechanisms of CA imply sulfoconjugation of unconjugated CA and deconjugation of CA sulfates, both substances entering the kidney by the arterial circulation. This introduces a new, unknown element in the calculation of the "fractional clearance" of NE, which should be used with caution. Only simultaneous measurements of arterial influx and renal venous and urinary outflow of free and conjugated CA make it possible to establish a renal balance for free and conjugated CA. The origin of conjugated CA in the urine and the role of the kidney in their excretion is not known. The urinary excretory mechanisms of NE result from a complicated interplay of an influx of unconjugated (free) and conjugated NE into the kidney, and an overflow of normally released NE neither metabolized nor reabsorbed by the axons but diffused into the renal venous circulation (101). It is not clear to what degree the NE newly released by nerve terminals in the kidney is subject to sulfoconjugation. We found that during the stress of surgery (123), there was no difference in the renal net balance of free NE, while the urinary excretion was too small to influence the balance. Conjugated NE was present in the renal outflow despite its absence from the arterial influx (probably due to stress), demonstrating that an endogenous conjugation mechanism exists for NE and E. This confirms earlier findings in the dog (110) in which, upon administration of labeled catechol to the renal artery, more than 60% of the activity was recovered as catechol sulfate in the urine. If the results of measurements of the total body clearance of NE can be applied to the kidney, it appears that approximately 25% of infused free NE remains unaccounted for by neuronal reabsorption, O-methylation, and action of MAO (24) and is thus probably due to conjugation. To complicate matters further, the kidney is a highly active organ not only in excretion but also in metabolism. Our study of renal handling of CA during stress suggests that part of the high influx of DA sulfate into the kidney, unaccounted for in the venous and urinary outflow, may serve as precursor material for the synthesis of free NE after being taken up by nerve terminals in the kidney and β-hydroxylated by DBH (124). The processes of conjugation and deconjugation of CA, and possible interconversions from DA sulfate to NE within the kidney, make the renal handling of NE much more complex than previously suspected. This will require new approaches in elucidating the kidney's role in generating and excreting circulating NE.

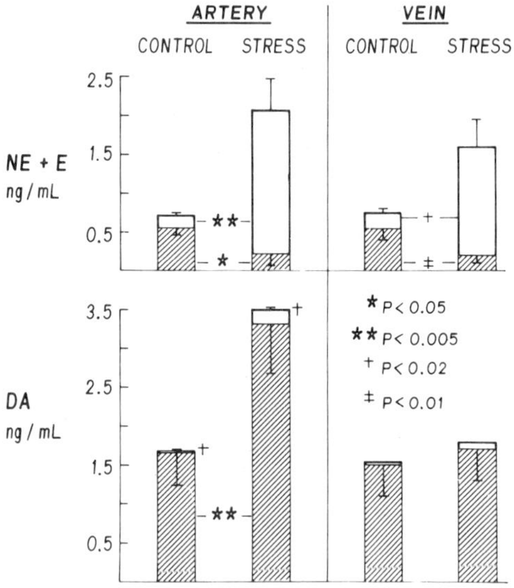

Figure 17.5. Effect of surgical stress on plasma-unconjugated (*open areas*) and conjugated (*hatched areas*) arterial and venous NE + E and DA. Mean ± SEM in 10 unilaterally nephrectomized dogs. (Reproduced with permission from P. S. Sever *et al.* (117) with the permission of the *Canadian Journal of Physiology and Pharmacology.*)

Conjugated CA and their Responses to Stress

The effects of stress on unconjugated CA have been reported (71). When conjugated E and NE were first discovered in human plasma (52), it was found that stressful stimuli (muscular work, insulin, hypoglycemia) increased free NE and E, but levels of conjugated CA did not change markedly. However, using more sensitive methods for the hydrolysis and determination of CA, we found that after surgical stress in dogs, the increase of free NE and E in arterial and venous blood was associated with a decrease in relative and absolute levels of conjugated NE and E (125) (Fig. 17.5). A similar phenomenon was observed in healthy subjects in response to bicycle exercise (65) and to a lesser degree when hypertensive subjects were stressed with venipuncture (86). These results were then compared with those from the same subjects when blood was sampled from an indwelling venous catheter (80). During stress, the increase of E and NE was concomitant with an increase of DA sulfate in the arterial blood. We have observed that several forms of experimental stress were associated with an increase in the plasma concentrations of conjugated DA and tissue content of DA (84). DA sulfate secretion by the adrenal was suggested from the higher concentration of DA sulfate in adrenal venous than in adrenal arterial blood and by the relatively high levels of conjugated DA in the granular fraction of the adrenal medulla (125). Whether the adrenomedullary occurrence of DA sulfate is primary or secondary, its increased concentration in the blood during stress suggests that DA sulfate may be not only a transport form of DA but a direct precursor of NE *in vivo*. Teleologically, this would be very advantageous since it offers an immediately available precursor for NE (after the DA sulfate is taken up in the granules) to compensate for the sluggish neuronal transport of DA. Much remains to be learned about the effects of the pituitary-adrenal system, activated during stress, on sulfoconjugating and deconjugating enzymes; perhaps they are analogous to its effects on the PNMT activity (71). The finding that rat kidney PST activity increased four to five times following dexamethasone in the presence or absence of the adrenals (89) suggests that a control system of conjugation may operate during stress.

CA Conjugation during Pregnancy and the Menstrual Cycle

Changes in free and conjugated NE during pregnancy and the menstrual cycle may vary with the sulfoconjugating process. It has recently been demonstrated (20) that during the third trimester of normal pregnancy, the excretion of conjugated NE and E increases more than 3-fold, and the proportion of conjugated NE and E rises from 42 to 75% of the total CA. It is not clear whether the extra urinary conjugates originate in the fetus, the fetoplacental unit, and/or result from their easier passage through the placenta than free CA. In rare cases of placental sulfatase deficiency (47) leading to the birth of boys with ichthyosis, some abnormalities in estrogens have been described, but CA conjugates have not yet been explored.

Preliminary findings indicate that during a normal menstrual cycle, there is a premenstrual decrease in the platelet PST activity (1). Whether this corresponds to a normally observed premenstrual increase in urinary excretion of CA and vanillylmandelic acid (VMA) (70) is yet unknown, since a clear correlation does not appear between the individual CA sulfates and platelet PST activity (74). Excessive premenstrual decrease of platelet PST activity may contribute to some hyperadrenergic manifestations observed during premenstrual tension.

Conjugation of Phenolic Drugs

Sulfoconjugation may affect the pharmacokinetics of some drugs. Although phenolic drugs such as propranolol (128) are conjugated mostly via glucuronoconjugation in the intestine and liver, such widely used drugs as acetaminophen and α-methyldopa are sulfoconjugated and excreted as sulfates. The oral and parenteral efficiency of some antiasthmatic drugs depends on the degree to which they are sulfoconjugated (42). It is evident that individual variations in human PST activity may have an effect on the sulfoconjugation of some drugs and of neurotransmitters now used as drugs. Drugs such as L-dopa may be conjugated upon absorption from the gut, the principal gut metabolite being DA glucuronide in the rat (119). L-dopa increases platelet PST activity (115), probably by enzyme induction, and thus accelerates its own sulfoconjugation and metabolism. The finding that human platelet PST activity can be assayed using such dissimilar substrates as DA, tyramine, serotonin (5-hydroxytryptamine; 5-HT), acetaminophen, and α-methyldopa with similar results (5) suggests that a wide variety of substances are metabolized via sulfoconjugation and that future pharmacokinetic studies should include measurements of PST activity. Genetic influences on platelet PST activity may contribute to variations in drug and neurotransmitter metabolism.

CA CONJUGATION AS A VARIABLE IN DISEASE STATES

The controversial relationship between circulating CA and human hypertension (10) (see Chapters 24 through 28) has led us to explore the possibility that some of the discrepant results derive from unobserved abnormalities in sulfoconjugation. This approach is based on two premises: (a) measurements of conjugated CA may prove to be a better means of assessing CA secretion rates since they circulate much longer than free CA by virtue of their longer half-life (five to ten times that of free CA); (b) the sulfoconjugation of CA resulting in up to 80% of NE or E circulating in the conjugated form is reversible in contrast to irreversible metabolites such as M which are themselves subjected to sulfoconjugation (69). The reversibility of the conjugation provides the potential for biological activity of NE or E sulfates. This potential may be either part of a labile complex from which free NE or E is split off, or a precursor such as DA sulfate in the generation of NE and E (30, 31, 33). Conjugation may enhance CA mobility to and from the cerebral and peripheral compartments so that CA sulfates represent a noncompartmentalized compound, either in the periphery (where they are usually determined) or in the CNS.

Pheochromocytoma is a rare but well characterized condition common to most discussions on the participation of CA in hypertension (see Chapter 12). This condition has been thought to have a clear-cut mechanism: a tumor producing and releasing excess NE and/or E causing excessive vasoconstriction and hypertension. However, there is no correlation between the concentration of NE and blood pressure (BP) (21). Examination of the role of conjugation in pheochromocytoma has shown that two populations of pheochromocytoma patients exist (Fig. 17.6). One has *normal* levels of conjugated NE and E and presents an expected "overflow" of NE or E with concomitant hypertension during crises. The other group has very high baseline-conjugated NE and E concentrations (10 to 15 times normal) in which the overflow of NE and E during crises materializes to a lesser degree while the proportion of conjugated NE and E, already high, increases further. Patients of this second group also have lower urinary VMA excretion than those of the first group (73, 78). Thus, the level of circulating sulfoconjugates in the blood modifies the biochemical (and possibly clinical) presentation of the disease. As a consequence of high conjugation, the metabolic pathway's alternative to conjugation is suppressed. These patients have lower urinary excretion of VMA than patients with normal

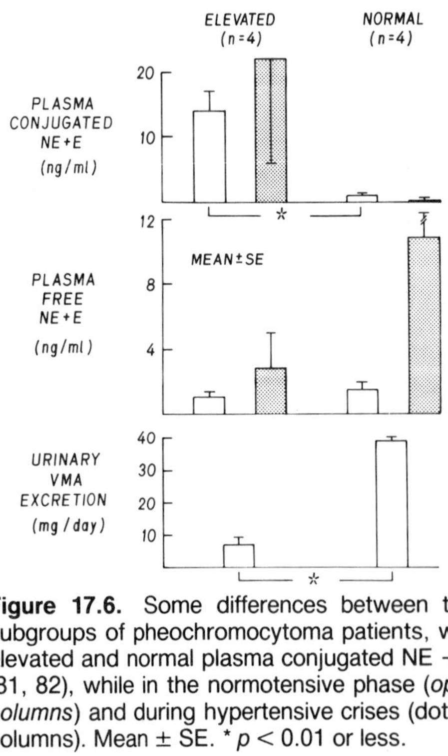

Figure 17.6. Some differences between two subgroups of pheochromocytoma patients, with elevated and normal plasma conjugated NE + E (81, 82), while in the normotensive phase (*open columns*) and during hypertensive crises (dotted columns). Mean ± SE. * $p < 0.01$ or less.

conjugated NE and E. In some patients, we have observed spontaneous crises that may relate not only to the type of CA released but also to the ratio of its unconjugated and conjugated fractions in plasma at the moment of crisis (78). Extremely high levels of the conjugated CA may also be related to BP, whereas no such relationship has been observed for the free CA (73, 78). This suggests that during an episode where secretion of CA is extreme, even conjugated CA may be biologically active. The mechanism of production of this high level of circulating sulfoconjugated CA is not yet clear. The high degree of sulfoconjugation may be due to an induction of PST activity by continuous rather than periodic release of CA by the tumor. In some cases, an elevated release of conjugated CA from the tumor itself has been observed. Further studies are necessary to evaluate the incidence of this high, conjugated CA "buffered" form of pheochromocytoma with its modifying effect on the clinical and biochemical aspects of the disease.

The effect of the degree of sulfoconjugation on the course of *essential hypertension* represents an even more complex and challenging problem. Uncertainty surrounds the role of the sympathetic nervous system (SNS) in this form of hypertension (10) (see Chapter 12). Most patients from a population with confirmed essential hypertension in whom pheochromocytoma had been strongly suspected on clinical grounds (including some who had been subjected to surgical exploration with negative results) had low or absent conjugated NE + E (75). These patients exhibit a syndrome similar to pheochromocytoma. When compared to patients with normal conjugated CA, these patients had: (a) higher baseline unconjugated CA that were hyperresponsive to direct venous puncture and (b) no conjugated NE or E in the regional samples obtained by catheterization. These patients have higher unconjugated NE + E, pulse pressure, and pulse rate than those with normal conjugated NE + E (Fig. 17.7). All patients combined have a maximum pulse rate that was correlated negatively with conjugated NE + E, but positively with free NE + E.

In the study mentioned, it was difficult to establish the respective roles of the defects in NE and E conjugation, since the two CA were not separately determined. We find that the hyperresponsiveness of "free NE + E" to glucagon in patients with low conjugated NE + E could be entirely ascribed to a higher response

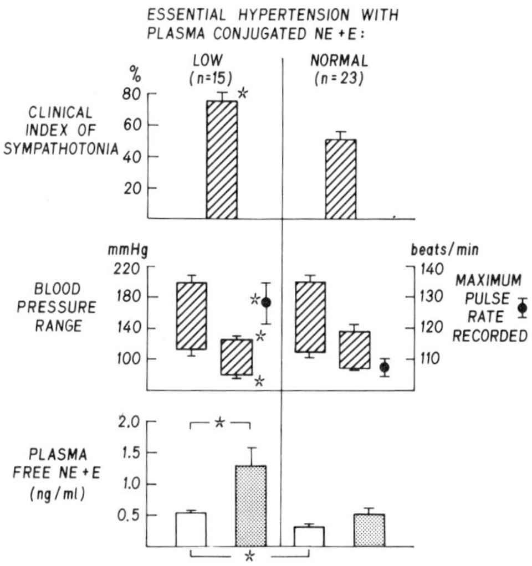

Figure 17.7. Some differences between two subgroups of essential hypertensive patients with low and normal plasma-conjugated NE, NE + E (80, 81). The clinical index of sympathotonia is based on 10 main symptoms in descending order of weighing (10 points attributed to first, 1 to last): palpitations, sweating, anxiety, weakness, loss of weight, dyspnea, headache, nausea, abdominal or lumbar pain, and tremor. The final individual score was expressed as a percentage of the maximum possible score of 55. The *middle panel* demonstrates the higher range between maximum and minimum blood pressure and the higher pulse rate in patients with low conjugated NE + E. In the *bottom panel* are results of sampling by indwelling catheter avoiding the stress of the immediate venous puncture (*open columns*) compared with sampling by direct venous puncture (*dotted columns*). Mean ± SE, * $p < 0.05$ or less.

of free E only (85). A follow-up study with separate determination of NE and E shows that a low degree of E conjugation results in an elevated unconjugated E level which correlates positively with pulse pressure and pulse rate, the latter also correlating negatively with the degree of E conjugation (74). Subdividing patients into one group with low and one with normal conjugated NE in plasma had no effect on the mean level of free NE. A positive correlation exists between diastolic BP and free NE (74) in accordance with a previous report (87), and a negative correlation exists between diastolic BP and the degree of NE conjugation. This study suggests that the conjugation defect affects E more than NE and results in a rise in the concentration of unconjugated E but not of unconjugated NE. The hemodynamic repercussions of the defect in E conjugation are also more evident and correspond with E concentrations that increase systolic BP and pulse pressure (39). In contrast, the degree of NE conjugation has no effect on the levels of unconjugated NE. Although the diastolic BP correlates positively with unconjugated NE and negatively with its degree of conjugation, this does not appear to be of biological significance, since the NE concentrations are far below the level of NE that raises diastolic BP (39). Plasma NE and its degree of conjugation thus reflects, rather than causes, the sympathetic neuronal activity involved in the rise of the diastolic pressure.

Despite differences between the impact of a conjugation defect on free E and NE and their actions on BP, there appears to be a common factor, as suggested by a correlation between the levels of NE and E sulfates ($r = 0.42$, $p < 0.005$), resulting in "unconjugated *hyperepinephrinemia.*" We explored the possibility that a deficiency in platelet PST, which appears to correlate well with some tissue PST activity (5), could be the common denominator of the defect in circulating E and NE conjugates. Although some preliminary results suggest low platelet PST activity in patients with low conjugated NE + E (1), no such difference was found in a more extensive series of patients (74). This suggests that *unconjugated hyperepinephrinemia* reflects more than just a defect in PST activity. It is conceivable that some interactions occur between DA, NE, and E competing for the same enzyme, PST. We find that conjugated plasma DA is elevated in *essential hypertension* (82), whatever the NE and E conjugation status (75), and that there are higher concentrations of unconjugated DA at several sites of regional sampling in patients with low or absent NE and E sulfate (77). We have observed that peaks in paroxysmic BP in some essential hypertensives are associated with huge surges of predominantly conjugated DA in plasma and urine (77). Perhaps the PST is saturated by DA, which has a high affinity for it, resulting in decreased NE and particularly E conjugation. The latter has the lowest affinity for PST and is therefore most affected by such a defect. E as a hormone is more dependent on sulfoconjugation than NE, a neurotransmitter leaking into the circulation and mostly inactivated by local mechanisms (39). This characteristic of E, when coupled with a high degree of PST occupancy by DA, may be a more plausible cause of unconjugated hyperepinephrinemia than an absolute decrease in PST activity. A hyper-β-adrenergic, *labile hypertension* is typical of the initial stages of hypertension, for instance, in younger patients.

It has been suggested that essential hypertension is an accelerated aging of the BP-regulating system (23). Free NE increases with age in healthy subjects (133) but not in patients with essential hypertension, since NE levels are already elevated in many younger patients (117). As a consequence of higher conjugated plasma DA progressively increasing with age (77), more NE and E may remain unconjugated. Only with advancing age, when the dopaminergic activity is known to decrease (14), do the conjugated NE and E increase parallel with the BP and age of the patient (81). A more equilibrated ratio of DA, NE, and E conjugation may possibly be found in longer-established hypertension, i.e., in older subjects. Despite these baseline tendencies, surges in conjugated DA were more frequently observed in older hypertensive subjects (77), a finding that remains unexplained.

In addition to this form of essential hypertension, there are other conditions in which a conjugation defect may be clinically significant. The best known is the benign form of *unconjugated hyperbilirubinemia* (*Gilbert's syndrome*), a hereditary defect of *glucuronyltransferase* that catalyzes conjugation of *bilirubin* (60). A similar mechanism is theorized for tyramine-sensitive *migraine*; patients with this disorder are unable to conjugate orally administered tyramine (132). More recently, patients with the lowest urinary output of *tyramine* (free and conjugated) after an oral tyramine load are more vulnerable to *depressive illness* than those with the highest tyramine output (16). Other relationships have been sought between the sulfoconjugating system (including platelet PST activity) and *Parkinson's disease* (115), with which it is suggested that there is enhanced tyramine conjugation (17). Urinary MHPG sulfate excretion was shown to correlate negatively with symptom severity in chronic *schizophrenic* patients (38) as well.

SUMMARY

All these data suggest that the effect of sulfoconjugation on the biological action of NE and other CA is still in an initial exploratory stage. Extensive work remains to be done to establish the factors, inherited and acquired, which modulate the sulfoconjugating process and so indirectly implicate NE and other CA in a number of diseases.

References

1. Abenhaim L, Romain Y, Kuchel O: Platelets, phenolsulfotransferase and catecholamines; physiological and pathological variations in humans. *Can J Physiol Pharmacol* 59:300–306, 1981.
2. Anderson RJ, Weinshilboum RM: Phenolsulphotransferase: Enzyme activity and endogenous inhibitors in the human erythrocyte. *J Lab Clin Med* 94:158–171, 1979.
3. Anderson RJ, Weinshilboum RM: Phenolsulphotransferase in human tissue: Radiochemical enzymatic assay and biochemical properties. *Clin Chim Acta* 103:79–90, 1980.
4. Anderson RJ, Weinshilboum R, Phillips S, Broughton D: Human platelet phenolsulphotransferase (PST) activity: Correlation with PST activity in the kidney and gut. *Pharmacologist* 22:774, 1980.
5. Anderson RJ, Weinshilboum RM, Phillips SF, Broughton D: Human platelet phenolsulphotransferase: Assay procedure, substrate and tissue correlations. *Clin Chim Acta* 110:157–167, 1981.
6. Andersson B, Berggren M, Moldeus P: Conjugation of various drugs in isolated hepatocytes. *Drug Metab Dispos* 6:611–616, 1978.
7. Anton AH, Serrano A, Tjandramaga TB, Goldberg LI: Dopa and dopamine in glusulase: Possible artefact in studies on catecholamine metabolism. *Science* 182:60–61, 1973.
8. Axelrod J: Metabolism of epinephrine and other sympathomimetic amines. *Physiol Rev* 39:751–776, 1959.
9. Axelrod J: Methylation reactions in the formation and metabolism of catecholamines and other biogenic amines. *Pharmacol Rev* 18:95–113, 1966.
10. Axelrod J: Catecholamines and hypertension. *Clin Sci Mol Med* 51:415s–421s, 1976.
11. Axelrod J, Tomchick R: Enzymatic O-methylation of epinephrine and other catechols. *J Biol Chem* 233:702–705, 1958.
12. Bannerjee RK, Roy AB: The sulfotransferases of guinea pig liver. *Mol Pharmacol* 2:56–66, 1966.
13. Bannerjee RK, Roy AB: Kinetic studies of the phenol sulphotransferase reaction. *Biochim Biophys Acta* 151:573–586, 1968.
14. Barbeau A: Aging and the extrapyramidal system. *J Am Geriatr Soc* 21:145–149, 1973.
15. Baulieu EE, Coperchot C, Draz F, Emiliozzi R, Lebeau MC, Mauvais-Jarvis P, Robel P: *Recent Prog Horm Res* 21:411–494, 1965.
16. Bonham-Carter SM, Revely MA, Sandler M, Dewhurst J, Little BC, Hayworth J, Priest RG: Decreased urinary output of conjugated tyramine is associated with lifetime vulnerability to depressive illness. *Psychiatry Res* 3:13–21, 1980.

17. Bonham-Carter SM, Youdim MBH, Sandler M, Hunter KR, Stern GM: Enhanced tyramine conjugation in Parkinson's disease. *Clin Chim Acta* 51:327–329, 1974.
18. Boobis AR, Jones DH, Murray S, Reid JL, Davies DS: MHPG and conjugates: Limitations in man. In *Workshop on NA Turnover*, London, Royal Postgraduate School of Medicine, 1978.
19. Boström H, Wengle B: Studies on ester sulphates. *Acta Endocrinol* 56:691–704, 1967.
20. Bourdon V, Demoulin A, Lecomte J: Conjugaison accrue des catécholamines durant la grossesse normale. *Arch Int Physiol Biochim* 86:1153–1155, 1978.
21. Bravo EL, Tarazi RC, Gifford RW, Steward BH: Circulating and urinary catecholamines in pheochromocytoma. *N Engl J Med* 301:682–687, 1979.
22. Bronaugh RL, Hattox SE, Hoehn MM, Murphy RC, Rutledge CO: The separation and identification of dopamine 3-O-sulfate and dopamine 4-O-sulfate in urine of Parkinsonian patients. *J Pharmacol Exp Ther* 195:441–451, 1975.
23. Brown JJ, Lever AF, Robertson JIS, Schalekamp MA: Pathogenesis of essential hypertension. *Lancet* 294:1217–1219, 1976.
24. Brown MH, Davies DS, Dollery CT: Quantitative analysis of noradrenaline clearance. *Br J Pharmacol* 67:452P–453P, 1979.
25. Bühler V, Da Prada M, Hefely NE, Picotti GB: Plasma adrenaline, noradrenaline and dopamine in man and different animal species. *J Physiol* 276:311–320, 1978.
26. Burns GRJ, Galanopoulou E, Wynn CH: Kinetic studies of the phenol sulphate-phenosulphotransferase and aspergillus oryzae. *Biochem J* 167:223–227, 1977.
27. Buu NT, Duhaime J, Kuchel O, Genest J: The convulsive effects of dopamine sulfate conjugates in rat brain. *Life Sci* 29:2311–2316, 1981.
28. Buu NT, Duhaime J, Savard C, Truong L, Kuchel O: Presence of conjugated catecholamines in rat brain: A new method of analysis of catecholamine sulfates. *J Neurochem* 36:769–772, 1981.
29. Buu NT, Kuchel O: A new method for the hydrolysis of conjugated catecholamines. *J Lab Clin Med* 90:680–685, 1977.
30. Buu NT, Kuchel O: The direct conversion of dopamine 3-O-sulfate to norepinephrine by dopamine-β-hydroxylase. *Life Sci* 24:783–790, 1979.
31. Buu NT, Kuchel O: Dopamine 4-O-sulfate: A possible precursor of free norepinephrine. *Can J Biochem* 57:1157–1162, 1979.
32. Buu NT, Kuchel O: Phenolsulfotransferase activity and levels of catecholamine sulfates in rat brain tissues. In Usdin E, Sandler M: *Phenolsulfotransferase in Mental Health Research*, New York, Macmillan, 1981, pp 145–151.
33. Buu NT, Nair G, Kuchel O: The extra-adrenal synthesis of epinephrine in rats: The possible involvement of dopamine sulfate. *J Lab Clin Med* 98:527–533, 1981.
34. Christensen NJ, Mathias CJ, Frankel HL: Plasma and urinary dopamine: Studies during fasting and exercise and in tetraplegic man. *Eur J Clin Invest* 6:403–409, 1976.
35. Cottet-Emard JM, Peyrin L, Bonnod J: Dietary induced changes in catecholamine metabolites in rat urine. *J Neural Transm* 48:189–201, 1980.
36. Coyle JT, Henry D: Catecholamines in fetal and newborn rat brain. *J Neurochem* 21:61–67, 1973.
37. Crout JR, Sjoerdsma A: The clinical and laboratory significance of serotonin and catechol amines in bananas. *N Engl J Med* 261:23–26, 1959.
38. Crow TJ: MHPG in brain and urine as an index of central noradrenaline turnover. In *Workshop on NA Turnover*. London, Royal Postgraduate School of Medicine, 1978.
39. Cryer PE: Physiology and pathophysiology of the human sympathoadrenal neuroendocrine system. *N Engl J Med* 303:436–444, 1980.
40. Da Prada M: Concentration, dynamics and functional meaning of catecholamines in plasma and urine. *Trends Pharmacol Sci* 1:157–159, 1980.
41. Davidson L, Vandongen R, Beilin LJ: Effect of eating bananas on plasma free and sulfate-conjugated catecholamines. *Life Sci* 29:1773–1778, 1981.
42. Davies DS: Pharmacokinetics of inhaled substances. *Postgrad Med J* 51:69–75, 1975.
43. Dulaney JT, Moser HW: Sulfatide lipidosis: Metachromatic leukodystrophy. In Stanbury JB, Wyngaarden JB, Frederickson DS: *The Metabolic Basis of Inherited Disease*, ed 4. New York, McGraw-Hill, 1978.
44. Elchisak MA: Determination of dopamine-O-sulfate and norepinephrine-O-sulfate isomers and serotonin-O-sulfate by high-performance liquid chromatography using dual-electrode electrochemical detection. *J Chromatogr* 264:119–127, 1983.
45. Eto Y, Rampini S, Wiesmann U, Hershkowitz NN: Enzymic studies of sulphatases in tissues of the normal human and in metachromatic leukodystrophy with multiple sulphatase deficiencies: Arylsulfatases A, B and C. Cerebroside sulphatase, psychosine sulphatase and steroid sulphatases. *J Neurochem* 23:1161–1170, 1974.
46. Euler US von, Floding I, Lishajho F: The presence of free and conjugated DOPAC in urine and blood plasma. *Acta Soc Med Upsal* 64:217–225, 1959.
47. Faroqui AA: Sulfatases, sulfate esters and their metabolic disorders. *Clin Chim Acta* 10:285–299, 1980.
48. Foldes A, Meek JL: Rat brain phenolsulfotransferase—partial purification and some properties. *Biochim Biophys Acta* 327:365–374, 1973.
49. Foldes A, Meek JL: Occurrence and localization of brain phenolsulfotransferase. *J Neurochem* 23:303–307, 1974.
50. Gilberg NJ, Itzkovitz HO: Renal functional status and patterns of catecholamine excretion, submitted for publication, 1984.
51. Goodal MC, Alton H: Metabolism of 3-4-dihydrophenylalanine (L-dopa) in human subjects. *Biochem Pharmacol* 21:2401–2408, 1972.
52. Haggendal J: The presence of conjugated adrenaline and noradrenaline in human blood plasma. *Acta Physiol Scand* 59:255–260, 1963.
53. Hart RF, Renskers KJ, Nelson EB, Roth JA:

Localization and characterization of phenol sulfotransferase in human platelets. *Life Sci* 24:125–130, 1979.
54. Hempel K, Lange HW, Kayser EF, Roger L, Hennemann H, Heidland A: Role of O-methylation in the renal excretion of catecholamines in dogs. *Arch Exp Pathol Pharmakol* 277:373–386, 1973.
55. Hoeldtke RD, Sloan JW: Acid hydrolysis of urinary catecholamines. *J Lab Clin Med* 75:159–165, 1970.
56. Hoeldtke RD, Wurtman RJ: Cereal ingestion and catecholamine excretion. *Metabolism* 23:33–41, 1974.
57. Houston JB, Levy G: Modification of drug biotransformation by vitamin C in man. *Nature* 255:78–79, 1975.
58. Imai K, Sugiura M, Tamura Z: The presence of conjugated dopamine in normal human plasma. *Chem Pharm Bull (Tokyo)* 18:2134, 1970.
59. Imai K, Wang MT, Shohei Y, Tamura Z: Determination of catecholamines in the plasma of patients with essential hypertension and of normal persons. *Clin Chim Acta* 43:145–149, 1973.
60. Isselbacher KI: Jaundice and hepatomegaly. In *Principles of Internal Medicine*, ed 7. New York, McGraw-Hill, 1974, p 221.
61. Jenner WN, Rose FA: Studies on the sulphation of 3-4-dihydroxyphenylethylamine (dopamine) and related compounds by rat tissues. *Biochem J* 135:109–115, 1973.
62. Jenner WN, Rose FA: Dopamine 3-O-sulphate, an end product of L-dopa metabolism in Parkinson patients. *Nature* 252:237–238, 1974.
63. Jenner WN, Rose FA: Dopamine 3- and 4-O-[^{35}S] sulphates and substrates for arylsulphatases in vitro and their metabolism by the rat in vivo. In Aito A: *Conjugation Reaction in Drug Biotransformation*. Amsterdam/New York, Elsevier/North Holland Biomedical Press, 1978, p 501.
64. Johnson GA, Baker CA, Smith RT: Radioenzymatic assay of sulfate conjugates of catecholamines and dopa in plasma. *Life Sci* 26:1591–1598, 1980.
65. Joyce DA, Beilin LJ, Vandongen R, Davidson L: Plasma free and sulfate conjugated catecholamine levels during acute physiological stimulation in man. *Life Sci* 30:447–454, 1982.
66. Kahane Z, Esser AH, Kline NS, Vestergaard P: Estimation of conjugated epinephrine and norepinephrine in urine. *J Lab Clin Med* 69:1042–1050, 1967.
67. Karoum F, Moyer-Schwing J, Potkin SG, Wyatt RJ: Presence of free, sulfate and glucuronide conjugated 3-methoxy-4-hydroxyphenylglycol (MHPG) in human brain, cerebrospinal fluid and plasma. *Brain Res* 125:333–339, 1977.
68. Kissinger PT, Riggin RM, Alcorn RL, Rau LD: Estimation of catecholamines in urine by high performance liquid chromatography with electrochemical detection. *Biochem Med* 13:299–306, 1975.
69. Kobayashi K, Koloch R, De Quattro V, Miano L: Increased plasma and urinary normetanephrine in young patients with primary hypertension. *Clin Sci* 57:173s–176s, 1979.
70. Kobus E, Wasilewska E, Bargiel Z: Urinary excretion of catecholamines and vanilmandelic acid during a normal menstrual cycle. *Bull Acad Pol Sci (Biol)* 27:71–74, 1979.
71. Kopin IJ: Catecholamines, adrenal hormones and stress. In Krieger DT, Hughes JC: *Neuroendocrinology*. Sunderland, Canada, Hospital Practice Book, 1980, pp 159–167.
72. Kuchel O, Buu NT: Dopamine, salt and blood pressure: The respective roles of free and conjugated catecholamines in hypertension. In Sambhi M: *The Fundamental Fault in Hypertension*. New York, Elsevier, in press, 1984.
73. Kuchel O, Buu NT, Fontaine A, Hamet P, Beroniade V, Larochelle P, Genest J: Free and conjugated plasma catecholamines in hypertensive patients with and without pheochromocytoma. *Hypertension* 2:177–186, 1980.
74. Kuchel O, Buu NT, Hamet P, Larochelle P, Bourque M: Catecholamine sulfates and platelet phenolsulfotransferase (PST) activity in human hypertension. In Usdin E, Sandler M: *Phenolsulfotransferase in Mental Health Research*, New York, MacMillan, 1981, pp 175–185.
75. Kuchel O, Buu NT, Hamet P, Larochelle P, Bourque M, Genest J: Essential hypertension with low conjugated catecholamines imitates pheochromocytoma. *Hypertension* 3:347–353, 1981.
76. Kuchel O, Buu NT, Hamet P, Larochelle P, Bourque M, Genest J: Unconjugated hyperepinephrinemia—a hallmark of pheochromocytoma imitating hypertension? *Hypertension* 3(Suppl II):II-129–II-133, 1981.
77. Kuchel O, Buu NT, Hamet P, Larochelle P, Bourque M, Genest J: Dopamine surges in hyperadrenergic essential hypertension. *Hypertension* 4:845–852, 1982.
78. Kuchel O, Buu NT, Hamet P, Larochelle P, Genest J: Conjugation of catecholamines, a variable to be accounted for in human hypertension. *Prog Biochem Pharmacol* 17:66–74, 1980.
79. Kuchel O, Buu NT, Hamet P, Nowaczynski W, Genest J: Free and conjugated dopamine in pheochromocytoma, primary aldosteronism and essential hypertension. *Hypertension* 1:267–273, 1979.
80. Kuchel O, Buu NT, Unger T: Free and conjugated dopamine: Physiological and clinical implications. In Imbs JL, Schwartz J: *Peripheral Dopaminergic Receptors*. Oxford, Pergamon Press, 1979, pp 15–27.
81. Kuchel O, Buu NT, Unger T, Genest J: Free and conjugated catecholamines in human hypertension. *Clin Sci Mol Med* 55:77–80, 1978.
82. Kuchel O, Buu NT, Unger T, Lis M, Genest J: Free and conjugated dopamine in human hypertension. *J Clin Endocrinol Metab* 48:425–429, 1979.
83. Kuchel O, Buu NT, Vecsei P, Bourque M, Hamet P, Genest J: Are plasma aldosterone surges in primary aldosteronism due to a loss of an

inhibitory dopaminergic control? *J Clin Endocrinol Metab* 51:337–344, 1980.
84. Kuchel O, Buu NT, Wang PC: Stress, catecholamines and hypertension: Respective roles of free and conjugated catecholamines. *Stress* 3:22–31, 1982.
85. Kuchel O, Hamet P, Buu NT, Larochelle P: Basis of false-positive glucagon tests for pheochromocytoma. *Clin Pharmacol Ther* 29:687–694, 1981.
86. Lake CR, Ziegler MG, Kopin IJ: Use of plasma norepinephrine for evaluation of sympathetic neuronal activity in man. *Life Sci* 18:1315–1326, 1976.
87. Louis WJ, Doyle AE, Anavekar S: Plasma norepinephrine levels in essential hypertension. *N Engl J Med* 288:599–601, 1973.
88. Mattok GL, Wilson DL, Heacock RA: Differential estimation of adrenaline, noradrenaline, dopamine, metanephrine and normetanephrine in urine. *Clin Chim Acta* 14:99–107, 1966.
89. Maus TP, Anderson RJ, Weinshilboum RM: Effect of dexamethasone on rat phenolsulphotransferase (PST) activity. *Pharmacologist* 22:773, 1980.
90. McKenna TJ, Island DP, Nicholson WE, Liddle GW: Dopamine inhibits angiotensin-stimulated aldosterone biosynthesis in bovine adrenal cells. *J Clin Invest* 64:287–291, 1979.
91. McKenna TJ, Island DP, Nicholson WE, Liddle GW: Stimulation of aldosterone production by catecholamines *in vitro*. In *Sixth International Congress of Endocrinology*. Melbourne, Australian Endocrinology Society, 1980, p 407.
92. Meek JL, Neff NH: Acidic and neutral metabolites of norepinephrine: Their metabolism and transport from brain. *J Pharmacol Exp Ther* 181:457–462, 1972.
93. Meek JL, Neff NH: Biogenic amines and their metabolites as substrates for phenol sulphotransferase (EC 2.8.2.1.) of brain and liver. *J Neurochem* 21:1–9, 1973.
94. Merits I: Formation of metabolism of [^{14}C] dopamine 3-O-sulfate in dog, rat and guinea pig. *Biochem Pharmacol* 25:829–833, 1976.
95. Merits I, Anderson DJ, Sonders RC: Species differences in the metabolism of orally administered dopamine-^{14}C and *N*-alanyl-dopamine-^{14}C. *Drug Metab Dispos* 1:691–697, 1973.
96. Michelot RJ, Lesko N, Stout RW, Cowan JK: Effect of S-adenosylhomocysteine and S-tubercidinylhomocystein on catecholamine methylation in neuroblastoma cells. *Mol Pharmacol* 13:368–373, 1977.
97. Moyer TP, Jing NS: Optimized isocratic conditions for analysis of catecholamines by high performance reversed-phase paired-ion chromatography with amperometric detection. *J Chromatogr* 153:365–372, 1978.
98. Mulder GJ, Krijgsheld KR: The availability of inorganic sulfate for sulfate conjugation in vivo. In Usdin E, Sandler M: *Phenolsulfotransferase and Mental Health Research*. New York, MacMillan, 1981, pp 127–144.
99. Mulder GJ, Scholtens E: Phenol sulfotransferase and uridine diphosphate glucuronyltransferase from rat liver in vivo and in vitro. *Biochem J* 165:553–559, 1977.
100. Nagel M, Schümann HJ: A sensitive method for determination of conjugated catecholamines in blood plasma. *J Clin Chem Clin Biochem* 18:431–432, 1980.
101. Oliver JA, Pinto J, Sciacca RR, Cannon PJ: Basal norepinephrine overflow into the renal vein: Effect of renal nerve stimulation. *Am J Physiol* 239:F371–F377, 1980.
102. Pearson RK, Anderson RJ, Weinshilboum RM: Rat tissue phenolsulfotransferase (PST): Assay and developmental changes. *Pharmacologist* 22:772, 1980.
103. Pennings EJ, van Kempen GM: Studies in vitro on the involvement of O-sulphate esters in the formation of O-methylated 3,4-dihydroxybenzoic acid by rat liver. *Biochem J* 191:133–138, 1980.
104. Pennings EJ, Vrielink R, Van Kempen GM: Anomalous effect of probenecid on rat brain phenolsulfotransferase. *Biochem Pharmacol* 25:1687–1690, 1976.
105. Pennings EJ, Vrielink R, Van Kempen GM: Kinetics and mechanism of the rat brain phenol sulphotransferase reaction. *Biochem J* 173:299–307, 1978.
106. Pennings EJ, Van Kempen GM: Studies on the meta and para-O-sulphation of the catechol compound 3,4-dihydroxybenzoic acid by rat liver sulphotransferase in vitro. *Biochem J* 193:869–874, 1981.
107. Picotti GB, Cesura AM, Galva MD, Mantegazza P, Kettler R, Da Prada M: Monoamine sulfoconjugation by human platelets. In Usdin E, Sandler M: *Phenolsulfotransferase in Mental Health Research*. New York, MacMillan, 1981, pp 44–54.
108. Raum WJ, Swerdloff RS: A radioimmunoassay for epinephrine and norepinephrine in tissues and plasma. *Life Sci* 28:2819–2827, 1981.
109. Rein G, Glover V, Sandler M: Phenolsulfotransferase in human tissue: Evidence for multiple forms. In Usdin E, Sandler M: *Phenolsulfotransferase in Mental Health Research*. New York, MacMillan, 1981, pp 98–126.
110. Rennick B, Quebbemann A: Site of excretion of catechol and catecholamines: Renal metabolism of catechol. *Am J Physiol* 213:1307–1312, 1970.
111. Renskers KH, Feor KD, Roth JA: Sulfation of dopamine and other biogenic amines by human brain phenol sulfotransferase. *J Neurochem* 34:1362–1368, 1980.
112. Richter D, MacIntosh FC: Adrenaline ester. *Am J Physiol* 135:1–5, 1941.
113. Roth JA, Young TG, Cohen DJ: Platelet monoamine oxidase activity in children and adolescents. *Life Sci* 18:919–924, 1976.
114. Roy AB: Sulphate conjugation enzymes. In Brodie BB, Gillette JR, Ackerman HS: *Handbook of Experimental Pharmacology, Concepts in Biochemical Pharmacology, Part 2*. Berlin, Springer-Verlag, 1971, pp 536–563.
115. Sandler M, Glover V, Lockwood J, Bonham

Carter S, Rein G: Clinical studies on platelet phenolsulphotransferase. In Usdin E, Sandler M: *Phenolsulfotransferase in Mental Health Research*. New York, MacMillan, 1981, pp 186–206.

116. Sekura RD, Jakoby WB: Phenolsulfotransferases. *J Biol Chem* 254:5658–5663, 1978.
117. Sever PS, Birch M, Osikowska B, Tunbridge RDG: Plasma noradrenaline in essential hypertension. *Lancet* 1:1078–1081, 1977.
118. Stöcker W, Hempel K: Inactivation and excretion of dopamine by the cat kidney in vivo. *Arch Exp Pathol Pharmakol* 295:123–126, 1976.
119. Stoff JS, Day B, Landsberg L: Bidirectional transport of H-L-dopa in the gut: Evidence for separate transport mechanisms. *Clin Res* 27:580A, 1979.
120. Subba Rao K, Seshadra Satry P, Ganguly J: Studies on metabolism of vitamin A. 2. Enzymic synthesis and hydrolysis of phenolic sulphates in vitamin-A-deficient rats. *Biochem J* 87:312–317, 1963.
121. Tycc GM, Sharpless NS, Kerr EWL, Muenter MD: Dopamine conjugate in cerebrospinal fluid. *J Neurochem* 34:210–213, 1980.
122. Udenfriend S, Lovenberg W, Sjoerdsma A: Physiologically active amines in common fruits and vegetables. *Arch Biochem Biophys* 85:487–490, 1959.
123. Unger T, Buu NT, Kuchel O: Renal handling of free and conjugated catecholamines following surgical stress in the dog. *Am J Physiol* 235:F542–F547, 1978.
124. Unger T, Buu NT, Kuchel O: Effects of surgical stress on free and conjugated catecholamines in the dog: Participation of dopamine sulfate in the stress response. In Usdin E, Kvetnansky R, Kopin IJ: *Stress and Catecholamines*. New York, Elsevier/North Holland, 1980, pp 205–210.
125. Unger T, Buu NT, Kuchel O, Schürch W: Conjugated dopamine: Peripheral origin, distribution, and response to acute stress in the dog. *Can J Physiol Pharmacol* 58:22–27, 1980.
126. Van Kempen GM, Jansen GS: Quantitative determination of phenolsulfotransferase using 4-methylumbelliferone. *Anal Biochem* 46:438–442, 1972.
127. Vignon F, Terqui M, Westley B, Derocq D, Rochefort H: Effects of plasma estrogen in mammary cancer cells. *Endocrinology* 106:1079–1086, 1980.
128. Walle T, Conradi ED, Walle UK, Fagan TC, Gaffney TE: Propranolol glucuronide cumulation during long-term propranolol therapy: A proposed storage mechanism for propranolol. *Clin Pharmacol Ther* 26:686–695, 1979.
129. Wettering JG, Krijgsheld KR, Mulder GJ: The availability of inorganic sulphate as a rate limiting factor in the sulphate conjugation of xenobiotics in the rat? Sulphation and glucuronidation of phenol. *Biochem Pharmacol* 28:757–762, 1979.
130. Wong KP: Species differences in the conjugation of 4-hydroxy-3-methoxyphenylethanol with glucuronic acid and sulfuric acid. *Biochem J* 158:33–37, 1976.
131. Wong KP: Measurement and kinetic study of the formation of adrenaline sulphate in vitro. *Biochem J* 174:777–782, 1978.
132. Youdim MBH, Carter SB, Sandler M, Hanington E, Wilkinson M: Conjugation defect in tyramine-sensitive migraine. *Nature* 230:127–128, 1971.
133. Ziegler MG, Lake CR, Kopin IJ: Plasma noradrenaline increases with age. *Nature* 261:331–335, 1977.

CHAPTER EIGHTEEN

Radioligand Binding to Adrenergic Receptors in Humans*

HARVEY J. MOTULSKY, M.D., and PAUL A. INSEL, M.D.

INTRODUCTION

A "lock and key" model is often used to describe the binding of a hormone or drug (key) to a receptor (lock). To carry the analogy further, locks are far more complex than keys: keys must only be shaped to fit the lock, but the lock must have both the appropriate shape to recognize the key and it must have the actual opening and closing mechanism. Similarly, whereas a drug or hormone need only have a structure that allows it to be recognized by a receptor, the receptor must recognize the drug or hormone (and discriminate between an agonist and an antagonist) and must have an effector mechanism to modulate biochemical events inside the cell. Thus receptors are more complex than drugs or hormones, and the study of receptors is critical for disciplines such as pharmacology, endocrinology, and neurobiology.

The concept of receptors as the initial recognition sites for neurotransmitters, hormones, and drugs (collectively known as ligands) has been well accepted for decades. During the last 10 to 15 years it has been possible to study the biochemical effects (such as ion fluxes and cAMP accumulation) of receptor activation. It has only been in the last 5 to 10 years, however, that the receptors themselves could be directly identified. This has

* Work in our laboratory is funded by grants from the National Science Foundation, National Institutes of Health, California Heart Association, and American Heart Association, a National Institutes of Health Postdoctoral Fellowship (HJM), and an Established Investigatorship of the American Heart Association (PAI). We thank Drs. R. J. Lefkowitz, A. J. J. Wood, J. M. Roberts, S. P. Galant, R. W. Alexander, and S. J. Shattil for giving us a scoop on unpublished results, and S. Dutky for preparing the manuscript.

occurred because of the development of radioactively labeled drugs (radioligands) of high specific activity. Radioligand binding has been a "growth industry" in recent years, and binding experiments have been used to study many types of receptors. Catecholamine (or adrenergic) receptors have been among the most extensively studied.

What advantages do radioligand binding methods offer to the study of the adrenergic system? Questions regarding the regulation of catecholamine response have baffled pharmacologists and physiologists for decades. The now classical studies of Ahlquist (4) were seminal in focusing attention on the differences between tissues in their response to catecholamines and in introducing the concept of discrete types of adrenergic receptors. This concept of adrenergic receptors has been the guiding force behind an abundance of pharmacological and physiological studies. In these studies the receptors served as conceptual models used to explain variations in tissue responses; the receptors themselves were not studied. The development of radioligand binding methods has enabled investigators to examine adrenergic receptors themselves, independent of any tissue response. These methods allow one to investigate questions regarding the distribution of receptors, coupling of receptors to biochemical response, mechanisms and circumstances by which receptors are regulated, and changes in adrenergic receptors in disease.

Most studies of adrenergic receptors have used cultured cells or animal tissues (for reviews see 28, 107, 140, 209, 214), but many studies, especially recently, have used human tissues. In this chapter we will first discuss the theoretical basis and practical problems of radioligand binding techniques and will then review the application of these techniques to the study of catecholamine receptors in humans. We have attempted to include studies published by August, 1981, and discussion is limited to α- and β-adrenergic receptors; dopamine receptors are not included.

RADIOLIGAND BINDING: METHODS AND PITFALLS

General Aspects

The theoretical basis for radioligand binding studies is straightforward: a radioactive agonist or antagonist ("radioligand") is incubated with a tissue, the unbound radioligand is removed, and the amount of radioactivity remaining with the tissue is used as a measure of the receptors within that tissue. Separation of "bound" and "free" radioactivity is generally accomplished either by filtering and washing the samples over inert (usually fiberglass) filters or by pelleting membrane-bound material by centrifugation. Filtration is the method used by most investigators. In most experiments, the radioactivity is quantitated by either liquid scintillation or gamma counting. Variations of the basic technique have included use of autoradiography or external nuclear scanning of tissues (29, 101, 161, 218). Although some investigators have studied receptors in intact target cells (for example, 24, 56, 73, 112, 195), most workers have undertaken radioligand-binding studies of adrenergic receptors using either homogenates or washed particulate ("membrane") preparations from receptor-bearing tissues. The methodology of what has been half-jokingly termed "grind and bind" experiments has been comprehensively reviewed in several monographs (e.g., refs. 140, 209), and we will therefore only briefly summarize some of the techniques and problems of radioligand binding to adrenergic receptors.

Radioactively labeled drugs bind not only to specific receptors but also to nonspecific sites on tissues, filters, and test tubes. This nonspecific binding can be determined by incubating the radioligand and tissue with an unlabeled drug that will bind to essentially all the receptors of interest. The choices of which drug and concentration to use to define nonspecific binding are important. One needs to select a receptor-specific drug (e.g., propranolol or isoproterenol for β-adrenergic receptors, phentolamine for α-adrenergic receptors) at a concentration high enough so that it will occupy more than 95% of the receptors and low enough so that it will not cause more general membrane perturbations. Problems can arise when either too low or too high a concentration is used (24, 154, 209). When all of the receptors are occupied by unlabeled drug, radioligand binding that remains is to nonspecific sites. Specific binding is derived by subtracting this nonspecific binding from the total binding.

The concept of receptors was developed from pharmacological experiments that measured a biochemical or physiological response to various drugs and was well accepted long before radioligand binding was possible. We can now easily measure radioligand binding sites, but it is very important to show that these binding sites are equivalent to receptors. A number of criteria have been developed (Table 18.1) to demonstrate that the radioactivity trapped on a filter truly represents radioligand binding to the pharmacologically defined receptor. Answering any one of these questions in the affirmative cannot prove that radioligand binding is to receptors, but fulfilling all of these criteria gives the investigators considerable confidence that they have identified the receptor of interest. The necessary experiments are straightforward, but misleading results have been obtained (and published) when some of these questions have been overlooked. Accordingly, it is important for investigators undertaking radioligand binding studies with new radioligands or with new tissues to do a variety of experiments to validate their assays.

Three main types of experiments—kinetic studies, saturation binding ("binding isotherms"), and competitive binding experiments—are used to validate an assay as well as to ask biological questions. In the sections that follow we will briefly describe the theory behind these experiments and the problems that can occur in performing them and in analyzing the results. These descriptions should be considered an introduction to make the reader aware of the types of radioligand binding data. Detailed descriptions (45, 209, 217, 222) are available for those who wish to perform and analyze such experiments.

Table 18.1
Questions That Should Be Answered in Evaluating a New Radioligand Assay

Is specific binding saturable (*i.e.*, are there a finite number of receptors)?
Does specific binding occur as rapidly and reversibly as does response to the unlabeled ligand?
Do unlabeled drugs compete for specific binding sites stereoselectively and in the rank order of potency predicted by pharmacological experiments?
Is specific binding proportional to tissue concentration?
Do independent methods of deriving the dissociation constant of the radioligand agree?
Does the radioligand remain unmetabolized during the experiment?
Has nonspecific binding been defined appropriately?

Kinetic Experiments

Radioligand binding reactions are assumed to be reversible and to demonstrate the law of mass action:

$$L^* + R \underset{k_2}{\overset{k_1}{\rightleftharpoons}} RL^*$$

where L^* is the radioligand, R is the receptor, and RL^* is the radioligand-receptor complex. The rate of association ("on rate") is measured in experiments where the radioligand is added to the tissue, and the extent of binding is measured at various time points (Fig. 18.1*A*). Such experiments permit calculation of the biomolecular association rate constant, k_1, which is expressed in units of time^{-1} M^{-1}. The rate of dissociation ("off rate") is measured by first allowing a radioligand to bind to the receptors, blocking the forward reaction of the equilibrium (either by increasing the volume to lower the concentration of receptor and radioligand or by adding an excess of an unlabeled receptor-specific drug to block all the receptors) and measuring the extent of binding at various time points thereafter (Fig. 18.1*B*). Such experiments allow one to calculate the monomolecular dissociation rate constant, k_2, which is expressed in units time^{-1}. Equilibrium in binding studies is reached (by definition) when the rate of association equals the rate of dissociation. The rate of association is $k_1[L^*][R]$; the rate of dissociation is $k_2[RL^*]$. Therefore at equilibrium, when the two rates are equal,

$$\frac{[L^*][R]}{[RL^*]} = \frac{k_2}{k_1}$$

$$= K_D \text{ (equilibrium dissociation constant).}$$

Some authors, especially those in Europe, prefer to use the inverse of this value, $1/K_D$, termed K_A, or equilibrium association constant. K_D is expressed in units of M and K_A in units of M^{-1}.

Kinetic experiments are useful in several ways. First, they allow a calculation of the equilibrium dissociation constant (K_D) as noted above. This K_D can also be determined by two

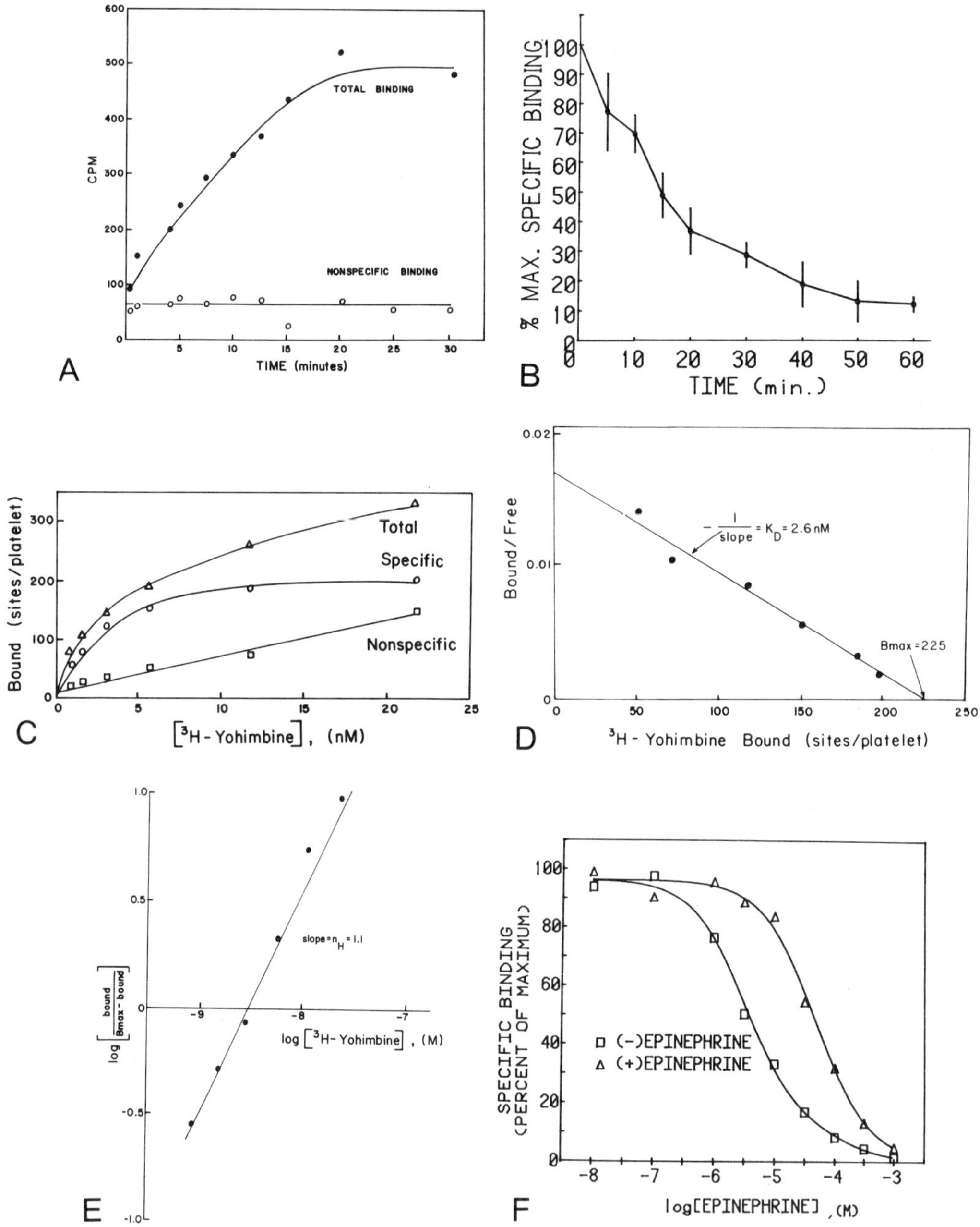

Figure 18.1. Experiments to characterize a radioligand binding assay. The example of [^{3}H]yohimbine binding to intact human platelets is used to illustrate the major types of binding experiments. (*A*) Rate of association ("on-rate" experiment). The platelets were incubated with 2 nM [^{3}H]yohimbine at 25°C and the total and nonspecific binding (expressed in counts per minute) determined at various times. Nonspecific binding (defined by 10 μM phentolamine) reaches equilibrium in less than a minute; total (and specific) binding reaches equilibrium in 20 to 30 minutes. The on-rate (K_1) is 10^8 min^{-1} M^{-1}. (*B*) Rate of dissociation ("off-rate" experiment). Platelets were incubated with 3 nM [^{3}H]yohimbine for 60 minutes. Phentolamine (to 10 μM) was then added and the binding of [^{3}H]yohimbine determined at various times. The results are the mean and standard deviation of three experiments. The off-rate is 0.04 min^{-1}. (*C*) Saturation binding isotherm. Various concentrations of [^{3}H]yohimbine were incubated with platelets and the total binding determined (△). Parallel incubations in the presence of 10 mM

other independent methods, and this redundancy allows one to check the internal consistency of the data. Second, the rate of association can be compared with the rate at which the pharmacological properties associated with receptor activation are known to occur. Clearly, radioligand binding that occurs over many hours may not be relevant to functional changes that occur in seconds to minutes. Third, the reversibility of the radioligand binding can be determined. Most hormones and radioligands bind to receptors reversibly. Irreversible binding can be experimentally useful (for example, in labeling receptors prior to solubilization and purification), but irreversibility of "binding" may in fact indicate that the radioligand is being transported, stored, or metabolized, rather than being bound to pharmacological receptors. Fourth, it is necessary to determine how long it takes for binding to reach equilibrium before one can set up other experiments that must be performed at equilibrium. Because the time required to reach equilibrium depends on the radioligand concentration used (the lowest concentrations require the longest time), it is preferable to do "on-rate" experiments with the lowest concentration of radioligand that is to be used in other experiments. Finally, it is important to check that dissociation occurs slowly enough that a negligible amount of specific binding dissociates during the time it takes to filter and wash the samples; this problem can often be minimized by using ice-cold buffer to wash the filters.

Binding Isotherms

Saturation binding studies ("binding isotherms") are performed by incubating various concentrations of the radioligand with a fixed concentration of tissue containing the receptor of interest and determining the radioligand bound at equilibrium. One generally incubates radioligand and tissue together in both the absence and presence of the compound chosen to define nonspecific binding. Figure 18.1C shows data for a typical equilibrium binding experiment of this type. Nonspecific binding increases linearly with increasing radioligand concentrations, and specific binding shows the typical rectangular hyperbolic pattern predicted from the equation (above) for formation of RL^*. Two important parameters can be estimated from this type of experiment: the equilibrium dissociation constant of the radioligand (K_D) and the maximum binding capacity of the receptor (B_{max}). These terms are analogous to the K_m and V_{max} used to characterize enzymes. As a measure of the affinity of the ligand for the receptor, K_D is defined as the concentration of radioligand needed to bind to half of the receptors. B_{max} is a measure of

---

phentolamine determined nonspecific (□) binding, and specific binding (○) was calculated as total minus nonspecific binding. (*D*) Scatchard plot. This graph is made from the specific binding data of *C*. The receptor number (B_{max}) is 225 [^{3}H]yohimbine binding sites/platelet, and the dissociation constant (K_D) of these sites is 2.6 nM. Why is the Scatchard plot linear? By definition

$$K_D = \frac{[R][L]}{[LR]} \tag{1}$$

In this equation $[L]$ is the concentration of free ligand ("free"), $[LR]$ is equal to the amount of ligand bound to receptor ("bound"), and R is the concentration of unbound receptor which is equal to the total concentration of receptors (B_{max}) minus the concentration of receptors bound to ligand ("bound"). Thus equation 1 can be rewritten as follows:

$$K_D = \frac{[\text{free}] \cdot (B_{max} - [\text{bound}])}{[\text{bound}]} \tag{2}$$

$$\frac{[\text{bound}]}{[\text{free}]} = \frac{B_{max}}{K_D} - \frac{1}{K_D} \cdot [\text{bound}] \tag{3}$$

A plot of bound/free *vs.* bound will therefore be linear with a slope of $-1/K_D$. (*E*) Hill plot. The specific binding data from *C* were used to create this Hill plot. The slope is 1.1, indicating binding to a single class of noninteracting sites. (*F*) Competitive binding experiment. Platelets were incubated with 2 nM [^{3}H]yohimbine and various concentrations of either (−) or (+)epinephrine, and the specific radioligand binding was determined. As the epinephrine concentration increases, less [^{3}H]yohimbine binds. (−)Epinephrine is over 10-fold more potent than (+)epinephrine. (Reproduced with permission from H. J. Motulsky *et al.* (151).)

receptor number and is usually expressed in units of receptors per cell, or moles of receptor per milligram of protein or per gram of tissue.

K_D and B_{max} can be determined from saturation binding data either using precise computerized nonlinear regression techniques, or using simpler graphical methods, most commonly that of Scatchard (179) (perhaps more accurately attributed to Rosenthal (172)). In a Scatchard analysis, one plots the amount of radioligand bound *vs.* the ratio of bound to free radioligand, and thereby linearizes the binding data. As shown in Figure 18.1*D*, a straight line on the Scatchard plot of the specific binding data indicates the presence of a single class of receptors. One estimates K_D from the negative inverse of the slope of this line and B_{max} from the intercept with the abscissa. More sophisticated nonlinear regression methods of analysis (155, 54) are needed to resolve curved Scatchard plots; such plots occur in several situations including site-site interaction, multiple classes of receptors, experimental artifacts, etc. (for discussion, see ref. 55). It is tempting to analyze curved Scatchard plots by hand, drawing two straight lines through the data points and using those lines to determine two values for K_D and B_{max}; that method gives results that are simply wrong (154, 159).

A second graphical method that can be used to analyze such binding studies is the Hill plot (134) which yields estimates of K_D and n_H, the Hill coefficient (Fig. 18.1*E*). This empirically derived value, which is the slope of the Hill plot, is an estimate of "cooperativity" in the binding of L^* to R. By convention, Hill slopes greater than 1.0 are said to represent positive "cooperativity," those less than 1.0 negative "cooperativity," and those equal to 1.0 (usually in the range of 0.9 to 1.1) no "cooperativity." Positive or negative "cooperativity," as estimated from Hill slopes, can occur for a variety of reasons, and one must perform additional experiments to figure out the molecular basis for these results. For most antagonist radioligands binding to adrenergic receptors the Hill coefficient is close to 1.0, reflecting the absence of cooperative interactions.

Competitive Binding Experiments

In competitive binding experiments the investigator measures the ability of various drugs to compete with the radioligand for receptor binding sites. These experiments are usually performed by incubating the tissue simultaneously with the radioligand and an unlabeled drug until equilibrium is reached, which takes longer than it does in saturation binding experiments (222). In such binding studies the unlabeled drug and the radioligand compete for binding sites, but some incorrectly term these studies "displacement" experiments. Usually, one uses a single concentration of the radioligand and varying concentrations of the unlabeled drug. As the concentration of the unlabeled drug increases, more receptors are occupied by that drug and fewer, therefore, are bound by the labeled ligand (Fig. 18.1*F*). The affinity, or dissociation constant, of the unlabeled drug for the radioligand binding sites can be determined from the concentration of that unlabeled drug needed to compete for 50% of the specific radioligand binding (IC_{50}). The IC_{50} is a measure of both the affinity of the unlabeled drug and the extent to which the receptors are occupied by the radioligand. To determine the absolute dissociation constant of the competitor, one must correct the IC_{50} for the affinity and quantity of the radioligand used. This correction is made using the following equation (33, 35):

$$K_D \text{ competitor} = \frac{IC_{50} \times K_D \text{ radioligand}}{S + K_D \text{ radioligand}}$$

where the K_D of the radioligand is determined in other (binding isotherm) experiments, and S refers to the concentration of that radioligand used in the competitive binding experiment.

A number of investigators have emphasized the value of more complex analyses of competitive binding studies. The slope of the competitive binding curve ("pseudo Hill slope" or "slope factor") indicates whether the unlabeled drug is binding to a single class of receptors (slope = 1.0) or possibly to multiple classes of receptors (slope <1.0). In effect, such slope factors are equal to 1.0 if a 10 to 90% competition for radioligand binding occurs over a 81-fold (~2 log units) concentration range of the unlabeled drug; slopes of <1.0 reflect more shallow competition curves that occur over a greater concentration range. More sophisticated analyses can be performed using computer programs that use nonlinear regression to fit binding data to mass-action binding equations for either one or several classes of binding sites (155). This method has been used to define adrenergic receptor subtypes (57, 94, 146).

A crucial step in characterizing radioligand binding to adrenergic receptors is to show that competition of radioligand binding by adrenergic agonists and antagonists demonstrates the expected rank order of potency. For β-adrenergic receptors, this means that the order of potency in competition for radioligand binding sites should be isoproterenol > epinephrine > norepinephrine, and for α-adrenergic receptors the rank order of agonist potency should be epinephrine ≥ norepinephrine ≫ isoproterenol. Similarly the affinity of adrenergic antagonists in competing for radioligand binding should correlate with their pharmacological potency; for example, propranolol should compete for radioligand binding to β-adrenergic receptors more potently than phentolamine. In addition, one can use a large number of available adrenergic compounds in competitive binding studies to help confirm the identity of a particular type or subtype of adrenergic receptor.

Adrenergic agents should also compete stereoselectively. Pharmacological stimulation of α- and β-adrenergic receptors occurs with (−) or (L)-isomers of epinephrine or norepinephrine about 100-fold more potently than with the (+) or (D)-isomers. Accordingly, the (−)isomers should compete for radioligand binding sites more potently than do (+)isomers (Fig. 18.1F).

Radioligands

A wide variety of radioligands are now available for the study of adrenergic receptors (Tables 18.2 and 18.3).

Both iodinated and tritiated radioligands are available for studying β-adrenergic receptors; none selects between β_1- and β_2-adrenergic receptors. Iodinated ligands have specific activities of ~2200 Ci/mmole, while tritiated ligands have a specific activity of less than 100 Ci/mmole. The iodinated ligands, such as [^{125}I] iodohydroxybenzylpindolol, are therefore particularly useful in situations where the availability of tissue is limiting (61, 140, 168). Two disadvantages of iodinated ligands are that (a) they decay much more rapidly than tritiated ligands (half-life of ^{125}I is 60 days *vs.* 13 years for ^{3}H) and thus the radioligand must be prepared more often, and (b) they are, for the most part, prepared as a racemic mixture rather than as a pure (−) stereoisomer (25). Iodination is a procedure that can be done in one's own laboratory, thus saving the expense of purchasing commercial preparations of tritiated ligands. Nevertheless, many investigators prefer to buy commercially prepared iodinated radioligands.

Most early studies of α-adrenergic receptors used [^{3}H]dihydroergocryptine (DHE), which does not select between α_1- and α_2-adrenergic receptors and which binds to other receptor types as well (30, 53). Selective radioligands for α_1- and α_2-receptors are now available. Since these have higher affinities for the respective receptors than does [^{3}H]DHE, less radioligand is needed, and thus the nonspecific binding is also decreased. [^{3}H]WB4101 was initially reported to be selective for α_1-receptors, but its selectivity has been questioned (95). [^{3}H]Prazosin has a high affinity (less than 1 nM in all systems studied) and high selectivity for α_1 receptors. In a recent report [^{125}I]BE2254 has been described as having high affinity and selectivity for α_1-adrenergic receptors (223).

Table 18.2
β-Adrenergic Radioligands[a]

Ligand	Specific Activity (Ci/mmole)	Comment
Antagonists		
(−)[^{3}H]Dihydroalprenolol ([^{3}H]DHA)	30–70	
(±)[^{125}I]Hydroxybenzylpindolol ([^{125}I]IHYP)	~2200	Often high nonspecific binding
(±)[^{125}I]Cyanopindolol ([^{125}I]ICYP)	~2200	Often low nonspecific binding; more stable than IHYP
(−)[^{125}I]Pindolol	~2200	Easily prepared as the active stereoisomer
[^{3}H]Propranolol	20	Affinity lower than other antagonists
Agonists		
(±)[^{3}H]Hydroxybenzylisoproterenol (HBI)	5–20	
(−)[^{3}H]Epinephrine	50	

[a]None of these ligands select between β_1- and β_2-receptors.

Table 18.3
α-Adrenergic Radioligands

Ligand	Pharmacology	Specific Activity (Ci/mmol)	Comment
Antagonists			
[^{3}H]Dihydroergocryptine (DHE)	Nonselective antagonist	20–50	Also binds to nonadrenergic sites
[^{3}H]Prazosin	α_1-Antagonist	10–30	High affinity, highly selective
[^{3}H]Yohimbine	α_2-Antagonist	~80	
[^{3}H]WB4101	"α_1-Antagonist"	15–30	Somewhat, but not absolutely, α_1-selective
[^{3}H]Phentolamine	Nonselective antagonist	23	Lower affinity than other ligands
[^{3}H]Rauwolscine	α_2-Antagonist	82	Relatively untested radioligand
[^{3}H]Phenoxybenzamine	Antagonist	3–15	Relatively α_1-selective, binds irreversibly
[^{125}I]BE2254	α_1-Antagonist	2175	Very new; only iodinated α-selective ligand
Agonists			
[^{3}H]Clonidine	α_2 Partial agonist	20–30	
[^{3}H]Paraaminoclonidine	α_2 Partial agonist	24–30	Lower nonspecific binding than with clonidine
[^{3}H]Epinephrine	Nonselective agonist	~50	
[^{3}H]Norepinephrine	Nonselective agonist	20–40	

[^{3}H]Yohimbine is similarly selective for α_2-adrenergic receptors, as is, presumably, the new ligand [^{3}H]rauwolscine, a stereoisomer of yohimbine.

CHARACTERIZATION OF ADRENERGIC RECEPTORS IN HUMANS

General

As noted above, the number of studies that characterize α- and β-adrenergic receptors in humans is rapidly increasing. In many of these reports, investigators have been careful to consider the details involved in validating the radioligand binding assays. Blood cells have been the most thoroughly studied "tissue" in humans for three reasons. First, they are the only conveniently accessible human "organ." Second, they can be easily separated to yield fairly pure cell types whose receptors and receptor-mediated biochemical and functional responses can be readily measured. Thus they can be used to examine coupling between receptor occupancy and response. Finally, receptors on blood cells have been used as possible mirrors of adrenergic receptors on more critical, but less easily studied, organs, such as heart, blood vessels, lung, or brain. An important question that is as yet poorly answered is the extent to which properties of adrenergic receptors on blood cells are representative of those of internal organs. Thus far, it is not possible to study both α- and β-adrenergic receptors on a single cell type. Blood platelets have been used to study α-adrenergic receptors, and leukocytes have been used to study β-adrenergic receptors. It would be useful if one could study these receptors in tandem in one cell type, but platelets do not appear to have β-adrenergic receptors detectable with radioligands (P. A. Insel, unpublished observations), and the existence of α-adrenergic receptors in leukocytes has been incompletely documented (46, 176).

Platelets

CHARACTERISTICS OF PLATELET α-ADRENERGIC RECEPTORS

When exposed to >1 μM epinephrine, platelets secrete their granular contents and aggregate (145). Exposure to lower concentrations of epinephrine will not aggregate platelets but will potentiate aggregation induced by other types of agents (*e.g.*, ADP, thrombin) (102). From the rank order of potency of adrenergic agonists and antagonists, one can readily conclude that these effects of epinephrine are mediated by an α-adrenergic receptor (102, 145). Biochemical events triggered by α-adrenergic agonists include inhibition of adenylate cyclase with a concomitant decrease in platelet cyclic AMP levels and enhancement of calcium influx (103, 114, 160, 191).

In other tissues, α-adrenergic receptors and responses have been classified into α_1 and α_2 subtypes (reviewed in ref. 96). This classification was devised to explain the different rank order of potencies for response to adrenergic agents in various tissues. This classification was initially thought to be equivalent to the distinction between presynaptic (α_2) and postsynaptic (α_1) adrenergic receptors (15, 132, 189). It is now clear, however, that this anatomic distinction is invalid; for example, α_2-receptors exist at both presynaptic and postsynaptic sites (15, 127). The nature of the receptor subtype mediating α-adrenergic response in platelets has been controversial. Some workers have suggested that all platelet α-receptors were of the α_2-subtype (102), while others thought that both α_1- and α_2-receptors were present (83, 84), or perhaps that platelets had a unique ("α_3") type of α-adrenergic receptor (113). As will be explained below, this controversy appears to have been resolved through use of radioligand binding studies.

Identification of α-adrenergic receptors on platelets using radioligand binding techniques was first accomplished using [^{3}H]dihydroergocryptine (DHE) binding to platelet particulate ("membrane") preparations. Kafka *et al.* (117) initially described [^{3}H]DHE binding to platelets, but there are several problems with this characterization: the K_D for [^{3}H]DHE was much higher than that subsequently observed in several laboratories, the affinities for several adrenergic compounds were unusual, and only about 50% of the specific [^{3}H]DHE binding was reversible. Two other groups almost simultaneously reported detailed characterizations of [^{3}H]DHE binding to platelets (5, 157). [^{3}H]DHE binding was shown to satisfy multiple criteria of receptor binding: it was rapid, reversible, of high affinity (~5 nM), stereoselective, and competed for by adrenergic drugs as expected for α-adrenergic receptors. The studies with platelet membranes indicated that there were about 200 fmole receptors/mg membrane protein. Similar results were later obtained with the antagonist [^{3}H]phentolamine (190). Compared to [^{3}H]DHE, [^{3}H]phentolamine had lower affinity, higher nonspecific binding, and more rapid dissociation. Most initial binding experiments with platelets were done with platelet membranes; attempts to use DHE with intact platelets yielded results of poorer quality and reproducibility (5, 157).

In an attempt to resolve the controversy regarding the subtype of α-adrenergic receptors on platelets, two groups examined the competition of [^{3}H]DHE binding to platelet membranes using the α_1-selective antagonist prazosin and the α_2-selective antagonist yohimbine (59, 94). The affinity of yohimbine was 1000-fold higher than that of prazosin. From the results of computerized nonlinear regression analysis, Hoffman *et al.* (94) concluded that [^{3}H]DHE bound to a single class of receptors and that these were exclusively of the α_2 subtype.

[^{3}H]Yohimbine is α_2-selective and has proven to be a very useful radioligand for identifying α_2-adrenergic receptors on both intact platelets (151) and platelet membranes (47, 151, 225). We have also found that [^{3}H]yohombine produces results that are of a higher consistency and reproducibility than [^{3}H]DHE, especially in studies with intact platelets. Direct comparison of saturation binding experiments using [^{3}H]DHE and [^{3}H]yohimbine indicates that [^{3}H]DHE binds to one and a half times as many sites as does [^{3}H]yohimbine (Fig. 18.2; 47, 149). Moreover, unlabeled yohimbine, a selective α_2-antagonist, competes for fewer [^{3}H]DHE sites than the nonselective antagonist phentolamine (which is usually used to define nonspecific [^{3}H]DHE binding) (149). Thus, both labeled and unlabeled yohimbine bind to fewer sites than [^{3}H]DHE. We find that [^{3}H] rauwolscine, a new α_2-adrenergic ligand, binds to the same number of sites as [^{3}H]yohimbine (149).

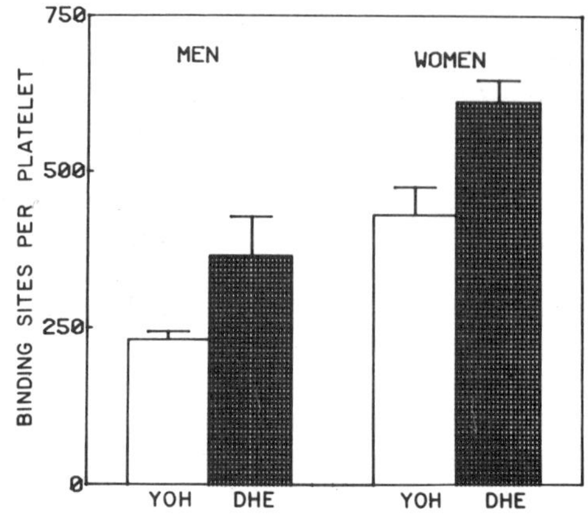

Figure 18.2. Comparison of number of [^{3}H]DHE and [^{3}H]yohimbine binding sites to intact platelets. Saturation binding isotherms were determined in parallel with both radioligands using 10 mM phentolamine to define nonspecific binding. Results are mean and standard error of data from five men and four women. (Data from Motulsky and Insel (149).)

Because both [^{3}H]yohimbine and [^{3}H]DHE were thought to bind only to α_2-adrenergic receptors on platelets, the discrepancy between the number of sites identified by these two ligands is surprising. To confirm that there are no α_1-adrenergic receptors on platelets, we have used the α_1-selective ligand [^{3}H]prazosin and found that its interaction with platelets is not compatible with binding to α_1-adrenergic receptors. Although it is possible that both [^{3}H] DHE and [^{3}H]yohimbine bind only to α_2-adrenergic sites but that some of these sites (perhaps those recently internalized or synthesized) are accessible only to [^{3}H]DHE, we believe that it is more likely that [^{3}H]DHE is binding to other (phentolamine-displaceable but non-α-adrenergic) sites in addition to binding to α_2-adrenergic receptors. Many of the clinical results summarized below are from experiments using [^{3}H]DHE and the conclusions, therefore, should be viewed with caution.

REGULATION OF BINDING TO PLATELET α-ADRENERGIC RECEPTORS BY NUCLEOTIDES AND IONS

Antagonists bind to receptors and prevent other drugs from binding; agonists bind to receptors and initiate a cascade of events leading to altered cellular function. α-Adrenergic agonists and antagonists appear to interact with a common population of membrane receptors (94, 203). However, binding of agonists is regulated by a variety of factors (including guanyl nucleotides, and monovalent and divalent cations), whereas binding of antagonists is relatively unaffected by such factors. Understanding how these factors modulate the binding of agonists to α-adrenergic receptors is likely to offer insight into molecular mechanisms by which these receptors work.

Guanyl nucleotides modulate binding of agonists to a variety of receptors (99), especially those that are linked to activation or inhibition of adenylate cyclase (170). Moreover, guanyl nucleotides are obligatory cofactors in the activation and inhibition of adenylate cyclase (140, 170, 173). Platelets are one tissue in which effects of guanyl nucleotides on agonist binding to α-adrenergic receptors have been examined in some detail. Lefkowitz and coworkers (99, 144, 199) studied effects of nucleotides on the ability of various drugs to compete for [^{3}H]DHE binding sites on platelet membranes. Whereas antagonists appeared to bind to a single class of α-adrenergic receptors, agonists appeared to interact with two classes of sites termed R_H (for high affinity) and R_L (for low affinity). Guanyl nucleotides had no effect on antagonist binding but decreased the affinity of agonists so that the agonists appeared to interact with a single class of receptors of low affinity (the R_L class). We have found similar results in our studies with [^{3}H]yohimbine and platelet membranes; moreover, we found that intact platelets behave similarly to platelet membranes that were assayed in the presence of guanyl nucleotides (151).

These findings have been confirmed more directly using labeled agonists in direct binding experiments with platelet membranes. Several laboratories have reported that labeled agonists (e.g., [^{3}H]epinephrine, [^{3}H]norepinephrine, and [^{3}H]clonidine) fulfill expected criteria for binding to platelet α-adrenergic receptors (98, 184, 138). These labeled agonists bind to washed platelet membranes with high affinity but to fewer sites than do labeled antagonists. Addition of GTP and other guanyl nucleotides markedly decreases the binding of these labeled agonists. These results have been interpreted to mean that labeled agonists preferentially bind to high affinity sites (R_H) in the membranes and that addition of nucleotides decreases the number of these high-affinity sites (presumably by converting them to low-affinity sites), thereby decreasing binding of the labeled agonists (98, 138, 184).

Recent studies have begun to explore the molecular basis for such nucleotide-mediated effects. [^{3}H]Yohimbine and [^{3}H]epinephrine binding can be detected in platelet membranes solubilized with the nonionic detergent digitonin (143, 187). When the receptors are solubilized after prelabeling with [^{3}H]epinephrine (in the absence of GTP) the sedimentation coefficient (and presumably the molecular weight) is greater than when the receptors are prelabeled with [^{3}H]yohimbine (143, 187). By analogy with similar results observed with solubilized β-adrenergic receptors (137), it has been suggested that R_H, the high affinity state of the receptor, may represent a macromolecular complex between the receptor and a guanyl nucleotide binding protein, while R_L may represent the receptor alone (136, 143). Addition of guanyl nucleotides would serve to dissociate receptor-nucleotide binding protein complexes and thereby lower agonist affinities and decrease the apparent molecular weight of the receptor. The existence of such a guanyl nucleotide binding protein (called N_s) is well documented for hormonal systems that stimulate adenylate cyclase, but only indirect evidence has been presented to prove the existence of a separate protein that mediates inhibition of adenylate cyclase (N_i) by receptors such as the α_2-adrenergic receptors of platelets (100, 170, 187).

What such findings tell us about α-adrenergic receptors in intact platelets is as yet unclear. Intracellular concentrations of GTP are likely to be high enough to decrease agonist affinities to a large extent. Accordingly, it is not surprising that we have observed that epinephrine has a low affinity ($K_D \backsim 1$ μM) in competing for [^{3}H]yohimbine binding sites on intact platelets (151). Membrane preparations contain variable amounts of the high affinity state depending, to some degree, on the concentration of GTP retained during homogenization and centrifugation of the tissue. Thus in adding back GTP to such membranes, one is, in part, recreating the "state" of the receptor in the intact cell.

Epinephrine potentiates platelet aggregation at concentrations (10 to 100 nM) far lower than those that influence equilibrium binding of [^{3}H]yohimbine to intact platelets. This apparent paradox between the low concentrations of epinephrine that are physiologically active (K_{ACT}) and the much higher concentrations required in equilibrium binding studies (K_D) has not been adequately explained. One possible explanation is that epinephrine initially binds to receptors in a high affinity state but that these are rapidly converted to low affinity sites. These transient, high affinity sites presumably mediate physiological responses, but we are currently unable to detect the nonequilibrium, intermediate states of the receptor in intact cells using equilibrium binding studies. Presumably the high affinity state is the receptor-N_i complex, and the low affinity state is the receptor alone. An alternative explanation is that there are many "spare" α_2-adrenergic receptors, and that only a small fraction of these need be occupied to produce biological responses. Further studies will be needed to test these and alternative hypotheses for the K_{ACT}/K_D discrepancy.

Monovalent and divalent cations also modulate the binding of agonists to platelet α-adrenergic receptors. Mg^{2+} increases the affinity of agonists for these receptors as well as potentiating the effect of GTP (184, 199). Conversely, monovalent ions, in particular Na^+, decrease agonist affinities (144). The mechanism by which this effect of Na^+ occurs is unknown, but it appears to operate physiologically because removal of Na^+ from the incubation buffer increases the affinity of an agonist for [^{3}H]yohimbine sites on intact platelets (151). Na^+ and GTP have synergistic effects in lowering the affinity of epinephrine for [^{3}H] yohimbine sites on platelet membranes (136, 151). Michel *et al.* (144) have recently reported similar results with [^{3}H]DHE and platelet membranes, and these workers have suggested that while GTP converts R_H to R_L, Na^+ decreases the affinity of both R_H and R_L without changing the relative proportion of high and low affinity states.

In summary, platelets are a useful tissue for studying α_2-adrenergic receptors in intact platelets, platelet membranes, and solubilized membranes. The factors involved in regulating agonist interaction with these receptors and in exploring the coupling of receptor binding to responses are under active investigation. In addition, as will be discussed below, alteration in α-adrenergic receptors on platelets has been investigated in a variety of diseases.

Leukocytes

As noted above, β-adrenergic receptors on peripheral blood leukocytes have been studied by several laboratories. Leukocytes respond to catecholamines in several ways, including β-adrenergic-mediated activation of adenylate cyclase, altered hemolytic plaque formation, mitogenesis of lymphocytes, and inhibition of release of hydrolytic enzymes by granulocytes (19, 104, 142, 221). Bourne *et al.* (18) proposed that the increase in cAMP produced by β-adrenergic agonists and other agents serves to inhibit immune and inflammatory response in a variety of settings. By contrast, there is only equivocal evidence of α-adrenergic-mediated responses of granulocytes or lymphocytes.

Williams *et al.* (210) first reported identification of β-adrenergic receptors on membranes prepared from leukocytes using radioligand methods. These workers used [^{3}H]dihydroalprenolol ([^{3}H]DHA) and showed that both lymphocytes and granulocytes contained β-adrenergic receptors. In their initial report, Williams *et al.* carefully characterized [^{3}H] DNA binding to lymphocytes and compared the radioligand binding results to parallel studies of activation of adenylate cyclase. Lymphocytes had about 75 fmole [^{3}H]DNA binding sites/mg protein, which corresponded to about 2000 receptors per lymphocyte. In preliminary studies, they found that granulocytes contained approximately similar numbers of β-adrenergic receptors, but that erythrocytes had virtually no [^{3}H]DHA binding sites. Based on the rank order of potency of adrenergic agents, lymphocytes appear to contain β_2-adrenergic receptors. Radioligand binding to β-adrenergic receptors on lymphocytes has also been demonstrated by other laboratories (1, 40, 129).

Lymphocytes are a heterogeneous population of several cell types including T-cells, B-cells, and null cells. This classification is a general one that has been subdivided based on surface antigens and antibody sites (e.g., ref. 166). Radioligand binding to "lymphocytes," therefore, is really to several distinct cell populations. Some investigators have found that the amount of cAMP accumulated after β-adrenergic stimulation varies among the several subpopulations, and the results can vary depending on the method used to separate the cells (68, 158). Does this variability in response between cell types result from variability in β-adrenergic receptors (either number or affinity)? Our impression is that inadequate data have been published to answer this question. Sheppard *et al.* (185) initially showed that T cells contain fewer β-adrenergic receptors than do B cells, but the number of patients studied was small. In a larger study, Bishopric *et al.* (16) isolated T and B cells (using adherence to nylon wool and ability to rosette with sheep red blood cells, respectively) and reported that both the binding of [^{3}H]DHA and β-adrenergic-mediated cAMP accumulation were virtually identical in the two cell populations. Because the methods used to separate T from B cells in these studies are based on differences in their membranes (e.g., adherence to nylon wool columns or to sheep red blood cells), it is perhaps not surprising that different investigators have arrived at different conclusions about expression of membrane-bound β-adrenergic receptors. Further studies using techniques that perturb membranes less (e.g., centrifugal elutriation (86)) may help resolve this issue. Clearly, if expression of β-adrenergic receptors differs among lymphocyte subpopulations, the interpretation of clinical studies will be murky. It will be difficult to know whether receptor abnormalities of lymphocytes from patients are really due to differences in receptor expression or are simply due to differences in the proportion of lymphocyte subpopulations present. Results with a homogeneous cell preparation (perhaps a specific lymphocyte subpopulation) may be much easier to interpret.

Granulocytes are usually presumed to be a fairly homogeneous cell population, comprised almost entirely of polymorphonuclear (PMN) cells. Some recent data, however, suggest that even these cells may exist in separate subpopulations, defined by the expression of F_c receptors (126). As noted above, β-adrenergic receptors of PMNs were identified in preliminary fashion by Williams *et al.* (210). Galant *et al.* (67) reported a more complete characterization of [^{3}H]DHA binding to these receptors in membranes prepared from PMNs. The receptors are of the β_2-subtype, and there are about 800 receptors per granulocyte. In further studies Galant and Allred (71) showed that GTP decreased the affinity of the receptors for agonists, that GTP was required for β-adrenergic activation of adenylate cyclase, and that there was a prominent (K_{ACT}/K_D) discrepancy between the low concentrations of agonists required to activate adenylate cyclase and the much larger concentrations necessary to compete for [^{3}H]DHA binding sites.

Two reports have appeared in which investigators have attempted to simultaneously compare β-adrenergic receptors on membranes from lymphocytes and PMNs (49, 177). In a preliminary report, Sano *et al.* (177) found that a single blood sample could be used to prepare lymphocytes and PMN membranes for studies of [^{125}I]IHYP binding; Davies and Lefkowitz (49) studied [^{3}H]DHA binding in lymphocyte and PMN membranes from a large number of control subjects (n = 70). They found 66 ± 21 (mean ± SD) fmoles [^{3}H]DHA bound/mg protein on the PMN membranes and 48 ± 21 fmoles [^{3}H]DHA bound/mg protein on the lymphocyte membranes. Although variations in the age or sex of the subjects, time of their last meal, and their consumption of tobacco, coffee, or alcohol did not influence the results, there was a wide interindividual and intraindividual variation in [^{3}H]DHA binding, a finding replicated by another group (79). As will be noted subsequently, β-adrenergic receptors on PMN and lymphocyte membranes appear to be regulated differently.

A problem with using [^{3}H]DHA is that, because of its low specific activity, a large amount of blood is required to obtain a small number of data points (Davies and Lefkowitz (49) drew 200 ml from their subjects). Sano *et al.* (177) and Aarons *et al.* (1) have shown that use of [^{125}I]IHYP allows one to study β-adrenergic receptors on peripheral blood cells using much less blood (25 to 50 ml). In preliminary studies, we have confirmed this observation using the radioligand [^{125}I]iodocyanopindolol ([^{125}I]ICYP) (P. A. Insel and M. D. Snavely, unpublished observations).

Although most studies of leukocyte β-adrenergic receptors have involved membrane preparations, a few reports suggest that one can also study these receptors on intact cells. Dullis and Wilson (58) reported that [^{3}H]DHA could be successfully used with intact PMN but only when chloroquine was included in the binding assay to block nonspecific uptake of the radioligand. These investigators confirmed most of the usual criteria for validation of a binding assay, and in addition suggested that microtubules may be involved in binding of agonists to β-adrenergic receptors in PMN. The latter observation is at variance with data obtained in other systems (108). Galant and Allred (73) have also reported [^{3}H]DHA binding to intact PMNs. Krall *et al.* (129) reported binding [^{3}H]DHA to intact lymphocytes, but their use of previously frozen cells in a hypotonic buffer makes it difficult to accept that these investigators were actually studying viable lymphocytes. Watanabe *et al.* (206) and Abrass and Scarpace (3) also demonstrated [^{3}H] DHA binding to intact lymphocytes (in saline) and showed that the affinity of agonists for these sites is much lower in intact cells than in membrane preparations. Furthermore, Watanabe *et al.* (206) also showed that addition of a cytoplasmic fraction decreases the affinity of agonists for β-receptors in membranes.

Adipocytes

Adipose tissue is the major source of energy storage in the human body and the breakdown (lipolysis) of adipocyte triglyceride is regulated by several hormones. Catecholamines are important regulators of lipolysis, but their effects are complex. Pharmacological studies have shown that adipocytes contain several types of adrenergic receptors, including β-adrenergic receptors (primarily β_1- and perhaps also β_2-receptors) that mediate stimulation of adenylate cyclase and an increase in lipolysis, and α_2-adrenergic receptors that mediate an inhibition of adenylate cyclase and a reduction in lipolysis (124, 131, 188). In addition, it has been suggested that some adipocytes contain α_1-adrenergic receptors that mediate an increase in phospholipid turnover; their effect on lipolysis is not clear (26, 75). Tissue for all these studies has been obtained either from abdominal tissue at laparotomy or from subcutaneous tissue obtained under local anesthesia.

Several characterizations of radioligand binding to adrenergic receptors on adipocyte membranes have been published. Berlan and LaFontan (14) showed that [^{3}H]clonidine binds to α_2-adrenergic receptors on these cells. Burns *et al.* (26) used three radioligands—[^{3}H]DHA, [^{3}H] p-aminoclonidine, and [^{3}H]WB4101—and interpreted the results as binding to β-, α_2- and α_1-receptors, respectively. These authors also showed that β-adrenergic stimulation (isoproterenol) increases, and α_2-stimulation (clonidine) decreases cAMP production, and that epinephrine alone (which binds to both β- and α_2-receptors) had little effect on cAMP. Epinephrine in the presence of either yohimbine or phentolamine (to block α-receptors), however, markedly stimulated cAMP. The choice of radioligands used in these studies weakens the conclusions. Because [^{3}H]clonidine is a partial agonist it may bind only to high affinity states of the α_2-receptor (184), and because [^{3}H] WB4101 may not be α_1-selective (95), it may bind to α_2-receptors. Tharp *et al.* (197) found that the α_2-selective antagonist [^{3}H]yohimbine and the nonselective antagonist [^{3}H]DHE bind to nearly the same number of sites (~145 fmoles/mg protein), thus casting doubt on the existence of α_1-receptors in adipose tissue. To directly address this point they used [^{3}H]prazosin and found no specific binding. These results are thus in conflict with the previously cited report of [^{3}H]WB4101 binding to α_1-receptors and the observed α_1-adrenergic mediated phosphoinositol hydrolysis.

It seems, then, that adipose tissue contains both β-receptors and α_2-receptors. Whether these are both present on the same individual cell and whether α_1-adrenergic receptors are found on these cells is not clear.

Placenta

Adrenergic stimulation of placenta tissue causes complex changes in synthesis and secretion of various hormones (207). Two groups have characterized the β-adrenergic receptor on human placentas using [^{3}H]DHA (207, 180). These studies were similar, but they reached opposite conclusions regarding the receptor subtype present. From competition experiments Shocken *et al.* (180) determined that the placental receptors are β_1, but Whitsett (207) determined that they are predominantly β_2.

The β-adrenergic receptor of placenta is linked to stimulation of adenylate cyclase, and receptors are found as early as 10 weeks gestation (207). Subcellular fractionation experiments suggest that the receptors are related to the fetal rather than the maternal circulation (207).

Because the placenta has many receptors (240 fmoles/mg protein) and because human placentas are readily available, the placenta may turn out to be a useful tissue from which to solubilize and purify β-adrenergic receptors. To this end Schocken *et al.* (180) found that the receptors could still be identified after solubilization with digitonin.

Other Human Tissues

Radioligand binding to adrenergic receptors on human lung, brain, and uterus have been reported. Binding of [^{125}I]ICYP to β-adrenergic receptors on human lung is very similar to results obtained with guinea pig lung (in terms of the ratio of β_1- to β_2-adrenergic receptors detectable) (61). Barnes *et al.* (11) performed detailed characterizations of [^{3}H]DHA and [^{3}H] prazosin binding to human lung tissue obtained at thoracotomy from patients with lung cancer or other lung disease. They attempted to use only relatively normal-appearing tissue and performed kinetic, saturation, and competition experiments. [^{3}H]DHA bound to similar numbers of sites in human lungs and guinea pig lungs, but [^{3}H]prazosin bound to 12-fold more sites in the human tissue. Maggi *et al.* (139) have examined [^{3}H]DHA binding to β-adrenergic receptors on the cerebellum and cerebral cortex of human brains obtained at autopsy. Finally, work from Roberts' laboratory has identified both α- ([^{3}H]DHE) and β-adrenergic ([^{3}H]DHA) receptors in human myometrium (89; personal communication).

PHYSIOLOGICAL AND PHARMACOLOGICAL CHANGES IN ADRENERGIC RECEPTORS

Work with animals and cultured cells documents that the number or affinity of adrenergic receptors can change as a function of intrauterine or postnatal development and maturation (85, 90, 174), circadian rhythm (213), hormonal status (reviewed in ref. 97), and levels of circulating or tissue catecholamines (97, 192). The influence of such factors on adrenergic receptors in humans has been evaluated to a much more limited extent; we summarize the fledgling efforts that have been undertaken thus far.

Age-Related Changes in Adrenergic Receptors

Although we are aware of no studies on developmental regulation of adrenergic receptors during fetal life or childhood in humans, recent work with animals in utero suggests that changes in β-adrenergic receptors in the lung may be related to the ability of pneumocytes to produce surfactant (32). Moreover, glucocorticoid treatment of fetal animals increases the number of these pulmonary β-adrenergic receptors (32). It will be important to determine if similar changes occur in humans, and if so, whether such changes explain the utility of glucocorticoids in treating the neonatal respiratory distress syndrome.

At the opposite end of the age spectrum, studies in animals indicate that changes in expression of β-adrenergic receptors occur with aging (174). The results in humans are inconclusive. Schocken and Roth (181) reported that the number of β-adrenergic receptors on mononuclear leukocytes from normal subjects was negatively correlated with the age of the subjects (Fig. 18.3). The reason for this negative correlation was not determined, although it is possible that differences in receptor number result from changes in the proportion of mononuclear cell subpopulations among subjects of different ages. Alternatively, it is possible that receptors of older subjects are "down regulated" by the higher plasma catecholamine levels found in the elderly (220). Others have found no age-related changes in β-adrenergic receptors on lymphocytes or on PMN (67, 49, 3). Lymphocytes from elderly subjects have been reported to have a decreased ability to stimulate adenylate cyclase (130), but this finding has been disputed (128). Maggi *et al.* (139) have reported that membranes prepared from the cerebellum of autopsy specimens of older patients have fewer [^{3}H]DHA sites than did those from younger patients. Similar changes were not observed with cerebral cortical membranes obtained from the same subjects. We have found no correlation between age and either the number or affinity of α_2-adrenergic receptors of intact platelets (150).

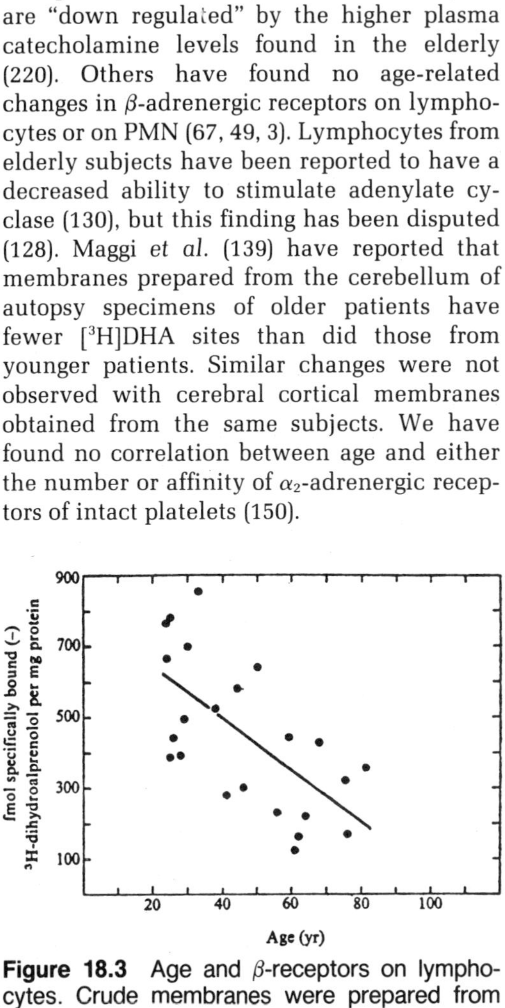

Figure 18.3 Age and β-receptors on lymphocytes. Crude membranes were prepared from mononuclear leukocytes and the B_{max} of [^{3}H]DHA binding determined. Each point represents a different donor. The correlation coefficient was 0.41 ($p < 0.001$). (Reproduced with permission from D. H. D. Schocken and G. S. Roth (181).)

Regulation of Adrenergic Receptors by Heterologous Hormones

Nonadrenergic hormones such as thyroxine, sex steroids, and glucocorticoids are well-recognized to regulate both α- and β-adrenergic receptors in several experimental systems (97). These hormones have been considered to be "gene-active," as opposed to the "membrane-active" hormones such as catecholamines, polypeptide hormones, and prostaglandins (107). Although detailed molecular studies have not yet been conducted, in at least one experimental system preliminary data indicated that glucocorticoids enhanced the number of β-adrenergic receptors by induction of receptor synthesis and not by altering receptor degradation (204). We anticipate numerous studies designed to explore mechanisms by which heterologous hormones regulate adrenergic receptors within the next several years.

THYROID HORMONE

Many of the clinical manifestations of hyperthyroidism resemble those of a hyperadrenergic state, but plasma levels of norepinephrine are decreased. Although considerable controversy exists as to whether hyperthyroidism is associated with enhanced adrenergic response, multiple reports have indicated that myocardial membranes from animals with hyperthyroidism or heart cells treated with triiodothyronine have decreased α-adrenergic receptors and increased β-receptors (reviewed in refs. 48, 200). However, β-adrenergic receptor number and isoproterenol-stimulated cAMP production are completely normal on lymphocytes from hyperthyroid patients (Fig. 18.4) (208). Scarpace and Abrass (178) investigated the contradiction between data derived in animal models of hyperthyroidism and in lymphocytes from hyperthyroid patients by simultaneously quantitating the β-adrenergic receptors on membranes prepared from the heart, lung, and lymphocytes of normal and hyperthyroid rats. These authors confirmed that myocardial β-adrenergic receptors are increased in hyperthyroidism, but they found that the receptors on lymphocyte and lung membranes were unaltered. These findings suggest that β-adrenergic receptors on lymphocyte membranes are regulated differently than those on the heart. Whether this is merely because the lymphocyte contains β_2-adrenergic receptors and the heart contains primarily

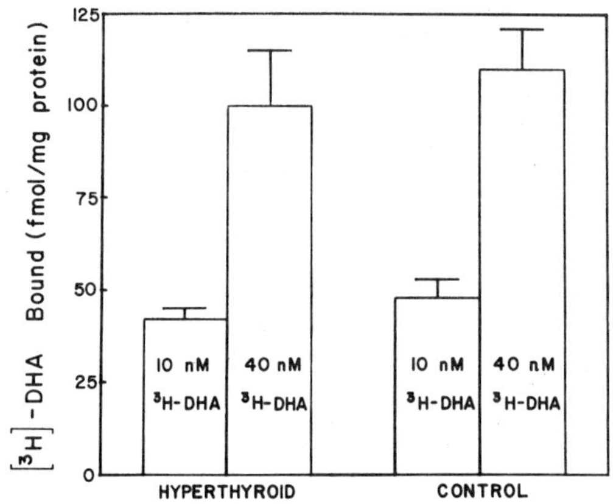

Figure 18.4. Binding of [^{3}H]DHA to lymphocyte membranes prepared from hyperthyroid patients and controls. Mean ± standard error of binding from 12 subjects in each group is shown. (Modified from R. S. Williams *et al.* 208).)

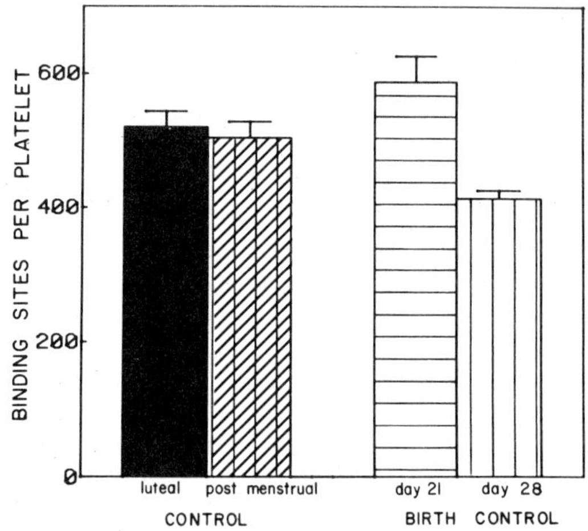

Figure 18.5. Platelet α_2-adrenergic receptors and oral contraceptives. The B_{max} for [^{3}H]DHE binding was determined in platelets from 15 women taking a combination contraceptive and from 8 controls. Then platelets of women taking birth control pills were assayed at the end of a 21-day cycle of treatment and again after 7 days withdrawal. The 15% increase in [^{3}H]DHE binding with birth control pills was significant ($p < 0.005$).

β_1-adrenergic receptors or whether there are really organ-specific changes produced by thyroid hormone is not clear. Moreover, whether such findings can be extrapolated to humans is still an open question. In a recent study, Ginsberg *et al.* (79) reported that administration of triiodothyronine (100 μg/day for 70 days) to 24 normal subjects produced laboratory evidence of hyperthyroidism and increased [^{3}H]DHA binding on lymphocyte membranes about 2-fold. Studies of α-adrenergic receptors in patients with thyroid disease have not been reported.

SEX STEROIDS

In animals, estrogen modulates expression of adrenergic receptors in the central nervous system and in peripheral tissues. For example, in rabbits treated with estrogen, the number of α-adrenergic receptors increases in myometrium in parallel with an enhancement of α-adrenergic response; the number of α-adrenergic receptors on platelets decreases from the same animals (89, 167). Thus, hormones such as estrogen may, to some degree, work through modulation of adrenergic receptors. An obvious clinical question follows: does the increased incidence of thromboembolic disease in estrogen-treated patients result from changes in α-adrenergic receptors on platelets? Peters *et al.* (164) studied the binding of [^{3}H] DHE to intact platelets of normal women and found no variation during the menstrual cycle but an increased number of [^{3}H]DHE sites in women taking oral contraceptives (Fig. 18.5). Other workers have been unable to find changes in [^{3}H]DHE binding in platelets from women taking oral contraceptives or from pregnant women (147; J. M. Roberts, personal communication). In our own studies, we noted that platelets from women generally have slightly more [^{3}H]yohimbine binding sites than do platelets from men (149; Fig. 18.2). To determine whether this is related to differences in sex steroids will require additional study. By contrast with these results for α-adrenergic receptors, no difference has been found in β-adrenergic receptor number in PMN and lymphocytes from men and women (49).

GLUCOCORTICOIDS

Adrenalectomy and glucocorticoid treatment change the expression in animals of adrenergic receptors in a tissue-specific and receptor-specific manner. For example, adrenalectomy increases hepatic β-adrenergic receptor number but decreases pulmonary β-adrenergic receptor number and has no effect on hepatic α-adrenergic receptors (87, 141). Four hours after people are given an oral dose of 100 mg cortisone, the number of [^{3}H]DHA binding sites is decreased 60% on membranes from mononuclear cells and is increased 40% on membranes from PMN (49). After 24 hours, the number of [³H]DHA binding sites on both PMN and mononuclear leukocyte membranes are about 20% higher than control (Fig. 18.6). The competition by isoproterenol for [³H]DHA sites on the PMN membranes is different in the steroid-treated subjects: the percentage of receptors in a high affinity state is somewhat higher, the affinity of the high affinity state is half normal, and the affinity of the low affinity state is reduced 6-fold (50). Since GTP is a critical determinant of the proportions of high affinity and low affinity states, it is unclear whether glucocorticoids change receptors on PMN membranes through a direct effect on membrane receptors or through indirect effects on GTP or other regulators of receptor binding. Other workers (72) have found nearly normal numbers of β-adrenergic receptors on PMN from patients receiving glucocorticoids.

Regulation of Adrenergic Receptors by Dietary Na⁺ and by Cholesterol

Little data have been presented on the role of diet in manipulating adrenergic receptors. Fraser *et al.* (66) have noted that subjects placed on 400 mEq Na^+ diets have about 50% more β-adrenergic receptors on PMN and lymphocyte membranes than do subjects consuming 10 mEq Na^+ daily (Fig. 18.7). Furthermore, the subjects on the 400 mEq Na^+ diet had lower plasma and urine catecholamines than did the subjects on the 10 mEq Na^+ diet. They also noted that the enhancement in β-adrenergic receptor number is accompanied by an increase in the sensitivity of the heart rate to infused isoproterenol in the same subjects. Similarly, Mulvihill-Wilson and Pettinger (153) found that the number of platelet α_2-adrenergic receptors of several patients with essential hypertension was twice as high when their Na^+ intake was 300 mEq/day than when it was 9 mEq/day.

Shattil and co-workers (182) have demonstrated that cholesterol content of platelet membranes can be manipulated *in vitro* by incubating platelets with liposomes having variable cholesterol/phospholipid (C/PL) ratios. Cholesterol-rich liposomes (which have high C/PL) can be used to enrich membrane cholesterol, and cholesterol-poor liposomes (low C/PL) can be used to deplete membrane cholesterol. Enrichment of membrane cholesterol strikingly enhances platelet aggregation in response to epinephrine, ADP, and other compounds, while depletion of membrane cholesterol decreases aggregation by such compounds (182). Such findings may offer information about mechanisms of enhanced platelet aggregation that have been reported in some hypercholesterolemic patients (31). [³H]DHE binds to α-receptors on both cholesterol-rich and cholesterol-depleted platelets with similar kinetics and affinities (110). Platelet α-

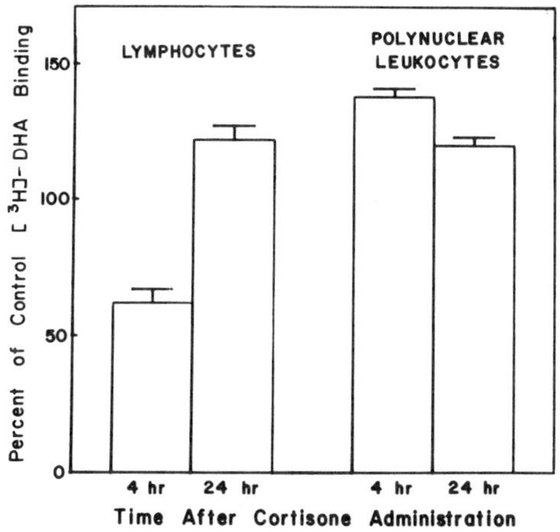

Figure 18.6. Binding of [³H]DHA to membranes prepared from lymphocytes and PMN obtained from normal individuals 4 and 24 hours after cortisone administration. The error bars are the standard error of the mean for 8 to 19 patients. The number of [³H]DHA sites was determined by Scatchard analysis. (Modified from A. O. Davies and R. J. Lefkowitz (49).)

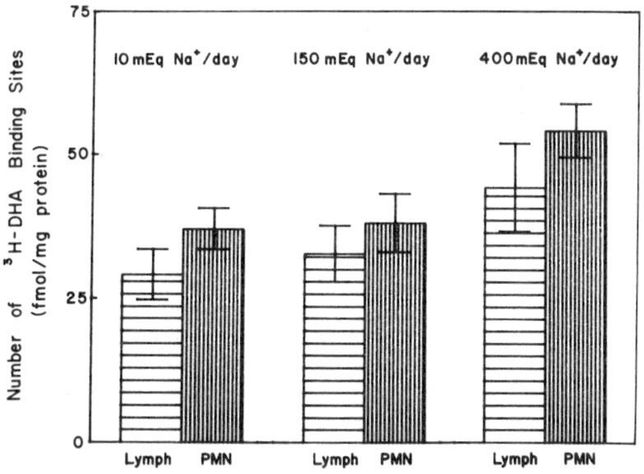

Figure 18.7. Effect of sodium intake on the number of leukocyte β-adrenergic receptors. Normotensive healthy volunteers were placed on diets containing 10, 150, or 400 mg/day for 10 days, and membranes prepared from their lymphocytes or monocytes were used in binding isotherm experiments with [³H]DHA. Data were analyzed using Scatchard plots. Results are the mean ± SE. (Data from Fraser *et al.* (66).)

adrenergic receptor number and α-adrenergic-mediated inhibition of adenylate cyclase are also not altered by cholesterol enrichment (110). Thus, alterations in α-adrenergic receptors do not appear to account for the altered platelet response accompanying cholesterol enrichment or depletion.

Regulation of Adrenergic Receptors by Catecholamines: Refractoriness (Desensitization)

Physiological responses of target cells exposed to a hormonal agonist often decline during continued exposure to agonist. This phenomenon has been variously termed refractoriness, desensitization, tolerance, or tachyphylaxis. Commonly, the decrease in response to agonist is accompanied by a decrease in the number of available receptors (termed "downregulation" of receptors) or by a decrease in the affinity of the receptors for agonists (termed "uncoupling"). Refractoriness to catecholamines, especially for β-adrenergic stimulation, has been well studied in a variety of experimental systems (reviewed in ref. 48), and some data have recently accrued regarding "downregulation" and "uncoupling" in studies of human tissues.

β-ADRENERGIC RECEPTORS

Galant *et al.* (69) were the first to report that therapy of humans with adrenergic agonists produced downregulation of β-adrenergic receptors on PMN membranes. These investigators found that treatment of normal subjects with 10 mg terbutaline daily decreased [^{3}H] DHA binding of PMN when binding was initially evaluated after 3 days of treatment. By 1 week of treatment, [^{3}H]DHA binding decreased about 75%, and this returned to control levels over several days after withdrawal of the β-agonist (Fig. 18.8). In a larger group of asthmatic patients receiving β-adrenergic agonists chronically, Galant *et al.* (72) found that [^{3}H]DHA binding to PMN was decreased about 70% compared to either controls or to asthmatics not receiving adrenergic agonists. Moreover, they showed that the decrease in [^{3}H]DHA binding reflected a decrease in β-adrenergic number without a change in receptor affinity for agonists or antagonists and that β-adrenergic-mediated adenylate cyclase activity was decreased in PMN membranes from patients who received the adrenergic agonists.

Tohmeh and Cryer (196) have studied β-adrenergic receptors on lymphocyte membranes during infusions of isoproterenol and found a biphasic response (Fig. 18.9). After infusions of 1 hour, the number of [^{3}H]DHA sites nearly doubled, but after a 4-hour infusion the number of receptors decreased to about 50% of control levels. Saturation binding curves indicated that the number, and not the affinity, of the [^{3}H]DHA sites was altered during both phases of the infusion. These authors also assessed adrenergic responsiveness of the heart by measuring the increase in pulse rate after a further bolus of isoproterenol. During the period when the number of receptors in the lymphocytes was increased, the isoproterenol-stimulated increase in pulse rate (17 beats/min) was about twice that seen before the infusion (9 beats/min).

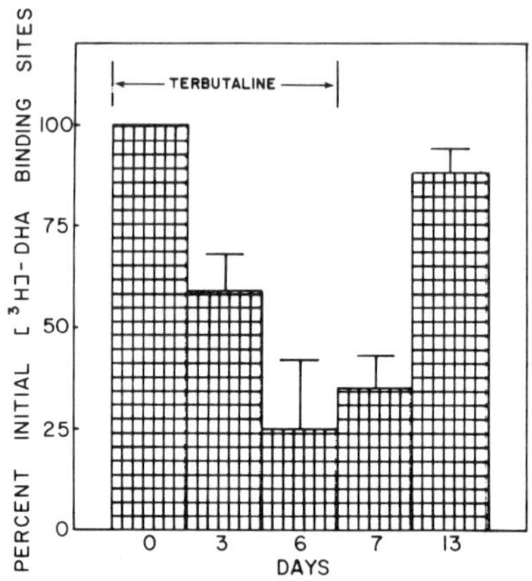

Figure 18.8. Effect of terbutaline administration on [^{3}H]DHA binding to PMN membranes. One asthmatic and two normal individuals were treated with terbutaline (2.5 mg four times a day) for 6 days, and specific [^{3}H]DHA (30 nM) binding to PMN was determined before, during, and after therapy. The data are expressed as percent of initial [^{3}H]DHA binding and are means ± standard error. (Data from Galant *et al.* (69).)

Krall *et al.* (129) studied refractoriness and downregulation of lymphocytes *in vivo* and *in vitro*. They confirmed that lymphocytes of subjects receiving isoproterenol infusions had decreased numbers of [^{3}H]DHA binding sites. They also exposed lymphocytes to isoproterenol *in vitro* and found that both cAMP production and the number of [^{3}H]DHA binding sites was decreased. The concentration of isoproterenol needed for these two effects was quite different, with as little as 0.01 nM isoproterenol decreasing subsequent cAMP production but with much greater concentrations of iso-

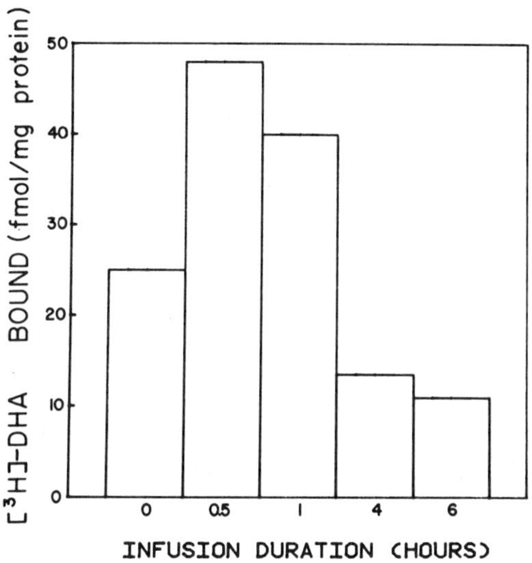

Figure 18.9. Specific [^{3}H]DHA binding to lymphocyte membranes during an infusion of isoproterenol (1 μg/min). Binding isotherms were performed with lymphocyte membranes obtained at various times during an isoproterenol infusion. Mean ± standard error of seven subjects is plotted. (Modified from J. F. Tohmek and P. E. Cryer (196).)

proterenol required to decrease the number of receptors. Thus, downregulation is not the only explanation for refractoriness in lymphocytes incubated *in vitro* with β-adrenergic agonists.

Fitzgerald *et al.* (63) have recently reported in preliminary fashion a different form of "downregulation," the decrease in binding to receptors that occurs with upright posture or exercise. These authors examined [^{3}H]DHA binding to lymphocyte membranes and compared exercise and posture-induced changes in plasma catecholamines with changes in binding to receptors. After 15 minutes of exercise, subjects had several-fold increases in plasma catecholamines and about a 40% decrease in receptor number. Although the catecholamines had nearly returned to resting levels by 1 hour, the decreases in receptor binding persisted. These findings of decreases in catecholamine receptors with a brief period of exercise contrast with reports of unchanged numbers of receptors on lymphocytes from patients who undertook chronic exercise (212).

Still another example of downregulation of β-adrenergic receptors is the study of Colucci *et al.* (38). They treated patients with congestive heart failure with pirbuterol, a β-adrenergic agonist. The patients initially responded well to the drug (as assessed clinically and by measuring the cardiac index), but after 1 month of therapy no clinical benefit was apparent. The number of β-adrenergic receptors on these patients' lymphocytes was about half normal, and the authors speculate that a similar downregulation of receptors on the heart or vasculature may account for the refractoriness to the drug.

α-ADRENERGIC RECEPTORS

Cooper *et al.* (42) reported that aggregation of platelets was diminished after platelets were incubated 4 hours with 100 μM epinephrine. They compared the binding isotherm of [^{3}H] DHE of epinephrine-preincubated and control platelets, and concluded that epinephrine had decreased the number of α-adrenergic receptors (*i.e.*, "downregulation" had occurred). We have recently attempted to confirm these observations (121). We examined the binding of [^{3}H]yohimbine to platelets washed 4 times after incubation with epinephrine for 4 hours *in vitro*, and found a decrease in binding when we used a single [^{3}H]yohimbine concentration. This decrease occurred only when the platelets had been incubated with >10 μM epinephrine, a concentration far in excess of physiological levels (≤0.01 μM). Binding isotherms indicated that the number of binding sites was

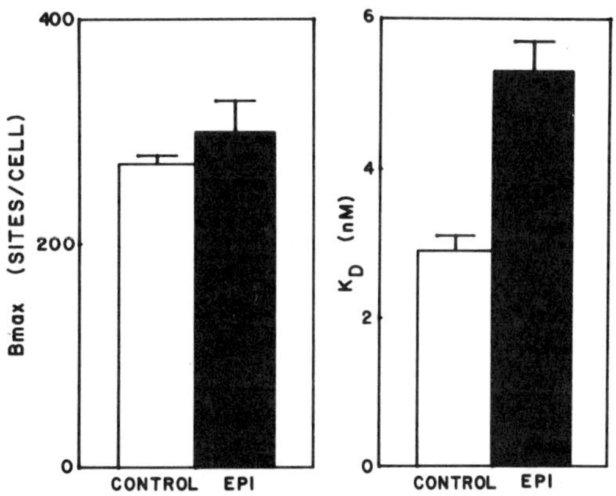

Figure 18.10. Lack of "downregulation" of α_2-adrenergic receptors of human platelets. Platelet-rich plasma was incubated with ("EPI") or without ("Control") 100 mM epinephrine for 4 hours, the platelets were washed four times, and [^{3}H]yohimbine binding isotherms were performed to determine the B_{max} (*left panel*) and the K_D (*right panel*). The results shown are the mean and standard error of four experiments. (Data from J. S. Karliner (121).)

similar in control and epinephrine-preincubated platelets but that the apparent K_D for [^{3}H]yohimbine was 2-fold higher in the epinephrine-preincubated platelets (Fig. 18.10). Further studies indicated that platelets accumulate enough epinephrine during these preincubations to account for the observed decrease in [^{3}H]yohimbine binding. Thus, retained epinephrine and not downregulation of α_2-adrenergic receptors explains our results. Some mechanism other than downregulation must therefore explain the refractoriness to epinephrine that platelets have been reported to demonstrate (42).

Our studies of possible downregulation of platelet α-adrenergic receptors illustrate the difficulty of distinguishing "true" downregulation from "pseudo" downregulation caused by retained agonist. Figure 18.11 shows that different binding isotherms will be obtained in these two settings. In "true" downregulation, one will see a decrease in maximal binding without a change in affinity (K_D). By contrast, in the setting of "pseudo" downregulation, the radioligand must compete with the retained agonist for binding sites. High concentrations of radioligand will bind to all the receptors, but the binding of lower concentrations of radioligand will be reduced by the retained agonist. Thus the B_{max} will be unaltered by retained agonist, but the apparent K_D will be increased. In addition to doing full binding isotherms (as opposed to single point determinations), investigators may undertake other maneuvers to help uncover "pseudo" downregulation. As noted above, guanyl nucleotides decrease the affinity of β-adrenergic receptors in most tissues (including lymphocyte and PMN membranes) and of β_2-adrenergic receptors in several tissues (including platelets), and similarly, Na^+ decreases affinity of agonists in binding to α_2-adrenergic receptors of platelets. Thus, if guanyl nucleotides (and, for α_2-receptors, Na^+) are included in binding studies, the affinity of the receptors for agonists is decreased, and the binding of any retained agonist is thereby minimized. This trick was utilized by us (121) to study [^{3}H]yohimbine binding to platelets preincubated with epinephrine and by Aarons *et al.* (2) to study refractoriness of lymphocyte β-adrenergic receptors produced by β-adrenergic agonists.

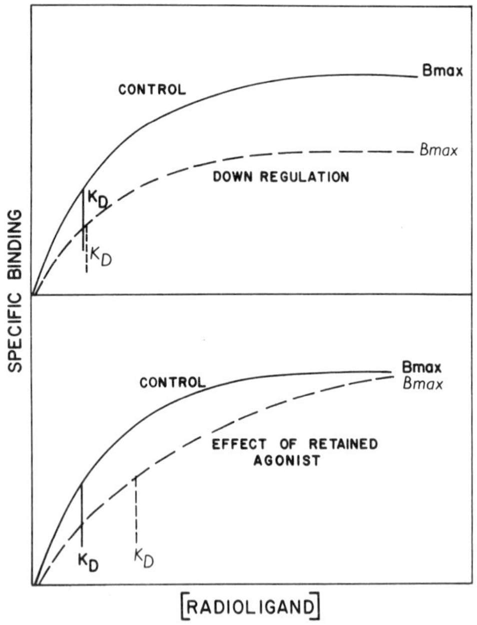

Figure 18.11. Theoretical curves for "true" down regulation of receptors (*top*) and "pseudo" down regulation resulting from retained agonist (*bottom*). The binding isotherms of down regulated receptors and receptors to which retained agonists are bound are quite different. With down regulation the B_{max} is decreased without a change in K_D; with retained agonist the B_{max} is unchanged and the K_D is increased. Binding with a single concentration of radioligand, however, may not differentiate these two effects.

The *in vitro* studies cited above lead to the conclusion that downregulation of platelet α_2-adrenergic receptors does not occur. We have also found that treatment of hypertensive patients with guanabenz, a partial α_2-adrenergic agonist in platelets (224), does not alter the number of α_2-receptors on the patients' platelets (150).

Regulation of Adrenergic Receptors by Catecholamines: Supersensitivity ("Upregulation")

Supersensitivity is the converse of the refractoriness following continued exposure of target tissues to increased concentrations of adrenergic agonists. Depletion of catecholamines results in a supersensitive response in the target tissue. Multiple mechanisms appear to be involved in this supersensitivity, including decreased catecholamine uptake, but recent attention has focused on "upregulation" of receptors—an increase in receptor number. In experimental animals, both α- and β-adrenergic receptors increase in number in settings of surgical or chemical sympathectomy (9, 202); studies of possible upregulation following receptor blockade have yielded more equivocal results (7, 80).

An important clinical setting, which has been invoked as possibly resulting from supersensitivity to catecholamines because of upregulation, is the "propranolol withdrawal syndrome." Rapid withdrawal of propranolol may be associated with enhanced risk of serious arrhythmias, myocardial ischemia, and possibly infarction (reviewed in ref. 78). Some, but not all, studies of animals treated with propranolol have indicated that therapy with the β-adrenergic antagonist increases the number of myocardial β-adrenergic receptors (7, 80). In studies of humans, Aarons *et al.* (1) gave healthy volunteers 160 mg propranolol for each of 7 days and found that the number of [^{125}I]IHYP binding sites in lymphocyte membranes increased (Fig. 18.12). Following withdrawal of propanolol, receptor number returned to control values within 4 days, but the plasma levels of propranolol decreased to zero within 24 hours. Thus, for several days following propranolol withdrawal, the number of receptors was increased while no propranolol was present. Moreover, during this period the orthostatic change in pulse and "double product" (pulse × blood pressure, used as an index of myocardial oxygen consumption) was increased over normal values, and the magnitude of these increases was correlated with the increase in β-adrenergic receptor number. Similar results have been observed by Fraser *et al.* (66), who also noted that the magnitude of the increase in receptor number during propranolol treatment was correlated with pretreatment levels of catecholamines. These findings are thus consistent with the possibility that increased numbers of β-adrenergic receptors may contribute to the heightened adrenergic state of patients during the first few days after withdrawal from propranolol. These findings conflict, however, with those of Goldstein *et al.* (82), who found a small decrease in [^{3}H]DHA binding to leukocytes during propranolol therapy, and found no rebound after the drug was withdrawn.

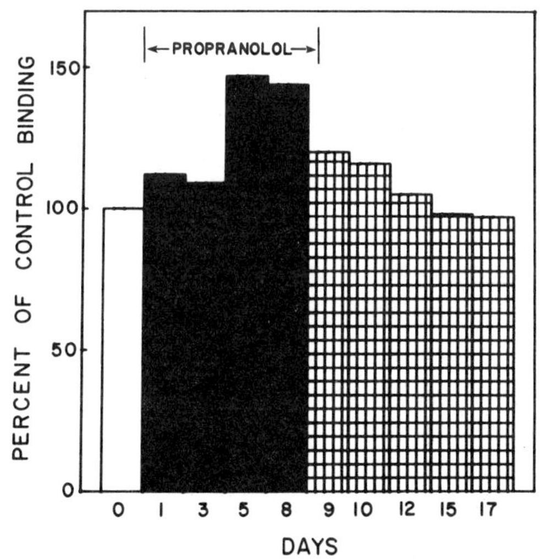

Figure 18.12. Effect of propranolol treatment on [^{3}H]DHA binding to lymphocyte membranes. Healthy volunteers were given propranolol (40 mg every 6 hours) for 8 days, and binding isotherms of [^{3}H]DHA binding to lymphocytes membranes were performed before, during, and after therapy. Shown are the mean data from 12 subjects. (Modified from R. D. Aarons *et al.* (1).)

Whether similar phenomena might be occurring when α-adrenergic antagonists are administered has not been documented. Nevertheless the possible existence of such syndromes has been suggested in patients receiving α-adrenergic agents, such as clonidine and prazosin (78).

Modulation of Adrenergic Receptors by Other Drugs

PENICILLINS

Carbenicillin or penicillin G can cause a bleeding diathesis as a result of platelet dysfunction. Shattil *et al.* (183) incubated platelets *in vitro* with high concentrations of both drugs and showed that platelet function, as assayed by aggregation and [^{14}C]serotonin release, was abnormal. The number of [^{3}H]DHE binding sites was normal, but the affinity of the receptors for either [^{3}H]DHE or epinephrine (in competition for [^{3}H]DHE binding) was decreased (Fig. 18.13). As the equilibrium dissociation constant for [^{3}H]DHE is the ratio of the rate of dissociation ("off rate") to the rate of association ("on rate"), decreased affinity for [^{3}H]DHE ought to be reflected in either or both rates. After preincubation of platelets with penicillin, the on rate of [^{3}H]DHE binding was decreased and the off rate was normal. Moreover, platelets that had been pretreated with antibiotics had several other abnormalities: decreased aggregation in response to ADP, decreased incorporation of an ADP affinity probe,

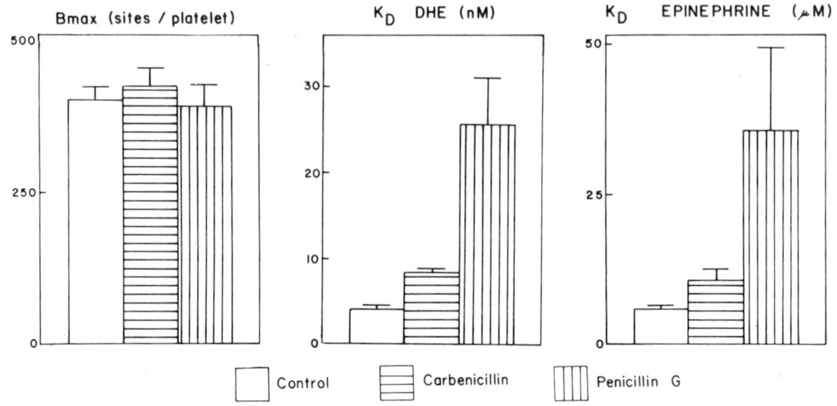

Figure 18.13. Effect of carbenicillin and penicillin on [^{3}H]DHE binding to intact platelets. Platelets were incubated with carbenicillin (36 mM), or penicillin G (31 mM) and the B_{max} for [^{3}H]DHE binding, and the dissociation constants of [^{3}H]DHE and for epinephrine competition of [^{3}H]DHE binding determined. (Data from Kaywin *et al.* (125).)

and altered interaction with von Willebrand factor. Thus several agonists (epinephrine, ADP, von Willebrand factor) interact with the platelets abnormally after the preincubation with the antibiotics. The common molecular mechanism is not known.

TRICYCLIC ANTIDEPRESSANTS

Tricyclic antidepressants such as imipramine block the reuptake of norepinephrine into nerve terminals. The ability of these drugs to block neuronal uptake, however, is poorly correlated with their clinical efficacy, so other mechanisms have been sought to explain their action. Some pharmacological studies indicate that tricyclic antidepressants block presynaptic α_2-receptors (37), although in radioligand binding studies imipramine has a low affinity for α_2-receptors (at least in platelets where the K_D of imipramine is 7 μM (151)). Treatment with imipramine decreases by ~35% the number of [^{3}H]clonidine binding sites present in rat brain and human platelets (77). Whether this is due to a direct or indirect effect of imiprimine on the α_2-receptors is not clear. It should be noted that both rat cerebral cortex (165) and human platelets (21, 162) also contain high affinity (K_D ~2 nM) binding sites for [^{3}H]imipramine, but because yohimbine competes poorly for these [^{3}H]imipramine sites (133) and because imipramine competes poorly for [^{3}H]yohimbine sites (151), it seems likely that the two sites are distinct.

An additional effect of tricyclic antidepressants is their ability to decrease the number of β-adrenergic, in particular β_1-adrenergic, receptors in the central nervous system (8). This effect, an apparent "downregulation" of receptors that is thought to result from increased synaptic concentrations of norepinephrine, has not thus far been studied in humans.

INDOMETHACIN

Indomethacin decreases prostaglandin synthesis. Güllner *et al.* (88) gave 2 mg/kg/day of indomethacin to normal volunteers and assessed [^{3}H]DHE binding to α-adrenergic receptors on platelets and [^{3}H]DHA binding to β-adrenergic receptors on leukocytes. The platelet binding did not change, but there was a 90% increase in [^{3}H]DHA binding to the leukocytes. The authors hypothesized that indomethacin decreased release of vasodilator prostaglandins, which then led to a decrease in sympathetic tone, and in turn to an "upregulation" of the β-receptors, but additional evidence will be required to substantiate this hypothesis.

ADRENERGIC RECEPTORS IN DISEASE

Disorders in catecholamine action have been hypothesized to occur in a variety of diseases, and accordingly, investigators have begun to explore possible abnormalities of adrenergic receptors in those settings. Because it has not been possible to assess adrenergic receptors on the involved internal tissues, blood cells have usually been used to examine changes in receptors. Little data are available to assess how well adrenergic receptors on blood cells resemble receptors on internal tissues. With other types of hormone receptors (and especially insulin receptors), changes in receptors on blood cells seem to be reasonably closely related to changes on the internal tissues of interest. In using receptors on blood cells as models of receptors elsewhere, one encounters an important difference between endocrine receptors (such as insulin receptors) and neuroendocrine receptors (such as adrenergic receptors). Endocrine receptors on blood cells and internal tissues are likely to be exposed to similar steady-state concentrations of the hormone (although under non-steady-state conditions differences may occur in hormonal levels in vascular and extravascular compartments (109)). Neuroendocrine receptors on blood cells, however, are exposed only to circulating neurotransmitter, while neuroendocrine receptors on internal tissues are exposed (even under steady-state conditions) to both circulating and neuronally released neurotransmitter. Thus studies on peripheral blood cells cannot assess the contributions of neuronal and other local factors that modulate receptors in catecholamine target tissues. Nevertheless, changes in blood cell adrenergic receptors are associated with several diseases.

Allergic Conditions and Pulmonary Disease

Szentivanyi (193) originally proposed that atopic diseases, in particular asthma, may be caused by an alteration in the balance between α- and β-adrenergic receptors in various target tissues, such as bronchial smooth muscle. Results of some, but not all, radioligand binding studies support this idea.

ASTHMA

Clinical studies have suggested that asthmatics have a relative decrease in β-adrenergic responsiveness (119, 39). Several recent studies have examined changes in β-adrenergic receptors in asthmatic patients and have attempted to separate the contribution of the disease from that of treatment. Galant *et al.* (69, 72) initially reported that therapy with adrenergic agonists, rather than asthma itself, appeared to be responsible for a decrease in the number of [^{3}H]DHA binding sites on PMN. Although these investigators could not absolutely rule out the possibility that severe asthmatics might have fewer β-adrenergic receptors than control subjects or mild asthmatics, the findings indicated that β-adrenergic receptors, as assessed on PMN, were essentially normal in asthmatics. These findings are at variance with several reports showing that lymphocytes of asthmatics have decreased numbers of β-adrenergic receptors. Brooks *et al.* (23) reported that the number of [^{3}H]DHA sites on lymphocytes from asthmatic patients correlates with the severity of the clinical disease, but these results were unusual in that the binding isotherm indicated a much lower affinity for [^{3}H] DHA (even in lymphocyte membranes from normal subjects) compared to previous reports from other laboratories. Kariman (119) found that [^{3}H]DHA binding was reduced to lymphocytes of both treated and untreated asthmatics, but he used only a single concentration (12 nM) of [^{3}H]DHA. In an abstract, Sano *et al.* (177) reported that lymphocytes, and not PMNs, from asthmatics had abnormally low numbers of [^{125}I]IHYP binding sites and that β-agonist therapy decreased the binding in both cell types. Thus, two studies show that β-adrenergic receptors on PMNs are normal in untreated asthmatics, and several studies show that asthmatics have reduced numbers of β-adrenergic receptors on lymphocytes (Fig. 18.14). Which cell type is likely to be a better model for the lung? Since both cell types contain predominantly β_2-adrenergic receptors, as does the lung, there is no obvious answer.

Szentivanyi *et al.* (194) have reported preliminary studies using both lymphocytes and pulmonary tissue from asthmatic patients; they found both decreased β-adrenergic receptors and increased α-adrenergic receptors in both sites. Similarly, Sano *et al.* (176) found increased numbers of α-adrenergic receptors and decreased numbers of β-adrenergic receptors ([^{3}H]DHA) on lymphocytes from asthmatics. (As noted above, it is unclear if, in fact, lymphocytes contain α-adrenergic receptors.) Similar findings were obtained with lungs from sensitized guinea pigs which have a 2-fold increase in the number of [^{3}H]prazosin (α_1) binding sites and a slight (~20%) decrease in [^{3}H]DHA sites (10). Interestingly these changes altered the ratio of α to β receptor number from 1:30 in controls to 1:12 in asthmatic animals. These findings are of interest because of the evidence that asthmatic patients may have enhanced α-adrenergic responsiveness (92).

A novel and potentially important observation regarding atopic disease was recently re-

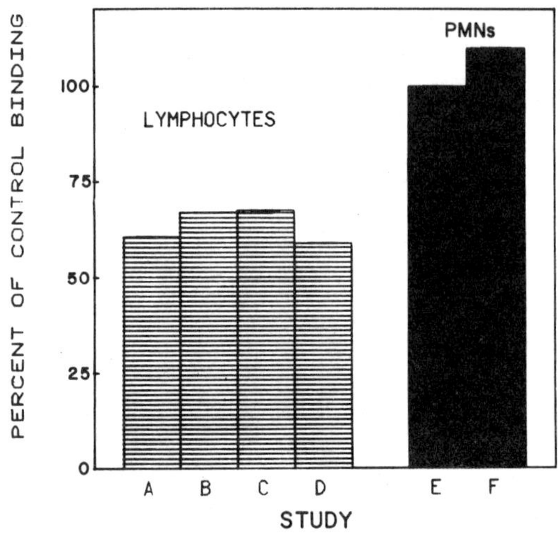

Figure 18.14. β-Adrenergic receptors on white blood cells from asthmatics. Results of six published studies of radioligand binding to white cells of untreated asthmatic patients are shown. In each study the results were compared to those of a control group, and the ratio is plotted here. Studies *A* to *C* were [^{3}H]DHA binding (at a single concentration) to membranes from lymphocytes; study *D* is [^{125}I]IHYP binding to membranes from lymphocytes; studies *E* and *F* are binding of [^{3}H] DHA to membranes from PMNs. (*A*, Data from Szentivanyi *et al.* (194); *B*, Data from Brooks *et al.* (23); *C*, Data from Kariman (119); *D*, Data from Galant *et al.* (12); *E*, Data from Galant *et al.* (69).)

ported by Venter *et al.* (205). These investigators demonstrated that serum from small numbers of patients with allergic rhinitis or asthma contained autoantibodies that appeared to be preferentially directed against β-adrenergic receptors. These autoantibodies were of reasonably high titers (1:60 to 1:130) and were able to precipitate solubilized β-adrenergic receptors from dog lung and to block [^{125}I]IHYP binding to membrane-bound receptors from lung and placenta, but not from heart. This latter finding suggested that the antibody was likely to be specifically directed at the β_2-adrenergic receptors and moreover, at the receptor binding site for [^{125}I]IHYP. These are intriguing observations that hint at a possible receptor-directed autoimmunity in atopic disease. In further studies with larger numbers of patients, however, the same investigators found autoantibodies in a similar proportion of normal and asthmatic subjects (65). Further studies are needed to clarify the situation.

ECZEMA

Atopic eczema is a second allergic condition in which decreased β-adrenergic response has been observed (175). Two groups have reported that the number of [^{3}H]DHA or [^{125}I] IHYP binding sites and their affinity for the respective ligands are unaltered in PMN from patients with atopic eczema (27, 70). Neither group reported the affinity of agonists for the radioligand binding sites.

CYSTIC FIBROSIS

Patients with cystic fibrosis have abnormal autonomic regulation. Davis *et al.* (51) showed that 50% more isoproterenol is required to raise the heart rate in patients with cystic fibrosis than in controls and that 50% less phenylephrine is needed to dilate the pupils of the patient. These results suggest that patients with cystic fibrosis may have a decreased response to β-adrenergic stimulation and an increased response to α-adrenergic stimulation. Furthermore, both lymphocytes and PMNs from cystic fibrosis patients produce only half the normal amount of cAMP when stimulated by isoproterenol (52). Galant *et al.* (74) have confirmed that cAMP production is low in PMNs from patients with cystic fibrosis and showed that this could be explained by a 50% decrease in the number of [^{3}H]DHA binding sites in the PMNs of these patients.

Psychiatric Disease

SCHIZOPHRENIA

The possibility that adrenergic receptors are altered in psychiatric patients has been examined in one study. Kafka *et al.* (118) studied [^{3}H]DHE binding and cAMP production in platelets from schizophrenic subjects and controls. Although the maximal cAMP production is somewhat lower in platelets from male schizophrenics, no overall pattern of differences was apparent in this study.

DEPRESSION

Although the "catecholamine hypothesis" for the etiology of depression is popular, we know of only one preliminary radioligand binding study assessing adrenergic receptors in depressed individuals. Garcia-Sevilla *et al.* (76) found that the number of [^{3}H]yohimbine binding sites on platelet membranes was 21% higher in patients with a "major depressive disorder" than in controls, and that this decreased to normal with lithium treatment. [^{3}H] Imipramine binding sites are decreased in the platelets of depressed patients (22), but the physiological role of these sites and their relationship, if any, to adrenergic receptors is not clear.

Cardiovascular Disease

Several publications have appeared on radioligand binding to tissues of animals with heart failure, ischemia, and hypertension. Hearts of animals with congestive failure produced by aortic constriction exhibit altered radioligand binding: increased [^{3}H]DHA binding in rats (135) and increased [^{3}H]DHA and [^{3}H]prazosin binding in guinea-pigs (120). Ischemic myocardial tissue (produced by ligating the left anterior descending artery) has twice the normal number of [^{3}H]prazosin binding sites, and this increase slowly reverses after reperfusion (43). Similar data were obtained by another group for [^{3}H]DHA binding in ischemic myocardium (152). Others have observed a decreased number and affinity of β-adrenergic receptors ([^{3}H]DHA sites) in nonischemic surviving right ventricular membranes several days following left ventricular infarction (13). Because hypertension may involve abnormalities in catecholamines and because drugs that alter the adrenergic system are the mainstay of antihypertensive therapy, many studies of radioligand binding to tissue from animals with experimental or genetic hypertension have been done. Although it is somewhat overly simplistic to generalize, in each study the receptor number is altered (in general, a decrease in β-adrenergic receptors, and an increase in α-adrenergic receptor number) in at least one "target tissue' (*i.e.*, kidney, heart, CNS, vascular tissue) likely to be involved in development or maintenance of the hypertensive state (*i.e.*, 215, 216). The findings tend to be dependent on the experimental model and species chosen for investigation. Radioligand binding experiments of humans with cardiovascular disease have been more limited and are summarized below.

HYPERTENSION

Kafka *et al.* (116) studied radioligand binding to blood cells of hypertensives and controls. No differences were found in [^{3}H]DHA binding to lymphocytes; binding of [^{3}H]DHE to α-adrenergic receptors on platelets was increased 60% in female, but not in male, hypertensives. We have performed a more extensive study of [^{3}H]yohimbine binding to platelets from untreated hypertensive men (Fig. 18.15). We compared 19 hypertensives and controls and

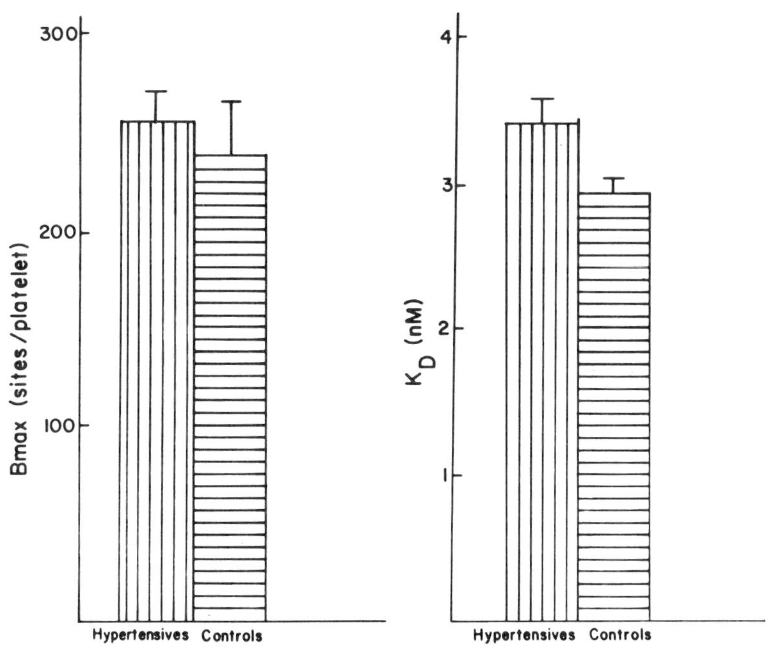

Figure 18.15. Platelet α_2-receptors in platelets of hypertensives. Saturation binding isotherms of [^{3}H]yohimbine were performed using platelets obtained from fasting supine men with essential hypertension or from fasting supine controls. There is no difference in K_D or B_{max} between hypertensives and controls. (Data from Motulsky *et al.* (150).)

found no differences in number of receptors or receptor affinity for [^{3}H]yohimbine or epinephrine (150). We followed many of these patients while they were treated with either the β-adrenergic antagonist propranolol or the α_2-adrenergic agonist Guanabenz and found no changes in the number or affinity of their receptors during therapy (150).

POSTURAL HYPOTENSION

Chobanian (34) has studied four patients with idiopathic orthostatic hypotension and found increased [^{3}H]DHE binding to platelet membranes and increased [^{3}H]DHA binding to PMNs. After treatment with ergotamine bitartrate the [^{3}H]DHE binding to platelets fell dramatically. Similarly, Hui and Connolly (104) found that lymphocytes from a patient with postural hypotension due to metastatic cancer had twice the normal number of β-adrenergic receptors.

CONGESTIVE HEART FAILURE

Colucci *et al.* measured [^{3}H]DHA binding to lymphocytes of patients with congestive heart failure and controls. Although there was a large amount of overlap between the two groups, the mean number of [^{3}H]DHA sites was 15% lower in the lymphocytes of patients than in those of controls (38). Conceivably, this small decrease might result from "downregulation" of receptors by the increased catecholamines found in patients with congestive heart failure (see Chapter 20).

Neoplasia

As we have shown, receptors are dynamic entities whose number and characteristics seem to be tightly regulated. The growth of neoplastic cells is not regulated normally. Is the expression of adrenergic receptors also abnormal? These are two examples of abnormal adrenergic receptors in human neoplastic disorders.

ESSENTIAL THROMBOCYTHEMIA

This is a myeloproliferative disorder characterized by an abnormally high number of platelets that do not function normally. Kaywin *et al.* (125) studied the platelets of several of these patients. In two patients whose platelets did not aggregate when exposed to epinephrine, the number of [^{3}H]DHE binding sites was reduced by about 50% on both intact platelets and platelet membranes (Fig. 18.16). The affinity of both [^{3}H]DHE and epinephrine for these sites was normal. The platelets of several other patients with essential thrombocythemia responded normally to epinephrine, and the number of [^{3}H]DHE binding sites on these platelets was normal. This was the first report to indicate that decreases in platelet α-adrenergic receptors could lead to decreased platelet aggregation. It was surprising that such a large loss of function could be caused by only a 50% decrease in receptor number. To prove that this is indeed possible, the authors looked at the aggregation of platelets in which half the receptors were bound to [^{3}H]DHE (by preincubation with 5 nM [^{3}H] DHE—which is the dissociation constant, or the amount needed to bind to half the receptors). The rate of aggregation was substantially diminished in these platelets when they were later exposed to epinephrine, lending credibility to the idea that a loss of receptors may be the cause of the loss of function in patients with essential thrombocythemia.

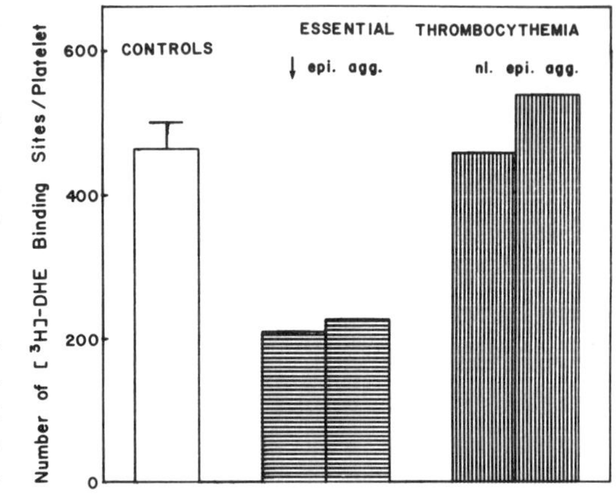

Figure 18.16. Specific binding of [^{3}H]DHE to platelets obtained from nine controls, two patients with essential thrombocythemia whose platelets exhibit decreased epinephrine aggregation, and two patients with essential thrombocythemia whose platelets exhibit normal epinephrine aggregation. Results were obtained from binding isotherms using intact platelets. (Data from Kaywin *et al.* (125).)

LEUKEMIA

The lymphocytes of patients with chronic lymphocyte leukemia were reported to have only one-third the number of [^{3}H]DHA binding sites as do normal lymphocytes (185). The loss in receptors was also reflected in a marked decrease in the isoproterenol-stimulated cAMP production of those lymphocytes. When adenylate cyclase was stimulated instead with sodium fluoride, which activates cyclase distal to β-adrenergic receptors, the cyclase activity was normal. Thus these neoplastic cells are deficient in β-adrenergic receptors, and this deficiency is reflected in blunted response to adrenergic stimulation.

OTHER USES FOR RADIOLIGAND BINDING

Radioligand binding is a technique devised to study receptors, and almost all studies using this method have been designed to characterize and ask questions about receptors. Radioligand binding assays can also be adapted to ask questions about drugs that bind to these receptors.

Radioreceptor Assay

In competitive binding experiments, known quantities of an unlabeled drug compete with a radioligand for binding to a receptor. From the shape of the binding curve the affinity of the receptor for the unlabeled drug can be determined. When the quantity of unlabeled drug is unknown but the affinity of the receptor for that drug is known, the same experiment becomes a radioreceptor assay. A standard tissue source and a standard amount of radioligand are incubated with the unknown sample, and the specific radioligand binding determined. The more drug present in the sample, the less the radioligand can bind; the amount of radioligand binding can thus be related to the amount of the drug in the sample. This method has been used to measure propranolol in human serum (106, 156, 171). The procedures used vary (plasma incubated with [^{3}H]DHA and bovine lung membranes (156), plasma dialysate with [^{3}H]DHA and calf cerebellar membranes (106), or extracted plasma with [^{125}I]IHYP and turkey erythrocyte membranes (171)), but all are reported to be sensitive and precise.

A feature of radioreceptor assays is that the property of the drug that is measured—its ability to bind to a receptor—is exactly the property that makes the drug clinically useful. This also means that active metabolites of the drug are measured as well. In the case of the propranolol assays, 4-hydroxypropranolol is also measured, unless the metabolite in the sample is first oxidized with hydrogen peroxide, in which case only propranolol itself is measured (171).

Because of their convenience, specificity, and precision, radioreceptor assays may come into general use for a variety of drugs. The principle is similar to that of radioimmunoassays, and the potential utility is as great.

Assessing a Drug's Mechanism of Action

Experiments based on radioligand binding can give new insights into the mechanism of drug action. An example is the recent findings with the "calcium channel-blocker" verapamil, which is used in the treatment of a variety of cardiovascular diseases (93). Work with several experimental systems has shown that verapamil, or its close methoxy derivative D600, competitively blocks radioligand binding to muscarinic, α_1-adrenergic, and α_2-adrenergic receptors (Fig. 18.17; 6, 12, 17, 62, 81, 122, 148). The dissociation constant for verapamil in these studies is ~1 μM, which is close to the plasma and myocardial levels of the drug achieved clinically (123). It is not clear whether the ability of verapamil to block receptors and its ability to block calcium channels are independent, coincidental effects or whether the receptors and calcium channels are somehow closely linked. It also is not clear whether blocking α-adrenergic and muscarinic receptors is important in terms of pharmacological responses to verapamil. Further studies will be required to answer these questions; without results from radioligand binding

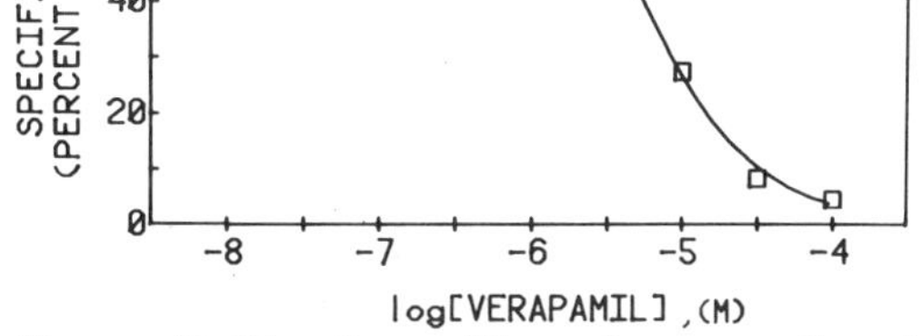

Figure 18.17. Competition of verapamil and [^{3}H]yohimbine for binding to platelet membranes. Platelet membranes were incubated with 4.7 nM [^{3}H]yohimbine and various concentrations of verapamil. The dissociation constant of verapamil is 1.2 μM; the "slope factor" is 1.1.

studies, however, they may never have been asked.

An additional example of the usefulness of radioligand binding methods for studying drug action is the report in which binding studies in several model systems were used to show that the drug dobutamine is a selective β_1-agonist that also interacts with α-adrenergic receptors (211).

Radioligand binding studies have also been used to investigate the mechanism of psychotropic drug action. U'Prichard *et al.* (201) have shown that the clinical efficacy of six antidepressants correlates well with their ability to compete for [^{3}H]WB4101 binding sites in rat brain. The authors conclude that blockade of α_1-adrenergic receptors may be a mechanism of antidepressant drug action. The same authors (111) show that the ability of antipsychotic drugs to compete for [^{3}H]WB4101 binding sites in rat brain correlates with their clinical sedative and hypnotic effect, but not with their antipsychotic effect. The affinity of the drugs for [^{3}H]haloperidol sites (dopamine receptors), on the other hand, correlates much more closely with the drugs' clinical usefulness as antipsychotics. The authors propose that the ratio of the two affinities can be used as a predictor of the therapeutic usefulness; an ideal has low affinity for [^{3}H]WB4101 sites (few side effects) and high affinity for [^{3}H]haloperidol (good efficacy). Thus radioligand binding experiments may be used to predict both the efficacy and toxicity of a new drug. This concept is more fully discussed, and data from other types of receptors are presented, in a review by Creese (44).

RADIOLIGAND BINDING IN HUMANS: PERSPECTIVES AND FUTURE

Early radioligand binding experiments focused on characterizing binding assays and proving that they actually described the interaction of radioligands with receptors. More recent investigations have begun to use radioligand binding to ask questions about the normal regulation of receptors, about adrenergic receptors in disease, about the molecular properties of receptors, and about the use of radioligand binding as a tool in clinical pharmacology (Table 18.4). Work in these areas has yielded interesting results and, as with any new scientific field, has raised many new questions (Table 18.5).

Table 18.4
Uses of Radioligand Binding Studies

1. To study the mechanism of drug or hormone action.
2. To study the normal regulation and variation of receptor number and of receptor binding properties.
3. To investigate the causes of altered tissue responses to drugs and to hormones.
4. To assay drug concentrations (radioreceptor assay).
5. To screen drugs for efficacy or toxicity.
6. To predict individual responses to drugs.

Table 18.5
Some Unanswered Questions Regarding Adrenergic Receptors

1. How accurately do properties of receptors on blood cells reflect those on internal tissues?
2. Under what conditions do altered tissue responses to drugs and hormones result from alterations in receptors?
3. How is the number of receptors regulated?
4. Does normal variation in receptor number and affinity have physiological significance?
5. What is the physiological significance of high and low affinity states of adrenergic receptors?
6. How does intracellular GTP act as a physiological regulator of the affinity of receptors for agonists?
7. Do antibodies directed against adrenergic receptors play a role in disease?
8. What are the "life cycles" of α- and β-adrenergic receptors?
9. Why is the concentration of an agonist needed to produce a physiological effect (K_{ACT}) different than the amount needed to bind to adrenergic receptors (K_D)?
10. How does Na^+ influence agonist binding to α_2-adrenergic receptors?

Almost all radioligand binding studies in humans have been studies of blood cells, and (as mentioned earlier) it is not clear whether the properties of receptors on these cells are accurate reflections of those of internal tissues. One approach to study this problem would be to simultaneously study adrenergic receptors of a variety of animal tissues, including blood cells, but this has not yet been done extensively. One problem with this approach is that interspecies differences may exist. For example, rat platelets, unlike human platelets, do not aggregate when exposed to epinephrine (219), and they do not have any α_2-adrenergic receptors detectable by [^{3}H]yohimbine binding (unpublished results). Another approach to circumvent the problem of only studying adrenergic receptors on blood cells is to develop noninvasive radioligand binding assays for internal organs. Conceivably, it will become possible to inject a small amount of a radioligand with high affinity and specific activity and visualize its distribution with nuclear isotope scanning. Perhaps future physicians will order such a "β-receptor scan" as routinely as they now order a liver-spleen scan.

Observation of changes in adrenergic receptor number and, to a lesser extent, in receptor affinity has been the focus of many studies, yet it is not clear how much physiological impact is caused by small changes in either receptor number or affinity. In studies of the adrenergic receptors on blood cells of normal individuals, a large variation in both number and affinity of receptors is seen; the coefficient of variation is often ~40%. Whether this variation reflects physiological differences in catecholamine response among individuals remains to be established. Moreover, the mechanism by which the number of adrenergic receptors is regulated is completely unknown. We have virtually no idea of the relative contribution of changes in the synthesis, degradation, or compartmentation of receptors in settings of altered receptor number (36, 91, 204). Many studies of the regulation of adrenergic receptors have looked at the changes in agonist affinities that occur in the presence of ions and nucleotides. These studies have been almost exclusively with membrane preparations, and the meaning of these results in terms of receptor function in intact cells is a mystery.

Although many studies of adrenergic receptors in disease have been reported, there is only a single report (of two patients with essential thrombocytosis, see Fig. 18.16) in which alterations in radioligand binding to adrenergic receptors on a diseased "organ" have clearly been demonstrated. For other diseases, such as asthma, the evidence that adrenergic receptors are abnormal is very suggestive, and adrenergic receptors are likely to be abnormal in other diseases for which no radioligand binding studies have yet been reported. Pheochromocytoma, which causes increases in plasma catecholamine levels, seems an obvious setting in which "down regulation" of adrenergic receptors has occurred (20, 41). Further examples of diseases with possible alterations in adrenergic receptors include diabetes mellitus, renal disease, neuropsychiatric disorders, obesity, and essential hypertension.

The intensive efforts that several laboratories have undertaken to solubilize and purify adrenergic receptors (e.g., ref. 64) may prove helpful in understanding molecular events involved in receptor function and in supplying material to develop antireceptor antibodies. Preliminary results suggest that such antibodies, including monoclonal antibodies, may offer powerful new tools for studying adrenergic receptors (64).

Perhaps the most exciting use of radioligand binding in patients is as a tool in clinical pharmacology. Several adrenergic agonists and antagonists are in wide use clinically, and in addition to working by binding to receptors, these agents may produce changes in these receptors. Currently, adrenergic agents are used without quantifying receptors in the patients receiving the drugs. Radioligand binding offers a new tool that perhaps will allow us (a) to measure receptors before initiation of drug treatment and thus predict the required dose of a drug; (b) to monitor changes in the receptors during treatment; and (c) to simplify determination of plasma levels of adrenergic drugs through the use of radioreceptor assays. A goal of clinical pharmacology is to predict individual responses to drugs, and radioligand binding may prove useful for this. Similarly, radioligand binding may be useful in predicting both the beneficial and harmful interaction of drugs with adrenergic receptors and thus may assist in drug design and development.

Radioligand binding has rapidly progressed from being a novel and unproven experimental technique to its new role as a powerful pharmacological tool. Results with this methodology have generated a large number of intriguing new questions about catecholamine action and have added sophistication and complexity to our previously naive and clumsy concepts regarding adrenergic receptors. The solutions of the increasingly difficult problems uncovered by radioligand binding techniques are likely to require novel approaches and the interaction of scientists working in different disciplines. Expertise in biochemistry, cell biology, pharmacology, and immunology will probably be necessary to resolve the questions listed in Table 18.5. We feel confident in predicting that these "second generation" radioligand binding studies will have a major impact on clinical concepts regarding catecholamines.

References

1. Aarons RD, Nies AS, Gal J, Hegstand LR, Molinoff PB: Elevation of beta-adrenergic receptor density in human lymphocytes after propranolol administration. *J Clin Invest* 65:949–957, 1980.
2. Aarons RD, Nies AS, Gerber JG, Molinoff PB: Decrease in lymphocyte beta-adrenergic receptor density during chronic agonist administration. *Clin Res* 29:80A, 1981.
3. Abrass IB, Scarpace PJ: Human lymphocyte beta-adrenergic receptors are unaltered with age. *J Gerontol* 36:298–301, 1981.
4. Ahlquist RP: A study of the adrenotropic receptors. *Am J Physiol* 53:586–600, 1948.
5. Alexander RW, Cooper B, Handin RI: Characterization of the human platelet alpha-adrenergic receptor: Correlation of [^{3}H]-dihydroergocryptine binding with aggregation and adenylate cyclase inhibition. *J Clin Invest* 61:1136–1144, 1978.
6. Atlas D, Adler M: Alpha-adrenergic antagonists as possible calcium channel inhibitors. *Proc Natl Acad Sci USA* 78:1237–1241, 1981.
7. Baker SP, Potter LT: Effect of propranolol on beta-adrenoceptors in rat hearts. *Br J Pharmacol* 68:8–10, 1980.
8. Banerjee SP, Kung LS: Development of beta-adrenergic receptor subsensitivity by antidepressants. *Nature* 268:455–456, 1977.
9. Banerjee SP, Sharma VK, Kung LS: Beta-adrenergic receptors in innervated and denervated skeletal muscle. *Biochim Biophys Acta* 470:123–127, 1977.
10. Barnes PJ, Dollery CT, MacDermot J: Increased pulmonary alpha-adrenergic and reduced beta-adrenergic receptor in experimental asthma. *Nature* 285:569–571, 1980.
11. Barnes PJ, Karliner JS, Dollery CT: Human lung adrenoceptors studied by radioligand binding. *Clin Sci* 58:447–461, 1980.
12. Barnathan E, Addonizio P, Shattil S: Interaction of verapamil with human platelet α-adrenergic receptors. *Am J Physiol* 242:H19–23, 1982.
13. Baumann G, Riess G, Erhardt WD, Felix SB, Ludwig L, Blümel G, Blomer H: Impaired beta-adrenergic stimulation in the uninvolved ventricle defect due to excessive circulating catecholamine-induced decline in number and affinity of beta-receptors. *Am Heart J* 101:569, 1981.
14. Berlan M, LaFontan M: Identification of alpha$_2$-adrenergic receptors in human fat cell membranes by [^{3}H]clonidine binding. *Eur J Pharmacol* 67:481–484, 1980.
15. Berthelsen S, Pettinger WA: A functional basis for classification of alpha-adrenergic receptors. *Life Sci* 22:595–606, 1977.
16. Bishopric NH, Cohen HJ, Lefkowitz RJ: Beta-adrenergic receptors in lymphocyte subpopulations. *J Allergy Clin Immunol* 65:29–33, 1980.
17. Blackmore PF, El-Refai MF, Exton JH: Alpha-adrenergic blockade and inhibition of A23187-mediated Ca^{++} uptake. *Mol Pharmacol* 15:598–606, 1979.
18. Bourne HR, Lichtenstein LM, Melmon KL, Henney CS, Weinstein Y, Shearer GM: Modulation of inflammation and immunity by cyclic AMP. *Science* 184:19–28, 1974.
19. Bourne HR, Melmon KL: Adenylate cyclase in human leukocytes: Evidence for activation by separate beta-adrenergic and prostaglandin receptors. *J Pharmacol Exp Ther* 178:1–7, 1971.
20. Bravo EL, Tarazi RC, Gifford RW, Steward BH: Circulating and urinary catecholamines in pheochromocytoma: diagnostic and pathophysiologic implications. *N Engl J Med* 301:682–686, 1979.
21. Briley MS, Raisman R, Langer SZ: Human platelets possess high-affinity binding sites for [^{3}H]imipramine. *Eur J Pharmacol* 58:347–348, 1979.
22. Briley MS, Langer SZ, Raisman R, Sechter D, Zarifian E: Tritiated imipramine binding sites are decreased in platelets of untreated depressed patients. *Science* 299:303–305, 1980.
23. Brooks SM, McGowan K, Bernstein K, Altenau P, Peagler J: Relationship between numbers of beta-adrenergic receptors in lymphocytes and disease severity in asthma. *J Allergy Clin Immunol* 63:401–406, 1979.
24. Brown EM, Gardner JD, Auerbach GD: Direct determination of ligand interactions with beta-adrenergic receptors on intact turkey erythrocytes: Correlation of binding with biological activity. *Endocrinology* 99:1370–1376, 1976.
25. Bürgisser E, Hancott AA, Lefkowitz RJ, DeLean A: Anomalous equilibrium binding properties of high-affinity racemic radioligands. *Mol Pharmacol* 19:205–216, 1981.
26. Burns TW, Langley PE, Terry BE, Bylund DB, Hoffman BB, Tharp MD, Lefkowitz RJ, Garcia-Sainz JA, Fain JB: Pharmacological characterizations of adrenergic receptors in human adipocytes. *J Clin Invest* 67:467–475, 1981.
27. Busse WW, Lee TP: Decreased adrenergic response in lymphocytes and granulocytes in atopic eczema. *J Allergy Clin Immunol* 58:586, 1976.
28. Bylund DB: Regulation of central adrenergic receptors. In Ehrlich VH, Volauka J, Davis LG, Brunngraber EG: *Modulators, Mediators, and Specifiers in Brain Function.* New York, Plenum Publishers, 1979.
29. Bylund DB, Charness ME, and Snyder SH: Beta-adrenergic receptor labelling intact animals with [^{125}I]hydroxybenzylpindolol. *J Pharmacol Exp Ther* 201:644–653, 1977.
30. Caron MG, Beaulieu M, Raymond V, Gagne B, Dronin J, Lefkowitz RJ, Labire F: Dopaminergic receptors in the anterior pituitary gland. *J Biol Chem* 253:2244–2250, 1978.
31. Carvalho ACA, Colman RW, Lees RS: Platelet function in hyperlipoproteinemia. *N Engl J Med* 290:434–438, 1974.
32. Chang JB, Goldfien A, Ballard PL, Roberts JR:

Glucocorticoids increase pulmonary beta-adrenergic receptors in fetal rabbit. *Endocrinology* 107:1646–1648, 1980.
33. Cheng Y, Prusoff WH: Relationship between the inhibition constant (K_I) and the concentration of an inhibitor that causes a 50% inhibition (I_{50}) of an enzymatic reaction. *Biochem Pharmacol* 22:3099–3108, 1973.
34. Chobanian AV: Alpha- and beta-adrenergic receptors in circulating blood cells in patients with idiopathic orthostatic hypotension and pheochromocytoma. *Clin Exp Hypertens A4*:793, 1982.
35. Chou TC: Relationships between inhibition constants and fractional inhibition in enzyme-catalyzed reactions with different numbers of reactants, different reaction mechanisms, and different types and mechanisms of inhibition. *Mol Pharmacol* 10:235–247, 1974.
36. Chuang D-M, Kinnier WJ, Farber L, Costa E: A biochemical study of receptor internalization during β-adrenergic receptor desensitization in frog erythrocytes. *Mol Pharmacol* 18:348–355, 1980.
37. Collis MG, Sheppard JT: Antidepressant drug action and presynaptic alpha-receptors. *Mayo Clin Proc* 55:567–572, 1980.
38. Colucci WS, Alexander RW, Williams GH, Rude RE, Holman BL, Konstam MA, Wyne J, Mudge GH, Brunwald E: Decreased lymphocyte beta-adrenergic receptor density in heart failure patients tolerant to the beta-adrenergic agonist pirbuterol. *N Engl J Med* 305:185–190, 1981.
39. Conolly ME, Greenacre JK: The lymphocyte beta-adrenoceptor in normal subjects and patients with bronchial asthma. *J Clin Invest* 58:1307–1316, 1976.
40. Conolly ME, Greenacre JK: The beta-adrenoceptor of the human lymphocyte and human lung parenchyma. *Br J Pharmacol* 59:17–23, 1977.
41. Conolly ME, Greenacre JK: Desensitization of beta-adrenergic receptors of lymphocytes from normal subjects and patients with pheochromocytoma: Studies *in vivo*. *Br J Clin Pharmacol* 5:191–197, 1978.
42. Cooper B, Handin RI, Young IH, Alexander RW: Agonist regulation of the human platelet alpha-adrenergic receptor. *Nature* 274:703–706, 1978.
43. Corr PB, Shayman JA, Kramer JB, Kipnis RJ: Increased alpha-adrenergic receptors in ischemic cat myocardium. *J Clin Invest* 76:1232–1236, 1981.
44. Creese I: Receptor binding as a primary drug screening device. In Yamamaura HI, Enna SJ, Kuhar MS: *Neurotransmitter Receptor Binding*. New York, Raven Press, 1979, pp 141–170.
45. Cuatrecasas P, Hollenberg MD: Membrane receptors and hormone action. *Adv Protein Chem* 70:261–286, 1975.
46. Darfler FJ, Marinetti GU, Condemi JJ, Leddy JP: Adrenergic receptors and adenylate cyclase of human mononuclear and polymorphonuclear leukocytes. *Fed Proc* 38:843A, 1979.
47. Daiguji M, Meltzer HY, U'Prichard DC: Human platelet alpha$_2$-adrenergic receptors: Labelling with [^{3}H]yohimbine, a selective antagonist ligand. *Life Sci* 28:2705–2717, 1981.
48. Davies AO, Lefkowitz RJ: Regulation of adrenergic receptors. In Lefkowitz RJ: *Receptor Regulation*. New York, Chapman and Hall, 1981.
49. Davies AO, Lefkowitz RJ: Corticosteroid-induced differential regulation of beta-adrenergic receptors in circulating human polymorphonuclear leukocytes and mononuclear leukocytes. *J Clin Endocrinol Metab* 51:599–605, 1980.
50. Davies AO, Lefkowitz RJ: Agonist-promoted high affinity state of the beta-adrenergic receptor in human neutrophils: Modulation by corticosteroids. *J Clin Endocrinol Metab* 53:703, 1981.
51. Davis PB, Shelhamer JR, Kaliner M: Abnormal adrenergic and cholinergic sensitivity in cystic fibrosis. *N Engl J Med* 302:1453–1456, 1980.
52. Davis PB, Braunstein M, Jay C: Decreased adenosine 3′,5′-monophosphate response to isoproterenol in cystic fibrosis leukocytes. *Pediatr Res* 12:703–707, 1978.
53. Davis JM, Strittmatter WJ, Hoyler E, Lefkowitz RJ: Dihydroergocryptine binding in rat brain. *Brain Res* 132:327–336, 1977.
54. DeLean A, Munson PJ, Rodbard D: Simultaneous analysis of families of sigmoidal curves: Application to bioassay, radioligand assay, and physiological dose-response data. *Am J Physiol* 235:E93–E102, 1978.
55. DeLean A, and Rodbard D: Kinetics of cooperative binding. In O'Brien RD: *Receptors: A Comprehensive Treatise*. New York, Plenum Press, 1979, pp 143–190.
56. Dibner MD, Wolfe RA, Insel PA: Replacement of serum with a defined medium increases beta-adrenergic receptor number of cultured glioma cells. *Exp Cell Res* 131:424–427, 1981.
57. Dickinson KP, Richardson A, Nahorski SR: Homogeneity of beta$_2$-adrenergic receptors on rat erythrocytes and reticulocytes: A comparison with heterogeneous rat lung beta-receptors. *Mol Pharmacol* 19:194–204, 1981.
58. Dullis BH, Wilson IB: The beta-adrenergic receptor of live human polymorphonuclear leukocytes. *J Biol Chem* 255:1043–1048, 1980.
59. Elliott JM, and Grahame-Smith DG: Evidence for a single type of alpha-receptor on human platelets. *Br J Pharmacol* 71:59P, 1980.
60. Elliott JM, Peters JR, Grahame-Smith DG: Oestrogen and progesterone change the binding characteristics of alpha-adrenergic and serotonin receptors on rabbit platelets. *Eur J Pharmacol* 66:21–30, 1980.
61. Engel G: Identification of different subgroups of beta-receptors by means of binding studies in guinea-pig and human lung. *Triangle* 19:69–76, 1980.
62. Fairhurst AS, Whittaker ML, Ehlert FJ: Interactions of D600 (methoxyverapamil) and local anesthetics with rat brain alpha-adrenergic and muscarinic receptors. *Biochem Pharmacol* 29:155–162, 1980.
63. Fitzgerald G, Robertson D, Feely J, Wood AJJ: Beta$_2$-adrenoreceptors are down-regulated by

dynamic exercise in man. *Clin Res* 29:564A, 1981.

64. Fraser CM, Venter JC: Monoclonal antibodies to beta-adrenergic receptors: Use in purification and molecular characterization of beta-receptors. *Proc Natl Acad Sci USA* 77:7034–7038, 1980.
65. Fraser CM, Harrison LC, Kaliner MC, Venter JC: Autoantibodies to the beta-adrenergic receptor are associated with adrenergic hyporesponsiveness. *Clin Res* 28:236A, 1980.
66. Fraser J, Nadeau J, Robertson D, Wood AJJ: Regulation of human leukocyte beta-receptors by endogenous catecholamines: Relationship of leukocyte beta-receptors to the cardiac sensitivity to isoproterenol. *J Clin Invest* 67:1777–1784, 1981.
67. Galant SP, Underwood S, Duriseti L, and Insel PA: Characterization of high-affinigy beta$_2$-adrenergic receptor binding of [^{3}H]dihydroalprenolol to human polymorphonuclear cell particulates. *J Lab Clin Med* 92:613–618, 1978.
68. Galant SP, Underwood SB, Lundak TC, Grancy CC, Mouratides DI: Heterogeneity of human lymphocyte subpopulation to pharmacological stimulation. 1. Lymphocyte responsiveness to beta-adrenergic agents. *J Allergy Clin Immunol* 62:349, 1978.
69. Galant SP, Duriseti L, Underwood S, Insel PA: Decreased beta-adrenergic receptors on polymorphonuclear leukocytes after adrenergic therapy. *N Engl J Med* 299:933–936, 1978.
70. Galant SP, Underwood S, Allred S, Hanifin JM: Beta-adrenergic receptor binding on polymorphonuclear leukocytes in atopic dermatitis. *J Invest Dermatol* 72:330, 1979.
71. Galant SP, Allred SJ: Demonstration of beta$_2$-adrenergic receptors of high coupling efficiency in human neutrophil sonicator. *J Lab Clin Med* 96:15–23, 1980.
72. Galant SP, Duriseti L, Underwood S, Allred S, Insel PA: Beta-adrenergic receptors of polymorphonuclear particulates in bronchial asthma. *J Clin Invest* 65:577–585, 1980.
73. Galant SP, Allred AJ: Binding and functional characteristics of beta-adrenergic receptors in the intact neutrophil. *J Lab Clin Med* 98:227–231, 1981.
74. Galant SP, Norton L, Herbst J, Wood C: Impaired beta-adrenergic receptor binding and function in cystic fibrosis neutrophils. *J Clin Invest* 68:253–258, 1981.
75. Garcia-Sainz JA, Hoffman BB, Li SY, Lefkowitz RJ, Fain JN: Role of alpha$_1$-adrenoceptors in the turnover of phosphatidyl-inositol in the regulation of cyclic AMP accumulation in hamster adipocytes. *Life Sci* 27:933–961, 1980.
76. Garcia-Sevilla JA, Zis AP, Hollingswort PJ, Greden JF, Smith CB: Platelet alpha$_2$-adrenoreceptors in major depressive disorder (abstr). *Pharmacologist* 23:216, 1981.
77. Garcia-Sevilla JA, Zis AP, Zelnik TC, Smith CB: Tricyclic antidepressant drug treatment decreases alpha$_2$-adrenoreceptors on human platelet membranes. *Eur J Pharmacol* 69:121–123, 1981.
78. Gerber JG, Nies AS: Abrupt withdrawal of cardiovascular drugs. *N Engl J Med* 301:1234–1235.
79. Ginsberg AM, Clutter WE, Shah SD, Cryer PE: Triiodothyronine-induced thyrotoxicosis increases mononuclear leukocyte β-adrenergic receptor density in man. *J Clin Invest* 67:1785–1791.
80. Glaubiger G, Lefkowitz RJ: Elevated beta-adrenergic receptor number after chronic propranolol treatment. *Biochem Biophys Res Commun* 78:720–725, 1977.
81. Glossman H, Hornung R: Calcium- and potassium-channel blockers interact with alpha-adrenoceptors. *Mol Cell Endocrinol* 19:243–251, 1980.
82. Goldstein RE, Corash LC, Tallman JF, Lake CR, Hyde J, Smith CC, Capurro NL, Anderson JC: Shortened platelet survival time and enhanced heart rate responses after abrupt withdrawal of propranolol from normal subjects. *Am J Cardiol* 47:1115–1123, 1981.
83. Grant JA, Scrutton MC: Novel alpha$_2$-adrenoreceptors primarily responsible for inducing human platelet aggregation. *Nature* 277:659–661, 1979.
84. Grant JA, Scrutton MC: Interaction of selective alpha-adrenoceptor agonists and antagonists with human and rabbit blood platelets. *Br J Pharmacol* 71:121–134, 1980.
85. Greenberg LH, Weiss B: Beta-adrenergic receptors in aged rat brain: Reduced number and capacity of pineal gland to develop supersensitivity. *Science* 201:61–63, 1978.
86. Griffith OM: Separation of T and B cells from human peripheral blood by centrifugal elutriation. *Anal Biochem* 87:97–107, 1978.
87. Guellaën G, Yates-Aggerbeck M, Vauquelin E, Strosberg D, Hanoune J: Characteristics with [^{3}H]DHE of the alpha-adrenergic receptor of the hepatic plasma membrane. *J Biol Chem* 253:1114–1120, 1978.
88. Güllner HG, Kafka MS, Bartter FC: Indomethacin increases leukocyte beta-adrenoceptors in man. *Clin Sci* 59:397–400, 1980.
89. Hayashida D, Leung R, Goldfien A, Roberts JM: Human myometrial adrenergic receptors: Identification of beta-adrenergic receptors with dihydroalprenolol, submitted for publication, 1983.
90. Harden TK, Wolfe BB, Sporn JR, Perkins JP, Molinoff PB: Ontogeny of beta-adrenergic receptors in rat cerebral cortex. *Brain Res* 125:99–108, 1977.
91. Harden TK, Cotton CU, Waldo GC, Cotton JK, Perkins JP: Catecholamine-induced alteration in sedimentation behavior of membrane bound β-adrenergic receptors. *Science* 210:441–443, 1980.
92. Henderson WR, Shelhamer JH, Reingold DB, Smith LJ, Evans R, Kaliner M: Alpha-adrenergic hyper-responsiveness in asthma: Analysis of vascular and pupillary response. *N Engl J Med* 300:642–647, 1979.
93. Henry PD: Comparative pharmacology of calcium antagonists: Nifedipine, verapamil and diltiazem. *Am J Cardiol* 46:1047–1058, 1980.

94. Hoffman BB, Delean A, Wood CL, Schoken DD, Lefkowitz RJ: Alpha-adrenergic receptor subtypes: Quantitative assessment by ligand binding. *Life Sci* 24:1739–1746, 1979.
95. Hoffman BB, Lefkowitz RJ: [^{3}H]WB4101: Caution about its role as an alpha-adrenergic subtype selective radioligand. *Biochem Pharmacol* 29:1537, 1979.
96. Hoffman BB, Lefkowitz RJ: Alpha-adrenergic receptor subtypes. *N Engl J Med* 302:1390–1396, 1980.
97. Hoffman BB, Lefkowitz RJ: Radioligand binding studies of adrenergic receptors: New insights into molecular and physiological regulation. *Ann Rev Pharmacol Toxicol* 20:581–608, 1980.
98. Hoffman BB, Michel T, Mullikin-Kilpatrick D, Lefkowitz RJ, Tolberg MEM, Gilman H, Fain JN: Agonist versus antagonist binding to alpha-adrenergic receptors. *Proc Natl Acad Sci USA* 77:4569–4573, 1980.
99. Hoffman BB, Mullikin-Kilpatrick D, Lefkowitz RJ: Heterogeneity of radioligand binding to alpha-adrenergic receptors: Analysis of guanine nucleotide regulation of agonist binding in relation to receptor subtypes. *J Biol Chem* 255:4645–4652, 1980.
100. Hoffman BB, Yim S, Tsai BS, Lefkowitz RJ: Preferential uncoupling by manganese of alpha-adrenergic receptor mediated inhibition of adenylate cyclase in human platelets. *Biochem Biophys Res Commun* 100:724–731, 1981.
101. Homcy CJ, Strauss HW, Kopiwoda S: Beta-receptor occupancy. Assessment in the intact animal. *J Clin Invest* 65:1111–1118, 1980.
102. Hsu CY, Knapp DR, Halushka PV: The effects of alpha-adrenergic agents on human platelet aggregation. *J Pharmacol Exp Ther* 208:366–370, 979.
103. Hughes RJ, Insel PA: Regulation of cyclic AMP levels in human platelets. *Fed Proc* 39:424, 1980.
104. Hui KKP, Conolly ME: Increased numbers of beta-adrenergic receptors in orthostatic hypotension, clue to autonomic dysfunction. *N Engl J Med* 304:1473–1476, 1981.
105. Ignarro LJ: Nonphagocytic release of neutral protease and beta-glucuronidase from human neutrophils. *Arthritis Rheum* 17:25, 1974.
106. Innis RB, Bylund OB, Snyder SH: A simple, sensitive, and specific radioreceptor assay for beta-adrenergic antagonist drugs. *Life Sci* 23:2031–2038, 1978.
107. Insel PA: Membrane active hormone receptors and receptor regulation. In Rickenberg HV: Biochemistry and Mode of Action of Hormones II. *International Review of Biochemistry*, Baltimore, University Park Press, 1978, vol 20, pp 1–43.
108. Insel PA, Kennedy MS: Colchicine potentiates beta-adrenoreceptor stimulated cyclic AMP by an action distal to the receptor. *Nature* 273:471–473, 1978.
109. Insel PA, Liljenquist JE, Tobin JD, Sherwin RS, Watkins P, Andres R, Berman M: Insulin control of glucose metabolism in man. A new kinetic analysis. *J Clin Invest* 55:1057–1066, 1975.
110. Insel PA, Nirenberg P, Turnbull J, Shattil SJ: Relationships between membrane cholesterol, alpha-adrenergic receptors, and platelet function. *Biochemistry* 17:5269–5274, 1978.
111. Insel PA, Snavely MDS: Catecholamines and the kidney: Receptors and renal function. *Ann Rev Physiol* 43:625–636, 1981.
112. Insel PA, Stoolman LM: Binding to beta-adrenergic receptors of intact cultured S49 lymphoma cells. *Mol Pharmacol* 14:549–561, 1978.
113. Jakobs KH: Synthetic alpha-adrenergic agonists are potent alpha-adrenergic blockers in human platelets. *Nature* 274:819–820, 1978.
114. Jakobs KH, Saur W, Schultz G: Reduction of adenylate cyclase activity in lysates of human platelets by the alpha-adrenergic component of epinephrine. *J Cyclic Nucl Res* 2:381–392, 1976.
115. Jakobs KH, Saur W, Schultz G: Inhibition of platelet adenylate cyclase by epinephrine requires GTP. *FEBS Lett* 85:167–168, 1978.
116. Kafka MS, Lake CR, Güllner HG, Tallman JF, Bartter FC, Fujita T: Adrenergic receptor function is different in male and female patients with essential hypertension. *Clin Exp Hypertens* 1:613–627, 1979.
117. Kafka MS, Tallman JF, Smith CC: Alpha-adrenergic receptors on human platelets. *Life Sci* 21:1429–38, 1977.
118. Kafka MS, Von Kammen DP, Bunney WE: Reduced cyclic AMP production in blood platelets from schizophrenic patients. *Am J Psychiatry* 136:685–687, 1979.
119. Kariman K: Beta-adrenergic receptor binding in lymphocytes from patients with asthma. *Lung* 158:41–51, 1980.
120. Karliner JS, Barnes P, Brown M, Dollery CT: Chronic heart failure in the guinea pig produces increases in cardiac alpha$_1$- and beta-adrenoceptors. *Eur J Pharmacol* 67:115–118, 1980.
121. Karliner JS, Motulsky HJ, Insel PA: Apparent down-regulation of human platelet α_2-adrenergic receptors is due to retained agonist. *Mol Pharmacol* 21:36–43, 1982.
122. Karliner JS, Motulsky HJ, Dunlap J, Brown JH, Insel PA: Verapamil is a competitive inhibitor of alpha$_1$-adrenergic and muscarinic, but not beta-adrenergic receptors in rat myocardium. *J Cardiovasc Pharmacol* 4:515–520, 1982.
123. Kates RE, Keefe D, Schwartz J, harapat S, Harrison D, Kirsten E, Johnson SM: Disposition kinetics of verapamil and norverapamil (abstr). *Clin Pharmacol Ther* 29:257.
124. Kather H, Pries J, Schrader V, Simon B: Inhibition of human fat cell adenylate cyclase mediated via alpha-adrenoreceptors. *Eur J Clin Invest* 10:345–358, 1980.
125. Kaywin P, McDonough M, Insel PA, Shattil SJ: Platelet function in essential thrombocythemia. *N Engl J Med* 299:505–509.
126. Klempner MS, Gallin JI, Balow JE, VanKammen DP: The effect of hemodialysis and C_5A on neutrophil subpopulations. *Blood* 55:777–783, 1980.
127. Kobinger W, Pichler L: Investigation into different types of post- and presynaptic alpha-

adrenoceptors at cardiovascular sites in rats. *Eur J Pharmacol* 65:393–402, 1980.
128. Kraft CA, Castlede CM: The effect of aging on beta-adrenoceptor-stimulated cyclic AMP formation in human lymphocytes. *Clin Sci* 60:587–589, 1981.
129. Krall JF, Connelly M, Tuck ML: Acute regulation of beta-adrenergic catecholamine sensitivity in human lymphocytes. *J Pharmacol Exp Ther* 214:554–560, 1980.
130. Krall JF, Connelly M, Tuck ML: Age-related loss of adenylate cyclase responsiveness to guanyl nucleotide and isoproterenol in the human lymphocyte. *The Endocrine Society 63rd Annual Meeting*, Abstract 202, 1981.
131. La Fontan M, Berlan M: Evidence for the alpha$_2$ nature of the alpha-adrenergic receptor inhibiting lipolysis in human fat cells. *Eur J Pharmacol* 66:87–93, 1980.
132. Langer SZ: Presynaptic regulation of catecholamine release. *Biochem Pharmacol* 23:1973–1800, 1974.
133. Langer SZ, Briley MS, Raisman R, Henry JF, Morselli PL: Specific [^{3}H]imipramine binding in human platelets: Influence of age and sex. *Naunyn Schmiedebergs Arch Pharmacol* 313:189–194, 1980.
134. Levitsky A: Quantitative aspects of allosteric mechanisms. *Molecular Biology, Biochemistry and Biophysics*. New York, Springer-Verlag, 1978, vol 28.
135. Limas CJ: Effect of phospholipid methylation on beta-adrenergic receptors in the normal and hypertrophied rat myocardium. *Circ Res* 47:536–541, 1980.
136. Limbird LE: The alpha-adrenergic receptor of the human platelet: Agonist receptor interactions in membrane and solubilized preparations. *Adv Cyclic Nucl Res* 14:189–191, 1981.
137. Limbird LE, Gill DM, Lefkowitz RJ: Agonist-promoted coupling of the beta-adrenergic receptor with guanine nucleotide regulatory protein of the adenylate cyclase system. *Proc Natl Acad Sci USA* 77:755, 1980.
138. Lynch CJ, Steer ML: Evidence for high and low affinity alpha$_2$-receptors: Comparison of [^{3}H] norepinephrine and [^{3}H]phentolamine binding to human platelet membranes. *J Biol Chem* 254:3298–3303, 1981.
139. Maggi A, Schmidt MJ, Ghetti B, Enna SJ: Effects of aging on neurotransmitter receptor binding in rat and human brain. *Life Sci* 24:367–374, 1979.
140. Maguire ME, Ross EM, Gilman AG: Beta-adrenergic receptor: Ligand binding properties and interaction with adenyl cyclase. *Adv Cyclic Nucleotide Res* 8:1–83, 1977.
141. Mano K, Akbarzadeh A, Townley RG: Effect of hydrocortisone on beta-adrenergic receptors in lung membranes. *Life Sci* 25:1925–1930, 1979.
142. Melmon KL, Bourne HR, Weinstein N, Shearer GM, Kram J, Bauminger S: Hemolytic plaque formation by leukocytes in vitro—control by vasoactive hormones. *J Clin Invest* 53:13–21, 1974.
143. Michel T, Hoffman BB, Lefkowitz RJ, Caron MG: Different sedimentation properties of agonist- and antagonist-labelled platelet alpha$_2$-adrenergic receptors. *Biochem Biophys Res Commun* 100:1131–1136, 1981.
144. Michel T, Hoffman BB, Lefkowitz RJ: Differential regulation of the alpha$_2$-adrenergic receptor by Na^+ and guanine nucleotides. *Nature* 288:709–711, 1980.
145. Mills DCB, Roberts GCK: Effects of adrenaline on human platelets. *J Physiol* 193:443–453, 1967.
146. Minneman KP, Pittman RN, Molinoff PB: Beta-adrenergic receptor subtypes: Properties, distribution, and regulation. *Ann Rev Neurosci* 4:419–461, 1981.
147. Mize N, Storr J, Goldfien A, Roberts JM: Platelet aggregation by epinephrine is increased during pregnancy by events distal to the alpha-adrenergic receptor (abstr). 1981 Annual meeting, St. Louis. *Soc Gynecol Invest*, in press, 1983.
148. Motulsky HJ, Hughes RJ, Snavely MD, Insel PA: Verapamil blocks binding to alpha-adrenergic receptors in platelets, kidney, and muscle cells. *Circ Res* 52:226–231, 1983.
149. Motulsky HJK, Insel PA: Binding of [^{3}H]dihydroergocryptine to alpha-adrenergic receptors on human platelets: A reassessment using the selective radioligands [^{3}H]yohimbine, [^{3}H]rauwolscine, and [^{3}H]prazosin, submitted for publication, 1983.
150. Motulsky HJ, O'Connor DT, Insel PA: The platelet α_2-adrenergic receptor of untreated and treated hypertensive men. *Clin Sci* 64:265–272, 1983.
151. Motulsky HJ, Shattil SJ, Insel PA: Characterization of alpha$_2$-adrenergic receptors on human platelet using [^{3}H]yohimbine. *Biochem Biophys Res Commun* 74:1562–1570, 1980.
152. Mukhergee AT, Wong TM, Buja LM, Lefkowitz RJ: Beta-adrenergic and muscarinic cholinergic receptors in canine myocardium. Effects of ischemia. *J Clin Invest* 64:1423–1428, 1979.
153. Mulvihill-Wilson J, Pettinger W: Can platelet alpha$_2$-receptors reflect a genetically-determined hypertensive mechanism? (abstr). *Pharmacologist* 23:216, 1981.
154. Munck A: General aspects of steroid hormone-receptor interactions, Part II. In Pasqualin JR: *Receptors and Mechanism of Action of Steroid Hormones*. New York, Marcel Dekker, 1976, pp 1–40.
155. Munson PJ, Rodbard D: Ligand: A versatile computerized approach for characterization of ligand-binding systems. *Anal Biochem* 107:220–239, 1980.
156. Nahorski SR, Batta MI, Barnett DB: Measurement of beta-adrenoceptor antagonists in biological fluids using a radioreceptor assay. *Eur J Pharmacol* 52:393–396, 1978.
157. Newman KD, Williams LT, Bishopric N, Lefkowitz RJ: Identification of alpha-adrenergic receptors in human platelets by [^{3}H]dihydroergocryptine binding. *J Clin Invest* 61:395–402, 1978.
158. Niaudet P, Beaurain G, Bach M: Differences in effect of isoproterenol stimulation on levels of cyclic AMP in human B and T lymphocytes. *Eur J Immunol* 6:834–836, 1976.
159. Norby JG, Ottolenghi P, Jenson J: Scatchard plot: Common misinterpretation of binding ex-

periments. *Anal Biochem* 102:318–320, 1980.
160. Owen NE, Feinberg A, LeBreton GC: Epinephrine induces Ca^{++} uptake in human blood platelets. *Am J Physiol* 8:H483–488, 1980.
161. Palacios JM, Kuhar MJ: Beta-adrenergic receptor localization by light microscopic autoradiography. *Science* 208:1378–1380, 1980.
162. Paul SM, Rehavi M, Skolnick P, Goodwin FK: Demonstration of specific "high affinity" binding sites for [^{3}H]imipramine on human platelets. *Life Sci* 26:953–959, 1980.
163. Peroutka SJ, U'Prichard DC, Greenberg DA, Snyder SH: Neuroleptic drug interactions with norepinephrine alpha-receptor binding sites in rat brain. *Neuropharmacology* 16:549–556, 1977.
164. Peters JR, Elliot JM, Grahame-Smith DG: Effects of oral contraceptives on platelet noradrenaline and 5-hydroxytryptamine receptors and aggregation. *Lancet* 2:933–936, 1979.
165. Raisman R, Briley M, Langer SZ: High-affinity [^{3}H]imipramine binding in rat cerebral cortex. *Eur J Pharmacol* 69:121–123, 1981.
166. Reinherz EL, Schlossman SF: Regulation of the immune response—inducer and suppressor T-lymphocyte subsets in human being. *N Engl J Med* 303:370–373, 1980.
167. Roberts JM, Goldfien RD, Tsuchiya A, Goldfien A, Insel PA: Estrogen treatment decreased alpha-adrenergic binding sites on rabbit platelets. *Endocrinology* 104:722–728, 1979.
168. Roberts JM, Insel PA, Goldfien RD, Goldfien A: Identification of beta-adrenergic binding sites in rabbit myometrium. *Endocrinology* 101:1839–1843, 1977.
169. Roberts JM, Insel PA, Goldfien A: Regulation of myometrial adrenoreceptor and adrenergic response by sex steroids. *Mol Pharmacol* 20:52–58, 1981.
170. Rodbell M: The role of hormone receptors and GTP-regulatory proteins in membrane transduction. *Nature* 284:17–22, 1980.
171. Rochester CL, Gammon DE, Shane E, Bilezikan JP: A radioreceptor assay for propranolol and 4-hydroxy propranolol. *Clin Pharmacol Ther* 28:32–39, 1980.
172. Rosenthal HE: A graphic method for the determination and presentation of binding parameters in complex systems. *Anal Biochem* 20:525–532, 1967.
173. Ross EM, Gilman AG: Biochemical properties of hormone-sensitive adenylate cyclase. *Ann Rev Biochem* 49:533–564, 1980.
174. Roth GS: Hormone receptor changes during adulthood and senescence: Significance for aging research. *Fed Proc* 38:1910–1914, 1979.
175. Ruoho AE, DeClerque JL, Busse WW: Characterization of granulocyte beta-adrenergic receptors in atopic eczema. *J Allergy Clin Immunol* 66:46–51, 1980.
176. Sano Y, Krakos K, Cheng JB, Bewtra A, Townley R: Comparison of alpha- and beta-adrenergic receptors in asthmatics and controls: Identification and characterization of alpha-adrenergic receptors in human lymphocytes. *Clin Res* 29:172A, 1981.
177. Sano Y, Ruprecht H, Mano K, Begley M, Bewtra A, Townley R: Leukocyte beta-adrenergic receptor assay in normals and asthmatics. *Clin Res* 27:403A, 1979.
178. Scarpace PJ, Abrass IB: Thyroid hormone regulation of rat heart, lymphocyte, and lung beta-adrenergic receptors. *Endocrinology 108*:1007–1011, 1981.
179. Scatchard G: The attractions of proteins for small molecules and ions. *Ann NY Acad Sci* 51:660–672, 1949.
180. Schocken DD, Caron MG, Lefkowitz RJ: The human placenta—a rich source of beta-adrenergic receptors: Characterization of the receptors in particulate and solubilized preparations. *J Clin Endocrinol Metab* 50:1082–1088, 1980.
181. Schocken DD, Roth GS: Reduced beta-adrenergic receptor concentrations in aging man. *Nature* 267:856–858, 1977.
182. Shattil SJ, Anaya-Galindo R, Bennett J, Colman RW, Cooper RA: Platelet hypersensitivity induced by cholesterol incorporation. *J Clin Invest* 55:636–640, 1975.
183. Shattil SJ, Bennett JS, McDonough M, Turnbull J: Carbenicillin and penicillin G inhibit platelet function in vitro by impairing the interaction of agonists with the platelet surface. *J Clin Invest* 65:329–337, 1980.
184. Shattil SJ, McDonough M, Turnball J, Insel PA: Characterization of alpha-adrenergic receptors in human platelets using [^{3}H]clonidine. *Mol Pharmacol* 19:179–183, 1981.
185. Sheppard JH, Gormus R, Moldow CF: Catecholamine hormone receptors are reduced on chronic lymphocytic leukaemic lymphocytes. *Nature* 269:693–695, 1977.
186. Shorr RGL, Lefkowitz RJ, Caron MC: Purification of the beta-adrenergic receptor. Identification of the hormone binding subunit. *J Biol Chem* 256:5820–5826, 1981.
187. Smith SK, Limbird LE: Solubilization of human platelet alpha-adrenergic receptors: Evidence that agonist occupancy of the receptor stabilizes receptor-effector interactions. *Proc Natl Acad Sci USA* 78:4026–4030, 1981.
188. Smith U: Adrenergic control of human adipose tissue lipolysis. *Eur J Clin Invest* 10:343–344, 1980.
189. Starke K: Regulation of noradrenaline release by presynaptic receptor systems. *Rev Physiol Biochem Pharmacol* 77:1–24, 1977.
190. Steer ML, Khorana J, Galgoci B: Quantitation and characterization of human platelet alpha-adrenergic receptors using [^{3}H]phentolamine. *Mol Pharmacol* 16:719–728, 1979.
191. Steer ML, Wood A: Regulation of human platelet adenylate cyclase by prostaglandin E_1 and guanine nucleotides. *J Biol Chem* 254:10791–10798, 1979.
192. Su Y, Harden TK, Perkins JP: Catecholamine-specific desensitization of adenylate cyclase. Evidence for a multistep process. *J Biol Chem* 255:7410–7419, 1980.
193. Szentivanyi A: The beta-adrenergic theory of the atopic abnormality in bronchial asthma. *J Allergy* 42:203–232, 1968.
194. Szentivanyi A, Heim O, Schultze P: Changes in adrenoceptor densities in membranes of lung tissue and lymphocytes from patients with atopic disease. *Ann NY Acad Sci* 332:295–298,

1979.

195. Terasaki WL, Brooker G: [^{125}I]iodohydroxybenzylpinolol binding sites on intact rat glioma cells: Evidence for beta-adrenergic receptors of high coupling affinity. *J Biol Chem* 253:5418–5425, 1978.

196. Tohmeh JF, Cryer PE: Biphasic adrenergic modulation of beta-adrenergic receptors in man. Agonist-induced early increment and late decrement in beta-adrenergic receptor number. *J Clin Invest* 65:836–840.

197. Tharp MD, Hoffman BB, Lefkowitz RJ: Alpha-adrenergic receptors in human adipocyte membranes: Direct determination by [^{3}H]yohimbine binding. *J Clin Endocrinol Metab* 52:709–714, 1981.

198. Triggle DJ, Moran JF: The identification and isolation of adrenergic receptors. In Kalsner S: *Trends in Autonomic Pharmacology.* Baltimore, Urban and Schwarzenberg, 1979, pp 387–409.

199. Tsai BS, Lefkowitz RJ: Agonist-specific effects of guanine nucleotides on alpha-adrenergic receptors in human platelets. *Mol Pharmacol* 16:61–68, 1979.

200. Tse J, Wrenn RW, Kuo JF: Thyroxine-induced changes in characteristics and activities of beta-adrenergic receptors and adenosine 3′,5′-monophosphate and guanosine 3′,5′-monophosphate systems in the heart may be related to reputed catecholamine supersensitivity in hyperthyroidism. *Endocrinology* 107:6–16, 1980.

201. U'Prichard DC, Greenberg DA, Sheenan PP, Snyder SH: Tricyclic antidepressants: Correlation of therapeutic properties with affinity for alpha-noradrenergic receptor binding sites. *Science* 199:197–198, 1977.

202. U'Prichard DC, Reisne TP, Mason ST, Fibiger HC, Yamamura HI: Modulation of rat brain alpha- and beta-adrenergic receptor populations by lesion of the dorsal noradrenergic bundle. *Brain Res* 187:143–154, 1980.

203. U'Prichard DC, Snyder SH: Distinct alpha-noradrenergic receptors differentiated by binding and physiological relationships. *Life Sci* 24:79–88, 1979.

204. Venter CJ, Fraser JC: The synthesis of beta-adrenergic receptors in cultured human lung cells: Induction by glucocorticoids. *Biophys Biochem Res Commun* 94:390–397, 1980.

205. Venter JC, Fraser CM, Harrison LC: Autoantibodies to beta$_2$-adrenergic receptors: A possible cause of adrenergic hyporesponsiveness in allergic rhinitis and asthma. *Science* 207:1361–1363, 1980.

206. Watanabe Y, Lai RT, Higuchi H, Maeda H, Yoshida H: Beta-adrenoceptors of human lymphocytes decrease in affinity of agonist for [^{3}H] dihydroalprenolol binding sites by cytoplasmic fractions. *Life Sci* 28:1399–1405, 1981.

207. Whitsett JA, Johnson CL, Noguchi A, Darovec-Beckerman C, Costello M: Beta-adrenergic receptors and catecholamine-sensitive adenylate cyclase of the human placenta. *J Clin Endocrinol Metab* 50:27, 1980.

208. Williams RS, Guthrow CE, Lefkowitz RJ: Beta-adrenergic receptors of human lymphocytes are unaltered by hyperthyroidism. *J Clin Endocrinol Metab* 48:503, 1979.

209. Williams LT, Lefkowitz RJ: *Receptor Binding Studies in Adrenergic Pharmacology.* New York, Raven Press, 1978.

210. Williams LT, Snyderman R, Lefkowitz RJ: Identification of beta-adrenergic receptors in human lymphocytes by [^{3}H]alprenolol binding. *J Clin Invest* 57:149–155, 1976.

211. Williams RS, Bishop T: Selectivity of dobutamine for adrenergic receptor subtypes. *J Clin Invest* 67:1703–1711, 1981.

212. Williams RS, Lefkowitz RJ: Physical conditioning and membrane receptors for cardioregulatory hormones. *Clin Res* 27:216A, 1979.

213. Wirz-Justice A, Kafka MS, Naber D, Wehr TA: Circadian rhythms in rat brain alpha- and beta-adrenergic receptors are modified by chronic imipramine. *Life Sci* 27:341–347, 1980.

214. Wolfe BB, Harden TK, Molinoff PB: In vitro study of beta-adrenergic receptors. *Ann Rev Pharmacol Toxicol* 17:575–604, 1977.

215. Woodcock EA, Johnston CI: Cardiovascular adrenergic receptors in experimental hypertension in the rat. *Circ Res* 46 (Suppl I): 145–146, 1980.

216. Yamada S, Yamamura HI, Roesce WR: Alterations in central and peripheral adrenergic receptors in deoxycorticosterone/salt hypertensive rats. *Life Sci* 27:2465–2416, 1980.

217. Yamamura H, Enna S, Kuhar MJ: *Neurotransmitter Receptor Binding.* New York, Raven Press, 1978.

218. Young WS, Kuhar MJ: Noradrenergic alpha$_1$ and alpha$_2$ receptors: Light microscopic autoradiographic localization. *Proc Natl Acad Sci USA* 77:1696–1700, 1980.

219. Yu SK, Latour JG: Potentiation by alpha- and inhibition by beta-adrenergic stimulations of rat platelet aggregation. A comparitive study with human and rat platelets. *Thromb Haemostasis* 37:413, 1977.

220. Ziegler MG, Lake CR, Kopin IJ: Plasma norepinephrine increases with age. *Nature* 261:333–335, 1976.

221. Zurier RB, Weissman G, Hoffstein S, Kammerman S, Tai HH: Mechanisms of lysosomal enzyme release. II. Effects of cAMP and cGMP, autonomic agonists and agents which affect microtubule function. *J Clin Invest* 53:297, 1974.

Added in Proof:

222. Weiland GA, Molinoff PB: Quantitative analysis of drug-receptor interactions: 1. Determination of kinetic and equilibrium properties. *Life Sci* 29:313–330, 1981.

223. Engel G, Hoyer D: [^{125}I]BE2254, a new high affinity radioligand for α_1-adrenoceptors. *Eur J Pharmacol* 73:221–224, 1981.

224. Glusa E, Markwardt F: Influence of clonidine-like hypotensive drugs on adrenergic platelet reactions. *Biochem Pharmacol* 30:1359–1360, 1981.

225. Mukherjee A: Characterization of α_2-adrenergic receptors in human platelets by binding of a radioactive ligand [^{3}H]yohimbine. *Biochim Biophys Acta* 676:141–148, 1981.

CHAPTER NINETEEN

α-Adrenoceptors in Cardiovascular Regulation

WALTER KOBINGER, M.D.

INTRODUCTION

One of the most important achievements in biological science was the discovery of chemical transmitters as mediators of information from and within the nervous system (67). Neurotransmitters stimulate receptors on target cells and trigger subsequent reactions. The target cell could be an effector organ, such as a muscle or secretory gland, or another neuron.

Neurotransmitters have been studied by investigating their storage, release, synthesis and degradation mechanisms. Receptors, however, cannot be studied by the same biochemical techniques. Instead, pharmacological methods, such as the interaction of drugs between antagonists and agonists, are used to characterize the receptor. Most of the substances used to study receptors are not natural transmitters, but are synthetic drugs of different chemical structures. During recent years, new drugs have been synthesized which can be used as tools in α-adrenoceptor research, and utilized therapeutically in hypertension and other cardiovascular diseases. Some of these substances are shown in Table 19.1.

Figure 19.1 illustrates our increasing knowledge of α-adrenoceptors involved in cardiovascular regulation. About 15 years ago, the only functions ascribed to peripheral vascular α-adrenoceptors were mediation of vasoconstriction and blood pressure increase following receptor stimulation and vasodilatation and pressure decrease following receptor blockade (Fig. 19.1, I.). Studies of the mode of action of the antihypertensive drugs clonidine and α-methyldopa subsequently revealed α-adrenoceptors in cardiovascular centers of the brain and led to the unexpected finding that stimulation of these receptors mediates a decrease in blood pressure (Fig. 19.1, II.).

The next step was the description of presynaptic α-adrenoceptors at the peripheral adrenergic nerve endings (Fig. 19.1, III.). Stimulation of these presynaptic receptors decreased the release of the neurotransmitter norepinephrine per nerve impulse, resulting in "adrenergic neuronal inhibition." Presynaptic α-adrenoceptors which inhibit the turnover and release of central norepinephrine were found in the CNS, but it was not known if cardiovascular depression triggered by central α-adrenoceptor stimulation was ultimately due to inhibition or activation of central adrenergic

Table 19.1
Chemical Structure of Some α-Adrenoceptor Agonists and Antagonists

1. α-ADRENOCEPTOR AGONISTS

1.1. AGONISTS LACKING HYPOTENSIVE EFFECT AFTER SYSTEMIC ADMINISTRATION.

CHEMICAL GROUP	STRUCTURE	NAME	R 1)
IMIDAZOLINE	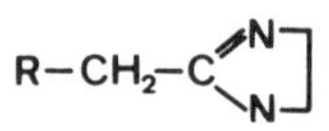	NAPHAZOLINE	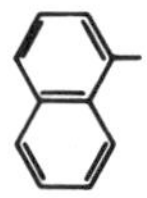
		OXYMETAZOLINE	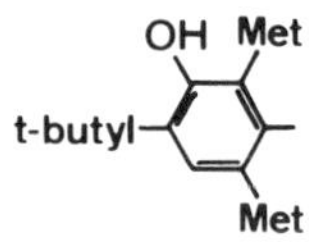
		ST 91	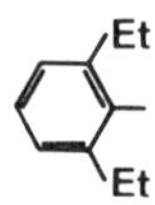
PHENYLETHYLAMINE	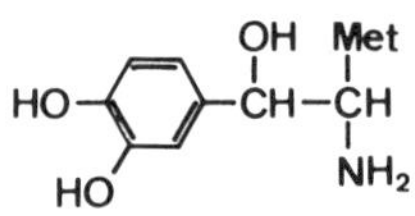	α-METHYLNOREPINEPHRINE	

1) MET,METHYL ; ET,ETHYL

1.2. CLONIDINE LIKE SUBSTANCES

CHEMICAL GROUP	STRUCTURE	NAME
IMINOIMIDAZOLIDINE	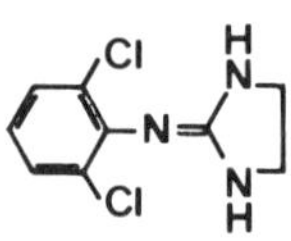	CLONIDINE
AMINO-DIHYDRO-THIAZINE	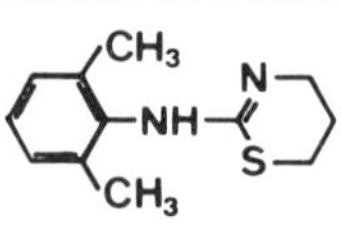	XYLAZINE
GUANIDINE	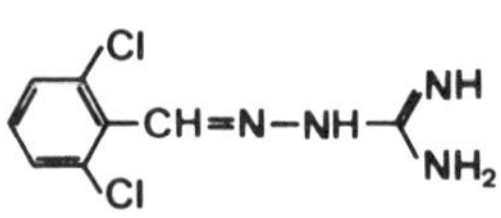	GUANABENZ
AZEPINES	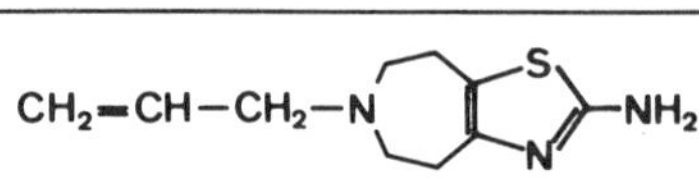	B-HT 920
	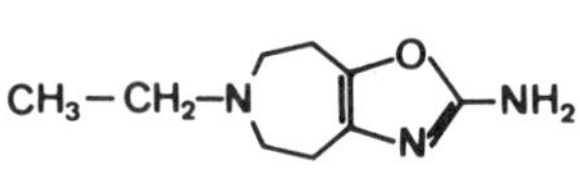	B-HT 933 (AZEPEXOLE)

TABLE 1 (CONTINUED 2)

2. α-ADRENOCEPTOR ANTAGONISTS.

CHEMICAL GROUP	STRUCTURE	NAME
IMIDAZOLINE	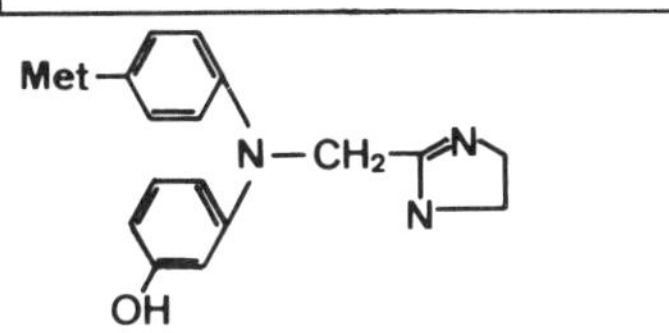	PHENTOLAMINE
PIPERAZINE	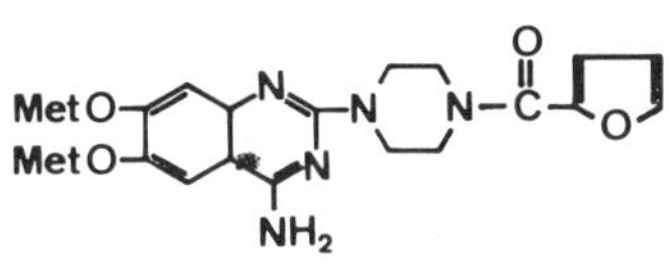	PRAZOSIN
INDOLEALKYLAMINE	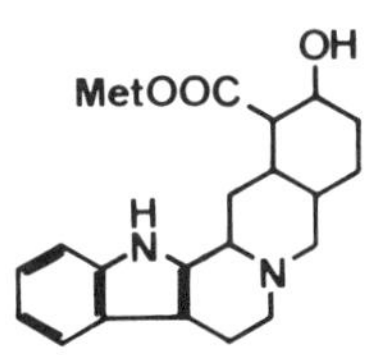	YOHIMBINE

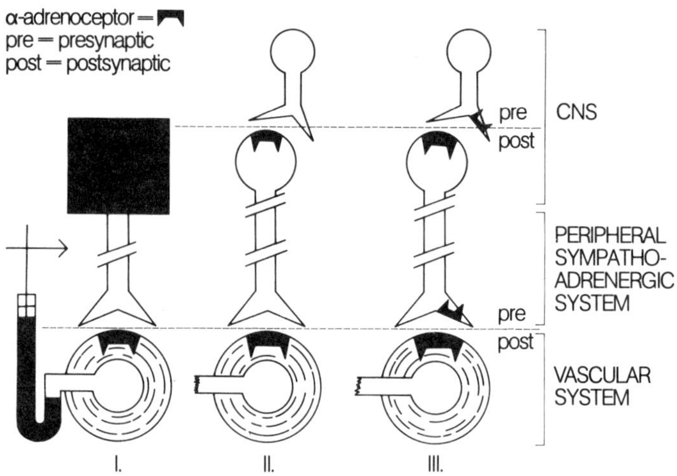

Figure 19.1. Schematic representation of our increasing information on α-adrenoceptors in cardiovascular regulation. (*I.*) Only postsynaptic (*post*) α-adrenoceptors were known. In the vascular system stimulation of these receptors leads to vasoconstriction and blood pressure increase. *Black box*, receptor systems in the CNS. (*II.*) Description of α-adrenoceptors in the CNS. Stimulation of these receptors mediate a decrease in peripheral sympathetic nerve activity and thereby a decrease in blood pressure and heart rate. (*III.*) Description of presynaptic (*pre*) α-adrenoceptors located on adrenergic neurons. Stimulation of these receptors decrease the activity of the presynaptic neuron and thereby decrease release of the adrenergic transmitter (norepinephrine, epinephrine).

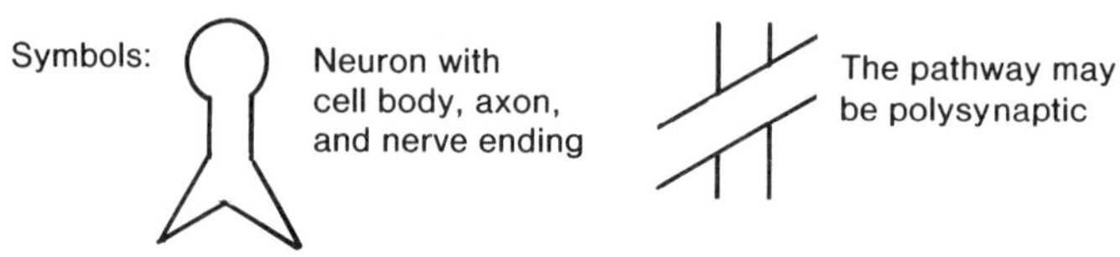

neurons. After peripheral adrenoceptors were classified into α_1 and α_2 subtypes (Fig. 19.2), the question arose as to whether central cardiovascular depression is mediated by either or both subtypes.

α-ADRENOCEPTOR SUBTYPES IN THE PERIPHERAL NERVOUS SYSTEM

Pre- and Postsynaptic α-Adrenoceptors

α-Adrenoceptors can be differentiated on a morphological basis into pre- and postsynaptic receptors (for reviews see refs. 63, 85, 87; Fig. 19.2). The "classical" postsynaptic receptor mediates responses of muscular, secretory, or other effector cells. The presynaptic receptor is located on or near adrenergic terminals and modifies the release of norepinephrine. Stimulation of presynaptic α-adrenoceptors was shown to decrease norepinephrine release for each stimulation impulse. Thus, clonidine decreased the outflow of norepinephrine into the perfusion fluid in isolated perfused rabbit hearts following electrical stimulation of the sympathetic cardiac nerve (90, 104). Analogous results were obtained using several α-adrenoceptor agonists of the clonidine type, imidazolines without hypotensive action, and phenylethylamines. These drugs decreased organ responses and/or norepinephrine release after electrical stimulation of the corresponding adrenergic nerve (4, 10, 70, 72, 83, 84, 104). This inhibitory effect of α-adrenoceptor agonists was shown to be dependent on the frequency of the electrical stimulation: the lower the frequency, the higher the inhibition. This effect was explained by the high transmitter concentration around the receptor during high-frequency stimulation, which leaves little room for any further inhibition by an exogenous agonist (84).

The important conclusion that pre- and postsynaptic α-adrenoceptors are structurally different was based on dissimilar activity ratios of various drugs. Starke *et al.* (88) used isolated pulmonary artery strips of rabbits to study the change in [^{3}H] overflow after preincubation with [^{3}H]norepinephrine (presynaptic re-

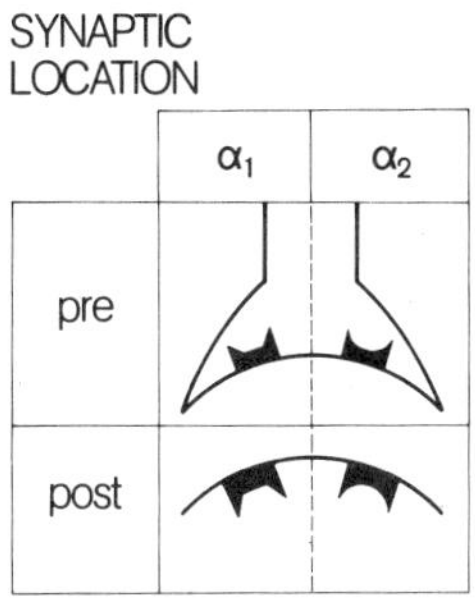

Figure 19.2. α-Adrenoceptors at adrenergic neuron terminals (presynaptic = pre) and target tissues (postsynaptic = post). The classification into the subtypes pre- and postsynaptic is morphological. The classification α_1-α_2 is pharmacological, based on different responses to different drugs (agonists and antagonists). As shown in the figure, the subtypes α_1 as well as α_2 might be present pre- as well as postsynaptically.

sponse), and in the same preparation, muscular contractions (postsynaptic response). In this study, 2 Hz of transmural electrical stimulation was applied. For nine agonists equieffective concentrations were evaluated and pre/postsynaptic activity ratios calculated. The drugs were classified into three groups: group 1 (preferentially postsynaptic agonists), methoxamine, phenylephrine; group 2 (similar pre- and postsynaptic potencies), norepinephrine, epinephrine, and naphazoline; group 3 (preferentially presynaptic agonists), clonidine, oxymetazoline, α-methylnorepinephrine, and tramazoline. Later, pre/postsynaptic activity ratios were determined for a great number of agonists, and their rank orders are in good agreement as determined by different methods and in various tissues *in vitro* and *in vivo* (review: ref. 87).

Evidence that pre- and postsynaptic α-adrenoceptors are not identical was also gained from experiments with α-adrenoceptor antagonists. In isolated perfused cat spleen, phenoxybenzamine was about 100 times more potent in blocking postsynaptic responses (increase in perfusion pressure), than in increasing presynaptically induced outflow of dopamine-β-hydroxylase activity by sympathetic nerve stimulation. On the other hand, the dissociation between pre- and postsynaptic potencies was much less apparent with phentolamine (9; for review see ref. 63). For the further development of α-adrenoceptor subtypes, it is important to note that yohimbine and rauwolscine appeared preferentially active at presynaptic sites while prazosin and phenoxybenzamine seemed preferentially active at postsynaptic sites (7, 103).

Although the theory of presynaptic α-adrenoceptors is widely accepted, results have been presented which cast some doubts as to whether the presynaptic sites regulating norepinephrine efflux are correctly designed as α-adrenergic (37, 38). Two *in vivo* methods that have been used repeatedly in many studies are outlined briefly: (a) In spinal (transection at C_2 and destruction of the brain) or pithed rats (destruction of brain and spinal canal by a steel rod), increases in blood pressure were recorded following i.v. injection of agonists as a parameter for *postsynaptic* α-adrenoceptor stimulation of arterial vasoconstriction. (b) In pithed rats electrical stimulation was applied to the steel rod in the spinal canal, which was isolated except at the level of C_7-Th_1 where the sympathetic roots emerge to the cardiac nerves (20). Electrical stimulation increases heart rate, and this increase is inhibited by α-adrenoceptor-stimulating drugs—provided that low-frequency stimulation is used (0.2 Hz). This inhibition of electrical stimulation is considered due to stimulation of *presynaptic* α-adrenoceptors: it is frequency dependent (see above), antagonized by several α-adrenoceptor blocking agents, and the rank order of activity of agonists and antagonists correspond well with more specific methods which directly measure transmitter release following postganglionic nerve stimulation (87). Some examples for post/presynaptic activity ratios shown by these two methods can be seen in Tabel 19.2.

Physiological and Pharmacological Meaning of Pre- and Postsynaptic α-Adrenoceptors

It has been suggested that presynaptic α-adrenoceptors operate a negative feedback system, regulating the release of endogenous norepinephrine (14, 73, 84). Norepinephrine released at nerve endings activates the presynaptic α-adrenoceptors and, above a threshold concentration, limits further norepinephrine release. Evidence for this hypothesis has been summarized by Langer (63) and Starke (85). Drugs with presynaptic agonistic activity should decrease blood pressure and heart rate, provided the postsynaptic (vasoconstrictor) effect does not outweigh this presynaptic mechanism.

Clonidine, a centrally acting antihyperten-

Table 19.2
α_1/α_2 Adrenoceptor Selectivity and Postsynaptic/Presynaptic α-Adrenoceptor Potency Ratio of Various Agonists[a]

Agonist	D_{10} (mg/kg i.v.)		$\frac{D_{10R}}{D_{10P}}$	$\frac{ID_{50}}{PD_{30}}$
	Rauwolscine	Prazosin		
B-HT 920	1.25	100.0	0.012	0.09
B-HT 933 (Azepexole)	0.7	9.0	0.08	0.22
α-*m*-Norepinephrine	1.9	10.0	0.19	
Xylazine	3.6	2.8	1.28	0.46
Clonidine	5.5	0.7	7.86	0.45
Norepinephrine	4.8	0.55	8.73	
Epinephrine	3.7	0.12	30.83	
St 91	14.0	0.36	38.89	1.45
Phenylephrine	9.5	0.06	158.33	
Methoxamine	15.0	0.07	214.28	5.88

[a] Reprinted with permission from W. Kobinger and L. Pichler (53). The blood pressure-increasing effect of the agonists was determined in pithed rats by means of dose-response curves. The α-adrenoceptor antagonists rauwolscine and prazosin shifted the dose-response curves to the right (see Fig. 19.3). From curves analogous to Figure 19.4 those antagonist doses were evaluated, which caused a 10-fold shift of agonist dose-response curves to the right (D_{10}). The ratio D_{10} rauwolscine/ D_{10} prazosin (D_{10R}/D_{10P}) was proposed as a measure of the α_1/α_2 selectivity for the agonists tested. In these experiments animals were treated with a β-adrenoceptor blocking agent. The post/presynaptic activity ratio is expressed by the ratio ID_{50}/PD_{30}, where ID_{50} = dose which inhibited electrically induced tachycardia in pithed rats by 50% (presynaptic effect); PD_{30} = dose which increased blood pressure in spinal rats by 30 mm Hg (postsynaptic effect). The ratios D_{10R}/D_{10P} and ID_{50}/PD_{30} are significantly correlated ($r = 0.998$; $p < 0.001$; $n = 6$).

sive drug, is a potent stimulator of presynaptic α-adrenoceptors. It has been argued that at least part of its action might be due to a mechanism at peripheral adrenergic nerves. Pichler and Kobinger (70) compared cardiovascular actions of clonidine with other imidazoline derivatives, oxymetazoline and St 91, which, in contrast to clonidine, do not penetrate well into the CNS. In isolated perfused hind limbs of cats all three drugs exerted approximately equal presynaptic "adrenergic nerve inhibition," demonstrated by a decreased response to the vasoconstriction elicited by low-frequency electrical stimulation of the lumbar sympathetic chain. The three drugs were also equally active as transient vasoconstrictors. However, only the centrally active clonidine decreased heart rate substantially and lowered blood pressure. These results therefore did not support a role of presynaptic α-adrenoceptors in the blood pressure-lowering effect of clonidine.

Electrical stimulation of sensory fibers in the tibial nerve of cats elicits a reflex increase in blood pressure and heart rate (101). This somatosympathetic reflex was inhibited by clonidine, oxymetazoline, and tramazoline (preferentially presynaptic α-adrenoceptor agonists) but not by phenylephrine (preferentially postsynaptic α-adrenoceptor agonist). The inhibitory effect was localized to peripheral α-adrenoceptors, an argument in favor of the participation of presynaptic α-adrenoceptors in cardiovascular regulation.

In urethane-anesthetized, vagotomized rats a great number of drugs with strong and preferential presynaptic α-adrenoceptor stimulating action (such as oxymetazoline, St 91, and naphazoline) did not reduce heart rate (45, 51). However, in pentobarbital-anesthetized, vagotomized rats, oxymetazoline caused bradycardia.

These differing results show the importance of experimental circumstances in demonstrating cardiovascular responses that may be due to peripheral inhibition of norepinephrine release mediated by presynaptic α-adrenoceptors. Recently Fitz-Gerald *et al.* (16) used six healthy humans to investigate the effect of two peripherally acting agonists with different pre/postsynaptic activity ratios (89, Table 19.2): α-methylnorepinephrine (preferential presynaptic) and phenylephrine (preferential postsynaptic). Blood pressure, heart rate, and norepinephrine plasma levels were measured during supine or upright positions and also during bicycle exercise. Both drugs were infused i.v. in equipressor doses (*i.e.*, comparable postsynaptic effects) and in nonpressor doses (subthreshold). There were no differences between the drugs with respect to the hemodynamic responses or increases in plasma norepinephrine in response to changes in posture and exercise. These results do not provide evidence for the importance of presynaptic α-adrenoceptors in regulation of hemodynamics and norepinephrine release under physiological conditions in man; they are supported by the reports of Reid *et al.* (75) on the failure of clonidine to lower blood pressure in tetraplegic patients (discussed later).

α_1 AND α_2 ADRENOCEPTOR CHARACTERIZATION

Because pre- and postsynaptic α-adrenoceptors do not appear identical, the suggestion was made to refer to the postsynaptic as α_1 and to the presynaptic as α_2 type (62). Berthelsen and Pettinger (65) pointed out that some drugs with high activity at presynaptic sites are also potent agonists at some postsynaptic effector cells (e.g., α-methyl-norepinephrine on frog skin melanocytes) and suggested the differentiation of α_1 and α_2 subtypes on functional rather than anatomical grounds. The rank order of potencies of various drugs can provide a pharmacological basis of differentiation. Wikberg (105) suggested that regardless of anatomical location or function, α_1 might denote an adrenoceptor where the potency of phenylephrine is higher than that of clonidine, whereas at α_2-receptors, the potency of clonidine is higher than that of phenylephrine. The same technique was used some years ago by Ahlquist (1) to differentiate adrenoceptors into the subtypes α and β. Further investigations led to the assumption that α_1 and α_2 adrenoceptors are present at pre- as well as at postsynaptic sites of adrenergic neurons and effector organs (52, 53; Fig. 19.2). Prior investigations, which indicated the existence of different α-adrenoceptor types at postsynaptic vascular sites, were recently reviewed by Timmermans and van Zwieten (96). Evidence for the coexistence of α_1 and α_2-adrenoceptors at postsynaptic vascular sites was presented by Drew and Whiting (13), Docherty *et al.* (11), Timmermans and van Zwieten (95), and Kobinger and Pichler (52). This evidence is based on the differential antagonism of one α-adrenoceptor blocker against two different agonists. The left curve in Figure 19.3*b* demonstrates the postsynaptic agonistic effect of B-HT 920 alone, which increased the blood pressure of pithed rats in a dose-dependent manner. After treatment with 1 mg/kg of the antagonist yohimbine, the dose-response curve of B-HT 920 is shifted more than 10 dose units to the right along the abscissa. Figure 19.3*c* shows dose-response curves for methoxamine, and also reveals the much weaker effect of yohimbine: 1 mg/kg of the antagonist shifts the agonist's curve only approximately two times to the right. Further inspection of Figure 19.3 shows the opposite action with prazosin, which was much more effective against methoxamine than against B-HT 920. In this study, dose-response curves of three agonists, B-HT 920, methoxamine, and clonidine were used, as well as different doses of the antagonists yohimbine and prazosin. The potency of the antagonists (doses plotted on abscissas in Fig. 19.4) were expressed by the shift of the agonist's dose-response curves to the right (ordinates in Fig. 19.4). The differential effects of each antagonist against the three agonists can be best explained by the assumption of two different receptor subtypes (or receptor sites): one receptor type (α_1) with high affinity to methoxamine and prazosin, and another type (α_2) with high affinity to B-HT 920 and yohimbine (52). Analogous results were obtained at presynaptic sites when the cardiac nerve was electrically stimulated from the spinal canal of pithed rats and the heart rate observed. Tachycardia was antagonized differentially by either yohimbine or prazosin. As revealed in Figures 19.4*c* and *d*, yohimbine was again more active as antagonist for B-HT 920 than for methoxamine, whereas prazosin showed the reverse potency. These results led to the same explanations offered for the postsynaptic vascular site: two receptor subtypes must be assumed, which exert the pharmacological characteristics of α_1 and α_2 at presynaptic adrenergic neuronal sites.

The following arguments are good evidence that pre- and postsynaptic α_2-adrenoceptors are similar structures: (a) the α_2-selective antagonist yohimbine antagonized the responses of the selective α_2-agonist B-HT 920 to approx-

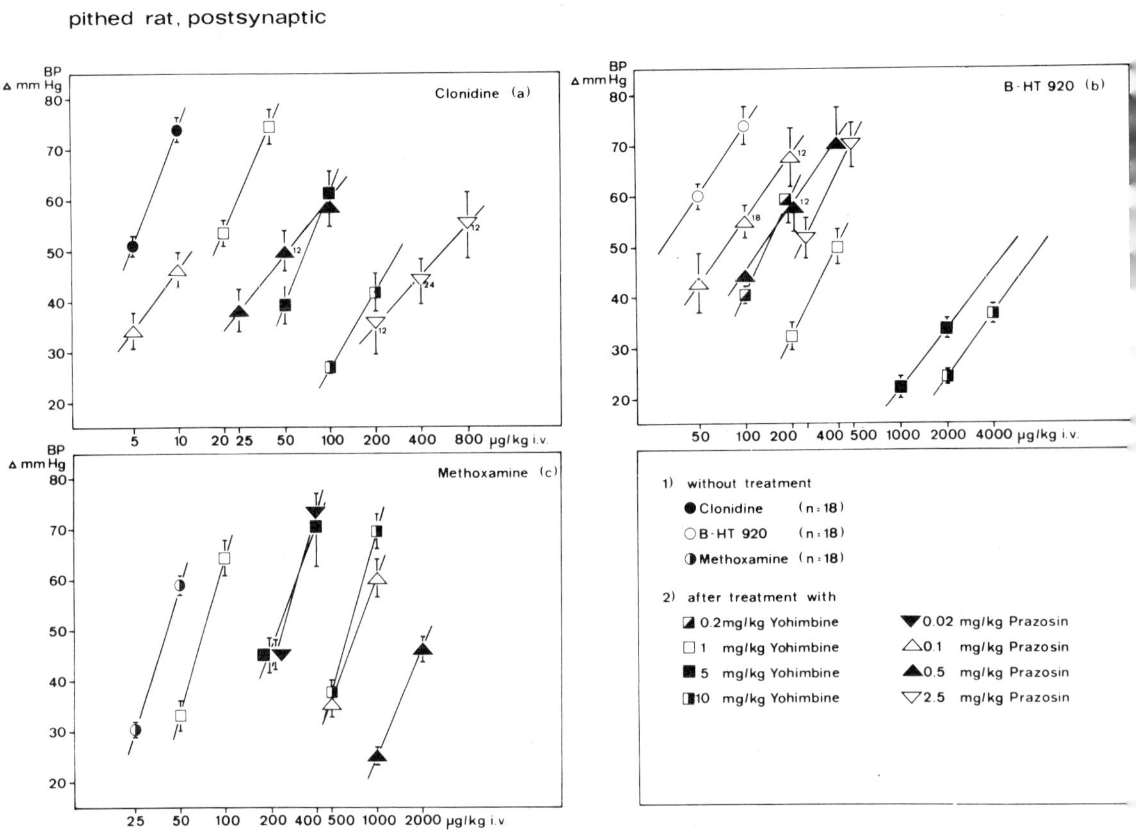

Figure 19.3. Postsynaptic activity of three agonists (clonidine, B-HT 920, methoxamine) without and after treatment with two antagonists (yohimbine, prazosin). Abcissae: doses of agonists. Ordinates: maximal blood pressure increase in pithed rats, following i.v. injection of the agonists. Note that a given dose of yohimbine shifts the dose-response curve of B-HT 920 much more to the right than of methoxamine, and that a given dose of prazosin shifts the dose-response curve of methoxamine much more to the right than of B-HT 920. (Reproduced with permission from W. Kobinger and L. Pichler (52).)

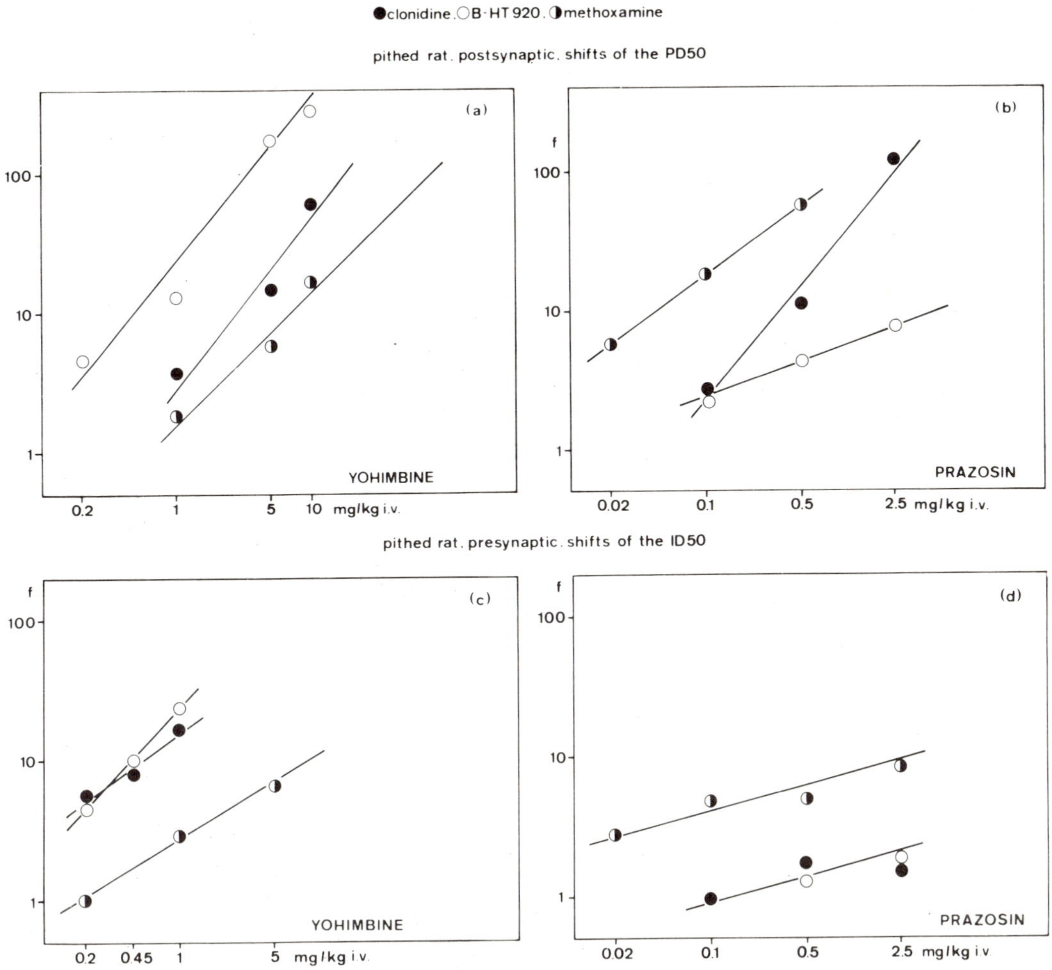

Figure 19.4. Antagonistic potency of yohimbine and prazosin against the α-adrenoceptor agonists clonidine (●), B-HT 920 (○), and methoxamine (◑). The upper part (*a* and *b*) was evaluated from the results presented in Figure 19.3 (postsynaptic activity, blood pressure increase in pithed rats after i.v. injection of agonists), the lower part (*c* and *d*) from analogous dose-response curves produced in pithed rats with electrically stimulated cardiac-sympathetic nerves. Thereby the heart rate was increased, and this was inhibited by i.v. injection of the agonists (presynaptic activity). Abscissae: dose of antagonists on logarithmic scale. Ordinates: ratios (*f*) of the agonists doses with/without antagonist (logarithmic scale), whereby doses were evaluated from dose-response curves at the level of blood pressure increase 50 mm Hg (postsynaptic effect, see Fig. 19.3) and 50% inhibition of the electrically induced tachycardia (presynaptic effect). (Reproduced with permission from W. Kobinger and L. Pichler (52).)

imately the same degree at both sites (52): a 10-fold shift of the dose-response curves was obtained with 0.45 mg/kg and 0.5 mg/kg yohimbine at the pre- and postsynaptic sites, respectively. (b) Following blockade of α_1-adrenoceptors by prazosin, the order of potencies of the agonists oxymetazoline, Guanabenz, clonidine, and xylazine at postsynaptic sites (blood pressure increase and contraction of anococcygeus muscle) was the same as their order of potency determined at the presynaptic site in the pithed rat (12). The presence of two α-adrenoceptor subtypes with the characteristics of α_1 and α_2 has also been demonstrated on the nicitating membrane of pithed rats, by the differential antagonism of prazosin and yohimbine, against the agonists methoxamine or B-HT 920 (53). Although these findings seem to contradict former research that sought to demonstrate differences between pre- and postsynaptic α-receptors, the discrepancies will be explained by the different selectivities of drugs used and different α_1/α_2 receptor ratios at pre- and postsynaptic sites (discussed further in following sections).

α_1/α_2 Selectivity Ratio of Various Substances

Figure 19.4 reveals the high affinity of B-HT 920 to the α_2 and that of methoxamine to the α_1 adrenoceptor subtype. The intermediate position of the interaction curves for clonidine indicates an affinity of clonidine for both receptor subtypes. The experimental approach shown in Figure 19.4*a* and *b* was then extended to a number of agonists (53) based on the findings that (a) both receptor types (α_1 and α_2) are present in the postsynaptic vascular preparation (blood pressure response in the pithed rat) and on (b) the selectivity of the two antagonists prazosin (α_1) and yohimbine or its stereoisomer rauwolscine (α_2) for the two receptor types (92). To include drugs with β-adrenoceptor-stimulating activities, the rats were pretreated with a β-adrenoceptor blocking agent (toliprolol).

In plottings similar to those in Figure 19.4*a* and *b*, those doses of the antagonist rauwolscine (R) and prazosin (P), respectively, were determined, which shifted the dose-response curve for an agonist 10-fold to the right along the abscissa (D_{10}); D_{10R} and D_{10P} are given in Table 19.2. The ratio D_{10R}/D_{10P} was then proposed as a measure of an agonist's α_1/α_2 selectivity ratio. These ratios for various agonists are presented in Table 19.2 and exerted a range over four orders of magnitude. B-HT 920 thereby appears as the agonist with the lowest, and methoxamine with the highest α_1/α_2 selectivity. The natural transmitters epinephrine and norepinephrine exerted affinities to both receptor subtypes as did clonidine.

Timmermans *et al.* (93) determined the selectivity of various antagonists in an analogous way, using the blood pressure increasing effect of the two highly selective agonists phenylephrine (α_1) and B-HT 933 (α_2) (Table 19.2) in the pithed rat preparations. The potencies of the antagonists to shift dose-response curves of the two agonists were expressed as K_B values (apparent dissociation constants); the ratios $K_B\alpha_2/K_B\alpha_1$ were calculated as a measure of selectivity of 11 antagonists which varied within four orders of magnitude. The highest selectivity at the subtype α_1 was revealed for prazosin, at α_2 for rauwolscine and yohimbine, while phentolamine had an intermediate position.

Relative Distribution of α_1 and α_2 Adrenoceptors in Various Tissues or Organs

To approach an estimate of the α_1/α_2 adrenoceptor ratio in a biological system, the selectivity of the agonists methoxamine (α_1) and B-HT 920 (α_2) was used (53, 55), and equieffective doses of both drugs were determined in different systems. In pithed rats the increase in blood pressure of 50 mm Hg (postsynaptic effect), the 50% inhibition of electrically induced tachycardia (presynaptic sympathoinhibitory effect), and the 50% contraction of the nictitating membrane (postsynaptic effect) were measured. In decapitated cats the increase of the blood pressure 50 mm Hg and in anesthetized cats the 50% contraction of the nictitating membrane (both postsynaptic effects) were used as parameters. Ratios of equieffective doses of B-HT 920/methoxamine are drawn in Figure 19.5 and were considered a measure of the relative "importance" of α_1/α_2 adrenoceptors in a tissue. This ratio is not necessarily identical with the ratio of the receptor numbers (density) because it also involves the steps between receptor interaction and physiological response. These *in vivo* experiments demonstrate that α_1 and α_2 subtypes are approximately equally "important" in the vascular system of both species; α_1 are relatively more "important" than α_2 in the nictitating membrane of both species, and the inverse relationship has been found at presynaptic receptor sites in rats. We must be aware that receptor "importance" depends on the experimental situation. *In vitro* preparations of the smooth muscles of vas deferens, pulmonary artery strips, or isolated perfused hindquarters of rats did not respond to selective α_2 agonists (43, 54), in contrast to their effectiveness *in vivo* or in isolated venous preparations (68).

Relation between Pre/Postsynaptic and α_1/α_2 Adrenoceptor Activity Ratio

From the two previous sections it appears that the response of a target system will depend on (a) the α_1/α_2-selectivity ratio of a given substance (agonist or antagonist) and (b) the α_1/α_2-receptor ratio within the target sys-

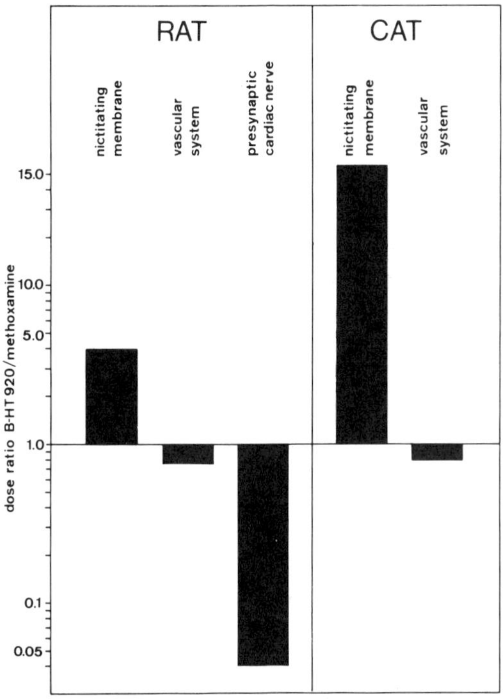

Figure 19.5. Ratios of equipotent doses (ordinate) of the two selective agonists B-HT 920 (α_2) and methoxamine (α_1), in three different test systems of two species. Postsynaptic systems: 50% contraction of the nictitating membrane in pithed rats or anesthetized cats; increase in blood pressure of 50 mm Hg in pithed rats or decapitated cats. Presynaptic system: 50% inhibition of tachycardia, as induced by electrical stimulation of the cardiac nerves in pithed rats. The dose ratio B-HT 920/methoxamine is proposed as an expression of the relative distribution of α_1/α_2 adrenoceptors in the different tissues ("receptor importance"). (Data from W. Kobinger and L. Pichler (53, 55).)

tem. Thus, the relative greater importance of α_2-adrenoceptors at presynaptic neuronal sites (small α_1/α_2 "importance") as compared with postsynaptic vascular sites, visualized in Figure 19.5, explains the reports of a high presynaptic/postsynaptic potency ratio for those drugs which are now defined by a high α_2/α_1 selectivity ratio. In the last column of Table 19.2 postsynaptic/presynaptic activity ratios for some agonists are given; they revealed a significant correlation with the corresponding α_1/α_2 selectivity ratio ($r = 0.99$; $p < 0.001$; $n = 6$; 59). Likewise, Timmermans *et al.* (93) found an excellent correlation between the selectivity ratio α_2/α_1 of 10 α-adrenoceptor antagonists and the presynaptic/postsynaptic activity ratio of these blockers as determined by other authors in isolated pulmonary artery strip preparations (7, 103).

Physiological and Pharmacological Meaning of α_1 and α_2 Adrenoceptor Subtypes

Antagonists such as phenoxybenzamine or prazosin antagonize the responses to sympathetic nerve stimulation more effectively than the responses to injected norepinephrine, and the reverse is known for other blockers such as piperoxan (for older literature, see ref. 65). Recently, these observations were explained by different locations of the α-adrenoceptor subtypes at the effector site of the vascular muscle (64, 106): the α_1-type, situated in the region of the neuroeffector junction (synaptic location) and easily reached by endogenously released transmitter, and the α_2-type, situated extrasynaptically and easily reached from the bloodstream. Different α-adrenoceptors in arteries and veins have been proposed since canine vascular rings showed different sensitivities to agonists and antagonists, indicating mainly α_2-adrenoceptors at the venous site (68).

Hirst and Neild (28) reported evidence for two different populations of excitatory sites in arteriolar smooth muscles. Both sites mediated excitatory responses to iontophoretically applied norepinephrine, but with different effects on the membrane potential and different sensitivities to α-adrenoceptor blockers, depending on the position of the micropipet. It will be interesting to find a possible correlation between these two sites and α_1 and α_2-adrenoceptors.

Structural Similarities between α_1 and α_2 Adrenoceptor Subtypes

As can be seen in the previous sections, great efforts have been made to differentiate the adrenoceptor subtypes α_1 and α_2. There are, however, some indications for structural similarities. The drugs B-HT 920 and B-HT 933 proved highly selective as α_2-agonists *in vivo*, but were ineffective as agonists in isolated perfused hind limbs of rats. In the same preparation, both drugs as well as clonidine were competitive antagonists of the vasoconstricting effect of norepinephrine and even methoxamine (43, 54). The high selectivity of methoxamine at α_1-adrenoceptors leaves little doubt that the interaction takes place at this receptor. The ranking order of the potency of these three drugs was the same when their α_2-agonistic effect *in vivo* was compared with their α_1-antagonistic effect *in vitro* (43). These results suggest that both α-adrenoceptor subtypes have common structural features. It will be a challenging task to analyze the molecular structure of such different compounds as methoxamine, clonidine, and the azepine derivative B-HT 920 to obtain information about the structure or dynamics of the α-receptors.

CENTRAL α-ADRENOCEPTORS AND THE CARDIOVASCULAR PATTERN OF CLONIDINE

There is little doubt that the antihypertensive drug clonidine played a key role in the discovery of central α-adrenoceptors. Pharmacological properties and effects of this drug on the cardiovascular system have been reviewed (41, 42, 77, 100, 108), but some of the drug's properties will be outlined briefly below, because they have not yet been used to characterize central α-adrenoceptors and their subtypes.

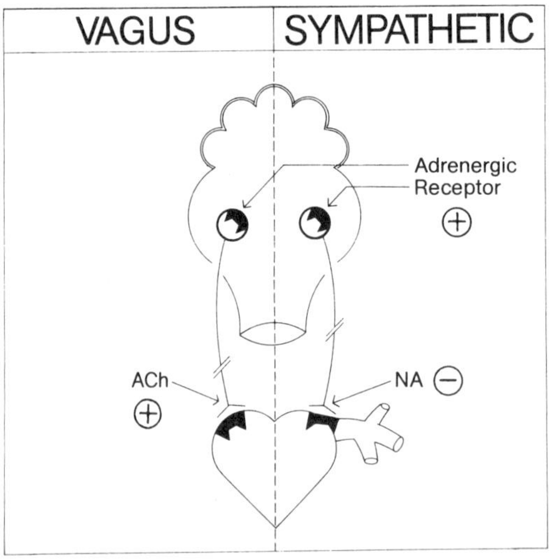

Figure 19.6. Schematic representation of the medullary autonomic nervous system and the cardiovascular system. The symbols denote receptors. *ACh*, acetylcholine, liberated at vagal nerve endings in the heart. *NA*, norepinephrine, liberated at sympathoadrenergic nerve endings in the heart and at vascular sites. Activation of central adrenoceptors (⊕) decreases the activity at peripheral adrenoceptors (⊖), and increases the activity at peripheral vagal cholinergic nerve endings. (Reproduced with permission from W. Kobinger (42).)

Rapid i.v. injection of clonidine initiates a brief increase in blood pressure followed by a long-lasting decrease. The hypotension is paralleled by a decrease in heart rate, cardiac output, and, depending on experimental conditions, decrease in total peripheral resistance (42). The initial increase in blood pressure is due to a peripheral vasoconstriction. The decrease in blood pressure, heart rate, and cardiac output is due to a central effect, as low doses of the drug (1 μg/kg) injected into the cisterna magna or into the vertebral artery of cats or dogs were effective, while the same doses given i.v. were ineffective (8, 40, 56, 76). Both parts of the autonomic nervous system are involved in the mediation of cardiovascular depression; sympathoinhibition was directly proved by a reduction of spontaneous electrical discharges in preganglionic sympathetic nerve fibers (34, 39, 79, 80). Central facilitation of the vagally mediated baroreceptor reflex was demonstrated in β-adrenoceptor-blocked dogs and rabbits after intracisternal or intraventricular injection of small doses of clonidine. The increase in blood pressure as provoked by i.v. angiotensin or aortic constriction elicited a greater bradycardia than in controls (57, 58, 61, 102). This pattern, sympathoinhibition + vagal reflex activation (Fig. 19.6), seems typical for clonidine and related central α-adrenoceptor-stimulating drugs and allows a differentiation from other drugs that lower blood pressure by different central mechanisms, such as morphine-like analgesics (50). Ablation experiments showed that the main cardiovascular effects of clonidine were localized to the medulla oblongata: cardiovascular depression, sympathoinhibition, and vagal activation were demonstrated after midbrain transection (34, 45, 46, 78).

Stimulation of Central α-Adrenoceptors Mediates Cardiovascular Depression

There are three arguments to support the hypothesis that central α-adrenoceptors mediate cardiovascular depression.

1. Clonidine causes a pattern of cardiovascular depression and activates various peripheral organs and tissues by α-adrenoceptor stimulation. The drug increased blood pressure in spinal animals (Fig. 19.3) and contracted the spleen, nictitating membrane, and various isolated vascular strips. These peripheral effects were antagonized by various α-

adrenergic blocking agents, such as phenoxybenzamine or phentolamine, and were demonstrated after surgical or pharmacological depletion of endogenous norepinephrine. These effects were thus assumed to be due to direct stimulation of peripheral α-adrenoceptors (for reviews: see refs. 42, 77). During recent years a number of "clonidine-like" drugs were described, and some of these were chemically quite different from clonidine, the azepine derivates B-HT 920 or B-HT 933 (azepexole; Table 19.1), for example. Despite chemical differences, these drugs have very similar pharmacological profiles, and the definition "clonidine-like" is based on their α-adrenoceptor-stimulating ability in combination with hypotensive properties (42).

2. Central effects of clonidine are antagonized by α-adrenoceptor blocking agents. Schmitt *et al.* (80, 82) reported that treatment with piperoxan or yohimbine antagonized the sympathoinhibitory effect of clonidine as demonstrated by registration of electrical activity of sympathetic nerve fibers in cats. Furthermore, treatment with phentolamine or piperoxan was shown to antagonize the facilitatory effect of clonidine on the vagally mediated baroreceptor reflex (45, 57, 59). The drugs act at the CNS level; 0.5 mg/kg phentolamine was effective after intracisternal injection, whereas i.v. injection of the same dose was ineffective.

3. Cardiovascular effects, similar to those following i.v. injection of clonidine, are achieved by central application of various sympathomimetic drugs. A variety of drugs which stimulate peripheral α-adrenoceptors and which increase blood pressure after systemic (i.v.) application have been shown to lower blood pressure, decrease sympathetic nerve discharges, and to facilitate the vagally mediated baroreceptor reflex after intracisternal, intracerebroventricular, vertebral arterial, or intracerebral injection. These include epinephrine (25), amphetamine (33), α-methylnorepinephrine, and norepinephrine (24, 35, 36), naphazoline, oxymetazoline, and St. 91 (46, 48). In the latter experiments (46, 48), piperoxan, yohimbine, or phentolamine antagonized the sympathomimetic drug effect.

Pre- or Postsynaptic α-Adrenoceptors as Mediators of Cardiovascular Depression?

In a previous section, the location of a presynaptic α-adrenoceptor was pinpointed as being on or near adrenergic nerve terminals. Within the CNS, presynaptic α-adrenoceptors might also be localized on cell bodies or dendrites of adrenergic neurons (somadendritic receptors (86), autoreceptors (91)). For the following considerations it is important to realize that the term "presynaptic" denotes an α-receptor, which modulates the activity of an adrenergic neuron—directly or via interneurons. Stimulation of these CNS receptors has been shown to decrease the activity of the adrenergic neuron. Liberation of norepinephrine from rat brain slices, as induced by electrical field stimulation, was inhibited by clonidine or oxymetazoline (15, 89), and the spontaneous firing rate of norepinephrine-containing cells in the locus coeruleus of rats was decreased by local application of clonidine (91).

The idea that presynaptic α-adrenoceptors in the CNS are involved in cardiovascular inhibition is based on the presence of a dense network of noradrenergic neurons in those areas of the brain which are known as centers of cardiovascular regulation (18, 97). This theory is supported by the observations that agonists (e.g., all "clonidine-like drugs") and antagonists (e.g., yohimbine), which were highly active at central cardiovascular sites, were preferentially active at peripheral presynaptic sites (89).

An agonistic action of a drug on a presynaptic receptor, as defined above, is expected to decrease the release of adrenergic transmitter from this "vasomotor center" neuron. A pharmacologic response to the agonist can only be expected if this neuron produces an effective level of the transmitter.

Kobinger and Walland (60) investigated the central pathway of the vagally mediated baroreceptor reflex after depleting catecholamine stores of dogs by pretreatment with reserpine. They described intracisternally injected clonidine as being just as effective in facilitating the reflex as in nonpretreated dogs. This finding indicated that endogenous norepinephrine is not essential for the central clonidine effect. The results from the vagal part of the response system were later extended using animals pretreated with reserpine and α-methyl-p-tyrosine (an inhibitor of catecholamine synthesis by inhibition of tyrosine hydroxylase) and using B-HT 933 (azepexole) as α-adrenoceptor agonist by intracisternal injection (44, 45, 72). Analogous experiments were carried out by recording the electrical activity of sympathetic nerves as a parameter of a central α-receptor stimulation. Hausler (22) pretreated cats with reserpine and α-methyl-p-tyrosine and observed a reduction in electrical nerve discharges after i.v. injection of clonidine. A sympathoinhibition in catecholamine-depleted cats was also seen after intracisternal injection of small doses of α-agonists, thereby avoiding reflex responses to the peripheral hypertensive action of these drugs: oxymetazoline, B-HT 933 (azepexole), and B-HT 920 (48, 72, 71).

These results practically exclude a hypotensive action of α-adrenoceptor agonists at presynaptic α-adrenoceptors on central adrenergic neurons. They suggest a direct action on effector α-adrenoceptors which operate in neurons that are not noradrenergic. It might be mentioned that the pretreatment schedule used with reserpine and α-methyl-p-tyrosine also widely excludes the participation of other catecholaminergic neurons (dopamine, epinephrine) as well as 5-hydroxytryptaminergic neurons. Furthermore, it excludes the possibility that the central cardiovascular action of clonidine-like drugs might be due to their α-adrenoceptor blocking properties (43, 54).

α_2-Adrenoceptors as Mediators of Cardiovascular Depression

After the relation between pre/post and α_1/α_2 adrenoceptors was revealed, and after the participation of presynaptic α-adrenoceptors was widely excluded, it seemed probable that the central cardiovascular receptors under discussion are of the α_2 type. Further arguments will be presented from experiments with agonists and antagonists.

All "clonidine-like" drugs as listed in Table 19.1 are agonists with high α_2/α_1 (low α_1/α_2) selectivity ratios (Table 19.2), although they differ considerably in their chemical structure. The two drugs with the highest selectivity towards α_2-adrenoceptors, B-HT 920 and B-HT 923 (azepexole), have been shown to exert the typical signs of central sympathoinhibition and vagal reflex facilitation which characterize the cardiovascular pattern of central α-adrenoceptor stimulation (49, 71). It is also of interest to note the high α_2/α_1 ratio of α-methylnorepinephrine (Table 19.2), which is structurally related more closely to phenylephrine than to any one of the clonidine-like compounds. This drug is probably the active component of the antihypertensive agent α-methyldopa from which it is derived as a metabolite and, within the CNS, has an action similar to clonidine's (26, 27; for review see ref. 108). A series of seven clonidine-like drugs from different chemical groups and with comparable brain penetrating abilities revealed a good correlation between central sympathoinhibition (vagotomized rats in urethane anesthesia, decrease in heart rate) and their potency at peripheral presynaptic sites. With respect to the "importance" of the α_2-receptor type at the peripheral presynaptic site, this correlation was used as another argument for α_2-adrenoceptors in cardiovascular centers (51).

Analogous suggestions came from observations with antagonists. Yohimbine, piperoxan, and phentolamine, now recognized as effective antagonists at peripheral α_2-adrenoceptors (93), were known to antagonize central cardiovascular actions of clonidine, whereas the more selective α_1 antagonist, phenoxybenzamine, was ineffective (59, 82).

Timmermans *et al.* (94) used the high selectivity of the stereoisomers rauwolscine and yohimbine for the α_2- and that of corynantine for the α_1-adrenoceptor type, to test their antagonism against clonidine-induced hypotension. All drugs were injected into the vertebral artery of cats, and rauwolscine and yohimbine were more effective than corynantine. Assuming similar physicochemical penetration abilities of the three isomers into the brain, these results strongly support the idea that central α_2-adrenoceptors do mediate hypotensive effects.

In recent years, specific binding sites in brain membrane preparations *in vitro* have been studied using radiolabeled α-adrenoceptor agonists and antagonists. Two distinct binding sites were recognized, one binding preferentially to prazosin and another one to clonidine, now classified as α_1 and α_2-adrenoceptors, respectively (21, 31, 32, 69, 98). Hammer *et al.* (23) studied the displacement of [^{3}H] clonidine from rat brain preparations by the nonradioactive compounds clonidine, B-HT 933 (azepexole), and B-HT 920. The order of potency of the three drugs was identical with their potencies in lowering heart rate in vagotomized urethane-anesthetized rats, a test for central sympathoinhibition. This result indicates the identity of the α_2-adrenoceptors mediating cardiovascular depression with those binding sites as determined biochemically.

Physiological and Pharmacological Meaning of α-Adrenoceptors in Central Cardiovascular Regulation

Neurons containing norepinephrine are present in brain centers concerned with autonomic nervous regulation. Dense networks of noradrenergic neurons were described in the nucleus tractus solitarius, the nucleus dorsalis nervi vagi, the reticular formation of the medulla oblongata and the hypothalamus (2, 18, 19, 97). On the other hand, α-adrenoceptor binding sites have recently been localized in the brain of rats by a light microscopic autoradiographic method (107). Following incubation of tissue sections with either [^{3}H]WB-4101 (α_1-antagonist) or [^{3}H]p-aminoclonidine (α_2-agonist), wide areas of the brain were labeled with these drugs, but with distinct differences. High density α_2-adrenoceptor labeling was found in parts of the limbic system, arcuate nucleus, locus coeruleus, and nucleus tractus solitarius. The α_1-receptor label was not concentrated in the same areas but in the olfactory bulb and the dentate gyrus of the hippocampus.

It seems logical to assume that under certain physiological conditions, endogenous norepinephrine (or epinephrine) is liberated from distinct neurons in the CNS and triggers similar responses as described for the α_2-adrenoceptor agonists of the clonidine type or others following intracisternal injection. This assumption seems even more acceptable if one considers the "physiological" response pattern to clonidine as shown by the reciprocal reaction of both parts of the autonomic nervous system (Fig. 19.6). A decrease in sympathetic and an increase in vagal activity are well known to occur during electrical stimulation of the "sympathoinhibitory" areas of the anterior hypothalamus, of the "depressor" areas of the anterior cingulate gyrus, of certain parts of the medulla ("depressor areas of the vasomotor center") and of the baroreceptor pathways (for reviews, see refs. 17, 66, 99). The observation that clonidine (and other α_2-adrenoceptor agonists) facilitated the cardiodepressor baroreceptor reflex by a central effect (57, 58) led to the hypothetical assumption that these drugs mimic endogenous norepinephrine, and that the latter is an essential link in the central part of the reflex loop. To prove this hypothesis, dogs were pretreated with a β-adrenoceptor blocking agent to exclude peripheral adrenergic influences upon the heart rate. Baroreceptor stimulation was provoked by i.v. injection of angiotensin, and the resulting bradycardia was supposed to be only mediated by the vagus, no adrenergic transmitter being involved in the peripheral part of the reflex (60). Intracisternal injection of the α-adrenoceptor blocking drug, phentolamine, markedly reduced the cardiodepressor reflex, a result in favor of the idea that norepinephrine is an essential central transmitter. However, dogs depleted of their endogenous catecholamine stores by pretreatment with reserpine still exerted a very pronounced reflex bradycardia, and this excluded the hypothesis of norepinephrine as a transmitter. In these reserpine-treated dogs, clonidine effectively facilitated the baroreceptor reflex, and this was antagonized by phentolamine. Essentially similar findings were obtained in a later study in rats pretreated with reserpine and α-methyl-p-tyrosine to exclude both synthesis and storage of catecholamines (44, 45); analogous results were obtained in dogs using the α_2 selective agonist B-HT 933 (azepexole) instead of clonidine (72). Thus, the α_2-adrenoceptors do not have an essential, but instead have a modulatory function within the baroreceptor reflex chain. Figure 19.7 proposes some sites of the central modulatory α-adrenoceptors. It might be pointed out that Figure 19.7c localizes the receptor morphologically into a presynaptic site, however, in a noncatecholamine neuron. This cannot be designated as presynaptic in the sense of an autoreceptor.

CONCLUSION

There is good evidence that clonidine-like substances act by the same mechanisms in human hypertension as in animal experiments. The decrease in sympathetic activity in hypertensive patients is accompanied by a reduction in urinary and plasma catecholamines in supine, upright, and exercising subjects (29, 30). In subjects with traumatic tetraplegia due to functional transection of the spinal cord, clonidine did not lower blood pressure (sympathetic bulbospinal pathways destroyed) but produced bradycardia (vagi intact)

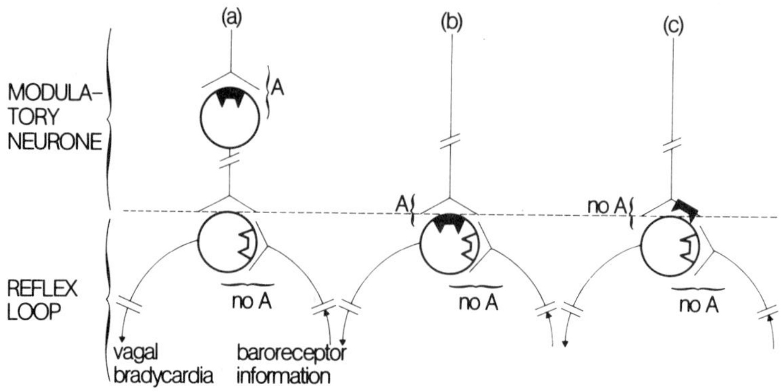

Figure 19.7. (*a* to *c*) Schematic representation of the central, medullary part of the baroreceptor reflex loop. No adrenergic link (*no A*) seems necessary for transmission in the reflex loop, as bradycardia can be elicited in spite of depletion of endogenous norepinephrine. Adrenoceptors may be incorporated in an additional neurone (or chain of neurones) which has a facilitatory influence upon the reflex loop (*a*), as an additional receptor at one of the neurones of the reflex loop (*b*) or as (presynaptic) inhibitory receptor at a modulatory nerve ending which (by means of a nonadrenergic transmitter) exerts a tonic inhibition upon the reflex (*c*). Under physiologic conditions, norepinephrine may act as a modulating transmitter (*A*) in (*a*) and (*b*). (Reproduced with permission from W. Kobinger (42).)

(74, 75). In those trials, effective clonidine plasma levels were proven.

These results show clearly that a central sympathoinhibitory effect parallels the hypotensive effect of clonidine in humans receiving therapeutic doses, and a causal link between these observations seems logical. The new compound B-HT 933 (azepexole) also decreased blood pressure in normal and hypertensive humans (5), suggesting a sympathoinhibitory function of the α_2-receptor subtype in the CNS of man, as in animals.

References

1. Ahlquist RP: A study of the adrenotropic receptors. *Am J Physiol* 153:586–600, 1948.
2. Andén NE, Dahström A, Fuxe K, Larsson K, Olson L, Ungerstedt U: Ascending monoamine neurons to the telencephalon and diencephalon. *Acta Physiol Scand* 67:313–326, 1966.
3. Armstrong JM, Cavero J, Lefèvre-Borg F: Some agents that stimulate presynaptic α_2-adrenoceptors produce bradycardia in pentobarbitone anaesthetized rats by an action on peripheral cardiac sympathetic nerves. *Br J Pharmacol* 70:169P, 1980.
4. Baum T, Shropshire AT, Rowless G, van Pelt R, Fernandez SP, Eckfeld D, Gluckmann MJ: Central pharmacological actions of the antihypertensive agent 2,6-dichlorbenzylidene aminoguanidine acetate (Wy-8678). *J Pharmacol Exp Ther* 171:276–287, 1970.
5. Benedikter L, Mey T, Middleton DA: Comparative study of the effect of azepexole and clonidine on blood pressure and pulse rate in normotensive subjects. *Eur J Clin Pharmacol* 20:321–328, 1981.
6. Berthelsen S, Pettinger WA: A functional basis for classification of adrenergic receptors. *Life Sci* 21:595–606, 1977.
7. Borowski E, Starke K, Ehrl H, Endo T: A comparison of pre- and postsynaptic effects of adrenolytic drugs in the pulmonary artery of the rabbit. *Neuroscience* 2:285–296, 1977.
8. Constantine JW, McShane WK: Analysis of the cardiovascular effects of 2-(2,6-dichlorphenylamino)-2-imidazoline hydrochloride (Catapres). *Eur J Pharmacol* 4:109–123, 1968.
9. Cubeddu LX, Barnes EM, Langer SZ, Weiner N: Release of norepinephrine and dopamine-β-hydroxylase by nerve stimulation. I. Role of neuronal and extraneuronal uptake and of alpha presynaptic receptors. *J Pharmacol Exp Ther* 190:431–450, 1974.
10. Deck R, Oberdorf A, Kroneberg G: Die Wirkung von 2-(2,6-dichlorphenylamino)-2-imidazolin Hydrochlorid (Clonidin) auf die Kontraktion und die Acetylcholinfreisetzung am isolierten koaxial elektrisch gereizten Meerschweinchenileum. *Arzneim Forsch* 21:1580–1584, 1971.
11. Docherty JR, McDonald A, McGrath JC: Further sub-classification of α-adrenoceptors in the car-

diovascular system, vas deferens and anococcygeus muscle of the rat. *Br J Pharmacol* 67:421P–422P, 1979.

12. Docherty JR, McGrath JC: A comparison of pre- and post-junctional potencies of several alpha-adrenoceptor agonists in the cardiovascular system and anococcygeus muscle of the rat. Evidence for two types of post-junctional alpha-adrenoceptor. Naunyn Schmiedebergs *Arch Pharmacol* 312:107–116, 1980.
13. Drew GM, Whiting SB: Evidence for 2 distinct types of postsynaptic α-adrenoceptors in vascular smooth muscle in vivo. *Br J Pharmacol* 67:207–215, 1979.
14. Enero MA, Langer SZ, Rothlin RP, Stefano FJE: Role of the α-adrenoceptor in regulating noradrenaline overflow by nerve stimulation. *Br J Pharmacol* 44:672–688, 1972.
15. Farneboe LD, Hamberger B: Drug-induced changes in the release of ^{3}H-monoamines from field stimulated rat brain slices. *Acta Physiol Scand Suppl* 371:35–44, 1971.
16. Fitz-Gerald GA, Watkins J, Dollery CT: Regulation of norepinephrine release by peripheral α-$_2$-receptor stimulation. *Clin Pharmacol Ther* 29:160–167, 1981.
17. Folkow B, Neil E: *Circulation*. New York, Oxford University Press, 1971.
18. Fuxe K: Evidence for the existence of monoamine neurons in the central nervous system. IV. The distribution of monoamine terminals in the central nervous system. *Acta Physiol Scand* 64(Suppl 247):37–85, 1965.
19. Fuxe K, Hökflet T, Bolme P, Goldstein M, Johansson O, Jonson G, Lidbrink P, Ljungdahl A, Sachs CH: The topography of central catecholamine pathways in relation to their possible role in blood pressure control. In Davies DS, Reid JL: *Central Action of Drugs in Blood Pressure Regulation*. Tunbridge Wells, England, Pitman Medical Publishing, 1975, pp 8–22.
20. Gillespie JS, Muir TC: A method of stimulating the complete sympathetic outflow from the spinal cord to blood vessels in the pithed rat. *Br J Pharmacol Chemother* 30:78–87, 1967.
21. Greengrass P, Bremner R: Binding characteristics of ^{3}H-prazosin to rat brain α-adrenergic receptors. *Eur J Pharmacol* 55:223–326, 1979.
22. Haeusler G: Clonidine induced inhibition of sympathetic nerve activity: No indication for a central pre-synaptic or an indirect sympathomimetic mode of action. *Naunyn Schmiedebergs Arch Pharmacol* 286:97–111, 1974.
23. Hammer R, Kobinger W, Pichler L: Binding of an imidazolidine (clonidine), an oxazoloazepin (B-HT 933) and a thiazoloazepin (B-HT 920) to rat brain α-adrenoceptors and relation to cardiovascular effects. *Eur J Pharmacol* 62:277–285, 1980.
24. Heise A, Kroneberg G. α-Sympathetic receptor stimulation in the brain and hypotensive action of α-methyl-DOPA. *Eur J Pharmacol* 17:315–317, 1972.
25. Heller, H: Über die zentrale Blutdruckwirkung des Adrenalins, *Naunyn Schmiedebergs Arch Pharmacol* 173:291–300, 1933.
26. Henning M, van Zwieten PA: Central hypotensive effect of α-methyl-DOPA. *J Pharm Pharmacol* 19:403–405, 1967.
27. Henning M, van Zwieten PA: Central hypotensive action of α-methyl-DOPA. *J Pharm Pharmacol* 20:409–417, 1968.
28. Hirst GDS, Neild TO: Evidence for two populations of excitatory receptors for noradrenaline on arteriolar smooth muscle. *Nature* 283:767–768, 1980.
29. Hökfelt B, Hansson BG, Manheim P: Effect of clonidine, metoprolol and penbutolol on plasma catecholamines and renin activity in response to sympathetic stimulation in hypertensive patients. In Fuxe K, Goldstein M, Hökfelt B, Hökfelt T: *Central Adrenaline Neurons*. Oxford, Pergamon Press, 1980, pp 317–325.
30. Hökfelt B, Hedeland H, Dymling J-F: Studies on catecholamines, renin and aldosterone following catapresan (2-(2,6-dichlor-phenylamine)-2-imidazoline hydrochloride) in hypertensive patients. *Eur J Pharmacol* 10:389–397, 1970.
31. Hoffman BB, Lefkowits RR: Alpha-adrenergic receptor subtypes. *N Engl J Med* 302:1390–1396, 1980.
32. Hoffman BB, Lefkowits RR: Radioligand binding studies of adrenergic receptors: New insights into molecular and physiological regulation. *Ann Rev Pharmacol Toxicol* 20:581–608, 1980.
33. Hoyer J, van Zwieten PA: The centrally induced fall in blood pressure after the infusion of amphetamine and related drugs into the vertebral artery of the cat. *J Pharm Pharmacol* 23:892–893, 1971.
34. Hukuhara T Jr, Otsuka Y, Takeda R, Sakai F: Die zentralen Wirkungen des 2-(2,6-dichlorphenylamino)-2-imidazolin-hydrochloride. *Arzneim Forsch* 18:1147–1153, 1968.
35. de Jong W: Noradrenaline: Central inhibitory control of blood pressure and heart rate. *Eur J Pharmacol* 29:179–181, 1974.
36. de Jong W, Nijkamp FB, Bohus B: Role of noradrenalin and serotonin in the central control of blood pressure in normotensive and spontaneously hypertensive rats. *Arch Int Pharmacodyn* 213:272–284, 1975.
37. Kalsner S, Chan CC: Adrenergic antagonists and the presynaptic receptor hypothesis in vascular tissue. *J Pharmacol Exp Ther* 211:257–264, 1979.
38. Kalsner S, Suleiman M, Dobson RE: Adrenergic presynaptic receptors: An overextended hypothesis? *J Pharm Pharmacol* 32:290–292, 1980.
39. Klupp H, Knappen F, Otsuka Y, Streller J, Teichmann H: Effects of clonidine on central sympathetic tone. *Eur J Pharmacol* 10:225–229, 1970.
40. Kobinger W: Über den Wirkungsmechanismus einer neuen antihypertensiven Substanz mit Imidazolinstruktur. *Naunyn Schmiedebergs Arch Pharmacol* 258:48–58, 1967.
41. Kobinger W: Pharmacologic basis of the cardiovascular actions of clonidine. In Onesti G, Kim KE, Moyer JH: *Hypertension, Mechanisms and Management*. New York, Grune & Stratten,

1973, pp 369–380.
42. Kobinger W: Central α-adrenergic systems as target for hypotensive drugs. *Rev Physiol Biochem Pharmacol* 81:39–100, 1978.
43. Kobinger W, Lillie C, Pichler L: Central cardiovascular α-adrenoceptors. Relation to peripheral receptors. *Circ Res* 46(Suppl 1):121–125, 1980.
44. Kobinger W, Pichler L: Evidence for direct α-adrenoceptor stimulation of effector neurons in cardiovascular centers by clonidine. *Eur J Pharmacol* 27:151–154, 1974.
45. Kobinger W, Pichler L: The central modulatory effect of clonidine on the cardiodepressor reflex after suppression of synthesis and storage of noradrenaline. *Eur J Pharmacol* 30:56–62, 1975.
46. Kobinger W, Pichler L: Localization in the CNS of adrenoceptors which facilitate a cardioinhibitory reflex. *Naunyn Schmiedebergs Arch Pharmacol* 286:371–377, 1975.
47. Kobinger W, Pichler L: Investigation into some imidazoline compounds, with respect to peripheral α-adrenoceptor stimulation and depression of cardiovascular centers. *Naunyn Schmiedebergs Arch Pharmacol* 291:175–191, 1975.
48. Kobinger W, Pichler L: Centrally induced reduction in sympathetic tone—a postsynaptic α-adrenoceptor stimulating action of imidazolines. *Eur J Pharmacol* 40:311–320, 1976.
49. Kobinger W, Pichler L: Pharmacological characterization of B-HT 933 (2-amino-6-ethyl-4,5,7,8-tetrahydro-6H-oxazolo-5,4-d-azepin-dihydrochloride) as a hypotensive agent of the "clonidine-type"). *Naunyn Schmiedebergs Arch Pharmacol* 300:39–46, 1977.
50. Kobinger W, Pichler L: Differentiation of drugs acting centrally upon the cardiovascular system by means of sympathetic and vagal responses. *Clin Exp Hypertens* 1:229–249, 1978.
51. Kobinger W, Pichler L: Relation between central sympathoinhibitory and peripheral pre- and postsynaptic α-adrenoceptors as evaluated by different clonidine-like substances in rats. *Naunyn Schmiedebergs Arch Pharmacol* 315:21–27, 1980.
52. Kobinger W, Pichler L: Investigation into different types of post- and presynaptic α-adrenoceptors at cardiovascular sites in rats. *Eur J Pharmacol* 65:393–402, 1980.
53. Kobinger W, Pichler L: α_1- and α_2-adrenoceptor subtypes: Selectivity of various agonists and relative distribution of receptors as determined in rats. *Eur J Pharmacol* 73:313–321, 1981.
54. Kobinger W, Pichler L: α_1-Adrenoceptor blockade by α_2-adrenoceptor agonists in the isolated perfused hindquarter of rats. *Eur J Pharmacol* 72:113–115, 1981.
55. Kobinger W, Pichler L. α-Adrenoceptor subtypes in cardiovascular regulation. *J Cardiovasc Pharmacol* 4:81–85, 1982.
56. Kobinger W, Walland A: Investigations into the mechanism of the hypotensive effect of 2-(2,6 dichlorphenylamino)-2-imidazoline HCl. *Eur J Pharmacol* 2:155–162, 1967.
57. Kobinger W, Walland A: Involvement of adrenergic receptors in central vagus activity. *Eur J Pharmacol* 16:120–122, 1971.
58. Kobinger W, Walland A: Evidence for a central activation of a vagal cardiodepressor reflex by clonidine. *Eur J Pharmacol* 19:203–209, 1972.
59. Kobinger W, Walland A: Facilitation of vagal reflex bradycardia by an action of clonidine on central α-receptors. *Eur J Pharmacol* 19:210–217, 1972.
60. Kobinger W, Walland A: Modulating effect of central adrenergic neurons on a vagally mediated cardioinhibitory reflex. *Eur J Pharmacol* 22:344–350, 1973.
61. Korner PJ, Oliver JR, Sleight P, Chalmers JP, Robinson JS: Effects of clonidine on the baroreceptor-heart rate reflex and on single aortic fibre discharge. *Eur J Pharmacol* 28:189–198, 1974.
62. Langer SZ: Presynaptic regulation of catecholamine release. *Biochem Pharmacol* 23:1773–1800, 1974.
63. Langer SZ: Presynaptic receptors and their role in the regulation of transmitter release. *Br J Pharmacol* 60:481–497, 1977.
64. Langer SZ, Massingham R, Shepperson NB: Differential sensitivity to prazosin blockade of endogenously released and exogenously administered noradrenaline: Possible relationship to the synaptic location of α_1- and the extrasynaptic location of α_2-adrenoceptors in dog vascular smooth muscle. *Br J Pharmacol* 72:123P, 1981.
65. Levin JA, Beck L: Selective reduction in neurogenically induced constriction by phenoxybenzamine. *J Pharmacol Exper Ther* 155:31–41, 1967.
66. Löfwing B: Cardiovascular adjustments induced from the rostral cingulate gyrus. With special reference to sympathoinhibitory mechanisms. *Acta Physiol Scand* 53(Suppl 184):1–82, 1961.
67. Loewi O: Über humorale Übertragbarkeit der Herznervenwirkung. I. Mitteilung. *Pfluegers Arch Physiol* 189:239, 1921.
68. De Mey JG, Vanhoutte PM: Differences in pharmacological properties of postjunctional α-adrenergic receptors among arteries and veins. *Arch Int Pharmacodyn* 244:328–329, 1980.
69. Miach PJ, Dausse JP, Meyer P: Direct biochemical demonstration of two types of α-adrenoceptors in rat brain. *Nature* 274:492–494, 1978.
70. Pichler L, Kobinger W: Presynaptic activity at peripheral adrenergic sites and blood pressure effect of α-adrenoceptor stimulating drugs. *Eur J Pharmacol* 52:287–295, 1978.
71. Pichler L, Kobinger W: Centrally mediated cardiovascular effects of B-HT 920 (6-allyl-2-amino-5, 6, 7, 8-tetrahydro-4H-thiazolo-(4,5-d)-azepine dihydrochloride), a hypotensive agent of the "clonidine type." *Cardiovasc Pharmacol* 3:269–277, 1981.
72. Pichler L, Placheta P, Kobinger W: Effect of azepexole (B-HT 933) on pre- and postsynaptic α-adrenoceptors at peripheral and central nervous sites. *Eur J Pharmacol* 65:233–241, 1980.
73. Rand MJ, Story DF, Allen GS, Glover AB, Mc Culloch MW: Pulse-to-pulse modulation of nor-

adrenaline release through a prejunctional α-receptor auto-inhibitory mechanism. In Ustin E, Snyder SH: *Frontiers in Catecholamine Research.* New York, Pergamon Press, 1973, pp 579–581.
74. Reid JL, Mathias JC, Jones DH, Wing LMH: The contribution of central and peripheral adrenoceptors to the actions of clonidine and alpha methyldopa in man. In Fuxe K, Goldstein M, Hökfelt B, Hökfelt T: *Central Adrenaline Neurons.* Oxford, Pergamon Press, 1980, pp 317–325.
75. Reid JL, Wing LMH, Mathias CJ, Frankel HL, Neill E: The central hypotensive effect of clonidine. Studies in tetraplegic subjects. *Clin Pharmacol Ther* 21:375–381, 1977.
76. Sattler RW, Van Zwieten PA: Acute hypotensive action of 2-(2,6-dichlorophenylamino)-2-imidazoline hydrochloride (St 155) after infusion into the cat's vertebral artery. *Eur J Pharmacol* 2:9–13, 1967.
77. Schmitt H: The pharmacology of clonidine and related products. In Gross F: *Antihypertensive Agents.* Berlin, Springer Verlag, 1977, pp 299–396.
78. Schmitt H, Schmitt H: Localization of the hypotensive effect of 2-(2,6-dichlorophenylamino)-2-imidazoline-hydrochloride (St 155, Catapresan). *Eur J Pharmacol* 6:8–12, 1969.
79. Schmitt H, Schmitt H, Boissier JR, Giudicelli JF: Centrally mediated decrease in sympathetic tone induced by 2-(2,6-dichlorophenylamino)-2-imidazoline (St 155, Catapresan). *Eur J Pharmacol* 2:147–148, 1967.
80. Schmitt H, Schmitt H, Boissier JR, Guidicelli JF, Fichelle J: Cardiovascular effects of 2-(2,6-dichlorophenylamino)-2-imidazoline hydrochloride (St 155). II. Central sympathetic structures. *Eur J Pharmacol* 2:340, 1968.
81. Schmitt H, Schmitt H, Fenard S: Evidence for an α-sympathomimetic component in the effects of Catapresan on vasomotor centres: Antagonism by piperoxan. *Eur J Pharmacol* 14:98–100, 1971.
82. Schmitt H, Schmitt H, Fenard S: Action of α-adrenergic blocking drugs on the sympathetic centres and their interactions with the central sympatho-inhibitory effect of clonidine. *Arzneim Forsch* 23:40–45, 1973.
83. Scholtysik G, Lauener H, Eichenberger E, Bürki H, Salzmann R, Müller-Schweinitzer E, Waite R: Pharmacological actions of the antihypertensive agent N-amino-2-(2,6-dichlorophenyl)-acetamide hydrochloride (BS 100-141). *Arzneim Forsch* 25:1483–1491, 1975.
84. Starke K: Alpha sympathomimetic inhibition of adrenergic and cholinergic transmission in the rabbit heart. *Naunyn Schmiedebergs Arch Pharmacol* 274:18–45, 1972.
85. Starke K: Regulation of noradrenaline release by presynaptic receptor systems. *Rev Physiol Biochem Pharmacol* 77:1–124, 1977.
86. Starke K: Presynaptic modulation of catecholamine release in the central nervous system: some open questions. In Langer SZ, Starke K, Dubocovich MC: *Presynaptic Receptors, Advances in the Biosciences.* Oxford, Pergamon Press, 1979, pp 129–136.
87. Starke K: α-Adrenoceptor subclassification. *Rev Physiol Biochem Pharmacol* 88:199–236, 1981.
88. Starke K, Endo T, Taube HD: Relative pre- and post-synaptic potencies of α-adrenoceptor agonists in the rabbit pulmonary artery. *Naunyn Schmiedebergs Arch Pharmacol* 291:55–78, 1975.
89. Starke K, Montel H: Involvement of alpha-receptors in clonidine-induced inhibition of transmitter release from central monoamine neurons. *Neuropharmacology* 12:1073–1080, 1973.
90. Starke K, Schümann HJ: Zur peripheren sympathicus-hemmenden Wirkung des clonidins. *Experientia* 27:70–71, 1971.
91. Svensson TH, Bunney BS, Aghajanian GK: Inhibition of both noradrenergic and serotonergic neurons in brain by the α-adrenergic agonist clonidine. *Brain Res* 92:291–306, 1975.
92. Tanaka T, Starke K: Binding of ^{3}H-clonidine to an α-adrenoceptor in membranes of guinea pig ileum. *Naunyn Schmiedebergs Arch Pharmacol* 309:207–215, 1979.
93. Timmermans PBMWM, Van Meel JCÄ, Van Zwieten PA: Evaluation of the selectivity of α-adrenoceptor blocking drugs for postsynaptic α_1- and α_2-adrenoceptor blocking drugs for postsynaptic α_1- and α_2-adrenoceptor in a simple animal model. *J Auton Pharmacol* 1:53–60, 1980.
94. Timmermanns PBMWM, Schoop AMC, Kwa HY, Van Zwieten PA: Characterization of α-adrenoceptors participating in the central hypotensive and sedative effects of clonidine using yohimbine, rauwolscine and corynantine. *Eur J Pharmacol* 70:7–15, 1981.
95. Timmermans PBMWM, Van Zwieten PA: Postsynaptic α_1- and α_2-adrenoceptors in the circulatory system of the pithed rat: Selective stimulation of the α_2-type by B-HT 933. *Eur J Pharmacol* 63:199–202, 1980.
96. Timmermans PBMWM, Van Zwieten PA: The postsynaptic α_2-adrenoceptor. *J Auton Pharmacol* 1:171–183, 1981.
97. Ungerstedt U: Stereotaxic mapping of the monoamine pathways in the rat brain. *Acta Physiol Scand* 82(Suppl 367):1–48, 1971.
98. U'Prichard DC, Charness ME, Robertson D, Snyder SH: Prazosin: Differential affinities for two populations of α-noradrenergic receptor binding sites. *Eur J Pharmacol* 50:87–89, 1978.
99. Uvnäs B: Central cardiovascular control. In *Handbook of Physiology, Section 1: Neurophysiology.* Washington, DC, American Physiological Society, 1960, Vol II, pp 1131–1162.
100. Walland A: Clonidine. In Goldberg ME, et al: *Drug Profiles: Pharmacological and Biochemical Properties of Drug Substances,* American Pharmaceutical Association, Acad Pharm Sci. New York, Academic Press, 1977, vol I, pp 67–107.
101. Walland A: Inhibition of a somato-sympathetic reflex via peripheral presynaptic α-adrenocep-

tors. *Eur J Pharmacol* 47:211–221, 1978.

102. Walland A, Kobinger W, Csongrady A: Action of clonidine on baroreceptor reflexes in conscious dogs. *Eur J Pharmacol* 26:184–190, 1974.
103. Weitzell R, Tanaka T, Starke K: Pre-, and postsynaptic effects of yohimbine stereoisomers on noradrenergic transmission in the pulmonary artery of the rabbit. *Naunyn Schmiedebergs Arch Pharmacol* 308:127–136, 1979.
104. Werner U, Starke K, Schümann HJ: Actions of clonidine and 2-[2-methyl-6-ethyl-cyclohexylamino]-2-oxazoline on postganglionic autonomic nerves. *Arch Int Pharmacodyn* 195:282–290, 1972.
105. Wikberg J: Differentiation between pre- and post-junctional α-receptors in guinea pig ileum and rabbit aorta. *Acta Physiol Scand* 103:225–239, 1978.
106. Yamaguchi J, Kopin IJ: Differential inhibition of alpha-1 and alpha-2 adrenoceptor-mediated pressor responses in pithed rats. *J Pharmacol Exp Ther* 214:275–281, 1980.
107. Young WS, Kuhar MJ: Noradrenergic α_1 and α_2 receptors: Autoradiographic visualization. *Eur J Pharmacol* 59:317–319, 1979.
108. Van Zwieten PA: Antihypertensive drugs with a central action. *Prog Pharmacol* 1:1–63, 1975.

SECTION 4

Clinical Aspects

Control of the Cardiovascular System

CHAPTER TWENTY

Effects of Catecholamines in Clinical Heart Disease

ROBERT E. GOLDSTEIN, M.D., and SOL I. RAFJER, M.D.

INTRODUCTION

Catecholamines (CA) play a crucial role in regulating mechanical and electrical performance of the heart, thereby modulating intrinsic properties of the myocardium to match changing demand for perfusion. The enormous influence that CA exert upon cardiac performance has important implications when defining pathogenic mechanisms or selecting proper therapy for many cardiac disorders. In this chapter we shall review the action of CA upon the normal cardiovascular system and discuss how CA relate to pathophysiology and treatment of myocardial ischemia, congestive heart failure, and cardiac arrhythmias.

CATECHOLAMINES AND THE NORMAL CARDIOVASCULAR SYSTEM

Sympathetic Regulation of the Circulation

Moment-to-moment adjustment of cardiovascular function is accomplished in part by responses of the heart and blood vessels to changing levels of autonomic stimulation. Heart rate, contractile function, arterial resistance, and venous capacitance are matched to physiologic requirements for blood pressure (BP) and flow by the rapid actions of the sympathetic nervous system (SNS) in concert with the reciprocally functioning parasympathetic nervous system and by the slower influence of the adrenal medulla.

Release of norepinephrine (NE) by postganglionic nerve endings in the heart and blood vessels and humoral delivery of NE and epinephrine (E) by the adrenal medulla are closely regulated by the central nervous system (CNS). Final control is exercised by sympathetic motor neurons in the intermediolateral nucleus of the thoracolumbar spinal cord. Those motor neurons specifically regulating cardiac activity are located in the first five to six thoracic segments of the spinal cord. Their fibers usually synapse in the stellate ganglia or other upper thoracic or cervical ganglia. Postganglionic fibers are distributed to the heart via multiple small cardiac nerves, which also contain afferent fibers and parasympathetic fibers. Experimental data suggest that stimulation from right-sided sympathetic outflow has greater influence on the sinoatrial node, while that from left-sided sympathetic outflow has greater influence on the atrioventricular node and left ventricular myocardium (82, 102, 117).

Neurons of the intermediolateral nucleus receive direct and indirect input from multiple centers within the neuraxis concerned with organization and integration of autonomic responses. Such centers are located in the cerebral cortex (particularly the cingulate gyrus), in the hypothalamus, and in the reticular formation of medulla, pons, and mesencephalon (78). Some centers control specific autonomic effectors, e.g., heart rate centers in the reticular formation of the medulla, while others, e.g., certain areas in the hypothalamus, induce more generalized cardiovascular responses. Medullary hypothalamic and cortical cardiovascular centers have direct connections with motor neurons of the intermediolateral nucleus. Cortical connections are probably important in coordinating cardiac and peripheral vascular responses with skeletal muscle activity. Other centers act only indirectly by modulating activity in centers with direct linkage. Regulation of sympathetic outflow is also influenced in a complex, nonlinear manner by feedback from autonomic afferents conveying information from arterial chemoreceptors and from an array of mechanoreceptors in arteries, great veins, cardiac chambers, and other locations in the thorax.

Adrenergic Receptors and the Circulation

Several different adrenergic receptor types are present within the cardiovascular system. β-adrenergic receptors mediate effects that characterize adrenergic stimulation of the heart: increase in heart rate due to accelerated depolarization within cells of the sinoatrial node (and within latent pacemaker cells elsewhere in the heart), decrease in conduction time through the atrioventricular node, and increase in the contractile state of atrial and ventricular myocardium. Many electrical effects of CA are thought to reflect enhancement of calcium conductance across sarcolemmal membranes resulting in greater "slow" inward current (14, 119, 120). This, in turn, results in more rapid diastolic depolarization of pacemaker cells and larger, more rapidly rising and more readily propagated action potentials in tissues of the atrioventricular node. The mechanical effects of CA are readily observed in either isolated myocardium or intact hearts when rate of depolarization and mechanical loading conditions are controlled (11, 14, 120, 130). CA cause myocardium with mild systolic loading to exhibit more complete and more rapid shortening. Myocardium that is held at a constant length by more severe systolic loading will develop greater peak tension and greater maximum rate of rise of tension after CA exposure. In addition, CA reduce the time span that tension is maintained and accelerate myocardial relaxation. These positive inotropic effects are thought to reflect enhancement of calcium delivery to contractile proteins with initiation of systole and acceleration in removal of calcium at the end of systole (35).

Other adrenergic receptors of the cardiovascular system are concerned with regulation of vascular tone. α_1-Adrenergic receptors constrict arterioles and veins (66). Distributed chiefly within vessels of the skin, kidney, and splanchnic bed, these receptors play an essential role in the maintenance of appropriate BP and central blood volume. α_2-Adrenergic receptors modulate adrenergic function by inhibiting release of NE from postganglionic fibers, thus completing a feedback loop. α_2-Adrenergic receptors are also present in certain vascular beds and cause vasoconstriction when activated. β_2-Adrenergic receptors cause vasodilation of arteries, arterioles, and veins. They are predominantly located in vessels supplying the myocardium, skeletal muscles, and digestive tract. Two dopamine (DA) receptors have been found in blood vessels (48, 49). DA_1 receptors cause vasodilation in renal, coronary, and cerebral vascular beds. DA_2 receptors, like α_2-adrenergic receptors, decrease NE release from postganglionic nerves. Unlike α_2-adrenergic receptors, DA_2 receptors are not present in vessels and do not produce vasoconstriction.

The mechanism by which β-adrenergic receptors stimulate myocardial contractility has been the subject of intensive study (137). These cell surface receptors activate adenylate cyclase to convert adenosine triphosphate to cyclic adenosine monophosphate (cAMP) (Fig. 20.1). Increased intracellular cAMP, in turn, activates protein kinases, leading to phosphorylation of one or more proteins concerned with regulation of contraction. Phosphorylation of proteins in the sarcoplasmic reticulum may enhance systolic release and diastolic reuptake of calcium from sequestered intracellular pools (72). Phosphorylation of the sarcolemma may enhance movement of calcium into the cell, thereby accelerating calcium-mediated calcium release from the sarcoplasmic reticulum (35). Alternatively, phosphorylation of the regulatory protein, troponin, may lead to greater calcium-induced activation of the contractile proteins, actin and myosin (63, 118). Adrenergic action is terminated with removal of CA from β-adrenergic receptors, breakdown of cAMP by phosphodiesterase, and dephosphorylation of proteins by phosphoprotein phosphatase. The capacity of phosphodiesterase inhibitors such as theophylline to interfere with breakdown of cAMP explains some of the

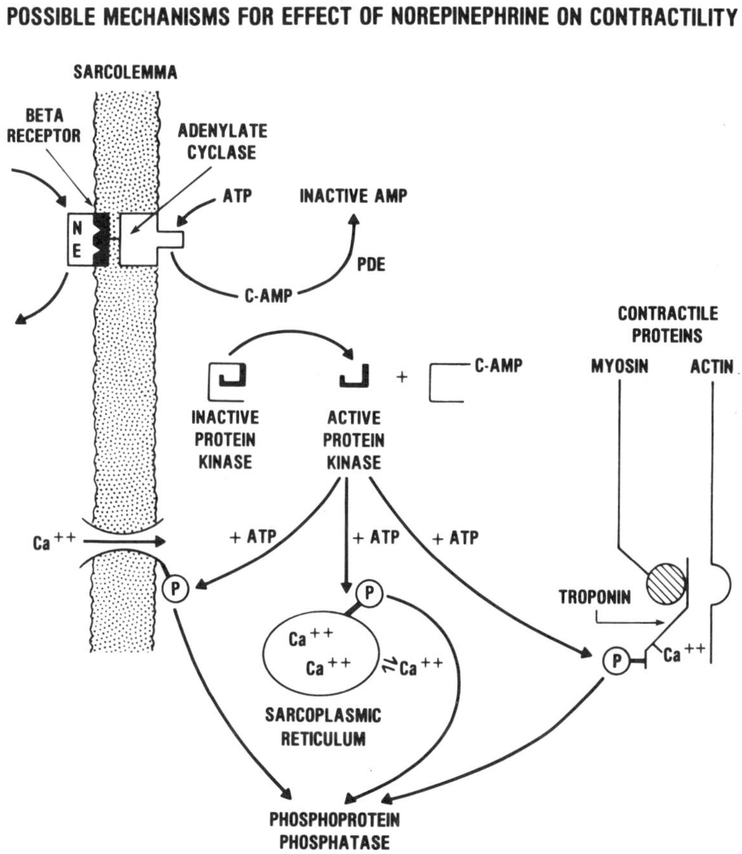

Figure 20.1. NE combines with beta receptors located on the sarcolemma, resulting in conversion of ATP to cAMP, which, in turn, activates protein kinase. The activated kinase may phosphorylate (*circled P*) proteins of the sarcolemma or sarcoplasmic reticulum, thus favoring Ca^{++} entry into the cell or calcium accumulation in stores that promote contractility when released in systole. In addition, activated kinase may phosphorylate the regulatory protein troponin, facilitating withdrawal of its inhibitory influence on contractile proteins with arrival of Ca^{++}. The inotropic process is limited by rapid dephosphorylation of proteins by phosphoprotein phosphatase and by cleavage of cAMP by phosphodiesterase (PDE).

sympathomimetic properties of these drugs (86).

β-Adrenergic stimulation may produce vasodilation by phosphorylation of myosin kinase. This reduces the capacity of myosin kinase to bind calmodulin, a calcium-sensitive "trigger" protein needed for phosphorylation of myosin and consequent activation of the contractile proteins of smooth muscle (137). Mediation of the electrical effects of β-adrenergic stimulation is less fully understood. Cyclic AMP production probably represents an intermediary step, since phosphodiesterase inhibitors potentiate the electrical as well as the mechanical actions of CA (142).

Catecholamine Effects in the Intact Circulation

The circulatory effects of a sympathomimetic drug are modified by a number of factors such as ambient level of sympathetic tone, reflex responses, and competing physiologic influences. The parasympathetic nervous system is a particularly important modifier (83). Parasympathetic stimulation can directly oppose sympathetic stimulation by causing bradycardia, increased atrioventricular conduction times, reduced contractility, and vasodilation. Moreover, acetylcholine released from parasympathetic fibers reduces NE release from sympathetic fibers. Stimulation of cholinergic receptors raises intracellular levels of cyclic guanine monophosphate, a nucleotide that may interfere with the production or activity of cAMP (137). The parasympathetic nervous system is not a symmetrical antagonist of the SNS. Due to differences in nerve distribution and efficacy of chemical mediators, parasympathetic influence is predominant for heart rate and atrioventricular conduction, minor for myocardial contractility, and minimal for regulation of tone in most vascular beds. In orchestrating cardiovascular responses, the CNS normally induces reciprocal changes in the two components of the autonomic nervous system. Thus, increasing sympathetic stimulation is accompanied by progressive parasympathetic withdrawal.

Sympathetic stimulation produces dramatic changes in the normal circulation. The nature of the change depends upon the intensity of stimulation of each type of adrenergic receptor (Table 20.1). If the stimulus is confined to the β_1-adrenergic receptor by administration of dobutamine, the rate and completeness of ventricular (and atrial) ejection is increased, ejection fractions rise, and end-systolic volumes fall. Heart rate tends to increase due to the direct influence of β_1-adrenergic stimulation, although baroreceptor reflexes initiated by the more rapid and extensive increase in systolic pressure may mute this tendency. Cardiac output rises only modestly, if at all (in contrast to the β_1-adrenergic response of the impaired circulation, as discussed subsequently). β_2-Adrenergic stimulation reduces peripheral vascular resistance. Isoproterenol stimulates β_2- as well as β_1-adrenergic receptors. In this context, β_2-adrenergic stimulation augments the β_1 effect on heart rate and contractile performance. This is caused by baroreceptor-mediated vagal withdrawal and reinforcement of sympathetic stimulation. Reduced aortic impedance also favors more complete left ventricular ejection. Pulse pressure and cardiac output rise markedly with isoproterenol. Agonists acting on both β_1- and α_1-adrenergic receptors, such as NE, have a different effect; the added α_1-adrenergic stimulation acts to raise peripheral resistance and oppose or reverse a tendency toward tachycardia. α_1-Mediated venoconstriction shifts blood toward the heart and increases cardiac output despite a slower heart rate and greater resistance to left ventricular emptying. Pure α_1-adrenergic agonists, e.g., methoxamine, cause hypertension and bradycardia. They reduce left ventricular performance by pressure and volume loading the myo-

Table 20.1
Change of Hemodynamic Performance in Response to Adrenergic Stimulation

	Contractile Performance	Heart Rate	Peripheral Vascular Resistance	Blood Pressure (S/D)	Cardiac Output
β_1 only	↑	↑ or 0	0	↑/0	↑ or 0
$\beta_1 + \beta_2$	↑↑	↑↑	↓	↑/↓	↑↑
$\beta_1 + \alpha_1$	↑	0 or ↓	↑	↑/↑	↑ or 0
α_1 only	↓	↓	↑↑	↑/↑	0 or ↓

cardium without increasing its intrinsic contractility.

Adrenergic stimulation has a profound effect on myocardial energy utilization and coronary blood flow. The heart is an aerobic organ; its energy use is directly related to myocardial oxygen consumption. Since the myocardium extracts more than 70% of oxygen arriving in arterial blood, coronary perfusion must rise in proportion to myocardial oxygen consumption. Autoregulation is the dominant influence governing coronary blood flow in the normal heart. The increases in heart rate, systolic BP, and contractility accompanying adrenergic stimulation each lead to large rises in myocardial oxygen consumption and coronary blood flow. The increase in myocardial oxygen consumption produced by CA is similar to the increase produced by similar hemodynamic changes achieved by other means (53, 54). The oxygen consumption of the quiescent potassium-arrested heart is unaltered by CA (75). α_1-Adrenergic stimulation tends to constrict coronary blood vessels, an effect that can overwhelm opposing autoregulatory influences in awake animals when the stimulus is sufficiently intense (148).

Neural and humoral delivery of CA play an essential role in making a wide variety of adjustments in response to changing demands. Sympathetic regulation of cardiovascular function is discussed elsewhere in this volume (see Chapters 1 and 14).

CATECHOLAMINES AND MYOCARDIAL ISCHEMIA

Catecholamines in the Pathogenesis of Ischemia and Its Sequelae

Myocardial ischemia occurs when local perfusion requirements outstrip the capacity of narrowed coronary vessels to deliver blood flow. CA augment myocardial ischemia by increasing myocardial demand for flow and may also interfere with flow delivery (Table 20.2). When flow deprivation is severe, local contractile activity ceases. Even though the number of β-adrenergic receptors and the level of adrenergically stimulated cAMP rise in ischemic tissue, CA are unlikely to influence contractile activity or the rate of tissue metabolism in the context of local ischemic arrest (24, 29, 95, 111). In regions with lesser degrees of flow deprivation, however, CA augmentation of contractile performance and heart rate may well promote ischemia by accelerating local myocardial oxygen consumption above the local oxygen supply, thereby extending the region of severe ischemia (88).

Table 20.2
Adverse Effects of Catecholamines on Myocardial Ischemia

I. Increased flow demands
   - ↑ Heart rate
   - ↑ Myocardial contractility
   - ↑ Systolic blood pressure

II. Decreased flow delivery
   - ↓ Duration of diastolic perfusion
   - ↑ Competition for perfusion—coronary "steal"
   - ↑ Platelet deposition in diseased vessels
   - ↑ Coronary constriction or spasm

III. Other potentially adverse effects
   - ↑ Electrical instability
   - ↑ Release of free fatty acids

In addition to augmenting demand for flow, CA may also interfere with coronary perfusion to regions with compromised flow by increasing heart rate, thereby abbreviating diastole, the time during which nearly all coronary collateral flow occurs (100, 110). By increasing metabolic activity in regions with relatively normal perfusion, CA can divert flow from ischemic tissue unable to compete for flow due to exhaustion of capacity for further vasodilation—a phenomenon described as "coronary steal" (36, 159).

CA may act on diseased arteries to exaggerate the effects of atherosclerotic narrowing. E may increase deposition of platelets within experimentally narrowed segments of coronary arteries, thus augmenting the obstruction to flow and possibly precipitating complete arterial occlusion (37). A similar rise in the obstructive influence of a preexisting stenosis may be caused by CA-induced arterial narrowing or focal spasm. Coronary spasm is frequently provoked as a clinical test by intracoronary ergonovine (62). Although this drug has serotonergic as well as adrenergic agonist properties, the resulting constriction of coronary arteries might be mimicked by NE.

CA may also have an adverse effect on ischemic hearts that is not immediately related to metabolic rate or effect on perfusion. The lipolytic influence of CA may raise plasma levels of free fatty acids and promote local production of free fatty acids from myocardial stores of neutral fat (105, 106). Ischemic myocardium is unable to metabolize fatty acids through oxidative pathways. Fatty acids, accumulating in excessive amounts for these reasons, may accelerate damage to cell membranes due to their detergent properties, thereby hastening cell death. In addition, CA reduce electrical stability of the heart (as explained subsequently). Fatty acids are also thought to favor arrhythmias in ischemic hearts (104, 122). Arrhythmias can increase ischemia by increasing metabolic demand and further compromising perfusion, thus initiating a vicious cycle. CA promote inhomogeneity of electrical conduction properties in ischemic hearts, lower the electrical threshold for ventricular fibrillation, and thus favor a fatal outcome.

The effects of CA are not all negative in the context of myocardial ischemia. When the remaining, nonischemic portion of myocardium is overburdened and failing due to extensive ischemic left ventricular damage, β-adrenergic stimulation may provide life-saving inotropic support. In addition, appropriate α-adrenergic stimulation may sustain myocardial and cerebral perfusion pressure when cardiac output is severely compromised by ischemia. α-Adrenergic stimulation may also reverse coronary steal by reducing excessive perfusion to normal myocardium.

β-Adrenergic Blockade

Considering all the negative effects of CA stimulation in the context of myocardial ischemia (Table 20.2), it is easy to see why β-adrenergic blockade has proven so useful in treatment of ischemic disorders. β-Blockers, in appropriate doses, reverse or attenuate CA-mediated metabolic stimulation by reducing tachycardia and rise in contractility caused by CA. Reduced contractility and actions on the renin-angiotensin system lower systolic BP, further attenuating the metabolic load. These same effects prolong diastole, favor collateral perfusion, and reverse coronary steal. β-Blockers also prevent the CA-induced rise in free fatty acids and decline in ventricular fibrillation threshold. Despite their great therapeutic potential in ischemia, the use of β-blockers is restricted. Patients with poor left ventricular function may develop worsened congestive heart failure due to the β-adrenergic blockade. The resultant increase in heart size may actually add to ischemia by augmenting left ventricular systolic wall stress, which is proportional to chamber dimensions. In addition, depressant effects on rhythm generation and conduction may cause intolerable bradycardia in susceptible patients.

Angina Pectoris Due to Coronary Occlusive Disease

In chronic, stable angina pectoris, coronary perfusion is adequate at rest but becomes insufficient when myocardial metabolism is stimulated by activity or emotion. In addition, constriction or spasm of arteries narrowed by atherosclerosis may transiently reduce blood flow to ischemic regions, particularly during spontaneous anginal episodes. CA release figures significantly in the hemodynamic changes that increase myocardial flow requirements during exercise or other circumstances precipitating myocardial ischemia. Thus, β-adrenergic blockers are very successful in preventing or forestalling ischemic symptoms and electrocardiographic evidence of myocardial ischemia during effort (2, 4, 156). Several factors can increase the extent of sympathetic response associated with a given level of activity, including low ambient temperature, a recent meal, and lack of physical conditioning. A physical conditioning program (particularly in a symptomatic individual who avoids activity) can reduce plasma NE and other evidence of sympathetic stimulation and, accordingly, raise the capacity of patients to perform exercise prior to onset of myocardial ischemia (23).

Experimental studies have suggested that myocardial ischemia may be a cause as well as a consequence of adrenergic stimulation (71, 111). Anginal episodes might stimulate a rise in plasma NE and predispose to further ischemia. Recent studies do not find changes in plasma NE with pacing-induced or effort-induced ischemic chest pain (23, 124). In the latter investigation, patients with coronary occlusive disease performing supine exercise had rises in plasma NE unrelated to the presence or absence of chest pain and the severity of exercise-induced mechanical dysfunction and ventricular arrhythmias. Thus, the transient ischemic episodes that characterize chronic, stable angina pectoris seem insufficient to elicit measurable changes in plasma NE content.

Individuals experiencing more prolonged,

frequent, or easily provoked anginal episodes are considered to have unstable angina. Seven to 15% of these patients will develop acute myocardial infarction within a few weeks (12). The status of CA in such patients has not been investigated to date, although it is apparent that β-adrenergic blockers are often useful in suppressing symptoms.

Acute Myocardial Infarction

Clinical acute myocardial infarction and acute coronary occlusion in conscious dogs produce rapid activation of the SNS as shown by sinus tachycardia and other tachyarrhythmias, hypertension, and elevation of plasma CA (24, 111). Although hemodynamic evidence of sympathetic activation is most frequently seen in anterior infarction in man, preponderant activation of parasympathetic nerves, associated with bradycardia, hypotension, and Mobitz type I atrioventricular block, is seen more often with inferior infarction (157). Combined autonomic activation can result in bradycardia which, when treated with parasympatholytic drugs, is converted to tachycardia (34). It should be emphasized that activation of the SNS is highly selective: vasoconstriction of skin and splanchnic beds is accompanied by vasodilation of renal beds. These responses very likely reflect a complex array of afferent input from ischemic tissue, dilated cardiac chambers and great veins, and altered stimulation of arterial baroreceptors and chemoreceptors as well as altered input from cerebral cortex and other sources within the neuraxis.

Elevations of plasma NE and, according to some studies, E, are variable and depend upon the severity of myocardial damage and circulatory dysfunction (5, 20, 40, 55, 58, 91, 93, 113, 128, 144, 150). Marked, sustained elevations in plasma CA are most often found in patients with cardiogenic shock or electrical instability incident to acute myocardial infarction. Little or no rise in plasma CA may occur in documented infarctions when accompanying hemodynamic stress is minimal. High levels of plasma NE are associated with substantial metabolic changes. In a series of 173 patients with acute transmural infarction, Mueller and Ayre (93) find that more than half have elevated plasma-free fatty acids (Fig. 20.2): 41% of patients have this rise along with increased myocardial utilization of free fatty acids and moderately elevated plasma CA; an

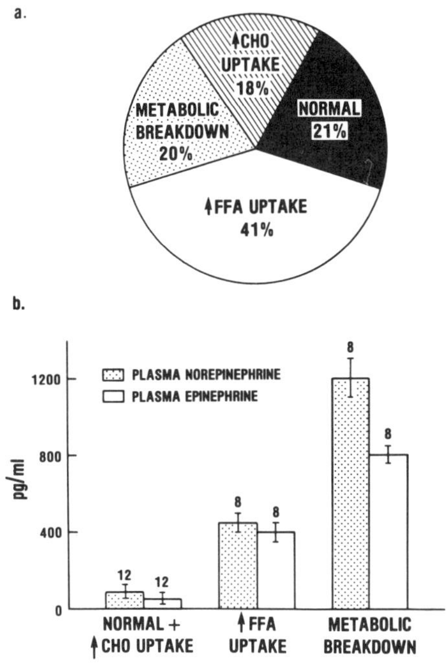

Figure 20.2. Metabolic responses to acute myocardial infarction, as described by Mueller and Ayres (93). Panel *a* shows relative number of 173 patients with increased free fatty acid (*FFA*) uptake, increased carbohydrate (*CHO*) uptake, metabolic breakdown, and normal metabolic patterns. Plasma CA associated with each pattern are shown in panel *b*. The numbers of patients evaluated appear over each bar. Standard errors are indicated by brackets. (Data adapted from Mueller and Ayres (93).)

additional 20% have increased plasma-free fatty acids, decreased myocardial utilization of both fatty acids and carbohydrates, and markedly elevated plasma CA (93). These latter two groups have a substantially higher incidence of simultaneous complications—arrhythmias, heart failure, and continuing chest pain—compared with infarct patients lacking these metabolic abnormalities. Other studies associate high risk of complications with elevated plasma-free fatty acid and CA (104, 106). These latter measurements also relate to decreased glucose tolerance and elevated plasma glucose in patients with acute myocardial infarction (106).

Although metabolic and hemodynamic responses to infarction are affected by numerous influences not directly mediated by CA, these findings demonstrate the close linkage of CA levels and complicated myocardial infarction and the potential for reversing deleterious CA-mediated influences when CA responses to infarction are in excess of circulatory requirements for inotropic and chronotropic support. Diminution of metabolic abnormalities, as well as relief of ischemic pain, electrocardiographic abnormalities, and tachyarrhythmias, are reported by many observers when β-blockers are given to selected patients with acute myocardial infarction (94). The capacity of propranolol to reduce infarct size is presently being evaluated in patients of a large, multicenter trial.

The actions of β-blockers to minimize ischemia and its sequelae may be of particular value in reducing the high mortality rates seen during the first year after myocardial infarction. Trials of practolol, alprenolol, timolol, metoprolol, and propranolol have demonstrated a reduction in postinfarction mortality in patients receiving these β-blocking drugs (6, 65, 96, 97, 138, 158).

CATECHOLAMINES AND HEART FAILURE

Cardiac dysfunction profoundly affects CA levels and the performance of the SNS: myocardial NE stores are severely depleted; certain autonomically mediated reflexes—sympathetic and parasympathetic—are markedly reduced, and plasma NE and E are substantially altered. The functional consequences of these changes in autonomic function contribute significantly to the clinical and physiologic manifestations of congestive heart failure.

Myocardial Norepinephrine

The heart normally contains large stores of NE within the abundant adrenergic nerve terminals that are distributed throughout the myocardium. Myocardial NE content is severely reduced in atria and ventricles of experimental animals with heart failure due to right ventricular (18, 131, 153) or left ventricular outflow tract obstruction (132) or primary myocardial disease (129). Similar results are observed in samples of human myocardium obtained at operation. Patients with heart failure have significantly lower myocardial NE content than patients with no evidence of cardiac decompensation (15, 19, 30, 79, 108). Left ventricular papillary muscles excised at the time of mitral valve replacement in patients with varying degrees of heart failure demonstrate a close connection between NE content and the functional state of the myocardium: NE content correlates directly with the contractile state of the papillary muscles, as assessed by length-tension relations (19).

Severe myocardial NE depletion, evident as a loss of normal NE fluorescence from adrenergic nerve terminals, is accompanied by reduced activity of tyrosine hydroxylase (TH) in failing myocardium (30). This enzyme is rate-limiting for the production of NE precursors and thus regulates biosynthesis of NE (see Chapter 1). Loss of TH is a specific manifestation of heart failure. It does not occur when NE depletion is due to denervation or reserpine administration and is not simply a consequence of dilution by increased muscle mass (112). The correlation between NE depletion and depression of TH activity in failing hearts suggests that depressed synthesis of neurotransmitter is an important aspect of the decrease in myocardial NE stores. In addition, impaired neuronal uptake of NE can be demonstrated in cardiac muscle of patients with heart failure of varying etiology (109). Hence, NE depletion in the failing heart is likely due to both deficient synthesis and impaired storage. In contrast to the marked reduction in myocardial NE stores, no significant depression is observed in renal (132) or splenic (129) NE stores in heart failure. Reduction in NE content is thus not uniformly present in all tissues innervated by the SNS and appears to be localized to the heart.

Although the exact reasons for depletion of myocardial NE stores with heart failure are not known, the functional consequences of this NE depletion are evident. Neurally mediated delivery of NE, normally an important mechanism augmenting contractility and heart rate during exercise and other circumstances requiring enhanced circulatory performance, is severely curtailed in heart failure. Positive inotropic and chronotropic responses to cardioaccelerator nerve stimulation are markedly diminished in animals with experimentally induced heart failure (25). For this reason, circulating NE and E provide a much greater proportion of the adrenergic stimulation to the heart in the presence of heart failure than in its absence. Relative lack of concomitant, neurally mediated adrenergic stimulation is one reason plasma levels of NE and E assume particular importance for the failing heart.

Even though myocardial NE depletion

causes impairment of neurally mediated adrenergic stimulation, it is not accompanied by decline in the intrinsic contractile state of the myocardium (133). Spann *et al.* (133) demonstrated that force-velocity and length-tension relations are not depressed in papillary muscles obtained from cats that have had surgical cardiac denervation or reserpine pretreatment. Thus, depletion of cardiac NE stores does not appear to be primarily responsible for the depression of the intrinsic myocardial contractile state in heart failure.

Autonomically Mediated Reflexes

Autonomically mediated circulatory responses to exercise and baroreceptor stimulation are significantly altered in heart failure. During exercise, animals with heart failure have a substantial, neurally mediated fall in renal and mesenteric perfusion. In contrast, animals without heart failure maintain perfusion to these beds at a constant level during exercise (64). This modification of the exercise response channels the limited blood flow provided by the failing heart toward the exercising muscles and away from beds with lesser metabolic demands; blood flow and BP are thus conserved. However, increased vasoconstriction during exercise also raises the aortic input impedance against which the failing left ventricular myocardium must pump.

Comparison of patients with heart failure and normal subjects at levels of exercise producing equivalent stress (assessed as intensity of pulmonary artery desaturation) has shown a marked reduction in heart rate response to exercise in the presence of heart failure (51). In addition, heart rate changes in response to upright tilt and drug-induced hypotension (51) or hypertension (17, 32) are severely diminished in patients with heart failure. Both sympathetic and parasympathetic components of baroreceptor-mediated heart rate responses are reduced. The diminution in responses mediated by the SNS does not appear related to alteration in quantity or sensitivity of β-adrenergic receptors, since heart rate and BP changes with graded doses of isoproterenol are the same in patients with heart failure and in normal subjects (51). Reduction in SNS-mediated changes in heart rate—and, presumably, contractility—very likely reflect depletion of myocardial stores of the sympathetic neurotransmitter, NE.

Plasma and Urinary Catecholamines

Early investigators were unable to document any failure-related changes in NE and E excretion, which were considered indices of sympathoadrenal activity. CA were assayed in these studies by colorimetric or bioassay techniques. Using a more sensitive fluorimetric method, however, Chidsey *et al.* (17) found that the augmentation in plasma NE concentrations induced by exercise in patients with cardiac decompensation exceeds that observed in normal volunteers. Thomas and Marks (139) employed a radioenzymatic method to document a direct correlation between plasma levels of NE and the degree of left ventricular decompensation as assessed by functional cardiac status and systolic time intervals.

A recent study of plasma NE and E (radioenzymatic method) compared values in normal subjects and patients with heart failure at rest, at equivalent submaximal exercise (matched total body oxygen consumption) and at maximal effort (38). At rest, patients with heart failure have higher levels of plasma NE than normal subjects, confirming earlier results. Patients with heart failure also have higher values of plasma NE than normal subjects after equivalent exercise. Normal subjects, however, are able to exercise at a much greater intensity than patients. In doing so, they attain far greater levels of plasma NE and E than patients performing maximal exercise. Thus, maximal, symptom-producing exercise causes far less elevation in circulating CA in patients with heart failure than vigorous effort causes in normal individuals.

The determinants of plasma NE and E remain somewhat unclear, particularly in the presence of heart failure, when the heart experiences NE depletion from adrenergic nerve terminals. It seems likely that elevation of plasma CA in patients with heart failure at rest and with mild exercise reflects increased sympathetic tone, at least in a portion of the SNS. An associated rise in urinary excretion of NE in patients with heart failure (16) confirms that elevated plasma NE levels are due to increased NE release rather than deficient removal of NE from the bloodstream. The smaller changes in plasma E relative to plasma NE suggest that the adrenal medulla plays a rather modest role in compensatory adjustments to heart failure (38).

Functional Consequences

Alterations in neural and humoral CA delivery figure in several important pathophysiologic consequences of heart failure. Increased circulating levels of NE and E and increased sympathetic tone in renal and other vascular beds are at least partially responsible for the vasoconstriction that is observed in many peripheral vessels in heart failure (92, 98, 162). Renal vasoconstriction favors salt and water retention, an important feature of congestive heart failure. Stimulation of β-adrenergic receptors of the kidney also leads to release of renin and increased blood levels of the vasoconstrictor angiotensin. Constriction of venous capacitance beds by adrenergic stimulation and other vasoconstrictor influences displaces blood toward the lungs and heart. The resultant increase in ventricular filling pressures may augment stroke volume through greater diastolic stretch of the ventricular myocardium (Frank-Starling mechanism). The benefits of this mode of compensation, however, are limited. Pulmonary and systemic venous congestion and tissue edema are, in part, consequences of the excessive central pooling of blood. Progressive increases in ventricular filling pressure produce less and less rise in stroke volume and place the ventricle under a steadily increasing mechanical and metabolic burden.

Adrenergic stimulation in heart failure causes constriction of resistance vessels as well as capacitance vessels. Although this compensatory influence directs blood flow to tissues with highest priority, it may also cause chronic flow deprivation in certain tissues and may augment the systolic burden on the left ventricle by increasing aortic input impedence. Moreover, the heart must deal with this added burden despite diminished inotropic and chronotropic support due to myocardial NE depletion.

The most important compensatory function of adrenergic stimulation in heart failure is direct inotropic, dromotropic, and chronotropic support of the heart. Increased myocardial contractility is beneficial for the patient with heart failure unless adrenergic stimulation produces concomitant myocardial ischemia or electrical instability. Increase in conduction and heart rate may be helpful in the presence of heart failure when stroke volume is limited by reduced systolic shortening of the myocardium, and intrinsic ventricular rate may be slow. The importance of adrenergic stimulation in sustaining the circulation is clearly demonstrated by the marked exacerbation of symptoms and hemodynamic difficulties frequently observed when patients with compensated heart failure are treated with β-adrenergic blocking drugs (152).

The effectiveness of elevated adrenergic stimulation may depend upon the etiology of the concomitant myocardial dysfunction. Papillary muscles, obtained from cats in which cardiac decompensation was induced by pulmonary artery constriction, responded to NE with an increment in contractile force that was greater than normal (131). However, the augmentation in contractile force of the left ventricle in response to NE is depressed in dogs with heart failure produced by an artificial aortocaval fistula (101, 164). These observations suggest that NE may exert a greater positive inotropic effect in heart failure due to pressure overload than in heart failure due to volume overload.

Vasodilator Therapy

Increased arteriolar and venous tone in heart failure represents an attempt to maintain a satisfactory BP in the face of an inadequate cardiac output. However, the elevated systemic vascular resistance results in an increased impedance to left ventricular ejection, which can further reduce cardiac output. This can lead to a deleterious cycle in which left ventricular function is progressively depressed. Vasodilators are used in the treatment of heart failure is an attempt to counteract this compensatory increase in peripheral vascular resistance (21). The improvement in left ventricular performance observed with the use of these agents supports the concept of an inappropriately elevated vascular resistance in heart failure (90).

Several investigators have attempted to evaluate the relationship between the hemodynamic effects of vasodilator agents and sympathetic neuronal activity in heart failure. Generally, a decrease in plasma NE levels is noted in patients who demonstrate clinical and hemodynamic improvement with vasodilator therapy (28, 103, 154). On the other hand, increases in plasma NE concentrations are reported in patients who have received long-term therapy with the α_1-adrenergic blocking drug, prazosin (22). This rise in NE levels suggests that augmented sympathetic neuronal activity may contribute to the attenuated hemodynamic responses to prazosin observed by several investigators during its chronic administration (3, 31, 33, 107). Further studies are needed to elucidate the precise role of the SNS in determining the effectiveness of vasodilator therapy in heart failure.

CATECHOLAMINES AND ARRHYTHMIAS

Effects on Rhythm and Conduction in Normal Hearts

β-Adrenergic stimulation exerts profound effects upon rhythm-generating and conduction mechanisms of normal and diseased hearts (8). These effects can be analyzed in terms of the influence of β-adrenergic stimulation on ion fluxes in cardiac tissues and resultant alterations in membrane potential across the sarcolemma (Fig. 20.3). Administration of NE or other β-agonists to isolated cardiac Purkinje fibers results in: (a) a significant rise in the calcium- and sodium-mediated "slow inward current" (i_{si}) that helps sustain the plateau (phase 2) of the cardiac action potential (11, 14, 119, 120, 130); (b) more rapid development of the potassium-mediated (outward) fast repolarizing current (i_{xi}) resulting in an abbreviation of the action potential (10, 142); (c) diminution and more rapid inactivation of the potassium-mediated (outward) transient diastolic current, i_{K2}, thereby speeding spontaneous diastolic depolarization (60, 61, 141); and (d) enhancement of electrogenic sodium pumping that can increase outward potassium current (145, 146). Acceleration of repolarization and diastolic depolarization figure in the NE-mediated decrement in effective refractory period and enhancement of Purkinje fiber automaticity (61). By bringing more calcium into the cell to trigger calcium-mediated calcium release, augmented i_{si} may be an important factor in inotropic response to NE (35). The changes just described are brought about by all β-agonists and are inhibited by β-blockers (47, 114). In addition, enhancement of these actions by theophylline and other phosphodiesterase inhibitors indicates intracellular mediation by cAMP (143). α-Adrenergic stimulation appears to have an effect opposite to the actions of β-adrenergic stimulation (27, 121). The reasons for this relatively weak α-adrenergic effect are not fully known.

Hoffman (67) suggests that adrenergic stimulation of certain tachyarrhythmias may be due to enhancement of triggered afterdepolarizations, erratic fluctuations in membrane potential following an action potential. NE can initiate early (systolic) afterdepolarization in otherwise normal Purkinje fibers and can increase the amplitude of delayed (diastolic) afterdepolarization, thereby increasing the likelihood that the latter afterdepolarization will give rise to a new action potential.

Although studies are not as complete in fibers of the sinoatrial node and atrioventricular junction, information from these rhythm-generating and conducting tissues indicates responses to adrenergic stimulation resembling

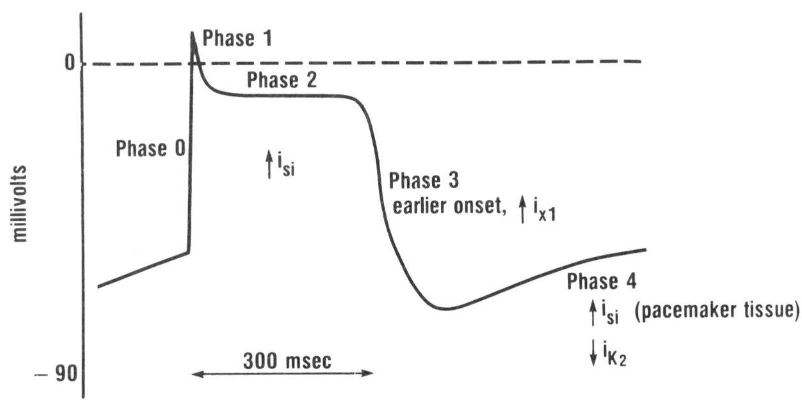

Figure 20.3. Action potential in a cardiac Purkinje fiber (schematic). Phases 2, 3, and 4 are influenced by β-adrenergic stimulation. Current changes are defined in text.

those in Purkinje fibers (10, 68, 89, 140, 147). However, i_{si} is more important in sustaining the action potential in fibers of the sinoatrial node and atrioventricular junction than in Purkinje fibers or fibers of ventricular or atrial myocardium (13). In addition, diastolic depolarization in pacemaker tissues is, in part, brought about by i_{si} that persists into diastole. Enhancement of i_{si} is responsible for CA-induced improvement in propagation of impulses through the atrioventricular junction and the sinoatrial node. Conduction of action potentials mediated by i_{si} is fragile—subject to decremental conduction and extinction—because these action potentials have low amplitude and rate of depolarization. By strengthening the action potential in these specialized tissues, β-adrenergic stimulation opposes conduction block within the atrioventricular junction and sinoatrial node (68, 89, 140, 147). Augmentation of diastolic i_{si} in these tissues also tends to increase their firing rate. Thus, increased i_{si} is an important aspect in the dromotropic, chronotropic, and possibly the inotropic effects of β-adrenergic stimulation.

β-Adrenergic stimulation can favor electrical instability even in normal hearts (57, 59, 85, 115, 149). Inhomogeneous action of either humoral or neural CA increases differences in rate of recovery from refractoriness ("temporal dispersion") in various portions of the Purkinje network (57). Temporal dispersion decreases the minimum stimulatory current necessary to induce repetitive ventricular extrasystoles and ventricular fibrillation in normal as well as diseased hearts (56, 57, 59, 74, 85, 115, 149). This effect can be exaggerated by imbalance of sympathetic stimulation. Unilateral, particularly left, stellate ganglion stimulation can lead to severe ventricular arrhythmias, including ventricular fibrillation (126, 127, 149).

Catecholamine Effects on Rhythm and Conduction in Diseased Hearts

Although i_{si} plays an essential role in the function of the normal sinoatrial node and atrioventricular junction, i_{si} may also figure prominently in the genesis of arrhythmias in diseased hearts (8). The i_{si} is propagated slowly over relatively long stretches of tissue with reduced resting potential. In addition, i_{si} persists, leaving tissues depolarized and, thus, refractory for relatively long periods. Each of these properties tends to favor development and maintenance of reentrant circuits in predisposed tissues. Diseased or damaged fibers with reduced resting potential that might fail to propagate action potentials in the absence of β-adrenergic stimulation of i_{si} will conduct impulses slowly in the presence of such stimulation. Slowly conducted wavefronts, upon encountering normally conducting tissue, cause reentrant extrasystoles and tachycardia (39, 123, 155, 161). Prolonged refractoriness in slowly conducting tissue tends to perpetuate and protect reentrant circuits by preventing synchronous depolarization.

Adrenergically induced speeding of phase four depolarization and enhancement of the capacity of injured, partially depolarized tissue to develop and propagate action potentials raises the firing of both normal and abnormal pacemakers (41, 141). Automaticity may also be increased in Purkinje fibers by CA-induced augmentation of delayed afterdepolarization (67). Increased automaticity results in early impulses that are particularly likely to trigger reentrant arrhythmias in tissues with appropriate anatomic and physiologic predisposition.

Although β-adrenergic stimulation of conditions favoring reentry and accelerated automaticity can increase the likelihood of tachyarrhythmia under many circumstances, adrenergic stimulation plays a special pathogenic role in the context of myocardial ischemia (67). Ischemic myocardium receives an abundance of neurally released NE through reflex action and local effects of ischemia upon adrenergic nerve terminals promoting NE release (87, 134). Ischemic impairment of NE reuptake may raise ambient NE levels in the vicinity of adrenergic receptors. Humoral delivery of CA may also play a role during ischemia, as mentioned in a previous section. β-Adrenergic stimulation is particularly significant in the presence of myocardial underperfusion because it can precipitate, intensify, or extend the ischemic process as well as foster electrophysiologic circumstances favoring arrhythmia. CA-induced rises in heart rate and contractility tend to increase myocardial metabolic activity, thereby precipitating or exaggerating local imbalance between coronary perfusion and metabolically mediated flow requirements in myocardium with perfusion limited by diseased arteries.

Although β-adrenergic stimulation can initiate lethal tachyarrhythmia, adrenergically mediated enhancement of conduction through the atrioventricular junction and stimulation of pacemaker activity within the distal Purkinje network can be life-sustaining when normal functioning of either the atrioventricular junction or the sinoatrial node is compromised, e.g., complete heart block or "sick sinus syndrome."

Catecholamines and Arrhythmias in the Clinical Setting

Neural or humoral delivery of CA or administration of sympathomimetic drugs can precipitate a variety of arrhythmias in appropriately predisposed individuals. Conversely, β-adrenergic blockade can be extremely useful in prevention or interruption of certain tachyarrhythmias (42, 45, 135). The importance of CA stimulation in a particular clinical setting depends upon the intensity of ambient adrenergic stimulation and the extent to which underlying predisposing factors render the heart vulnerable to arrhythmogenic influences (including CA). Vagal withdrawal, which often accompanies sympathetic stimulation, may reinforce arrhythmogenic effects, since the two reciprocal autonomic changes have similar electrophysiologic influence (76, 77).

The adrenergic stimulation and vagal withdrawal accompanying exercise or emotional upset commonly precipitate either supraventricular or ventricular tachyarrhythmias in several different conditions: coronary occlusive disease, mitral prolapse, syndromes featuring QT interval prolongation, preexcitation syndromes (e.g., Wolff-Parkinson-White) and hypertrophic cardiomyopathy (1, 43, 44, 52, 70, 80, 151, 160, 163). In coronary occlusive disease (and in certain cases of hypertrophic cardiomyopathy) adrenergic stimulation may initiate ischemia and hemodynamic deterioration as well as favor ectopic automaticity, reentry, and ventricular fibrillation. Partial removal of adrenergic stimulation by β-blocking drugs is often helpful in control or prevention of arrhythmias in the disease states just mentioned. Arrhythmias precipitated by exercise are particularly likely to respond to treatment with β-blocking drugs (42, 45).

Drugs and disease states associated with enhanced or imbalanced adrenergic activity frequently lead to tachyarrhythmias that may be controlled with adequate doses of β-blockers (26, 81). Pheochromocytoma, hyperthyroidism, certain anesthetic agents (e.g., chloroform and cyclopropane), other drugs releasing NE (e.g., amphetamines), drugs interfering with NE reuptake (e.g., cocaine and tricyclic antidepressants), and inhibitors of NE breakdown (e.g., MAO inhibitors), as well as direct sympathomimetic agents, can all precipitate atrial or ventricular arrhythmias through increased adrenergic stimulation (26, 48, 69, 73, 81, 84, 116). Prolonged QT interval syndromes, in which patients frequently die of ventricular fibrillation, have been attributed to excessive sympathetic stimulation via the left stellate ganglion. Marked decrease in incidence of ventricular fibrillation is reported in such patients after left stellectomy or β-blockade (125).

Digitalis toxicity in the clinical setting is probably mediated, at least in part, by adrenergic stimulation, although the exact nature of this mediation has not been fully defined (46). β-Adrenergic activity may augment the amplitude of delayed (triggered) afterdepolarizations stimulated by toxic doses of cardiac glycosides so that these afterdepolarizations initiate ectopic rhythms (67). β-Adrenergic blockers can be effective treatment for tachyarrhythmias associated with digitalis toxicity (42, 45).

Plasma NE levels are associated with increased incidence of ventricular arrhythmias in certain specific circumstances (50). Early in the course of acute myocardial infarction high plasma NE levels have been correlated with ventricular ectopy (7, 55, 91, 99). High plasma NE levels and other measures of adrenergic activity, however, may simply reflect the severity of the ischemic process and resulting hemodynamic and metabolic impairment, factors which may be closely connected with genesis of arrhythmias (8). During the later stages of myocardial infarction, plasma NE appears to have little predictive power (9, 128, 136, 150). Under these conditions plasma NE probably represents only one of many factors that influence the initiation and perpetuation of serious ventricular arrhythmias.

SUMMARY

CA play a crucial role in regulating circulatory function. Neural and humoral delivery of CA supports myocardial contractility and rhythm-generating activity, particularly in situations calling for enhanced cardiac performance such as exercise. CA also mediate regulation of resistance and capacitance vessels that control loading conditions for the heart and perfusion parameters for peripheral tissues. The pathogenesis and treatment of many clinical disorders are influenced by CA. In the presence of coronary occlusive disease, myocardial ischemia is often precipitated by increased CA stimulation and alleviated by β-adrenergic blockade. Heart failure from any cause is generally characterized by profound depletion of NE from adrenergic nerve terminals of the myocardium. Thus SNS-mediated compensation for heart failure is mediated largely by elevation of plasma CA and vasomotor changes in peripheral vessels. In all cardiac tissues, CA can markedly increase the calcium-mediated slow inward current, thereby speeding action potential initiation and propagation in certain tissues. Although the sinoatrial node and atrioventricular junction are particularly sensitive to influences affecting the slow inward current, the same sensitivity is also thought to occur in diseased Purkinje fibers. Augmentation of the slow, inward current very likely accounts for the well-documented arrhythmogenic influence of β-adrenergic stimulation.

Plasma NE generally reflects the extent of circulatory stress present in patients with heart disease, although the variability of individual values is quite considerable. In heart failure, plasma NE is usually elevated both at rest and during exercise. In ischemic heart disease unaccompanied by heart failure, plasma NE is generally not elevated unless the patient is experiencing a particularly severe, acute myocardial infarction. In the early postinfarction period, patients with increased plasma NE are especially vulnerable to ventricular arrhythmias. With this exception, plasma NE values appear to have little demonstrable relation to occurrence of arrhythmias, despite the arrhythmogenic properties of CA.

References

1. Allen JD, Pantridge JF, Shanks RG: The effects of practolol on the dysrhythmias complicating acute ischemic heart disease. *Am J Med* 58:199–208, 1975.
2. Amsterdam EA, Gorlin R, Wolfson S: Evaluation of long-term use of propranolol in angina pectoris. *JAMA* 210:103–106, 1969.
3. Arnold SB, Williams RL, Ports TA, Baughman RA, Benet LZ, Parmley WW, Chatterjee K: Attenuation of prazosin effect on cardiac output in chronic heart failure. *Ann Intern Med* 91:345–349, 1979.
4. Aronow WS, Turbow M, Van Camp S, Lurie M, Wittaker K: The effect of timolol vs placebo on angina pectoris. *Circulation* 61:66–69, 1980.
5. Benedict CR, Grahame-Smith DG: Plasma adrenaline and noradrenaline concentrations and dopamine-β-hydroxylase activity in myocardial infarction with and without cardiogenic shock. *Br Heart J* 42:214–220, 1979.
6. Beta-blocker Heart Attack Trial Research Group. A randomized trial of propranolol in patients with acute myocardial infarction. *JAMA* 247:1707–1714, 1982.
7. Bhat PS, Udupa KN, Vaish SK: Plasma catecholamines during arrhythmias following acute myocardial infarction. *Indian Heart J* 31:345–348, 1979.
8. Bigger JT Jr: Mechanisms and diagnosis of arrhythmias. In Braunwald E: *Heart Disease: A Textbook of Cardiovascular Medicine*. Philadelphia, WB Saunders, 1980, pp 630–690.
9. Borer JS, Rosing DR, Miller RH, Stark RM, Kent KM, Bacharach SL, Green MV, Lake CR, Cohen H, Holmes D, Donohue D, Baker W, Epstein SE: Natural history of left ventricular functioning during 1 year after acute myocardial infarction: Comparison with clinical, electrocardiographic, and biochemical determinations. *Am J Cardiol* 46:1–12, 1980.
10. Brown HF, Noble SJ: Effects of adrenaline on membrane currents underlying pacemaker activity in frog atrial muscle. *J Physiol* 238:51P–53P, 1974.
11. Burns JW, Covell JW, Ross J Jr: Mechanics of isotonic left ventricular contractions. *Am J Physiol* 224:725–732, 1973.
12. Cairns JA, Fantus IG, Klassen GA: Unstable angina pectoris. *Am Heart J* 92:373–386, 1976.
13. Carmeliet E, Vereecke J: Electrogenesis of the action potential and automaticity. In Berne RM, Sperelakis N, Geiger SR: *Handbook of Physiology, Section 2: The Cardiovascular System*. Baltimore, Williams & Wilkins, 1979, vol 1, pp 269–334.
14. Carmeliet EE, Vereecke J: Adrenaline and the plateau phase of the cardiac action potential. *Pfluegers Arch* 313:303–315, 1969.
15. Chidsey CA, Braunwald E, Morrow AG, Mason DT: Myocardial norepinephrine concentration in man. *N Engl J Med* 269:653–658, 1963.
16. Chidsey CA, Braunwald E, Morrow AG: Catecholamine excretion and cardiac stores of norepinephrine in congestive heart failure. *Am J Med* 39:442–450, 1965.
17. Chidsey CA, Harrison DC, Braunwald E: Augmentation of the plasma norepinephrine response to exercise in patients with congestive heart failure. *N Engl J Med* 267:650–654, 1962.
18. Chidsey CA, Kaiser GA, Sonnenblick EH, Spann JF, Braunwald E: Cardiac norepi-

nephrine stores in experimental heart failure in the dog. *J Clin Invest* 43:2386–2393, 1964.

19. Chidsey CA, Sonnenblick EH, Morrow AG, Braunwald E: Norepinephrine stores and contractile force of papillary muscle from the failing human heart. *Circulation* 33:43–51, 1966.
20. Christensen NJ, Videbaek J: Plasma catecholamines and carbohydrate metabolism in patients with acute myocardial infarction. *J Clin Invest* 54:278–286, 1974.
21. Cohn JN: Vasodilator therapy for heart failure. *Circulation* 48:5–8, 1973.
22. Colucci WS, Williams GH, Braunwald E: Increased plasma norepinephrine levels during prazosin therapy for severe congestive heart failure. *Ann Intern Med* 93:452–453, 1980.
23. Cooksey JD, Reilly P, Brown S, Bomze H, Cryer PE: Exercise training and plasma catecholamines in patients with ischemic heart disease. *Am J Cardiol* 42:372–375, 1978.
24. Corr PB, Witkowski FX, Sobel BE: Mechanisms contributing to malignant dysrhythmias induced by ischemia in the cat. *J Clin Invest* 61:109–119, 1978.
25. Covell JW, Chidsey CA, Braunwald E: Reduction of the cardiac response to postganglionic sympathetic nerve stimulation in experimental heart failure. *Circ Res* 19:51–56, 1966.
26. Crago RM, Eckoldt JW, Wiswell JG: Pheochromocytoma: Treatment with alpha and beta adrenergic blocking drugs. *JAMA* 202:870–874, 1967.
27. Cranefield PF, Hoffman BF, Wit AL: Block of conduction in partially depolarized cardiac Purkinje fibers induced by an alpha-adrenergic agent. *Nature* 234:159–160, 1971.
28. Curtiss C, Cohn JN, Vrobel T, Franciosa JA: Role of the renin-angiotensin system in the systemic vasoconstriction of chronic congestive heart failure. *Circulation* 58:763–770, 1978.
29. Davidson S, Maroko PR, Braunwald E: Effect of isoproterenol on the contractile function of the ischemic and anoxic heart. *Am J Physiol* 227:439–443, 1974.
30. DeQuattro V, Nagatsu T, Mendez A, Verska J: Determinants of cardiac noradrenaline depletion in human congestive failure. *Cardiovasc Res* 7:344–350, 1973.
31. Desch CE, Magorien RD, Triffon DW, Blanford MF, Unverferth DV, Leier CV: Development of pharmacodynamic tolerance to prazosin in congestive heart failure. *Am J Cardiol* 44:1178–1182, 1979.
32. Eckberg DL, Drabinsky M, Braunwald E: Defective cardiac parasympathetic control in patients with heart disease. *N Engl J Med* 285:877–883, 1971.
33. Elkayam U, Lejemtel TH, Mathur M, Ribner HS, Frishman WH, Strom J, Sonnenblick EH: Marked early attenuation of hemodynamic effects of oral prazosin therapy in chronic congestive heart failure. *Am J Cardiol* 44:540–545, 1979.
34. Epstein SE, Goldstein RE, Redwood DR, Kent KM, Smith ER: The early phase of acute myocardial infarction: Pharmacological aspects of therapy. *Ann Intern Med* 78:918–936, 1973.
35. Fabiato A, Fabiato F: Calcium and cardiac excitation-contraction coupling. *Ann Rev Physiol* 41:473–484, 1979.
36. Fam WM, McGregor M: Effect of nitroglycerin and dipyridamole on regional coronary resistance. *Circ Res* 22:649–659, 1968.
37. Folts JD, Crowell ER, Rowe GG: Platelet aggregation in partially obstructed vessels and its elimination with aspirin. *Circulation* 54:365–370, 1976.
38. Francis GS, Goldsmith SR, Ziesche SM, Cohn JN: Response of plasma norepinephrine and epinephrine to dynamic exercise in patients with congestive heart failure. *Am J Cardiol* 1152–1156, 1982.
39. Friedman PL, Stewart JR, Wit AL: Spontaneous and induced cardiac arrhythmias in subendocardial Purkinje fibers surviving extensive myocardial infarction in dogs. *Circ Res* 33:612–626, 1973.
40. Gazes PC, Richardson JA, Woods EF: Plasma catecholamine concentrations in myocardial infarction and angina pectoris. *Circulation* 29:657–661, 1959.
41. Gelband H, Rosen MR, Myerburg RJ, Bush HL, Bassett AL, Hoffman FG: Restorative effect of epinephrine on electrical and mechanical properties of diseased atrial tissue. *J Electrocardiol* 10:313–320, 1977.
42. Gettes LS, Chen C-M: Mechanisms and efficacy of beta-blocking agents in treatment of arrhythmias. In Schwartz PJ, Brown AM, Malliani A, Zanchetti A: *Neural Mechanisms in Cardiac Arrhythmias*. New York, Raven Press, 1978, pp 377–400.
43. Gettes LS, Yoshonis KF: Rapidly recurring supraventricular tachycardia: A manifestation of reciprocating tachycardia and an indication for propranolol therapy. *Circulation* 41:689–700, 1970.
44. Gettes LS: Electrophysiologic basis of arrhythmias in acute myocardial ischemia. In Oliver MF: *Modern Trends in Cardiology*. London, Butterworths, 1974, pp 219–246.
45. Gibson D, Sowton E: The use of beta adrenergic blocking drugs in dysrhythmias. *Prog Cardiovasc Dis* 12:16–32, 1969.
46. Gillis RA: Cardiac sympathetic nerve activity: Changes induced by ouabain and propranolol. *Science* 166:508–510, 1969.
47. Giotti A, Ledda F, Maunaioni PF: Effects of noradrenaline and isoprenaline, in combination with alpha- and beta-receptor blocking substances on the action potential of cardiac Purkinje fibers. *J Physiol*

229:99–113, 1973.
48. Goldberg LI: Cardiovascular and renal actions of dopamine: Potential clinical applications. *Pharmacol Rev* 24:1–29, 1972.
49. Goldberg LI: Dopamine: Clinical uses of an endogenous catecholamine. *N Engl J Med* 291:707–710, 1974.
50. Goldstein DS: Plasma norepinephrine as an indicator of sympathetic neural activity in clinical cardiology. *Am J Cardiol* 48:1147–1154, 1981.
51. Goldstein RE, Beiser GD, Stampfer M, Epstein SE: Impairment of autonomically mediated heart rate control in patients with cardiac dysfunction. *Circ Res* 36:571–578, 1975.
52. Goodwin JF: Congestive and hypertrophic cardiomyopathies. A decade of study. *Lancet* 1:731–739, 1970.
53. Graham TP Jr, Covell JW, Sonnenblick EH, Ross J Jr, Braunwald E: The control of myocardial oxygen consumption: Relative influence of contractile state and tension development. *J Clin Invest* 47:375–385, 1968.
54. Graham TP Jr, Ross J Jr, Covell JW, Sonnenblick EH, Clancy RL: Myocardial oxygen consumption in acute experimental cardiac depression. *Circ Res* 21:123–138, 1967.
55. Griffiths J, Leung A: The sequential estimation of plasma catecholamines and whole blood histamine in myocardial infarction. *Am Heart J* 82:171–179, 1971.
56. Han J, Moe GK: Nonuniform recovery of excitability of ventricular muscle. *Circ Res* 14:44–60, 1964.
57. Han J, Garcia de Jalon P, Moe GK: Adrenergic effects on cardiac vulnerability. *Circ Res* 14:516–524, 1964.
58. Hanson JF, Christensen NJ, Hesse B: Determinants of coronary sinus noradrenaline in patients with ischaemic heart disease: Coronary sinus catecholamine concentration in relation to arterial catecholamine concentration, pulmonary artery oxygen saturation and left ventricular end-diastolic pressure. *Cardiovasc Res* 12:415–421, 1978.
59. Harris AS, Otero H, Bocage A: The induction of arrhythmias by sympathetic activity before and after occlusion of a coronary artery in the canine heart. *J Electrocardiol* 4:34–43, 1971.
60. Hauswirth O, McAllister RE, Nobel D, Tsien RW: Reconstruction of actions of adrenaline and calcium on cardiac pacemaker potentials. *J Physiol* 204:126P–128P, 1969.
61. Hauswirth O, Nobel D, Tsien RW: Adrenaline: Mechanism of action on the pacemaker potential in cardiac Purkinje fibers. *Science* 162:916–917, 1968.
62. Heupler FA, Proudfit WL, Razoni M, Shirley EK, Greenstreet R, Sheldon WC: Ergonovine maleate provocative test for coronary arterial spasm. *Am J Cardiol* 41:631–640, 1978.
63. Hicks MJ, Shigekowa M, Katz AM: Mechanism by which cyclic adenosine 3′:5′-monophosphate-dependent protein kinase stimulates calcium transport in cardiac sarcoplasmic reticulum. *Circ Res* 44:384–391, 1979.
64. Higgins CB, Vatner SF, Millard RW, Franklin D, Braunwald E: Alterations in regional hemodynamics in experimental heart failure in conscious dog. *Trans Assoc Am Physicians* 85:267–278, 1972.
65. Hjalmarson Å, Elmfeldt D, Herlitz J, Holmberg S, Málek I, Nyberg G, Rydén L, Swedberg K, Vedin A, Waagstein F, Waldenström A, Waldenström J, Wedel H, Wilhelmsen L, Wilhelmsson C: Effect on mortality of metoprolol in acute myocardial infarction. *Lancet* 2:823–827, 1981.
66. Hoffman BB, Lefkowitz RJ: Alpha-adrenergic receptor subtypes. *N Engl J Med* 302:1390–1396, 1980.
67. Hoffman BF: Role of the sympathetic nervous system in arrhythmias occurring after coronary artery occlusion and myocardial infarction. In Schwartz PJ, Brown AM, Mallani A, Zanchetti A: *Neural Mechanisms in Cardiac Arrhythmias.* New York, Raven Press, 1978, pp 155–165.
68. Hutter OF, Trautwein W: Vagal and sympathetic effects on the pacemaker fibers in the sinus venosus of the heart. *J Gen Physiol* 39:715–733, 1956.
69. Jefferson JW: A review of the cardiovascular effects and toxicity of tricyclic antidepressants. *Psychosom Med* 37:160–179, 1975.
70. Jeresaty RM: Sudden death in the mitral valve prolapse-click syndrome. *Am J Cardiol* 37:317–318, 1976.
71. Karlsberg RP, Penkoske PA, Cryer PE, Corr PB, Roberts R: Rapid activation of the sympathetic nervous system following coronary artery occlusion: Relationship to infarct size, site, and haemodynamic impact. *Cardiovasc Res* 13:523–531, 1979.
72. Katz AM, Bailin G, Kirchberger MA, Today M: Regulation of myocardial cell function by agents that increase cyclic AMP production in the heart. In Fishman AP: *Heart Failure.* Washington DC, Hemisphere Publishing Corp, 1978, pp 11–28.
73. Katz RL, Bigger JT Jr: Cardiac arrhythmias during anesthesia and operation. *Anesthesiology* 33:193–213, 1970.
74. Kliks BR, Burgess MM, Abildskov JA: Influence of sympathetic tone on ventricular fibrillation threshold during experimental coronary occlusion. *Am J Cardiol* 36:45–49, 1975.
75. Klocke FJ, Kaiser GA, Ross J Jr, Braunwald E: Mechanism of increase of myocardial oxygen uptake produced by catecholamines. *Am J Physiol* 209:913–918, 1965.
76. Kolman BS, Verrier RL, Lown B: The effect

of vagus nerve stimulation upon excitability of the canine ventricle: Role of sympathetic-parasympathetic interactions. *Am J Cardiol* 37:1041–1045, 1975.
77. Kolman BS, Verrier RL, Lown B: The effect of vagus nerve stimulation upon vulnerability of the canine ventricle: Role of sympathetic-parasympathetic interactions. *Circulation* 52:578–585, 1975.
78. Korner PI: Central nervous control of autonomic cardiovascular function. In Berne RM: *Handbook of Physiology.* Baltimore, Williams & Wilkins, 1979, pp 691–740.
79. Kramer RS, Mason DT, Braunwald E: Augmented sympathetic neurotransmitter activity in the peripheral vascular bed of patients with congestive heart failure and cardiac norepinephrine depletion. *Circulation* 38:629–634, 1968.
80. Kupersmith J, Shiang H, Litwak RS, Herman MV: Electrophysiological and antiarrhythmic effects of propranolol in canine acute myocardial ischemia. *Circ Res* 38:302–307, 1976.
81. Levey GS: The heart and hyperthyroidism. *Med Clin North Am* 59:1193–1201, 1975.
82. Levy MN, Ng JL, Zieske H: Functional distribution of the peripheral cardiac sympathetic pathways. *Circ Res* 19:650–661, 1966.
83. Levy MN: Sympathetic-parasympathetic interactions in the heart. *Circ Res* 29:437–445, 1971.
84. Loeb HS, Khan M, Klodnycky ML, Sinno MZ, Towne WD, Gunner RM: Haemodynamic effects of dobutamine in man. *Anesthesiology* 44:281–286, 1976.
85. Lown B, Verrier RL: Neural activity and ventricular fibrillation. *N Engl J Med* 294:1165–1170, 1976.
86. Luttgau HC: Caffeine, calcium, and the activation of contraction. In Cuthbert AW: *Calcium and Cellular Function.* New York, St. Martins Press, 1970, pp 241–248.
87. Malliani A, Schwartz PJ, Zanchetti A: A sympathetic reflex elicited by experimental coronary occlusion. *Am J Physiol* 217:703–709, 1969.
88. Maroko PR, Kjekshus JK, Sobel BE, Covell JW, Ross J Jr, Braunwald E: Factors influencing infarct size following experimental coronary artery occlusions. *Circulation* 43:67–82, 1971.
89. Matsuda K, Hoski T, Kameyama S: Action of acetylcholine and adrenaline upon the membrane potential of the atrioventricular node. *Tohoku J Exp Med* 68:16, 1958.
90. Mason DT: Afterload reduction and cardiac performance. *Am J Med* 65:106–125, 1978.
91. McDonald L, Baker C, Bray C, McDonald A, Restieaux N: Plasma catecholamines after cardiac infarction. *Lancet* 2:1021–1023, 1969.
92. Merrill AJ: Edema and decreased renal blood flow in patients with chronic congestive heart failure: Evidence of "forward failure" as a primary cause of edema. *J Clin Invest* 27:272–277, 1948.
93. Mueller HS, Ayres SM: Metabolic responses of the heart in acute myocardial infarction in man. *Am J Cardiol* 42:363–371, 1978.
94. Mueller HS, Ayres SM: The role of propranolol in treatment of acute myocardial infarction. *Prog Cardiovasc Dis* 19:405–412, 1977.
95. Mukherjee A, Wong TM, Buja M, Lefkowitz RJ, Willerson JT: Beta adrenergic and muscarinic cholinergic receptors in canine myocardium. *J Clin Invest* 64:1423–1428, 1979.
96. Multicentre International Study. Improvement in prognosis of myocardial infarction by long-term beta-adrenoceptor blockade using practolol: A multicentre international study. *Br Med J* 3:735–740, 1975.
97. Multicentre International Study. Reduction in mortality with long-term beta-adrenoceptor blockade: A multicentre international study. *Br Med J* 2:419–421, 1977.
98. Muth HAV, Wormald PH, Bishop JM, Donald KW: Further studies of blood flow in the resting arm during supine leg exercise. *Clin Sci* 17:603–610, 1958.
99. Nadeau RA, DeChamplain J: Plasma catecholamines in acute myocardial infarction. *Am Heart J* 98:548–554, 1979.
100. Neill WA, Phelps NC, Oxendine JM, Mahler DJ, Sun DN: Effect of heart rate on coronary blood flow distribution in dogs. *Am J Cardiol* 32:306–312, 1973.
101. Newman WH: Volume overload heart failure: Length-tension curves, and response to beta-agonists, Ca^{+2}, and glucagon. *Am J Physiol* 735:H690–H700, 1978.
102. Norris JE, Foreman RD, Wurster RD: Responses of the canine heart to stimulation of the first five ventral thoracic roots. *Am J Physiol* 227:9–12, 1974.
103. Ogasawara B, Ogawa K, Sassa H: Effects of nitroglycerin ointment on plasma norepinephrine and cyclic nucleotides in congestive heart failure. *J Cardiovasc Pharmacol* 3:867–875, 1981.
104. Oliver MF, Kurlen VA, Greenwood TW: Relation between serum-free-fatty-acid and arrhythmias and death after acute myocardial infarction. *Lancet* 1:710–714, 1968.
105. Oliver MF: Metabolic response during impending myocardial infarction. II. Clinical implications. *Circulation* 45:491–500, 1972.
106. Opie LH: Metabolism of free fatty acids, glucose, and catecholamines in acute myocardial infarction: Relation to myocardial ischemia and infarct size. *Am J Cardiol* 36:938–953, 1975.
107. Packer M, Meller J, Gorlin R, Herman MV: Hemodynamic and clinical tachyphylaxis to prazosin-mediated afterload reduction

in severe chronic congestive heart failure. *Circulation* 59:531–539, 1979.
108. Petch MC, Nayler WG: Concentration of catecholamines in human cardiac muscle. *Br Heart J* 41:340–344, 1979.
109. Petch MC, Nayler WG: Uptake of catecholamines by human cardiac muscle in vitro. *Br Heart J* 41:336–339, 1979.
110. Pitt B, Gregg DE: Coronary hemodynamic effects of increasing ventricular rate in the unanesthetized dog. *Circ Res* 22:753–761, 1968.
111. Podzuweit T, Dalby AJ, Cherry GW, Opie LH: Tissue levels of cyclic AMP in ischemic and nonischemic myocardium following coronary artery ligation. *J Mol Cell Cardiol* 10:81–94, 1978.
112. Pool PE, Covell JW, Levitt M, Gibb J, Braunwald E: Reduction of cardiac tyrosine hydroxylase activity in experimental congestive heart failure: Its role in depletion of cardiac norepinephrine stores. *Circ Res* 20:349–353, 1967.
113. Prakash R, Parmley WW, Horvat M, Swan HJC: Serum cortisol, plasma free fatty acids, and urinary catecholamines as indicators of complications in acute myocardial infarction. *Circulation* 45:736–745, 1972.
114. Quadbeck J, Reiter M: Adrenoceptors in cardiac ventricular muscle and changes in duration of action potential caused by noradrenaline and isoprenaline. *Naunyn Schmiedebergs Arch Pharmacol* 288:403–414, 1975.
115. Rabinowitz SH, Verrier RL, Lown B: Muscarinic effects of vasosympathetic trunk stimulation on the repetitive extrasystole threshold. *Circulation* 53:622–627, 1976.
116. Raisfeld IH: Cardiovascular complications of antidepressant therapy: Interactions at the adrenergic neuron. *Am Heart J* 83:129–133, 1972.
117. Randall WC, McNally H, Cowan J, Caliquiri L, Rohse WG: Functional analysis of the cardioaugmentor and cardioaccelerator pathways in the dog. *Am J Physiol 191:* 213–217, 1957.
118. Ray KP, England PJ: Phosphorylation of the inhibitory subunit of troponin and its effect on the calcium dependence of the cardiac myofibril adenosine triphosphatase. *FEBS Lett* 70:11–16, 1976.
119. Reuter H, Scholz H: The regulation of the calcium conductance of cardiac muscle by adrenaline. *J Physiol* 264:49–62, 1977.
120. Reuter H: Localization of beta-adrenergic receptors and effects of noradrenaline and cyclic nucleotides on action potentials, ionic currents and tension in mammalian cardiac muscle. *J Physiol* 242:429–451, 1974.
121. Rosen MR, Hordof AJ, Ilvento JP, Danilo P: Effects of adrenergic amines on electrophysiological properties and automaticity of neonatal and adult canine Purkinje fibers. Evidence for alpha- and beta-adrenergic actions. *Circ Res* 40:390–400, 1977.
122. Rutenberg HL, Pamintuah JC, Soloff LA: Serum-free-fatty-acids and their relation to complications after acute myocardial infarction. *Lancet* 2:559–564, 1969.
123. Scherlag BJ, El-Sherif N, Hope R, Lazzara R: Characterization and localization of ventricular arrhythmias resulting from myocardial ischemia and infarction. *Circ Res* 35:372–383, 1974.
124. Schwartz L, Sole MJ, Vaughan-Neil EF, Hussain NM: Catecholamines in coronary sinus and peripheral plasma during pacing-induced angina in man. *Circulation* 59:37–42, 1979.
125. Schwartz PJ, Periti M, Malliani A: The long Q-T syndrome. *Am Heart J* 89:378–390, 1975.
126. Schwartz PJ, Stone HL, Brown AM: Effects of unilateral stellate ganglion blockade on the arrhythmias associated with coronary artery occlusion. *Am Heart J* 92:589–599, 1976.
127. Schwartz PJ, Verrier RL, Lown B: Effect of stellectomy and vagotomy on ventricular refractoriness. *Circ Res 40:* 536–540, 1977.
128. Sigger DCM, Salter C, Fluck DC: Serial plasma adrenaline and noradrenaline levels in myocardial infarction using a new double isotope technique. *Br Heart J* 33:878–883, 1971.
129. Sole MJ, Lo CM, Laird CW, Sonnenblick EH, Wurtman RJ: Norepinephrine turnover in the heart and spleen of the cardiomyopathic syrian hamster. *Circ Res* 37:855–862, 1975.
130. Sonnenblick EH: Active state in heart muscle: Its delayed onset and modification by inotropic agents. *J Gen Physiol* 50:661–676, 1967.
131. Spann JF Jr, Buccino RA, Sonnenblick EH, Braunwald E: Contractile state of cardiac muscle obtained from cats with experimentally produced ventricular hypertrophy and heart failure. *Circ Res* 21:341–353, 1967.
132. Spann JF Jr, Chidsey CA, Pool PE, Braunwald E: Mechanism of norepinephrine depletion in experimental heart failure produced by aortic constriction in the guinea pig. *Circ Res* 17:312–321, 1965.
133. Spann JF Jr, Sonnenblick EH, Cooper T, Chidsey CA, Willman VL, Braunwald E: Cardiac norepinephrine stores and the contractile state of heart muscle. *Circ Res* 19:317–325, 1966.
134. Staszewska-Barczak J: The reflex stimulation of catecholamine secretion during the acute stage of myocardial infarction in the dog. *Clin Sci* 41:419–439, 1971.
135. Stock JPP, Dale N: Beta-adrenergic receptor blockade in cardiac arrhythmias. *Br Med J* 2:1230–1233, 1963.
136. Strange RC, Rowe MJ, Oliver MF: Lack of relation between venous plasma total catecholamine concentrations and ventricular arrhythmias after acute myocardial in-

farction. *Br Med J* 2:921–922, 1978.

137. Stull JT, Mayer SE: Biochemical mechanisms of adrenergic and cholinergic regulation of myocardial contractility. In Berne RM, Sperelakis N, Geiger SR: *Handbook of Physiology, Section 2: The Cardiovascular System*. Baltimore, Williams & Wilkins, 1979, vol 1, pp 741–774.
138. The Norwegian Multicenter Study Group. Timolol-induced reduction in mortality and reinfarction in patients surviving acute myocardial infarction. *N Engl J Med* 304:801–807, 1981.
139. Thomas JA, Marks BH: Plasma norepinephrine in congestive heart failure. *Am J Cardiol* 41:233–243, 1978.
140. Toda N: Electrophysiological effects of potassium and calcium ions in the sino-atrial node in response to sympathetic stimulation. *Pfluegers Arch* 310:45–63, 1969.
141. Tsien RW: Effects of epinephrine on the pacemaker potassium current of cardiac Purkinje fibers. *J Gen Physiol* 64:293–319, 1974.
142. Tsien RW, Giles W, Greengard P: Cyclic AMP mediates the effects of adrenaline on cardiac Purkinje fibers. *Nature* 240:181–183, 1972.
143. Tsien RW: Mode of action of chronotropic agents in cardiac Purkinje fibers. Does epinephrine act by directly modifying the external surface change? *J Gen Physiol* 64:320–342, 1974.
144. Valori C, Thomas M, Shillingford J: Free noradrenaline and adrenaline excretion in relation to clinical syndromes following myocardial infarction. *Am J Cardiol* 20:605–617, 1967.
145. Vassalle M, Barnabei O: Norepinephrine and potassium fluxes in cardiac Purkinje fibers. *Pfluegers Arch* 322:287–303, 1971.
146. Vassalle M: Electrogenic suppression of automaticity in sheep and dog Purkinje fibers. *Circ Res* 27:361–377, 1970.
147. Vassort GO, Rougier D, Garnier MP, Sauviat E, Coraboeuf, Gargouil Y-M: Effects of adrenaline on membrane inward currents during the cardiac action potential. *Pfluegers Arch* 309:70–81, 1969.
148. Vatner SF, Higgins CB, Braunwald E: Effects of norepinephrine on coronary circulation and left ventricular dynamics in the conscious dog. *Circ Res* 34:812–823, 1974.
149. Verrier RL, Thompson P, Lown B: Ventricular vulnerability during sympathetic stimulation: Role of heart rate and blood pressure. *Cardiovasc Res* 8:602–610, 1974.
150. Videbaek J, Christensen NJ, Sterndorff B: Serial determination of plasma catecholamines in myocardial infarction. *Circulation* 46:846–855, 1972.
151. Vincent GM, Abildskov JA, Burgess MJ: Q-T interval syndromes. *Prog Cardiovasc Dis* 26:523–530, 1974.
152. Vogel JHK, Chidsey CA: Cardiac adrenergic activity in experimental heart failure assessed with beta receptor blockade. *Am J Cardiol* 24:198–208, 1969.
153. Vogel JHK, Jacobowitz D, Chidsey CA: Distribution of norepinephrine in the failing bovine heart. Correlation of chemical analysis and fluorescence microscopy. *Circ Res* 24:71–83, 1969.
154. Vrobel TR, Cohn JN: Comparative hemodynamic effects of converting enzyme inhibitor and sodium nitroprusside in severe heart failure. *Am J Cardiol* 45:331–336, 1980.
155. Waldo AL, Kaiser GK: Study of ventricular arrhythmias associated with acute myocardial infarction in the canine heart. *Circulation* 47:1222–1228, 1973.
156. Warren SG, Bremer DL, Orgain ES: Long-term propranolol therapy for angina pectoris. *Am J Cardiol* 37:420–426, 1976.
157. Webb SW, Adgey AAJ, Pantridge JF: Autonomic disturbance at onset of acute myocardial infarction. *Br Med J* 3:89–92, 1972.
158. Wilhelmsson C, Vedin JA, Wilhelmsen L, Tibblin G, Werkö L: Reduction of sudden deaths after myocardial infarction with alprenolol: Preliminary results. *Lancet* 2:1157–1159, 1974.
159. Winbury MM, Howe BB, Hefner MA: Effect of nitrates and other coronary dilators on large and small coronary vessels: An hypothesis for the mechanism of action of nitrates. *J Pharmacol Exp Ther* 176:184–199, 1971.
160. Winkle RA, Lopes MG, Goodman DJ, Fitzgerald JW, Schroeder JS, Harrison DC: Propranolol for patients with mitral valve prolapse. *Am Heart J* 93:422–427, 1977.
161. Wit AL, Cranefield PF, Hoffman BF: Slow conduction and reentry in the ventricular conducting system. II. Single and sustained circus movement in networks of canine and bovine Purkinje fibers. *Circ Res* 30:11–22, 1972.
162. Wood JE, Litter J, Wilkins RW: Peripheral venoconstriction in human congestive heart failure. *Circulation* 13:524–527, 1956.
163. Wu D, Denes P, Dhingra R, Kahn A, Rosen KM: The effects of propranolol on induction of A-V nodal re-entrant paroxysmal tachycardia. *Circulation* 50:665–677, 1974.
164. Zucker IH, Waltke E, Gilmore JP: Cardiac responses to beta-adrenergic stimulation in anesthetized dogs with chronic congestive heart failure. *Basic Res Cardiol* 75:697–711, 1980.

CHAPTER TWENTY-ONE

Type A Behavior and Catecholamines: A Critical Review

DAVID G. GLASS, Ph.D., and RICHARD J. CONTRADA, A.B.

INTRODUCTION

Coronary heart disease, the major cause of death in this country, is believed to result from damage to the coronary arteries—a thickening of the arterial walls called *atherosclerosis*. The major clinical manifestations of the disease include *myocardial infarction*, *angina pectoris*, and sudden cardiac death. Many factors enhancing the risk of coronary heart disease have been identified by epidemiological research during the past several decades (6, 56), including age, hypertension (see Chapter 24), elevated levels of cholesterol and related lipids in the blood, cigarette smoking, family history of coronary heart disease, and the presence of diabetes mellitus. While the likelihood of heart disease is greater when more of the risk factors are present, the best combination of these factors still fails to identify most new cases of the disease (48).

Abbreviations used in Chapter 21 are: BP, blood pressure; SNS, sympathetic nervous system; E, epinephrine; NE, norepinephrine; CA, catecholamines; VMA, vanillylmandelic acid; COMT, catechol-O-methyltransferase.

TYPE A BEHAVIOR AND ITS CONCEPTUAL FOUNDATIONS

In recent years, considerable evidence has implicated psychosocial factors in the etiology and pathogenesis of coronary heart disease (38, 48, 50). Psychological stress and coping are among the more promising variables (38), with a specific coping style called the *Type A coronary-prone behavior pattern* receiving particular attention (74, 86). Type A behavior is characterized by competitive achievement striving, excessive impatience and time urgency, hostility, and aggressiveness. A contrasting Type B pattern is described as the relative absence of these characteristics. Classification of individuals as A or B ranges along a continuum, with extreme A and B points as polar opposites.*

The Type A pattern is conceptualized as the outcome of a person-situation interaction, in which the defining behaviors are observable when a susceptible individual is confronted by appropriately challenging and/or stressful circumstances (38, 83, 86). A major difficulty with this definition is the meaning of "appropriately challenging and/or stressful circumstances." Some effort has been made to specify these terms (15, 38), but it often remains unclear whether a given experimental protocol or a naturally occurring set of events will elicit Type A behavior.

* Issues of Type A measurement are beyond the purpose and scope of this chapter. Suffice it to note here that the major diagnostic tools are a structured interview developed by Friedman and Rosenman (83, 87) and a self-administering questionnaire called the Jenkins Activity Survey for Health Prediction (51–53). More detailed considerations of measurement can be found elsewhere (20, 74, 107). It should be noted here, however, that the structured interview may be a better predictor of episodic changes (reactivity) in sympathetic-adrenal medullary arousal than the Jenkins Activity Survey (19, 74). This point needs to be kept in mind when evaluating the CA research discussed in this chapter.

This lack of clarity may partly stem from definitions of psychological stress and challenge. The former is regarded as an individual's internal state when he is faced with physical and/or psychic threats (65). The salient feature of a stressor stimulus is the individual's perception of its potential for actual harm and his belief that his own resources are inadequate to cope with the threat. This leads to the experience of stress, and to the stress response, which may be measured in behavioral or physiologic terms (12). In like manner, challenge may be defined perceptually. A major difference is that, unlike a stressor, the salient feature of a challenge is the potential for reward rather than harm (2, 66). In both cases, however, perception-cognition is central in eliciting behavioral and physiologic responses, including those subsumed under the rubric of Type A behavior. The parameters for determining when a stimulus will be perceived as a stressor or as a challenge have yet to be specified; therefore, it is difficult to predict, *a priori*, when a cirumstance will produce Type A responses in a susceptible individual.

While the situational component of the person-situation equation requires greater conceptual clarity, recent research has been more successful in isolating facets of behavior pattern A that enhance the risk of coronary heart disease (75) and potentiate cardiovascular responses that may be implicated in disease end points (19). It appears that hostility and competitive drive are the features of the behavior pattern most likely to correlate with clinical coronary heart disease as well as with experimentally induced elevations in systolic blood pressure (BP) and heart rate. Researchers are beginning to examine the hypothesis that among psychosocial variables, Type A behavior alone is insufficient to account for physiologic changes related to coronary disease. The possibility exists that heightened cardiovascular and neuroendocrine reactivity to stress and challenge are determined by some combination of Type A characteristics and other psychological factors, including intrapsychic defenses (39).

TYPE A BEHAVIOR AND CLINICAL CORONARY HEART DISEASE

It would be difficult to predict the outcome of varied efforts to resolve the theoretical issues discussed in the previous section. The fact remains, however, that Type A behavior is an established risk factor for clinical coronary heart disease (107). The strongest available evidence for this association comes from the prospective *Western Collaborative Group Study*. The results indicate that Type A men experienced about twice the incidence of acute clinical events over an 8½-year follow-up period compared to Type B men (85). This difference occurred independently of other risk factors, including total serum cholesterol, systolic BP, and daily cigarette smoking. Still other research, using prospective data from the *Framingham Heart Study*, found that Type A was related to the incidence of coronary heart disease even when other risk factors were controlled (45).

An association between behavior pattern A and coronary atherosclerosis may also exist. Several studies, using angiographic procedures, have examined this relationship. Three report positive associations between severity of occlusion of the coronary arteries and Type A behavior (5, 27, 114). Other investigations, however, have failed to confirm this finding (21). It remains to be seen, therefore, what conditions determine a positive or null relationship between atherosclerosis and Type A behavior.

SUGGESTED MECHANISMS LINKING TYPE A BEHAVIOR TO CORONARY HEART DISEASE

Accepting the existence of a relationship between the behavior pattern and clinical coronary heart disease leads to the question of physiologic mechanisms underlying the association. Some evidence indicates that fully developed (*i.e.*, extreme) Type A individuals have higher lipid and related hormonal levels than Type Bs (86). However, there is little indication of a correlation between Pattern A and still another potential mediator, hypertension (91), nor is there a consistent relationship between Type A and elevated basal BP levels (39). On the other hand, a growing body of research has documented greater pressor reactivity in As than Bs when both types of subjects are confronted with stressful and/or challenging experimental situations (18, 40, 71). Still other studies have shown increased heart rate reactivity in Type A subjects (18, 40, 102), although several investigators have reported unsuccessful replications (41, 71).

Increased BP reactivity to stress and challenge in Type As may help explain their enhanced risk for coronary heart disease. It is by no means clear, however, that episodic elevations in cardiovascular variables are related to coronary heart disease. BP reactivity, though, is at least in part mediated by enhanced activity of the sympathetic nervous system (SNS) and consequent discharge of catecholamines (CA), such as epinephrine (E) and norepinephrine (NE) (78, 106). This relationship is exceedingly complex. In the case of plasma NE, the data do not suggest a direct association with BP elevations (62). Indeed, some studies report that NE levels in excess of 1800 pg/ml are required to produce hemodynamic effects—a level usually encountered only during vigorous exercise or major acute illness (92). Other research with exogenous infusions of E indicates that it may be 10 times more potent than NE in producing hemodynamic actions, in which case the relationship with BP may be stronger.

Aside from BP, CA potentiate other physiologic processes (e.g., lipid mobilization and atherogenesis) implicated in coronary disease (9, 23, 44, 46). There are also studies showing that psychological stress and certain forms of coping produce increments in E and NE (28, 77, 95, 98, 100). These results are more significant when viewed in the light of research suggesting a link between stress and coronary heart disease (38). It appears, then, that CA reactivity may be significant for understanding the mechanisms whereby Type A behavior leads to cardiovascular pathology.

TYPE A BEHAVIOR AND CATECHOLAMINES

General and Methodological Issues

The remainder of this chapter is devoted to a consideration of the evidence for an association between Type A behavior and CA. However, several methodological issues must be considered at the outset. First, with occasional exceptions (32, 35), little difference has been reported in CA levels between A and B subjects under conditions of relaxation; baseline values are approximately the same for the two types of individuals. As with cardiovascular measures such as BP, A and B discrimination is made primarily in terms of the magnitude of reactivity to stressful and/or challenging stimuli. In view of this finding, particular attention must be given to establishing stable baseline levels prior to the introduction of experimental manipulations.

Experimental results on A-B differences in catecholaminergic reactivity can be classified into those that derive their measurements from urine samples and those that rely on plasma. Urinary CA represent estimates of SNS activity integrated over extended periods of time, for example, 1 to 3 hours (28). They reflect all CA-producing activities that have occurred since the subject last voided. Plasma CA, by contrast, reflect momentary or episodic changes in response to discrete stimuli, including psychosocial manipulations in the laboratory (72). Recent research suggests that urinary NE is an approximate index of plasma NE (at least under various baseline conditions), but urinary and plasma indices of E appear to be dissociated (88). While the reasons for this dissociation are not well understood, the lack of relationship nevertheless has implications for interpreting discrepancies between E findings in Type A studies using the two types of measurements of this hormone.

CA levels are sensitive to a number of confounding factors including: the intake of caffeine and alcohol; tobacco (since nicotine stimulates CA release); age; physical activity; weight; body position; extreme cold; and various medications related to sympathetic-adrenal medullary activity (28, 67, 72, 111) (see Chapters 1 and 24). Since diurnal variations can have profound effects, particularly on E, it is advisable to obtain measurements under basal conditions covering the same period of the day as the experimental conditions (28). Menstrual cycle may also affect CA variability, although the data here are less clear than for the other factors (55).

Human experiments with plasma CA measurements present additional difficulties (68, 72). E and NE are thought to be released intermittently, and major changes in these substances occur in very short periods of time. Moreover, NE is subject to rapid neuronal reuptake at the sympathetic nerve endings, with studies reporting mean half-times of about 2.5 minutes (92). There is also evidence that E may be subject to half-times as short as 1.2 minutes (59). Blood samples must, therefore, be collected expeditiously after an experimental intervention.

The problem of drawing serial blood samples introduces another difficulty; repeated venipuncture is itself a stressor that cannot be ignored in behavioral research (29) (see Chapter 1). One approach to this problem has been the use of indwelling intravenous catheter techniques followed by a 25- to 30-minute "resting period" before blood samples are drawn (40, 64). Recent technical advances in continuous blood withdrawal may also ease the logistical problems of serial sampling (14, 22, 105).

Plasma concentrations of NE increase substantially when the SNS is stimulated by acute postural change or physical exercise such as a hand-grip task (64, 112) (see Chapter 14). Standing up from a supine position, for example, elevates NE by about 90% (62). This phenomenon introduces special problems in the interpretation of NE data collected in psychophysiological studies involving upright posture and the forearm exercise entailed in many psychomotor tasks.

Still another issue warranting comment pertains both to urine and plasma research. While E is a product of the adrenal medulla, NE is an adrenal medullary product as well as a neurotransmitter. This difference may account for the fact that NE levels are typically about four to five times greater than E levels. In addition, NE is more sensitive to physical effort and physical stressors than are E responses (22, 28). E, on the other hand, may have greater sensitivity to psychological aspects of the situation (28, 73).

The psychological significance of different E-NE ratios remains unclear. It may be a question of individual response-stereotypy or stimulus-response specificity (60, 96)—phenomena that have appeared repeatedly in psychophysiological experimentation involving cardiovascular and related autonomic channels. A longstanding hypothesis is the association of E with fear and anxiety, whereas anger and aggressiveness are associated with elevations in NE (1, 25, 37, 93). More recent studies, however, do not provide uniform support for the assumption that E and NE are selectively released in different emotional states, or in individuals (e.g., Type As) predisposed to reacting differentially to affect-laden events (28).

Most of the Type A research reviewed in this chapter has attempted to take into account the sources of error, technical problems, and interpretational difficulties described in the preceding paragraphs. There have, of course, been exceptions, in which case the data may be flawed or misinterpreted. These studies will be identified as we proceed, and due caution will be placed upon generalization of the results.† Note, also, that many of the studies measured the impact of Type A behavior upon physiological processes other than CA, including cortisol, BP, and heart rate. Since the principal focus of this chapter is on catecholaminergic activity, these other measurements will not be discussed unless they have special relevance for interpreting E and NE data.

Type A Behavior and Urinary Catecholamines

One of the first studies of Type A behavior and CA was conducted by Friedman and his colleagues (36). They examined urinary excretion of E and NE in a sample of American middle-class men who exhibited either behavior pattern A or B in extreme form. Nocturnal CA excretion was the same in both groups. When urine formed during more active working hours (pooled over 4 consecutive days) was analyzed, however, the investigators found a significantly greater NE increase in A than B subjects. The relevant percentage increments were 173 and 64, respectively. Both groups showed an increase in E, but the A-B difference was not statistically significant. A subsequent analysis of urinary samples in essentially the same subjects (8) revealed that during working hours, Type As exhibited almost twice as much vanillylmandelic acid (VMA) as that found in Type Bs.

The researchers interpret their data as support for a concomitance between aggressive and time-urgent behavior of Type As and elevated excretion of NE—a finding reminiscent of some literature on aggression and CA (25, 104). As previously mentioned, such a hypoth-

† We will not make extensive comments upon the nature and reliability of collection and assay procedures. Technical discussions of this kind are treated in Chapter 1 of this book. It should be noted, however, that the vast majority of studies considered later in this chapter used acceptable protocols for collection and assay—at least within limitations imposed by the "state-of-the-art" at the time a given experiment was conducted.

esis has received only weak support. Moreover, the Friedman *et al.* research contained a number of methodological deficiencies that cast doubt upon their observations. The experimental groups were exceedingly small (12 As and 11 Bs), cigarette smoking and alcohol consumption were greater in the Type A group, and a variety of selection criteria (other than the structured interview) may well have produced a unique set of cases that limit generalization of the results.

In view of these limitations in design and protocol, it is not surprising that the results could not be replicated in a study of Belgian white collar workers (13). This study included considerably more cases (50) than the original research, introduced more careful controls over various sources of error (e.g., there were about an equal number of cigarette smokers in the A and B groups), and used less potentially biasing criteria in subject selection. In all fairness, however, four differences between the Belgian and American studies should be emphasized: (1) Only subjects at the extreme ends of the A-B continuum were accepted in the American study, whereas most of the subjects in the replication represented more intermediate forms of A and B types; (2) Cultural differences in speech stylistics, for example, may reduce the validity of the structured interview; (3) The working day of Belgian white-collar workers may be less challenging and stressful than that of their American counterparts; (4) The Belgian study examined CA levels based upon samples collected during a single working day, which may have yielded less reliable estimates of E and NE than the 4-day collections used in the American study. Nevertheless, the failure to replicate differences in excretion of urinary NE and its metabolites during the working day period raises questions about the reliability of the original results.

Two other nonlaboratory studies of Type A behavior and urinary CA have reported null effects, although in at least one case, urine samples were collected during what may be considered benign periods of the day. This study examined VMA excretion in 27 Type A and 13 Type B postinfarction, male outpatients (89). Urine specimens were collected by each subject immediately after awakening on alternate mornings during a 2-week period. For each case, VMA determinations were made for two randomly selected weekdays and one randomly selected day during the weekend. There were no differences between A and B subjects for the average of these VMA values.

The other study (97) is essentially a replication of a Swedish investigation (99) aimed at documenting a relationship between stressful life events (e.g., children leaving home, death of a close friend) and excretion of urinary CA over a period of several weeks. A measure of Type A behavior, based on the structured interview, was included in the American study. Thirty As and 28 Bs were used in the analysis. Contrary to the Swedish results, the American data did not yield significant correlations between weighted life event scores (*i.e.*, life change units) and E excretion in urine collected at the time of each of four life event assessments. (Neither study observed associations with NE). Despite initial expectations, the American investigators also failed to find stronger correlations between CA and life change units within the A compared to the B group.

More systematic laboratory experiments of Type A and urinary CA have also been conducted, mostly by Marianne Frankenhaeuser and her colleagues at the University of Stockholm and the Karolinska Institute in Sweden. Frankenhaeuser typically uses college student subjects who are classified as A or B on the basis of a Swedish questionnaire modelled on the *Jenkin's Activity Survey* (69). Extreme groups (e.g., upper and lower quartiles) on the A-B score distribution of this test are selected for study. Urine collection for CA assessments are made as follows: (a) Baseline values are ascertained in a separate session, between 8:30 a.m. and 12 noon, a few days after the experimental session(s), with subjects relaxing, reading and listening to music; (b) experimental levels are determined from urine samples taken after each experimental condition, also between 8:30 a.m. and 12 noon. Subjects are asked to void before each session in order to facilitate inferences about the effects of experimental manipulations without contamination by preexperimental stress and activity. Along with attention to diurnal variation factors, Frankenhaeuser includes a number of other controls in her work, such as elimination of smokers, and she issues careful instructions to subjects to abstain from caffeine- and alcohol-containing beverages and drug intake prior to experimental and baseline sessions.

In one study (70), 48 male and female university students were studied while doing mental work under conditions described as understimulation (a monotonous vigilance task) and overstimulation (color-word conflict task and a nonnumerical reasoning test). Each condition lasted about 60 minutes. Baseline levels of E and NE were about the same for A and B subjects. E output was more pronounced during over- than understimulation, but there was no evidence of differential levels of excretion in Type A compared to Type B subjects of either sex. NE excretion was unaffected by either experimental conditions or the A/B variable. Cortisol excretion was also assayed in this study, with only Type As showing significantly greater levels in under- than overstimulation.

The latter result tends to agree with a previous study which compared reactions of A and B subjects during inactivity and while performing a mental arithmetic task under noise distraction (32). Assuming these conditions were approximately equivalent to understimulation and overstimulation in the earlier study, the similarity of results is reflected in the tendency of As to secrete more cortisol than Bs in the inactivity conditions. During work, no cortisol differences approaching statistical significance were found between A and B groups. Of greater interest to us here, however, is the finding that Bs secreted more E and tended to secrete more NE during work than inactivity. For the As, the corresponding differences approached significance only for NE. The study also reports reliably greater E secretion in As than Bs within the inactivity condition. Reference is made in this study to an unpublished Master's thesis by Forsman (26) which found that the difference in E excretion between work at an experimental task and a 5-hour inactivity period tended to be smaller for Type A than for Type B subjects.

The authors conclude that inactivity is about as stressful as work for Type As, whereas for Type Bs work is more stressful. This observation accords well with descriptions of the Type A individual as hard-driving and impatient. The inactivity condition may have been incongruent with these Type A characteristics, since it asked subjects to relax, read, and listen to music. On the other hand, such a condition can be construed as simply a baseline procedure, which raises the question of why there were A-B differences in urinary E. Lundberg and Forsman (70) did not report such differences, and the Forsman thesis (26) reports only a nonsignificant effect with a somewhat longer period of inactivity. The discrepancy remains something of an enigma, particularly in view of still another study (reported below) which presents baseline data suggesting that Type As do not excrete more E and NE than Type Bs.

Frankenhaeuser *et al.* (33) had A and B subjects perform a choice-reaction time task that allowed each person to perform at his/her preferred pace and to maintain "control" over stimulus rate throughout the experimental period. A "no control" condition was not, unfortunately, included in the design. While Type As performed better than Type Bs—both in terms of reaction time and number of correct responses—there were no differences in E and NE excretion between subject groups. Indeed, the general picture was the same for each: an increase in CA excretion and a decrease in cortisol during the experiment.

These and other related data have been interpreted by Frankenhaeuser (29, 31) in terms of a two-component approach to psychological arousal. When effort and distress are evoked by a given situation, both sympathetic-adrenal medullary and pituitary-adrenal cortical systems are activated, as reflected in increased secretion of CA and cortisol. The experience of effort without distress—as reported by subjects in the experiment under consideration—typifies, according to Frankenhaeuser, situations involving controllability. Under these circumstances, the sympathetic-adrenal medullary system is aroused, whereas the pituitary-adrenal cortical system may be depressed. While interesting and seminal, it is difficult to see how these ideas could account for the absence of A-B differences in CA excretion.

A more likely possibility is the way in which subjects were classified as A or B. Frankenhaeuser relies upon a Swedish version of the Jenkin's Activity Survey, whereas the structured interview tends to be a better predictor of experimentally induced SNS arousal (see footnote*). Indeed, the degree of agreement between A-B classifications based upon the Jenkin's Activity Survey and the structured interview is not very impressive (74). Even by taking only subjects from the extreme ends of the Jenkin's Activity Survey distribution, Frankenhaeuser and her colleagues may well have included misclassified cases. The null effects may, therefore, be attributable to unreliable A-B diagnosis.

OTHER RELEVENT URINARY CA STUDIES

The preceding section of this chapter was limited to a consideration of CA research that involved explicit measurements of Type A. It is of interest to discuss urinary studies that measure or manipulate personality and related variables that bear a similarity to one or more components of Type A behavior. Only a sample of studies in this group will be reviewed here. Moreover, an attempt has been made to cite research that does not contain flaws in design and other forms of unreliable methodology. References to studies comparing CA levels and reactions of coronary heart disease and noncoronary heart disease cases, such as the type reported by Nestel *et al.* (79), have also been ignored unless there were explicit assessments of personality variables closely related to pattern A. It would be risky, at best, to assume an absence of Type A cases in noncoronary heart disease groups (39, 85).

McClelland *et al.* (76) examined the urinary CA output of 27 Harvard sophomores who completed a version of the Picture-Story Test, a form of the *Thematic Apperception Test.* Analysis of the content of the scores permitted delineation of a subcategory of individuals who were high both in need for power and in activity inhibition (*i.e.*, they thought of inhibiting their actions as indicated by frequent use of the word "not" in their stories). These individuals may be likened to Type As insofar as they are assertive and hard-driving but carefully controlled. A measure of life stress was used in the study, thereby allowing further differentiation of subjects into those who experienced greater or lesser stress related to the achievement of power and control. Compared to other subjects in the sample, more of those who showed the inhibited power motive syndrome and reported high power stresses had above average E excretion rates in urine pooled from samples collected before and after a 2½-hour session of perceptual and learning tests. Although the significance level was marginal at .07, the results are in accord with the idea that Type As, in contrast to their B counterparts, manifest heightened sympathetic adrenal medullary activity when exposed to stressful events that threaten their sense of control.

Other research has studied the relationship between CA excretion and measures of various descriptive facets of Type A behavior, including *achievement motivation*, *aggressiveness*, and *time urgency*. While at least one study failed to detect significant associations (101), others have reported positive effects. For example, increased output of E has been found in a variety of achievement situations, such as a 6-hour matriculation examination (82), a mental arithmetic test for 13-year-old children who received instructions emphasizing the importance of performing well (3), and during the defense of a doctoral dissertation (54). Still other research indicates a relationship between competitiveness (*e.g.*, in professional baseball players) and CA excretion (24), and between aggressive activity (*e.g.*, in hockey players) and increases in NE output (24).

The time-urgency component of pattern A has also been related to CA levels. Frankenhaeuser and Kareby (30), for example, administered a battery of conventional intelligence subtests to medical students who were harassed and rushed during performance. Both E and NE elevations were observed in the experimental period relative to values derived from baseline samples taken on a different day. A similar study without the time-urgency manipulations produced weaker effects (34).

SUMMARY

The most salient feature of the studies reviewed in this section is that urinary indices of CA were usually unrelated to explicit measures of Type A behavior. Aside from the nonreplicated results of Friedman *et al.* (36), extant field studies do not suggest that Type A individuals excrete more CA and their metabolites than do Type B individuals. The picture is much the same for laboratory experiments. Other than the finding of an A-B difference in E under what amounts to a baseline condition (what was called "inactivity") in one of Frankenhaeuser's studies, there was little evidence that As reacted to stress and challenge with greater CA output.

The conclusion appears to be that urinary CA research is not a promising approach to understanding the mechanisms whereby Type A behavior leads to cardiovascular pathology; however, such a conclusion might be premature. Many of the studies without direct measures of Type A reported associations between CA output and various components of the behavior pattern. Admittedly, these findings must be interpreted with caution. The situational variations and individual differences used in these studies may appear to be facets of the behavior pattern, but, at best, we can only infer from their presence that A and B behaviors were, in fact, being measured. On the other hand, the studies do suggest that factors such as competitiveness are related to elevations in E and NE. To the extent that these behaviors reflect the Type A pattern, the results are sufficiently encouraging to warrant additional research with urinary CA measurements.

This line of research must, of course, include both field and laboratory designs. The collection of urine is particularly suitable for field studies where the subject's customary routines cannot be altered, and where large numbers of cases are involved. The costs of urine collection are relatively low and cost-effective (see Chapter 3). Morever, urine samples show a slower rate of change and can, therefore, be used to assess long-term changes in CA and their metabolites. This feature allows for examination of catecholaminergic consequences of the psychosocial influences in the everyday life of A and B individuals. Such research might well prove to be a fruitful way of establishing the mechanisms by which A-like behaviors culminate in disease end points.

Type A Behavior and Plasma Catecholamines

Plasma measurements of CA, as mentioned earlier, are well-suited for determining the impact of acute stressors, those used in laboratory experimentation, for example. The use of plasma samples allows for detection of rapid fluctuations in CA levels that often occur when subjects are exposed to systematic stress and challenge manipulations. The research reviewed in this section is based primarily upon such laboratory designs. It is interesting to note, therefore, that an often cited study in the Type A literature entailed plasma measurements in a field setting. Taggart and Carruthers (98) examined, among other neurochemical variables, the CA levels of professional racing car drivers. They report sizeable increments in NE values (about 180%) several minutes after a race, described as a highly aggressive competition, relative to values observed immediately before. Recall that this result is remarkably similar to urine data reported with professional hockey players (24). Unfortunately, neither study measures Type A behavior, either by standard techniques or via variables that had seeming relevance to the components of the behavior pattern.

Assessment of the relationship between explicit measures of Type A and plasma CA begins with a study by Friedman and his colleagues (35). Fifteen pairs of middle-aged Type A and B males were classified by the structured interview and recruited into an experiment involving competition. Sessions began at 8:00 a.m. Subjects were required to work against each other on mental puzzles for 15 minutes, with the winner receiving a prize of a bottle of French wine. The puzzles were virtually insolvable without preliminary hints, and these hints were never given. Subjects were led to believe the puzzles were solvable, and that the average person required only 10 minutes to discover the solution, but they would be given 15 minutes. After these instructions, a 10 ml blood sample was obtained, and subjects were told to begin. Two radios tuned to different rock-'n'-roll stations were turned on. As each minute passed, the number of minutes remaining was announced by an experimenter who remained in the room throughout the contest. Subjects were seated during the task and for 30 minutes after the end of the task. Subjects were bled again by repeated venipuncture 10, 15, and 30 minutes after they started work on the puzzle, although the contest was stopped at the end of 15 minutes. All subjects returned to the laboratory 1 week later (presumably at 8:00 a.m.) to give a baseline blood sample. They were assured that no experiment would precede or follow the taking of this final sample.

The NE results indicated: (a) no significant differences between As and Bs in baseline values; and (b) a seemingly nonsignificant trend toward higher values among As at the start of the contest. (No significance test was reported by the investigators.) An averaged NE level during the contest was significantly different from baseline for Type As, but not for Type Bs. The mean increase during the task period was also reliably greater in Type A subjects and remained significantly higher after the end of the contest. Plasma E did not show corresponding effects.‡ The results indi-

‡ While not reported in the 1975 paper, Rosenman

cate heightened sympathetic-adrenal medullary responses in Type As who are confronted by challenging and, perhaps, stressful conditions.

This study represents a landmark effort to examine the relationship between plasma CA and Type A reactions to challenge and stress. There are, however, several difficulties with the design and analysis that severely limit interpretation of the results. Among the more important flaws are: a confounding of competition with stressful manipulations, such as loud noise and instructions designed to arouse impatience and time urgency; repeated venipuncture which, as noted earlier, is a stressor that can affect behavior-CA associations; and the absence of a noncompetition control group against which to evaluate experimental manipulations. In addition, there is a strong possibility that baseline and experimental plasma samples were analyzed in separate assay runs because CA analysis was reported as having occurred within 72 hours of blood collection, and baseline samples were said to have been drawn one week after collection of experimental samples. The assay of baseline samples in separate runs from those which analyze experimental samples can only increase variability in NE and E values, in which case individual changes from baseline may have been inflated or deflated. In either case, the procedure was ill-advised and raises questions about reliability of the findings.

Another early study of plasma CA and Type A behavior was conducted by Simpson and colleagues of Boston University (94). Forty-eight male college students were classified into A and B groups by dividing them at the median of the Jenkin's Activity Survey distribution for these cases. A similar procedure was used to classify subjects in terms of their scores on three factor analytically derived subscales of the survey. Following a 30-minute rest in a horizontal position on the test day, a baseline blood sample was drawn. A modified Balke treadmill test was then performed, with each subject continuing the task until his pulse reached 180. A second blood sample was obtained immediately after termination of exercise, and a third 15 minutes later.

There were no differences between A and B subjects in plasma NE levels at rest, but after exercise, Type As had somewhat higher levels than Type Bs ($p = .10$). Although no statistical test is presented, the reported mean NE elevations are substantially higher after exercise compared to baseline in both A and B cases. It was also found that "fitness" (indicated by the amount of work done until pulse rate reached 180) was a significant predictor of the increase in NE ($p = .05$). These findings are in accord with other data, discussed earlier, showing NE increments as a function of postural change and physical exercise (64). Comparable effects were not observed for plasma E, nor did A-B differences in E appear in the study.

The only other CA result worth noting here involved measurements of job involvement and dedication to career in the Jenkin's Activity Survey. Subjects scoring high on this scale showed significantly higher E increments than subjects scoring below the median ($p = .05$). Platelet aggregation tests were performed in this study, including those that measured percent aggregation in the *in vitro* challenge of E and NE. Type As had no reduced intensity of platelet aggregation to NE after exercise, whereas Type Bs did ($p < .05$). Similar findings were reported for subjects scoring high and low on a subscale measuring speed and impatience.

CA results in the Simpson study achieved only marginal significance; hence, the only conclusion that may be drawn is that NE reactivity to physical challenge and stress is somewhat greater among As than Bs. The relative unreliability of these data may be due to crudity of behavior pattern assessment and classification. The investigators did, after all, rely upon the Jenkin's Activity Survey and a "median-split" to categorize their cases. One study described below, however, used a similar procedure and obtained substantial results.

A more serious problem with Simpson's pro-

(84) has published a bar graph (identified as figure 3) showing essentially the same NE data, along with NE increases in cases identified as having coronary heart disease. The point being made is that there are no differences between A, B, and coronary heart disease subjects during the resting period, but heart disease and Type A subjects show similar elevations of NE immediately before, during, and after the contest. No statistical tests are reported for comparisons of coronary heart disease cases with the A and B groups, nor is there any mention of the number of cases in the heart disease group, who they were, and from what source they were recruited. Equally puzzling is a failure to indicate when the data on coronary cases were collected. No mention of coronary subjects is made in the original Friedman *et al.* study (35).

tocol is the way in which blood was collected. Individuals were subjected to repeated venipuncture which, as mentioned, is a stressor that can elevate CA levels irrespective of the A-B dimension and could have attenuated exercise-induced differences between A and B subjects. A related problem is with the CA assay. No data are given on intra- and interassay coefficients of variation, thereby making it difficult to judge reliability. On balance, then, the results must be considered equivocal and, at best, exploratory.

About 4 years ago, Glass and his colleagues (40) began a program of research aimed at examining the association between psychological stress, Type A behavior, and plasma CA. The first experiment in the series studied 22 Type A and 22 Type B men (approximately 35 to 50 years of age) working for the New York City Transit Authority. All subjects were paid volunteers and free of the major risk factors of coronary heart disease which include BP of 145/90 mm Hg or greater, serum cholesterol levels of more than 260 mg/100 ml, electrocardiogram abnormalities, diabetes mellitus, and smoking more than a pack of cigarettes per day. Five cases reported using cigarettes, but none of them had smoked for 4 hours prior to the experiment. No subject had taken any medication or drug nor drunk caffeine or alcohol-containing beverages during the same time period. Subjects were classified as A or B on the basis of the structured interview and randomly assigned to the two experimental conditions described below. Two cases were lost from each group because of technical difficulties connected with assay and blood collection procedures.

A major purpose of the study was to assess the effects upon plasma CA of simple competition in a game *vs.* competition with a hostile and harassing opponent. A 25-minute relaxation period during which the seated subjects listened to music and read magazines followed venipuncture for insertion of an indwelling venous catheter in the preferred arm. Each subject then played a series of nine games of Pong, a computerized television game similar to tennis, against a confederate who had been trained at the game to be virtually unbeatable. All cases remained seated throughout baseline and experimental periods. During the "tournament," the subject's opponent was either nonharassing (No Harass condition), or expressed hostility toward the subject by delivering a series of preprogrammed comments at designated times (Harass condition), such as "Come on, can't you keep your eye on the ball" "Damn, you're not even trying," and "I don't understand why you're having so much trouble hitting the ball." As in the Friedman *et al.* study (35), the winner of the nine-game competition (the confederate) received a prize—in this case a $25 gift certificate to a major department store in New York City.

At the end of the 25-minute relaxation period, a blood sample was taken to determine baseline E and NE levels. Samples were taken again after the third, sixth, and ninth (final) games. Plasma E and NE were measured by radioassay using catechol-*O*-methyltransferase (COMT) procedures (40). Interassay coefficients of variation were 31 and 24% for E and NE, respectively. Intraassay variation of duplicates was 10% for E and 9% for NE.

There were no differences between As and Bs or between the two experimental conditions in mean baseline values for E and NE ($Fs < 1$). Increments of CA over baseline values were statistically significant ($p < .05$) in both groups. Of greater interest, however, are the change scores computed by subtracting each subject's baseline E from the average E levels during the contest. Analysis of variance of these data revealed a Harass-No Harass main effect ($p = .04$), as well as an interaction between this variable and the A-B classification ($p = .04$). The increase in plasma E was greater for Harassed As than for any of the other experimental groups. Change scores for plasma NE, while in the same direction, did not achieve statistical significance.

The conclusions of this experiment are as follows: (a) Competition elicits significant and similar increases in plasma CA in Type A and B men; (b) A hostile opponent causes no reliable increase in plasma CA responses of Type Bs; and (c) The presence of a harassing opponent potentiates plasma E (but not NE) during competition in As. It would appear that Type As are selectively predisposed to having enhanced E reactivity to hostile interactions, but competition alone does not produce differential E elevations in As and Bs.

The null NE effects could be attributed to subjects being seated and actively manipulating a response lever with the arm containing the catheter throughout the experimental task. Recall that substantial elevations in plasma NE occur in individuals who are in an upright position and/or are engaged in a hand-grip exercise (62, 64, 112). The NE increments produced by these factors might have obscured the differential magnitude of change in Harassed As compared to their Type B counterparts. On the other hand, two studies, discussed below, report differential NE elevations as a function of psychological stress in A and B subjects who were seated upright throughout the experiment. One of these experiments also entailed forearm activity as vigorous as that observed in the Glass *et al.* experiment (40, 41).

The results of the Pong study do not agree with those reported by Friedman *et al.* (35) who found NE (but not E) changes in As compared to Bs. As previously mentioned, uniform support does not exist for the idea that E and NE are released selectively in different situations, or in different types of individuals confronted by these situations. This relative lack of consistency in predominance of E and NE responses may reflect variations of assay technique from laboratory to laboratory, with some using procedures that are more sensitive in detecting E than NE (or vice-versa). Another, perhaps more viable explanation is the failure to conceptualize and specify the precise parameters of stress and challenge that elicit different patterns of neuroendocrine and related responses.

The fact that A and B subjects did not show differential elevations in plasma E in the No Harass condition was contrary to initial expectations. Greater E arousal might be expected in As under conditions of simple competition because they are defined as highly competitive. Two explanations may be offered for failing to confirm this expectation. The presence of a nonhostile competitor could have depressed Type A reactivity, or alternatively, Type Bs could have risen to the challenge of competition. Indeed, the Bs may have reached an asymptote of responsiveness in No Harass, since the hostile opponent caused no significant increase in E over the level reached with the nonhostile opponent (40). There is, in fact, some evidence to suggest that Bs may struggle as compulsively as prototypic As in a situation that encourages competitive striving and the attainment of concrete goals (7).

In a second experiment, 20 Type A and B subjects worked alone on an individualized version of Pong similar to basketball (40). All aspects of subject selection and classification, measurement, and procedure were virtually identical to the first experiment, except that this study was conducted in two parts—one with no gift certificate for success (No Incentive), and the other in which the subject could win the prize if he attained a designated criterion of performance (Incentive). As before, subjects were seated during the testing session.

There were no significant differences between A and B cases in either baseline E or NE values. Experimental effects were assessed on the mean of the E (or NE) concentration in two blood samples taken during the No Incentive part of the study, and on the E (or NE) concentration in the one blood sample drawn during the Incentive treatment. As in the previous experiment, baseline levels were subtracted from these samples to yield E and NE change scores. Analysis of E data revealed an A-B main effect ($p = .06$), with no other terms (including those involving the incentive manipulation) approaching statistical reliability. Comparable analysis of NE changes also failed to produce significant effects. The A-B result suggests—tentatively because of marginal significance—that the presence of a nonharassing competitor in the first study may have attenuated differences between As and Bs in E reactivity. In any event, it documents somewhat greater E reactions in As than Bs when both are confronted by the challenge of task performance.

Still another experiment in this series (41) recruited 30 Type A and B paid male volunteers to work in a task-overload situation. The structured interview was the basis for A-B classification. Two-thirds of the cases were students in an introductory psychology course, and one-third was from the screened pool of Transit Authority workers described earlier. Therefore, all data analyses included a worker-student factor, along with the A-B variable, as between terms in the design. Instructions regarding smoking, the intake of caffeine- and alcohol-containing beverages, and use of drugs prior to the experimental session were the same as those used in earlier studies.

Overload was induced by requiring subjects to work on primary (visual tracking) and subsidiary (digit recall) tasks simultaneously. On two of the four trials, subjects worked while listening to high-intensity white noise. Noise-no noise trials were counterbalanced by randomly assigning A and B cases either to N $\bar{N}$ $\bar{N}$ N or $\bar{N}$ N N $\bar{N}$ sequences, where N designates the presence of noise and $\bar{N}$ the absence of noise. Blood samples were drawn as follows: (a) immediately following a resting period (baseline sample), which lasted about 30 minutes after venipuncture; (b) one sample after each of the 5-minute experimental trials. Subjects were seated throughout the entire experiment. CA assay procedures were identical to those used in Glass' previous work.

Contrary to previous studies, Type As showed reliably higher baseline E concentrations than Type Bs ($p < .05$). The difference for NE was nonsignificant ($p > .10$). The usual change scores were computed and submitted to analysis of covariance, where baseline levels were the covariate. In the case of E, As showed greater elevations on each of the four trials ($p < .02$). There was also a worker-student effect, with older subjects exhibiting greater increments irrespective of the A-B factor ($p < .01$). The NE data were parallel to E, although the A-B main effect was only marginally significant at the .07 level. The addition of noise on two of the experimental trials did not potentiate effects for either CA.

These results show that Type As experience greater E reactivity during an appropriate level of challenge and stress. Unlike the experiment using an individualized version of Pong, the demanding task of performing two dissimilar activities was sufficiently arousing to produce reliable A-B differences in plasma E elevations. It is interesting to note that NE increments also were affected (albeit marginally) by the joint influence of pattern A and the magnitude of task demands. It is time, therefore, to examine more closely the impact of differing challenges and stresses on CA responses in A and B subjects.

SITUATIONAL SPECIFICITY AND CATECHOLAMINE RESPONSES

It was suggested earlier in this chapter that variability in obtaining positive as well as null CA effects in the Type A literature may be due to the wide variety of eliciting situations used in the experiments. Different types of challenging and stressful situations may elicit different patterns of catecholaminergic responses, or no response at all. The NE-aggression and E-fear associations represent one effort at specifying discriminative situational parameters, but as we noted previously, research does not fully support this hypothesis. A-B differences in catecholaminergic reactivity seem to emerge only under particular circumstances, including those that are "relevant" to Type A characteristics (e.g., harassing competition) and entail a high degree of challenge or stress (e.g., performing two tasks simultaneously). Situational specificity of physiologic response has appeared in other contexts, as mentioned earlier in this chapter, as well as in Type A studies of BP reactivity (16, 43, 81). Most of this research, however, has only pointed to descriptive differences between situations that do and do not elicit A-B effects. Little effort has been devoted to conceptualizing the dimensions that discriminate the situations.

Williams' group is an exception that bases its approach on the theories of the Laceys (61) and on previous work showing muscle vasodilatation during mental work behavior and muscle vasoconstriction during sensory intake behavior (108). These two behaviors are proposed to covary with qualitatively distinct patterns of cardiovascular and neuroendocrine response; enhanced expression of these patterns (hyperresponsivity) may be associated with Type A behavior. It is suggested that the hyperresponsivity of Type A subjects is specific to certain neuroendocrine response parameters, as well as to the two types of task behaviors.

These ideas were tested in an experiment with 21 male undergraduates classified as A or B by the somewhat unusual procedure of using both Jenkin's Activity Survey and structured interview assessments (109). There were 11 As and 10 Bs in the analysis. After venipuncture for insertion of a butterfly needle in a superficial arm vein, subjects rested for 1 hour in a seated position; a 20-minute baseline period followed relaxation. Continuous, integrated blood samples were obtained throughout baseline and subsequent task periods using a Cormed withdrawal pump. CA were averaged over 7-minute intervals—three for baseline and three for task periods. A and B subjects were compared using the mean of the three baseline samples *vs.* the mean level obtained across the three task samples. Covariance procedures were used on these change scores with baseline levels as the covariate.

(Reliable differences between A and B cases in baseline levels of E and NE are not reported in the study.)

The assay protocol for CA was similar to that reported by Glass *et al.* (40). Three other neuroendocrine parameters—cortisol, prolactin, and testosterone—were assayed in this study. Cortisol is known to stimulate CA-synthesizing enzymes, inhibit COMT, and increase the sensitivity of adrenergic receptors to a given level of neurotransmitter (58). Among Type A cases, therefore, enhanced adrenocortical arousal could potentiate physiologic and metabolic effects of any given level of sympathetic activation.

The mental work task entailed serial subtractions of 13 from 7683, with a small prize at the end of the study (6 months later) to the subject with the highest subtraction rate. A choice reaction time task was used to induce sensory intake behavior. Each subject was tested at the same time of day on two separate occasions—1 week apart. The tasks were counterbalanced so that half of the subjects performed the mental work task first.

All subjects showed significant increases in E during mental arithmetic tasks, with Type As exhibiting greater increases than Type Bs ($p = .05$). A similar effect was observed for NE ($p = .05$). The difference in cortisol response of A and B subjects also was statistically significant ($p = .006$), whereas prolactin and testosterone levels showed no reliable differences. During the reaction time task, significant differences between Type A and B subjects did not appear on any of the neuroendocrine measures, except testosterone, with As showing greater elevations than Bs ($p = .05$).

Williams *et al.* (109) conclude that during mental work Type A subjects show selective hyperresponsivity in E, NE, and cortisol but not prolactin. During sensory intake behavior, by contrast, Type A subjects were not more reactive than Type Bs. The sole exception for testosterone remains something of an enigma. Williams suggests that previous studies may have failed to find A-B differences as uniform as his because of "a greater expression of individual differences in response resulting from the use of complex tasks with mixed requirements to attend to visual stimuli, plan strategies and cope with various forms of harassment. . . ." In his own study, Williams argues that there was the relatively specific requirement either for mental work or sensory intake.

These conclusions are reasonable and are perhaps even correct interpretations of the results. The arithmetic task, however, may well differ from reaction time on a number of dimensions other than sensory intake *vs.* mental work. For example, serial subtractions are probably more difficult for college subjects than rapid response to stimulus lights. Indeed, NE and E increments during reaction time were smaller than during mental work. A task difficulty dimension could, therefore, account for differential A-B effects. In other words, Type As were challenged and/or stressed by calculational demands but not by the requirements of fast reaction time.

Alternative explanations of this sort always abound when qualitatively different tasks are used to manipulate a conceptual variable. Moreover, there are, in fact, some studies that have obtained A-B differences in physiological responses to sensory-intake tasks such as choice reaction time (74), and Williams did, after all, obtain an A-B effect for testosterone. Nevertheless, this general approach represents a rare effort to identify theoretical parameters of the situation that elicit differential CA responsiveness in A and B individuals. Paradigms of this sort may help reconcile seeming discrepancies between studies reporting one rather than another CA elevation.

Another effort to specify such conditions comes from the work of Glass and his colleagues (38). A series of studies conducted between 1972 and 1976 suggested that Type A behaviors may reflect an attempt to assert and maintain control over stressful aspects of the environment. The accelerated pace and aggressive stance of the Type A individual may reflect efforts to avoid the negative affect associated with the threat of failing to cope with situations by rapidly gaining control over them. In short, Type As increase their efforts to master the environment when confronted by threats to their ability to control. If these efforts meet with repeated failure, however, Type As may give up responding and act helpless. Brief exposure to threatened loss of control thus accelerates control efforts, whereas prolonged exposure leads to a decrement in these behaviors. This response pattern has been described as hyperresponsiveness followed by hyporesponsiveness (38). When efforts at control fail—that is, when hyporesponsiveness sets in—the theory predicts a shift from sympathetic to parasympathetic dominance, hence a decrement in CA elevations (see Fig. 21.1).

A recent experiment by Contrada *et al.* (11) examined the CA hyperresponsiveness part of the model. Eighty-seven paid male volunteers (29 to 53 years of age) from the New York City Police Department served as subjects. All cases were screened for both coronary heart disease risk factors and CA-contaminating practices, as in earlier studies from this laboratory. Subjects were classified as A or B based on the structured interview and were randomly assigned to four experimental conditions. Seven As and three Bs were unavailable for CA analysis because of difficulties with assay and blood collection procedures.

Following venipuncture for insertion of the forearm venous catheter, a 30-minute resting period ensued, and a baseline blood sample was drawn at the end of this period. Three experimental blood samples were taken after each of three blocks of 26 experimental trials on a reaction time task (see below). Measurement of plasma E and NE followed the assay procedure used in previous studies (40). NE and E change scores were computed by subtracting the subject's baseline from each of his experimental values. Covariance analysis of these change scores was the principal statistical procedure used in the study.

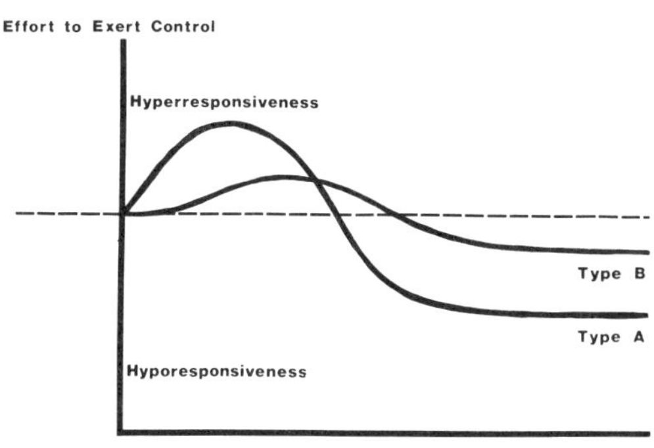

Figure 21.1. The relationship between amount of exposure of uncontrollable threat and the A-B variable in determining a person's motivation to exert control is shown. The curves are, of course, idealized and theoretical, and the "amount of exposure" parameter is, as yet, imprecisely defined. Nevertheless, experimental results have provided some behavioral support for the model (38). Note, also, that a central feature of this approach is the idea that active efforts to control are accompanied by sympathetic activation and elevated CA levels.

Subjects performed a complex choice reaction time task during which they received loud noise bursts and/or electric shocks on designated trials. About half the cases were told they could avoid noise and shock by attaining a predetermined criterion of reaction time speed (Contingency condition). The remaining half were instructed that noise and shock would be delivered randomly, irrespective of their performance (No Contingency condition). Half the cases in each group received a high frequency of aversive stimulation, whereas the remaining cases received a low frequency of noise and shock. Within the Contingency condition, high and low frequencies were designed to convey to the subject that his coping efforts were unsuccessful and successful, respectively. It is under unsuccessful, *i.e.*, failure, conditions that threatened loss of control is likely to be maximal, thereby potentiating A-B differences in CA reactivity. Aversive stimulation in the No Contingency condition was gratuitous and, therefore, did not convey information about performance. Differences between A and B subjects should be minimal under these conditions.

There were no differences between conditions, or between A and B cases in either baseline NE or E values ($p > .15$). Analysis of NE change scores revealed a significant interaction of the A-B variable, the two experimental treatments, and the three blocks of reaction time trials ($p < .05$). Further analysis indicated that Contingency/High frequency elicited a significant linear increase in NE for Type As but not for Type Bs. Reliable increments were not observed for As in contingency/Low frequency, nor for As and Bs in the two No Contingency conditions. Contrary to expectations, however, there was a significant linear increment in NE for Bs in the Contingency/Low frequency condition.

The results for E are more puzzling. Type As showed higher E elevations than Type Bs in all conditions save Contingency/High frequency, where Bs reliably exceeded As. This effect was reflected in a triple-order interaction involving A-B and the two experimental treatments ($p = .05$). We would have expected, as with NE, to find greater reactivity among As in Contingency/High frequency.

The Contrada experiment does not fully support the catecholaminergic-hyperresponsivity hypothesis. Greater increases in NE among As in Contingency/High frequency were consistent with theoretical expectations, but the greater NE reactivity of Bs in Contingency/Low frequency is not easily explicable in terms of the original model. The E results are even more inconsistent with theory.

Many reasons undoubtedly exist for such inconsistencies. Complex laboratory experiments of the kind conducted by Contrada attempt to create a psychological state in the subject by means of instructions, reinforcements, and induced expectations. Imprecision in calibrating various features of the experimental situation (e.g., the intensity of noise and shock, or the frequencies of aversive stimulation used to induce successful and unsuccessful coping), as well as other factors such as subjects' attitudes toward the research, may influence the manipulations and consequently produce what appear to be anomalous results.

The uncontrollable threat model may, nevertheless, have the potential for specifying at least one dimension of the situation necessary for eliciting A-B differences in CA reactivity. Definitive statements cannot be made at this time, but better controlled future research may prove more conclusive. Such experimentation should, of course, address both the hyperresponsive and hyporesponsive parts of the model.

STUDIES SHOWING REVERSE A-B EFFECTS

The idea of situational specificity implies the possible existence of conditions under which Bs will exceed As in catecholaminergic reactivity. We have seen such effects, albeit unanticipated, in the Contrada experiment. Behavioral studies of pattern A have also reported greater responsiveness in Bs under specific conditions of stress and challenge (38). The research reviewed next has a similar, common feature: CA results indicate greater elevations in Type Bs.

Consider, first, an experiment by Hilton *et al.* (47). Forty-two paid male volunteers from the New York City Transit Authority and the New York City Police Department participated. Subject selection and classification followed procedures used in previous studies (40). Blood collection and CA assay protocols were also the same.

The rationale for the study derives from: (a) findings suggesting that Type As, in contrast to Type Bs, suppress or publicly deny subjective fatigue in the course of performing challenging physical tasks (10); and (b) observations of the Type A individual as someone who expresses overt bravado, struggles to inhibit the expression of strong affect, and generally presents himself as being in control and capable of handling any crisis (49). It may be, therefore, that Type As would attempt to conceal emotional reactions when relating incidents in their lives that gave rise to, for example, fear or distress. While describing the affect-laden nature of the situation, they may show few outward signs of internal upheaval. To the extent that they are effective in concealing such affect, we might expect a concomitant dampening of physiological arousal, although it is possible that suppression may be targeted only at expressive behaviors.

To explore these possibilities, Hilton and his colleagues (47) asked subjects, via tape-recorded instructions, to "think of a situation that made you (angry, fearful, distressed) in the past . . . Imagine . . . the events contributing to increasing (anger, fear, distress) . . ." The order of occurrence of the three emotions was counterbalanced in order to control for possible sequence effects. Subjects were told, after 3 minutes of imaging, ". . . . to describe the situation . . . Tell . . . how you reacted. Pretend you are telling this to a sympathetic friend who is listening carefully. Describe the situation to your friend in such a way that he can also begin to feel (angry, fearful, distress) . . . ." Three minutes were allotted for this "relating" phase of the study.

Blood samples were drawn after each emotional incident was related. Plasma E and NE values in these samples were taken as deviations from baseline values obtained from a sample drawn at the end of a 30-minute resting period following venipuncture. The resulting change scores were analyzed by the usual covariance procedures. A similar protocol was used for the blood drawn after each of the 3 minutes of imaging.

There were no differences between A and B cases in baseline values of E and NE ($t < 1$). There were no A-B differences, or interactions with type of emotion, for E in both the imaging and relating phases of the study ($p > .20$). NE, by contrast, showed an A-B main effect for imaging ($p < .04$), as well as for relating ($p < .01$). Type Bs showed higher elevations than As for both phases of the study.

The results indicate that Bs exhibit greater CA reactivity under conditions that entail thinking about and revealing the experience of strong affect. Neurochemical effects alone, however, do not permit us to conclude that dampened physiological responsiveness was accompanied by inhibited affective expressiveness. Behavioral data are necessary to provide support for the theoretical rationale underlying the experiment. Consequently, video tapes were made as each subject thought and talked about his emotional experiences, and these tapes were rated, by naive observers, along several expressive dimensions (e.g., "pleasantness-unpleasantness") using scales and procedures similar to those discussed by Osgood (80). During both the imaging and relating phases of study, unpleasant affect, as measured by ratings of facial displays, was greater in Type B than Type A subjects ($p <$ .05). It appears, then, that lower physiological reactivity in As was paralleled by fewer signs of affective expressiveness when they were asked to recount "real-life" emotional situations.

The idea of psychosocial situations that depress hyperresponsivity is not unlike Glass' notion of hyporesponsiveness. The Hilton *et al.* (47) experiment did not test this hypothesis, but instead suggested another set of conditions for greater reactivity in Bs. Whether there are common themes that will bring both sets of data under a single theoretical rubric remains a question for future investigation; it is, without question, a complex issue. Adding to that complexity is a recent study by Zahn *et al.* (110).

Twenty upper middle-class, male volunteers (31 to 59 years of age) were classified as A or B on the basis of the structured interview (110). Fourteen were As and six were Bs. After venipuncture for insertion of the needle of a heparin lock, the subject reclined to a prone position and rested for 15 minutes. Two experimenters were in the room throughout this time and during subsequent stress periods. The testing session consisted of two 5-minute stress periods (standing, mental arithmetic) and one 8- to 9-minute period (structured interview and a history quiz). Each period was preceded by a 15-minute rest.

After the first rest period, the subject was asked to stand for 5 minutes. This procedure constituted the first task of the study. A second 15-minute rest period ensued. Mental arithmetic came next, in which a 5-minute series of multiplication problems of increasing difficulty were given orally. Each answer was followed by "Right" or "Wrong" flashed on a screen. Speed and accuracy were emphasized in the instructions for this test. A third 15-minute rest period then occurred, followed by the structured interview and ending with a 2-minute American history quiz (17).

Blood samples were drawn as follows: (a) after venipuncture; (b) at the end of each of the three rest periods; (c) after standing; (d) after 3 minutes and 5 minutes of the arithmetic problems; (e) after 3 minutes of the structured interview; (f) at the end of the history quiz. These samples were analyzed for plasma NE only by a radioenzymatic assay (64) (see Chapter 2). The major statistical analyses involved Groups × Periods analysis of variance for the rest periods only and for three change scores (stress minus preceding rest). Subsequent contrasts on the change scores were used to assess the effects of individual stress procedures.

Type B subjects had significantly higher supine baseline NE concentrations across all three rest periods ($p < .03$). For A and B groups combined, NE showed the usual marked increase from supine to standing (about a 2-fold increase) but did not increase reliably during the arithmetic task and showed only a small (albeit significant) increase during the structured interview/history quiz procedure ($p <$ .05). The analyses also revealed that Bs increased their plasma NE levels more than As in response to the combined stressors ($p < .01$). A significant Groups × Periods interaction (p $< .03$), along with individual comparisons, indicated that the effect was due primarily to the stand procedure ($p < .01$). There were no A-B differences for the other two stress periods.

These data indicate that the minimally challenging task of standing upright for 5 minutes produced greater increments in plasma NE in Bs than in As. The psychological significance of the stand procedure is not clear. It cannot be interpreted by the rationale used to predict the B-A reversal in the Hilton *et al.* experiment, nor is it explicable as an instance of hyporesponsiveness to uncontrollable threat. It is important to note, therefore, that Bs had higher NE levels during supine rest periods, including the one at the beginning of the experiment. This result suggests that Bs may have entered the experiment in a more aroused state than As, in which case the issue becomes one of accounting for initial differences in NE.

Several possibilities suggest themselves. First, the results may be due to sampling error given the small number of Type B cases (n = 6). Second, misclassification along the A-B dimension may have occurred, since the structured interview was administered as just one more task in the sequence of experimental procedures. Indeed, the structured interview came after the stressful arithmetic task, which may have affected the subjects' responses to the interview, despite an intervening 15-minute rest period. Moreover, the interview was not administered under customary conditions (83): two experimenters were in the room—one of whom acted as interviewer; the interview itself was interrupted for a blood sample; periodic measurements were made of autonomic variables; and the subjects had electrodes and a heparin lock attached to them throughout the interview. Still another reason for A-B differences in baseline NE levels may be that blood samples were obtained after the subjects had been supine for at least 15 minutes, whereas in most previous studies the samples were taken while subjects were sitting.

The Zahn *et al.* experiment (110), then, has produced a type of B-A reversal that may be a result of methodological artifacts. Yet, it does suggest the possibility of another set of conditions leading to greater NE reactivity in Type B individuals, since the Bs showed higher elevations in response to the standing procedure. The higher baseline values for these cases were accompanied by accentuated responsivity to this task. Further research based upon the more precise conceptualization of situational parameters is required in order to clarify and interpret this result.

SUMMARY

The studies reviewed in this section do not generally document baseline differences between A and B individuals for either E or NE. There were only two exceptions, one for E, where As showed higher bases (41), and one for NE, where Bs exceeded As (110). A-B differences in catecholaminergic reactivity, though, were observed in all the studies. One (35) reported differences in NE elevations under conditions of stress and competition, whereas another (40) observed E effects only when the competition involved explicitly hostile interactions. Somewhat greater consistency appeared in experiments using challenging and stressful tasks with subjects working alone. Glass *et al.* (40, 41) reported higher elevations in E among As, although one of these studies yielded only a marginal effect. The second study also found marginal NE increments, as did Simpson and his colleagues (94) for a treadmill task. E differences between As and Bs in this latter study did not even approach statistical significance.

In view of the flaws in the Friedman *et al.* (35) experiment, there are actually three reliable studies reporting greater E responses in As and one documenting an A-B difference for NE. Since previous research suggests greater sensitivity of NE than E to physical effort (22), it is not surprising that Simpson and his coworkers (94) obtained a specific NE result. The remaining studies used psychological stressors and/or challenges and produced primarily E results. There is some evidence that this hormone has greater sensitivity to psychological aspects of the situation (73), although precisely which aspects are still unclear.

Two other experiments reviewed in this section focused on specifying situational parameters that elicit catecholaminergic hyperresponsivity in Type A individuals. Williams *et al.* (109) suggested that the nature of the task determines such hyperresponsivity, with mental work (in contrast to sensory-intake tasks) producing greater CA elevations in As than Bs. While their results seem to confirm this hypothesis, an interpretation emphasizing the greater challenge inherent in the mental work task is a viable alternative.

The Contrada *et al.* (11) study also reported greater CA elevations in Type As, but only under certain psychological conditions. Uncontrollable threat, as predicted, potentiated NE reactivity, but a similar effect was observed for Bs in a condition which presumably minimized such threat. Moreover, Bs exceeded As in E elevations under conditions of uncontrollable threat, whereas the reverse occurred in the other experimental conditions.

Both studies represent potentially useful approaches to specifying the conditions for catecholaminergic hyperresponsivity in Type

As. However, neither one provides clear-cut data indicating precisely when As will elevate more than Bs, or, for that matter, when NE or E will predominate. The situation is still such that As elevate more than Bs when appropriately stressful and/or challenging circumstances occur. The nature of these circumstances has yet to be clarified.

The idea that Bs will show greater CA elevations than As under certain conditions received a systematic test in the Hilton *et al.* (47) experiment. Greater NE levels in Bs (as well as more unpleasant facial affect) were observed when the task involved revealing affect-laden expriences. The psychological significance, if any, of the B-A reversal reported by Zahn *et al.* (110) is unclear, for it indicates only that the act of standing upright produces greater NE increments in Bs.

The picture given by this summary of findings is more encouraging than the one projected by the urinary CA research. The plasma studies have provided some replicated data points indicating that under a number of conditions As show catecholaminergic hyperresponsivity. Specification of the underlying psychological dimension of those conditions is a major challenge for future research.

It is also necessary to develop a more differentiated approach to the relationship between facets of Type A behavior and physiological reactivity. Very recent work suggests that the magnitude of catecholaminergic responses to a psychological stressor (*i.e.*, an audio visual conflict task) was predictable on the basis of BP responses to a different stressor (*i.e.*, a serial mental arithmetic task) administered several months earlier (42). The question now becomes one of describing, in behavioral terms, the nature of high and low BP reactors. While preliminary data suggest that part of the answer lies in selected A-B characteristics (e.g., anger, hostility), other factors are almost certain to emerge from a fuller analysis of the results.

CONCLUDING REMARKS

This chapter has examined the evidence for an association between Type A behavior and catecholaminergic responses to psychological stressors and challenges. The review covers virtually all of the published and many of the unpublished field and laboratory studies in the area. Consideration was given primarily to urinary and plasma CA data. Many of the studies included other physiological measures (e.g., hemodynamic variables) that may, in the future, help to clarify A-B differences in E and NE reactivity. A systematic treatment of cardiovascular data was simply beyond the scope of this chapter. However, a recent review (74), taken in conjunction with the material presented here, could serve as the basis for specifying promising research directions.

At the risk of redundancy, let us emphasize once again that a major aspect of this research should be concerned with delineating the psychological dimensions of the situations that elicit hyperresponsiveness in type A individuals. In doing so, attention also must be given to the nature of type B behavior. Research is needed to specify those conditions in which Bs respond with higher CA levels than As and to evaluate the significance of catecholaminergic responses in behavior pattern B. Such efforts are especially important from a clinical point of view, for epidemiological data have documented coronary heart disease in Bs as well as As (85).

It remains to be seen whether studies of urinary and plasma CA will help explain how situational parameters elicit A and B behaviors. These responses are, in part, determined by perceptual-cognitive processes, thus giving importance to the cerebral cortex. While some evidence suggests that epinephrine may cross the blood-brain barrier, little is known about intracerebral concentrations needed to achieve effects upon the reticular formation and hypothalamus—structures implicated in the expression of affective behavior (28). Still other data indicate that urinary 3-methoxy-4-hydroxy-phenylglycol (MHPG)—the major metabolite of NE—cannot be used as a valid index of brain norepinephrine metabolism (4). Moreover, plasma NE and NE in cerebrospinal fluid (an index of central norepinephrine activity) are dissociated in certain populations, presumably because this catecholamine penetrates the blood-brain barrier poorly or not at all (63, 113). These data raise questions about the interpretation of associations between Type A behavior and peripheral catecholamine levels. Nevertheless, they do not alter whatever empirical value urinary and plasma CA measurements may have in understanding the mechanisms whereby Type A behavior potentiates cardiovascular disease.

Indeed, recent work with β-blocking agents in modifying Type A behavior and its physiologic correlates has yielded promising results (57, 84, 90). Alterations in CA production or metabolism may occur in peripheral tissues as well as in the brain. As noted above, changes in urinary and plasma CA may not provide a positive basis for supporting or discarding hypotheses regarding the role of these substances in Type A behavior. However, there seems to be a growing conviction concerning their significance in the pathogenesis and treatment of cardiovascular disease. New research designs are indicated, in which concurrent measures are taken of behavior patterns, psychological stress, and peripheral as well as central CA responses. Such biobehavioral investigations are, of course, fraught with technical, logistical, and even ethical difficulties. They will, nonetheless, contribute to a fuller understanding of the relationship between behavior, CA, and cardiovascular disease.

References

1. Ax AF: The physiological differentiation between fear and anger in humans. *Psychosom Med* 15:433–442, 1953.
2. Baum A, Singer JE, Baum CS: Stress and the environment. *J Social Issues* 37:4–35, 1981.
3. Bergman LR, Magnusson D: Overachievement and catecholamine excretion in an achievement-demanding situation. *Psychosom Med* 41:181–188, 1979.
4. Blombery PA, Kopin IJ, Gordon EK, Markey SP, Ebert MH: Conversion of MHPG to vanillylmandelic acid. *Arch Gen Psychiatry* 37:1095–1098, 1980.
5. Blumenthal JA, Williams RB, Kong Y, Schanberg SM, Thompson LW: Type A behavior and angiographically documented coronary disease. *Circulation* 58:634–639, 1978.
6. Brand RJ, Rosenman RH, Sholtz RI, Friedman M: Multivariate prediction of coronary heart disease in the Western Collaborative Group Study compared to the findings of the Framingham study. *Circulation* 53:348–355, 1976.
7. Burnam MA, Pennebaker JW, Glass DC: Time consciousness, achievement striving, and the Type A coronary-prone behavior pattern. *J Abnorm Psychol* 84:76–79, 1973.
8. Byers SO, Friedman M, Rosenman RH, Freed SC: Excretion of 3-methoxy-4-hydroxymandelic acid in men with behavior pattern associated with high incidence of coronary artery disease. *Fed Proc* 21:99–101, 1962.
9. Carruthers ME: Aggression and atheroma. *Lancet* 2:1170–1171, 1969.
10. Carver CS, Coleman AE, Glass DC: The coronary-prone behavior pattern and the suppression of fatigue on a treadmill test. *J Pers Soc Psychol* 33:460–466, 1976.
11. Contrada RJ, Glass DC, Krakoff LR, Krantz DS, Kehoe K, Isecke W, Collins C, Elting E: Effects of control over aversive stimulation and Type A behavior on cardiovascular and plasma catecholamine responses. *Psychophysiology* 19:408–419, 1982.
12. Cox T: *Stress*. Baltimore, University Park Press, 1980.
13. DeBacker G, Kornitzer M, Kittel F, Bogaert M, Van Durme JP, Vincke J, Rustin RM, Degre C, De Schaepdrijver A: Relation between coronary-prone behavior pattern, excretion of urinary catecholamines, heart rate, and heart rhythm. *Prev Med* 8:14–22, 1979.
14. De Lacerda L, Kowarski A, Migeon CJ: Integrated concentration and diurnal variation in plasma cortisol. *J Clin Endocinol Metab* 36:227–238, 1973.
15. Dembroski TM, MacDougall JM, Eliot RS, Buell JC: Effects of competitive challenge on cardiovascular reactions of pattern A and B subjects. Omaha, Nebraska, University of Nebraska Medical Center, unpublished manuscript, 1982.
16. Dembroski TM, MacDougall JM, Herd JA, Shields JL: Effect of level of challenge on pressor and heart rate responses in Type A and B subjects. *J Appl Soc Psychol* 9:209–228, 1979.
17. Dembroski TM, MacDougall JM, Lushene R: Interpersonal interaction and cardiovascular response in Type A subjects and coronary patients. *J Hum Stress* 5:28–36, 1979.
18. Dembroski TM, MacDougall JM, Shields JL: Physiologic reactions to social challenge in persons evidencing the Type A coronary-prone behavior pattern. *J Hum Stress* 3:2–9, 1977.
19. Dembroski TM, MacDougall JM, Shields JL, Petitto J, Lushene R: Components of the Type A coronary-prone behavior pattern and cardiovascular responses to psychomotor performance challenge. *J Behav Med* 1:159–176, 1978.
20. Dembroski TM, Weiss SM, Shields JL, Haynes SG, Feinleib M (eds): *Coronary-Prone Behavior*. New York, Springer-Verlag, 1978.
21. Dimsdale JE, Hackett TP, Hutter AM, Block PC: The risk of Type A mediated coronary artery disease in different populations. *Psychosom Med* 42:55–62, 1980.
22. Dimsdale JE, Moss J: Plasma catecholamines in stress and exercise. *JAMA* 243:340–342, 1980.
23. Eliot RS: *Stress and the Major Cardiovascular Diseases*. New York, Futura, 1979.
24. Elmadjian F: Excretion and metabolism of epinephrine and norepinephrine in various emotional states. *Proceedings of the 5th Pan American Congress on Endocrinology*. Lima, Peru, November, 1963, pp 341–370.
25. Elmadjian F, Hope JM, Lamson CT: Excretion of epinephrine and norepinephrine under stress. *Recent Prog Horm Res* 14:513–553, 1957.
26. Forsman L: Interindividual differences in psychophysiological reactions during work and inactivity. Unpublished Master's thesis, Department of Psychology, University of Stockholm, Sweden, 1977.
27. Frank KA, Heller SS, Kornfield DS, Sporn AA, Weiss MB: *JAMA* 240:761–763, 1978.

28. Frankenhaeuser M: Experimental approaches to the study of catecholamines and emotion. In Levi L: *Emotions: Their Parameters and Measurement*. New York, Raven Press, 1975, pp 209–234.
29. Frankenhaeuser M: Psychoneuroendocrine approaches to the study of emotion as related to stress and coping. In Howe HE, Dienstbier RA: *Nebraska Symposium on Motivation* 1978. Lincoln, University of Nebraska Press, 1979, pp 123–161.
30. Frankenhaeuser M, Kareby S: Effect of meprobamate on catecholamine excretion during mental stress. *Percept Mot Skills* 15:571–577, 1962.
31. Frankenhaeuser M, Lundberg U: Psychoneuroendocrine aspects of effort and distress as modified by personal control. In Bachmann W, Udris I, Daniel J: *Mental Load and Stress in Activity*. Berlin, Publishing House Verlag der Wissenschaften der DDR, in press, 1983.
32. Frankenhaeuser M, Lundberg U, Forsman L: Note on arousing Type A persons by depriving them of work. *J Psychosom Res* 24:45–47, 1980.
33. Frankenhaeuser M, Lundberg U, Forsman L: Dissociation between sympathetic-adrenal and pituitary-adrenal responses to an achievement situation characterized by high controllability: Comparison between Type A and Type B males and females. *Biol Psychol* 10:79–91, 1980.
34. Frankenhaeuser M, Post B: Catecholamine excretion during mental work as modified by centrally acting drugs. *Acta Physiol Scand* 55:74–81, 1962.
35. Friedman M, Byers SO, Diamant J, Rosenman RH: Plasma catecholamine response of coronary-prone subjects (Type A) to a specific challenge. *Metabolism* 24:205–210, 1975.
36. Friedman M, St George S, Byers SO, Rosenman RH: Excretion of catecholamines, 17-ketosteroids, 17-hydroxy-corticoids and 5-hydroxyindole in men exhibiting a particular behavior pattern (A) associated with high incidence of clinical coronary artery disease. *J Clin Invest* 39:758–764, 1960.
37. Funkenstein DH: Nor-epinephrine-like and epinephrine-like substances in relation to human behavior. *J Nerve Ment Dis* 124:58–68, 1956.
38. Glass DC: *Behavior Patterns, Stress, and Coronary Disease*. Hillsdale, New Jersey, Lawrence Erlbaum Associates, 1977.
39. Glass DC: Psychological and physiological responses of individuals displaying Type A behavior. *Acta Med Scand* [Suppl] 660:193–202, 1982.
40. Glass DC, Krakoff LR, Contrada R, Hilton WF, Kehoe K, Mannucci EG, Collins C, Snow B, Elting E: Effect of harassment and competition upon cardiovascular and plasma catecholamine response in Type A and B individuals. *Psychophysiology* 17:453–463, 1980.
41. Glass DC, Krakoff LR, Finkelman J, Snow B, Contrada R, Kehoe K, Mannucci EG, Isecke W, Collins C, Hilton WF, Elting E: Effect of task overload upon cardiovascular and plasma catecholamine responses in Type A and B individuals. *Basic Appl Soc Psychol* 1:199–218, 1980.
42. Glass DC, Lake CR, Contrada RJ, Kehoe K, Erlanger LR: Stability of individual differences in physiological responses to stress. *Health Psychol* 2:317–341, 1983.
43. Goldband S: Stimulus specificity of physiological response to stress and the Type A behavior pattern. *J Pers Soc Psychol* 39:670–679, 1980.
44. Haft JI: Cardiovascular injury induced by sympathetic catecholamines. *Prog Cardiovasc Dis* 17:73–86, 1974.
45. Haynes SG, Feinleib M, Kannel WB: The relationship of psychosocial factors to coronary heart disease in the Framingham study III. Eight-year incidence of coronary heart disease. *Am J Epidemiol* 111:37–58, 1980.
46. Herd JA: Physiological correlates of coronary-prone behavior. In Dembroski TM, Weiss SM, Shields JL, Haynes SG, Feinleib M: *Coronary-prone Behavior*. New York, Springer-Verlag, 1978, pp 129–136.
47. Hilton WF, Glass DC, Lake CR, Krakoff LR, Kehoe K, Contrada RC: Effects of emotional imagery on cardiovascular and plasma catecholamine responses in Type A and Type B individuals. Graduate Center, City University of New York, manuscript in preparation, 1983.
48. Jenkins CD: Psychologic and social precursors of coronary disease. *N Engl J Med* 284:244–255, 307–317, 1971.
49. Jenkins CD: The coronary-prone personality. In Gentry WD, Williams RB: *Psychological Aspects of Myocardial Infarction*. Saint Louis, CV Mosby, 1975, pp 5–23.
50. Jenkins CD: Recent evidence supporting psychologic and social risk factors for coronary disease. *N Engl J Med* 294:987–994, 1033–1038, 1976.
51. Jenkins CD: A comparative review of the interview and questionnaire methods in the assessment of the coronary-prone behavior pattern. In Dembroski TM, Weiss SM, Shields JL, Haynes SG, Feinleib M: New York, Springer-Verlag, 1978, pp 71–88.
52. Jenkins DC, Rosenman RH, Friedman M: Development of an objective psychological test for the determination of the coronary-prone behavior pattern in employed men. *J Chronic Dis* 20:371–379, 1967.
53. Jenkins CD, Zyzanski SJ, Rosenman RH: *Manual for the Jenkins Activity Survey*. New York, The Psychological Corporation, 1979.
54. Johansson G: Activation, adjustment and sympathetic-adrenal medullary activity: Field and laboratory studies of adults and children. Reports from the Psychological Laboratories, University of Stockholm, 1973, Supplement 21.
55. Johansson G, Post B: Catecholamine output of males and females over a one-year period. *Acta Physiol Scand* 92:557–565, 1974.
56. Kannel WB, McGee D, Gordon T: A general cardiovascular risk profile: The Framingham Study. *Am J Cardiol* 38:46–51, 1976.
57. Krantz DS, Durel LA, Davia JE, Shaffer RT, Arbian JM, Dembroski TM, MacDougall JM: Propranolol medication among coronary patients: Relationship to Type A behavior and

cardiovascular stress. *J Hum Stress* 8:4–12, 1982.

58. Kvetnansky R: Recent progress in catecholamines under stress: In Usdin E, Kvetnansky R, Kopin IJ: *Catecholamines and Stress: Recent Advances*. Amsterdam, Elsevier/North Holland, 1980, pp 7–18.
59. LaBrosse EH, Mann JD, Kety SS: The physiological and psychological effects of intravenously administered epinephrine and its metabolites in normal and schizophrenic man-III. *J Psychiatr Res* 1:68–75, 1961.
60. Lacey JI: Somatic response patterning and stress: Some revisions of activation theory. In Appley MH, Trumbull R: *Psychological Stress*. New York, Appleton-Century-Crofts, 1967, pp 14–42.
61. Lacey BC, Lacey JI: Studies of heart rate and other bodily processes in sensorimotor behavior. In Obrist PA, Black AH, Brener J, DiCara LV: *Cardiovascular Psychophysiology—Current Issues in Response Mechanisms, Biofeedback, and Methodology*. Chicago, Aldine, 1974, pp 538–564.
62. Lake CR: Relationship of sympathetic nervous system tone and blood pressure. *Nephron* 23:84–90, 1979.
63. Lake CR, Gullner HG, Polinsky RJ, Ebert MH, Ziegler MG, Bartter FC: Essential hypertension: Central and peripheral norepinephrine. *Science* 211:955–957, 1981.
64. Lake CR, Ziegler MG, Kopin IJ: Use of plasma norepinephrine for evaluation of sympathetic neuronal function in man. *Life Sci* 18:1315–1326, 1976.
65. Lazarus RS: *Psychological Stress and the Coping Process*. New York, McGraw-Hill, 1966.
66. Lazarus RS: The stress and coping paradigm. In Eisdorfer C, Cohen D, Kleinman A, Maxim P: *Theoretical Bases for Psychopathology*. New York, Spectrum, 1980, pp 173–209.
67. Levi L: Some principles and sources of error in psychophysiological research. *Foervarsmedicin* 3(Suppl 2):72–90, 1967.
68. Levi L: Stress and distress in response to psychosocial stimuli: Laboratory and real life studies on sympathoadrenomedullary and related reactions. *Acta Med Scand* [Suppl 528], 1972.
69. Lundberg U: Type A behavior and its relation to personality variables in Swedish male and female university students. *Scand J Psychol* 21:133–138, 1980.
70. Lundberg U, Forsman L: Adrenal-medullary and adrenal-cortical responses to understimulation and overstimulation. Comparison between Type A and Type B persons. *Biol Psychol* 9:79–89, 1979.
71. Manuck SB, Craft S, Gold KJ: Coronary-prone behavior pattern and cardiovascular response. *Psychophysiology* 15:403–411, 1978.
72. Mason JW: Organization of psychoendocrine mechanisms: A review and reconsideration of research. In Greenfield NS, Sternbach RA: New York, Holt, Rinehart and Winston, 1972, pp 3–91.
73. Mason JW: Emotion as reflected in patterns of endocrine integration. In Levi L: *Emotions: Their Parameters and Measurement*. New York, Raven Press, 1975, pp 143–181.
74. Matthews KA: Psychological perspectives on the Type A behavior pattern. *Psychol Bull* 91:293–323, 1982.
75. Matthews KA, Glass DC, Rosenman RH, Bortner R: Competitive drive, pattern A, and coronary heart disease: A further analysis of some data from the Western Collaborative Group Study. *J Chronic Dis* 30:489–498, 1977.
76. McClelland DC, Floor E, Davidson RJ, Saron C: Stressed power motivation, sympathetic activation, immune function, and illness. *J Hum Stress* 6:11–19, 1980.
77. McCubbin JA, Richardson J, Obrist PA, Kizer JS, Langer AW: Catecholaminergic and hemodynamic responses to behavioral stress in young adult males (abstr). *Psychophysiology* 18:183, 1981.
78. McGeer PI, McGeer EG: Chemistry of mood and emotion. *Ann Rev Psychol* 31:273–307, 1980.
79. Nestel PJ, Verghese A, Lovell RRH: Catecholamine secretion and sympathetic nervous responses to emotion in men with and without angina pectoris. *Am Heart J* 73:227–234, 1967.
80. Osgood CE: Dimensionality of the semantic space for communication via facial expressions. *Scand J Psychol* 7:1–30, 1966.
81. Pittner MS, Houston BK: Response to stress, cognitive coping strategies, and the Type A behavior pattern. *J Pers Soc Psychol* 39:147–157, 1980.
82. Rauste-von Wright M, von Wright J, Frankenhaeuser M: Relationships between sex-related psychological characteristics during adolescence and catecholamine excretion during achievement stress. *Psychophysiology* 18:362–370, 1981.
83. Rosenman RH: The interview method of assessment of the coronary-prone behavior pattern. In Dembroski TM, Weiss SM, Shields JL, Haynes SG, Feinleib M: New York, Springer-Verlag, 1978, pp 55–69.
84. Rosenman RH: The role of the Type A behavior pattern in ischaemic heart disease: Modification of its effects by beta-blocking agents. *Br J Clin Pract* 32(Suppl 1):58–66, 1978.
86. Rosenman RH, Brand RJ, Jenkins CD, Friedman M, Straus R, Wurm M: Coronary heart disease in the Western Collaborative Group Study: Final follow-up experience of 8½ years. *JAMA* 233:872–877, 1975.
86. Rosenman RH, Friedman M: Neurogenic factors in pathogenesis of coronary heart disease. *Med Clin North Am* 58:269–279, 1974.
87. Rosenman RH, Friedman M, Straus R, Wurm M, Kositchek R, Hahn W, Werthessen NT: A predictive study of coronary heart disease. *JAMA* 189:103–110, 1964.
88. Schiffl H, Weidmann P, Meier A, Ziegler WH: Relationship between plasma catecholamines and urinary catecholamine excretion rates in normal subjects and certain diseased states. *Klin Wochenschr* 59:837–844, 1981.
89. Schlegel RP, Wellwood JK, Copps BE, Gruchow

WH, Sharratt MT: The relationship between perceived challenge and daily symptom reporting in Type A vs. Type B postinfarct subjects. *J Behav Med* 3:191–204, 1980.

90. Schmieder RS, Friedrich G, Neus H, Rüdel H, von Eiff AW: The influence of β-blockers on cardiovascular reactivity and Type A behavior pattern in hypertensives. *Psychosom Med* 45:417–423, 1983.
91. Schekelle RB, Schoenberger JA, Stamler J: Correlates of the JAS Type A behavior pattern score. *J Chronic Dis* 29:381–394, 1976.
92. Silverberg AB, Shah SD, Haymond MW, Cryer PE: Norepinephrine: Hormone and neurotransmitter in man. *Am J Physiol* 234:E252–E256, 1978.
93. Silverman AJ, Cohen SI: Affect and vascular correlates of catecholamines. *Psychiatr Res Rep* 12:16–30, 1960.
94. Simpson MT, Olewine DA, Jenkins CS, Ramsey FH, Zyzanski SJ, Thomas G, Hames CG: Exercise-induced catecholamines and platelet aggregation in the coronary-prone behavior pattern. *Psychsom Med* 36:476–487, 1974.
95. Steptoe A: *Psychological Factors in Cardiovascular Disorders*. London, Academic Press, 1981.
96. Sternback RA (ed): *Principles of Psychophysiology: An Introductory Text and Readings*. New York, Academic Press, 1966.
97. Swan GE, Black GW, Chesney MA, Ward MM, Rosenman RH: Life-event-physiological relationships: Mediated by coronary-prone behavior? Paper presented to the American Psychological Association, New York, September, 1979.
98. Taggart P, Carruthers M: Endogenous hyperlipidaemia induced by emotional stress of racing driving. *Lancet* 1:363–366, 1971.
99. Theorell T, Lind E, Froberg J, Karlsson CG, Levi L: A longitudinal study of 21 subjects with coronary heart disease: Life changes, catecholamine excretion and related biochemical reactions. *Psychosom Med* 34:505–516, 1972.
100. Ursin H, Baade E, Levine S (eds): *Psychobiology of Stress: A Study of Coping Men*. New York, Academic Press, 1978.
101. van Dooren LJP: Physiological reactivity to stress as a function of personality variables. Paper presented at the International Symposium on Psychophysiological Risk Factors of Cardiovascular Diseases, Karlovy Vary (Carlsbad) Czechoslovakia, September, 1981.
102. Van Egeren LF: Cardiovascular changes during social competition of a mixed motive game. *J Pers Soc Psychol* 37:858–864, 1979.
103. Vlachakis ND, Ribeiro AB, Krakoff LR: Effect of *saralasin* upon plasma catecholamines in hypertensive patients. *Am Heart J* 95:78–80, 1978.
104. von Euler US, Gemzell CA, Levi L, Strom G: Cortical and medullary adrenal activity in emotional stress. *Acta Endocrinol (Kbh)* 30:567–573, 1959.
105. Ward MM, Mefford IN, Parker SD, Chesney MA, Taylor CB, Keegan DL, Barchas JD: Behavioral Medicine Program, SRI International, Menlo Park, California, unpublished manuscript, 1982.
106. Weiner H: *Psychobiology of Essential Hypertension*. New York, Elsevier, 1979.
107. Weiss SM, Cooper T, Detre T: Coronary-prone behavior and coronary heart disease: A critical review. *Circulation* 63:1199–1215, 1981.
108. Williams RB, Bittker TE, Buchsbaum MS, Wynne LC: Cardiovascular and neurophysiologic correlates of sensory intake and rejection. I. Effect of cognitive tasks. *Psychophysiology* 12:427–433, 1975.
109. Williams RB, Lane JD, Kuhn CM, Melosh W, White A, Schanberg SM: Type A behavior and elevated physiological and neuroendocrine responses to cognitive tasks. *Science* 218:483–485, 1982.
110. Zahn TP, Haynes SG, Mikkelsen ES, Lake CR: Effects of postural and psychological stress on plasma norepinephrine and autonomic variables in Type A and Type B men. Laboratory of psychology and psychopathology, National Institutes of Mental Health, Bethesda, Maryland, unpublished manuscript, 1982.
111. Ziegler MG, Lake CR, Kopin IJ: Plasma noradrenaline increases with age. *Nature* 261:333–335, 1976.
112. Ziegler MG, Lake CR, Kopin IJ: The sympathetic-nervous-system defect in primary orthostatic hypotension. *N Engl J Med* 296:293–297, 1977.
113. Ziegler MG, Lake CR, Wood JH, Brooks BR, Ebert MH: Relationship between norepinephrine in blood and cerebrospinal fluid in the presence of a blood-cerebrospinal fluid barrier for norepinephrine. *J Neurochem* 28:677–679, 1977.
114. Zyzanski SJ, Jenkins CO, Ryan TJ, Flessas A, Everest M: Psychological correlates of coronary angiographic findings. *Arch Intern Med* 136:1234–1237, 1976.

CHAPTER TWENTY-TWO

Catecholamines in Critical Care Medicine

BART CHERNOW, M.D., THOMAS G. RAINEY, M.D., and
C. RAYMOND LAKE, M.D., Ph.D.

INTRODUCTION

Critical care medicine is a new medical specialty concerned with the care of seriously ill patients (104). Since the endogenous and synthetic catecholamines (CA) (norepinephrine (NE), epinephrine (E) and dopamine (DA)) play an important role in the response to and recovery from serious illness or injury, the critical care specialist requires a thorough understanding of CA metabolism and the therapeutic application of the synthetic forms of these biogenic amines. In this chapter we will first review what is known about the endogenous CA response to critical illness and then discuss the therapeutic considerations involved in the uses of the synthetic CA.

THE ENDOGENOUS CATECHOLAMINE RESPONSE TO SERIOUS ILLNESS OR INJURY

E, NE, and DA are hormones (and neurotransmitters) released in injury, stress, or illness to maintain homeostasis (31). The vasoregulation resulting from augmented plasma concentrations of these CA is a key element in the body's defense against problems such as sepsis, hypoxia, and hypotension. CA-induced vasoregulatory changes have been observed in a variety of clinical settings, (24, 85, 164, 165) and serve as protection against circulatory failure. Most of these conditions are characterized by intravascular depletion (141) or respiratory failure. Therefore, the discussion will focus on CA alterations in these specific states.

Hemorrhagic Shock

Hemorrhagic shock activates the sympathoadrenal system, resulting in elevated plasma concentrations of E (50-fold increases from baseline) and NE (10-fold increments from baseline) (24, 85, 141, 158). Fredholm *et al.* (47) have shown that the adrenal medulla is the major source of plasma CA in hemorrhagic states. With prolonged (4-hour) hemorrhages in rats, peak plasma E concentrations are achieved in the first hour, with only minimal further increments when hemorrhage continues for 3 additional hours (47). Unlike the rapid rise to a plateau in plasma E levels, plasma NE concentrations rise continuously in hemorrhage during the 4-hour period. The E response to hemorrhage can be eliminated with adrenal demedullation but is unaffected by sympathectomy (47) . The rise in circulating NE in hemorrhagic shock is blunted by adrenal demedullation and sympathectomy (47, 110).

The rapid rise in plasma E and delayed but steady elevation in NE concentrations is repeatedly observed after hemorrhagic shock (44, 55, 109). One study suggests that the most marked increase in plasma E may occur in the first 5 minutes of hemorrhage. Plasma DA levels rise minimally in the hemorrhaging animal (44). Farnebo *et al.* (44) analyzed tissue CA content in the setting of 4-hour hemorrhagic hypotension, finding significant increases in the content of cardiac E with reductions in adrenal gland E. These data are consistent with the hypothesis that E is released from the adrenal medullae when hemorrhage occurs and that the circulating E engages β-adrenergic receptors in the myocardium. With the coupling of this hormone-receptor complex, an inotropic and chronotropic effect occurs, resulting in an increased cardiac output. NE content in the heart and adrenal gland is unaffected by hemorrhage, suggesting that NE is released, acts and undergoes reuptake locally, while E travels to target tissues.

If the acidosis of hemorrhagic shock is prevented by the intravenous (i.v.) administration of sodium bicarbonate, the plasma E response to hemorrhage is blunted (37). This implies that the mechanism for E release is related to arterial pH. Acidosis is a characteristic of shock. When cardiac output is inadequate, hypoxic cells function anaerobically and convert pyruvic acid to lactic acid. This relationship between acidosis and E release would explain the clinical observation that the heart rate decreases as an acidic arterial pH is corrected to normal.

The sympathetic nervous system (SNS) maintains blood pressure (BP) during acute blood loss. Lake and Ziegler (84) have shown that moderate volume depletion increases plasma NE concentration by 50% in both normotensive and hypertensive human subjects without a correlated increase in plasma dopamine-β-hydroxylase (DBH) concentration. Although NE output increases during volume depletion, there is no decrease in NE with volume expansion (84). In contrast, another group of investigators identified a genetic influence upon the plasma CA response to volume expansion (114). No one has shown a hereditable component in the SNS response to volume contraction.

Gastrointestinal Bleeding

In clinical practice, hemorrhagic shock is often a consequence of a bleeding peptic ulcer. Christensen (25) demonstrated that plasma E mediates stress-induced increases in serum gastrin concentrations. Indeed, patients suffering from peptic ulcer disease have markedly increased gastrin levels after an E infusion and are noted to have elevations in supine resting plasma NE (25). The reason for augmentation of the SNS in peptic ulcer disease is unclear. It is not related to abdominal pain, since plasma NE levels remain elevated even after successful vagotomy (26). Other possibilities are that patients with peptic ulcer disease drink and smoke more or that the increased NE comes from other gastrointestinal or non-gastrointestinal sources.

Ironically, in its approach to maintaining homeostasis, the body increases NE and E release with hemorrhage, and these CA may in themselves lead to gastrointestinal bleeding. CA shunt blood preferentially away from the gut, leading to mucosal ischemia and at times extensive and severe ulceration (7, 14, 43, 97, 115, 132). High plasma concentrations of either E (98) or NE can cause these lesions (15). This suggests that the best therapy for acute gastrointestinal bleeding is volume replacement and *not* the intravenous administration of synthetic CA. However, local use of CA may be an effective adjunct to fluid resuscitation. NE may be administered via a nasogastric tube as therapy for bleeding gastric lesions (38, 39, 91, 92) or intraperitoneally for more diffuse gastrointestinal hemorrhage (see therapy section of this chapter). Plasma concentrations of CA achieved after these types of local therapy have not been measured.

Other Models of Shock

The CA response to hemorrhagic shock is well described and other forms of shock have also been studied. It has been known for more than 50 years that regardless of the cause of shock, the adrenal medullae respond with an outpouring of E into the circulation (4). The hypotension observed in shock probably stimulates baroreceptors, which in turn trigger a reflex loop involving the central nervous system (CNS) and SNS as well as the adrenal medullae (69).

Marked elevations in both plasma E and NE are observed in shock due to anaphylaxis (137), trauma (130), poisoning (157), and endotoxin (133), in addition to hemorrhagic shock. In all of these, increased SNS activity unfortunately leads to ischemia of visceral organs and marked restriction of blood flow to the intestinal mucosa. As noted earlier, ischemic necrosis, mucosal ulceration, and frank gastrointestinal bleeding may then occur. Shock compounds this problem, causing elevations in endogenous plasma vasopressin and angiotensin levels, both of which actively contribute to intestinal vasoconstriction (107).

Intestinal ileus often accompanies shock. Although ileus may occur as a consequence of a number of variables, (*i.e.*, hypokalemia, hypoalbuminemia, etc.), at least one study shows that removal of the adrenal medullae could abolish ileus after induction of hemorrhagic shock (63). The decrease in gut peristaltic activity with shock usually occurs throughout the intestinal tract (63, 95, 156) or may involve only the upper gastrointestinal tract with hypermotility in the lower tract (122).

Shock shunts related blood flow away from the gut and from the adrenal cortex to the adrenal medulla (101). Herman *et al.* (68) show that for adrenal cortical secretion to continue, a blood flow of at least 1 ml/min must be maintained. When this critical minimum blood flow is not achieved, corticosteroid secretion decreases, accompanied by a further drop in BP (68). For this and other reasons, some physicians have advocated the use of exogenous corticosteroid therapy in the treatment of shock, especially septic shock (141).

Stress Hyperglycemia: Is It a Complication of the Catecholamine Response to Critical Illness?

Hyperglycemia almost invariably accompanies the acute onset of critical illness. Trauma and shock impair glucose tolerance (71), so it is common for nondiabetic patients in critical care units to have blood glucose concentrations between 150 and 300 mg/dl (normal range 60 to 120 mg/dl). This impairment in glucose tolerance is a consequence of elevations in the body's counterregulatory hormones. These hormones elevate blood glucose, thus preventing hypoglycemia, and serve as a balance to insulin which lowers blood glucose. These "antiinsulin" hormones include cortisol, glucagon, E, NE, and growth hormone, all of which are elevated in shock-trauma patients (19, 20, 57, 108, 117). Of these five hormones, it has recently been shown that E, NE, and glucagon induce the highest blood glucose elevations (49). Halter and Pflug (59) demonstrated impairment of insulin secretion in conjunction with the endogenous E and NE release induced by surgical stress (59). α-Adrenergic receptors on islet cells probably mediate the impaired insulin release.

Although the etiology of hyperglycemia in the critically ill subject may be multifactorial in origin, the heightened plasma E and NE levels play a central role. The CA alterations in glucose homeostasis are particularly pronounced in diabetic subjects admitted to shock trauma units (139). Blood glucose levels in this subgroup are frequently greater than 400 mg/dl and require aggressive insulin therapy via continuous intravenous infusion.

An intriguing hypothesis concerning "stress hyperglycemia" has been proposed by Baum and Porte (3). They suggest that hyperglycemia may be protective, especially in acute hypoxemia (a state almost synonymous with critical illness; see section on pulmonary considerations in this chapter). This would ensure the provision of adequate nutrition to the brain and other vital organs. Their experiments demonstrate that in hypoxemia, β-adrenergic receptors dysfunction in the pancreatic islet cells, thus allowing for the predominance of α-adrenergic activity. The relatively unopposed α-adrenergic input promotes glucagon release while inhibiting insulin secretion (3). In addition, the already elevated plasma levels of E, NE, and glucagon increase hepatic glycogenolysis, and it seems to be this combination of events that results in "stress hyperglycemia."

Although the hyperglycemia of critical illness may have a homeostatic basis, it is often quite detrimental in the clinical setting. If the blood glucose concentration exceeds the renal threshold for glucose, glucosuria occurs which can create an osmotic diuresis and dehydration. In addition, hyperglycemia raises the serum osmolality and may contribute to hyperosmolar coma. Hyperglycemia may also impair leukocyte function and depress serum sodium concentrations. Therefore, we feel that "stress hyperglycemia" may be another adverse effect of the CA response to critical illness.

Catecholamine Fluctuations with Fluid Resuscitation

There is a great deal of controversy concerning the proper fluid resuscitation in shock (141). Many clinicians advocate the use of crystalloid therapy (normal saline solution, lactated Ringer's solution), while others argue for colloid (albumin, plasma, dextran) as the best fluid for resuscitating a patient in shock. Regardless of the type of fluid used, it is clear that acute volume expansion (e.g. several liters of normal saline over 2 hours) does not significantly affect plasma NE levels (84, 146). On the other hand, if the fluid used contains glucose, plasma NE concentrations may ironically increase (161). Chronic expansion of intravascular fluid volume suppresses the SNS with resulting decreases in plasma NE and DBH concentrations (146).

Anesthesia As It Affects Catecholamine Homeostasis

Quite often, the critically ill patient will require surgical and/or anesthetic intervention. Almost from its inception, the administration of anesthetic agents has recognizably altered the SNS (see Chapter 16). For example, in 1912 it was shown that urethane, ether, and chloroform all caused adrenal medullary release of CA (18, 42). Similarly, surgical procedures by themselves result in elevations in plasma CA concentrations (60).

Diethyl or divinyl ether sharply increases both plasma NE and E levels (131). Chloroform results in similar CA changes, whereas thiopentone use does not significantly alter the plasma CA levels (18). Much of the information concerning anesthetic-CA interactions comes from animal experiments. Millar and Morris (113) discovered that although ether causes significant increases in both plasma NE and E in the dog, the increments in E were smaller in humans. Thus, the effects of anesthesia on plasma CA concentrations in a dog model may not represent what occurs in man. Ether-induced changes in CA content do, however, seem to be dose-responsive regardless of species (8).

Volatile anesthetics have variable effects on BP. Cyclopropane may raise BP by CA augmentation, and halothane diminishes BP (perhaps by a CA depressive effect). Stokke *et al.* (147) showed that halothane anesthesia does not alter plasma NE or E levels, but BP falls, indicating a direct depressive effect on vascular smooth muscle and the myocardium. Cyclopropane anesthesia, on the other hand, provokes a significant rise in plasma NE without altering plasma E concentrations. This SNS stimulation results in a proportionate elevation in BP (147). The increment in plasma NE observed with cyclopropane anesthesia parallels the depth of anesthesia achieved (129). Methoxyflurane (112), trichloroethylene (153), thiopental (129), and barbiturates (131) have minimal impact on plasma CA levels. Urethane, on the other hand, probably augments plasma CA concentrations (138).

Halter and Pflug (58) have recently demonstrated that with spinal anesthesia there are no changes intraoperatively or postoperatively in plasma NE or E when compared to preanesthesia concentrations. Minimizing changes in endogenous CA levels is probably beneficial in the anesthetic management of the critically ill patient with ischemic cardiac disease.

A rare but fascinating problem encountered in anesthesiology is malignant hyperthermia (see Chapter 23). Patients with this disorder develop extreme pyrexia and become critically ill when given anesthesia. Gronert *et al.* (56) report that the increase in myocardial oxygen consumption and decrease in myocardial efficiency in these patients are consequent to increased β-adrenergic stimulation. The implications of this finding are unclear. However, from work in hyperthermic cancer therapy, evidence has evolved that alterations in SNS activity (NE) modulate the hemodynamic changes accompanying hyperthermia (81).

Pulmonary Considerations in Catecholamine Metabolism

The clinical findings that characterize the critically ill patient most frequently include intravascular depletion, shock and/or respiratory failure. The two major components of the latter are inadequate oxygenation (hypoxia) and inadequate ventilation (hypercapnia). In this section we will discuss CA alterations observed with hypoxia, hypercapnia, bronchoconstriction and pulmonary artery hypertension.

Hypoxia, regardless of etiology, increases plasma CA levels in concert with the degree of deoxygenation (99). Changes in plasma NE concentration are even observed with subtle decrements in arterial oxygen tension, such as those found when subjecs travel to high altitudes (35). While lowering arterial oxygen tension raises CA levels, oxygen administration does not change NE levels at rest but decreases NE levels during exercise (30). In neonatal rats, short periods of hypoxia have been shown to decrease levels of tyrosine hydroxylase (TH) (64, 65), the first enzyme in CA biosynthesis. In acute hypoxia, CA synthesis, as well as release, appears to be stimulated. With prolonged hypoxia in adults, no enzyme decrement is observed. This suggests an adaptive process and a loss of CA responsiveness to the development of a chronic state.

Hypoxia alters a number of hemodynamic variables, presumably through its effect on CA. With arterial pH and carbon dioxide tension kept constant, hypoxia causes increases in cardiac output and stroke volume (148). Total peripheral resistance and mean arterial BP may decrease as a result of unopposed β-adrenergic activity or depressed α-adrenergic activity, especially if the hypoxia is a result of carbon monoxide poisoning (148).

The development of hypercapnia is also a potent stimulus for CA release (111, 149). It may not be the hypercapnia but the resulting respiratory acidosis that serves as the precipitant for CA discharge in humans (120). In the fetal lamb, hypercapnia and acidemia have little effect on CA release from the adrenal gland, but this animal model may not reflect the human situation (29).

The i.v. infusion of NE increases the body's ventilatory response to hypoxia and hypercapnia (33), as long as the fraction of inspired oxygen is not very high (34). In addition, endogenous circulating CA can stimulate respiration. What is unclear is whether CA actually contribute much to the respiratory effect of the critically ill patient in shock or with asphyxia.

ASTHMA

Reversible obstructive airway disease (asthma) is a model for the clinical application of our knowledge about adrenergic receptor physiology. Agonists and antagonists for α, β_1, β_2, and DA receptors are well-defined (see Table 22.1). The sensitivity of these receptors to endogenous or exogenous amines may vary considerably. Henderson *et al.* (67) have shown that in asthmatics a hyperresponsiveness exists in the α-adrenergic receptors. Similarly, some have argued that chronic adrenergic therapy, as in asthmatics, may lead to receptor desensitization with attenuation of drug effect (89). This subject remains debatable, however, as other investigations have shown no receptor resistance with the chronic administration of adrenergic receptor agonists (86).

Table 22.1
Agonists and Antagonists for Adrenergic Receptors*

RECEPTOR	AGONISTS‡	ANTAGONISTS
Alpha	NE	phenoxybenzamine
	E	phentolamine
	Dopamine	ergot alkaloids
	Isoproterenol	
$Beta_1$	Isoproterenol	propranolol
	E	alprenolol
	NE	practolol
	Dopamine	dichlorisoproterenol
$Beta_2$	Isoproterenol	propranolol
	E	dichlorisoproterenol
	NE	alprenolol
	Dopamine	butoxamine
Dopamine	Dopamine	haloperidol
	epinine	phenothiazines
	apomorphine	pimozide
		bulbocapnine

*reference- Steer ML: Adrenergic Receptors Clinics Endocrinol Metab 6:577-598, 1977
‡agonists are listed in order of relative potency.

Agonists engaging pulmonary α-adrenergic receptors cause bronchoconstriction of isolated airways (142). β-Adrenergic receptors are of two types: β_1, which primarily mediate the inotropic and chronotropic actions of adrenergic nerves on the heart, and β_2 which primarily control effects on bronchial smooth muscle and peripheral vascular beds (162). The desired CA effect in asthmatic patients is to cause bronchodilation without inducing tachycardia, that is, β_2-adrenergic receptor stimulation with minimal β_1 effect. Metaproterenol, isoetharine, salbutomol, and terbutaline are relatively "β_2-selective" agonists (83). Isoproterenol, E, and ephedrine have both β_1- and β_2-agonistic properties. Although clinicians hoped that the therapy of status asthmaticus might be managed best with β_2-specific agonists (in addition to aminophylline derivatives, corticosteroids etc.), isoproterenol and E remain as drugs of choice. This reflects the diverse potency of various CA and sympathomimetic agents at the level of the adrenergic receptor. Isoproterenol is 10 times more potent than E as a β-agonist (83, 162). Another reason the "nonspecific" β-agonists are often preferred may be that they offer other favorable pulmonary actions besides bronchodilation. For example, isoproterenol decreases airway resistance considerably more than other β-adrenergic agents (163). Additional information about adrenergic therapy appears under Catecholamines as Therapeutic Agents in Critical Care Medicine.

Circulating CA vary in concentration diurnally. Barnes *et al.* (2) show that plasma E levels fall considerably in the early morning hours (midnight to 4 a.m.). They offer evidence that this circadian drop in E values may explain the high frequency of nocturnal wheezing in asthmatics. The lowered plasma E concentrations may allow sensitized mast cells to degranulate, releasing histamine with resultant bronchial-constriction (2). Their data imply that patients suffering from nocturnal asthma should benefit from β_2-agonist administration at the hour of sleep.

PULMONARY HYPERTENSION

Primary pulmonary hypertension has an interesting relationship to CA metabolism. Primary pulmonary hypertension is a progressive disease of unknown cause characterized by a symptom complex which includes dyspnea, chest pain, and syncope. Therapy has been disappointing at best, with the exception of isoproterenol, which is beneficial even when administered chronically (140). Why this β-adrenergic agonist is effective is unclear, but likely involves its ability to decrease pulmonary vascular resistance, which is very high in this condition.

Normally, the lungs serve in part to extract and metabolize circulating NE (61). The lungs normally extract circulating NE with a clearance rate of about 270 ng/min but do not contribute to the metabolism of E or DA. In pulmonary hypertension, this metabolic function of the lungs is lost (143). Whether or not isoproterenol restores this function is unknown. Investigations concerning the metabolic functions of the lungs in health and disease are underway in a number of laboratories, including our own.

Interaction between Catecholamines and Opioid Peptides: Implications for Critical Care Medicine

CA play a central role in the physiologic response to stress. Evidence is rapidly accumulating that the endogenous opioid peptides are also important contributors to the maintenance of homeostasis during stressful events (78, 79, 94, 121). One report has already shown that antagonism of the endogenous opioid receptor may be useful in the treatment of shock (155).

The endogenous opioid peptides and CA share many common features, suggesting interaction between the two groups of chemicals. They both modulate vasopressin release (96), anterior pituitary hormone release (116), ventilation (118), as well as a number of cardiovascular effects (45). In addition, some evidence suggests that like CA, the opioid peptides are involved in thermoregulation (22, 70).

Initial research concerning the opioid peptides was directed toward understanding their role in analgesia, addiction, and psychiatric disorders (16). Attention is now drawn towards clinical applications in critical illness.

CATECHOLAMINES AS THERAPEUTIC AGENTS IN CRITICAL CARE MEDICINE

The synthetic CA are routinely used in the therapy of critically ill patients. In this section we will first review the chemistry of these CA and then detail the indications for their individual administration in the seriously ill patient.

Catecholamines *vs.* Sympathomimetic Agents

The sympathomimetic amines include the CA as well as other drugs that mimic their actions. Whereas all CA are sympathomimetic agents, not all sympathomimetic agents are CA. The difference between these and other agents lies in the biochemical structure of the various compounds (151). All of the sympathomimetic agents have a core structure consisting of a benzene ring and an ethylamine side chain (Fig. 22.1). The CA (NE, E, DA) and isoproterenol all have OH groups in the 3 and 4 positions of the benzene ring. This dihydroxybenzene is called a "catechol," and therefore sympathomimetic amines with a "catechol" structure are known as CA (Fig. 22.1). In addition to the CA listed above, a number of synthetic ones have been created to improve receptor specificity. Such compounds include the β_2 specific agonists isoetharine, metaproterenol, terbutaline, and salbutomol.

The non-CA sympathomimetic agents (those lacking the "catechol" structure) frequently exert their effect by an indirect mechanism, such as being taken up into the presynaptic neuron and causing release of NE. Included in this group of agents are phenylephrine, methoxamine, amphetamine, tyramine, ephedrine, and many others. To review their specific actions is beyond the scope of this chapter. The remainder of the discussion will comprise a review of the specific CA used therapeutically in critical care.

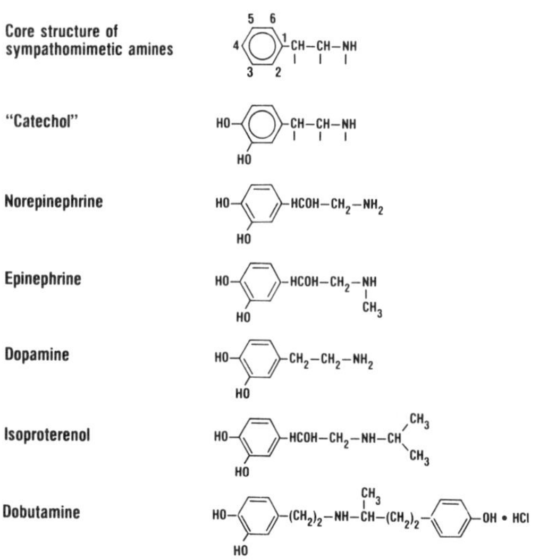

Figure 22.1. Chemical structure of the catecholamines.

Dopamine

DA is the immediate precursor of NE in the endogenous biosynthetic pathway (Fig. 22.2). Interest in DA as a therapeutic agent began in the mid-1960s, when evidence started to accumulate that it had vasodilatory effects on certain vascular beds, especially the renal vasculature (102, 103, 105, 106). Besides the renal vascular bed, vasodilation also takes place in the mesenteric, coronary, and intracerebral vasculature in response to DA infusion (51).

DA is a fascinating agent since three different types of receptors mediate its actions, depending on the dosage employed. When DA hydrochloride is infused at rates between 2 and 10 μg/kg/min, it acts on β-adrenergic receptors, producing an increase in myocardial contractility. Within this dosage range, one sees a rise in cardiac output without an increase in heart rate or BP (128). β-Adrenergic blockade reverses these effects of DA, supporting its action on the β-receptor (50). When DA is administered in dosages above 20 μg/kg/minute, α-adrenergic stimulation predominates, with vasoconstriction in all vascular beds (51, 128). Between 10 and 20 μg/kg/min, there is both α and β-adrenergic agonist activity. In addition to engaging α and β-adrenergic receptors, DA acts at a third receptor, known as the DA or dopaminergic receptor (51). This receptor mediates DA's unique ability to selec-

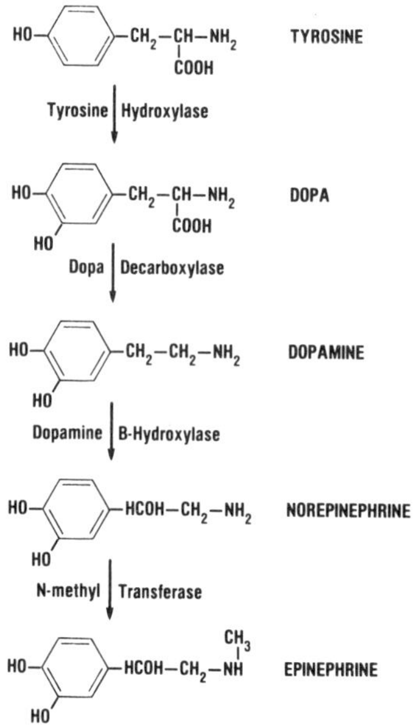

Figure 22.2. Biosynthetic pathway of the endogenous catecholamines.

tively dilate renal arterioles and is activated most when DA is infused at very low dosages such as 0.5 to 2 μg/kg/min. The receptor's actions are not antagonized by α- or β-blocking agents, but are reversible with DA receptor antagonists such as haloperidol (51).

THERAPEUTIC USES

DA is indicated in the treatment of shock, especially in patients with oliguria and/or those patients with a low peripheral vascular resistance (e.g., septic shock) (141). In conditions where the peripheral vascular resistance is elevated, DA may be used in combinations with a vasodilating agent, such as nitroprusside, in order to maintain normal systemic perfusion. DA reduces pulmonary capillary pressure and increases cardiac index in patients with cardiac pump failure. Forrester and Waters (46) recommend this medication for heart failure patients in subset IV (*i.e.*, patients with pulmonary congestion and peripheral hypoperfusion). Weil *et al.* (134) suggest that one may predict how successful DA will be in reversing circulatory shock by pretherapy measurement of arterial blood lactate concentration. They found that in patients with markedly elevated blood lactate levels, DA administration failed to increase peripheral perfusion, improve survival, or normalize the blood lactate level.

Although DA is frequently administered to patients in shock from myocardial infarction, its positive inotropic effect may be achieved at the expense of increased myocardial oxygen demand (119). Its usage in this case remains controversial, as does its use in hepatorenal syndrome, where some investigators have shown it to be beneficial (5).

PRECAUTIONS AND OTHER COMMENTS

DA will correct hypotension in most cases, but is not a substitute for fluid resuscitation in patients with hypovolemia. Using DA in the presence of tachyarrhythmias is dangerous and contraindicated in patients with pheochromocytomas. Alkaline media inactivates the drug, and therefore it should not be added to any alkaline solutions. Dopamine can cause nausea, emesis, tachycardia, and headaches, but in most cases it is very well tolerated and free of major side effects. However, when α-adrenergic dosages are required, peripheral ischemia and even gangrene may occur (52).

A number of hormonal alterations occur as a result of DA administration. Insulin secretion is inhibited and may thus contribute to the hyperglycemia commonly found in the critically ill patient (166). Aldosterone secretion is under dopaminergic control; its release from the zona glomerulosa is tonically suppressed by stimulation of dopaminergic receptors (124) and augmented with consequent elevations of plasma aldosterone by DA antagonists (21, 123). This may in part explain the increased urine output after DA administration. Its use also alters the release of a number of hormones from the anterior pituitary gland. Short-term or several hours infusion of DA results in diminished secretion of prolactin, thyrotropin, luteinizing hormone (LH), with increased release of growth hormone (6, 17, 75, 87, 88). During prolonged infusion (48 hours) of DA, there is sustained suppression of prolactin and TSH release, with a rebound increase in both hormones (prolactin > TSH) upon discontinuation of the DA (76). The gonadotropins, LH and follicle-stimulating hormone (FSH), are initially inhibited in their release from the pituitary, but with time "escape" the suppressive effect of DA (76). The clinician needs to be aware of these drug-induced changes in order to properly interpret the results of endocrine testing in the critically ill patient. A more detailed discussion of CA-induced changes in endocrine function appears elsewhere in this book (see Chapter 29).

Dobutamine

Dobutamine is a synthetic CA developed by modification of the chemical structure of isoproterenol (Fig. 22.1) (145). Dobutamine infusion produces a positive cardiac inotropic effect with minimal changes in heart rate or vascular resistance (152). Like DA, dobutamine exerts its inotropic effect via direct stimulation of β_1-adrenergic receptors in the myocardium. However, unlike DA, which causes release of myocardial NE stores, dobutamine lacks the ability to indirectly stimulate the myocardium by NE release (145, 152).

THERAPEUTIC USES

Dobutamine has some clear advantages over other CA agents in both adults and children (40). It lacks the chronotropic effect of isoproterenol, the arrhythmias associated with DA and the peripheral vasoconstriction seen with NE. It is of value in patients with acute or chronic congestive heart failure from ischemic heart disease (73). Indeed, in congestive heart failure from myocardial infarction, dobutamine has proven as effective alone as DA and nitroprusside in combination (80). As therapy for cardiac failure, dobutamine is superior to digoxin (53). Dobutamine decreases left ventricular end-diastolic filling pressure, increases cardiac index, and relieves pulmonary congestion. Digoxin, on the other hand, has no effect on filling pressure and only slightly elevates cardiac index (53).

Dobutamine is the drug of choice for patients in congestive heart failure after myocardial infarction, especially if the heart rate and/or peripheral vascular resistance are elevated (160). It also seems particularly useful in patients with depressed myocardial function after cardiopulmonary bypass. In this setting, the combination of dobutamine and nitroprusside is advantageous (160).

PRECAUTIONS AND OTHER COMMENTS

Dobutamine administration is contraindicated in patients with idiopathic hypertrophic subaortic stenosis, since it may augment the obstruction of the aortic outflow tract. Like DA, it is inactivated by alkaline solutions. Positive inotropic effects are seen with doses as low as 0.5 μg/kg/min, though the usual dosage range is 2.5 to 10 μg/kg/minute. Higher infusion rates seem to predispose the patient to the development of tachycardia or other arrhythmias. However, dobutamine has a lower incidence of induced arrhythmias than either DA (90) or isoproterenol (136). Since dobutamine increases A-V conduction in the heart, it must be used with great caution in patients with atrial fibrillation.

As to the pharmacokinetics of dobutamine, it is known that steady-state levels of the drug increase in plasma in proportion to the infusion rate (77). The drug has been infused for days in some patients without ill effect. However, there is evidence that pharmacodynamic tolerance to dobutamine may develop when it is given for more than 72 hours (154).

Epinephrine

E activates both α- and β-adrenergic receptors. It is a potent vasopressor agent that increases BP by its β_1 agonist action and to a lesser degree by its α-agonist stimulation. The β_1 stimulation offers a positive inotropic and chronotopic effect on the heart, and the α stimulation results in peripheral vessel vasoconstriction (151).

THERAPEUTIC USES

E is used in critical care medicine primarily in the treatment of status asthmaticus, hypersensitivity reaction to drugs, in some cases of gastrointestinal bleeding, in cardiac arrest, and in cardiogenic shock. It is generally administered subcutaneously since i.v. administration may result in death from ventricular arrhythmias or cerebral hemorrhage (159). These serious side effects are probably due to the rapid rise in BP with the i.v. route.

The half-life of parenterally administered E is measured in minutes due to the rapid metabolism of E by catechol-O-methyltransferase (COMT) and monoamine oxidase (MAO) in the body. It is this short half-life that allows the intensivist, caring for an asthmatic patient, to give subcutaneous E every 20 minutes (127). Although this is frequently a very effective form of therapy in status asthmaticus, some patients may be refractory to this approach. This is especially true when the bronchospasm is associated with the presence of viscous, mucous secretions. In this setting, E may aggravate the situation by further drying bronchial secretions and thereby making existing mucous plugs more viscous and more difficult to dislodge. In such cases, a combination of chest physical therapy, fluid hydration, and pharmacologic therapy with aminophylline, a β_2-agonist, and corticosteroids will be beneficial. The pregnant woman with asthma presents another special problem in terms of E therapy. In such patients, the therapeutic usage of E has been associated with an increase in fetal malformations, perhaps due to transient decreases in uterine perfusion (150). If the mother's life is threatened by bronchospasm, however, E therapy should not be withheld if the clinician feels it will be helpful (150).

Since locally applied E causes a marked reduction in mucosal blood flow, some have advocated its instillation into the stomach in cases of acute upper gastrointestinal bleeding. We have used NE in such patients with only limited success (see below). If E or NE is administered in this fashion, systemic blood levels of these CA remain unchanged, since both are rapidly conjugated and oxidized in the gut and liver. The use of E as a topical hemostatic agent requires further investigation.

In the intensive care unit, clinicians give many medications to patients, often without knowing whether they are allergic to the drugs. For this reason it is not uncommon to observe a hypersensitivity reaction to a medication. The subcutaneous administration of E offers prompt and effective relief in these conditions.

A condition in which the i.v. administration of E is appropriate is in treatment of the cardiac arrest victim. Such a patient may have "fine" ventricular fibrillation which has been refractory to repeated attempts at electrical defibrillation allowing for successful cardioversion (126). In asystole, the i.v. or intracardiac administration of E will frequently be followed by the initiation of an intrinsic rhythm (72). Parenteral E may also be useful in the treatment of syncope secondary to atrioventricular block (Stokes-Adams attacks) (10).

PRECAUTIONS AND OTHER COMMENTS

We mentioned earlier that CA may contribute to the hyperglycemia frequently seen in critically ill patients. E is a major contributor to this effect, since even small elevations in plasma E concentrations cause marked insulin antagonism (62). Another mechanism for the hyperglycemia seen with E administration is diminished glucose clearance. In humans, E infusion at a rate of 1.2 $\mu g/m^2/min$ results in a 25% decrease in glucose clearance from the circulation (144). The E-induced hyperglycemia may be responsible for the hypophosphatemia seen with E therapy (74, 82). That is, the hyperglycemia and insulin antagonism of E combine with augmented plasma insulin levels, and the latter cause phosphate to move from the circulation into cells. Hypophosphatemia can cause myocardial and respiratory depression, neurologic symptoms, and hematological problems, none of which benefit the critically ill patient (82).

Another potential adverse effect of E therapy is its alteration of white blood cell function. MacGregor has shown that E reduces granulocyte adherence by 25% and causes a marked demargination granulocytosis (100). For the patient in the intensive care unit with a serious bacterial infection, the altered leukocyte function can prove fatal. This may be particularly true in the setting of lung infection, where bacteria tend to be especially resistant to antimicrobial agents. Another possible hindrance to the recovery of the pneumonia patient is that E may serve as a respiratory depressant at the bulbar level (23).

Norepinephrine

Although frequently considered by clinicians as a "pure" α-adrenergic agonist, NE in low dosages (<0.03 $\mu g/kg/min$) does stimulate β-adrenergic receptors (11). NE is used less commonly in critical care than the other agents discussed. NE is effective in the therapy of hypotensive states where myocardial contractility has decreased in association with diminished total peripheral vascular resistance. Indeed, if peripheral vascular resistance is already elevated, NE may worsen the situation by increasing cardiac "afterload," resulting in heart failure (11). Unlike E, NE must *only* be given i.v. since it causes marked tissue necrosis with extravasations outside the vein (11).

THERAPEUTIC USES

To properly use NE, a pulmonary artery catheter should be inserted to allow for serial measurement of cardiac and pulmonary artery wedge pressure, as well as to provide data for the calculation of peripheral and pulmonary vascular resistance. When such information is available, the decision whether NE is a suitable agent becomes more rational.

In congestive heart failure, cardiac stores of NE may be depleted (13). This can be important for shock treatment due to cardiac pump failure, where a response to i.v. NE infusion may reflect this endogenous CA depletion. On the other hand, when dosages of NE above 15 $\mu g/min$ result in no improvement, it is fair to assume that cardiac stores of NE are adequate and that the myocardium is receiving maximal CA stimulation (12).

NE is a potent agent in the therapy of hypotension. It elevates arterial BP predominantly by augmenting peripheral vascular resistance, although cardiac output is also increased (46). The vasoconstriction causes a reflex increase in vagal tone, and consequently the heart rate may remain stable or decrease. If used, NE should be considered a short-term inotropic agent, as opposed to DA, which may be used for days to weeks.

The use of NE in the treatment of gastrointestinal bleeding is becoming more popular. The addition of NE to ice-cold normal saline (16 mg in 200 ml) provides a lavage solution given via a nasogastric tube that can be quite effective in the management of bleeding gastric lesions (38, 39, 91, 92). NE can also be administered by the peritoneal route as a therapeutic approach to severe gastrointestinal hemorrhage. In patients with variceal hemorrhage as a consequence of portal vein hypotension, such therapy can reduce splanchnic blood flow by 50% (91).

PRECAUTIONS AND OTHER COMMENTS

NE should preferably be given through a central vein to avoid any chance of extravasation and tissue necrosis. E, and to a lesser degree NE, can cause the patient to be anxious, fearful, restless, and aware of a pounding heart beat. Metabolically, NE causes hyperglycemia (135), although less frequently than E. NE can cause hypocalcemia (36), although human data to support this adverse effect are lacking.

Plasma NE levels may be altered by the simultaneous administration of vasodilator agents. For example, prazosin, a vasodilator used in the therapy of heart failure (1), has recently been shown to increase plasma NE levels (28). Such an augmentation in plasma NE may offer the myocardium maximum inotropic stimulation and hence, if NE is then given, no response is observed. More data are needed concerning the interaction of vasodilator agents with the SNS.

Patients receiving bretylium tolysate can have an exaggerated response to NE therapy (66). Bretylium is an antiarrhythmic agent taken up by postganglionic adrenergic nerve terminals where it first causes NE release and later prevents NE release (that is, sympathetic blockade) (1). Like many of the CA-drug interactions, the mechanism for these observations remains unexplained. We speculate, however, that the enhanced response to NE infusions results from the bretylium-induced block of CA release from the nerve terminal with consequent "upregulation" of the adrenergic receptors for NE.

Isoproterenol

Isoproterenol acts almost exclusively on β-adrenergic receptors (β_1 and β_2 types). It is a potent β-agonist. Although it can increase cardiac output by augmenting "preload," it is *not* useful in shock, since by its cutaneous and muscle vasodilation, it redistributes blood to nonessential areas (141).

THERAPEUTIC USES

Isoproterenol helps in states of cardiac failure associated with bradycardia, asthma, and cor pulmonale. It is also used as a cardiac stimulant in patients with heart block, but rarely in septic shock. It should be given parenterally, although unlike the other CA, it is available in sublingual tablets. However, its absorption by this route is unreliable. As a bronchodilator, its usefulness is often limited by the development of tachycardia. As a cardiac stimulant its usefulness is limited by its vasodilatation of nonessential vascular beds. For these reasons it is used less frequently than DA, dobutamine, or E.

We are currently investigating the use of isoproterenol in the treatment of pulmonary hypertension. Our preliminary, unpublished results suggest that this might be an effective agent in both primary (idiopathic) and secondary (cor pulmonale) pulmonary hypertension.

PRECAUTIONS AND OTHER COMMENTS

Unlike the other CA, isoproterenol does not cause hyperglycemia and actually increases insulin release (125). An area of interest requiring further investigation involves β-adrenergic mediation of antidiuretic hormone activity (93). This is important for the clinician to know, since isoproterenol administration may produce an antidiuresis. The resultant hyponatremia is frequently attributed to the syndrome of inappropriate antidiuretic hormone (SIADH) secretion, when in fact it is iatrogenic in etiology.

Although isoproterenol is quite effective in the treatment of severe asthma, its use is not without risk. Its potent cardiac stimulation causes tachycardia, and sometimes even myocardial necrosis. Interestingly, with isoproterenol-induced improvements in pulmonary gas exchange, one may find significant decrements in arterial oxygen tension (48). To obviate this problem we recommend that asthmatic patients receiving isoproterenol be given supplemental oxygen.

A final note concerning β-agonist therapy involves alterations of such agents on other CA. Cryer *et al.* (32) recently presented early evidence that E and NE may be cleared through β-adrenergic mechanisms. This is an area of active investigation at present.

SUMMARY

Knowledge of CA metabolism affords the clinician a physiologic basis for choosing exogenous CA therapy in the treatment of shock.

A great deal of our understanding about CA derives from animal studies, not always reflecting the situation in man. In addition, many of the early studies used urinary CA measurements and less-than-optimal plasma assays to determine CA responsiveness to a given variable. Today's assay techniques allow a more sophisticated analysis of the CA response to critical illness.

More clinical information is needed. Which subsets of patients in shock do best with α-adrenergic stimulation? Which will improve with β stimulation? What is the clinical significance of the interactions being described between CA and opioid peptides? What clinical states alter CA metabolism, and what are the hormonal alterations observed as a consequence of CA therapy? Clearly, a great deal of work has yet to be done.

References

1. Abramowicz M (ed): Bretylium for ventricular arrhythmias. *Med Lett* 20:105–106, 1978.
2. Barnes P, Fitzgerald G, Brown M, Dollery C: Nocturnal asthma and changes in circulating epinephrine, histamine and cortisol. *N Engl J Med* 303:263–267, 1980.
3. Baum D, Porte D Jr: Stress hyperglycemia and the adrenergic regulation of pancreatic hormones in hypoxia. *Metabolism* 29(Suppl 1):1176–1185, 1980.
4. Bedford EA, Jackson HC: The epinephrine content of the blood in conditions of low blood pressure and shock. *Proc Soc Exp Biol Med* 13:85–87, 1916.
5. Bennett WM, Keege E, Melnyk C, Mahler D, Rosch J, Porter GA: Response to dopamine hydrochloride in the hepatorenal syndrome. *Arch Intern Med* 135:964–971, 1975.
6. Besses GS, Burrow GN, Spaulding SW, Donabedian RK: Dopamine infusion acutely inhibits the TSH and prolactin response to TRH. *J Clin Endocrinol Metab* 41:985, 1975.
7. Blalock A: Shock: Further studies with particular reference to the effects of hemorrhage. *Arch Surg* 29:837–857, 1945.
8. Black GW, McArdle L, McCullough H, Unni VKN: Circulating catecholamines and some cardiovascular, respiratory, metabolic and pupillary responses during diethyl ether anesthesia. *Anesthesia* 24:168–179, 1969.
9. Bolme P, Fuxe K, Agnati LF, Bradley R, Smythies J: Cardiovascular effects of morphine and opioid peptides following intracisternal administration in chloralose-anesthetized rats. *Eur J Pharmacol* 48:319–324, 1978.
10. Boucher M, Duchene-Marullaz P, Lavarenne J: Catecholamines and cardiac rhythms in the unanesthetized dog with chronic AV block. *Am J Physiol* 237:H10–H17, 1979.
11. Braunwald E: Pharmacologic treatment of cardiovascular disorders. In Isselbacher KJ, Adams KD, Braunwald E, Petersdorf RG, Wilson JD: *Harrison's Principles of Internal Medicine.* New York, McGraw Hill, 1980, pp 1067–1068.
12. Braunwald E, Alpert JS, Ross RS: Acute myocardial infarction. In Isselbacher KJ, Adams KD, Braunwald E, Petersdorf RG, Wilson JD: *Harrison's Principles of Internal Medicine.* New York, McGraw Hill, 1980, pp 1125–1136.
13. Braunwald E, Ross J, Sonneblick EH: Disorders of myocardial function. In Isselbacher KJ, Adams KD, Braunwald E, Petersdorf RG, Wilson JD: *Harrison's Principles of Internal Medicine.* New York, McGraw Hill, 1980, pp 1025–1035.
14. Brown RB, Borawsky M: Further observation on intestinal lesions associated with pheochromocytoma. *Ann Surg* 151:683–692, 1960.
15. Brown RB, Rica BH, Szakas JE: Intestinal bleeding and perforation complicating treatment with vasoconstrictors. *Ann Surg* 150:790–798, 1959.
16. Bunney WE Jr, Pert CB, Klee W, Costa E, Pert A, Davis GC: Basic and clinical studies of endorphins. *Ann Intern Med* 91:239–250, 1979.
17. Burrow GN, May PB, Spaulding SW, Donabedian RK: TRH and dopamine interactions affecting pituitary hormone secretion. *J Clin Endocrinol Metab* 45:65, 1977.
18. Callingham BA: Catecholamines in blood. In Geiger SR: *Handbook of Physiology.* Bethesda, Md, American Physiological Society, 1975, Sect 7, pp 427–445.
19. Carey LC, Cloutier CT, Lowery DB: Growth hormone and adrenal cortical response to shock and trauma in the human. *Ann Surg* 174:451–460, 1971.
20. Carey LC, Lowery BD, Cloutier CT: Blood sugar and insulin response of humans in shock. *Ann Surg* 172:342–350, 1970.
21. Carey RM, Thorner MO, Ortt EM: Effects of metoclopramide and bromocriptine on the renin-angiotensin aldosterone system in man. *J Clin Invest* 63:727, 1979.
22. Carr D: Endogenous opioids and fever: A hypothesis. *Perspect Biol Med* 23:1–6, 1979.
23. Champagnat J, Danavi-Saubie, Henry JL, Leviel V: Catecholaminergic depressant effects on bulbar respiratory mechansims. *Brain Res* 160:57–68, 1979.
24. Chein S: Role of the sympathetic nervous system in hemorrhage. *Physiol Rev* 47:214–288, 1967.
25. Christensen NJ: Adrenergic mechanisms in selected diseases: Arterial hypertension, duodenal ulcer, primary depressive illness, malignant tumors and ketotic hypoglycemia. *Metabolism* 29(Suppl 1):1190–1197, 1980.
26. Christensen NJ, Brandsborg O, Lovgreen NA, Bransborg M: Elevated plasma noradrenaline concentrations in duodenal ulcer patients are not normalized by vagotomy. *J Clin Endocrinol Metab* 49:331–334, 1979.
27. Cohn JN, Franciosa JA: Selection of vasodilator, inotropic or combined therapy for the management of heart failure. *Am J Med* 65:181–188,

1978.
28. Colucci WS, Williams GH, Braunwald E: Increased plasma norepinephrine levels during prazosin therapy for severe congestive heart failure. *Ann Intern Med* 93:452–453, 1980.
29. Comline RS, Silver IA, Silver M: Factors responsible for the stimulation of the adrenal medulla during asphyxia in the fetal lamb. *J Physiol (Lond)* 178:211–238, 1965.
30. Cross MH, Houlihan RT: Sympathoadrenal medullary response of the rat to high oxygen exposures. *J Appl Physiol* 27:523–527, 1969.
31. Cryer PE: Physiology and pathophysiology of the human sympathoadrenal neuroendocrine system. *N Engl J Med* 303:436–444, 1980.
32. Cryer PE, Rizze RA, Haymond MW, Gerich JE: Epinephrine and norepinephrine are cleared through beta-adrenergic, but not alpha-adrenergic, mechanisms in man. *Metabolism* 29(Suppl 1):1114–1118, 1980.
33. Cunningham DJC, Hey EN, Patrick JM, Lloyd BB: The effect of noradrenaline infusion on the relation between pulmonary ventilation and the alveolar PO_2 and PCO_2 in man. *Ann NY Acad Sci* 109:756–771, 1963.
34. Cunningham DJC, Lloyd BB, Patrick JM: The respiratory effect of infused noradrenaline at raised partial pressures of oxygen in man. *J Physiol (Lond)* 165:45–46, 1963.
35. Cunningham WL, Becker EJ, Kreuzer F: Catecholamines in plasma and urine at high altitudes. *J Appl Physiol* 20:607–610, 1965.
36. Cuthman J: Norepinephrine-induced hypocalcemia in sheep. *Acta Vet Scand* 11:327–328, 1980.
37. Darby TD, Watts DT: Acidosis and blood epinephrine levels of hemorrhagic hypotension. *Am J Physiol* 206:1281–1284, 1964.
38. DelGuercio LRM: Diagnosis and control of gastrointestinal hemorrhage. In Weil MH, Henning RJ: *Critical Care Medicine*. Chicago, Year Book Medical Publishers, 1979, pp 373–379.
39. Douglass HO Jr: Levarterenol irrigation: Control of massive gastrointestinal bleeding in poor risk patients. *JAMA* 230:653–1657, 1974.
40. Driscoll DJ, Gillette PC, Duff DF, Nihill MR, Gutgesell HP, Vargo TA, Mullins CE, McNamara DG: Hemodynamic effects of dobutamine in children. *Am J Cardiol* 43:581–585, 1979.
41. Drugs for asthma. *Med Lett* 20(16):69–72, 1978.
42. Elliott TR: The control of the suprarenal glands by the splanchnic nerves. *J Physiol (Lond)* 44:374–409, 1912.
43. Erlanger J, Gasser HS: Studies in secondary traumatic shock II. Shock due to mechanical limitation of blood flow. *Am J Physiol* 49:151–173, 1919.
44. Farnebo LO, Hallman H, Hamberger B, Jonsson G: Catecholamines and hemorrhagic shock in awake and anesthetized rats. *Circ Shock* 6:109–118, 1979.
45. Florez J, Mediavilla A: Respiratory and cardiovascular effects of Met-enkephalin applied to the central surface of the brain stem. *Brain Res* 138:585–590, 1977.
46. Forrester JS, Waters DD: Hospital treatment of congestive heart failure—management according to hemodynamic profile. *Am J Med* 65:173–180, 1978.
47. Fredholm BB, Farnebo LO, Hamberger B: Plasma catecholamines, cyclic AMP, and metabolic substrates in hemorrhagic shock in the rat. The effect of adrenal demedullation and 6-OH-dopamine treatment. *Acta Physiol Scand* 105:481–485, 1979.
48. Gazioglu K, Condemi JJ, Hyde RW, Kaltreider NL: Effect of isoproterenol on gas exchange during air and oxygen breathing in patients with asthma. *Am J Med* 50:185–190, 1971.
49. Gerich J, Cryer P, Rizza R: Hormonal mechanisms in acute glucose counterregulation: The relative role of glucagon, epinephrine, norepinephrine, growth hormone and cortisol. *Metabolism* 29(Suppl 1):1164–1175, 1980.
50. Goldberg LI: Cardiovascular and renal actions of dopamine: Potential clinical applications. *Pharmacol Rev* 24:1–29, 1972.
51. Goldberg LI: Dopamine-clinical uses of an endogenous catecholamine. *N Engl J Med* 291:707–710, 1974.
52. Goldbranson FL, Lurie L, Vance RM, Vandell RF: Multiple extremity amputations in hypotensive patients treated with dopamine. *JAMA* 243:1145–1146, 1980.
53. Goldstein RA, Passaman ER, Roberts R: A comparison of digoxin and dobutamine in patients with acute infarction and cardiac failure. *N Engl J Med* 303:846–850, 1980.
54. Graham RM, Pettinger WA: Prazosin. *N Engl J Med* 300:232–236, 1979.
55. Greever CJ, Watts DT: Epinephrine levels in peripheral blood during irreversible hemorrhagic shock in dogs. *Circ Res* 7:192–195, 1959.
56. Gronert GA, Theye RA, Milde JH, Tinker JH: Catecholamine stimulation of myocardial oxygen consumption in porcine malignant hyperthermia. *Anesthesiology* 49:330–337, 1978.
57. Halmagyi DFJ, Neering IR, Lazarus L, Young JD, Pullin J: Plasma glucagon in experimental post-hemorrhagic shock. *J Trauma* 9:320–326, 1969.
58. Halter JB, Pflug AE: Effects of anesthesia and surgical stress on insulin secretion in man. *Metabolism* 29(Suppl 1):1124–1127, 1980.
59. Halter JB, Pflug AE: Relationship of impaired insulin secretion during surgical stress to anesthesia and catecholamine release. *Clin Endocrinol Metab* 51:1093–1098, 1980.
60. Halter JB, Pflug AE, Porte D Jr: Mechanism of plasma catecholamine increases during surgical stress in man. *J Clin Endocrinol Metab* 45:936–941, 1977.
61. Halter JB, Pflug AE, Tolas AG: Arterial-venous differences of plasma catecholamines in man. *Metabolism* 29:9–12, 1980.
62. Hamburg S, Hendler R, Sherwin RS: Influence of small increments of epinephrine on glucose tolerance in normal humans. *Ann Intern Med* 93:566–568, 1980.
63. Hamilton AS, Collins DA, Oppenheimer MJ: Effects of blood pressure on intestinal motility. *Fed Proc* 3:17, 1944.

64. Hedner T, Lundborg P: The effect of hypoxia on monoamine synthesis in brains of developing rats. *Biol Neonate* 31:122–126, 1977.
65. Hedner T, Lundborg P, Engel J: The effect of hypoxia on monoamine synthesis in brains of developing rats. II. Different length of exposure. *Biol Neonate* 32:229–236, 1977.
66. Heissenbuttel RH, Bigges JT: Bretylium tosylate: A newly available antiarrhythmic drug for ventricular arrhythmias. *Ann Intern Med* 91:229–238, 1979.
67. Henderson WR, Shelhamer JH, Reingold DB, Smith LJ, Evans R, Kaliner M: Alpha-adrenergic hyper-responsiveness in asthma. *N Engl J Med* 300:642–647, 1979.
68. Herman AH, Mack E, Egdahl RH: The relationship of adrenal perfusion to corticosteroid secretion in prolonged hemorrhagic shock. *Surg Gynecol Obstet* 131:795–802, 1971.
69. Heymans C: Le sinus carotidien, zone reflexogene regulatrice de ton vs vagal cardiaque du tonus neurovasculaire et de l'adrenalosecretion. *Arch Intern Pharmacodyn Ther* 35:269–306, 1928.
70. Holaday JW, Loh HH, Li CH: Unique behavioral effects of beta endorphins and their relationship to thermo-regulation and hypothalamic function. *Life Sci* 22:1525–1536, 1978.
71. Howard HM: Studies of the absorption and metabolism of glucose following injury. *Ann Surg* 141:321–326, 1955.
72. Hurst JW: *The Heart*, ed 3. New York, McGraw Hill, 1974, pp 596.
73. Jewitt D, Jennings K, Jackson PG: Efficacy of new inotropic drugs in clinical coronary heart failure. *Am J Med* 65:197–202, 1978.
74. Juan D, Elrazak MA: Hypophosphatemia in hospitalized patients. *JAMA* 242:163–164, 1979.
75. Judd SJ, Rakoff JS, Yen SSC: Inhibition of gonadotropin and prolactin release by dopamine: Effect of endogenous estradiol levels. *J Clin Endocrinol Metab* 47:494, 1978.
76. Kaptein EM, Kletzy DA, Spencer CA, Nicoloff JT: Effects of prolonged dopamine infusion on anterior pituitary function in normal males. *J Clin Endocrinol Metab* 51:488–491, 1980.
77. Kates RE, Leier CV: Dobutamine pharmacokinetics in severe heart failure. *Clin Pharmacol Ther* 24:537–541, 1978.
78. Katz RJ: Stress-induced tail elevation and further behavioral evidence in rats for the involvement of endorphins in stress. *Neurosci Lett* 13:249–252, 1979.
79. Katz RJ, Roth KA: Stress-induced grooming in the rat: An endorphin mediated syndrome. *Neurosci Lett* 13:282–292, 1979.
80. Keung ECH, Siskind SJ, Sonneblick EH, Ribner HS, Schwartz WJ, Lejemtel TH: Dobutamine therapy in acute myocardial infarction. *JAMA* 245:144–146, 1981.
81. Kim YD, Lake CR, Lees DE, Schuette WA, Bull JM, Weise V, Kopin IJ: Hemodynamic and plasma catecholamine responses to hyperthermic cancer therapy in humans. *Am J Physiol* 237:570–574, 1979.
82. Knochel JP: The pathophysiology and clinical characteristics of severe hypophosphatemia. *Arch Intern Med* 137:203–220, 1977.
83. Konzett H: Neue broncholytische hochwirksame Korper der Adrenalinreihe. *Arch Exp Pathol Pharmakol* 197:27–40, 1940.
84. Lake CR, Ziegler MG: Effect of acute volume alterations on norepinephrine and dopamine-β-hydroxylase in normotensive and hypertensive subjects. *Circulation* 57:774–778, 1978.
85. Landsberg L, Young JB: Catecholamines and the adrenal medulla. In Bondy PK, Rosenberg LE: *Metabolic Control and Disease.* Philadelphia, WB Saunders, 1980, pp 1621–1693.
86. Larsson S, Svedmyr N, Thiringer G: Lack of bronchial beta adrenoceptor resistance in asthmatics during long term treatment with terbutaline. *J Allergy Clin Immunol* 59:93–100, 1977.
87. Leblanc H, Lachelin GCL, Abu-Fadil S, Yen SSC: Effects of dopamine infusion on pituitary hormone secretion in humans. *J Clin Endocrinol Metab* 43:668–674, 1976.
88. Leebaw WF, Lee LA, Woolf PD: Dopamine affects basal and augmented pituitary hormone secretion. *J Clin Endocrinol Metab* 47:480–487, 1978.
89. Lefkowitz RJ: Direct binding studies of adrenergic receptors: Biochemical, physiologic and clinical implications. *Ann Intern Med* 91:450–458, 1979.
90. Leier CV, Heban PT, Huss P, Bush CA, Lewis RP: Comparative systemic and regional hemodynamic effects of dopamine and dobutamine in patients with cardiomyopathic heart failure. *Circulation* 58:466–475, 1978.
91. LeVeen HH: Intraperitoneal and intragastric use of vasopressors. *J St Barnabas Med Ctr* 10:7–11, 1973.
92. LeVeen HH, Falk G, Diaz C, Wynkoop BJ, Piccone VA, Yarnoz MD, Langsam AA, Nelson JH, Pedowitz WJ, Belfasky RB: Control of gastrointestinal bleed. *Am J Surg* 123:154–159, 1972.
93. Levi J, Coburn J, Kleeman CR: Mechanism of the antidiuretic effect of beta adrenergic stimulation in man. *Arch Intern Med* 136:25–29, 1976.
94. Lewis JW, Cannon JT, Liebeskind JC: Opioid and nonopioid mechanisms of stress and analgesia. *Science* 208:623–625, 1980.
95. Liere EJV, Northup D, Stickney JC: The effects of anemic anoxia on the motility of the small and large intestine. *Am J Physiol* 142:260–264, 1944.
96. Lightman SL, Forsling ML: Evidence for endogenous opioid control of vasopressin release in man. *J Clin Endocrinol Metab* 50:569–571, 1980.
97. Lillehei CW, Dixon JL, Wangensteen OH: Relation of anemia and hemorrhagic shock to experimental ulcer production. *Proc Soc Exp Biol Med* 68:125–128, 1948.
98. Lillehei RC, Longerbeam JK, Rosenberg JC: The nature of irreversible shock: Its relationship to intestinal changes. In Bock KD: *Shock: Pathogenesis and Therapy.* Berlin, Springer, 1962, pp 106–129.
99. Ludeman HH, Filbert MG, Cornblath M: Application of a fluorometric method for adrenaline-

like substances in peripheral plasma. *J Appl Physiol* 8:59–66, 1955.
100. MacGregor RR: Granulocyte adherence changes induced by hemodialysis, endotoxin, epinephrine and glucocorticoids. *Ann Intern Med* 86:35–39, 1977.
101. Mack E, Wyler DJ, Egdahl RH: Adrenal microcirculation in hemorrhagic shock. *Surg Gynecol Obstet* 128:511–518, 1969.
102. McDonald RH Jr, Goldberg LI, McNay JL, Tuttle EP Jr: Augmentation of sodium excretion and blood flow by dopamine in man. *Clin Res* 11:248, 1963.
103. McDonald RH Jr, Goldberg LI, McNay JL, Tuttle EP Jr: Effect of dopamine in man: Augmentation of sodium excretion, glomerular filtration rate and renal plasma flow. *J Clin Invest* 43:1124, 1964.
104. McLees BD: Critical care medicine. In Beeson PB, McDermott W, Wyngaarden JB: *Textbook of Medicine.* Philadelphia, WB Saunders, 1979, pp 1031–1058.
105. McNay JL, Goldberg LI: Direct renal vasodilation produced by dopamine in the dog. *Circ Res* 16:510–517, 1965.
106. McNay JL, McDonald RH Jr, Goldberg LI: Comparative effects of dopamine on renal and femoral blood flows. *Pharmacologist* 5:269, 1963.
107. McNeill JR, Stark RD, Greenway CV: Intestinal vasoconstriction after hemorrhage: Roles of vasopressin and angiotension. *Am J Physiol* 219:1342–1347, 1970.
108. Meyer V, Khobil E: Growth hormone secretion in the unanesthetized rhesus monkey in response to noxious stimuli. *Endocrinology* 80:163–172, 1967.
109. Millar RA, Bentey BG: The fluorimetric estimation of adrenaline and noradrenaline during hemorrhagic hypotension. *Br J Anaesthesia* 30:158–165, 1958.
110. Millar RA, Keener EB, Benfey BG: Plasma adrenaline and noradrenaline after phenoxybenzamine administration and during hemorrhagic hypotension in normal and adrenalectomized dogs. *Br J Pharmacol Chemother* 14:9–13, 1959.
111. Millar RA, Morris ME: Apneic oxygenation in adrenalectomized dogs. *Anesthesiology* 22:433–439, 1961.
112. Millar RA, Morris ME: A study of methoxyflurane anesthesia. *Can Anesth Soc J* 8:210–215, 1961.
113. Millar RA, Morris ME: Sympatho-adrenal responses during general anesthesia in the dog and man. *Can Anaesth Soc J* 8:356–386, 1961.
114. Miller JZ, Luft FC, Grim CF, Henry DP, Christian JC, Weinberger MH: Genetic influences on plasma and urinary norephinephrine after volume expansion and contraction in normal men. *J Clin Endocrinol Metab* 50:219–222, 1980.
115. Ming SC, Leviton R: Acute hemorrhagic necrosis of the gastrointestinal tract. *N Engl J Med* 263:59–65, 1960.
116. Morley JE: The endocrinology of the opiates and opioid peptides. *Metabolism* 30:195–209, 1981.
117. Moss GS, Cerchio GM, Siegel DC, Popovich PA, Butler E: Serum insulin response in hemorrhagic shock in baboons. *Surgery* 68:34–39, 1970.
118. Moss IR, Friedman E: β-endorphin: Effects on respiratory regulation. *Life Sci* 23:1271–1276, 1978.
119. Mueller HS, Evans R, Ayres JM: Effect of dopamine on hemodynamics and myocardial infarction in man. *Circulation* 57:361–365, 1978.
120. Nahas GG, Ligou JC, Mehlman B: Effects of pH changes on O_2 uptake and plasma catecholamine levels in the dog. *Am J Physiol* 198:60–66, 1960.
121. Nakao K, Nakai Y, Jingami H, Oki S, Fukata J, Imura H: Substantial rise of plasma β-endorphin levels after insulin induced hypoglycemia in human subjects. *J Clin Endocrinol Metab* 49:838–841, 1979.
122. Nechells H,Walker L, Olson W: A gradient of gastrointestinal motility following hemorrhage. *Fed Proc* 5:75–76, 1946.
123. Norbiato G, Bevilacqua M, Raggi U, Micossi P, Moroni C: Metoclopramide increases plasma aldosterone concentration in man. *J Clin Endocrinol Metab* 45:1313–1316, 1977.
124. Noth RH, McCallum RW, Contino C, Havelick J: Tonic dopaminergic suppression of plasma aldosterone. *J Clin Endocrinol Metab* 51:64–69, 1980.
125. Palmer JP, Halter J, Werner PL: Differential effect of isoproterenol on acute glucagon and insulin release in man. *Metabolism* 28:237–240, 1979.
126. Pearson JW, Redding JS: Epinephrine in cardiac resuscitation. *Am Heart J* 66:210–214, 1963.
127. Pekkarinen A: Studies on the chemical determination, occurrence and metabolism of adrenaline in the animal organism. *Acta Physiol Scand [Suppl]* 54:1–111, 1948.
128. Petersdorf RG, Dale DC: Gram negative bacteremia and septic shock. In Isselbacher KJ, Adams RD, Braunwald E, Petersdorf RG, Wilson JD: *Harrison's Principles of Internal Medicine*, ed 9. New York, McGraw Hill, 1980, pp 561–566.
129. Price HL, Llnde HW, Jones RE, Black GW, Price ML: Sympatho-adrenal responses to general anesthesia in man and their relation to hemodynamics. *Anesthesiology* 20:563–575, 1959.
130. Rapport D: Studies in experimental traumatic shock. VI. The liberation of epinephrine in traumatic shock. *Am J Physiol* 60:461–475, 1922.
131. Richardson JA, Woods EF, Richardson AK: Plasma concentrations of epinephrine and norephinephrine during anesthesia. *J Pharmacol Exp Ther* 119:378–384, 1957.
132. Roach PJ: Gastrointestinal bleeding in pheochromocytoma and following the administration of norepinephrine. *Arch Intern Med* 104:175–177, 1959.
133. Rosenberg JC, Lillehei RC, Longerbeam JK, Zimmerman B: Studies on hemorrhagic and endotoxin shock in relation to vasomotor changes and endogenous circulating epinephrine, norepinephrine and serotonin. *Ann Surg* 154:611–627, 1961.

134. Ruiz CE, Weil MH, Carlson RW: Treatment of circulatory shock with dopamine. *JAMA* 242:165–168, 1979.
135. Sacca L, Morrone G, Cicala M, Corso G, Ungaro B: Influence of epinephrine, norepinephrine and isoproterenol on glucose homeostasis in normal man. *J Clin Endocrinol Metab* 50:680–684, 1980.
136. Sakamokto T, Yamade T: Hemodynamic effects of dobutamine in patients following open heart surgery. *Circulation* 55:515–533, 1977.
137. Sato H, Ohguri M, Wada M: Epinephrine discharge, blood sugar and blood pressure in anaphylactic shock of dogs, non-anesthetized, non-fasted. *Tohoku J Exp Med* 28:504–519, 1935.
138. Satke Y: The amount of epinephrine secreted from the suprarenal glands in dogs in hemorrhage, and poisoning with guanidine, peptone, caffeine urethane, camphor etc. *Tohoku J Exp Med* 17:333–344, 1937.
139. Sherwin RS, Shamoon H, Hendler R, Sacca L, Eigler N, Walesky M: Epinephrine and the regulation of glucose metabolism: Effect of diabetes and hormonal interactions. *Metabolism* 29(Suppl 1):1146–1154, 1980.
140. Shettigar UR, Hultgren HN, Spector M, Martin R, Davies DH: Primary pulmonary hypertension: Favorable effect of isoproterenol. *N Engl J Med* 295:1414–1415, 1976.
141. Shine KI, Kuhn M, Young LS, Tillisch JHT: Aspects of the management of shock. *Ann Intern Med* 93:723–734, 1980.
142. Simonsson BG, Anderson R, Bergh NP, Skoogh BE, Svedmyr N: In-vivo and in-vitro studies of pharmacological effects on different receptors regulating bronchial tone in man. In Orie NGM, VanderLende R: *Bronchitis III*. Forest Grove, Ore, International Scholarly Book Service, 1970, pp 334–342.
143. Sole MJ, Drobac M, Schwartz L, Hussain MN, Vaughan-Neil EF: The extraction of circulating catecholamines by the lungs in normal man and in patients with pulmonary hypertension. *Circulation* 60:160–163, 1979.
144. Soman VR, Shamoon H, Sherwin RS: Effects of physiological infusion of epinephrine in normal humans: Relationship between the metabolic response and beta-adrenergic binding. *J Clin Endocrinol Metab* 50:294–297, 1980.
145. Sonnenblick EH, Frishman WA, LeJemtel TH: Dobutamine: A new synthetic cardioactive sympathetic amine. *N Engl J Med* 300:17–22, 1979.
146. Stene M, Panagiotis N, Tuck ML, Sowers JR, Mayes D, Berg G: Plasma norepinephrine levels are influenced by sodium intake. Glucocorticoid administration, and circadian changes in normal man. *J Clin Endocrinol Metab* 51:1340–1345, 1980.
147. Stokke DB, Christensen NH, Hole P, Andersen PK, Juhl B: Plasma catecholamines during equipotent anaesthesia with cyclopropane and halothane-N_2O in man. *Anaesthetist* 27:469–474, 1978.
148. Sylvester JT, Scharf SM, Gilbert RD, Fitzgerald RS, Traystman RJ: Hypoxic and carbon monoxide hypoxia in dogs: Hemodynamics, carotid reflexes and catecholamines. *Am J Physiol* 235:22–28, 1978.
149. Tenney SM: Sympathoadrenal stimulation by carbon dioxide and the inhibitory effect of carbonic acid on epinephrine response. *Am J Physiol* 187:341–346, 1956.
150. Turner ES, Greenberg PA, Patterson R: Management of the pregnant asthmatic patient. *Ann Intern Med* 6:905–918, 1980.
151. Turner P: *Clinical Aspects of Autonomic Pharmacology*. Philadelphia, JB Lippincott, 1969, pp 21–41.
152. Tuttle RR, Mills J: Dobutamine: Development of a new catecholamine to selectively increase cardiac contractility. *Circ Res* 36:185–196, 1975.
153. Unni VKN, McArdle L, Black GW: Sympathoadrenal, respiratory and metabolic changes during trichloroethylene anaesthesia. *Br J Anaesth* 42:429–433, 1970.
154. Unverferth DV, Blanford M, Kates RE, Leier CV: Tolerance to dobutamine after a 72 hour continuous infusion. *Am J Med* 69:262–266, 1980.
155. Vargish T, Reynolds DG, Gurll NJ, Lecher RB, Holaday JW, Faden AL: Naloxone reversal of hypovolemic shock in dogs. *Circ Shock* 7:31–38, 1980.
156. Wakim KG, Mason JW: The influence of hemorrhage and of depletion of plasma proteins on intestinal activity. *Gastroenterology* 4:92–101, 1945.
157. Watanabe M: Simultaneous determination of the blood sugar content and epinephrine output from the supra-renal gland in the non-anesthetized non-fasted dog during peptone poisoning. *Tohoku J Exp Med* 9:412–453, 1927.
158. Watts DT: Adrenergic mechanisms in hypovolemic shock. In Mills LC, Moyer JK: *Shock and Hypotension*. New York, Plenum Press, 1965, pp 385–391.
159. Webb-Johnson DC, Andrews JL Jr: Bronchodilator therapy. *J Engl J Med* 297:476–483, 1977.
160. Weiner N: NE, E and the sympatho-mimetic amines. In Gilman AG, Goodman LS, Gilman A: *The Pharmacologic Basis of Therapeutics*, ed 6. New York, MacMillan, 1980, pp 138–175.
161. Welle S, Lilavivathana V, Campbell RG: Increased plasma norepinephrine concentrations and metabolic rates following glucose ingestion in man. *Metabolism* 29:806–809, 1980.
162. Widdicombe JG: Action of catecholamines on bronchial smooth muscle. In Geiger S: *Handbook of Physiology*. Bethesda, Md, American & Physiological Society, 1975, pp 507–513.
163. Widdicombe JG: Regulation of tracheobronchial smooth muscle. *Physiol Rev* 43:1–37, 1963.
164. Wilmore DW: Carbohydrate metabolism in trauma. *Clin Endocrinol Metab* 5:731–741, 1976.
165. Wilmore DW, Cong JM, Mason AD Jr, Skreen RW, Pruitt BA Jr: Catecholamines: Mediators of the hypermetabolic response to thermal injury. *Ann Surg* 180:653–668, 1974.
166. Zern RT, Foster LB, Blalock JA, Feldman JM: Characteristics of the dopaminergic and noradrenergic systems of the pancreatic islets. *Diabetes* 28:185–189, 1979.

CHAPTER TWENTY-THREE

Malignant Hyperthermia: The Possible Role of Norepinephrine

NGUYEN BICH THOA, Ph.D., and GERTRUDE D. MAENGWYN-DAVIES, Ph.D.

INTRODUCTION

Malignant hyperthermia is a complex pharmacogenetic disorder described in both humans and pigs (8–10, 15, 16, 27, 29), as well as in several other mammalian species (14, 39, 49). Afflicted subjects are prone to sudden death upon exposure to some chemicals, including inhalation anesthetics, e.g., halothane and ether, and the depolarizing skeletal muscle relaxant, succinylcholine (9, 15, 16, 25, 27, 29, 30). In genetically susceptible humans and animals, the syndrome of malignant hyperthermia can also be readily elicited by stress (24, 35, 53, 55). In the majority of human cases there is a rapid and irreversible increase in body temperature (2°C every 5 minutes up to 47.5°C just prior to death) (35). Hyperthermia is frequently, but not necessarily, accompanied by excessive contracture of the skeletal musculature leading to rigidity. Biochemically, a pronounced increase in the skeletal muscle oxygen consumption and a severe lactacidosis are standard features of this syndrome. Many potent inhalation anesthetics trigger malignant hyperthermia in humans. Halothane is the agent most frequently implicated and documented (2, 3, 9, 28, 31, 35, 58).

The rate of occurrence of malignant hyperthermia is about 1:10,000 in a normal U.S. hospital population but may be much higher in geographical areas with a high concentration of malignant hyperthermia-susceptible families (46). Malignant hyperthermia has a worldwide distribution and can affect both sexes and all age groups, although it seems to occur more frequently in males between the ages of 16 and 30 (9).

Investigations of the primary defect in malignant hyperthermia concern the skeletal muscle; this is because: (a) calculations of the rate of heat production based on measurements of skeletal muscle *vs.* total body oxygen consumption show that the center of heat production in malignant hyperthermia is the skeletal muscle (23, 26); and (b) skeletal muscles from malignant hyperthermia patients and pigs usually respond to the application of caffeine, halothane, and succinylcholine with greater contractures than those observed in normal muscles (6, 36, 45, 47).

THE ROLE OF CATECHOLAMINES IN MALIGNANT HYPERTHERMIA

Little is known about the participation of the catecholamines (CA) in either the etiology of malignant hyperthermia or its clinical manifestations. Both NE and epinephrine (E) may be contributors since: (a) abnormally high amounts of these transmitters are found in the plasma of malignant hyperthermia susceptible (MHS) pigs, and the concentrations increase further after a challenge with succinylcholine and/or halothane (42); and (b) both humans and pigs susceptible to malignant hyperthermia develop the syndrome when subjected to stress (53, 55–58), an event in which CA have a central role (40).

Dopamine (DA) levels are lower than normal in both the urine and caudate nucleus of the stress-susceptible pigs (1). A role for DA in malignant hyperthermia is possible for two reasons: first, striatal DA has a central role in the control of motor activity (32, 33); and second, there is evidence that heat loss may be regulated by central DA (18). Investigations of central and/or peripheral adrenergic mechanism in malignant hyperthermia may help to elucidate this disorder.

Skeletal muscles isolated from both humans and pigs susceptible to malignant hyperthermia contract more in response to caffeine than those obtained from normal subjects (36, 37, 47). This observation is the basis for using caffeine either alone or in combination with halothane in *in vitro* tests to screen humans and pigs for their susceptibility to the disorder (37, 38). Combined therapy with a tricyclic antidepressant, imipramine, and a monoamine oxidase inhibitor (MAO-I), such as phenelzine or tranylcypromine, also induces a severe hyperthermia in humans which may result in death (5, 12, 13, 50).

In pigs, when this disorder is induced by succinylcholine followed by halothane, levels of plasma E and NE increase several fold (42). A good correlation exists between plasma CA and lactic acid levels (42). Lactacidosis is a constant feature of malignant hyperthermia (22, 41, 42). Following administration of halothane to MHS pigs, blood acidity, E and NE plasma levels, temperature of the blood draining the animals' limbs, and heart rate increase and simultaneously reach significantly higher levels 30 minutes after halothane induction (22). Intravenous infusion of NE (5 μg/kg/min for 50 minutes) into pigs pretreated with the β-adrenergic receptor antagonist, propranolol, produces no significant body temperature change in normal animals but induces a fatal hyperthermia in MHS animals (41). In MHS pigs in which reserpine decreased the plasma NE content from 1,230 ± 130 pg/ml to 140 ± 40 pg/ml, succinylcholine does not induce fatal hyperthermia (41). Thus, at least in the pig, CA and specifically NE seem to be very important contributing factors in the hyperthermic syndrome.

In vitro studies show that CA can affect the mechanical properties of both normal and MHS porcine skeletal muscle. At a dose of 100 mg/kg, both E and NE potentiate the combined effect of succinylcholine (50 μg/ml) and halothane (2% v/v) in muscles of MHS animals. In normal muscles, CA, succinylcholine, and halothane together produce only a moderate depression of tetanic tension. In MHS muscles, however, this drug combination dramatically decreases tetanic tension and initiates muscle contractures (20). A smaller stimulus is needed to induce contraction in MHS pig muscle since its mechanical threshold is lower than that of normal muscle (11). Halothane causes depolarization of the MHS muscle but not of the normal muscle (19). In humans and animals given the combination of succinylcholine-halothane, the increase in NE and E release, beginning in the early phase of the hyperthermic crisis, may have an additive role in the perpetuation of the syndrome. E increases the response of the cat soleus muscle to electrical stimulation of its motor nerve (7). Since the potentiating effect is observed with a supramaximal stimulus, *i.e.*, when every muscle cell is activated, the amine effect is thought to be due to a potentiation of the contractile force of the activated muscle cells. This response would suggest a pattern of repetitive activation of the muscle, an event caused by repetitive firing of the motor nerve terminals. Therefore, it is suggested that E causes its potentiation by increasing the release of acetylcholine from the motor nerve endings (7). If so, the excess of plasma CA observed during the malignant hyperthermia syndrome may influence the effect of succinylcholine/halothane-like combinations through both a direct synergism of these drugs at postsynaptic sites and/or an indirect stimulation of acetylcholine release from presynaptic stores.

CA, and more specifically DA, may participate in the etiology of malignant hyperthermia. DA levels in the caudate nuclei of MHS pigs (5,961 ng/gm) are significantly lower than in control pigs (10,878 ng/gm) (1). Pretreatment of humans susceptible to malignant hyperthermia with the DA receptor antagonist haloperidol prevents the development of hyperthermia during surgery and alleviates symptoms related to the syndrome in the postoperative period (58). Dantrolene sodium, the drug of choice for malignant hyperthermia, is also effective in several motor disorders associated with DA in the central nervous system (CNS) (17). Fatal hyperthermia, elicited in rats with lithium and the MAO inhibitor, tranylcypromine, is prevented by various DA receptor antagonists, but not by receptor antagonists of other central neurotransmitters (48). Denervation of the caudate nucleus is associated with a decrease of its DA content and an increase in the number of postsynaptic receptors. The low levels of DA observed in the caudate nucleus of MHS pigs and the blocking effect of hyperthermia by haloperidol in both humans and animals suggest that malignant hyperthermia may be associated with a denervation-like state of the caudate nucleus and perhaps of other central dopaminergic structures. For example, DA in the hypothalamus may have a role in the mediation of heat loss (21). In pigs susceptible to malignant hyperthermia, measurements of heat loss by radiation through a peripheral vasodilation process shows that heat loss increases during the first 20 to 25 minutes of the hyperthermic syndrome, then decreases to a negligible level after peripheral vasoconstriction (54). A very intense peripheral vasoconstriction is observed in both humans and pigs undergoing the malignant hyperthermia syndrome (53, 54, 55, 58). Animals treated with Arlidin, a potent peripheral vasodilator, survive the induction of malignant hyperthermia (53).

The availability of techniques for measuring the number of central pre- and postsynaptic DA receptors, as well as their affinities for agonists and antagonists, will permit studies which might clarify the role of DA in the etiology of malignant hyperthermia.

STRESS AND MALIGNANT HYPERTHERMIA

Humans and pigs susceptible to malignant hyperthermia tolerate stress very poorly. In humans, physical exercise, heat, automobile travel and emotional excitement can cause sudden death (4, 35, 44, 56, 58). The relationship between stress and malignant hyperthermia has been more thoroughly investigated in pigs. In a group of MHS pigs in which halothane was shown to produce fetal hyperthermia in 42% of the animals, 5 minutes of treadmill exercise brought the incidence of fatal hyperthermia to 100% (52). The stress of being put into a metabolic chamber may also induce fatal hyperthermia (43). Emotional excitement, as elicited by hearing a littermate's cries of distress, was fatal for one pig (43). The stress of heat, generated by cramped transportation, is known to cause many deaths in MHS pigs (51, 55). Because of the important involvement of NE in the body's response to stress (40), the abnormal susceptibility of MHS animals and humans is indirect evidence for the participation of NE in malignant hyperthermia.

Humans and pigs susceptible to malignant hyperthermia frequently have other abnormalities. In humans, cardiac disorders (34), emotional problems, nervousness, and hypertension are observed (24, 56, 58). In pigs, abnormally high metabolic rates and high blood pressure (BP) are frequently seen (55).

SUMMARY

Malignant hyperthermia may not only involve an abnormality of the skeletal musculature but may involve more generalized defects. The remedy for it awaits further investigation of the CNS and the periphery. Because of the obvious involvement of CA in the symptomatic manifestations of the syndrome, studies of CA dynamics in discrete brain areas may help to broaden knowledge of CNS adrenergic mechanisms and may further clarify the still essentially unknown etiology of this genetic disease.

References

1. Altrogge DM, Topel DG, Cooper MA, Halberg JW, Draper DD: Urinary and caudate nuclei catecholamine levels in stress-susceptible and normal swine. *J Anim Sci* 57:74–77, 1980.
2. Arens JF, McKinnon WMP: Malignant hyperpyrexia during anesthesia. *JAMA* 215:919–922, 1971.
3. Auerback VH, DiGeorge AM, Mayer BW, Hayden M, Carpenter GG, Krumperman LW, Trutten TR: Rhabdomyolysis and hyperpyrexia in children after administration of succinylcholine. In Gordon RA, Britt BA, Kalow W: *International Symposium of Malignant Hyperthemia.* Springfield, Ill, Charles C Thomas, 1973, pp 31–51.
4. Berman MC: Role of preoperative management in the development of malignant hyperthermia. *Lancet* 2:743, 1973.
5. Brachfeld J, Wirtshafter A, Wolfe S: Imipramine-tranylcypromine incompatibility. Near fatal reaction. *JAMA* 186:1172–1173, 1963
6. Bradley WG, Murchison D: Screening for malignant hyperpyrexia. *Br Med J* 4:108–109, 1972.
7. Breckenridge BM, Burn JH, Matschinski FH: Theophylline, epinephrine and neostigmine facilitation of neuromuscular transmission. *Proc Natl Acad Sci USA* 57:1893–1897, 1967.
8. Britt BA, Locher WG, Kalow W: Hereditary aspects of malignant hyperthermia. *Can Anaesth Soc J* 16:89–98, 1969.
9. Britt BA, Kalow W: Malignant hyperthermia. A statistical review. *Can Anaesth Soc J* 17:293–315, 1970.
10. Britt BA, Kwong FHK, Endrenyi L: The clinical and laboratory features of malignant hyperthermia management. A review. In Henschel EO: *Malignant Hyperthermia. Current Concepts.* New York, Appleton-Century-Crofts, 1977, pp 9–46.
11. Bryant SH, Anderson IL: Mechanical activation

and electrophysiological properties of intercostal muscle fibers from malignant hyperthermia-susceptible (MHS) pigs. *Neurosci Abstr* 3:213, 1977.

12. Byck R: Drugs and the treatment of psychiatric disorders. In Goodman LS, Gilman A: *The Pharmacological Basic of Therapeutics*. New York, McMillan, 1975, pp 152–200.
13. Carlsson A: Effect of drugs on amine uptake mechanisms in the brain. In Schümann HJ, Kroneberg G: *New Aspects of Storage and Release Mechanisms of Catecholamines*. New York, Springer-Verlag, 1970, pp 223–233.
14. DeJongh RH, Heavner JE, Amory DW: Malignant hyperthermia in the cat. *Anesthesiology* 41:608–609, 1974.
15. Denborough MA, Lovell RRH: Anesthetic deaths in a family. *Lancet* 2:45, 1960.
16. Denborough MA, Forster JFA, Lovell RRH, Maplestone PA, Villier JD: Anesthetic deaths in a family. *Br J Anaesth* 34:395–396, 1962.
17. Franz DA: Drugs for Parkinson's disease: Centrally acting muscle relaxants. In Goodman LS, Gilman A: *The Pharmacological Basis of Therapeutics*. New York, McMillian, 1975, pp 227–244.
18. Frens J: Neurotransmitter mapping in central thermoregulation. In Cox B, Lomax P, Milton AS, Schönbaum E: *Thermoregulatory Mechanisms and Their Therapeutic Implications*. Basel, Karger, 1979, pp 1–5.
19. Gallant EM, Godt R, Gronert GA: Role of plasma membrane defect of skeletal muscle in malignant hyperthermia. *Muscle Nerve* 2:491–494, 1979.
20. Gallant EM, Godt R, Gronert GA: Mechanical properties of normal and malignant hyperthermia susceptible porcine muscle: Effects of halothane and other drugs. *J Pharmacol Exp Ther* 213:91–96, 1980.
21. Gisolfi CV, Francisco M, Beattie M, Wall PT, Brown S, Magnes S: The role of dopamine in temperature regulation during exercise stress. In Cox B, Lomax P, Milton AS, Schönbaum E: *Thermoregulatory Mechanisms and Their Therapeutic Implications*. Basel, Karger, 1979, pp 31–32.
22. Gronert GA, Theye RA: Halothane induced porcine malignant hyperthermia: Metabolic and hemodynamic changes. *Anesthesiology* 44:36–43, 1976.
23. Gronert GA, Heffron JJA, Milde JH, Theye RA: Porcine malignant hyperthermia: Role of skeletal muscle in increased oxygen consumption. *Can Anaesth Soc J* 24:103–109, 1977.
24. Gronert GA, Thompson RL, Onofrio BM: Human malignant hyperthermia: Awake episodes and correction by dantrolene. *Anesth Analg* (Cleve) 59:377–378, 1980.
25. Guedel: *Inhalation Anesthesia*. New York, McMillan, 1937.
26. Hall GM, Bendell JR, Lucke JN, Lister D: Porcine malignant hyperthermia. II. Heat production. *Br J Anaesth* 48:305–308, 1976.
27. Hall LW, Woolf N, Bradley JWP, Jolly DW: Unusual reaction to suxamethonium chloride. *Br Med J* 4:1305, 1966.
28. Halsall PJ, Cain PA, Ellis FR: Retrospective analysis of anesthetics received by patients before susceptibility to malignant hyperpyrexia was recognized. *Br J Anaesth* 51:949–954, 1979.
29. Harrisson GG, Biebuyk JF, Terblanche J, Dent DM, Hickman R, Saunders JJ: Hyperpyrexia during anesthesia. *Br Med J* 3:594–595, 1968.
30. Harrisson GG, Saunders S, Biebuyck JF, Hickman R, Dent DM, Weaver V, Terblanche J: Anesthetic induced malignant hyperpyrexia and a method for its prediction. *Br J Anaesth* 41:844–854, 1969.
31. Henschel EO, Koh TB: Malignant hyperthermia in a 14 month old infant. In Henschel EO: *Malignant Hyperthermia: Current Concepts*. New York, Appleton-Century-Crofts, 1977, pp 3–7.
32. Hornykiewicz O: Dopamine (3-hydroxytyramine) and brain functions. *Pharmacol Rev* 18:925–964, 1966.
33. Hornykiewicz O: Parkinson's disease: From brain homogenates to treatment. *Fed Proc* 32:183–190, 1973.
34. Hickell VF, Staniloff HM, McLaughlin PR, Britt BA, Morch JE: Cardiovascular manifestations of normothermic malignant hyperthermia. In Aldrete JA, Britt BA: *The Second International Symposium on Malignant Hyperthermia*. New York, Grune & Stratton, 1978, pp 373–377.
35. Jardon OM, Wingard DW, Barak AJ, Connolly JF: Malignant hyperthermia. A potentially fatal syndrome in orthopedic patients. *J Bone Joint Surg* 61A:1064–1070, 1979.
36. Kalow W, Britt BA, Terreau ME, Haist E: Metabolic error of muscle metabolism after recovery from malignant hyperthermia. *Lancet* 2:895–898, 1970.
37. Kalow W, Britt BA, Richter A: The caffeine test of isolated human muscle in relation to malignant hyperthermia. *Can Anaesth Soc J* 24:678–694, 1977.
38. Kalow W, Britt BA, Peters P: Rapid, simplified techniques for measuring caffeine contraction for patients with malignant hyperthermia. In Aldrete JA, Britt BA: *Second International Symposium on Malignant Hyperthermia*. New York, Grune & Stratton, 1978, pp 339–350.
39. Klein LV: Case report: A hot horse. *Vet Anesth* 2:41–42, 1975.
40. Kopin IJ: Catecholamines and stress. In Usdin E, Kvetnansky R, Kopin IJ: *Catecholamines and Stress*. New York, Pergamon Press, 1976, pp 1–5.
41. Lister D, Hall GM, Lucke JN: Porcine malignant hyperthermia. III. Adrenergic blockade. *Br J Anaesth* 48:831–837, 1976.
42. Lucke JN, Hall GM, Lister D: Porcine malignant hyperthermia. I. Metabolic and physiological changes. *Br J Anaesth* 48:297–304, 1976.
43. Lucke JN, Hall GM, Lister D: Malignant hyperthermia in the pig and the role of stress. *Ann NY Acad Sci* 317:326–337, 1979.
44. Mogensen JV, Misfeldt BB, Hanel HK: Pre-operative excitement and malignant hyperthermia. *Lancet* 1:461, 1974.
45. Moulds RFW, Denborough MA: Biochemical basis of malignant hyperpyrexia. *Br Med J* 2:241–244, 1974.

46. National Institute of General Medicine and Science Research Conference Report. *Anesthesiology: Its Expanding Role in Medicine.* Washington, DC, DHEW, October, 1974.
47. Nelson TE: Excitation-contraction coupling: A common etiologic pathway for malignant hyperthermia susceptible muscle. In Aldrete JA, Britt BA: *Second International Symposium on Malignant Hyperthermia.* New York, Grune & Stratton, 1978, pp 23–36.
48. Shimomura K, Tomoi M, Mori J: The role of caudate nucleus dopamine and cyclic AMP in the hyperpyrexia induced by LiCl plus tranylcypromine in rats. *Jpn J Pharmacol* 29:447–456, 1979.
49. Short CE, Paddleford RR: Malignant hyperthermia in the dog. *Anesthesiology* 39:462–463, 1973.
50. Stanley B, Pal NR: Fatal hyperpyrexia with phenelzine and imipramine. *Br Med J* 2:1011, 1964.
51. Topel DG, Bricknell EJ, Preston KS: Porcine stress syndrome. *Mod Vet Pract* 49:40–41, 1968.
52. Van den Hende C, Lister D, Muylle E, Oams L, Oyaert W: Malignant hyperthermia in Belgian Landrace pigs rested or exercised before exposure to halothane. *Br J Anaesth* 48:821–828, 1976.
53. Williams CH: The development of an animal model for the fulminant hyperthermia porcine stress syndrome. In Henschel EO: *Malignant Hyperthermia: Current Concepts.* New York, Appleton-Century-Crofts, 1977, pp 117–140.
54. Williams CH, Shanklin MD, Houchins C: Energy metabolism in fulminant hyperthermia stress syndrome in swine. In Henschel EO: *Malignant Hyperthermia: Current Concepts.* New York, Appleton-Century-Crofts, 1977, pp 149–156.
55. Williams CH, Shanklin MD, Hendrix HB, Muhrer ME, Stubbs DH, Krause GF, Payne CG, Benedict JD, Hutcheson DP, Lasley JF: The fulminant hyperthermia-stress syndrome: Genetic aspects, hemodynamic and metabolic measurements in susceptible and normal pigs. In Aldrete JA, Britt BA, *Second International Symposium on Malignant Hyperthermia.* New York, Grune & Stratton, 1978, pp 113–140.
56. Wingard DW: Malignant hyperthermia: A human stress syndrome. *Lancet* 2:1450–1451, 1974.
57. Wingard DW: Malignant hyperthermia—acute stress syndrome of man? In Henschel EO: *Malignant Hyperthermia: Current Concepts.* New York, Appleton-Century-Crofts, 1977, pp 79–95.
58. Wingard DW, Gatz EE: Some observations on stress susceptible patients. In Aldrete AJ, Britt BA: *Second International Symposium on Malignant Hyperthermia.* New York, Grune & Stratton, 1978, pp 363–372.

CHAPTER TWENTY-FOUR

Plasma Norepinephrine in Essential Hypertension*

DAVID S. GOLDSTEIN, M.D., Ph.D., MICHAEL G. ZIEGLER, M.D., and C. RAYMOND LAKE, M.D., Ph.D.

INTRODUCTION

The introduction about a decade ago of sensitive, specific techniques for measuring plasma norepinephrine (NE) (40), and indications that those levels reflect sympathetic neural activity (39), promised to help answer a perennial question in hypertension research: Does sympathetic hyperactivity cause essential hypertension? Despite the subsequent publication of at least 45 comparative studies in hypertensive patients and in normotensive controls, controversy persists about whether patients with hypertension exhibit elevated levels of NE.

So many studies have compared NE concentrations in hypertensive and normotensive groups that a statistical approach can be undertaken in which the mean hypertensive and normotensive values from each study provide single data points. This approach can identify factors differentiating "positive" studies (those reporting a statistically significant hypertensive-normotensive difference in NE) from "negative" studies (those reporting no significant difference).

A recent review (22) found that three factors differentiated positive from negative studies: the type of assay technique used to measure NE levels; the normotensive control NE levels; and the age of the population studied. Studies were much more likely to be positive when a fluorometric assay technique was used than when a radioenzymatic technique was used. Normotensive control NE levels were significantly lower in the positive than in the negative studies. Also, studies were much more likely to be positive when they involved subjects less than about 40 years of age.

This chapter extends the analysis of resting hypertensive and normotensive NE levels to 46 studies and considers the effects of stress on NE in these groups.

METHODS

With a few exceptions discussed later, the studies considered in this review satisfied all of the following criteria: (a) They were published since 1970, in English; (b) They were not merely abstracts, but primary data sources independent of data from other sources; (c) They specifically involved plasma NE, not "total catecholamines (CA)"; (d) They used a sensitive and reliable fluorometric technique, such as that of Renzini *et al.* (50), a radioenzymatic assay (39), or high pressure liquid chromatography (HPLC) with electrochemical detection (23); and (e) They reported plasma NE levels in both normotensive control group

* Since the drafting of this chapter, several additional reports of plasma norepinephrine in essential hypertension have appeared. The conclusions of this review remain the same.

and an entire hypertensive group (as opposed to a subgroup selected for certain characteristics).

To locate these studies, we conducted several MEDLINE searches of the published literature for interactions among hypertension, CA, NE, stress, cold, hypoglycemia, emotions, hypoxia, and pain. We culled additional articles by inspection of the bibliographies in the publications listed in MEDLINE.

When necessary, mean group NE values were derived from figures or from the weighted contributions of listed subgroups. In the few allowed cases where "total CA" were reported, the NE concentration was assumed to be 80% of the total.

Statistical testing used independent and dependent-means "*t*"-tests, Pearson correlation coefficients, and chi-square analyses (13).

NOREPINEPHRINE AT SUPINE REST

Table 24.1 displays hypertensive and normotensive mean NE levels from the 46 studies. Of the 46, 40 (87%) reported higher levels in the hypertensives, by an average of 53 pg/ml ($t = 6.06$, $p < 0.001$). Only 18 (39%) reported statistically significant hypertensive-normotensive differences in NE. Since about two-fifths of the studies are positive and about three-fifths negative, one can appreciate that any generalizations from the literature might be questioned.

What made the positive studies positive, and the negative ones negative? If the cause were a variably represented "*hyperadrenergic*" subgroup of hypertensives, then studies with a preponderance of these patients should have been positive due to the excessive mean NE levels in the hypertensive groups. Across the 46 studies, the mean hypertensive NE levels in the hypertensive studies were similar to those of the negative studies (287 *vs.* 273 pg/ml). In contrast, across the 46 studies, normotensive control levels were lower in the positive studies (182 *vs.* 254 pg/ml; $t = 3.86$, $p < 0.001$). The average standard deviation of normotensive NE levels was also much lower in the positive studies (73 *vs.* 139 pg/ml; $t = 3.30$, $p < 0.005$). These results emphasize the need for close attention not only to the hypertensive groups but also to characteristics of the normotensive controls and to sources of variability in obtained NE levels. Several of these factors are considered below.

Number of Subjects

It would be inappropriate to give equal weight to studies with different numbers of patients if small studies, perhaps involving selected series of patients and therefore smaller standard deviations, tended to be positive while larger studies, involving more generalized patient populations and larger standard deviations, tended to be negative. In fact, 10 of 20 (50%) studies with 50 subjects or less were positive, and eight of 26 (31%) with more than 50 subjects were positive, a nonsignificant difference in proportions. Average standard deviations of NE values were also similar in the two types of studies. Hence, the size of the study was unrelated to the frequency of positive results.

Type of Assay

Of 10 studies using a fluorometric assay technique, eight (80%) were positive, while only nine of 34 (26%) using a radioenzymatic assay technique were positive (chi-square = 6.89, $p < 0.001$). The average standard deviation of NE levels was significantly greater with radioenzymatic studies for both hypertensive (154 *vs.* 97 pg/ml; $t = 2.99$, $p < 0.01$) and normotensive (135 *vs.* 44 pg/ml; $t = 3.79$, $p < 0.001$) groups.

The definition of a "normal" NE level depended on the type of assay. For the fluorometric type, the mean normotensive level was 147 pg/ml, compared with 246 pg/ml for the radioenzymatic type ($t = 4.88$, $p < 0.001$). Further, for the phenylethanolamine-*N*-methyltransferase (PNMT) radioenzymatic assay, the mean normotensive NE level tended to be greater than for the catechol-*O*-methyltransferase (COMT) radioenzymatic assay (270 *vs.* 229 pg/ml; $t = 1.93$, $0.05 < p < 0.10$).

Because positive studies were weighted disproportionately by the results derived from fluorometric assay techniques, we also compared positive and negative studies within assay techniques. Among fluorometric studies, mean hypertensive NE levels were higher in positive than in negative studies (212 *vs.* 165 pg/ml; $t = 2.58$, $p < 0.05$). Similarly, among radioenzymatic studies, mean hypertensive NE levels were higher in the positive studies (349 *vs.* 279 pg/ml; $t = 2.97$, $p < 0.01$). This positive-negative difference was especially pronounced for COMT radioenzymatic studies, in which the mean hypertensive NE level

Table 24.1
Plasma Norepinephrine Levels in Patients with Essential Hypertension and in Normotensive Controls during Supine Rest[a]

Author	H/N	Age H/N	Assay	MPA H/N	NE H/N	SD H/N
Beretta-Piccoli (1)	34/25	39/41	Rc	109/88	238/230	115/86
Beretta-Piccoli (2)	45/26	42/39	Rc	116/89	262/221	138/94
Bertel (3)	59/15	41/40	Rc	112/93	263/250	98/121
Brecht (4)	59/15	?	F	?	257/135[b]	147/59
Brecht (5)	125/107	40/42	F	?	201/128[b]	?
Corea (6)	19/7	38/34	F	128/85	172/147	96/32
Cousineau (8)	46/28	41/39	Rp	?	332/226[b]	198/106
DeLeeuw (10)	69/22	45/41	Rp	120/95	270/240	150/100
DeQuattro (11)	11/12	?	F	?	157/148	99/45
Eide (14)	28/12	38/?	F	109/?	193/148[b]	101/45
Eide (15)	7/7	40/36	F	109/89	240/167[b]	99/31
Eng (16)	20/17	47/34	Rc	114/90	390/250[b]	175/62
Elser (18)	21/11	?	F	?	193/138[b]	77/36
Esler (17)	41/24	37/34	?	?	323/196[b]	118/71
Franco-Morselli (20)	19/11	43/45	Rc	116/90	269/248	122/186
Franco-Morselli (19)	27/12	?	Rc	104/87	277/250	286/364
Geffen (21)	20/8	?	Rc	?	400/160[b]	223/113
Henquet (24)	25/25	?	Rc	108/89	409/370	254/254
Henry (25)	32/93	41/27	Rp	119/88	151/144	107/105
Hjemdahl (26)	7/7	35/35	H	101/87	436/353	156/94
Hofman (27)	18/18	19/18	Rp	97/90	351/248[b]	110/123
Ibsen (28)	33/31	40/40	Rc	115/91	170–180	?
Jones (29)	31/28	47/38	Rp	?	410/354	223/160
Kafka (30)	15/18	50/50	Rp	112/91	265/289	194/136
Kiowski (31)	44/34	43/43	Rc	?	289/283	98/115
Kobayashi (32)	27/21	42/42	Rc	111/81	291/224[b]	94/128
Kolloch (33)	7/9	29/31	Rc	109/?	410/235[b]	108/111
Lake (38)	67/84	44/33	Rp	112/86	339/304	188/183
Lake (39)	24/44	43/35	Rp	?	306/287	127/146
Lake (35)	151/117	43/45	Rp	112/88	297/294	143/159
Lake (37)	56/29	46–40	Rp	112/89	249/253	120/129
Lake (36)	11/9	48/44	Rp	115/87	222/172	129/120
Louis (40)	24/7	?	Rc	?	390/160[b]	?
Miura (42)	120/30	35/30	F	112/87	188/130[b]	88/60
Miura (43)	120/49	35/33	F	112/83	209/160[b]	?
Parfrey (45)	16/8	36/31	Rp	113/80	296/191[b]	111/31
Pedersen (46)	19/32	41/40	Rc	142/97	242/254	?
Philipp (48)	29/29	38/33	F	?	216/173[b]	72/45
Robertson (52)	9/10	25/27	Rc	101/86	226/196	?
Sever (53)	56/59	46/46	Rp	124/92	411/403	197/184
Sever (54)	100/48	45/47	Rp	121/95	352/372	178/171
Taylor (55)	51/26	46/40	Rc	119/89	240/260	230/257
Vlachakis (56)	38/14	48/49	Rc	119/92	256/205	139/60
Vlachakis (57)	22/13	50/44	Rc	?	277/234	89/90
Vlachakis (58)	60/23	48/46	Rc	116/88	282/206[b]	155/72
Weidmann (60)	79/90	46/37	Rc	123/89	202/169	128/91
Sum or mean	1941/1364	41/38		114/89	279–226[b]	142/114

[a] H, hypertensive; N, normotensive; Rc, COMT radioenzymatic; Rp, PNMT radioenzymatic; F, fluorimetric; H, HPLC.
[b] Significant H-N difference with $p < 0.05$.

in the positive studies was 361 pg/ml, compared to 259 pg/ml in the negative studies ($t = 3.80$, $p < 0.005$). Thus, when analyzed separately by assay type, the data remain consistent with the hypothesis that a variably represented hyperadrenergic subgroup exists among the population of hypertensive patients.

Separate analysis by assay type was also applied to the normotensive control values in positive and negative studies. Among fluorometric studies, the mean normotensive control NE levels were identical in the positive and negative studies (147 pg/ml). Among radioenzymatic studies, mean normotensive levels were significantly lower in the positive studies (211 vs. 258 pg/ml; $t = 2.03$, $p < 0.05$); within the COMT and PNMT groups of studies, no significant differences appeared between the normotensive control NE levels of the positive and negative studies.

Why did the fluorometric studies tend to yield positive results? We have mentioned that standard deviations of plasma NE were lower in fluorometric than radioenzymatic studies. When standard deviations were expressed as a percent of the mean, however, this difference from the radioenzymatic studies diminished dramatically. On the other hand, the ages of the hypertensive patients differentiated the fluorometric and radioenzymatic studies: Among eight fluorometric studies listing the hypertensives' mean age, all eight reported a mean age of 40 years old or less; among 30 radioenzymatic studies, only six (20%) reported a mean hypertensive age of 40 years old or less ($t = 14.1$, $p < 0.001$). Inclusion of relatively young hypertensives appears to have contributed to the high frequency of positive results associated with the fluorometric type of assay.

Patient Age

Lake *et al.* (38) have suggested that systematic differences in group mean age may have introduced artifactual effects in earlier positive studies, because plasma NE varied with age and because hypertensives were older than normotensives. Across the 37 studies that listed normotensive group ages between 20 and 50 years, mean group NE levels did correlate significantly with mean group age ($r = 0.39$, $p < 0.01$), even when only radioenzymatic studies were considered ($r = 0.36$, $p < 0.05$). Across the hypertensive groups, however, no significant correlation was obtained between mean group age and mean group NE. The hypertensives were older than the normotensives by an average of 3 years ($t = 4.26$, $p < 0.001$).

This artifactual effect of *age* would require that positive studies include older hypertensives than normotensives and negative studies include groups of similar age. Instead, the 0.08 correlation between the difference in mean group ages and the difference in mean group NE levels were negligible. Further, of 21 studies where the hypertensive and normotensive groups differed by less than 3 years in average age, seven were positive (33%), while of 17 studies with hypertensive-normotensive differences in age of 3 years or more, six were positive (35%). Hypertensive-normotensive NE differences in positive studies did not, therefore, derive from the artifactual effects of poor age matching.

In contrast, hypertensive-normotensive NE differences correlated negatively with the mean age of the hypertensive patients ($r = -0.40$, $p < 0.01$). Among 20 studies where the mean hypertensive group age was more than 41 years, only three (15%) were positive, while among 19 studies where the mean hypertensive group age was 41 or less, 11 were positive (58%; chi-square = 6.12, $p < 0.02$). When hypertensive-normotensive NE differences were expressed as a percent of the normotensive values, in order to take into account the effects of fluorometric studies, the percent of the hypertensive-normotensive difference in NE still correlated negatively with the hypertensive mean age ($r = 0.40$, $p < 0.01$). When the hypertensive group mean age was less than 35 years, the average hypertensive-normotensive difference was 44%; when the hypertensive group mean age was between 35 and 39 years, the average hypertensive-normotensive difference was 33%; between 40 and 44 years, 16%; between 45 and 49 years, 17%; and 50 years or more, 5%.

Taken together, the findings that only normotensive group mean NE levels correlated with age and that hypertensive-normotensive differences correlated negatively with the mean age of the hypertensive groups, suggest that a subgroup of young patients with essential hypertension show elevated NE levels during supine rest.

Kopin (34) has suggested that positive studies tend to report no significant correlation between normotensive NE and age, while negative studies tend to report significant correlations. If this were true, one might reasonably conclude that positive studies include normotensive groups where the older subjects have relatively low NE levels. This might occur if the investigators attempted to achieve age-matching with the usually older hypertensive groups by sampling laboratory or hospital personnel, in whom NE levels are known to be lower than in patients unaccustomed to a medical environment (29). Table 24.2 displays the reported correlation coefficients relating NE and age in 19 studies. Of these 19 studies, 18 reported the significance of age-NE relationships in the normotensive groups. Of the six positive studies, four reported significant age-NE relationships in the normotensives; of the 12 negative studies, eight reported significant age-NE relationships in the normotensives. Although Kopin's hypothesis may apply in isolated cases, overall it appears that *both* positive and negative studies have reported significant age-NE relationships among normotensives.

Severity of Hypertension

If NE levels varied with the severity of hypertension, then positive results could have derived from the inclusion of hypertensives with excessively high blood pressures (BP). Across the 53 groups in which a radioenzymatic assay was used and in which mean pressures were reported or calculable, only a small, statistically nonsignificant correlation of 0.21 was obtained between mean pressures and mean NE levels. Across the nine groups in which a fluorometric assay was used, the correlation was 0.60 ($p < 0.05$). However, the mean pressures of the hypertensives in the positive studies were not greater than in the negative studies (110 *vs.* 116 mm Hg).

Sources and Treatment of Control Groups

The available literature provides inadequate information about the controls: Were they inpatients, outpatients, laboratory workers, or newcomers to a hospital setting? Had they previously undergone phlebotomy? Had they smoked, drunk alcohol or coffee, or eaten recently? How long had they been off medications? What was their sodium intake? Were they paid? How were they recruited? In fact, none of the 46 studies answered all these questions.

The sources of control subjects ranged from patients with "various neurological disorders" (15) to patients recovering from uterine hemorrhage, respiratory infection, or malnutrition (20), to laboratory workers (16), patients detected on population screening (53), and clinic frequenters with minor complaints (42).

Table 24.2
Correlation Coefficients between Plasma Norepinephrine and Age[a]

Author	*r*(H)	*r*(N)	Study Positive?	Notes
Bertel (3)	NS	0.63[b]	No	
Brecht (5)	0.27[b]	0.1	Yes	
DeLeeuw (10)	NS	NS	No	
Eide (15)		0.89[b]	Yes	
Eng (16)		NS	Yes	
Franco-Morselli (19)	NS	0.81[b]	No	
Franco-Morselli (20)	NS	0.81[b]	No	
Henry (25)	NS		No	
Hofman (27)	0.14	0.30[b]	Yes	
Lake (39)		0.33[b]	No	
Lake (36)	0.65[b]		No	Implied NS *r*(N)
Miura (42)	NS		Yes	Implied S *r*(N)
Pedersen (46)	S	S	No	
Sever (53)	0.25	0.49[b]	No	
Sever (54)	0.01	0.47[b]	No	*r*(H) = 0.63[b] in Blacks
Vlachakis (56)		0.37	No	0.31 (labile, 0.80[b] estab'd
Vlachakis (57)	0.71[b]	0.48	No	
Vlachakis (58)	0.59[b]	0.48[b]	Yes	
Weidmann (60)	0.43[b]	0.32[b]	No	

[a] S, significant; NS, nonsignificant.
[b] Significant H-N difference with $p < 0.05$.

Laboratory workers are known to have lower NE levels than outpatient normotensives (29). Inpatient normotensives have lower levels than outpatients (35). Smoking, salt deprivation, and caffeine increase NE levels (41, 51), as does fasting 15 hours (47). Inadequate consideration of the sources and treatment of the normotensive controls undoubtedly helped to produce the wide range of standard deviations of NE levels across studies, from 16 to 146% of the mean! Standard deviations of the normotensive NE levels, in turn, contributed to the likelihood of positive or negative results: When the normotensive standard deviations were less than or equal to 30% of the mean, 65% (four to seven) of the studies were positive; when they were between 30 and 39% of the mean, 50% (three of six) were positive; between 40 and 49% of the mean, 33% (four of 12) were positive; between 50 and 59% of the mean, 33% (two of six) were positive; and greater than or equal to 60% of the mean, 13% (one of eight) were positive.

NOREPINEPHRINE WHILE STANDING

Table 24.3 summarizes results from 29 comparative studies of plasma NE during *standing* in patients with essential hypertension and in

Table 24.3
Plasma Norepinephrine in Patients with Essential Hypertension and in Normotensives during Standing[a]

Author	No. H/N	Age H/N	Time (min)	Assay	Basal MAP	Stand MAP	Basal NE	Stand NE
							H/N	H/N
Beretta-Piccoli (2)	45/26	43/39	60	Rc	116/89	108/84	262/221	572/500
Bertel (3)	24/15			Rc			260/250	505/406[b]
Brecht (4)	59/15		7	F			257/135	498/359[b]
Brecht (5)	87/87		7	F			201/128	405/278[b]
DeChamplain (9)	14/10		20	Rc (TC)			345/182	666/408[b]
DeQuattro (12)	27/25		5	Rc (TC)	117/76	119/77	283/218	418/386
DeQuattro (11)	11/12		60	F			157/148	382/419
Eide (15)	7/7	40/36	60	F	109/89	110/92	240–167	494/529
Eide (14)	28/12	38/?	60	F	109/?	114/?	193/148	359/419
Eng (16)	20/17	47/34	5	Rc	114/90	123/102	390/250	654/405[b]
Franco-Morselli (19)	27/12		10	Rc	103/87	124/92	252/248	472/529
Franco-Morselli (20)	19/11	43/35	10	Rc	116–90		269/248	491/559
Henry (25)	73/100		120	Rp	119/88		151/144	263/246
Hjemdahl (26)	7/7	35/35	10	H	101/87	115/95	436/353	776/533
Jones (29)	31/28	47/37	3	Rp	124/88	134/98	409/354	639/596
Kafka (30)	15/18	50/50	5	Rp	112/91		265/289	497/524
Lake (35)	151/117	43/45	5	Rp			297/294	509/590
Lake (39)	24/44	43/35	10/5	Rp			306/287	588–538
Lake (38)	67/84	44/33	5	Rp			339/304	595/543
Lake (37)	56/29	46/40	5	Rp	112/89	113/87	249/253	588/673
Lake (36)	11/9	48/44	5	Rp	115/87		222/172	388/379
Miura (42)	60/18		60	F	120/87		189/130	364/287
Parfrey (45)	16/8	36/31	5	Rp	113/80	119/92	296/191	573/465
Sever (54)	100/48		5	Rp	121/95	124/98	352/372	628/643
Sever (53)	56/59		5	Rp	124/92	126–94	411/403	737/639
Taylor (55)	51/26	46/40	5	Rc	119/89	115/87	240/260	435/456
Vlachakis (56)	38/14	48/49	10	Rc			256/205	540/429
Vlachakis (58)	60/23	48/46	10	Rc	116/88	127/95	282/206	564/458[b]
Weidmann (60)	79/90		60	Rc	123/89		202/169	302/344
Mean					115/88	119/92	276/232	514/467[b]

[a] Abbreviations the same as for previous tables; IC, total catecholamines.
[b] Significant hypertensive-normotensive difference during standing, with $p < 0.05$.

normotensive controls. Basal mean NE levels were higher in the hypertensive than the normotensive groups in 25 of the 29 studies, with a mean basal hypertensive-normotensive difference of 44 pg/ml ($t = 5.10$, $p < 0.001$). During standing, the mean hypertensive-normotensive difference remained statistically significant (47 pg/ml; $t = 2.60$, $p < 0.05$), but was much less consistent than with the supine condition.

Standing did not significantly increase the mean hypertensive-normotensive difference in NE, and the increment in NE with standing for the hypertensives resembled that for the normotensives (241 *vs.* 235 pg/ml). Basal hypertensive-normotensive differences correlated strongly with hypertensive-normotensive differences during standing ($r = 0.77$, $p < 0.001$).

When expressed in percents, plasma NE during standing actually increased more in normotensives than in hypertensives (110 *vs.* 88%; $t = 3.49$, $p < 0.005$).

Six studies reported significantly higher hypertensive levels during standing, as well as an increase in the hypertensive-normotensive difference. The reason for these positive results was not excessive mean NE levels in the hypertensives during standing, since mean standing hypertensive levels were similar in the positive and negative studies (549 *vs.* 505 pg/ml). Rather, normotensive control levels during standing were significantly lower in the positive studies (386 *vs.* 488 pg/ml; $t = 2.13$, $p < 0.05$). One of the positive studies (16) included a normotensive control group comprised mainly of members of the laboratory staff, and this type of group has a lower resting level of NE than other normotensives (29). The other five positive studies did not discuss the constitution of the control groups beyond mentioning that the normotensives were healthy.

Across the hypertensive groups, the absolute change in NE from baseline with standing correlated ($r = 0.61$, $p < 0.01$) with the resting NE level; in contrast, across the normotensive groups, the change in NE from baseline was unrelated to the rest level ($r = 0.17$). This last finding is consistent with the existence within the hypertensive population of a subgroup of patients with elevated NE levels at rest and excessive sympathetic responsiveness to standing. However, the lack of a significant increment in overall hypertensive-normotensive differences with standing suggests that, when considered as a single population, hypertensives do not show exaggerated plasma NE responses to this stimulus.

NOREPINEPHRINE DURING EXERCISE

Table 24.4 summarizes the results of 10 studies comparing plasma NE responses to *exercise* in hypertensive patients and normotensive controls. In order to analyze the maximum number of studies, the table includes studies in which total CA were reported.

Overall, mean NE levels during or after exercise were higher in the hypertensive than in the normotensive groups (1169 *vs.* 758 pg/ml; $t = 2.49$, $p < 0.05$), and the average hypertensive-normotensive difference increased with exercise from 127 to 412 pg/ml ($t = 2.69$, $p < 0.05$). With *isotonic exercise*, the mean hypertensive NE level was 1584 pg/ml, compared with 947 in the normotensives, and the average hypertensive-normotensive difference increased from 172 to 637 pg/ml ($t < 3.73$, $p < 0.05$). In contrast, with *isometric* exercise, the mean hypertensive NE level was 548 pg/ml, not notably different from the normotensive level (474 pg/ml). The average hypertensive-normotensive difference did not increase significantly (73 *vs.* 59 pg/ml). Although too few studies were available to apply "t"-statistics, it appears that hypertensive-normotensive differences during isotonic exercise were most pronounced when the same workload was applied to each subject, as opposed to workloads calculated individually to result in a percentage of maximum work capacity or heart rate.

As with standing, hypertensive-normotensive differences during exercise correlated significantly with hypertensive-normotensive differences at rest ($r = 0.84$, $p < 0.01$). Indeed, for isotonic exercise the absolute change in NE from baseline correlated significantly with the baseline level across the hypertensive groups ($r = 0.66$, $p < 0.01$) but not across the normotensive groups ($r = 0.18$).

Since resting NE levels as well as increments in NE with stress increase with age (44), poor age-matching could have produced greater hypertensive-normotensive differences during exercise than at rest. However, in four of the studies, the groups were age-matched, and positive results were still obtained.

OTHER STRESSES

Only one study has compared plasma CA responses to the cold pressor test in hypertensives

Table 24.4
Plasma Norepinephrine in Patients with Essential Hypertension and in Normotensives during Exercise[a]

Author	No. H/N	Age H/N	Type/Intensity/Time (min)	Basal MAP	Exercise MAP	Basal NE	Exercise NE
Bertel (3)	24/20		Bicycle 75% of PWC_{max} 5			H/N 263/250	H/N 1393/1190
Henquet (24)	25/25		Bicycle 75% of PWC_{max} 5	108/89	123/103	409/370	1230/1210
Lake (36)	11/9	48–44	Handgrip 30% of max 5	115/87		222/172	487/626
Philipp (48)	29/29	38/33	Bicycle 200 w 2			216/173	1213/563[b]
Planz (49)	8/8	31/30	Bicycle 150 w 5	119/97	141/115	552/240	1792/776[b]
Robertson (52)	9/10	25/27	Treadmill 4 mph 3	101/86		510/400	1610/890[b]
Vlachakis (57)	22/13	50/44	Handgrip ⅔ max 3	113/95	145/135	277/234	675/570
Vlachakis (56)	38/14	48/49	Handgrip ⅔ max 5			265/205	501/350[b]
Vlachakis (58)	37/23	47/49	Handgrip ⅔ max 5	116/88	144/117	282–206	527/351[b]
Watson (59)	6/5	46/35	Bicycle 85% HR_{max} 8	130/102		938/429	2265/1050[b]
Mean				115/92	138/118	393/268	1169/758[b]

[a] PWC_{max}, predicted maximum work capacity; w, watts; HR_{max}, maximum predicted heart rate. Basal norepinephrine values not necessarily supine.
[b] Significant hypertensive-normotensive difference during exercise, with $p < 0.05$.

and normotensives (7), and this study used measurement of total CA rather than NE alone. The level of plasma CA during the test was significantly higher in the hypertensives than in the normotensives (619 *vs.* 467 pg/ml), with an increase in the mean hypertensive-normotensive difference from 61 to 152 pg/ml.

Similarly, only one study has compared hypertensive and normotensive responses to a psychological stress, in this case, a color-word conflict test (26). The stress increased NE only slightly in both hypertensives (436 to 479 pg/ml) and normotensives (353 to 369 pg/ml), with an increase in the hypertensive-normotensive differences from 83 to 110 pg/ml.

No studies have compared plasma NE responses of hypertensives and normotensives to hypoxia, hypoglycemia or pain.

EPINEPHRINE

Of the 46 studies in this review, 19 included data about plasma epinephrine (E) (Table 24.5). Eight of the 19 studies showed a significant hypertensive-normotensive difference in E. Comparison with Table 24.1 indicates that there was little, if any, relationship between the likelihood of positive findings regarding E and positive findings regarding NE. Perhaps subgroups of hypertensives show *either* elevated E or NE. If so, this would help to explain why studies reporting "total CA" have usually been positive.

Table 24.5
Plasma Epinephrine in Patients with Essential Hypertension and in Normotensive Control during Supine Rest

Author	E H/N	SD H/N
Beretta-Piccoli (1)	45/31[a]	28/21
Beretta-Piccoli (2)	44/35	26/17
Bertel (3)	54/35[a]	39/18
Corea (6)	32/15[a]	16/3
Cousineau (8)	160/89[a]	128/74
Eide (15)	38/46	32/45
Eng (16)	50/30[a]	26/18
Esler (18)	66/69	38/43
Franco-Morselli (19)	87/45[a]	87/58
Franco-Morselli (20)	83/41[a]	34/18
Henquet (24)	70/90	101/146
Hjemdahl (26)	123/68	
Ibsen (28)	30/40	
Pedersen (46)	34/47	
Robertson (52)	51/36[a]	24/20
Vlachakis (56)	47/40	20/19
Vlachakis (57)	45/35	32/10
Vlachakis (58)	49/40	29/23
Weidmann (60)	58/62	69/62

[a]Significant H-N difference with $p < 0.05$.

DISCUSSION

Most studies measuring NE in patients with essential hypertension and in normotensive controls have reported higher mean levels in the hypertensives. To the extent that plasma NE reflects sympathetic activity, this finding is consistent with the hypothesis of increased sympathetic nervous system (SNS) activity in a proportion of patients with essential hypertension. Only about two-fifths of the studies, however, reported *statistically* significant hypertensive-normotensive differences, leading to an effort to identify factors differentiating positive from negative studies.

When the results were analyzed within assay techniques, hypertensive NE levels were higher in the positive studies. This argues for the existence of a variably-represented hyperadrenergic subgroup among the hypertensive population. However, normotensive control NE levels were lower overall and less variable in the positive studies. Unfortunately, the published literature has virtually ignored the sources, characteristics, and treatment of the normotensive controls.

The likelihood of positive results also depended on the type of assay used to measure plasma NE. The radioenzymatic assay techniques were associated both with a lower frequency of positive results and greater intrastudy standard deviations than the fluorometric. "Normal" NE levels varied by up to 100% across assay types. The available literature provides no clues about why the results should have been so assay-dependent.

The age factor has obscured the role of plasma NE, since NE tends to increase with age and since hypertensives have been older than normotensives. Hypertensive-normotensive differences in NE did not derive from the artifactual effects of poor age-matching. On the other hand, the largest hypertensive-normotensive differences occurred when the patients were relatively young. In addition, the age-NE relationship was statistically significant across the normotensive but not the hypertensive groups. These two findings are consistent with the hypothesis that elevated resting plasma NE levels characterize a proportion of *young patients* with essential hypertension.

It is possible that such a hyperadrenergic group merely reacts excessively to environmental situations eliciting increased SNS activity and that the data reflect a nonresting state in these individuals without any pathophysiologic significance of the increased NE. To examine this possibility further, the available literature was reviewed to test the hypothesis that during stress, hypertensive-normotensive differences in NE become more apparent. Of the several stresses known to stimulate the SNS, only orthostasis and exercise have been extensively studied. Increments in NE in response to standing were similar in hypertensives and normotensives. In contrast, hypertensives showed larger increments in NE in response to exercise—in particular, isotonic exercise—resulting in large hypertensive-normotensive differences averaging 412 pg/ml. For both standing and isotonic exercise, the absolute change in NE from baseline correlated positively with the baseline level for hypertensive groups but not for normotensive groups.

Why should there have been such a difference between the results obtained during standing and those obtained in response to exercise? The simplest explanation is that orthostasis is not as potent a stimulus for NE release as is exercise. A prediction from this is that the extent of hypertensive-normotensive differences in NE should vary with the intensity of the stress. In the two studies with graded levels of exercise, the hypertensive-normotensive differences in

NE increased progressively from the resting to the orthostatic positions and then across the several intensities of exercise.

An alternate explanation for the discrepancy in results between standing and exercise is that only relatively younger subjects would be exercised severely, and it has been previously shown that hypertensive-normotensive differences in NE vary inversely with age. Too few studies are available to analyze the relevant results statistically. In the one study in which NE responses to isotonic exercise were analyzed separately by age group (3), hypertensive-normotensive differences in response to isotonic exercise *increased* with the age of the group studied.

It is also possible that exercise differentiates hypertensive and normotensive groups for reasons specific to exercise. For instance, if the hypertensives were simply less physically trained, one might predict that they would show greater NE responses to exercise—especially submaximal isotonic exercise—than to other stresses (61). One finding consistent with this interpretation is that during isotonic exercise, hypertensive-normotensive differences in NE were most pronounced when the same workload was applied to each subject, as opposed to workloads calculated individually to result in a percentage of maximum work capacity or heart rate.

The available exercise data have been difficult to analyze because of (a) variability from study to study in the type, intensity, and duration of exercise; (b) testing of only poorly described—and possibly biased—subgroups from among the hypertensives and normotensives; (c) poor age-matching in a situation where both resting and stress-related increments in NE are known to vary with age; (d) small group sizes; and (e) reporting total CA levels where the relative contributions of NE and E to the total may change during the testing.

A rather surprising finding, derived from studies of orthostatic stress and of isotonic exercise, was that absolute changes in NE from baseline during stress correlated with baseline NE levels across the hypertensive but not the normotensive groups. This implies the existence of a subgroup of hypertensive patients with elevated basal NE levels and excessive SNS responsiveness to stress, while among the normotensive population, variations in basal NE represent sampling error of no physiological significance. Unfortunately, none of the analyzed studies examined the NE responses of the normotensives with relatively high or low basal NE.

SUMMARY

In summary, during supine rest, a proportion of patients with essential hypertension—particularly among the younger age groups—show excessive overall SNS activity, as indicated by plasma NE levels. Hypertensives show excessive NE responses to isotonic exercise at the same workloads as applied to normotensives but do not show excessive responsiveness to standing.

We suspect that the combination of the typically small hypertensive-normotensive differences in NE, the notorious difficulty in measuring plasma CA reliably, and inadequate attention to the control groups and to the wide variety of stimuli known to increase SNS activity have resulted in a plethora of conflicting data and in continuing controversy. Since positive studies have tended to include younger hypertensives, studies targeted to evaluate neural circulatory control in these patients may yield more definitive evidence than the hypertensive population as a whole.

References

1. Beretta-Piccoli C, Weidmann P, Meier A, Grimm M, Keusch G, Gluck Z: Effects of short-term norepinephrine infusion on plasma catecholamines, renin, and aldosterone in normal and hypertensive man. *Hypertension* 2:623–629, 1980.
2. Beretta-Piccoli C, Weidmann P, Keusch G, Grimm M, Meier A, Gluck Z, Ziegler WH: Renin-hyporesponsiveness in essential hypertension dissociation between plasma renin and catecholamines or aldosterone following furosemide. *Klin Wochenschr* 58:457–465, 1980.
3. Bertel O, Buhler FR, Kiowski W, Lutold BE: Decreased beta-adrenoreceptor responsiveness as related to age, blood pressure, and plasma catecholamines in patients with essential hypertension. *Hypertension* 2:130–138, 1980.
4. Brecht HM, Banthien F, Ernst W, Schoeppe W: Increased plasma noradrenaline concentrations in essential hypertension and their decrease after long-term treatment with a beta-receptor-blocking agent (prindolol). *Clin Sci Mol Med* 51:485s–488s, 1976.
5. Brecht HM, Schoeppe W: Relation of plasma noradrenaline to blood pressure, age, sex, and sodium balance in patients with stable essential hypertension and in normotensive subjects. *Clin Sci Mol Med* 55:81s–82s, 1978.
6. Corea L, Miele N, Bentivoglio M, Boschetti E, Agabitirosei E, Muiesan G: Acute and chronic effects of nifedipine on plasma renin activity and plasma adrenaline and noradrenaline in controls and hypertensive patients. *Clin Sci Mol Med* 57:115s–117s, 1979.
7. Chobanian AV, Gavras H, Gavras I, Breshnahan M, Sullivan P, Melby JC: Studies on the activity of the sympathetic nervous system in essential hypertension. *J Hum Stress* 4:22–28, 1978.

8. Cousineau D, DeChamplain J, Lapointe L: Circulating catecholamines and systolic time intervals in labile and sustained hypertension. *Clin Sci Mol Med* 55:65s–68s, 1978.
9. DeChamplain J, Farley L, Cousineau D, Van Ameringen M-R: Circulating catecholamine levels in human and experimental hypertension. *Circ Res* 38:109–114, 1976.
10. DeLeeuw PW, Wester A, Punt R, Falke HE, Birkenhager WH: Noradrenaline levels in essential hypertensives and normotensive controls. *Neth J Med* 22:145–147, 1979.
11. DeQuattro V, Campese V, Lurvey A, Yen G, Kypridakis G: Low response of serum dopamine-beta-hydroxylase to stimuli in primary hypertension. *Biochem Med* 15:1–9, 1976.
12. DeQuattro V, Chan S: Raised plasma-catecholamines in some patients with primary hypertension. *Lancet* 1:806–809, 1972.
13. Edwards AL: *Statistical Methods.* New York, Holt, Rinehart, and Winston, 1967.
14. Eide I, Campese V, Stein D, Eide K, DeQuattro V: Clinical assessment of sympathetic tone: Orthostatic blood pressure responses in borderline primary hypertension. *Clin Exp Hypertens* 1:51–65, 1978.
15. Eide I, Kolloch R, DeQuattro V, Miano L, Dugger R, Van der Meulen J: Raised cerebrospinal fluid norepinephrine in some patients with primary hypertension. *Hypertension* 1:255–269, 1979.
16. Eng FWHT, Huber-Smith M, McCann DS: The role of sympathetic activity in normal renin essential hypertension. *Hypertension* 2:14–19, 1980.
17. Esler M, Jackman G, Leonard P, Bobik A, Skews H, Jennings G, Kelleher D, Korner P: Determination of noradrenaline uptake, spillover to plasma and plasma concentration in patients with essential hypertension. *Clin Sci* 59:311s–313s, 1980.
18. Esler M, Zweifler A, Randall O, Julius S, DeQuattro V: Agreement among three different indices of sympathetic nervous system activity in essential hypertension. *Mayo Clin Proc* 52:379–382, 1977.
19. Franco-Morselli R, Baudouin-Legros M, Meyer P: Plasma adrenaline and noradrenaline in essential hypertension and after long-term treatment with beta-adrenergic-blocking agents. *Clin Sci Mol Med* 55:97s–99s, 1978.
20. Franco-Morselli R, Elghozi JL, Joly E, DiGiuliu S, Meyer P: Increased plasma adrenaline concentrations in benign essential hypertension. *Br Med J* 2:1251–1254, 1977.
21. Geffen LB, Rush RA, Louis WJ, Doyle AE: Plasma dopamine-beta-hydroxylase and noradrenaline amounts in essential hypertension. *Clin Sci* 44:617–620, 1973.
22. Goldstein DS: Plasma norepinephrine in essential hypertension: A study of the studies. *Hypertension* 3:48–52, 1981.
23. Goldstein DS, Feuerstein GZ, Izzo JL Jr, Kopin IJ, Keiser HR: Validity and reliability of liquid chromotagraphy with electrochemical detection for measuring plasma levels of norepinephrine and epinephrine in man. *Life Sci* 28:467–475, 1981.
24. Henquet JW, Kho T, Schols M, Thijssen H, Rahn KH: The sympathetic nervous system and the renin-angiotensin system in borderline hypertension. *Clin Sci* 60:25–31, 1981.
25. Henry DP, Luft FC, Weinberger MH, Fineberg NS, Grim CE: Norepinephrine in urine and plasma following provocative maneuvers in normal and hypertensive subjects. *Hypertension* 2:20–28, 1980.
26. Hjemdahl P, Eliasson K: Sympatho-adrenal and cardiovascular responses to mental stress and orthostatic provocation in latent hypertension. *Clin Sci* 57:189s–191s, 1979.
27. Hofman A, Boomsa F, Schalekamp MADH, Valkenburg HA: Raised blood pressure and plasma noradrenaline concentration in teenagers and young adults selected from an open population. *Br Med J* 1:1536–1539, 1979.
28. Ibsen H, Christensen NJ, Hollnagel H, Leth A, Kappelgaard AM, Giese J: Plasma noradrenaline concentration in hypertensive and normotensive forty-year-old individuals: Relationship to plasma renin concentration. *Scand J Clin Lab Invest* 40:333–339, 1980.
29. Jones DH, Hamilton CA, Reid JL: Choice of control groups in the appraisal of sympathetic nervous system activity in essential hypertension. *Clin Sci* 57:339–344, 1979.
30. Kafka MS, Lake CR, Gullner H-G, Tallman JF, Bartter FC, Fujita T: Adrenergic receptor function is different in male and female patients with essential hypertension. *Clin Exp Hypertens* 1:613–627, 1979.
31. Kiowski W, Van Brummelen P, Buhler FR: Plasma noradrenaline correlates with alpha-adrenoceptor-mediated vasoconstriction and blood pressure in patients with essential hypertension. *Clin Sci* 57:177s–180s, 1979.
32. Kobayashi K, Kolloch R, DeQuattro V, Miano L: Increased plasma and urinary normetanephrine in young patients with primary hypertension. *Clin Sci* 57:173s–176s, 1979.
33. Kolloch R, Kobayaski K, DeQuattro V: Dopaminergic control of sympathetic tone and blood pressure: Evidence in primary hypertension. *Hypertension* 2:390–394, 1980.
34. Kopin IJ: Plasma catecholamines in human and experimental hypertension. In Meyer P, Schmitt H: *Nervous System and Hypertension.* New York, Wiley, 1979, pp 267–276.
35. Lake CR: Relationship of sympathetic nervous system tone and blood pressure. *Nephron* 23:84–90, 1979.
36. Lake CR, Polinsky RJ, Gullner H-G, Ebert MH, Ziegler MG, Bartter FC, Kopin IJ: Dissociation between central and peripheral noradrenergic activity in essential hypertension. *Clin Sci* 59:229s–233s, 1980.
37. Lake CR, Ziegler MG: Effect of acute volume alterations on norepinephrine and dopamine-beta-hydroxylase in normotensive and hypertensive subjects. *Circulation* 57:774–777, 1978.
38. Lake CR, Ziegler MG, Coleman MD, Kopin IJ: Age-adjusted plasma norepinephrine levels are similar in normotensive and hypertensive subjects. *N Engl J Med* 296:208–209, 1977.
39. Lake CR, Ziegler MG, Kopin IJ: Use of plasma norepinephrine for evaluation of sympathetic neuronal function in man. *Life Sci* 18:1315–1326,

1976.

40. Louis WJ, Doyle AE, Anavekar S: Plasma norepinephrine levels in essential hypertension. *N Engl J Med* 288:599–601, 1973.
41. Luft FC, Rankin LI, Henry DP, Bloch P, Grim CE, Weyman AE, Murray RH, Weinberger MH: Plasma and urinary norepinephrine values at extremes of sodium intake in normal man. *Hypertension* 1:261–266, 1979.
42. Miura Y, Kobayashi K, Sakuma H, Tomioka H, Adachi M, Yoshinaga K: Plasma norepinephrine levels and hemodynamics in young patients with essential hypertension. *Jpn Circ J* 42:609–612, 1978.
43. Miura Y, Kobayashi K, Sakuma H, Tomioka H, Adachi M, Yoshinaga K: Plasma noradrenaline concentrations and haemodynamics in the early stage of essential hypertension. *Clin Sci Mol Med* 55:69s–71s, 1978.
44. Palmer GJ, Ziegler MG, Lake CR: Response of norepinephrine and blood pressure to stress increases with age. *J Gerontol* 33:482–487, 1978.
45. Parfrey PS, Wright P, Goodwin FJ, Vandenburg MJ, Holly JMP, Evans SJW, Ledingham JM: Blood pressure and hormonal changes following alteration in dietary sodium and potassium in mild essential hypertension. *Lancet* 1:59–63, 1981.
46. Pedersen EB, Christensen NJ: Catecholamines in plasma and urine in patients with essential hypertension determined by double-isotope derivative techniques. *Acta Med Scand* 198:373–377, 1975.
47. Pequignot JM, Peyrin L, Peres G: Catecholamine-fuel interrelationships during exercise in fasting men. *J Appl Physiol* 48:109–113, 1980.
48. Philipp T, Distler A, Cordes A: Sympathetic nervous system and blood-pressure control in essential hypertension. *Lancet* 2:959–963, 1978.
49. Planz G, Gierlichs HW, Hawlina A, Planz R, Stephany W, Rahn KH: A comparison of catecholamine concentrations and dopamine-beta-hydroxylase activities in plasma from normotensive subjects and from patients with essential hypertension at rest and during exercise. *Klin Wochenschr* 54:561–565, 1976.
50. Renzini V, Brunori CA, Valori C: A sensitive and specific fluorimetric method for the determination of noradrenalin and adrenalin in human plasma. *Clin Chim Acta* 30:587–594, 1970.
51. Robertson D, Johnson GA, Robertson RM, Nies AS, Shand DG, Oates JA: Comparative assessment of stimuli that release neuronal and adrenomedullary catecholamines in man. *Circulation* 59:637–643, 1979.
52. Robertson D, Shand DG, Hollifield JW, Nies AS, Frolich JC, Oates JA: Alterations in the response of the sympathetic nervous system and renin in borderline hypertension. *Hypertension* 1:118–124, 1979.
53. Sever PS, Osikowska B, Birch M, Tunbridge RDG: Plasma-noradrenaline in essential hypertension. *Lancet* 1:1078–1081, 1977.
54. Sever PS, Peart HS, Davies IB, Tunbridge RDG, Gordon D: Ethnic differences in blood pressure with observations on noradrenaline and renin. *Clin Exp Hypertens* 1:745–760, 1979.
55. Taylor AA, Pool JL, Lake CR, Ziegler MG, Rosen RA, Rollins DE, Mitchell JR: Plasma norepinephrine concentrations: No differences among normal volunteers and low, high, or normal renin hypertensive patients. *Life Sci* 22:1499–1510, 1978.
56. Vlachakis ND: Blood pressure and catecholamine responses to sympathetic stimulation in normotensive and hypertensive subjects. *J Clin Pharmacol* 19:458–469, 1979.
57. Vlachakis ND, Aledort L: Platelet aggregation in relationship to plasma catecholamines in patients with hypertension. *Atherosclerosis* 32:451–460, 1979.
58. Vlachakis ND, Mendlowitz M: Plasma catecholamines in primary hypertension. *Biochem Med* 23:35–46, 1980.
59. Watson RDS, Hamilton CA, Jones DH, Reid JL, Stallard TJ, Littler WA: Sequential changes in plasma noradrenaline during bicycle exercise. *Clin Sci* 58:37–43, 1980.
60. Weidmann P, Beretta-Piccoli C, Ziegler WH, Keusch G, Gluck Z, Reubi FC: Age versus urinary sodium for judging renin, aldosterone, and catecholamine levels: Studies in normal subjects and patients with essential hypertension. *Kidney Int* 14:619–627, 1978.
61. Winder WW, Hagberg J, Hickson RC, Ehsani AA, McLane JA: Time course of sympathoadrenal adaptation to endurance exercise training in man. *J Appl Physiol* 45:370–374, 1978.

CHAPTER TWENTY-FIVE

Hemodynamic and Pharmacologic Correlates of Plasma Norepinephrine in Hypertension

STEVO JULIUS, M.D., Sc.D., HANS IBSEN, M.D., and HARRY T. COLFER, M.D.

INTRODUCTION

Plasma norepinephrine (NE) levels do not always adequately reflect prevailing sympathetic tone (see Chapter 14). Plasma NE reflects secretion, clearance, and neuronal reuptake in a complex neuronal system (16), so interpretation is particularly difficult when one attempts to relate cardiovascular sympathetic tone to NE levels. What proportion of NE reflects gastrointestinal, bronchial, metabolic, and endocrine as opposed to cardiovascular tone? The cardiac output (14), vascular resistance (21), and venous tone (51) are altered to various degrees in different forms of hypertension. If all these interpretative problems were somehow solved, the final effect of the sympathetic tone on the maintenance of elevated blood pressure (BP) in pathologic states could not be understood without additional information about receptor sensitivity and organ responsiveness.

A good example of such complexities is the increase of plasma NE with age (34, 43). In spite of this increase, the overall sympathetic drive to the cardiovascular system does not appear to increase; cardiac output is lower, heart rate is not increased, and there is no evidence of a sympathetic increase in resting oxygen consumption (28). During exercise, old people do not have a hypersympathetic state; cardiac output, heart rate, and vascular resistance responses at submaximal exercise are similar in old and young subjects (28). In elderly subjects, the increased catecholamine (CA) levels either do not represent an increased sympathetic tone, or if the tone is increased, the overall sympathetic responsiveness must be decreased. In the case of aging, then, a simple measurement of plasma NE is not sufficient. However, additional information can be gained by pharmacologic means to assess both the sympathetic cardiovascular drive and the responsiveness of sympathetic receptors.

The relationship of CA and aging is only one instance illustrating the thesis of this chapter. To understand the role of the sympathetic nervous system (SNS) in hypertension, it is useful to combine biochemical measurements with hemodynamic and pharmacologic observations. For further discussion of the relationship of NE to age see Chapters 1, 8, 14, and 28.

RELATIONSHIP BETWEEN NE, BP, AND ITS HEMODYNAMIC COMPONENTS

BP is a product of cardiac output and vascular resistance. In early phases of hypertension a substantial proportion of patients with borderline and mild hypertension have increased cardiac output, whereas in established hypertension the vascular resistance is elevated. To understand the role of NE as a measure of sympathetic cardiovascular tone in hypertension and normotension, one must analyze how NE relates to BP and to its components.

Plasma NE may more readily reflect the cardiac pool of sympathetic neurons (53) than the NE release from other vascular areas. Mancia *et al.* (38) elicited increased vascular resistance in humans without significantly elevating plasma NE values. However, recent data indicate that NE released from kidneys and legs contributes substantially to plasma NE in systemic venous blood (5). We will review the existing evidence about the relationship between plasma NE, BP, and systemic hemodynamics. The following questions will be addressed: (a) Is there a relationship between plasma NE and BP, and is the relationship different in normotension and hypertension? (b) How do plasma NE values correlate with cardiac output and vascular resistance in hypertension? Such an analysis should provide a better understanding of the role of the SNS in the pathophysiology of hypertension.

Louis *et al.* (36) reported the strongest correlation found between plasma NE and BP in hypertensive patients. The correlation coefficient value between diastolic BP and plasma NE was 0.73. In other studies (1, 3, 15, 50), the correlation varied from 0.69 to 0.29, whereas others (8, 18, 24, 44) did not find a significant correlation. The wide range of correlation coefficients between plasma NE and BP in hypertension is not surprising, since patients with hypertension are not a homogeneous population. Different severities of hypertension, differences in hormonal profiles, and responsiveness to antihypertensive agents in patients with similar severity of hypertension suggest different pathophysiologic mechanisms and, conceivably, even different etiologies among patients with essential hypertension. Some authors attempted to account for these differences when analyzing the relationship between plasma NE and BP. Vlachakis (49) finds a significant correlation only in patients with sustained and severe hypertension. This result is surprising, as evidence for an increased sympathetic drive in many patients with mild and labile hypertension is good. A review by Goldstein (22) recently suggested that young, mild hypertensives are the only subgroup of patients in whom there is real evidence of elevated plasma NE (also see Chapter 24). However, Esler *et al.* (17), while generally finding increased plasma NE in such patients, found a correlation between plasma NE and heart rate but not between plasma NE and BP. These patients are characterized by an unusual personality pattern (26), which may render them overresponsive to mental stress. For example, increased BP and NE responsiveness to emotional stimuli (20, 42) and a faster heart rate response invariably accompany excessive BP responses to emotion in such patients (23, 35). It is conceivable that plasma NE elevation in these patients predominantly reflects cardiac sympathetic drive. There is no correlation of plasma NE with BP in patients with mild hypertension (18, 49) in whom NE may be elevated (17) or normal (49). On the other hand, a correlation was found in patients whose plasma NE values were not elevated (49). In a similar vein, Chobanian *et al.* (6) and DeQuattro *et al.* (9) did not find a correlation between BP and NE in patients with elevated plasma NE values, but did find one among patients whose NE levels were normal. Finally, Eng *et al.* (15) found an equal degree of correlation between plasma NE and BP in their "high" and "low" plasma NE subset of hypertensive patients.

Esler *et al.* (17) report that patients who have elevated plasma renin values also have elevated plasma NE values, and their BPs fall to normal values after autonomic blockade. Their study promises to provide a new tool for stratifying the heterogenous pool of patients with essential hypertension into more homogeneous subsets of patients with "neurogenic" and "nonneurogenic" hypertension. As indicated earlier, we could not find a correlation between BP and plasma NE in these patients (17). Although not all investigators have been able to find a relationship between NE and renin in hypertensives (48), Chobanian *et al.* (6) and DeQuattro *et al.* (9) confirmed the finding that patients with high renin have a higher plasma NE, but again no correlation was found between plasma NE and BP. Such classification may delineate those who apparently have neurogenic hypertension.

Other factors, such as age, can influence the plasma NE levels (6, 33, 43). When these are taken into account, our understanding of the relationship of plasma NE and BP in hypertension is not improved. Chobanian *et al.* (6) find a correlation of NE and BP only in male patients, but the pathophysiologic implications of these findings are not clear. Similarly, Vlachakis *et al.* (50) attempted to unravel the relationship between plasma NE, age, and BP. They found that plasma NE correlated with both age and BP in hypertension. In normotension, the NE correlated with age, but not with BP. Since the slope of NE increase with age is steeper in hypertensive than in normotensive subjects, it is inferred that another factor, possibly BP (also correlated with age), accounts for the difference in the slopes. Others find the age-NE correlation only in normotensive subjects, not in hypertensive patients (see Chapter 24).

The failure to define the subset of hypertensive patients who show a strong correlation between the NE value and BP does not in itself preclude the possibility of a SNS role in the development of hypertension. In experimental renovascular hypertension, plasma renin is initially elevated and directly related to the degree of BP elevation. In later phases, plasma renin decreases, and a dependence of BP on renin levels is more difficult to prove (4). With plasma NE, a similar course may be even more obscured as the hypertensive's hemodynamics change from a high heart rate-high output "normal" resistance state to normal heart rate-normal output high resistance state. It is of interest to analyze the relationship of plasma NE to hemodynamic components of BP, but very little data on this important topic are available. We (18) found no correlation between vascular resistance and plasma NE levels in a group of patients with hypertension. These patients did not have an elevated cardiac output nor was there a correlation between cardiac output and plasma NE. A correlation with heart rate was found ($r = 0.44$), and the preejection period correlates strongly with plasma NE values ($r = -0.75$). It is thus conceivable that plasma NE levels in these patients reflect sympathetic cardiac drive (25). Two groups found a correlation between total peripheral vascular resistance and plasma NE. Patients in these two studies (1, 41) had a more established hypertension than patients in our study (18). The observations are consistent with the concept that, in the evolution of hypertension, sympathetic drive is first more evident at the heart and later at the blood vessels and that the relationship of plasma NE to hemodynamics reflects such a shift. Such changes of the sympathetic drive during hypertension appear to result from downregulation of cardiac β-responsiveness to the increased sympathetic tone (25, 30) and the simultaneous occurrence of increased responsiveness of arterioles due to structural changes (27). Even though a correlation to vascular resistance was found in two studies, it is weak and explains less than 25% of the variance. Thus, plasma NE levels do not support a strong role of the SNS in the maintenance of the major hemodynamic abnormality in hypertension—the elevation of vascular resistance. For this reason, some authors are combining pharmacologic measures of sympathetic tone with NE measurement. Results of such studies are evaluated in the next section.

There does not appear to be a correlation between plasma NE and BP in normotensive subjects (24, 45, 49). DeQuattro and Miura (11) studied 64 subjects awaiting vasectomy, many of whom were normotensive, and found a correlation of $r = 0.39$ between total plasma CA and systolic BP. This may, however, reflect the emotional variability of both systolic BP and plasma NE in subjects expecting minor surgery since emotional stimuli elevate both BP and NE levels (20, 42). Jones *et al.* (24) report a wider variability of NE in their small group of normotensive patients than in another group of well-trained volunteers. Situational anxiety, then, may account for some correlation between BP and NE in the general population.

RELATIONSHIP BETWEEN PLASMA NE AND HEMODYNAMIC RESPONSE TO ADRENERGIC INHIBITION OR STIMULATION

As discussed above, measurements of plasma NE or plasma epinephrine (E) alone are not sufficient to elucidate the role of SNS on BP homeostasis. Specific antagonists or agonists of α- or β-adrenoceptors may be more useful to determine SNS effects on BP and other hemodynamic parameters. A combination of both methods, precise measurement of plasma NE and plasma E with pharmacologic blockade or stimulation, is bound to provide more useful information.

An early study of ganglionic blockade (13) notes a more pronounced fall in BP in hypertensives than in controls. An exaggerated response to sympatholytic agents in patients with hypertension might not only be due to increased neurogenic activity, but could conceivably be related to structural vascular change in fixed essential hypertension. In the course of hypertension, arteriolar wall thickens, and the lumen decreases. Folkow (19) has shown that vasoconstrictive or vasodilating stimuli in such vessels elicit exaggerated changes of vascular resistance. Therefore, the effect of adrenolytic agents in early hypertension, where structural changes are minimal, might be of great interest.

Julius *et al.* (29) found that cardiac blockade

with atropine and propranolol normalized the increased heart rate and cardiac output in the high-output subgroup of borderline hypertensives but did not normalize BP. As cardiac output decreased, the total peripheral resistance increased. They thought this might be a reflex adjustment. Subsequently, they administered phentolamine but normalized the increased total peripheral resistance and BP in only about 30% of the investigated group (29). The group consisted predominantly of patients with normal cardiac output. Unfortunately, similar experiments were not carried out in the high-output subgroup where one would expect the largest prevalence of neurogenic hypertension. These studies suggest that in a substantial proportion of patients with borderline hypertension, BP elevation is not neurogenic. The only patients in whom blockade normalized BP were those with high-renin borderline hypertension (29). Similar results were found in patients with mild established high-renin hypertension (17).

In more recent studies, the outcome of adrenergic blockade was related to levels of plasma NE or plasma E. Esler *et al.* (18) found that the decrease in heart rate with β-adrenergic blockade correlates with higher levels of plasma NE. They also note a positive correlation between the decrease in cardiac index and the increase in preejection period with plasma NE levels. Similarly, another study (17) in which high-renin patients had high plasma NE levels shows a more marked decrease in cardiac output and heart rate during β-blockade compared to normotensive controls and normal renin hypertension. However, the high-renin patients did not have elevated resting cardiac output.

Louis *et al.* (36) report that the fall in BP during pentolinium infusion correlates with resting plasma NE. During treatment with tiamenidine, a clonidine-like drug (10), BP fall is most pronounced in patients with high plasma NE.

Cousineau *et al.* (7) found that the BP response to atenolol is more pronounced in patients with high plasma NE. Similarly, Distler *et al.* (12) report that the BP fall on atenolol treatment correlates closely to an index of sympathetic nerve activity (plasma NE during exercise/plasma NE at rest) prior to treatment.

Agabiti-Rosei *et al.* (1) report that phentolamine does not decrease mean arterial pressure in hypertension due to reflex increase in cardiac output, but the fall in total peripheral resistance correlates to plasma CA at rest. Labetalol, a drug possessing both α- and β-blocking activity, decreases BP and total peripheral resistance, which correlates with the resting level of plasma NE plus E prior to treatment.

Overall, the picture is consistent with increased BP responsiveness to adrenergic inhibition in patients who have high plasma NE levels. Before these results are accepted as proof of the neurogenic maintenance of BP elevation in these patients, one must consider methodological factors that may becloud such interpretation. Of foremost interest are the adequacy of baroreflex responses and the already mentioned problem of arteriolar structural wall to lumen ratio changes. It is, therefore, important to analyze whether NE responses in these studies reflect BP responses. Louis *et al.* (36, 37) show a strong correlation between the BP fall in plasma NE during ganglionic blockade with pentolinium or acute clonidine administration.

On the other hand, Weidmann *et al.* (52), using debrisoquine, note that the fall in plasma NE in normotension, borderline hypertension and established hypertension is identical, but the fall in BP is only significant in patients with hypertension and, more markedly, in established hypertension. This finding raises the question of whether the observed BP fall during adrenergic inhibition reflects other factors in addition to sympathetic tone. This question was systematically investigated by Bühler's group (2, 32). They infused small doses of an α-adrenergic antagonist into the forearm and measured forearm blood flow changes. (The small doses do not cause systemic hemodynamic alterations.) Kiowski *et al.* (32) found similar resting plasma NE levels and similar increases in forearm blood flow to phentolamine infusion in hypertensive and normotensive individuals, but the hypertensives showed a close correlation between increase in forearm blood flow and plasma NE at rest prior to infusion. Mean arterial pressure correlates to plasma NE and to the increase in flow during phentolamine infusion. Consequently, there is no evidence of overall increased adrenergic activity in hypertension. Plasma NE in hypertension, however, is a better marker of the α-adrenergic vasoconstriction than in normotension. This tighter relationship between plasma

NE and sympathetic vasoconstriction in hypertension, according to the authors, cannot be attributed to an increased wall to lumen ratio of the resistance vessels, since the overall decrease of forearm vascular resistance is similar in hypertensive and normotensive subjects (32). The same group attempted to take morphological changes into account by another approach: by comparing the changes in vascular resistance during intraarterial infusion of prazosin to infusion of the nonspecific vasodilator nitroprusside (2). They found that while the increase in flow due to nitroprusside was similar in hypertensives and normotensives, prazosin elicits larger forearm blood flow increases in hypertensive subjects. Surprisingly, this study did not find a relationship between plasma NE and the changes in flow but did find a close correlation between plasma E and forearm blood flow changes. The results also indicate that plasma CA are more closely related to α-adrenergic vasoconstriction in hypertension than in normotension. The authors view their results as indicative of a specific increase of α-receptor responsiveness in hypertension. This conclusion is tempting but cannot be accepted unless the comparison of the nonspecific vasodilator and prazosin is carried out across a wide range of dose-response curves for both agents.

Whereas the Basel group used α-adrenergic antagonists, other authors utilized agonists to investigate whether sympathetic tone, as reflected by plasma NE or receptor sensitivity, is altered in hypertension. Philipp *et al.* (44) report an inverse correlation between reactivity to NE infusion and plasma NE levels during physical exercise in normotensive individuals. A similar relationship exists in hypertensives, but the response depends on BP levels. The hyperbolic curve is shifted to the right in mild hypertension and even further to the right in severe hypertension. It follows that for a given level of plasma NE, the reactivity to NE infusion is higher in patients with hypertension.

The authors interpret their findings as a demonstration of a specific increase in vascular reactivity to NE hypertension (44). Central to their argument is the finding that reactivity to angiotensin is also increased in hypertension, but the curve describing the relationship of plasma renin and angiotensin reactivity is not shifted to the right. It could be argued, however, that the use of plasma renin measurements obtained after stimulation with furosemide plus orthostasis might not be the proper frame of reference; the authors carried out the infusion of angiotensin II in the resting position, and the patient's sodium balance was not affected with the diuretic.

In similar studies, Weidmann *et al.* (52) and Meier *et al.* (39, 40) also report an increased reactivity to exogenous NE as well as to angiotensin II. The correlation between plasma NE and NE pressor dose is significant in both normotensive and hypertensive individuals, but for a given level of plasma NE the pressor response is exaggerated in hypertensives. Weidmann *et al.* (52) also note a decrease in the threshold dose for NE in hypertension compared to normotension. A decrease in threshold dose might favor a mechanism related to increased vascular sensitivity rather than to morphological vascular changes. However, as pointed out by the authors, the response to exogenous NE is not only dependent on the peripheral vascular tone, but also on the responsiveness of the integrated cardiovascular system.

These studies utilize an infusion of exogenous NE. It is not clear whether reactivity to exogenous NE quantitatively reflects the sensitivity, reactivity, and available receptor mass to endogenously released NE. Krakoff *et al.* (31) found an exaggerated increase in mean arterial pressure during NE infusion in hypertensives compared to normotensives, but the level of plasma NE before and during the infusion was identical in the two groups. During isometric handgrip exercises, the increase in NE was more pronounced in the hypertensives, although the increase in mean arterial pressure was the same as in normotensives. Based on their results, one must question whether an increased responsiveness to exogenous NE infusion in hypertension in fact reflects increased reactivity to the endogenous release of NE. The observed differences in responsiveness to exogenous NE between hypertensives and normotensives depend upon differences in reflex regulation of the overall circulation (46). Furthermore, recent animal data suggest that the pressor response to exogenous NE and to endogenously released NE is mediated by activation of two different populations of α-adrenoceptors (54).

Although experiments with infusions of α-adrenergic antagonists and agonists appear to yield very diverse results, some major findings are remarkably consistent. Compared to normotension, a different relationship between α-adrenergic responses and plasma NE in hypertension is observed by all investigators. The data strongly suggest that BP in hypertension is more dependent on sympathetic drive if vascular responsiveness, NE levels, and BP are taken into account. The disagreement is in interpretation. Suggestions that there is a specific hypersensitivity of α-adrenoreceptors in hypertension must, at present, be considered speculative. However, the most important yield of all these studies, and particularly of the study by Philipp *et al.* (44), is an understanding that NE values in hypertension can be better interpreted if combined with measures of NE responsiveness and if the prevailing BP is taken into account. Philipp *et al.* (44) used a multiregression analysis to obtain a formula allowing prediction of mean arterial pressure from plasma NE and reactivity to NE. Next, computer plots defining the relationship between plasma NE and NE reactivity for different levels of mean arterial pressure were generated. With a given mean arterial pressure value, the variation in NE reactivity could be accounted for by variation in plasma NE, and vice versa. This means that both factors are equally important determinants for BP in hypertension and normotension.

Whereas vascular NE responsiveness is enhanced in hypertension, β-adrenergic responsiveness is uniformly decreased. Julius *et al.* (29) report a decreased chronotropic responsiveness to isoproterenol infusion in patients with borderline hypertension who, based on their cardiac response to propranolol, were presumed to have increased sympathetic tone. However, they also found a decreased responsiveness in patients with normal cardiac output and no increased sympathetic drive as assessed by pharmacologic blockade (30). Bertel *et al.* (3) note an inverse relationship between plasma NE during exercise and sensitivity to isoproterenol. Furthermore, the sensitivity decreases with age, especially in patients with hypertension.

The apparent discrepancy between α- and β-adrenergic responsiveness in hypertension is confusing and hard to explain in terms of presently accepted physiologic knowledge. If the work of the Basel group (2, 3, 32) is interpreted in its entirety, one must presume that exposure to a sustained "sympathetic overactivity" in hypertension leads to β-receptor downregulation and α-receptor upregulation. In that respect, the data by Philipp *et al.* (44) also present difficulties. The expected negative relationship between the sympathetic "tone" (expressed by plasma NE) and reactivity to NE is present, indicating a normal relationship between agonist and receptor. Another factor, then, intervenes and causes a BP-dependent "hypersensitivity." The nature of this relationship between increased BP and increased α-adrenergic responsiveness is not elucidated. In the heart, structural changes have no effect on chronotropic responses, and subjects with higher plasma NE respond less to isoproterenol infusion. In view of the complexity of α-adrenergic responses, it is prudent to reserve judgment on whether a specific increase in α-adrenergic sensitivity is present in hypertension.

A promising new approach, combining stimulation and blockade procedures, has recently been proposed by Tanaka *et al.* (47). The major advantage of this method is that it utilizes sequential autonomic blockade to decrease the reflex influence on pharmacologic responses. An indirect measure of adrenergic tone, called "α-secretion," has been defined from the response to infusion phenylephrine and phentolamine; vascular α-adrenergic tone is expressed as an amount of exogenous phenylephrine, which is believed to maintain vascular tone equal to the amount of NE released for a given degree of α-adrenergic tone. Similarly, "β-secretion," a measure of cardiac β-adrenergic tone, is expressed as an amount of isoproterenol. Many questions could be raised regarding the value of expressing adrenergic tone as an amount of infused phenylephrine or isoproterenol. Nevertheless, Tanaka *et al.* (47) found a close correlation between plasma NE and α and β secretion over a wide range of plasma NE levels, suggesting the validity of the method.

SUMMARY

Development of precise methods to determine plasma NE was an important step in the assessment of SNS activity in hypertension. Comparisons of plasma NE in hypertension and normotension, however, have not led to any clear-cut understanding of the importance of the SNS in the pathogenesis or maintenance of hypertension. Plasma NE is not elevated in the total population of patients with essential hypertension, but subgroups with increased levels may exist, most likely in younger age groups (see Chapter 24). An increased level of plasma NE does not necessarily imply increased SNS activity, but changes in neuronal reuptake or clearance may also be responsible. Furthermore, plasma NE in systemic venous blood represents only a small fraction of the amount secreted, and NE from various organs contributes to the level of plasma NE. Even when CA are elevated it is not known whether these higher levels reflect increased vascular tone or stem from other organs.

Consequently, understanding the relationship between plasma NE levels and hemodynamics in hypertension may provide new insights. A relationship between plasma NE and BP exists in hypertension, but the correlation is generally weak and has not been found in all studies. Such a correlation has never been found in normotensive subjects. Consequently, plasma NE seems to be a better marker for the BP in hypertension than in normotension. Data on the relationship between plasma NE and the hemodynamic components of BP are scanty but suggest that plasma NE predominantly reflects sympathetic drive to the heart early in the course of hypertension, and in later, more advanced stages of the disease, reflects the sympathetic drive to the blood vessels. It is not clear whether this pattern reflects increased sympathetic nerve activity or enhancement of the influence of normal sympathetic nerve activity due to morphological changes in hypertensive vessels. It should be emphasized that the correlation of NE to hemodynamic parameters, initially to cardiac function and later to vascular resistance, is weak.

Measurement of plasma NE in combination with specific adrenoceptor blockade is a more specific approach for determination of the influence of sympathetic nerve activity on BP and other hemodynamic parameters. In mild hypertension, the decreased heart rate and cardiac output after β-adrenoceptor blockade correlate with the level of plasma NE. In established hypertension, the fall in BP or in calculated total peripheral resistance after α-adrenoceptor blockade is related to plasma NE levels. The response to systemic α-adrenergic inhibition not only depends on the effect of peripheral vascular adrenergic tone, but also on reflex-mediated adjustments in the integrated cardiovascular system. Therefore, local α-adrenergic blockade in a well-defined vascular bed, such as the forearm, is a more specific approach for determination of the neurogenic component in vascular resistance. Changes in forearm vascular resistance induced by local α-adrenergic blockade correlate with plasma NE or with plasma E in hypertension, but not in normotension, suggesting that plasma NE is a better marker for α-adrenergic vasoconstriction in hypertension than in normotension.

The number of subjects studied with adrenergic blocking agents is small, the methodology is complex and the correlation found is not strong. More work is needed in this promising field for evaluation of sympathetic activity in human hypertension.

Receptor sensitivity in hypertension has been studied by relating plasma NE with the response to infusion of adrenergic agonists. An inverse correlation between plasma NE and response to β-receptor agonists as well as to α-receptor agonists has been shown. The response to cardiac β-receptor stimulation decreases during the course of hypertension, while vascular responsiveness to exogenous NE apparently increases with progression of the hypertensive disease. The interpretation of response to exogenous NE is very complex. The response depends not only on receptor sensitivity, affinity, and amount, but also on many other factors such as plasma NE level, baroreflex sensitivity, and morphological vascular changes. If data from recent animal studies suggesting that exogenous and endogenous NE interact with different populations of α-receptors also apply to humans, the value of infusion of exogenous NE as a measure of synaptic receptor sensitivity could be seriously questioned.

The available literature allow some tentative conclusions about the importance of the SNS and the pathogenesis and maintenance of hypertension. There is evidence to suggest that in some patients with early or borderline hypertension, an increased sympathetic tone is involved in the maintenance of the elevated BP. In established essential hypertension there is no definite indication of increased sympathetic tone, but sympathetic activity plays a more prominent role in the maintenance of BP. A vascular hyperreactivity to adrenergic stimulation is characteristically associated with hypertension. It is not clear whether this is due to specific increased sensitivity of vascular α-adrenoceptors or whether it is a consequence of the hypertension. In our opinion, morphological vascular changes in hypertension may well be a single most important factor for the observed vascular hyperresponsiveness.

References

1. Agabiti-Rosei E, Alicandri C, Fariello R, Muiesan G: Catecholamines and haemodynamics in fixed essential hypertension. *Clin Sci* 57:193s–196s, 1979.
2. Amann FW, Bolli P, Kiowski W, Bühler FR: Enhanced alpha-adrenoreceptor-mediated vasoconstriction in essential hypertension. *Hypertension* 3(Suppl 1):I119–I123, 1981.
3. Bertel O, Bühler FR, Kiowski W, Lütold BE: Decreased beta-adrenoreceptor responsiveness as related to age, blood pressure, and plasma catecholamines in patients with essential hypertension. *Hypertension* 2:130–138, 1980.
4. Brown JJ, Davis DL, Morton JJ, Robertson JIS, Cuesta V, Lever AF, Padfield PL, Trust P: Mechanism of renal hypertension. *Lancet* 1:1219–1221, 1976.
5. Brown MJ, Jenner DA, Allison DJ, Dollery CT: Variations in individual organ release of noradrenaline measured by an improved radioenzymatic technique; limitations of peripheral venous measurements in the assessment of sympathetic nervous activity. *Clin Sci* 61:585–590, 1981.
6. Chobanian AV, Gavras H, Melby JC, Gavras I, Jick H: Relationship of basal plasma norepinephrine to blood pressure, plasma renin activity, mineralocorticoids, and plasma volume in essential hypertension. *Trans Assoc Am Phys* 91:368–380, 1978.
7. Cousineau D, deChamplain J, LaPointe L: Circulating catecholamines and systolic time intervals in labile and sustained hypertension. *Clin Sci Mol Med* 55:65s–68s, 1978.
8. DeLeeuw PW, Wester A, Willemse PJ, Birkenhäger WH: Patterns of noradrenaline and renin in essential hypertension—effects of stress and therapy. In Philipp TH, Distler A: *Hypertension: Mechanisms and Management.* New York, Springer-Verlag, 1980, pp 49–54.
9. DeQuattro V, Campese V, Miura Y, Meijer D: Increased plasma catecholamines in high renin hypertension. *Am J Cardiol* 38:801–804, 1976.
10. DeQuattro V, Kolloch R, Kobayashi K, Johnston W, Miano L: Circulatory neurohormones in primary hypertension: Probes in etiology and treatment. In Philipp T, Distler A: *Hypertension: Mechanisms and Management.* New York, Springer-Verlag, 1980, pp 3–10.
11. DeQuattro V, Miura Y: Plasma and tissue catecholamines in primary hypertension. In Julius S, Esler MD: *The Nervous System in Arterial Hypertension.* Springfield, Ill, Charles C Thomas, 1976, pp 355–374.
12. Distler A, Keim HJ, Cordes U, Philipp T, Wolf HP: Sympathetic responsiveness and antihypertensive effect of beta-receptor blockade in essential hypertension. *Am J Med* 64:446–451, 1978.
13. Doyle AE, Smirk FH: The neurogenic component in hypertension. *Circulation* 12:543–552, 1955.
14. Eich RH, Cuddy RP, Smulyan H, Lyons RH: Hemodynamics in labile hypertension: A follow-up study. *Circulation* 34:299–307, 1966.
15. Eng FWHT, Huber-Smith M, McCann DS: The role of sympathetic activity in normal renin essential hypertension. *Hypertension* 2:14–19, 1980.
16. Esler M, Jackman G, Bobik A, Leonard P, Kelleher D, Skews H, Jennings G, Korner P: Norepinephrine kinetics in essential hypertension. Defective neuronal uptake of norepinephrine in some patients. *Hypertension* 3:149–156, 1981.
17. Esler M, Julius S, Zweifler A, Randall O, Harburg E, Gardiner H, DeQuattro V: Mild high-renin essential hypertension. Neurogenic human hypertension? *N Engl J Med* 296:405–411, 1977.
18. Esler M, Zweifler A, Randall O, Julius S, DeQuattro V: Agreement among three different indices of sympathetic nervous system activity in essential hypertension. *Mayo Clin Proc* 52:379–382, 1977.
19. Folkow B: Cardiovascular structural adaptation; its role in the initiation and maintenance of primary hypertension. *Clin Sci Mol Med* 55:3s–22s, 1978.
20. Falkner B, Onesti G, Angelakos ET, Fernandes M, Langman C: Cardiovascular response to mental stress in normal adolescents with hypertensive parents. Hemodynamics and mental stress in adolescents. *Hypertension* 1:23–30, 1979.
21. Freis ED: Hemodynamics of hypertension. *Physiol Rev* 40:27–54, 1960.
22. Goldstein DS: Plasma norepinephrine in essential hypertension. A study of the studies. *Hypertension* 3:48–52, 1981.
23. Hollenberg NK, Williams GH, Adams DF: Essential hypertension: Abnormal renal vascular and endocrine responses to a mild psychological stimulus. *Hypertension* 3:11–17, 1981.
24. Jones DH, Hamilton CA, Reid JL: Plasma noradrenaline, age and blood pressure: A population study. *Clin Sci Mol Med* 55:73s–75s, 1978.
25. Julius S: Neurogenic component in borderline hypertension. In Julius S, Esler MD: *The Nervous System in Arterial Hypertension.* Springfield, Ill, Charles C Thomas, 1976, pp 301–330.
26. Julius S: The psychophysiology of borderline hypertension. In Weiner H, Hofer MA, Stunkard AJ: *Brain, Behavior and Bodily Disease.* New York, Raven Press, 1981, pp 293–303.
27. Julius S: Psychophysiologic evidence for the role of the nervous system in hypertension. In Amery A, Lijnen P, Fagard R, Stalssen J: *Hypertensive Cardiovascular Disease: Pathophysiology and Treatment.* The Hague, The Netherlands, Nijhoff Publishing Company, 1982, pp 217–230.

28. Julius S, Amery A, Whitlock LS, Conway J: Influence of age on the hemodynamic response to exercise. *Circulation* 36:231–237, 1967.
29. Julius S, Esler MD, Randall OS: Role of the autonomic nervous system in mild human hypertension. *Clin Sci Mol Med* 48:243s–252s, 1975.
30. Julius S, Randall OS, Esler MD, Kashima T, Ellis C, Bennett J: Altered cardiac responsiveness and regulation in the normal cardiac output type of borderline hypertension. *Circ Res* 36–37 (Suppl I):I199–I207, 1975.
31. Krakoff LR, Vlachakis ND, Mendlowitz M: Effect of posture, isometric hand-grip exercise and norepinephrine infusion in normal renin hypertensive patients. In Laragh JH, Bühler FR, Seldin DW: *Frontiers in Hypertension Research.* New York, Springer-Verlag, 1981, pp 290–296.
32. Kiowski W, Bühler FR, van Brummelen P, Amann FW: Plasma noradrenaline concentration and α-adrenoceptor-mediated vasoconstriction in normotensive and hypertensive man. *Clin Sci* 60:483–489, 1981.
33. Lake CR, Ziegler MG, Coleman MD, Kopin IJ: Age-adjusted plasma norepinephrine levels are similar in normotensive and hypertensive subjects. *N Engl J Med* 296:208–209, 1977.
34. Lake CR, Ziegler MG, Kopin IJ: Use of plasma norepinephrine for evaluation of sympathetic neuronal function in man. *Life Sci* 18:1315–1326, 1976.
35. Light KC, Obrist PA: Cardiovascular reactivity to behavioral stress in young males with and without marginally elevated casual systolic presures. Comparison of clinic, home, and laboratory measures. *Hypertension* 2:802–808, 1980.
36. Louis WJ, Doyle AE, Anavekar S: Plasma norepinephrine levels in essential hypertension. *N Engl J Med* 288:599–601, 1973.
37. Louis WJ, Doyle AE, Anavekar SN, Johnston CI, Geffen LB, Rush R: Plasma catecholamine, dopamine-beta-hydroxylase, and renin levels in essential hypertension. *Circ Res* 34–35(Suppl I):I57–I61, 1974.
38. Mancia G, Leonetti G, Picotti GB, Ferrari A, Galva MD, Gregorini L, Parati G, Pomidossi G, Ravazzani C, Sala C, Zanchetti A: Plasma catecholamines and blood pressure responses to the carotid baroreceptor reflex in essential hypertension. *Clin Sci* 57:165s–167s, 1979.
39. Meier A, Gübelin U, Weidmann P, Grimm M, Keusch G, Glück Z, Minder I, Beretta-Piccoli C: Age-related profile of cardiovascular reactivity to norepinephrine and angiotensin II in normal and hypertensive man. *Klin Wochenschr* 58:1183–1188, 1980.
40. Meier A, Weidmann P, Grimm M, Keusch G, Glück Z, Minder I, Ziegler WH: Pressor factors and cardiovascular pressor responsiveness in borderline hypertension. *Hypertension* 3:367–372, 1981.
41. Miura Y, Kobayashi K, Sakuma H, Tomioka H, Adachi M, Yoshinaga K: Plasma noradrenaline concentrations and haemodynamics in the early stage of essential hypertension. *Clin Sci Mol Med* 55:69s–71s, 1978.
42. Nestel PJ: Blood pressure and catecholamine excretion after mental stress in labile hypertension. *Lancet* 1:692–694, 1969.
43. Pedersen EB, Christensen NJ: Catecholamines in plasma and urine in patients with essential hypertension determined by double-isotope derivative techniques. *Acta Med Scand* 198:373–377, 1975.
44. Philipp TH, Distler A, Cordes U: Sympathetic nervous system and blood-pressure control in essential hypertension. *Lancet* 2:959–963, 1978.
45. Sever PS, Osikowska B, Birch M, Tunbridge RDG: Plasma-noradrenaline in essential hypertension. *Lancet* 1:1078–1081, 1977.
46. Sleight P: Reflex control of the heart. *Am J Cardiol* 44:889–894, 1979.
47. Tanaka N, Kudo A, Kawahira K, Niimura T, Tanaka H, Kanahisa T: A new method in the quantitative determination of the sympathetic nervous tone of the cardiovascular system in man. Proceedings of the 18th International Congress of Neurovegetative Research, 1977, pp 202–204.
48. Taylor AA, Pool JL, Lake CR, Ziegler MG, Rosen RA, Rollins DE, Mitchell JR: Plasma norepinephrine concentrations: No differences among normal volunteers and low, high or normal renin hypertensive patients. *Life Sci* 22:1499–1510, 1978.
49. Vlachakis ND: Blood pressure and catecholamine responses to sympathetic stimulation in normotensive and hypertensive subjects. *J Clin Pharmacol* 19:458–466, 1979.
50. Vlachakis ND, Mendlowitz M: Plasma catecholamines in primary hypertension. *Biochem Med* 23:35–46, 1980.
51. Walsh JA, Hyman C, Maroude RF: Venous distensibility in essential hypertension. *Cardiovasc Res* 3:338–349, 1969.
52. Weidmann P, Grimm M, Meier A, Glück A, Keusch G, Minder I, Beretta-Piccoli C: Cardiovascular pressor reactivity as related to plasma catecholamines: Role in the pathogenesis of essential hypertension and in the antihypertensive mechanism of diuretic treatment. In Philipp T, Distler A: *Hypertension: Mechanisms and Management.* New York, Springer-Verlag, 1980, pp 23–36.
53. Yamaguchi N, de Champlain J, Nadeau R: Correlation between the response of the heart to sympathetic stimulation and the release of endogenous catecholamines into the coronary sinus of the dog. *Circ Res* 36:662–668, 1975.
54. Yamaguchi I, Kopin IJ: Differential inhibition of alpha-1 and alpha-2 adrenoceptor-mediated pressor responses in pithed rats. *J Pharmacol Exp Ther* 214:275–281, 1980.

CHAPTER TWENTY-SIX

The Role of Norepinephrine in the Regulation of Sodium and Water Excretion by the Kidney

HANS-GEORG GÜLLNER, M.D.

INTRODUCTION: HISTORICAL AND ANATOMICAL CONSIDERATIONS

A link between the sympathetic nervous system (SNS) and salt and water excretion was first recognized by Claude Bernard (7) more than a century ago. He found that section of the greater splanchnic nerve in anesthetized dogs increases urinary volume on the affected side, whereas electrical stimulation of the distal portion of the interrupted nerve corrected these changes. Many similar studies have since been reported, and these changes associated with section of the splanchnic nerve are now known as denervation diuresis and natriuresis. Some studies have noted an increase in glomerular filtration rate following acute denervation, and the subsequent diuresis has been attributed to the increase in filtered load (16). Denervation natriuresis occurs in anesthetized animals but has not yet been clearly demonstrated in conscious animals (17).

Anatomic, clearance, and micropuncture experiments confirm that renal nerves can increase tubular reabsorption of sodium by a direct effect on the tubule. Varicosities of sympathetic nerve fibers are in contact with the basement membranes of proximal and distal tubular cells (2, 19). The kidney is well innervated with sympathetic fibers (8), and some directly supply renal tubular cells of the rat and monkey. Adrenergic nerve endings make direct contact with the basement membrane of proximal and distal tubular cells similar to the innervation of smooth muscle cells (2, 19).

EFFECT OF SECTION OF RENAL NERVES ON SODIUM AND WATER EXCRETION

An increase in salt and water excretion without a change in filtered load or renal blood flow following section of the renal nerves was first demonstrated in 1919 (18). Electrical stimulation of renal nerves decreases sodium excretion (21). Micropuncture experiments have confirmed that autonomic neurons control sodium and water reabsorption by the proximal tubule through a direct mechanism independent of changes in systemic hemodynamics, glomerular filtration rate, or renal blood flow (3, 4, 6). Renal nerves in the rat are usually destroyed by stripping the renal artery of its adventitia and by applying phenol in alcohol to the artery. This procedure causes depression of sodium and water reabsorption by the proximal tubules. Sodium excretion increases 5-fold in the treated kidney, despite an increase in sodium reabsorption in distal segments of the nephron that partially compensates for the decrease in proximal reabsorption (3). Section of the renal nerves in anesthetized hypophysectomized dogs undergoing a water diuresis produced an increase in urine flow, sodium excretion, and free water clearance, indicating a decrease in proximal reabsorption (15), and similar results were obtained in rats expanded 10% above their body weight by the infusion of isotonic saline (4). Denervation did not change single nephron glomerular filtration rate or renal plasma flow but increased urine flow and sodium excretion by 45%. Absolute water reabsorption in the proximal tubule was markedly reduced, and absolute sodium reabsorption was increased in the loop of Henle, the distal convoluted tubules, and collecting ducts. Denervation increases sodium excretion for at least 2 weeks in rats (25) and dogs (26) and chronically impairs other proximal tubular transport systems for glucose and phosphate. These results can be reproduced by the α-adrenergic antagonist phenoxybenzamine (12), suggesting that there are α-receptors on proximal tubular cells.

SODIUM REABSORPTION INCREASE BY RENAL NERVE STIMULATION AND NE

Electrical stimulation of renal sympathetic nerves in volume expanded rats increased proximal water and sodium reabsorption by 25% without changing systemic or intrarenal hemodynamics (5). Similarly, renal nerve stimulation in dogs decreased sodium excretion without affecting glomerular filtration rate, renal blood flow, or the distribution of radiolabeled microspheres (23). The effect of nerve stimulation on sodium reabsorption could be prevented by infusion of an α-adrenergic antagonist (29). Conversely, the effects of renal nerve stimulation on the handling of sodium and water by the proximal tubule can be mimicked by infusion of the α-adrenergic agonist norepinephrine (NE) (13). In anesthetized, hypophysectomized dogs infusion of NE into the left renal artery in the presence of β-receptor blockade produced an ipsilateral decrease in urine flow and in free water clearance without significant changes in systemic and intrarenal hemodynamics and without changes in distal fractional sodium reabsorption, suggesting that an increase in proximal tubular reabsorption was responsible for the observed effects (13).

MECHANISM OF ACTION OF ENHANCED SODIUM REABSORPTION BY INCREASED ADRENERGIC ACTIVITY

Little is known about the nature of the changes in the renal tubular cells which are responsible for the enhanced sodium and water transport induced by increased adrenergic activity. There is some indirect evidence that qualitative and quantitative changes in renal adrenergic activity may contribute to the natriuresis that occurs when the extracellular fluid volume is expanded. Regulation of SNS activity in the kidney is controlled by low-pressure receptors located in the cardiopulmonary region of the circulation and by high-pressure receptors such as the carotid sinus baroreceptors (9, 20, 28). A reduction in blood volume by hemorrhage, through a reflex-mediated increase in renal sympathetic nerve activity, increases the tubular reabsorption of sodium (14). The effects of hemorrhage on sodium excretion can be blocked by renal denervation or by infusion of the adrenergic antagonists guanethidine or phenoxybenzamine into the renal artery (14, 28).

The splanchnic nerve also mediates constriction of the thoracic inferior vena cava, another potent stimulus to the renal reabsorption of sodium in the dog. As with other adrenergic stimuli to renal tubular sodium reabsorption, the effects of constriction of the thoracic inferior vena cava can be attenuated by pharmacological blockade of autonomic transmission or by renal denervation (1, 11) or can be prevented by pretreatment with hexamethonium or phenoxybenzamine (22). Similarly, surgical denervation of the left kidney and infusion of guanethidine or phenoxybenzamine into the renal artery were equally effective in decreasing proximal sodium reabsorption during infusion of saline in dogs with constriction of the thoracic inferior vena cava without changing renal hemodynamics (24).

The effects of NE on tubular transport might be caused directly or by other hormones such as angiotensin or prostaglandins. The available evidence implicates a direct effect of NE. The renal response to nerve stimulation is not altered by administration of an angiotensin II antagonist or by indomethacin, an inhibitor of prostaglandin synthesis (10, 27).

Since it is not clear whether or not a physiologically significant basal nerve activity in the kidney exists, it is difficult to assess whether an increase in blood volume can suppress renal adrenergic activity sufficiently to cause natriuresis. In anesthetized rabbits, expansion of blood volume decreased renal nerve activity (9). In anesthetized rats, volume expansion with subsequent renal denervation resulted in a further increase in sodium excretion (4). The effects of blood volume expansion and renal denervation in anesthetized animals may represent attenuation of increased renal adrenergic activity induced by the "stress" of anesthesia and surgery. In one of the few experiments in conscious animals reported to date, volume expansion in cats caused a decrease in efferent renal sympathetic activity (21). Additional experiments in conscious animals are needed to clarify this question.

SUMMARY

The physiological role of the adrenergic nervous system in the regulation of salt and water excretion remains to be established. The rapidly responding SNS may be complementary to the more slowly activated renin-angiotensin-aldosterone system or may have an independent role in the control of sodium excretion.

References

1. Azer M, Gannon R, Kaloyanides GJ: Effect of renal denervation on the antinatriuresis of caval constriction. *Am J Physiol* 22:611–616, 1972.
2. Barajas L, Müller J: The innervation of the juxtaglomerular apparatus and surrounding tubules: A quantitative analysis by serial section electron microscopy. J Ultrastruct Res 43:107–132, 1973.
3. Bello-Reuss E, Colindres RE, Pastoriza-Munoz E, Mueller RA, Gottschalk CW: Effects of acute unilateral renal denervation in the rat. *J Clin Invest* 56:208–217, 1975.
4. Bello-Reuss E, Pastoriza-Munoz E, Colindres RE: Acute unilateral renal denervation in rats with extracellular volume expansion. *Am J Physiol* 232:F26–F32, 1977.
5. Bello-Reuss E, Trevino DL, Gottschalk CW: Effect of renal sympathetic nerve stimulation on proximal water and sodium reabsorption. *J Clin Invest* 57:1104–1107, 1976.
6. Bencsath P, Bonvalet J-P, deRouffignac C: Tubular factors in denervation diuresis and natriuresis. In Wirz H, Spinelli F: *Recent Advances In Renal Physiology. International Symposium on Renal Handling of Sodium*, Basel, Karger, 1972, pp 96–106.
7. Bernard C: *Lecons sur les Proprietes Physiologique des Liquides de l'Organisme.* Paris, Bailliere, 1859, pp 172–173.
8. Bradford JR: The innervation of the renal blood vessels. *J Physiol* 10:358–407, 1889.
9. Clement DL, Pelletier CL, Shepherd JT: Role of vagal afferents in the control of renal sympathetic nerve activity in the rabbit. *Circ Res* 31:824–830, 1972.
10. DiBona GB, Zambraski EJ, Aquilera AJ, Kaloyanides GJ: Neurogenic control of renal tubular sodium reabsorption in the dog. *Circ Res* 40(Suppl I):I127–I130, 1977.
11. Gill JR Jr, Carr AA, Fleischman LE, Casper AGT, Bartter FC: Effects of pentolinium on sodium excretion in dogs with constriction of the vena cava. *Am J Physiol* 212:191–196, 1967.
12. Gill JR Jr, Casper AGT: Depression of proximal tubular sodium reabsorption in the dog in response to renal beta-adrenergic stimulation by isoproterenol. *J Clin Invest* 50:112–118, 1971.
13. Gill JR Jr, Casper AGT: Effect of renal alpha-adrenergic stimulation on proximal tubular sodium reabsorption. *Am J Physiol* 223:1201–1205, 1972.
14. Gill JR Jr, Casper AGT: Role of the sympathetic nervous system in the renal response to hemorrhage. *J Clin Invest* 48:915–922, 1969.
15. Hatori N: Effect of acute unilateral renal denervation on tubular sodium reabsorption in the dog. *Am J Physiol* 232:F16–F19, 1977.
16. Kamm DE, Levinsky NG: The mechanism of denervation natriuresis. *J Clin Invest* 44:93–102, 1965.
17. Lifschitz MD: Lack of a role for the renal nerves in renal sodium reabsorption in conscious dogs. *Clin Sci Mol Med* 54:567–572, 1978.
18. Marshall EK Jr, Kolls AC: Studies on the nervous control of the kidney in relation to diuresis and urinary secretion. I. The effect of unilateral excision of the adrenal, section of the splanchnic nerve and section of the renal nerves on the secretion of the kidney. *Am J Physiol* 49:302–316, 1919.
19. Müller J, Barajas L: Electron microscopic and histochemical evidence for a tubular innervation in the renal cortex of the monkey. *J Ultrastruct Res* 41:533–549, 1972.
20. Öberg B, White S: The role of vagal cardiac nerves and arterial baroreceptors in the circulatory adjustments to hemorrhage in the cat. *Acta Physiol Scand* 80:395–403, 1970.
21. Schad H, Seller H: Reduction of renal nerve activity by volume expansion in conscious cats. *Pfluegers Arch* 363:155–159, 1976.
22. Schrier RW, Humphrey MH, Ufferman RC: Role of cardiac output and the autonomic nervous system in the antinatriuretic response to acute constriction of the thoracic superior vena cava. *Circ Res* 29:490–498, 1971.
23. Slick GL, Aquilera AJ, Zambraski EJ, DiBona GF, Kaloyanides GJ: Renal neuroadrenergic transmission. *Am J Physiol* 229:60–65, 1975.
24. Slick GL, DiBona GF, Kaloyanides GJ: Renal sympathetic nerve activity in sodium retention of acute caval constriction. *Am J Physiol* 226:925–932, 1974.
25. Spielman WS, Colindres RE, Gottschalk CW: Role of the renal nerves in the adaptive response of the kidney to contralateral renal denervation (abstr). *VIIth International Congress on Nephrology.* Basel, Karger, 1978, pp. C–11.
26. Takacs L, Bencsath P, Szalay L: Decreased proximal tubular transport capacity after renal sympathectomy. *Proceedings of the VIIth International Congress on Nephrology* Basel, Karger, 1978, pp 553–558.

27. Zambraski EJ, DiBona GF: Angiotensin II in antinatriuresis of low-level nerve stimulation. *Am J Physiol* 231:1105–1110, 1976.
28. Zambraski EJ, DiBona GF, Kaloyanides GJ: Effect of sympathetic blocking agents on the antinatriuresis of reflex nerve stimulation. *J Pharmacol Exp Ther* 198:464–472, 1976.
29. Zambraski EJ, DiBona GF, Kaloyanides GJ: Specificity of neural effect on renal tubular sodium reabsorption. *Proc Soc Exp Biol Med* 151:543–546, 1976.

CHAPTER TWENTY-SEVEN

Adrenergic-Opioid Interactions and the Central Control of Blood Pressure in Hypertension*

GEORGE KUNOS, M.D., Ph.D., MARIA DOLORES RAMIREZ-GONZALEZ, Ph.D., and CSABA FARSANG, M.D.

CARDIOVASCULAR EFFECTS OF OPIATES

The identification of stereoselective receptors for opiates in the brain (68, 88, 93) and the subsequent discovery of endogenous opioid peptides (45) have stimulated intensive research into the possible physiological role of these substances. The involvement of endogenous opioid peptides in the control of pain sensation (26, 35), behavioral (50) and endocrine functins (61) is well documented. Although a role for the endogenous opioids in cardiovascular regulation is less clear, two lines of evidence are suggestive. The first is the well-documented cardiovascular effectiveness of various opiates in mediating changes in central sympathetic outflow, which indicates the existence of specific opiate receptors in the brain; the second consists of observations that the opiate antagonist naloxone, an otherwise pharmacologically inert compound, causes changes in cardiovascular parameters (see below).

Morphine and morphine-like opiates are known to cause hypotension and bradycardia (16). In addition to peripheral mechanisms (16), central inhibition of sympathetic and potentiation of parasympathetic outflow via opiate receptors in the medulla oblongata are involved in these effects (22, 27, 56). The signs of increased sympathetic activity during opiate withdrawal (1, 33) indirectly suggest a sympathoinhibitory action of narcotic opiates. Centrally administered vasopressor responses to morphine are observed when the compound is administered into the lateral cerebral ventricles (22). An opiate-mediated facilitation of the central pressor response to angiotensin is also described (91). Opioid peptides also have centrally mediated cardiovascular effects. These vary greatly with the type of peptide, dose, anesthesia, species, and the central site of administration (4). β-Endorphin and some enkephalins administered intracisternally or into the fourth cerebral ventricle have naloxone-reversible depressor actions (23, 27), whereas peptides injected into the lateral ventricles usually have pressor effects (23, 81, 83, 89). Both the pressor and depressor effects of centrally administered opioid peptides appear

* C. Farsang, M.D., was supported by a postdoctoral fellowship of the S.S.H.R.C. of Canada, and M.D. Ramirez-Gonzalez, Ph.D., by a fellowship of the Consejo Nacional de Ciencia 7 Tecnologia, Mexico. Studies from the authors' laboratory were supported by grants from the Medical Research Council of Canada, the Quebec Heart Foundaton, and Boehringer/ Ingelheim of Canada, Ltd.

to be mediated by opiate receptors (7). Most evidence suggests that delta-receptors are involved in pressor responses (72a), although the nature of the receptor subtype involved may vary with the site of injection (23a). Intravenous injections of enkephalins into normotensive rats increases blood pressure (89); intravenous β-endorphin has a hypotensive effect (59).

Endogenous opioid peptides released from the brain may produce similar effects (11, 12). Studies by Dashwood and Feldberg (11, 12) show that naloxone, which does not significantly influence blood pressure and heart rate in intact animals, causes acute rises in blood pressure and heart rate after various surgical procedures that reduce visceral input into the brain (12). The results of several recent studies indicate that naloxone can reverse hypotension associated with various physiological and pathological conditions: such observations suggest that release of endogenous opioid peptides mediate the hypotension in various forms of shock (9, 17, 42, 47), spinal injury (18), and in inhalational anesthesia (2).

CARDIOVASCULAR EFFECTS OF CENTRAL α-ADRENOCEPTOR STIMULATION

Clonidine is the first compound shown to have a centrally mediated antihypertensive action. Clonidine is an α-adrenergic agonist with a high degree of selectivity for α_2-adrenoceptors (101). At higher concentrations it also binds to α_1-receptors, on which it has a weak stimulatory (101) as well as a competitive inhibitory effect (78). Stimulation of peripheral vascular α-receptors accounts for the acute, transient rise in blood pressure after intravenous administration (52). A central α_1-receptor-mediated pressor mechanism has been recently described (43) which may also contribute to the transient hypertensive response. The prolonged decrease in blood pressure that follows the initial pressor effect of clonidine is believed to be mediated by α_2-adrenoceptors in the central nervous system (CNS) (71), although histamine H_2 (49) and muscarinic cholinoceptors (90) are also implicated. These α_2-receptors are probably located in the pontomedullary region with additional sites in the hypothalamus (99). Their activation results in a decrease in central sympathetic and increase in parasympathetic outflow. Clonidine also produces bradycardia; the central α_2-receptors mediating this effect may be located in the nucleus of the solitary tract (NTS) (80), and their activation would facilitate the baroreceptor reflex. Decreased norepinephrine (NE) release caused by peripheral presynaptic α_2-receptor stimulation does not seem to contribute to the hypotensive action of clonidine, as other potent α_2-selective agonists that do not readily cross the blood-brain barrier are devoid of any hypotensive effect (40). The exact sites of action of clonidine in the brain and their possible pre- or postsynaptic localizations have been extensively discussed in recent reviews (46, 52, 85, 99).

The antihypertensive action of α-methyldopa involves a mechanism similar to that of clonidine. α-Methyldopa is first taken up into sympathetic nerve terminals where it is converted into α-methylnorepinephrine (α-MeNE). The α-MeNE released by nerve impulses in the brain stimulates α_2-receptors similar or identical to those involved in the effects of clonidine, although some evidence points to different sites of action for the two drugs in the brain (99). The requirement for intraneuronal transformation of α-methyldopa accounts for the onset of its hypotensive action being slower than that of clonidine.

SIMILARITIES BETWEEN THE EFFECTS OF OPIATES AND CLONIDINE

Although clonidine and opiates act on distinct receptors in the CNS (1, 19), there are remarkable similarities and overlaps in their effects. Both clonidine and opiates can reduce central sympathetic outflow and, consequently, blood pressure and heart rate. Activation of receptors in the pontomedullary region are involved in these effects, although the localization of the receptors for the two types of compounds is not strictly overlapping (85). The naloxone-reversible analgesic effect of morphine-like compounds and endogenous opioids is well-documented (26); noradrenergic mechanisms are implicated in these effects, and α-receptor antagonists inhibit opiate analgesia (54, 102). Clonidine is also a potent analgesic (84), and this effect in rats could be partially inhibited by naloxone (5, 60), although lack of antagonism has also been reported (24). Withdrawal symptoms after discontinuation of either opiate (32, 33) or clonidine treatment (29) include signs of increased sympathetic activity. Moreover, clonidine was found effective in reversing symptoms of opiate withdrawal in animals (95) and man (32, 33), and morphine reverses symptoms of clonidine withdrawal (94). Some endocrine effects of opiates and clonidine are similar: both compounds increase growth hormone release (55, 61) and suppress the release of the neurotransmitters acetylcholine and NE (14, 44). These negative feedback effects are mediated by α_2 and opiate receptors, possibly located on nerve terminals. The coexistence of independent α_2 and opiate receptors on central (1) or peripheral neurons (14) may explain many of the above overlaps in the effects of the two compounds. For example, either morphine or clonidine suppresses the activity of noradrenergic neurons in the locus coeruleus (1). A possible explanation of the reported reversal of symptoms of opiate withdrawal by clonidine (33, 71) is: morphine withdrawal in addicts leads to increased activity of these neurons, which are then suppressed by clonidine acting on α_2-receptors located on the same cells (32). A more direct interaction between the opiate and adrenergic receptor systems is suggested by the findings of Kosterlitz and co-workers (30), who report that isolated ilea from guinea pigs that had been implanted with morphine pellets *in vivo* show a reduced presynaptic response to clonidine when morphine is absent from the tissue bath ("withdrawn" state), and that this reduction can be reversed by the *in vitro* addition of opiate agonists. Facilitation of the α_2-adrenergic response by opiate receptor activation at a yet undefined postreceptor site may account for these observations (30). The relationship between central adrenergic and opiate mechanisms involved in blood pressure regulation is discussed in detail below.

ADRENERGIC-OPIATE INTERACTION IN THE CENTRAL CONTROL OF BLOOD PRESSURE IN SPONTANEOUSLY HYPERTENSIVE RATS (SHR)

The above similarities between the central effects of clonidine and opiates led us to test the possibility that activation of opiate receptors may be involved in the antihypertensive action of clonidine. As the hypotensive effect of clonidine is much greater in hypertensive than in normotensive rats (3, 65), genetically hypertensive male rats of the Okamoto-Aoki strain (SHR) were selected for this study. Detailed results of most of these experiments are published (19, 20, 21, 21a, 53, 75, 76). Blood pressure and heart rate of the unanesthetized animals are measured by the tail cuff technique. Intravenous injection of various doses of clonidine cause a dose-dependent fall in blood pressure and heart rate, as illustrated in Figure 27.1. When the opiate-receptor antagonist naloxone is injected 5 minutes before clonidine, both responses decrease significantly in a dose-related fashion, the lowest dose of naloxone that causes inhibition being 0.2 mg/kg i.p. Thus, the inhibitory potency of naloxone is somewhat lower than its potency to block the effects of morphine. However, naloxone is 10 to 15 times less potent on δ- than on μ-receptors (25, 31), and its potency in inhibiting the cardiovascular effects of clonidine is compatible with an interaction with δ-receptors, which are thought to mediate the actions of opioid peptides.

When naloxone is given after clonidine or α-methyldopa, it causes an acute reversal of the hypotension and for clonidine, also the bradycardia (Fig. 27.2). Note that the reversal caused by naloxone given during the plateau phase of the α-methyldopa response lasts for about 1 hour, which is consistent with the short half-life of parenterally administered naloxone (62). A single i.p. injection of 2 mg/kg naltrexone, an opiate antagonist with a prolonged duration of action (62), inhibits responses to clonidine or α-methyldopa for up to 12 hours (20). When given alone, neither naloxone nor naltrexone has any effect on blood pressure or heart rate (20, 76).

Since anesthesia interferes with cardiovascular regulation (73), the interaction of naloxone and clonidine was also tested in SHR anesthetized with Nembutal. Careful dosage of Nembutal, 40 mg/kg i.p., is necessary in these experiments, because higher doses significantly suppress blood pressure in SHR, close to levels in anesthetized normotensive control rats. Blood pressure and heart rate are continuously monitored through an intraarterial catheter connected to a pressure transducer and polygraph. Intravenous injection of 5 μg/kg clonidine causes a transient increase in blood pressure followed by prolonged hypotension and bradycardia. Naloxone, 2 mg/kg given 5 minutes before the administration of clonidine, attenuates the hypotension and bradycardia, but does not affect the initial rise in pressure (20). This observation supports the possibility that naloxone inhibits the action of clonidine in SHR at a central site. Further experiments show that naloxone reverses the antihypertensive effect of chronic clonidine treatment in SHR (20).

As clonidine and opiates act at distinct receptor sites in the CNS, a concern is the selectivity and specificity of their interaction. The possibility that naloxone inhibits hypotensive drug effects in SHR nonspecifically was tested by using the direct smooth muscle relaxant sodium nitrite. Sodium nitrite, 10 mg/kg i.v., lowers systolic blood pressure in unanesthe-

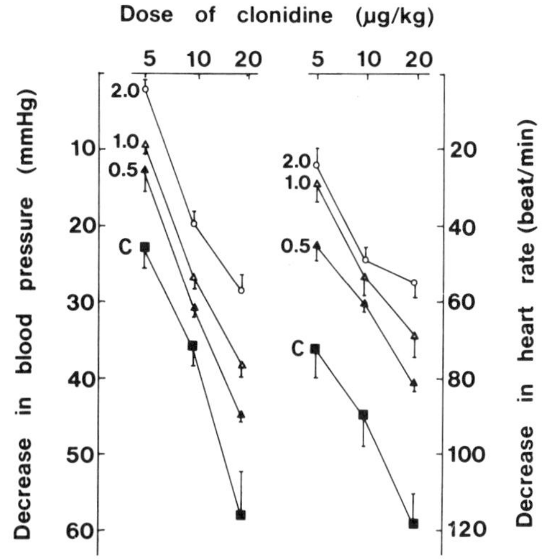

Figure 27.1. Naloxone antagonism of the hypotensive and bradycardic effects of clonidine in unanesthetized SHR. Clonidine, 5 to 20 μg/kg, was injected i.v. in the absence of naloxone (*C*, ■) or 5 minutes after the i.p. injection of 0.5 (▲), 1.0 (△) or 2.0 mg/kg of naloxone (○). Points and vertical bars represent means and S.E.s as obtained in six different animals. The mean systolic blood pressure and heart rate were 196.0 ± 2.9 mm Hg and 435.9 ± 5.2 beats/min before naloxone and 194.8 ± 1.7 mm Hg and 431.4 ± 6.3 beats/min 5 minutes after 2 mg/kg of naloxone.

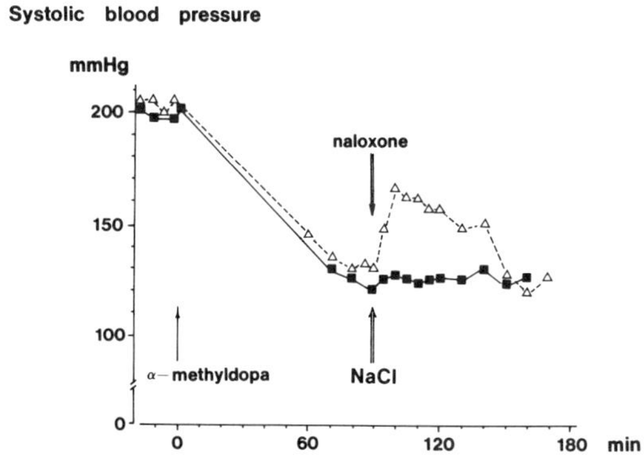

Figure 27.2. Reversal of the hypotensive effect of α-methyldopa by naloxone in a spontaneously hypertensive rat. α-Methyldopa, 100 mg/kg, was injected i.p. at 0 minute on two different occasions. At 90 minutes, the animal was injected i.p. with either 0.6 ml of NaCl or 2 mg/kg of naloxone in the same volume of saline. Naloxone increased blood pressure from 130 to 166 mm Hg (△), whereas NaCl only increased blood pressure from 122 to 130 mm Hg (■).

tized animals from 192 ± 2 to 169 ± 3 mm Hg and causes a reflex increase in heart rate from 424 ± 3 to 476 ± 6 beats/min. Naloxone, 2 mg/kg i.p., does not alter blood pressure or heart rate, and the effect of subsequently administered sodium nitrite is also largely unaffected: blood pressure decreased from 195 ± 2 to 174 ± 4 mm Hg, and heart rate rose from 431 ± 6 to 484 ± 7 beats/min. This observation supports the selectivity of the effect of naloxone in reversing clonidine-induced hypotension. Naloxone is known to have effects other than the antagonism of opiate receptors (82). However, the finding that different kinds of opiate antagonists were equally effective reduces the possibility that such nonspecific actions account for the reversal of the response to central α-receptor agonists. The possibility remains that clonidine may nonspecifically interact with central opiate receptors, or that opiate antagonists inhibit central α-adrenergic receptors. These possibilities were tested by direct binding studies using brain membrane preparations from SHR (19). Figure 27.3*A* shows that binding of the α-adrenoceptor binding probe, [^{3}H]dihydroergocryptine, is significantly suppressed by low concentrations of clonidine, probably as a result of the interaction of the two drugs at α_2-receptors, whereas naloxone is completely ineffective. Because the concentration of naloxone in the cerebrospinal fluid of rats after an i.v. dose of 2 mg/kg is less than 10^{-6} M (28), block of central α-receptors by naloxone is unlikely. Figure 27.3*B* shows that the stereoselective binding of [^{3}H] naloxone was suppressed by clonidine only at the very high concentration of 10^{-4} M, which is more than two orders of magnitude higher than the peak plasma or brain concentration calculated to occur after the highest clonidine dose used in our experiments (48). These observations confirm that, at the hypotensive

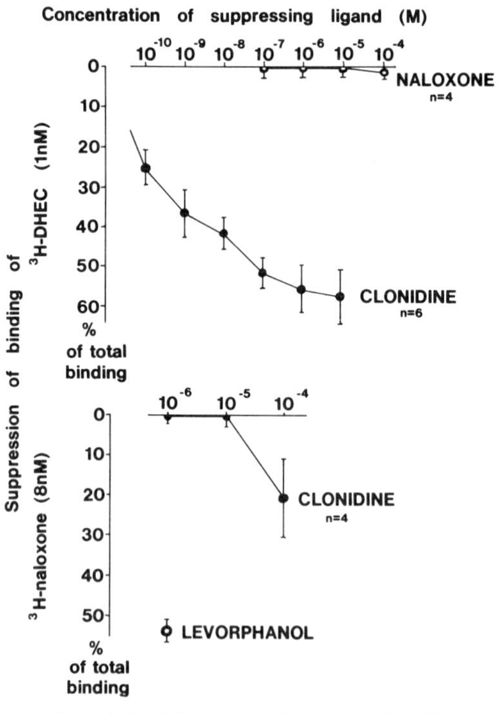

Figure 27.3. Interaction of clonidine and naloxone with opiate and α-adrenergic receptor binding sites in brain membranes from SHR. Crude membranes from whole brains of SHR were prepared as in ref. 64. α-Adrenoceptors (*top panel*) were identified by ^{3}H-dihydroergocryptine (*^{3}H-DHEC*) binding according to ref. 37. Competition-displacement studies with clonidine and naloxone were done as decribed in ref. 18. Opiate receptors (*bottom panel*) were identified by ^{3}H-naloxone binding according to ref. 64. Total binding of ^{3}H-naloxone represents binding in the presence of 1 μM dextrorphan. *Vertical bars* indicate 2 × S.E.

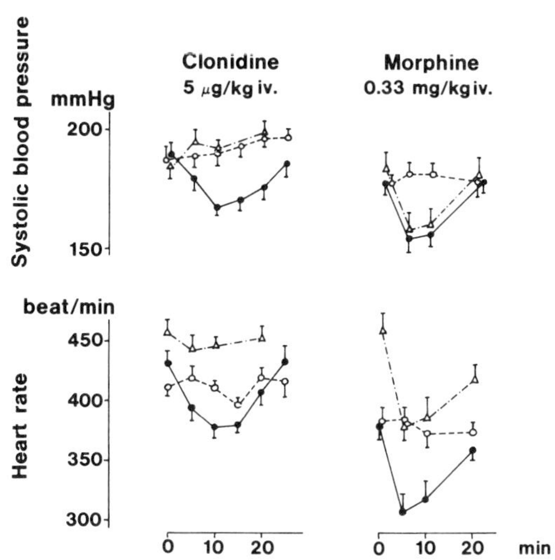

Figure 27.4 The effects of yohimbine and naloxone on the hypotension and bradycardia produced by clonidine or morphine in SHR. Clonidine (*left panels*) or morphine (*right panels*) were injected i.v. at 0 minute under control conditions (O), 5 minutes after the i.p. injection of 2 mg/kg of naloxone (O), or 30 minutes after the i.p. injection of 1 mg/kg of yohimbine (Δ). *Vertical bars* with horizontal lines represent means and S.E.s from four to eight experiments. Note that the hypotension and bradycardia produced by clonidine were abolished by either naloxone or yohimbine, whereas the similar effects of morphine were only inhibited by naloxone.

doses used, clonidine does not interact with opiate receptors, and that the inhibition of its *in vivo* effects by naloxone in SHR cannot be due to block of central α_2-receptors by naloxone.

Without a direct interaction between the two types of drugs, the possible mechanism of their interaction remains to be established. Adrenergic α_2 and opiate receptors involved in central depressor mechanisms may be arranged in parallel or in series. To distinguish between these two possibilities we studied the hypotensive and bradycardic actions of clonidine and of an equiactive dose of morphine in SHR, and tested the sensitivity of these responses to naloxone and the selective α_2-receptor antagonist yohimbine. The dose of yohimbine used (1 mg/kg) is sufficient to inhibit α_2-receptor-mediated responses in rats without influencing the α_1-receptor-mediated vasoconstrictor tone (36), the presence of which is necessary to monitor changes in central sympathetic outflow. The results in Figure 27.4 show that whereas the effects of i.v. clonidine are inhibited by either naloxone or yohimbine, similar effects of morphine are only inhibited by naloxone and not yohimbine. The increased basal heart rate after yohimbine is due probably to inhibition of cardiac presynaptic α_2-receptors, whereas the lack of change in basal blood pressure after yohimbine confirms the absence of a significant block of postsynaptic vascular α_1-receptors. The finding that opiate antagonists inhibit the hypotensive action of clonidine, but α_1-receptor blockade does not inhibit the similar action of opiate agonists, suggests that activation of opiate receptors is secondary to activation of α_2-receptors by clonidine and occurs at sites distal from the site of α_2-receptors in the sympathetic effector pathway.

Although the above experiments do not distinguish between central and possible peripheral localization of these opiate receptors, experiments with centrally administered clonidine and naltrexone favor the former. In these latter experiments (75) a chronic cannula is implanted in SHR so that the tip is wedged in the aquaeductus Sylvii. Drugs administered through the cannula reach the floor of the fourth cerebral ventricle. Clonidine, 10 μg given intracerebroventricularly (i.c.v.) produces prolonged hypotension and bradycardia, confirming the central origin of these effects (52, 85, 99). In the same rats the effects of i.c.v. administered naltrexone (100 μg) on the hypotensive and bradycardic response to i.v. administered clonidine was also tested. Centrally given naltrexone inhibits blood pressure but not heart rate responses to clonidine, suggesting that both α_2 and opiate receptors that are involved in the hypotensive effects of clonidine are located in the medulla, close to the fourth ventricle, whereas the opiate receptors involved in the bradycardic effects of clonidine in SHR may have a different location. These latter receptors may be located at a central site not readily reached by naltrexone via diffusion from the fourth ventricle. Alternatively, they may be located in the periphery; cardiac presynaptic α_2-receptors appear to participate in the bradycardic effect of clonidine (13), and a possible adrenergic-opiate interaction at a peripheral site has not yet been tested.

Although some of the i.c.v. administered naltrexone may have leaked out into the peripheral circulation, the plasma concentrations thus achieved would be much lower than concentrations following systemic administratin of the higher dose required to inhibit the actions of clonidine (20). However, the most compelling evidence for a central action of a β-endorphin-like opioid comes from experiments with β-endorphin antiserum. In four SHR implanted with a chronic cannula for i.c.v. injections, the effects of 10 μg/kg i.v. clonidine were tested 90 minutes after the i.c.v. injection of either control rabbit serum (2 μl) or a rabbit antiserum to human β-endorphin, kindly provided by Dr. M. Chretien (also 2 μl). The data in Table 27.1 indicate that β-endorphin antiserum significantly attenuates the hypotensive and bradycardic responses to clonidine (75).

The involvement of opiate receptors in the hypotensive and bradycardic actions of clonidine is further supported by the results of experiments in which SHR are treated with naltrexone chronically (2 mg/kg/day for 14 days, stopped 2 days before the experiment), to cause a supersensitivity of opiate receptors (86). The potentiation of the hypotensive and bradycardic effects of clonidine by chronic naltrexone treatment, shown in Figure 27.5, can be explained by the sensitization of opiate receptors and the indirect action of clonidine via these receptors.

LACK OF ADRENERGIC-OPIOID INTERACTION IN NORMOTENSIVE RATS

The effects of antihypertensive drugs may be different in normotensive and hypertensive animals (6). The adrenergic-opiate interaction observed in SHR has been tested in genetically matched Kyoto Wistar normotensive rats (WKY) (20). Figure 27.6 shows that the hypotension caused by clonidine in WKY is unaffected by naloxone, in striking contrast to SHR. The hypotensive effect of α-methyldopa is also unaffected by opiate antagonists (20). A similar lack of naloxone effect was found in unilaterally nephrectomized Sprague-Dawley rats used as normotensive controls of steroid salt and renal hypertensive rats (see below). Thus, it appears that the adrenergic-opiate interaction identified in SHR is absent or inactive in normotensive rats. This raises the question of whether the adrenergic-opiate interaction in SHR is related to the hypertensive process itself or whether it is genetically determined

Table 27.1
The Effects of i.c.v. Administered β-Endorphin Antiserum on the Cardiovascular Response to Clonidine in Unanesthetized SHR[a]

	Blood Pressure (mm Hg)		Heart Rate (beats/min)	
Pretreatment (i.c.v.)	Basal	Change by Clonidine	Basal	Change by Clonidine
Control rabbit serum	214 ± 9	−41 ± 12	405 ± 18	−63 ± 9
β-Endorphin antiserum	213 ± 7	−15 ± 4[b]	402 ± 17	−40 ± 8[b]

[a] Systolic blood pressure and heart rate were measured by the tail cuff technique (75). Clonidine, 10 μg/kg, was injected i.v. 90 minutes following the i.c.v. injection of 2 μl of control rabbit serum or 2 μl of β-endorphin antiserum. Means and SE of four experiments are shown.
[b] Significant difference from value after control rabbit serum treatment. For further details see ref. 75.

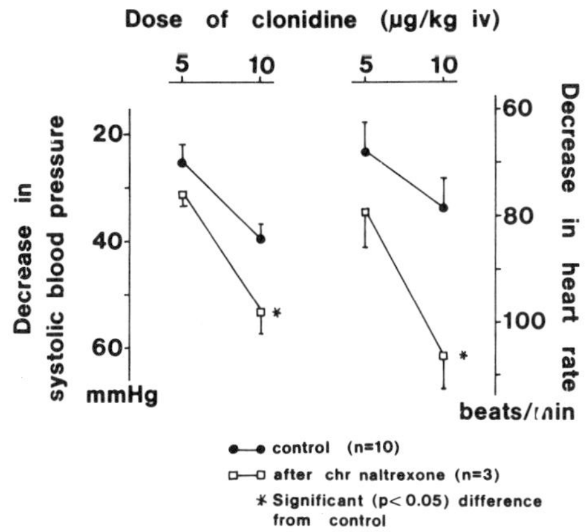

Figure 27.5. Potentiation of the hypotensive and bradycardic effects of clonidine by chronic naltrexone treatment of SHR. Naltrexone treatment was 2 mg/kg/day ip. for 14 days. Measurements of blood pressure and heart rate by the tail cuff technique were done 2 days after the last dose of naltrexone. *Vertical bars* indicate S.E. of means.

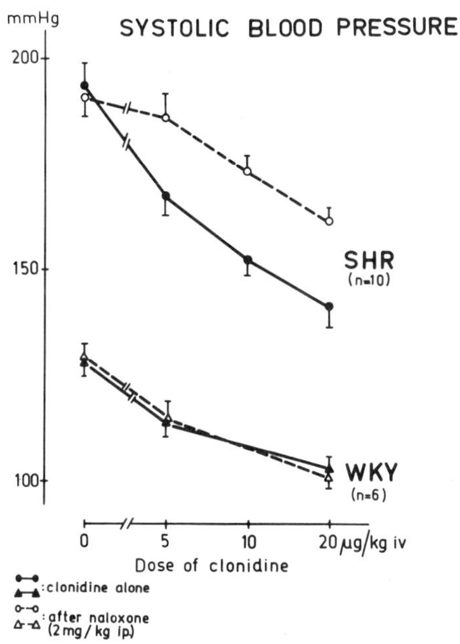

Figure 27.6. The effect of naloxone on clonidine-induced hypotension in SHR and WKY. Single intravenous doses of clonidine were tested before and 5 minutes after the i.p. injection of 2 mg/kg naloxone. *Vertical bars* indicate S.E. of means.

and limited to this strain. Preliminary studies providing an answer to this question are discussed under Adrenergic-Opioid Interaction in Different Models of Experimental Hypertension.

Although the reason for the above difference between SHR an WKY is not yet clear, it is noteworthy that the hypotensive and bradycardic actions of clonidine are significantly smaller in WKY than in SHR (20). This is not simply due to the lower baseline values because (a) the hypotension and reflex tachycardia caused by sodium nitrite are similar in the two strains (19) and (b) basal heart rate is not lower but even higher in WKY (20). Although changes at the level of central α_2-receptors could conceivably account for the different clonidine sensitivity, studies in our laboratory so far have failed to reveal any significant difference in the properties of [^{3}H]clonidine binding sites in membranes obtained from whole brainstem of SHR and WKY. Localized differences in α_2-receptors limited to certain brainstem nuclei are possible, however, as illustrated by a selective increase in the density of [^{3}H]clonidine binding sites in the hypothalamus of young SHR compared to WKY rats (63a). Facilitation of the release of an endogenous opioid involved in the effects of clonidine is an alternative possibility.

RELEASE OF AN ENDORPHIN-LIKE OPIOID BY CENTRAL α_2-RECEPTOR STIMULATION IN SHR

The experiments discussed so far demonstrate that in SHR inhibition of opiate receptors or induction of their supersensitivity can antagonize or potentiate responses to clonidine, respectively. A possible explanation is that in SHR direct activation of central α_2-receptors is associated with the release of an endogenous opioid. β-Endorphin is suspected for several reasons. In physiological studies β-endorphin is far more potent and longer acting than enkephalins (39). β-Endorphin is present in the hypothalamus and pontomedullary region (37, 64), and its calcium-dependent release suggests that it is a neurotransmitter (66). β-Endorphin has potent hypotensive and bradycardic actions (57, 59), whereas the enkephalins often have pressor effects (83, 89), potentiated even in SHR (81). Furthermore, an α-receptor-mediated increase in β-endorphin release from rat pituitary cells has been demonstrated (97).

Slices of the brainstem that included the medulla, pons, and hypothalamus have been obtained from SHR and WKY and perfused *in vitro*. β-Endorphin immunoreactivity (β-EI) was measured in the effluent buffer by radioimmunoassay (53). In slices from WKY, clonidine (10^{-8} M) does not significantly influence the amount of β-EI released (53), in contrast to preparations from SHR where the same concentration of clonidine causes a 3-fold increase in release of β-EI (Fig. 27.7). One micromolar of yohimbine, a selective α_2-receptor antagonist, reduces the release and abolishes the effect of added clonidine, suggesting that this effect of clonidine is mediated by α_2-receptors. A specific receptor interaction is further supported by the finding that *l*-α-MeNE, but not its biologically inactive *d*-stereoisomer, significantly increases β-EI release (Fig. 27.7). The data in this figure also show that naloxone (10^{-6} M) does not influence the effect of clonidine upon release, suggesting that naloxone antagonizes the effect of clonidine *in vivo* by blocking the postsynaptic response to the endogenous opioid rather than by inhibiting its release.

The above observations strongly suggest that release of an endorphin-like opioid from the brain is involved in the antihypertensive action of central α-receptor stimulation in SHR. The ability of clonidine to induce release of β-EI was recently confirmed. In normotensive Sprague-Dawley rats, clonidine, in doses

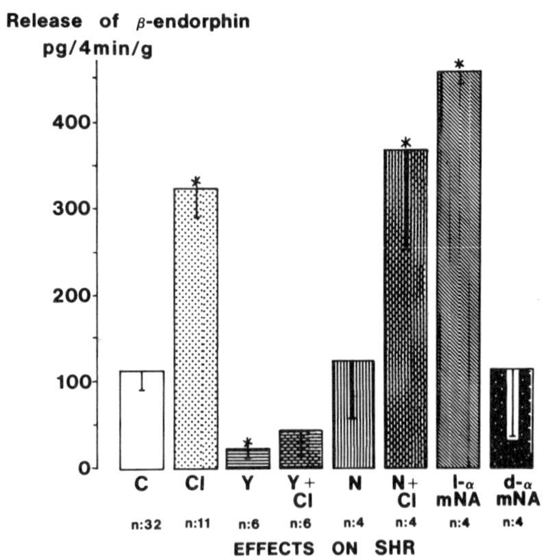

Figure 27.7. Release of β-endorphin immunoreactive material from brainstem slices of SHR (for details see ref. 76). *C*, control period; *Cl*, clonidine, 10^{-8} M; *Y*, yohimbine, 10^{-6} M; *N*, naloxone, 10^{-6} M; *l-αmNA*, *l*-α-methylnoradrenaline, 10^{-7} M; *d-αmNA*, *d*-α-methylnoradrenaline, 10^{-7} M. *Vertical bars* indicate S.E. of means from 4 to 32 experiments. Values are expressed as picograms of human β-endorphin equivalent per gram of wet brain tissue, released over a period of 4 minutes.

higher than those required to produce hypotension in SHR (0.1 to 1 mg/kg s.c.), was found to increase serum levels of β-EI, and α-receptor antagonists prevented this effect (69). The source of this β-EI is probably the pituitary, as hypophysectomy eliminated the increase, and clonidine also increased the release of β-EI *in vitro*, in cultured cells of rat anterior pituitary (70). Again, the concentrations of clonidine required for release were 10 to 100 times higher (70) than those found effective in brainstem slices of SHR (53). This could suggest that if release mechanisms in the brainstem and pituitary are similar, the absence of release from the brainstem of WKY by low concentrations of clonidine reflects a quantitative rather than qualitative difference.

Further studies are required to elucidate the nature of the opioid released from the brainstem of SHR (53). The antibody used (53) does not cross-react with the enkephalins, α-MSH or γ-endorphin, but does cross-react with β-lipotropin (53). Although in the rat brain β-lipotropin represents a very small fraction of β-EI (64), reports indicate that compounds with molecular weights smaller than that of β-endorphin can cross-react with the β-endorphin antibody (8, 41). Since the effluent fluid was not fractionated (53), undefined compounds may have contributed to the β-EI detected. It is also possible that release of β-EI by clonidine *in vitro* and naloxone antagonism of its effect *in vivo* are parallel but not necessarily cause and effect events.

ADRENERGIC-OPIOID INTERACTION IN DIFFERENT MODELS OF EXPERIMENTAL HYPERTENSION

To determine whether the adrenergic-opiate interaction detected in SHR is related to the hypertensive process itself, it was useful to examine the possibility of a similar interaction in other models of experimental hypertension (74, 75) . Weekly injections of deoxycorticosterone pyvalate (10 mg s.c.) and 1% saline in the drinking water induces steroid-salt hypertension in unilaterally nephrectomized male Sprague-Dawley rats within 4 to 6 weeks. In these animals α-methyldopa, 100 mg/kg i.p., has a transient hypertensive effect followed by marked and sustained hypotension. Naloxone given during the plateau phase of the hypotensive response acutely increases blood pressure, as shown in Table 27.2. In unilaterally neph-

Table 27.2
Naloxone Reversal of the Antihypertensive Action of Clonidine but not Propranolol in Hypertensive Patients[a]

	Clonidine Treatment			Propranolol Treatment		
	Control	Clonidine	Clon. + Naloxone	Control	Propranolol	Propr. + Naloxone
Blood pressure (mm Hg, systolic/diastolic)	164 ± 6/101 ± 2	131 ± 10/82 ± 1[b]	169 ± 8/99 ± 1	165 ± 5/100 ± 3	131 ± 8/83 ± 4[b]	132 ± 7/85 ± 4[b]
Heart rate (beats/min)	78 ± 2	64 ± 1[b]	80 ± 4	80 ± 3	65 ± 3[b]	67 ± 3[b]

[a] Five hypertensive patients of the "reacting" group were treated with clonidine, 3 × 0.1 mg/day for 3 days. Several weeks later and after a second drug-free period of 2 weeks, the same patients were treated with propranolol, 3 × 40 mg/day for 6 days. On the last day of either drug treatment period the patients were given a single i.v. injection of 0.4 mg naloxone, the blood pressure and heart rate were monitored at frequent intervals, as described in ref. 21. Values after naloxone represent peak effects. Means and standard errors are shown.

[b] Significant difference ($p < 0.05$) from control value.

rectomized, normotensive control rats, α-methyldopa is ineffective, and naloxone given 90 minutes after α-methyldopa also fails to influence blood pressure.

Renal hypertension is induced in male Sprague-Dawley rats by unilateral nephrectomy and partial obstruction of the arterial blood flow to the contralateral kidney, either by a silver clip or by ligation of one of the two branches of the renal artery (Goldblatt one-kidney model). Control animals have a unilateral nephrectomy only. In renal hypertensive rats, α-methyldopa (100 mg/kg i.p.) causes significant reduction in blood pressure. When the same dose of α-methyldopa is given 30 minutes after the i.p. injection of 2 mg/kg naltrexone, the decrease in blood pressure is significantly less than in the absence of naltrexone.

Clonidine was also tested in steroid and renal hypertensive rats. However, responses to the usual i.v. doses of clonidine (5 to 20 μg/kg) vary widely: the initial hypertensive phase is significantly greater in these models than in SHR, and it often masks the subsequent hypotensive response, which makes the interaction with naloxone difficult to test. Similar findings are reported and are attributed to a sensitization of the postsynaptic vascular α-receptor in hypertension, which is greater in steroid-salt and renal hypertension than in SHR (10).

ADRENERGIC-OPIOID INTERACTION IN HUMAN ESSENTIAL HYPERTENSION

The experiments described above suggest that the observed adrenergic-opioid interaction is related to the development of hypertension in the animals used. It is of interest whether a similar interaction can also be detected in human hypertensive disease. The effect of naloxone on the cardiovascular actions of clonidine was tested in 27 hospitalized patients with uncomplicated, mild to moderate essential hypertension (21). In a double-blind crossover study, clonidine is given orally in a dose of 0.1 mg three times a day for 3 days and significantly reduces systolic and diastolic blood pressure and heart rate in all patients, whereas placebo is ineffective. A 0.4-mg dose of naloxone given i.v. on the third day of clonidine treatment causes a rapid increase in blood pressure and heart rate in 14 patients (reacting group), whereas it was ineffective in the remaining 13 patients (nonreacting group). Peak changes in blood pressure and heart rate are shown in Table 27.2. Naloxone given during the placebo period is ineffective in all patients. Patients in the reacting and nonreacting groups are similar in terms of average age or previous antihypertensive medication. However, both the hypotension induced by clonidine and the rebound increase in blood pressure after stopping clonidine treatment are greater in the reacting than in the nonreacting group. Thus, it appears that in a subset of patients with essential hypertension, an endogenous opiate is involved in the antihypertensive action of clonidine.

A recent report of the failure of naloxone (Rodgers JF, Cubeddu LX: *Clin Pharmacol Ther* 34:68, 1983) to inhibit the hypotensive response to clonidine in six hypertensive patients may be related to a preponderance of nonrespondents in that small group. Also, differences in the way drugs were administered (continuous infusion of naloxone followed by a single large dose of clonidine in that study) may influence their possible interaction, as we found that the hypotensive response to a single dose of clonidine is more resistant to block by naloxone than the effect of chronic treatment of patients (unpublished).

ENDOGENOUS OPIOIDS AND THE ANTIHYPERTENSIVE ACTION OF PROPRANOLOL

Propranolol is becoming one of the most widely used drugs in the treatment of hypertension. Despite numerous hypotheses put forth to account for the antihypertensive action of β-receptor antagonists, their mechanism of action is still unclear (87). Propranolol causes an acute reduction of sympathetic tone by an action in the CNS (77), and this effect is believed to have a role in its chronic antihypertensive action in man (96). Increasing evidence indicates that the central sympathoinhibitory effect of propranolol is mediated indirectly, by release of NE in the CNS and subsequent stimulation of central α_2-receptors: centrally administered propranolol causes hypotension and bradycardia and a parallel increase in NE concentration in the cerebrospinal fluid (92). Furthermore, the hypotensive response to propranolol can be inhibited by destruction of noradrenergic nerve terminals in the CNS by 6-hydroxydopamine (63), depletion of their NE stores by centrally administered reserpine (51), or block of the effects of the released NE at central α_2-receptors by phentolamine (92) or rauwolscine (98). Thus, the centrally mediated hypotensive action of propranolol and clonidine appear to be similar, which raises the possibility of endogenous opioid involvement in the action of propranolol. This possibility has recently been tested in SHR (76). In pentobarbital-anesthetized, 14-week-old SHR intraperitoneally administered naltrexone, 2 mg/kg, inhibited the acute hypotension and bradycardia caused by the intracisternal injection of 50 μg ($\pm$) propranolol (76). When young SHR were treated with propranolol chronically, 60 mg/kg/day orally through the drinking water from age 4 to 14 weeks, systolic blood pressure and heart rate of the unanesthetized animals were significantly lower than in untreated littermates. When the propranolol-treated animals were given a single intraperitoneal injection of 2 mg/kg of naloxone or naltrexone 2 days after the last dose of propranolol, blood pressure and heart rate acutely rose to or close to levels in the untreated littermates. In these latter animals the injection of a similar dose of the opiate antagonists did not significantly influence blood pressure and only caused a slight and transient bradycardia.

These findings are interpreted to indicate that in SHR activation of opiate receptors, possibly through the release of an endogenous opioid, is involved in the centrally mediated hypotensive and bradycardic actions of propranolol. They further suggest that this mechanism is also active in the chronic antihypertensive action of propranolol, which supports the idea that this effect has an important CNS component.

In hypertensive humans, however, the mechanism of action of propranolol may be different: five hypertensive patients of the "reacting" group, tested earlier with clonidine (see text under previous heading), were treated with oral propranolol, 120 mg/day, for 6 days. This treatment caused a reduction of blood pressure and heart rate similar to that seen in the same patients with clonidine. However, unlike during clonidine treatment, the intravenous injection of 0.4 mg naloxone on the 6th day of propranolol treatment failed to increase either blood pressure or heart rate (Table 27.2).

DISCUSSION AND CONCLUSIONS

The centrally mediated cardiovascular effects of clonidine and α-methyldopa are inhibited or reversed by opiate antagonists in rats with various forms of hypertension (19, 20, 75), as well as in a subgroup of patients with essential hypertension (21, 21a). This interaction indicates that the antihypertensive action of central α-receptor stimulation involves the release of an endogenous opioid. The inhibition of the centrally mediated hypotensive action of propranolol by opiate antagonists may involve a similar mechanism (76), although this latter interaction appears to be limited to SHR.

Several lines of evidence support this conclusion. First, there is no direct interaction between clonidine and opiate receptors or naloxone and α_2-adrenoceptors that could explain the *in vivo* interaction between the two types of drugs. Second, while the hypotension and bradycardia caused by clonidine are inhibited either by naloxone or by yohimbine, only naloxone inhibits the similar actions of morphine. This suggests that opiate receptors mediating inhibition of sympathetic and potentiation of parasympathetic outflow from the brain are located distally from the medullary α_2-receptors mediating similar effects. Finally, *in vitro* experiments showed that central α_2-receptor stimulation (*i.e.*, clonidine) in the brainstem of SHR, but not WKY, results in an increased release of β-endorphin-immunoreactive material (53). As β-endorphin has potent hypotensive actions of its own, this suggests that the opioid involved in the effects of clonidine is an endorphin-like compound. This possibility is strongly supported by the inhibition of the cardiovascular actions of clonidine by i.c.v. injection of β-endorphin antiserum in rats with three different models of hypertension (75). A similar antagonism has been reported between β-endorphin antiserum and α-methylnorepinephrine injected into the NTS.

An important aspect of the observations presented here is the absence of an adrenergic opioid interaction in normotensive rats. A similar lack of interaction between clonidine and naloxone has been found in normotensive Wistar rats (16a) and in normotensive humans (100), although failure of these authors to consider differences between normotensive and hypertensive conditions led them to incorrect generalizations. Interestingly, in some recent studies in normotensive Sprague-Dawley rats (4a, 40a) or normotensive human opiate addicts (77a), opiate receptor antagonists did reverse clonidine-induced hypotension.

Whereas such differences may be due to differences in resting sympathetic tone resulting in variable responses to clonidine, they also indicate that hypertension potentiates an existing regulatory mechanism, rather than triggers its *de novo* development.

It is interesting to note that the results of several recent studies suggest a relationship between central mechanism controlling blood pressure and those involved in reactions to pain and noxious stimuli (15, 103–105). Activation of baroreceptors in rats by phenylephrine infusion acutely reduces sensitivity to aversive, painful stimuli, an effect that is abolished by baroreceptor denervation (15). Other studies show that pain sensitivity is reduced in three different forms of experimental hypertension in rats, but can revert to normal sensitivity with naloxone or by allowing blood pressure to return to normal after removing the cause of hypertension (103). The decreased pain sensitivity of the hypertensive animals is subsequently shown to be associated with increased brain levels of endogenous opioids (105), and decreased pain sensitivity is also found in a group of patients with essential hypertension (104).

The observed difference between hypertensive and normotensive rats also raises the question of the mechanism and central site of the cardiovascular actions of clonidine. A recent study indicates that destruction of the nucleus of the solitary tract (NTS) abolishes the hypotensive and bradycardic effects of clonidine in SHR but does not significantly influence these effects in WKY (80). Several sites within the pontomedullary region are implicated in the cardiovascular actions of clonidine (42, 85, 99), and the above findings indicate that the relative importance of various sites may change in hypertension. It is possible that in SHR, release of an endorphin-like opiate occurs either in the NTS or in a pathway involving the NTS.

Alternatively, clonidine can release β-endorphin from the hypothalamus (66), which is also the possible source of β-endorphin released from our brainstem preparation of SHR (53). Clonidine can also release β-endorphin from the anterior pituitary of rats (70), and the clonidine-induced increase in plasma β-endorphin in hypertensive humans (21a) is probably also of pituitary origin. Peptides released from either site can reach brainstem sites involved in cardiovascular regulation (5a , 76a). Further studies are required to distinguish between these possibilities.

While opiate antagonists reverse the hypotensive effect of central α-receptor stimulation, when given alone they do not influence blood pressure and heart rate, either in normotensive or in hypertensive conditions. This indicates that, as suggested by Goldstein for endorphin-systems in general (34), the endorphin mechanism involved in central blood pressure control is in a quiescent, standby state, so blockade by naloxone can only be demonstrated if the system is first activated. The reduced pain sensitivity in hypertension discussed above is not contrary to such a possibility, because testing pain sensitivity involves activation by a pain stimulus. Alternatively, the absence of naloxone effects on blood pressure in the unstimulated state may reflect a balance between opioid-mediated pressor and depressor mechanisms. Enkephalins preferentially produce pressor responses when applied in certain brain areas (4, 23, 83), and such effects are potentiated in SHR (89). An enkephalin-mediated inhibition of baroreflex sensitivity is postulated (72), which is opposite to the baroreflex facilitation by narcotic analgesics and some endogenous opiates (58). In agreement with the possible involvement of endogenous opioids in both pressor and depressor mechanisms, we recently reported that naloxone inhibits clonidine bradycardia, but potentiates baroreflex bradycardia, and both effects are present only in SHR and not in WKY (75). Thus, endogenous opioid peptides may be involved in the development of hypertension as well as in compensatory mechanisms aimed at maintaining blood pressure within normal limits. Further studies are needed to clearly establish the nature of the opioids and the localization of their central pathways in the regulation of cardiovascular homeostasis.

References

1. Aghajanian GK: Tolerance to locus coeruleus neurons to morphine and suppression of withdrawal response by clonidine. *Nature* 276:186, 1978.
2. Arndt JO, Freye E: Opiate antagonist reverses the cardiovascular effects of inhalational anesthesia. *Nature* 277:399, 1979.
3. Atkinson J, Boillat N, Pera-Bally R, Peters-Haefeli L, Kirchertz EJ: Effect of clonidine treat-

ment and its abrupt cessation on mean blood pressure of rats with normal and elevated blood pressure. *Clin Sci* 57:195, 1979.

4. Bellet M, Elghozi JL, Meyer P, Pernollet MG, Schmitt H: Central cardiovascular effects of narcotic analgesics and enkephalins in rats. *Br J Pharmacol* 71:365, 1980.

4a. Bennett DA, DeFeo JJ, Elko EE, Lal H: Naloxone-induced reversal of clonidine but not hydralazine, hypotension. *Drug Dev Res* 2:175, 1982.

5. Bentley GA, Starr J, Newton S: Evidence for a peripheral component in the antinociceptive action of morphine and α-agonist drugs. *Proc Aust Physiol Pharmacol Soc* 11:37, 1980.

5a. Bergland R, Blume H, Hamilton A, Monica P, Paterson R: Adrenocorticotropic hormone may be transported directly from the pituitary to the brain. *Science* 210:541, 1980.

6. Bhagat BD: Possible mechanisms of action of propranolol in hypertension. *Biochem Pharmacol* 27:1698, 1978.

7. Bolme P, Fuxe K, Agnati LF, Bradley R, Smythies J: Cardiovascular effects of morphine and opioid peptides following intracisternal administration in chloralose-anesthetized rats. *Eur J Pharmacol* 48:319, 1978.

8. Borvendeg J, Graf L, Hermann I, Palkovits M, Meretey K: Radioimmunoassay of β-endorphin: Immunoreactive substances in the rat brain and the pituitary. In Graf L, Palkovits M, Ronai AZ: *Endorphins '78*. Amsterdam, Excerpta Medica, 1978, p 177.

9. Curtis MT, Lefer AM: Protective actions of naloxone in hemorrhagic shock. *Am J Physiol* 239:H416, 1980.

10. Dadkar NK, Aroskar VA, Dohadwalla AN: Differential antihypertensive effects of clonidine in different models of experimental hypertension in rats. *J Pharm Pharmacol* 31:264, 1979.

11. Dashwood MR, Feldberg W: A pressor response to naloxone. Evidence for release of endogenous opioid peptides. *J Physiol* 281:30P, 1978.

12. Dashwood MR, Feldberg W: Release of opioid peptides in anesthetized cats. *Br J Pharmacol* 68:697, 1980.

13. De Jonge A, Timmermans PB, MWM, van Zwieten PA: Participation of cardiac presynaptic α_2-adrenoceptors in the bradycardine effects of clonidine and analogues. *Nauyn Schmiedebergs Arch Pharmacol* 317:8, 1981.

14. Dubocovich ML, Langer SZ: Pharmacological differentiation of presynaptic inhibitory α-adrenoceptors and opiate receptors in the cat nictitating membrane. *Br J Pharmacol* 70:383, 1980.

15. Dworkin BR, Filevich RJ, Miller NE, Craigmyle N, Pickering TG: Baroreceptor activation reduces reactivity to noxious stimulation: Implications for hypertension. *Science* 205:1299, 1979.

16. Eckenhoff JE, Oech SR: The effects of narcotics and antagonists upon respiration and circulation. *Clin Pharmacol Ther* 1:483, 1960.

16a. Elghozi JL, Bellet M, Meyer P: Central pressor action of enkephalins in rats. In Buckley JP, Ferrario CM: *Central Nervous System Mechanisms in Hypertension*. New York, Raven Press, 1981, p 249.

17. Faden AI, Holaday JW: Opiate antagonists: A role in the treatment of hypovolemic shock. *Science* 205:317, 1979.

18. Faden AI, Jacobs TP, Holaday JW: Opiate antagonist improves recovery after spinal injury. *Science* 211:493, 1981.

19. Farsang C, Kunos G: Naloxone reverses the antihypertensive effect of clonidine. *Br J Pharmacol* 67:161, 1979.

20. Farsang C, Ramirez-Gonzalez MD, Mucci L, Kunos G: Possible role of an endogenous opiate in the cardiovascular effects of central alpha adrenoceptor stimulation in spontaneously hypertensive rats. *J Pharmacol Exp Ther* 214:203, 1980.

21. Farsang C, Kapocsi J, Juhasz I, Kunos G: Possible involvement of an endogenous opiate in the antihypertensive effect of clonidine in patients with essential hypertension. *Circulation* 66:1268, 1982.

21a. Farsang C, Vajda L, Kapocsi J, Malisak Z, Alfoldi S, Varga K, Juhasz I, Kunos G: Diurnal rhythm of beta endorphin in normotensive and hypertensive patients: The effect of clonidine. *J Clin Endocrinol Metab* 56:865, 1983.

22. Feldberg W, Wei E: The central origins and mechanisms of cardiovascular effects of morphine as revealed by naloxone in cats. *J Physiol* 272:99P, 1977.

23. Feldberg W, Wei E: Central cardiovascular effects of enkephalins and C fragment of lipotropin. *J Physiol* 280:18P, 1978.

23a. Feurstein G, Faden AI: Differential cardiovascular effects of μ, δ, κ opiate agonists at discrete hypothalamic sites in the anesthetized rat. *Life Sci* 31:2197, 1982.

24. Fielding S, Wilker J, Hynes M, Szewczak M, Novick WJ, Lal H: A comparison of clonidine with morphine for antinociceptive and antiwithdrawal actions. *J Pharmacol Exp Ther* 207:899, 1978.

25. Fields HL, Emson PC, Leigh Gilbert R, Iversen LL: Multiple opiate receptor sites on primary afferent fibers. *Nature* 284:351, 1980.

26. Foley KM, Kourides IA, Inturrisi CE, Kaiko RF, Zaroulis CG, Posner JB, Houde RW, Li CH: β-Endorphin: Analgesic and hormonal effects in humans. *Proc Natl Acad Sci USA* 76:5377, 1979.

27. Freye E, Arndt JO: Perfusion of the fourth cerebral ventricle with fentanyl induces naloxone reversible bradycardia, hypotension, and EEG synchronization in conscious dogs. *Naunyn Schmiedebergs Arch Pharmacol* 307:123, 1979.

28. Fu TC, Halenda S, Lawrence L, Chau-Pham TT, Martin BR, Dewey WL: Evidence for the blockade of the release of the morphine-induced endogenous opiates by naloxone. *Fed Proc* 38:364A, 1979.

29. Geyskes GG, Boer P, Durhout Mees EJ: Clonidine withdrawal. Mechanism and frequency of rebound hypertension. *Br J Clin Pharmacol* 7:55, 1979.

30. Gillan MGC, Kosterlitz HW, Robson LE, Water-

field AA: The inhibitory effects of presynaptic α-adrenoceptor agonists on contraction of guinea-pig ileum and mouse vas deferens in the morphine-dependent and withdrawn states produced *in vitro*. *Br J Pharmacol* 66:601, 1979.

31. Gillan MGC, Kosterlitz HW, Paterson SJ: Comparison of the binding characteristics of tritiated opiates and opioid peptides. *Br J Pharmacol* 70:481, 1980.
32. Gold MS, Redmond DE Jr, Kleber HD: Clonidine blocks acute opiate withdrawal symptoms. *Lancet* 2:599, 1978.
33. Gold MS, Redmond DE, Kleber HD: Noradrenergic hyperactivity in opiate withdrawal supported by clonidine reversal of opiate withdrawal. *Am J Psychol* 136:100, 1979.
34. Goldstein A: Opioid peptides (endorphins) in pituitary and brain. *Science* 193:1081, 1976.
35. Graf L, Szekely JI, Ronai AZ, Dunai-Kovacs Z, Bajusz S: Comparative study on analgesic effect of Met[5]-enkephalin and related lipotropin fragments. *Nature* 263:240, 1976.
36. Graham RM, Stephenson WH, Pettinger WA: Pharmacological evidence for a functional role of the prejunctional alpha-adrenoreceptor in noradrenergic neurotransmission in the conscious rat. *Naunyn Schmiedebergs Arch Pharmacol* 311:129, 1980.
37. Gramsch C, Kleber G, Hollt V, Pasi A, Mehraein P, Herz A: Pro-opiocortin fragments in human and rat brain: β-endorphin and α-MSH are the predominant peptides. *Brain Res* 192:109, 1980.
38. Greenberg DA, U'Prichard DC, Sheehan P, Snyder SH: α-Noradrenergic receptors in the brain: Differential effects of sodium on binding of ^{3}H-antagonists. *Brain Res* 140:378, 1978.
39. Guillemin R: Peptides in the brain: The new endocrinology of the neuron. *Science* 202:390, 1978.
40. Haeusler G: Studies on the possible contribution of a peripheral presynaptic action of clonidine and dopamine to their vascular effects under *in vivo* conditions. *Naunyn Schmiedebergs Arch Pharmacol* 295:191, 1976.

40a. Hamilton TC, Longman SD: A comparison of the cardiovascular and sedative actions of alpha-adrenoceptor agonists FLA-136 and clonidine in the rat. *Br J Pharmacol* 75:13, 1982.

40b. Hassen AH, Feuerstein GZ, Faden AI: Cardiovascular responses to opioid agonists injected into the nucleus of tractus solitarius of anesthetized cats. *Life Sci* 31:2193, 1982.

41. Ho W, Kwok KY, Lam S: Characterization of a plasma factor having opiate and immunoactivity like beta-endorphin. *Biochem Biophys Res Commun* 87:448, 1979.
42. Holaday JW, Faden AI: Naloxone reversal of endotoxin hypotension suggests role of endorphins in shock. *Nature* 275:450, 1978.

42a. Holaday JW, Ruvio BA, Robles LE, Johnson CE, d'Amato RJ: M 154,129, a putative delta antagonist, reverses endotoxic shock without altering morphine analgesia. *Life Sci* 31:2209, 1982.

43. Huchet AM, Chelly J, Schmitt H: Role of α_1 and α_2-adrenoceptors in the modulation of the baroreflex vagal bradycardia. *Eur J Pharmacol* 71:455, 1981.
44. Hughes J, Kosferlitz HW, Leslie FM: Effect of morphine on adrenergic transmission in the mouse vas deferens. Assessment of agonist and antagonist potencies of narcotic analgesics. *Br J Pharmacol* 41:239, 1975.
45. Hughes J, Smith TW, Kosterlitz HW, Fothergill LA, Morgan BA, Morris HR: Identification of two related pentapeptides from the brain with potent opiate agonist activity. *Nature* 258:577, 1975.
46. Isaac L: Clonidine in the central nervous system: Site and mechanism of hypotensive action. *J Cardiovasc Pharmacol* 2:S5, 1980.
47. Janssen HJ, Lutherer LO: Ventriculocisternal administration of naloxone protects against severe hypotension during endotoxin shock. *Brain Res* 194:608, 1980.
48. Jarrott B, Spector S: Disposition of clonidine in rats as determined by radioimmunoassay. *J Pharmacol Exp Ther* 207:195, 1978.
49. Karppanen H, Paakkari I, Paakkari P, Huotari R, Orma A-L: Possible involvement of central histamine H_2-receptors in the hypotensive effect of clonidine. *Nature* 259:587, 1976.
50. Kastin AJ, Olson RD, Schally AV, Coy DH: CNS effects of peripherally administered brain peptides. *Life Sci* 25:401, 1979.
51. Kelliher GJ, Buckley JP: Central hypotensive activity of dl- and d-propranolol. *J Pharm Sci* 59:1276, 1970.
52. Kobinger W: Central α-adrenergic systems as targets for hypotensive drugs. *Rev Physiol Biochem Pharmacol* 81:39, 1978.
53. Kunos G, Farsang C, Ramirez-Gonzalez MD: β-Endorphin: Possible involvement in the antihypertensive effect of central α-receptor activation. *Science* 211:82, 1981.
54. Kuraishi Y, Harada Y, Satah M, Takagi H: Antagonism by phenoxybenzamine of the analgesic effect of morphine injected into the nucleus gigantocellularis of the rat. *Neuropharmacology* 18:107, 1979.
55. Lal S, Tolis G, Martin JB, Brown GM, Guyda H: Effect of clonidine on growth hormone, and thyroid stimulating hormone in the serum of normal man. *J Clin Endocrinol Metab* 41:827, 1975.
56. Laubie M, Schmitt H, Canellas J, Roquebert J, Desmichel P: Centrally mediated bradycardia and hypotension induced by narcotic analgesics: dextromoramide and fentanyl. *Eur J Pharmacol* 28:66, 1974.
57. Laubie M, Schmitt H, Vincent M, Remond G: Central cardiovascular effects of morphinomimetic peptides in dogs. *Eur J Pharmacol* 46:67, 1977.
58. Laubie M, Schmitt H: Action of the morphinomimetic agent, fentanyl, on the nucleus tractus solitarii and the nucleus ambiguus cardiovascular neurons. *Eur J Pharmacol* 67:403, 1980.
59. Lemaire I, Tseng R, Lemaire S: Systemic administration of endorphin: Potent hypotensive effect involving a serotonergic pathway. *Proc Natl Acad Sci USA* 75:6240, 1978.

60. Lin MT, Chi ML, Chandra A, Tsay BL: Serotonergic mechanisms of beta-endorphin and clonidine-induced analgesia in rats. *Pharmacology* 20:323, 1980.
61. Meites J, Bruni JF, Van Vugt DA, Smith AF: Relation of endogenous opioid peptides and morphine to neuroendocrine functions. *Life Sci* 24:1325, 1979.
62. Misra AL: Metabolism of opiates. In Adler ML, Manara L, Samanin S: *Factors Affecting the Action of Narcotics.* New York, Raven Press, 1978, p. 297.
63. Montastruc JL, Montastruc P: Effect of intracisternal application of 6-hydroxydopamine on the antihypertensive action of propranolol in the dog. *Eur J Pharmacol* 63:103, 1980.
63a. Morris MJ, Devynck M-A, Woodcock EA, Johnston CI, Meyer P: Specific changes in hypothalamic alpha-adrenoceptors in young spontaneously hypertensive rats. *Hypertension* 3:516, 1981.
64. Ogawa N, Panerai AE, Lee S, Forsbach G, Havlicek V, Friesen HG: β-Endorphin concentration in the brain of intact and hypophysectomized rats. *Life Sci* 25:317, 1979.
65. Ohkusu Y: Susceptibility of the hypertensive rats to a new hypotensive agent, ST-155 (Catapres): 2(2,6-dichlorophenylamino)-2-imidazoline hydrochloride. *Jpn Circ J* 35:1175, 1971.
66. Osborne H, Przewlocki R, Holt V, Herz A: Release of β-endorphin from rat hypothalamus *in vitro*. *Eur J Pharmacol* 55:425, 1979.
67. Pert CB, Snyder SH: Properties of opiate-receptor binding in rat brain. *Proc Natl Acad Sci USA* 70:2243, 1973.
68. Pert CB, Snyder SH: Opiate receptor: Demonstration in nervous tissue. *Science* 179:1011, 1973.
69. Pettibone DJ, Mueller GD: α-Adrenergic stimulation by clonidine increases plasma concentration of immunoreactive β-endorphin in rats. *Endocrinology* 109:758, 1981.
70. Pettibone DJ, Mueller GP: Clonidine releases immunoreactive β-endorphin from rat pars distalis. *Brain Res* 221:409, 1981.
71. Pettinger WA: Pharmacology of clonidine. *J Cardiovasc Pharmacol* 2:S21, 1980.
71a. Petty MA, de Jong W: Does β-endorphin contribute to the central antihypertensive action of α-methyldopa? *Clin Sci* 63:293S, 1982.
72. Petty MA, Reid JL: The effects of opiates on baroreflex sensitivity in the normotensive rabbit. *Br J Pharmacol* 70:64P, 1980.
72a. Pfeiffer A, Feuerstein GZ, Faden AI, Kopin IJ: Evidence for an involvement of mu-, but not delta- or kappa-opiate receptors in cardiovascular responses to opiates upon intrahypothalamic microinjection. *Life Sci.* 31:1279, 1982.
73. Price HL: General anesthesia and general homeostasis. *Physiol Rev* 40:187, 1960.
74. Ramirez-Gonzalez MD, Kunos G: Endogenous opiates and the central adrenergic control of blood pressure in hypertension. *Proc Can Fed Biol Sci* 24:128A, 1981.
75. Ramirez-Gonzalez MD, Tchakarov L, Mosqueda-Garcia R, Kunos G: β-Endorphin acting on the brainstem is involved in the antihypertensive action of clonidine and α-methyldopa in rats. *Circ Res* 53:150, 1983.
76. Ramirez-Gonzalez MD, Farsang C, Tchakarov L, Kunos G: Opiate antagonists reverse the centrally mediated antihypertensive action of propranolol in spontaneously hypertensive rats. *Eur J Pharmacol* 81:167, 1982.
76a. Rapoport SI, Klee WA, Pettigrew KD, Ohno K: Entry of opioid peptides in the central nervous system. *Science* 207:84, 1980.
77. Reid JL, Lewis PJ, Myers MG, Dollery CT: *J Pharmacol Exp* 168:394, 1974.
77a. Resnick RB, Washton AM, Lal H: Reversal of clonidine-induced hypotension by opioid antagonists in man. In Way EL: *Endogenous and Exogenous Opiate Agonists and Antagonists.* New York, Pergamon Press, 1980, p. 565.
78. Ress RJ, Field FP, Lockley OE, Fregly MJ: Effect of clonidine on the contractile responsiveness of aortic smooth muscle to norepinephrine. *Pharmacology* 18:149, 1979.
79. Rockhold RW, Caldwell RW: Effect of lesions on the nucleus tractus solitarii on the cardiovascular actions of clonidine in conscious rats. *Neuropharmacology* 18:347, 1979.
80. Rockhold RW, Caldwell RW: Cardiovascular effects following clonidine microinjection into the nucleus tractus solitarii of the rat. *Neuropharmacology* 19:919, 1980.
81. Rockhold RW, Crofton JT, Share L: Increased pressor responsiveness to enkephalin in spontaneously hypertensive rats: The role of vasopressin. *Clin Sci* 59:235s, 1980.
82. Sawynok J, Pinsky C, LaBella FS: Minireview on the specificity of naloxone as an opiate antagonist. *Life Sci* 25:1621, 1979.
83. Schaz K, Stock G, Simon G, Schlor KH, Unger T, Rockhold R, Ganten D: Enkephalin effects on blood pressure, heart rate, and baroreceptor reflex. *Hypertension* 2:395, 1980.
84. Schmitt H, Le Douarec JC, Petillot N: Antinociceptive effects of some α-sympathomimetic agents. *Neuropharmacology* 13:289, 1974.
85. Schmitt H, Schmitt-Jubeau H, Daskalopoulos NT: Central mechanisms of clonidine. *Trends Pharmacol Sci* 1:71, 1979.
86. Schutz R, Wuster M, Herz A: Supersensitivity to opioids following the chronic blockade of endorphin action of naloxone. *Naunyn Schmiedebergs Arch Pharmacol* 306:93, 1979.
87. Scriabine A: β-Adrenoceptor blocking drugs in hypertension. *Annu Rev Pharmacol Toxicol* 19:269, 1979.
88. Simon EJ, Hiller JM, Edelman I: Stereospecific binding of the potent narcotic analgesic [^{3}H] etorphine to rat-brain homogenate. *Proc Natl Acad Sci USA* 70:1947, 1973.
89. Simon W, Schaz K, Ganten U, Stock G, Schlor KH, Ganten D: Effects of enkephalins on arterial blood pressure are reduced by propranolol. *Clin Sci Mol Med* 55:2375, 1978.
90. Srimal RC, Gulati K, Dhawan BN: On the mechanism of central hypotensive action of clonidine. *Can J Physiol Pharmacol* 55:1007, 1977.
91. Szilagyi JE, Ferrario CM: The role of endoge-

nous opiates in the area postrema pressor pathway. *Clin Sci* 59:267s, 1980.
92. Tackett RL, Webb JG, Privitera PJ: Cerebroventricular propranolol elevates cerebrospinal fluid norepinephrine and lowers blood pressure. *Science* 213:911, 1981.
93. Terenius L: Stereospecific interaction between narcotic analgesics and a synaptic plasma membrane fraction of rat cerebral cortex. *Acta Pharmacol Toxicol (Kbh)* 32:317, 1973.
94. Thoolen M, Timmermans P, Van Zwieten PA: Morphine suppresses the blood pressure responses to clonidine withdrawal in the spontaneously hypertensive rat. *Eur J Pharmacol* 71:351, 1981.
95. Tsenge LF, Loh HH, Wei ET: Effects of clonidine on morphine withdrawal signs in the rat. *Eur J Pharmacol* 30:93, 1975.
96. Turner P: Central nervous actions of β-adrenoceptor blocking drugs in man. *Trends Pharmacol Sci* 1:49, 1979.
97. Vale W, Rivier C, Yang L, Minick S, Guillemin R: Effects of purified hypothalamic corticotropin-releasing factor and other substances on the secretion of adrenocorticotropin and β-endorphin-like immunoactivities *in vitro*. *Endocrinology* 103:1910, 1978.
98. Van den Bosch RA, Nykamp FP, De Jong W: The cardiovascular effects of propranolol in adrenalectomized, spontaneously hypertensive rats and the opposing influence of rauwolscine. *Eur J Pharmacol* 71:505, 1981.
99. Van Zwieten PA, Timmermans P: The role of central α-adrenoceptors in the mode of action of hypotensive drugs. *Trends Pharmacol Sci* 1:39, 1979.
100. Watkins J, Fitzgerald G, Zamboulis C, Brown MJ, Dollery CT: Absence of opiate and histamine H_2 receptor-mediated effects of clonidine. *Clin Pharmacol Ther* 28:605, 1980.
101. Wikberg JES: Pharmacological classification of adrenergic α-receptors in the guinea pig. *Nature* 273:164, 1978.
102. Yaksh TL: Direct evidence that spinal serotonin and noradrenaline terminals mediate the spinal antinociceptive effects of morphine in the periaqueductal gray. *Brain Res* 160:180, 1979.
103. Zamir N, Segal M: Hypertension-induced analgesia: Changes in pain-sensitivity in experimental hypertensive rats. *Brain Res* 160:170, 1979.
104. Zamir N, Shuber E: Altered pain perception in hypertensive humans. *Brain Res* 201:471, 1980.
105. Zamir N, Simantov R, Segal M: Pain sensitivity and opioid activity in genetically and experimentally hypertensive rats. *Brain Res* 184:299, 1980.

CHAPTER TWENTY-EIGHT

Sympathetic Activity in Human Extremity Nerves and Its Relationship to Plasma Norepinephrine*

B. GUNNAR WALLIN, M.D.

INTRODUCTION

The recent development of new techniques for studying sympathetic nervous activity in man has widened the horizons of our investigations into this system. Previously, conclusions could only be drawn from indirect measures of sympathetic nervous activity, such as blood pressure and heart rate, but new methods allow us to study sympathetic nervous activity itself, levels of sympathetic neurotransmitter release from nerve terminals, and the interaction of this neurotransmitter with various adrenergic receptors. This chapter will review results obtained with one of these new techniques, microelectrode recordings of sympathetic action potentials in nerves supplying skin and muscles in man. Although visceral sympathetic nerves are still inaccessible for recordings in man, the technique provides several new insights into specific sympathetic reflex mechanisms and into the relationship between neuronal activity in the sympathetic nervous system and plasma concentrations of norepinephrine.

METHODS

Sympathetic action potentials are recorded in peripheral nerves with tungsten microelectrodes with a tip diameter of a few micrometers. The electrode is inserted manually through the skin into an underlying nerve and a reference electrode is placed subcutaneously a few centimeters away. The usual record obtained by this method gives multiunit signals, but occasionally single units can be detected. Most recordings are made in large nerves, such as the peroneal, median, or tibial nerve, but it is possible to record sympathetic nervous activity in smaller branches supplying the skin. The peripheral nerves in man are composed of several fascicles, and each fascicle is surrounded by connective tissue. The tungsten microelectrode records action potentials only when the tip has penetrated the fascicle, which prevents cross-talk from adjacent fascicles. Fascicles in the proximal parts of an extremity consist of a mixture of fibers which separate distally into individual fascicles to skin and to muscle. Nerves are impaled as far distally as possible to record from relatively "pure" muscle or skin nerve fascicles. Simultaneous measures may be made of blood pressure, blood flow, skin electrical resistance, and finger or toe photoelectric plethysmograms. Detailed descriptions of these techniques have been previously published (3, 8, 19, 22).

SYMPATHETIC OUTFLOW TO SKIN

Blood vessels and sweat glands in the human skin are innervated by separate sympathetic fibers with different conduction velocities (12).

* This work was supported by Swedish Medical Research Council, Grant No. B81-14X-03546-10C.

In nerves supplying the feet and the palms of the hands, vasoconstrictor and sudomotor impulses can be recorded from the same intrafascicular electrode site, suggesting that the fibers are intermingled. When a person is at rest at normal room temperature, spontaneous skin sympathetic activity (SSA) consists of bursts of impulses of irregular strength and duration (14). Some individual bursts of SSA are followed by transient changes in skin resistance and by vasoconstriction, indicating a simultaneous activation of vasoconstrictor and sudomotor impulses. Other bursts contain only vasoconstrictor impulses, whereas pure sudomotor bursts are unusual at normal room temperature (3). The arterial baroreceptors apparently do not modulate the outflow of vasoconstrictor impulses to the skin, but there is a loose coupling between sympathetic activity and respiratory rhythms, and a sudden deep breath will regularly evoke a strong burst of impulses (14).

Effect of Thermoregulatory and Mental Stimuli

Since the skin is important in thermoregulation, one may expect SSA to be affected by thermal stimuli. Body cooling leads to an increased outflow of vasoconstrictor impulses while moderate warming reduces the strength of activity to a minimum (3, 10). When body heating is associated with sweating, there is also an increase of SSA which is due to activation of sudomotor rather than vasoconstrictor impulses (3). Thus, by changing environmental temperature, selective activation or suppression of either the vasoconstrictor or sudomotor neural system can be obtained.

In contrast to thermoregulatory reflexes which cause opposite responses in sudomotor and vasomotor fibers, arousal and stress responses usually involve simultaneous activation of sudomotor and vasoconstrictor fibers. This simultaneous activation constitutes the basis for "cold sweat." As illustrated in Figure 28.1, unexpected arousal stimuli (such as sudden sound, sudden touch or sudden pain) regularly evoke a single burst of impulses after 0.5 to 1.0 seconds, depending on the recording site and the subject's height (12, 14). Emotional excitement caused by a stressful conversation (Fig. 28.1) or a request to solve an arithmetic problem usually leads to an increase of SSA which is of longer duration than that caused by arousal stimuli (10).

SYMPATHETIC OUTFLOW TO MUSCLES

Arterial Baroreflex Influence

Sympathetic bursts in muscle nerve fascicles are composed of vasoconstrictor impulses which are controlled by arterial baroreflexes. Figure 28.2 illustrates that the bursts are grouped in the heart rhythm, occur most frequently during transient blood pressure reductions, and disappear with rising blood pressure (8, 20, 23, 24, 27). An atrioventricular block with a sudden fall of diastolic blood pressure in a single heart beat is regularly followed by a strong sympathetic burst (Fig. 28.2) (24), indicating that a fall in pressure is sensed by the baroreceptors and counteracted by increasing vasoconstriction. When recording in the peroneal nerve at the knee, the latency between pressure fall and burst is 1.2 to 1.55 seconds (again depending on subject's height), which corresponds to the delay in the baroreflex arc (8, 12, 20).

A known reflex latency makes it possible to determine which heart beat corresponds to a given burst, and therefore which pressure parameters control the occurrence of activity. It was found that variations of muscle sympathetic activity (MSA) are determined by fluctuations of diastolic blood pressure (20). Strength of activity also depends on the direction of an ongoing blood pressure change, and Figure 28.2 shows that more bursts occur during falling than during rising diastolic blood pressure. In contrast to the relationship between dynamic variations of blood pressure and sympathetic activity, no systematic correlation was found when subjects were compared with regard to their mean levels of diastolic blood pressure and sympathetic activity. This suggests that MSA is important for buffering acute changes in blood pressure, but has little importance in setting long-term blood pressure levels.

Other evidence supports the idea that muscle sympathetic activity is affected primarily by *changes* in blood pressure. Application of subatmospheric pressure in a collar around the neck (neck suction) leads to increased transmural carotid sinus pressure and increased arterial baroreceptor firing. It was expected that sympathetic outflow would be inhibited, but continuous neck suction at −40 mm Hg led to only an initial inhibition corresponding to the application of suction with only minor

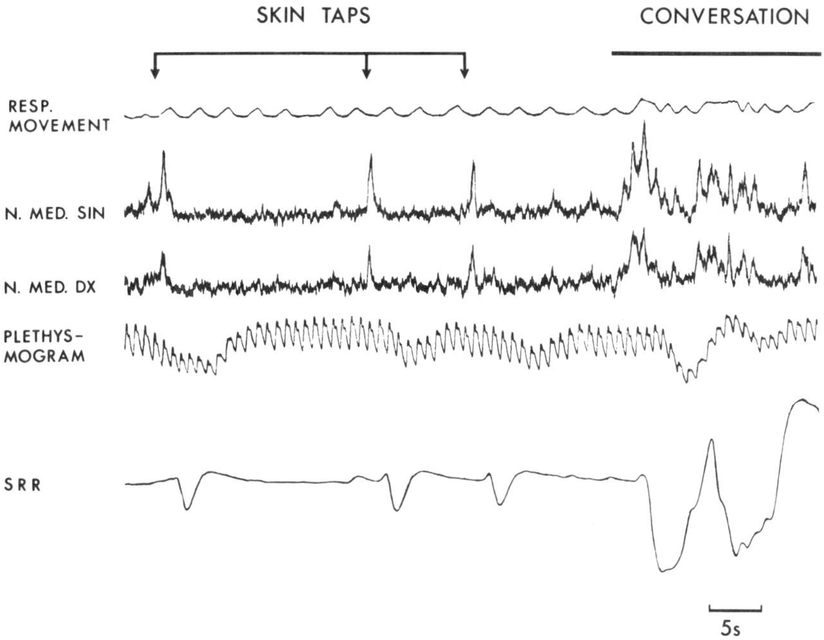

Figure 28.1. Sympathetic responses to weak touch stimuli (*left*) and stressful conversation (*right*) recorded in cutaneous branches of both median nerves in healthy subject. Nerve activities are shown as mean voltage neurograms (time constants 0.1 seconds). Skin resistance responses (*SRR*) and photoelectric finger pulse plethysmogram were recorded from the palm and third finger of the left hand, respectively. Respiratory movements are shown with inspiration upwards. Note the similarity between neural records. (Reproduced with permission from L. Lidberg and B. G. Wallin (16). Copyright 1981, The Society for Psychophysiological Research.)

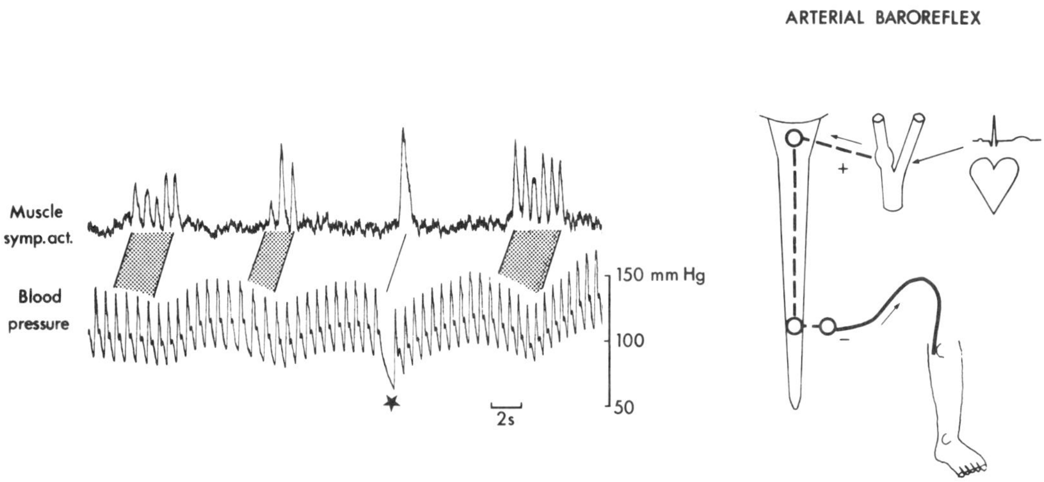

Figure 28.2. Relationship between spontaneous fluctuations of blood pressure and muscle nerve sympathetic activity (*left*) recorded in right peroneal nerve. Baroreflex accounts for the pulse synchrony of nerve activity and the inverse relationship to blood pressure fluctuations. †, diastolic blood pressure fall due to sudden AV block. *Stippled areas* indicate corresponding sequences of bursts and heart rates. (Reproduced with permission from B. G. Wallin *et al.* (28) and Oxford University Press.)

sustained effects (2). In another study (30a), neck suction or neck pressure was applied very rapidly and maintained for 5 seconds. There were clear transient changes of nerve activity, but adaption was almost complete within 1 to 2 seconds. Cardiac responses, on the other hand, showed much less adaptation, suggesting that the stimuli induced maintained changes of afferent baroreceptor traffic. Presumably, therefore, adaptation of muscle sympathetic outflow was mainly of central origin. Only carotid baroreceptors are affected directly by this type of stimulus, so we do not know whether there would be an equally pronounced adaptation if firing of all high pressure receptors were changed at the same time.

To summarize, available evidence indicates that: (a) rapid changes of arterial baroreceptor activity effectively modulate sympathetic outflow to muscles; (b) it is uncertain to what extent static changes of baroreceptor activity lasting more than a few seconds induce static changes of sympathetic activity in muscle nerves; (c) interindividual differences in arterial blood pressure are not related to interindividual differences in the level of muscle nerve sympathetic activity.

Influence from "Low-Pressure" Receptors

Apart from arterial baroreceptors, "low-pressure receptors" also influence muscle nerve sympathetic activity. These volume receptors are located primarily in the heart and large vessels entering the heart, but have also been found in hepatic vessels. A reduction of intrathoracic blood volume, brought about by applying subatmospheric pressure around the lower body, caused an increase in MSA which cannot be explained as an arterial baroreceptor effect and therefore was presumably elicited from the "low-pressure" receptors (21). This increase in sympathetic activity did not wear off for several minutes, so these receptors may exert a more static influence on muscle sympathetic activity than arterial baroreceptors.

Relationship to Heart Rate

If mean levels of heart rate and muscle sympathetic activity are compared in recumbent subjects, subjects with low heart rates tend to have more sympathetic activity than those with high heart rates (2, 29). These findings suggest that in recumbent subjects a low level of resting sympathetic activity is balanced by a high heart rate and vice versa. There is also evidence that baroreceptor-induced changes of heart rate and muscle sympathetic activity vary with preexisting tone in cardiac and muscle autonomic nerves.

Subjects with a low level of sympathetic activity in muscle nerves while recumbent had larger increases in sympathetic activity, and smaller changes in heart rate on transition to sitting or standing than subjects with many bursts while lying (6). There was also an inverse linear relationship between heart rate while lying down and the change in heart rate induced by a standardized orthostatic test (17).

Effect of Various Maneuvers

In contrast to skin sympathetic activity, the outflow of sympathetic impulses to muscles is not easily affected by minor disturbances such as arousal stimuli, which do not themselves affect blood pressure. Any situation associated with a change in blood pressure and/or central blood volume is likely to produce a change of muscle sympathetic activity, however. Figure 28.3 illustrates the response to the Valsalva maneuver, which causes transient reductions of both blood pressure and central blood volume, and reliably brings about an increase in muscle sympathetic activity (9). Changes in posture from reclining to sitting to standing induce progressive increases of muscle sympathetic activity (6). Drug-induced blood pressure reductions affect muscle sympathetic activity in different ways, depending on the mechanism of action of the drug. The vasodilator sodium nitroprusside regularly induces an increase of activity (B. G. Wallin *et al.*, unpublished observations), but clonidine, which has many modes of action, may lead either to a decrease or an increase of sympathetic outflow to muscles (30).

Emotional stress elicited by a request to solve an arithmetic problem and muscle work (strong handgrips) are examples of complex maneuvers usually associated with an increase in blood pressure. Although no systematic study of the response to these maneuvers has been made, both increases and decreases of muscle sympathetic activity have been observed (9, 23). Sympathetic outflow to muscles has been recorded during two instances of

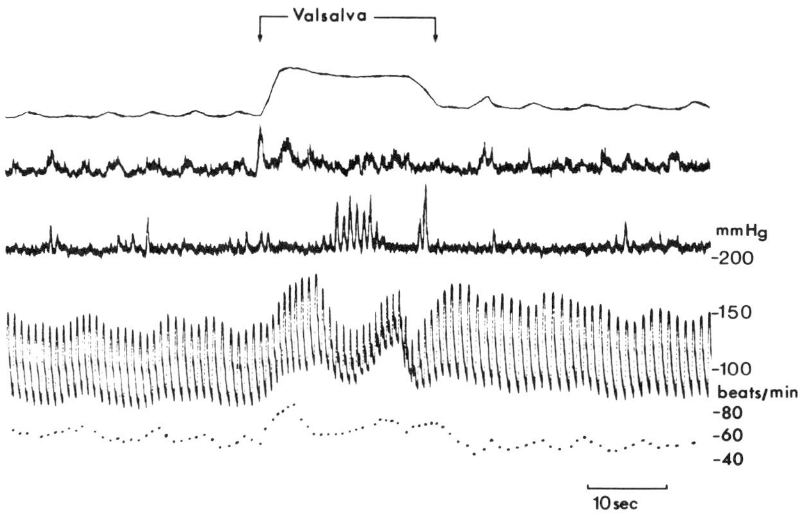

Figure 28.3. Effects of the Valsalva maneuver on sympathetic outflow to skin of left foot (*second trace*) and muscle of right lower leg (*third trace*) recorded simultaneously in left and right peroneal nerves of healthy subject. *Upper trace*, respiratory movements (inspiration upwards); *fourth trace*, blood pressure; *bottom trace*, instantaneous heart rate. Note different temporal patterns of activity at rest and different responses to maneuver in the two nerve types. (Reproduced with permission from B. G. Wallin *et al.* (25) and Hippokrates Verlag.)

vasovagal syncope (6, 30b). In both cases the syncope was associated with disappearance of sympathetic bursts at about the same time as bradycardia developed and blood pressure fell. It has been suggested that muscle vasodilation during vasovagal syncope is due to activation of vasodilator fibers (1), but our results show that there is inhibition of vasoconstrictor activity.

ESSENTIAL HYPERTENSION

Excess sympathetic nervous system activity has been considered an important factor in the pathogenesis of essential hypertension, but circulatory studies have failed to produce conclusive evidence supporting this hypothesis. Microelectrode recordings of sympathetic outflow at rest and during various mild stresses revealed no qualitative differences in muscle or skin sympathetic activity between normotensive and hypertensive subjects (23). This suggests that the sympathetic activity recorded in hypertensives is made up of impulses from fiber populations similar to those in normotensive subjects and that the outflow of these impulses is controlled by similar mechanisms in the two groups.

A recent study quantitated sympathetic outflow to muscles at rest in 12 hypertensive patients (27). As in normal subjects, the impulses were grouped in pulse-synchronous bursts which showed the usual inverse relationship to transient blood pressure variations. In the hypertensives, higher blood pressure levels were needed to inhibit the sympathetic bursts, indicating an elevated operating range of the baroreflex. This was probably a secondary effect of the hypertension caused by resetting of the arterial baroreceptors (18). As in normal subjects, the strength of the activity in hypertensives correlated best to diastolic blood pressure variations. For a given pressure value there was more activity if blood pressure was in a falling phase rather than a rising phase, and this directional sensitivity was at least as great as in normal subjects. Spontaneous variation of diastolic blood pressure was the same as in normal subjects.

When the mean diastolic blood pressure *level* was related to the mean *level* of sympathetic activity, there was a slight tendency for higher sympathetic activity in patients than in controls. However, since muscle sympathetic activity increases with age and since our patients were about 10 years older than controls, this difference was probably an effect of age. Taken together, the results suggest that there is no difference between sympathetic outflow to muscles in normotensive and hypertensive subjects that is likely to be of pathogenetic importance. The finding is not surprising, since vascular resistance of skeletal muscle is normal or even decreased in hypertensives (5). Consequently, studies of both effector organs and sympathetic nerves indicate that sympathetic outflow to the vascular bed of skeletal muscle does not contribute to the development of essential hypertension. It must be recognized, however, that this conclusion cannot be generalized to include other sympathetic nerves and other vascular beds. Since sympathetic outflow is controlled regionally, the possibility that there is an increased outflow of sympathetic impulses to some other cardiovascular region cannot be excluded. In view of the dynamic characteristics of the system and its way of adapting to sustained stimuli, it is also highly unlikely that arterial baroreflex mechanisms exerted via sympathetic outflow to muscles can protect man against the development of hypertension, irrespective of the underlying cause.

"SYMPATHETIC TONE"

Functional Organization of Sympathetic Outflow

Traditionally, sympathetic reactions were thought to be slow and protracted and to occur in parallel in different parts of the body. This view of a diffusely acting system led to the term "sympathetic tone" to describe the strength of a presumed global level of activity in sympathetic nerves. The concept of clinically describing some subjects as having "high sympathetic tone" is not tenable. When recordings are made selectively from sympathetic nerves supplying well defined effector organs, there are clear differences between the activity in one nerve and that in another (22, 25, 26). For example, when comparing skin and muscle sympathetic activity, the temporal patterns of activity are different even at rest. In addition, various maneuvers affect these activities differently. For example, body cooling increases skin sympathetic activity without changing muscle sympathetic activity. Figure 28.3 shows that the Valsalva maneuver always causes an increase in muscle sympathetic activity, but it does not consistently change skin sympathetic activity. In contrast to such evidence for differences between sympathetic outflows to skin and muscle, there is a remarkable parallelism between the two neurograms when sympathetic activity is recorded simultaneously in different muscle nerves (19, 23). Double recordings of sympathetic activity from skin nerves innervating the palm of the hands and the feet reveal a similar close parallelism between both neurograms (Fig. 28.1) (4, 23). These findings show that there are different populations of sympathetic neurons, each of which is subjected to its own homogenous supraspinal drive, which may be different from that of other populations. Consequently *there is no common "sympathetic tone."* If the term is to be retained, it should be used only when considering the sympathetic drive in specified nerves supplying well defined effector organs.

Relationship between Sympathetic Muscle Nerve Activity and Plasma Norepinephrine

Plasma levels of norepinephrine are often used as an index of sympathetic activity. This is partly based on the fact that norepinephrine is the transmitter released from postganglionic sympathetic neurons and partly on physiological studies showing that plasma levels of norepinephrine increase when sympathetic nerves to specific organs are stimulated electrically (13, 32) or when human subjects perform maneuvers known to increase activity in some sympathetic nerves (7, 11, 15).

Plasma levels of norepinephrine are known to vary markedly between human subjects, but in a given individual the level is fairly reproducible (7, 15). Older subjects tend to have higher levels of plasma norepinephrine than young subjects. Similarly, the level of sympathetic activity in muscle nerves in a given individual is remarkably constant from day to day over many months, but there are wide interindividual differences, with a tendency for increasing levels with increasing age (19, 20). Recently, a link between these two sets of observations was obtained (29) when a positive correlation was found between the frequency of sympathetic bursts in muscle nerves and plasma concentrations of norepinephrine in forearm venous blood in supine normotensive subjects (Fig. 28.4). This corre-

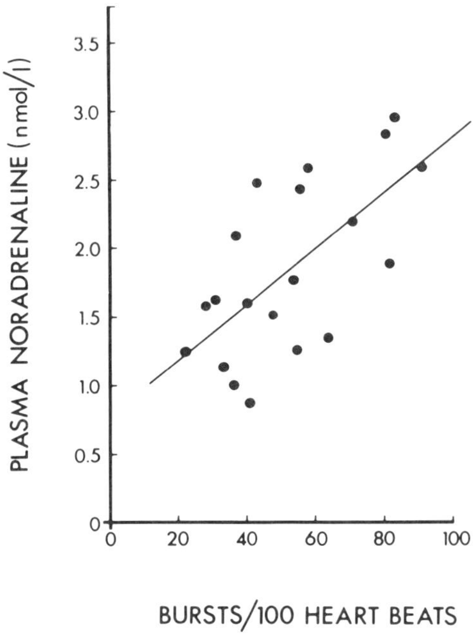

Figure 28.4. Relationship between level of sympathetic activity in muscle nerves (expressed as number of sympathetic bursts/100 heart beats) and plasma concentration of norepinephrine in forearm venous blood. Each point represents one healthy subject. (Reproduced with permission from B. G. Wallin *et al.* (29) and *Acta Physiologica Scandinavica*.)

lation was shown despite the fact that nerve recordings and blood samples were obtained on different occasions, 4 to 5 months apart.

One possible explanation for this finding is that the level of sympathetic activity in muscle nerves simply reflects or correlates with the level of activity in other sympathetic nerves. However, sympathetic outflows to different effector organs are controlled by different mechanisms, which makes it unlikely that the level of activity should be similar in all sympathetic nerves. A second and more likely possibility is that muscles, comprising 40% of total body weight, contain such a large number of adrenergic terminals that spillover from these is a major determinant of the plasma concentration of norepinephrine. The finding by Vendsalu (31) of a higher concentration of norepinephrine in the antecubital vein than in the brachial artery is in agreement with this hypothesis. Since norepinephrine is metabolized in the liver, concentration in the hepatic vein is low (31), and therefore the contribution to plasma from the splanchnic vascular bed may be less than would be expected on the basis of its size.

SUMMARY

Muscle sympathetic activity appears to be a major determinant of plasma norepinephrine levels, so these studies clarify why many investigators fail to find a relationship between plasma norepinephrine and blood pressure in subjects at rest. We find no relationship between the mean level of muscle sympathetic activity and blood pressure in normal or in hypertensive subjects (27). Thus, this major source of plasma norepinephrine has a level of activity unrelated to long-term blood pressure levels.

References

1. Barcroft H, Edholm OG: On the vasodilatation in human skeletal muscle during posthaemorrhagic fainting. *J Physiol (Lond)* 104:161–175, 1945.
2. Båth E, Lindblad LE, Wallin BG: Effects of dynamic and static neck suction on muscle nerve sympathetic activity, heart rate and blood pressure in man. *J Physiol (Lond)* 311:551–564, 1981.
3. Bini G, Hagbarth KE, Hynninen P, Wallin BG: Thermoregulatory and rhythm-generating mechanisms governing the sudomotor and vasoconstrictor outflow in human cutaneous nerves. *J Physiol (Lond)* 306:537–552, 1980.
4. Bini G, Hagbarth KE, Hynninen P, Wallin BG: Regional similarities and differences in thermoregulatory vaso- and sudomotor tone. *J Physiol (Lond)* 306:553–565, 1980.
5. Brod J, Fencl V, Hejl Z, Jirka J, Ullrych M: General and regional haemodynamic pattern underlying essential hypertension *Clin Sci* 23:339–349, 1962.
6. Burke D, Sundlöf G, Wallin BG: Postural effects on muscle nerve sympathetic activity in man. *J Physiol (Lond)* 272:399–414, 1977.
7. deChamplain J, Farley L, Cousineau D, van Ameringen MR: Circulating catecholamine levels in human and experimental hypertension. *Circ Res* 38:109–114, 1976.
8. Delius W, Hagbarth KE, Hongell A, Wallin BG: General characteristics of sympathetic activity in human muscle nerves. *Acta Physiol Scand* 84:65–81, 1972.
9. Delius W, Hagbarth KE, Hongell A, Wallin BG: Manoeuvres affecting sympathetic outflow in human muscle nerves. *Acta Physiol Scand* 84:82–94, 1972.
10. Delius W, Hagbarth KE, Hongell A, Wallin BG: Manoeuvres affecting sympathetic outflow in human skin nerves. *Acta Physiol Scand* 84:177–186, 1972.

11. de Quattro V, Chan S: Raised plasma catecholamines in some patients with primary hypertension. *Lancet* 1:806–809, 1972.
12. Fagius J, Wallin BG: Sympathetic reflex latencies and conduction velocities in normal man. *J Neurol Sci* 47:433–488, 1980.
13. Haefely W, Hürlimann A, Thoenen H: Relation between the rate of stimulation and the quantity of noradrenaline liberated from sympathetic nerve endings in the isolated perfused spleen of the cat. *J Physiol (Lond)* 181:48–58, 1965.
14. Hagbarth KE, Hallin RG, Hongell A, Torebjörk HE, Wallin BG: General characteristics of sympathetic activity in human skin nerves. *Acta Physiol Scand* 84:164–176, 1972.
15. Lake CR, Ziegler MG, Kopin IJ: Use of plasma norepinephrine for evaluation of sympathetic neuronal function in man. *Life Sci* 18:1315–1326, 1976.
16. Lidberg L, Wallin BG: Sympathetic skin nerve discharges in relation to amplitude of skin resistance responses. *Psychophysiology* 18:340–346, 1981.
17. Lindblad LE, Atterhög JH, Wallin BG: Sympathetic activity in muscle nerves—a factor influencing the postural heart rate increase? *Acta Physiol Scand* 111:509–510, 1981.
18. McCubbin JW, Green JH, Page IH: Baroreceptor function in chronic renal hypertension. *Circ Res* 4:205–210, 1956.
19. Sundlöf G, Wallin BG: The variability of muscle nerve sympathetic activity in resting recumbent man. *J Physiol (Lond)* 272:383–397, 1977.
20. Sundlöf G, Wallin BG: Human muscle nerve sympathetic activity at rest. Relationship to blood pressure and age. *J Physiol (Lond)* 274:621–637, 1978.
21. Sundlöf G, Wallin BG: Effect of lower body negative pressure on human muscle nerve sympathetic activity. *J Physiol (Lond)* 278:525–532, 1978.
22. Vallbo ÅB, Hagbarth KE, Torebjörk HE, Wallin BG: Somatosensory, proprioceptive and sympathetic activity in human peripheral nerves. *Physiol Rev* 59:919–957, 1979.
23. Wallin BG, Delius W, Hagbarth KE: Comparison of sympathetic nerve activity in normotensive and hypertensive subjects. *Circ Res* 33:9–21, 1973.
24. Wallin BG, Delius W, Sundlöf G: Human muscle nerve sympathetic activity in cardiac arrhythmias. *Scand J Clin Lab Invest* 34:293–300, 1974.
25. Wallin BG, Delius W, Hagbarth KE: Regional control of sympathetic outflow in human skin and muscle nerves. In Umbach W, Koepchen HP: *Central Rhythmic and Regulation*. Stuttgart, Hippokrates Verlag, 1974, pp 190–195.
26. Wallin BG: Recordings of impulses in unmyelinated nerve fibres in man: Sympathetic activity. *Acta Anaesthesiol Scand [Suppl]* 70:130–136, 1978.
27. Wallin BG, Sundlöf G: A quantitative study of muscle nerve sympathetic activity in resting normotensive and hypertensive subjects. *Hypertension* 1:67–77, 1979.
28. Wallin BG, Sundlöf G, Lindblad LE: Baroreflex mechanisms controlling sympathetic outflow to the muscles in man. In Sleight P: *Arterial Baroreceptors and Hypertension*, Oxford, Oxford University Press, 1980, pp 101–107.
29. Wallin BG, Sundlöf G, Eriksson BM, Dominiak P, Grobecker H, Lindblad LE: Plasma noradrenaline correlates to sympathetic muscle nerve activity in normotensive man. *Acta Physiol Scand* 111:69–73, 1981.
30. Wallin BG, Frisk-Holmberg M: On the antihypertensive mechanism of clonidine in man—evidence against a generalized reduction of sympathetic activity. *Hypertension* 3:340–346, 1981.
30a. Wallin BG, Eckberg DL: Sympathetic transients caused by abrupt alterations of carotid baroreceptor activity in man. *Am J Physiol* 242:H185–190, 1982.
30b. Wallin BG, Sundlöf G: Sympathetic outflow to muscles during vasovagal syncope. *J Auton Nerv Syst* 6:287–291, 1982.
31. Vendsalu A: Studies on adrenaline and noradrenaline in human plasma. *Acta Physiol Scand* 49, Suppl 173, 1960.
32. Yamaguchi N, de Champlain J, Nadeau R: Correlation between the response of the heart to sympathetic stimulation and the release of endogenous catecholamines into the coronary sinus of the dog. *Circ Res* 36:662–668, 1975.

CHAPTER TWENTY-NINE

Overview of Catecholamines in Selected Endocrine Systems

BART CHERNOW, M.D., and JOHN T. O'BRIAN, M.D.

INTRODUCTION

Epinephrine (E) and norepinephrine (NE) have hemodynamic actions and also help regulate hormone secretion from a variety of endocrine organs. These catecholamines (CA) are also hormones. For example, E is secreted from the adrenal medulla into the circulation and travels to target tissues. The CA act at adrenergic receptors (α, β, dopaminergic and presynaptic) and can "downregulate" receptors by decreasing their number and affinity (64). This chapter will detail the impact of these CA hormones on selected endocrine systems.

CATECHOLAMINE MODULATION OF CALCIUM HOMEOSTASIS

The plasma calcium concentration is kept within a narrow range by the action of parathyroid hormone (PTH) and to a lesser degree by thyrocalcitonin. The prevailing serum levels of calcium, phosphate, and magnesium modulate the secretion rate of PTH and thyrocalcitonin. In addition, these two hormones are influenced by adrenergic input. *In vitro* (132) and *in vivo* (55) experiments show that β-adrenergic stimulation with E increases PTH release. This effect can be reproduced with the use of the β-adrenergic agonist, isoproterenol, and inhibited by β-blockade with propranolol (55). Stimulation of α-receptors with agonists such as phenylephrine or α-blockade with phentolamine produce no change in PTH secretion or release.

The β-adrenergic influence on PTH release has some important clinical implications. It has been noted that asthmatics have an increased frequency of hyperparathyroidism (1). The β-agonists administered therapeutically to such patients may contribute to a hyperparathyroid state in the asthmatic. Acute administration of β-agonists clearly augments plasma PTH levels (28), but whether chronic CA excess has this same effect is unknown (80). Among patients with CA-secreting tumors, only those with a genetic predisposition to the development of hyperparathyroidism (those with the multiple endocrine neoplasia syndrome) are hypercalcemic (80). Occasionally, however, removal of a pheochromocytoma will correct existing hypercalcemia (56).

In vivo studies show that NE (as well as E and isoproterenol) has a positive effect on PTH release (96). This is not surprising as NE stimulates β-adrenergic receptors (14). Surprisingly, dopamine (DA), a β-agonist at high doses, does not stimulate PTH release. Yet, when administered by intravenous infusion, DA may inhibit phosphate reabsorption (22). Explanations for these observations are lacking.

Stimulation of the parathyroid glands' β-adrenergic receptors activates membrane-bound adenylate cyclase (96), which in turn catalyzes the conversion of ATP to cyclic AMP (cAMP) (16). CA exert their effect on PTH release by this "second messenger." Non-β-agonists, such as glucagon, histamine, and prostaglandin E_2, also provoke PTH release (16). Receptors other than adrenergic ones are, therefore, probably also involved.

The sympathetic nervous system (SNS) modulation of thyrocalcitonin secretion is less clear. Metz *et al.* (78) demonstrated that α- and β-adrenergic agonists are equally potent secretagogues of calcitonin. β-Adrenergic blockade decreases basal serum calcitonin levels (124). NE increases calcitonin secretion *in vitro*; yet, α-adrenergic blockade augments the calcitonin response to E infusion (137). β-Adrenergic receptors are probably important in calcitonin release, whereas α-receptors may have an inhibitory role. CA regulation of calcium homeostasis rarely becomes a clinical issue, perhaps with the exceptions of asthmatics, patients with pheochromocytomas, and patients with hyperthyroidism.

CATECHOLAMINE THYROID INTERACTIONS

Effect of Thyroid Hormones on CA

A relationship between thyroid gland function and the SNS has long been known but poorly understood. The clinical findings associated with excessive (or inadequate) plasma concentrations of thyroid hormone resemble those observed with hyper-(or hypo-) adrenergic tone (Table 29.1). Originally ascribed to changes in CA "sensitivity," investigators now know that the relationship between these two groups of hormones is much more complex.

Table 29.1
Adrenergic Findings in Patients with Thyroid Disorders

HYPERTHYROIDISM	HYPOTHYROIDISM
· Hyperadrenergic	· Hypoadrenergic
Tremor	Bradycardia
Tachycardia	Constipation
Diaphoresis	Decreased Sweating
Anxiety/Nervousness	Cold Intolerance
Heat Intolerance	Mental Sluggishness
Hyperdefecation	Fatigue
Palpitations	Weight Gain
Weight Loss	

In hyperthyroidism, plasma and urinary levels of E and NE are normal or even low (62). Indeed, Stoffer *et al.* (114) have shown an inverse relationship between thyroid hormone levels and plasma CA concentration. Although reports conflict (37, 39, 59) concerning this negative association, most laboratories now agree that as plasma thyroid hormone levels rise, plasma E and NE fall, and vice versa. A recent study provides evidence that changes in plasma CA levels occur slowly and, in hypothyroidism, for example, plasma NE levels may not rise for 6 to 9 months after the patient is clinically hypothyroid (125).

If thyroid hormone excess does *not* elevate CA levels, why then do patients with thyrotoxicosis manifest signs and symptoms of a hyperadrenergic state? Investigations performed in the last decade suggest an answer related to changes in adrenergic receptors. A number of laboratories show an increase in the number of β-receptor sites per cell in cardiac tissue from hyperthyroid animals (5, 20, 133). The increase in β-receptors as a consequence of thyroid hormone excess correlates well with the thyroxine-dependent augmentation in the adenylate cyclase response to isoproterenol infusion (19). Although the concentration of β-receptors varies directly with the serum thyroxine concentration, the affinity of these receptors remains unaltered (19).

Although myocardial cells and adipocytes demonstrate an increase in β-receptor concentration in hyperthyroidism, β-adrenergic receptor concentration (sites/cell) remains unaltered in other tissues such as lymphocytes (134). Therefore, an alternative explanation of adrenergic changes in thyroid disorders may apply. One such theory is that changes in thyroid homeostasis may alter the "amplification" of the β-adrenergic signal (71, 113). Experimental evidence supporting this theory is noted by Bilezikian and Loeb (11, 12). In their studies, thyroid hormone excess resulted in an increased cAMP response to CA without any detectable change in β-receptor concentration or affinity. Although this is an attractive theory, it remains to be confirmed by other investigators.

Thyroid status also seems capable of modulating α-adrenergic receptors (107). Indeed, a reciprocal relationship exists between thyroid-dependent changes in α- and β-adrenergic receptors (107). The affinity of the α-receptor decreases as the plasma thyroxine level increases, although the affinity of these receptors is also decreased in hypothyroidism (19). α-Receptor affinity is highest in the euthyroid subject. The reciprocal nature of α- and β-adrenergic receptors in altered thyroid states has led to the hypothesis that an ability may exist for adrenergic receptors to convert from one form to another (e.g., α to β) (58, 62, 65). Such a balance might be modulated by the amount of circulating thyroid hormone.

Although earlier studies indicated that in thyrotoxicosis an increased sensitivity of the cardiovascular system to CA exists (38, 83, 104), more recent investigations do not support this concept (3, 4). On the other hand, some of the metabolic effects of E and NE are potentiated by thyroid hormone. These include glycogenolysis, lipolysis, and heat production (nonshivering thermogenesis) (62).

In hyperthyroidism, in addition to low normal plasma NE levels, plasma dopamine-β-hydroxylase (DBH) concentrations are depressed (86). In hypothyroidism the opposite is true; that is, both plasma NE and DBH concentrations are elevated (87).

EFFECTS OF CATECHOLAMINES ON THYROID FUNCTION

Sympathetic nerves, which innervate the thyroid gland (74, 117), release thyroid hormone when they are stimulated (76). Similarly, infusion of either E or NE causes release of thyroxine (27). Adrenergic blocking agents (β-blockers more than α-blockers) or sympathectomy inhibit thyroid hormone secretion (21, 75). Propranolol, in addition to transiently reducing thyroxine secretion from the thyroid gland, also blocks 5′-deiodinase, the enzyme that catalyzes the conversion of T_4 to T_3 (67). In addition to their effects on thyroid hormone release, the CA also affect thyroid hormone biosynthesis. *In vitro* experiments demonstrate that CA can increase uptake and organification of iodine (an early step in T_4 biosynthesis) by the thyroid gland (68, 77).

The literature suggests that CA may augment thyroxine breakdown (31); however, this theory has not been substantiated (41, 85). It is interesting, though, in view of the findings in the "sick euthyroid" syndrome (often associated with high plasma CA levels). Serum triiodothyronine (T_3) and thyroxine (T_4) concentrations are often decreased, yet the patient (with a nonthyroidal illness) is clinically euthyroid. If CA did increase deiodination (*i.e.*, thyroid hormone catabolism), this could explain the low T_3 levels. Some "sick euthyroid" patients receive DA hydrochloride infusions which inhibit TSH release, and impair the serum TSH response to the administration of thyrotropin-releasing hormone (TRH) (52). These effects of DA may therefore help to explain a low serum thyroxine level in the sick patients receiving DA.

The interactions between CA and thyroid hormone remain to be fully delineated. It is clear that a relationship *does* exist, and this knowledge is clinically useful in thyrotoxic states, where β-adrenergic blockers and other antiadrenergic medications are effective therapeutic agents (69, 73).

CATECHOLAMINES AND ADRENAL CORTICAL FUNCTION

Effects of ACTH and Cortisol on CA

The hypothalamic-pituitary-adrenal axis is important in the biosynthesis of CA. If this axis is interrupted, by hypophysectomy for example, there is a marked reduction in the medullary content of tyrosine hydroxylase (TH) (82), DBH (127), and phenylethanolamine-N-methyltransferase (PNMT) (108, 135) and thus of adrenal E (2). Replacement of adrenocorticotropic hormone (ACTH) or glucocorticoid administration normalizes the concentrations of these enzymes (2).

A link has been suggested between the release of CA and the release of ACTH (110). Both exhibit circadian variations, and both are secreted in stress situations. Preliminary studies suggest that corticotropin-releasing factor (CRF) is the common secretagogue. Stene *et al.* (112) have shown that dexamethasone administration suppresses supine and upright plasma NE levels, whereas ACTH administration for 3 days does not. These investigators speculate that the glucocorticoid suppression of NE is either related to suppression of ACTH or is a direct inhibitor of dexamethasone on NE release. Suppression of CRF by dexamethasone may inhibit both ACTH and NE release.

Glucocorticoids cause a potentiation of the metabolic actions of CA. Bilateral adrenalectomy markedly blunts the lipolytic and glycogenolytic responses to exogenous E administration (25). This abnormality is corrected with glucocorticoid therapy. Similarly, CA-induced thermogenesis is virtually absent in adrenalectomized rats and restored with glucocorticoid replacement (25). These findings may reflect altered adrenergic receptor function in the absence of glucocorticoids (15).

Glucocorticoid administration potentiates the response of various tissues to E and NE (9), and, in the absence of glucocorticoids, a blunted CA response is observed in most tissues (18, 93), although conflicting data have been reported (98). This steroid-induced potentiation of CA requires both glucocorticoid and mineralocorticoid activity (18). Several reports suggest that the potentiation is more significant for E than NE (36, 136). Therefore, adrenal cortical function is important in the biosynthesis, release, and the action of the CA.

Effect of CA on Adrenal Cortical Function

CA probably play an inhibitory role in the hypothalamic-pituitary-adrenal (cortical) axis (Fig. 29.1). Administration of NE in small doses to the rat hypothalamus *in vitro* inhibits the release of CRF and, in turn, ACTH (34, 47). When CA stores are depleted by reserpine, an augmentation of CRF and ACTH release is observed (10, 43). An inverse relationship exists between the NE content in hypothalamus and plasma corticosterone concentration (34, 47), at least in the animal model.

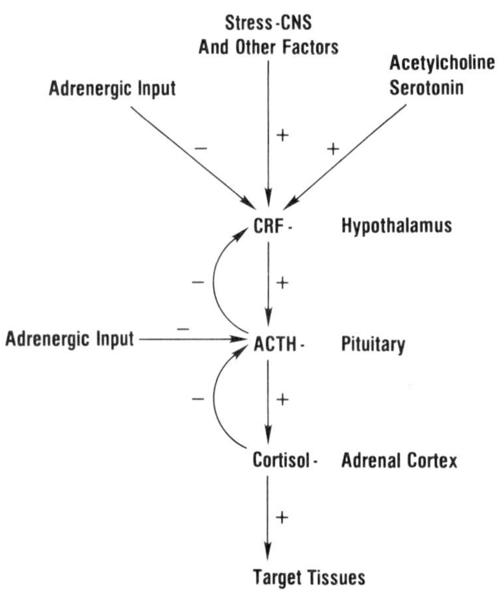

Figure 29.1. The hypothalamic-pituitary-adrenal axis.

The inhibitory effect of the CA on CRF release seems to be mediated by α-adrenergic receptors. Multifaceted evidence supports this concept. Ganong has shown that α-adrenergic receptor stimulation with clonidine (an α-agonist) inhibits the adrenal cortical response to surgical stress (33). Along these same lines, when guanfacine (a centrally acting α-adrenergic agonist) was administered to normal volunteers, there was a significant reduction in the normal ACTH response to insulin-induced hypoglycemia (61). The inhibitory effect of the α-agonists is dose-dependent and blocked by α-adrenergic blockade with phentolamine but not by β-blockade with propranolol (42, 103).

NE is probably responsible for adrenergic control of CRF and ACTH release. Fuxe *et al.* (29) have shown that after adrenalectomy in animals, NE turnover increases in both the cerebral cortex and hypothalamus. Others have shown that in rats, administration of α-methyl-*p*-tyrosine (AMPT) results in a decrease in hypothalamic NE content, along with an increase in plasma ACTH concentration (102). A very minimal amount of exogenous NE is capable of inhibiting CRF release *in vitro* (61). By comparison, the hypothalamus contains more NE, which allows the tonic inhibition of CRF release. Acetylcholine and serotonin (5-hydroxytryptamine, 5-HT) are capable of stimulating CRF release, and NE can inhibit cholinergic stimulation of CRF (46).

Although some investigators (84, 95, 131) have reported divergent results, the bulk of the evidence supports the concept that noradrenergic input helps control the CRF and ACTH responses to stress. This exciting area of neuroendocrinology has important clinical implications in the care of seriously ill patients.

CATECHOLAMINE INFLUENCE ON GONADAL FUNCTION (Fig. 29.2)

The CA are involved in the regulation of male (72, 109, 115) and female sexual function (30, 88). Adrenergic innervation of gonadal tissues is present in several species, including humans (17, 88, 89). Fuxe and others (30) have shown that regions of the hypothalamus con-

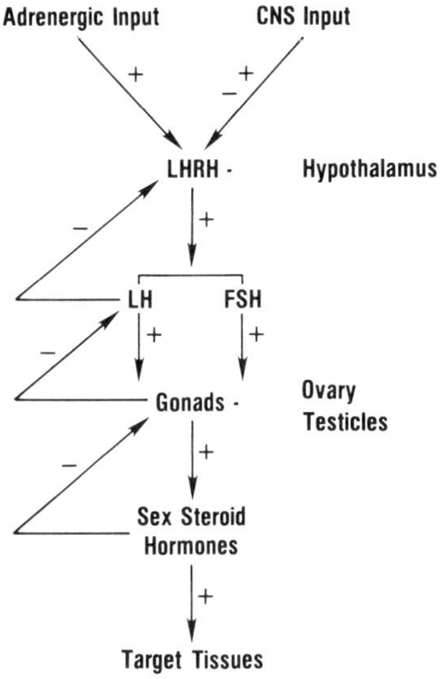

Figure 29.2. The hypothalamic-pituitary-gonadal axis.

taining luteinizing hormone-releasing hormone (LHRH) are invested with NE nerve terminal plexuses.

Female Sexual Function

Sawyer (99, 101) pioneered studies relating the SNS to ovulation. He and his colleagues demonstrated that ovulation could be produced by infusing NE into the cerebrospinal fluid (CSF) (via the ventricles) and that ovulation could be prevented by the administation of α-adrenergic blocking agents (99, 101). These findings have subsequently been confirmed by a number of other laboratories (57, 79, 118).

The relationship between the human menstrual cycle and the *peripheral* SNS is not so clear. At least two groups have shown that plasma NE levels rise before, or at the same time as, the midcycle gonadotropin surge (97, 141). However, these findings were not substantiated by a third group, which found no association between the plasma NE levels and the midcycle gonadotropin surge (139). In addition, studies in rhesus monkeys showed a perimenstrual peak in DBH activity with a nadir at midcycle (94), whereas the opposite DBH cycle (*i.e.*, peak at midcycle, nadir with menstruation) was found in humans (60). Similar conflict exists concerning the cyclical pattern of blood monoamine oxidase (MAO) levels during the menstrual cycle (8, 35).

DA inhibits secretion of gonadotropins. DA inhibits the secretion of LHRH and blocks the gonadotropin surge induced by NE (100). Most agree that DA inhibits luteinizing hormone (LH) and follicle-stimulating hormone (FSH) release (48, 49, 63). However, a number of investigators have reported that DA does *not* inhibit LHRH but rather causes its release (51, 54, 105). Kaptein *et al.* (53) showed that although serum LH and FSH levels decrease with short-term DA administration, with prolonged infusion of DA the two "escape" suppression. Although CA input is present and probably important in mediating gonadotropin release, the mechanisms involved require further definition.

Women with hypogonadotropic hypogonadism (low serum LH and FSH levels), may have increased dopaminergic activity. Quigley *et al.* (90) found that 50% of such patients had a significant rise in LH and FHS levels in response to the DA receptor antagonist metoclopramide. The same authors have also provided evidence that in polycystic ovary syndrome, there may be a decreased amount of hypothalamic DA activity (92). In this entity, there is an increase in LH secretion and increased LH sensitivity to DA inhibition. These data have important therapeutic implications, since they offer new avenues of therapy in the treatment of sexual disorders, namely adrenergic rather than sex steroid hormone treatment.

Male Sexual Function

The SNS is essential for normal erectile and ejaculatory function (128). Drugs that affect the autonomic nerve system, such as propranolol, α-methyldopa, and reserpine, may cause impotence. This is especially frequent with guanethidine (32).

The Leydig cells within the testicles are adrenergically innervated (6). Scrotal sweating, important to the thermoregulatory aspect of spermatogenesis, is under sympathetic control as well (81). Sympathetic input is essential in the movement of sperm within the male genital tract (7), and normal ejaculation cannot occur without normal SNS function.

Arousal and erection are accompanied by marked increases in NE (129). With orgasm, plasma NE levels reach their peak, declining rapidly afterwards (130).

In summary, the CA are important in the function of the pituitary-gonadal axis in men and women. Much remains to be done in defining the mechanism of adrenergic control and clinical implications of the CA-gonadal association.

CATECHOLAMINES AND THE PITUITARY GLAND

DA plays a leading role in neurotransmitter control of anterior pituitary function. DA inhibits prolactin release, and DA agonists cause sustained suppression of serum prolactin levels (70). DA receptor antagonists, such as chlorpromazine, elevate serum prolactin levels (119). Opiates can also increase prolactin, probably by inducing DA receptor blockade (119).

Dopaminergic inhibition of prolactin secretion has important therapeutic implications for patients suffering from the galactorrhea-amenorrhea syndrome. This syndrome usually results from the development of a prolactin-secreting pituitary adenoma, although it may be seen in any hyperprolactinemic state. Bromocriptine, an ergot alkaloid that activates DA receptors, has been effective in normalizing serum prolactin levels in these patients (24) and can also reduce the size of prolactin-secreting pituitary tumors (116).

Bromocriptine causes CA alterations in addition to its effect on prolactin secretion. It inhibits NE release by stimulating presynaptic receptors (140). In normotensive patients, it depresses both CSF and plasma NE levels by 50%. It can lower blood pressure (BP), has been used as an antihypertensive agent, and can cause postural hypotension. Patients with hyperprolactinemia may have psychological dysfunction (12), which normalizes with bromocriptine therapy and its resultant lowering of plasma NE concentration (138).

DA administration inhibits thyroid-stimulating hormone (TSH), LH, and FSH release (23, 49) and stimulates growth hormone release (63). Like DA, the opioid peptides affect many of the pituitary hormones. Administration of these agents results in decreases in the release of TSH, LH, and FSH and increases in ACTH, growth hormone, prolactin, melanocyte-stimulating hormone (MSH), and antidiuretic hormone (ADH) (44). The opiates may exert these effects via changes in CA. It has been shown that the opioid peptides cause an increase in the hypothalamic concentration of DA and a decrease in hypothalamic NE (26). As we noted earlier, NE exerts an inhibitory influence on ACTH release, and DA affects most of the pituitary hormones in a way similar to the opiates.

Van Loon (121) has recently begun studying variations in plasma CA between normal subjects and those with pituitary disorders. He began his studies with patients suffering from growth hormone or prolactin-secreting pituitary tumors, since CA are so important in the regulation of these two pituitary hormones. Basal plasma levels of E, NE, and DA did not differ in the acromegalics or prolactinoma patients as compared to controls. However, when the plasma CA response to bromocriptine administration was determined, marked differences occurred. In normal male and female subjects, a single dose (2.5 mg) of bromocriptine consistently decreased plasma concentrations of E, NE, and DA. Neither acromegalics (122) nor patients with prolactinomas (123) decreased the three CA after bromocriptine. In addition, Van Loon studied a patient with Cushing's disease in whom NE and DA failed to drop in response to bromocriptine therapy.

Thus, CA are important in the normal functioning of the anterior pituitary gland. Elsewhere in this text, posterior pituitary function is discussed in relation to CA. CA not only play a role in hormonal synthesis and release, but may also be of diagnostic and therapeutic importance in dealing with patients with pituitary disorders.

CATECHOLAMINES AND GASTRIN SECRETION

Adrenergic input is but one of many factors involved in gastrin production. Gastrin-secreting cells have adrenergic innervation (45). β-Adrenergic agonists increase gastrin levels (40), and this effect is blocked by propranolol (137). Stressful situations such as hypoglycemia, burns, and conditions associated with acidosis are also associated with hypergastrinemia (50, 137). In hyperthyroidism, a CA-sensitive state, hypergastrinemia has also been reported (106), although controversy surrounds this finding (66). Increased serum gastrin levels have been found in patients with pheochromocytoma, and these augmented gastrin levels may be normalized by α-adrenergic blockade (126).

These data support a relationship between gastrin secretion and plasma CA concentration. This association has a number of clinical implications in critically ill patients receiving mechanical ventilatory support; hypergastrinemia and "stress peptic ulcers" are, unfortunately, frequent. Furthermore, patients with peptic ulcer disease may hypersecrete gastrin in response to stress (111) and /or β-adrenergic stimulation (13). This information suggests that CA may be involved in the etiology of ulcer disease.

SUMMARY

NE, E, and DA are important regulators of synthesis and release of many hormones. Hypothalamic CA regulate releasing factors, pituitary DA regulates prolactin secretion, and peripheral NE alters thyroid, parathyroid, and gastrin levels. Not only do CA regulate hormone levels, but hormones such as the glucocorticoids alter CA release and receptor response. This closely interrelated system is complex, but understanding the system has led to advances such as bromocriptine therapy of hyperprolactinemia. Recent studies, such as those concerning the effects of CRF on CA levels, promise further important advances.

References

1. Aberg H, Johansson H, Werner L: Hyperparathyroidism and asthma. *Lancet* 2:381, 1972.
2. Axelrod J: Catecholamines: Effects of ACTH and adrenal corticoids. *Ann NY Acad Sci* 297:275–283, 1977.
3. Aoki VS, Wilson WR, Theilen EO: Studies of the reputed augmentation of the cardiovascular effects of catecholamines in patients with spontaneous hyperthyroidism. *J Pharmacol Exp Ther* 181:362–368, 1972.
4. Aoki VS, Wilson WR, Theilen EO, Lukensmeyer WW, Leaverton PE: The effects of triiodothyronine on hemodynamic responses to epinephrine and norepinephrine in man. *J Pharmacol Exp Ther* 157:62–68, 1967.
5. Banerjie SP, King LS: Beta-adrenergic receptors in rat heart: Effects of thyroidectomy. *Eur J Pharmacol* 43:207–208, 1977.
6. Baumgarten HG, Holstein AF: Ultrastructure, innervation and noradrenaline content of smooth muscle layers of the human epididymis and ductus deferens. *J Anat* 109:348–349, 1971.
7. Bell C: Autonomic nervous control of reproduction: Circulatory and other factors. *Pharmacol Rev* 24:657–736, 1972.
8. Belmaker RH, Murphy DL, Wyatt RJ, Loriaux DL: Human platelet monamine oxidase changes during the menstrual cycle. *Arch Gen Psychiatry* 31:553–556, 1974.
9. Beese JC, Bass AD: Potentiation by hydrocortisone of responses to catecholamines in vascular smooth muscle. *J Pharmacol Exp Ther* 154:224–238, 1966.
10. Bhattacharya AN, Marks BH: Reserpine and chlorpromazine induced changes in hypothalamo-hypophyseal-adrenal system in rats in the presence and absence of hypothermia. *J Pharmacol Exp Ther* 165:108–116, 1969.
11. Bilezikian JP, Loeb JN: Thyroid status and the beta-adrenergic responsiveness of the turkey erythrocyte. Presented at the 60th Annual Meeting of the Endocrine Society, Miami, Florida, June 14–16, 1978.
12. Bilezikian JP, Loeb JN, Gammon DE: The influence of hyperthyroidism and hypothyroidism in the beta-adrenergic response of the turkey erythrocyte. *J Clin Invest* 63:207–217, 1979.
13. Brandsborg O, Brandsborg M, Christensen NJ: The role of the beta-adrenergic receptor in the secretion of gastrin: Studies in normal subjects in patients with duodenal ulcers. *Eur J Clin Invest* 6:395–401, 1976.
14. Braunwald E: Pharmacologic treatment of cardiovascular disorders. In Isselbacher KJ, Adams RD, Braunwald E, Petersdorf RG, Wilson JD: *Harrison's Principles of Internal Medicine.* New York, McGraw Hill, 1980, pp 1067–1068.
15. Brodie BB, Davies JI, Dynie S, Krishna G, Weiss B: Interrelationships of catecholamines with other endocrine systems. *Pharmacol Rev* 18:273–289, 1966.
16. Brown EM, Gardner DG, Windeck RA, Hurwitz S, Brennan MF, Aurbach GD: β-adrenergically stimulated adenosine 3′,5′-monophosphate accumulation in and parathyroid hormone release from dispersed human parathyroid cells. *J Clin Endocrinol Metab* 48:618–626, 1979.
17. Brundin T: Distribution and function of adrenergic nerves in the rabbit fallopian tube. *Acta Physiol Scand* 66(Suppl 259):1–57, 1965.
18. Chopae CR, Brahmankar DM, Sheorey RV, Udhoji AG, Dorle AK: The effect of acute bilateral adrenalectomy on vasopressor responses to catecholamines in dogs. *J Pharm Pharmacol* 27:262–267, 1975.
19. Ciaraldi TP, Marinetti GV: Hormone action at the membrane level. VIII. Adrenergic receptors in rat heart and adipocytes and their modulation by thyroxine. *Biochim Biophys Acta* 541:334–346, 1978.
20. Ciaraldi T, Marinetti GV: Thyroxine and propylthiouracil effects *in vivo* in alpha and beta-adrenergic receptors in rat heart. *Biochem Biophys Res Commun* 74:984–991, 1977.
21. Coleon AH: Effects of the administration of catecholamine depleting drugs on the thyroid function of the rat. *Pharmacology* 8:300–310, 1972.
22. Cuche J, Marchand GR, Greger RS, Lang SC, Knox FG: Phosphaturic effect of dopamine on dogs: Possible role of intrarenally-produced dopamine in phosphate regulation. *J Clin Invest* 58:71, 1976.
23. Dilitala G: Dopamine and TSH secretion in man. *Lancet* 2:760, 1977.
24. delPozo E, Varga L, Wyss H, Tolis G, Friesen H, Wenner R, Vetter L, Uettwiler A: Clinical and hormonal response to bromocriptine (CB-154) in the galactorrhea syndromes. *J Clin Endocrinol Metab* 39:18, 1974.

25. DeQuattro V, Campese VM: Functional components of the sympathetic nervous system: Regulation of organ systems. In DeGroot LJ, Odell L, Martin JT, Potts DH, Nelson GF, Cahill WD, Steinberger E, Winegard AI: New York, Grune & Stratton, 1979, vol 11, pp 1261–1278.
26. deWied D, van Ree JM, deJong W: Narcotic analgesia and the neuroendocrine control of anterior pituitary function. In Zimmerman E, George R: *Narcotics and the Hypothalamus.* New York, Raven Press, 1974, pp 251–266.
27. Ericson LE, Melander A, Owman C, Sundler F: Endocytosis of thyroglobulin and release of thyroid hormone in mice by catecholamines and 5-hydroxytryptamine. *Endocrinology* 87:915–923, 1970.
28. Fisher JA, Blum JW, Binswanger U: Acute parathyroid hormone response to epinephrine *in vivo. J Clin Invest* 52:2434–2440, 1973.
29. Fuxe K, Hökfelt T, Jonsson G, Levine S, Lidbrink P, Lofstrom A: Brain and pituitary-adrenal interaction studies on central monoamine neurons. In Brodish A, Redgate ES: *Brain-Pituitary-Adrenal Interrelationships.* Basel, Switzerland, 1973, pp 239.
30. Fuxe K, Lofstrom A, Hökfelt T, Ferland L, Andersson K, Agnati L, Eneroth P, Gustafsson J, Sketh P: Influence of central catecholamines on LHRH-containing pathways. *Clin Obstet Gynecol* 5:251–263, 1978.
31. Galton VA: Thyroid hormone-catecholamine interrelationships. *Endocrinology* 77:278–284, 1965.
32. Gannon BJ, Iwayama T, Burnstock G: Prolonged effects of chronic guanethidine treatment on the sympathetic innervation of the genitalia of male rats. *Med J Aust* 2:207–208, 1976.
33. Ganong WF: Brain amines and the control of ACTH and growth hormone secretion. In Motta VE, Crosignami PG, Martini L: *Physiology, Pharmacology and Clinical Uses.* New York, Academic Press, 1975, p 239.
34. Ganong WF: Evidence for a central noradrenergic system that inhibits ACTH secretion. In *The First International Symposium on Brain-Endocrine Interactions-Median Eminence.* Basal, Switzerland, 1972, pp 239–269.
35. Gilmore NJ, Robinson DS, Nies A, Sylvester D, Ravaris CL: Blood monoamine oxidase levels in pregnancy and during the menstrual cycle. *J Psychosom Res* 15:215–219, 1971.
36. Ginsburg J, Duff RS: Influence of intraarterial hydrocortisone on adrenergic responses in the hand. *Br Med J* 2:424–427, 1958.
37. Goldfein A, Zileli S, Goodman D, Thorn G: The estimation of epinephrine and norepinephrine in human plasma. *J Clin Endocrinol Metab* 21:281, 1961.
38. Groetsch E: Newer methods in the diagnosis of thyroid disorders: Pathological and clinical. *NY State J Med* 18:259–267, 1918.
39. Harrison TS: Adrenal medullary and thyroid relationship. *Physiol Rev* 44:161, 1964.
40. Hayes JR, Adrill J, Kennedy TL, Shanks RG, Buchanan KD: Stimulation of gastrin release by catecholamines. *Lancet* 1:819–821, 1972.
41. Hays MT, Solomon DH: Effect of epinephrine on the peripheral metabolism of thyroxine. *J Clin Invest* 48:1114–1123, 1969.
42. Hillhouse EW, Burden JL, Jones MT: The effect of various putative neurotransmitters on the release of corticotropin-releasing hormone from the hypothalamus of the rat *in vitro. Neuroendocrinology* 17:1–11, 1975.
43. Hodges JR, Vellucci SV: The effect of reserpine on hypothalamo-pituitary-adrenocortical function in the rat. *Br J Pharmacol* 53:555–561, 1975.
44. Holaday JW, Loh HH: Endorphin-opiate interactions with neuroendocrine systems. In Loh HH, Russ DH: *Neurochemical Mechanisms of Opiates and Endorphins.* New York, Raven Press, 1979, pp 227–258.
45. Jacobowitz D: Histochemical studies of the autonomic innervation of the gut. *J Pharmacol Exp Ther* 149:358–364, 1965.
46. Jones MT, Hillhouse EW, Burden J: Effect of various putative neurotransmitters on the secretion of corticotrophin-releasing hormone from the rat hypothalamus *in vitro*; a model of the neurotransmitters involved. *J Endocrinol* 69:1–9, 1976.
47. Jones MT, Hillhouse E, Burden J: Secretion of corticotropin-releasing hormone *in vitro.* In Martini L, Ganong WF: *Frontiers in Neuroendocrinology.* New York, Raven Press, vol 4, pp 195–226.
48. Judd SJ, Rakoff JS, Yen SSC: Inhibition of gonadotropin and prolactin release of dopamine: Effect of endogenous estradiol levels. *J Clin Endocrinol Metab* 47:494, 1978.
49. Judd SJ, Rigg LA, Yen SSC: The effects of ovariectomy and estrogen treatment of the dopamine inhibition of gonadotropin and prolactin release. *J Clin Endocrinol Metab* 49:182, 1979.
50. Kaess H, Utz G, Teckentrup U, Hauck AM, Dorner M: The effect of propranolol and phentolamine on serum gastrin concentration in response to respiratory acidosis in normal man. *Eur J Clin Invest* 5:401–408, 1975.
51. Kamberi IA, Mical RS, Porter JL: Effect of anterior pituitary perfusion and intraventricular injection of catecholamines and indolamines on LH release. *Endocrinology* 87:1–12, 1980.
52. Kaptein EM, Kamiel M, Spencer C, Hall T, Nicoloff JT: Suppression of thyroid function with severe illness (abstr 74). 60th Meeting of the Endocrine Society.
53. Kaptein EM, Kletzky OA, Spencer CA, Nicoloff JT: Effects of prolonged dopamine infusion on anterior pituitary function in normal males. *J Clin Endocrinol Metab* 51:488–491, 1980.
54. Kordon C, Glowinski J: Selective inhibition of superovulation by blockade of dopamine synthesis during the "critical period" in the immature rat. *Endocrinology* 85:924, 1969.
55. Kukreja SC, Hargis GK, Bowser EN, Henderson WJ, Fisherman EN: Role of adrenergic stimuli in parathyroid hormone secretion in man. *J Clin Endocrinol Metab* 46:478, 1975.
56. Kukreja SC, Hargis GK, Rosenthal IM, Williams GA: Pheochromocytoma causing excessive parathyroid hormone production and hypercalcemia. *Ann Intern Med* 79:838, 1973.

57. Kun I, Feszt G, Cuparencu B, Orban I: The different influence of predominantly antiadrenergic and antidopaminergic neuroleptics on the estrous cycle in rats. *Pharmacology* 15:104–111, 1977.
58. Kunos G, Kunos IV, Nickerson M: Effects of thyroid state on adrenoceptor properties. *Nature* 250:779–781, 1974.
59. Kuschke HJ, Warnzl H, Becker G: Sympathoadrenal activity in thyrotoxicosis. *Br Med J* 1:1656, 1960.
60. Lamprecht F, Matta RJ, Little B, Zahn TP: Plasma dopamine-beta-hydroxylase (DBH) activity during the menstrual cycle. *Psychosom Med* 36:304–310, 1974.
61. Lancranjan I, Ohnhaus E, Girard J: The alpha-adrenoceptor control of adrenocorticotropin secretion in man. *J Clin Endocrinol Metab* 49:227–230, 1979.
62. Landsberg L: Catecholamines and hyperthyroidism. *Clin Endocrinol Metab* 6:697–718, 1977.
63. Leblanc H, Lachelin GCL, Aub-Fadil S, Yen SSC: Effects of dopamine infusion on pituitary hormone secretion in humans. *J Clin Endocrinol Metab* 43:668–674, 1976.
64. Lefkowitz RJ: Direct binding studies of adrenergic receptors: Biochemical, physiologic and clinical implications. *Ann Intern Med* 91:450–458, 1979.
65. Levitzki A: Catecholamine receptors. *Rev Physiol Biochem Pharmacol* 82:1–26, 1978.
66. Lipson A, Hus T-H, Hickoloff EL: Blood gastrin levels in hyperthyroidism. *J Clin Endocrinol Metab* 50:176–178, 1980.
67. Lumholtz IB, Nielson-Siersboek K, Faber J, Kirkegaard C, and Friis T: The effect of propranolol on the extrathyroidal metabolism of thyroxine and 3,3′,5 triiodothyronine evaluated by noncompartmental kinetics. *J Clin Endocrinol Metab* 47:587–591, 1978.
68. Maayan ML, Ingbar SH: Effect of epinephrine on iodine and intermediary metabolism in isolated thyroid cells. *Endocrinology* 87:588–595, 1970.
69. Mackin JF, Canary JJ, Pitman CS: Thyroid storm and its management. *N Engl Med* 291:1396–1398, 1974.
70. MacLeod RM, Lehmeyer JE: Studies on the mechanisms of the dopamine mediated inhibition of prolactin secretion. *Endocrinology* 94:1077, 1979.
71. Malbon CC, Moreno FJ, Cabelli RJ, Fain JN: Fat cell adenylate cyclase and beta-adrenergic receptors in altered thyroid states. *J Biol Chem* 253:671–678, 1978.
72. Malmnas CO: Monoaminergic influence on testosterone activated copulatory behavior in the castrated male rat. *Acta Physiol Scand (Suppl)* 395:1–128, 1973.
73. Mazzaferri EL, Reynolds JC, Young RL, Thomas CN, Parisi AF: Propranolol as primary therapy for thyrotoxicosis. *Arch Intern Med* 136:50–56, 1976.
74. Melander A, Ericson LE, Ljunggren JG, Norberg KA, Persson B, Sundler F, Tibblin S, Westgren U: Sympathetic innervation of the normal human thyroid. *J Clin Endocrinol Metab* 39:713–718, 1974.
75. Melander A, Ericson LE, Sundler F, Ingbar SH: Sympathetic innervation of the mouse thyroid and its significance in thyroid hormone secretion. *Endocrinology* 94:959–966, 1974.
76. Melander A, Sundler F: Interactions between catecholamines, 5-hydroxytryptamine and TSH on the secretion of thyroid hormone. *Endocrinology* 90:188–193, 1972.
77. Melander A, Sundler F, Westgren U: Intrathyroidal amines and the synthesis of thyroid hormone. *Endocrinology* 93:193–200, 1973.
78. Metz SA, Deftos LJ, Baylink DJ, Robertson RP: Neuroendocrine modulation of calcitonin and parathyroid hormone in man. *J Clin Endocrinol Metab* 47:151–159, 1978.
79. McCann SM, Moss RL: Putative neurotransmitters involved in discharging gonadotropin-releasing neurohormones and the action of LH-releasing hormone on the CNS. *Life Sci* 16:833–852, 1975.
80. Miller SS, Sizemore GW, Sheps SG, Type GM: Parathyroid function in patients with pheochromocytoma. *Ann Intern Med* 81:372–375, 1975.
81. Monro PAG: *Sympathectomy*. London, Oxford University, 1959.
82. Mueller RA, Thoenen H, Axelrod J: Effect of pituitary and ACTH on the maintenance of basal tyrosine hydroxylase activity in the rat adrenal gland. *Endocrinology* 86:751–755, 1970.
83. Murray JF, Kelly JJ Jr: The relation of thyroidal hormone level to epinephrine response: A diagnostic test for hyperthyroidism. *Ann Intern Med* 51:309–321, 1959.
84. Nakai Y, Imura H, Yoshimi T, Matsukura S: Adrenergic control mechanism for ACTH secretion in man. *Acta Endocrinol* 74:263, 1973.
85. Nicoloff JT: A new method for the measurement of acute alterations in thyroxine deiodination rate in man. *J Clin Invest* 49:267–273, 1970.
86. Noth RH, Spaulding SW: Decreased serum dopamine-beta-hydroxylase in hyperthyroidism. *J Clin Endocrinol Metab* 39:614–617, 1974.
87. Nishizawa Y, Hamada N, Fujiis C, Morii H, Okuda K, Wada M: Serum dopamine-beta-hydroxylase activity in thyroid disorders. *J Clin Endocrinol Metab* 39:599–602, 1974.
88. Owman C, Rosengran E, Sjoberg N-O: Adrenergic innervation of the human female reproductive organs: A historical and chemical investigation. *Obstet Gynecol* 30:763–773, 1967.
89. Owman C, Sjoberg N-O: Adrenergic nerves in the female genital tract of the rabbit. *Z Zellforsch* 74:182, 1966.
90. Perlman RL, Chaltie M: Catecholamine release from the adrenal medulla. *Clin Endocrinol Metab* 6:551–576, 1977.
91. Quigley ME, Rakoff JS, Yen SSC: Increased LH sensitivity to dopamine inhibition in polycystic ovary syndrome. *J Clin Endocrinol Metab* 52:231–234, 1981.
92. Quigley ME, Sheehan KL, Casper RF, Yen SSC: Evidence for increased dopaminergic and opioid activity in patients with hypothalamic

hypogonadotropic amenorrhea. *J Clin Endocrinol Metab* 50:949–954, 1980.

93. Ramey ER, Goldstein M: The adrenal cortex and the sympathetic nervous system. *Physiol Rev* 37:155–195, 1957.
94. Redmond DE Jr, Murphy DL, Banlu J, Ziegler MG, Lake CR: Menstrual cycle and ovarian hormone effects on plasma and platelet monoamine oxidase and plasma dopamine-beta-hydroxylase activities in the rhesus monkey. *Psychosom Med* 37:417–428, 1975.
95. Rees LP, Bulter WP, Gosling C, Besser GM: Adrenergic blockade and the corticosteroid and growth hormone responses to methylamphetamine. *Nature* 228:565, 1970.
96. Rodrigues HJ, Morrison A, Slatopolsky E, Klahr S: Adenylate cyclase of human parathyroid gland. *J Clin Endocrinol Metab* 47:319–325, 1978.
97. Rosner JM, Nagle CA, deLaborde NP, Pedroze E, Badand A, Figueroa Casas PR, Carril M: Plasma levels of NE during the periovulatory period and after LHRH stimulation in women. *Am J Obstet Gynecol* 124:567, 1976.
98. Salmoiraghi GC, McCubbin JW: Effect of adrenalectomy on pressor responsiveness to angiotensin and renin. *Circ Res* 2:280–284, 1954.
99. Sawyer CH: Stimulation of ovulation in the rabbit by the intraventricular injection of epinephrine or norepinephrine. *Anat Rec* 112:385–387, 1952.
100. Sawyer CH, Hilliard J, Kanematsu S, Scaramuzzi R, Blake CA: Effects of intraventricular infusion of neorepinephrine and dopamine on LH release and ovulation in the rabbit. *Neuroendocrinology* 15:328–337, 1974.
101. Sawyer CH, Markee JE, Townsend BF: Cholinergic and adrenergic components in the neurohumoral control of the release of LH in the rabbit. *Endocrinology* 44:18–37, 1949.
102. Scapagnini V, Annuziate L, Lombardi G, Oliver CB, Preziosi P: Time-course of the effect of alpha-methyl-p-tyrosine on ACTH secretion. *Neuroendocrinology* 18:272–275, 1975.
103. Scapagnini V, Preziosi P: Receptor involvement in the control of ACTH secretion. *Neuropharmacology* 12:32–38, 1973.
104. Schneckloth RE, Kurland GS, Freedberg AS: Effect of variation in thyroid function on the pressor response to norepinephrine in man. *Metabolism* 2:546–555, 1953.
105. Schneider HPG, McCann SM: Mono- and indolamines and control of LH secretion. *Endocrinology* 86:1127–1133, 1970.
106. Seino Y, Miyamoto Y, Moridera K, Taminato T, Matsukura Sand Immura H: The role of the beta-adrenergic mechanism in the hypergastrinemia of hyperthyroidism. *J Clin Endocrinol Metab* 50:368–370, 1980.
107. Sharma VK, Vannerjee SP: Alpha-adrenergic receptor in rat heart—effects of thyroidectomy. *J Biol Chem* 253:5277–5279, 1978.
108. Sheperd DM, West GB: Noradrenaline and the suprarenal medulla. *Br J Pharmacol* 6:665–674, 1951.
109. Soulairac ML, Soulairac A: Monoaminergic and cholinergic control of sexual behavior in the male rat. In Sandler M, Gessa GL; *Sexual Behavior: Pharmacology and Biochemistry.* New York, Raven Press, 1975, pp 99–116.
110. Spark RF, Connolly PB, Gluckin DS, White R, Sacks B, Landsberg L: ACTH secretion from a functioning pheochromocytoma. *N Engl J Med* 301:416–418, 1979.
111. Stadil F, Rehfeld JF: Effect of insulin injection on serum gastrin concentrations in duodenal ulcer patients and normal patients. *Scand J Gastroenterol* 9:143–147, 1974.
112. Stene M, Panagiotis N, Tuck ML, Sowers JR, Mayes D, Berg G: Plasma norepinephrine levels are influenced by sodium intake, glucocorticoid administration, and circadian changes in normal man. *J Clin Endocrinol Metab* 51:1340–1345, 1980.
113. Sterling K: Thyroid hormone action at the cell level. *N Engl J Med* 300:173–177, 1979.
114. Stoffer SS, Siang NS, Gorman CA, Pikler GM: Plasma catecholamines in hypothyroidism and hyperthyroidism. *J Clin Endocrinol Metab* 36:587–589, 1973.
115. Tagliamonte A, Fratta W, Del Fiacco M, Gessa GL: Possible stimulatory role of brain dopamine in the copulatory behavior of male rats. *Pharmacol Biochem Behav* 2:257–260, 1974.
116. Thorner MO, Martin WH, Rogol AD, Morris JL, Perryman RL, Conway BP, Howards SS, Wolfmann MG, MacLeod RM: Rapid regression of pituitary prolactinomas during bromocriptine treatment. *J Clin Endocrinol Metab* 51:438–445, 1980.
117. Tice LW, Creveling CR: Electron microscopic identification of adrenergic nerve endings on thyroid epithelial cells. *Endocrinology* 97:1123–1129, 1975.
118. Tima L, Flerko B: Ovulation induced by NE in rats made anovulatory by various experimental procedures. *Neuroendocrinology* 15:346–354, 1974.
119. Tolis G, Dent R, Guyda H: Opiates, prolactin and the dopamine receptor. *J Clin Endocrinol Metab* 47:200–203, 1978.
120. Tyson JE, Andreassen B, Huth J, Smith B, Zacur H: Neuroendocrine dysfunction with psychogenic implications in post-pill galactorrhea-amenorrhea. *Obstet Gynecol* 46:1–10, 1975.
121. Van Loon GR: Abnormal catecholamine mechanisms in hypothalamic-pituitary disorders. *Metabolism* 29(Suppl 1):1198–1202, 1980.
122. Van Loon GR: Abnormal plasma catecholamine responses in acromegalics. *J Clin Endocrinol Metab* 48:784–789, 1979.
123. Van Loon GR: A defect in catecholamine neurons in patients with prolactin secreting pituitary adenoma. *Lancet* 2:868–871, 1978.
124. Vora NM, Williams GA, Hargis GK, Bowser EN, Kawahara W, Jackson B, Henderson WJ, Kukneja SL: Comparative effect of calcium and of the β-adrenergic system on calcitonin secretion in man (abstr 494). 59th Meeting of the Endocrine Society, Chicago, June 8–10, 1977.
125. Wahl RS, Chansovria JPN, Udupa KN: Normal levels of norepinephrine in short term hypothyroidism. *J Clin Endocrinol Metab* 51:934–958, 1980.

126. Walsh JH, Grossman MI: Gastrin. *N Engl J Med* 292:1377–1384, 1975.
127. Weinshilboum R, Axelrod J: Dopamine-beta-hydroxylase activity in the rat after hypophysectomy. *Endocrinology* 87:894–899, 1970.
128. Whitelaw GP, Smithwick RH: Some secondary effects of sympathectomy with particular reference to disturbance of sexual function. *N Engl J Med* 235:121–130, 1951.
129. Wiedeking C, Lake CR, Ziegler M, Kowarski AA, Money J: Plasma NE and DBH during sexual activity. *Psychosom Med* 39:143–148, 1977.
130. Wiedeking C, Ziegler MG, Lake CR: Plasma NE and DBH during human sexual activity. *J Psychiatr Res* 15:139–145, 1979.
131. Wilcox CS, Aminoff MJ, Millar JGV, Kennan J, Kremer M: Circulating levels of corticotrophin and cortisol after infusions of L-dopa, dopamine and noradrenaline in man. *Clin Endocrinol* 4:191–194, 1975.
132. Williams GA, Hargis GK, Bowser EN, Henderson WJ, Martinez N: Evidence for a role of adenosyine 3′,5′-monophosphate in parathyroid hormone release. *Endocrinoogy* 92:687–689, 1973.
133. Williams LT, Lefkowitz RJ, Watanabe AM, Hathaway DR, Besch HR Jr: Thyroid hormone regulation of beta-adrenergic receptor number. *J Biol Chem* 252:2787–2789, 1977.
134. Williams RS, Guthrow CE, Lefkowitz RJ: Beta-adrenergic receptors of human lymphocytes are unaltered by hyperthyroidism. *J Clin Endocrinol Metab* 48:503–505, 1979.
135. Wurtman RJ, Axelrod J: Control of enzymatic synthesis of adrenaline in the adrenal medulla by adrenal cortical steroids. *J Biol Chem* 241:2301–2305, 1966.
136. Yard AC, Kadowitz PJ: Studies on the mechanism of hydrocortisone potentiation of vasoconstrictor responses to epinephrine in the anesthetized animal. *Eur J Pharmacol* 20:1–9, 1972.
137. Young JB, Landsberg L: Catecholamines and the regulation of hormone secretion. *Clin Endocrinol Metab* 6:657–695, 1977.
138. Zacur HA, Chapanis NP, Lake CR, Ziegler MG, Tyson JE: Galactorrhea-amenorrhea: Psychological interaction with neuroendocrine function. *Am J Obstet Gynecol* 125:859–862, 1976.
139. Zacur HA, Tyson JE, Ziegler MG, Lake CR: Plasma dopamine-beta-hydroxylase activity and NE levels during the human menstrual cycle. *Am J Obstet Gynecol* 130:148–157, 1978.
140. Ziegler MG, Lake CR, Williams AC, Teychenne PF, Shoulson I, Steinsland O: Bromocriptine inhibits norepinephrine release. *Clin Pharmacol Ther* 25:137–142, 1979.
141. Zuspan FP, Zuspan KJ: Ovulatory plasma amine (E and NE) surge in the women. *Am J Obstet Gynecol* 117:654–656, 1973.

CHAPTER THIRTY

Thyroid Hormones and Catecholamines in Burn Patients: A Hypermetabolic Low T_3 Syndrome

GEORGE M. VAUGHAN, M.D., AND RICHARD A. BECKER, M.D.*

INTRODUCTION

A correspondence between the thyroid and the sympathoadrenal systems has been suspected in several pathologic conditions in which circulating concentrations of thyroid hormones and catecholamines become altered. For example, since many of the signs and symptoms of thyrotoxicosis resemble excess sympathetic autonomic activity, it was assumed for many years that excess sympathetic tone was the effective functional anomaly in thyrotoxic states (39, 64). Amelioration of some symptoms and signs of thyrotoxicosis by antisympathetic drugs has further solidified the nexus between thyroid hormones and catecholamines, even though the elusive scientific basis for the wide use of drugs such as propranolol in thyrotoxicosis is still being sought. Because of these considerations, as well as observations of altered thyroid hormones and catecholamines in other pathologic conditions to be discussed, potential interaction between thyroid hormones and catecholamines has been investigated at every level of the brain-pituitary-thyroid-end organ axis.

* The authors wish to express their appreciation to Dr. Michael G. Ziegler for performing the catecholamine assays, to Dr. A. D. Mason, Jr., for his excellent advice and assistance in the statistical analyses, to Leonard Seraile for performing the thyroid hormone assays, to Jennifer Tucker for technical assistance, to Edwin W. Hander and Steven G. Fehrman for their computer analyses, and to Sandy Hale for typing, technical, editorial, and administrative assistance.

The opinions or assertions contained herein are the private views of the authors and are not to be construed as official or as reflecting the views of the Department of the Army or the Department of Defense. Human subjects participated in these studies after giving their free and informed voluntary consent. Investigators adhered to AR 70-25 and USAMRDC Reg 70-25 on Use of Volunteers in Research.

Sympathetic-Thyroid Interdependence

CONTROL OF TSH

The hypothalamus is believed to have a neurochemical coding system for controlling each anterior pituitary hormone. This is envisioned as an aminergic system with certain neurotransmitters as coding elements for stimulation or inhibition of the peptidergic system. A releasing or release-inhibiting hormone of the hypothalamic peptidergic system in turn stimulates or inhibits secretion of an anterior pituitary hormone. Thus, the pituitary hormone response is partly dependent upon integration of central nervous system aminergic signals (62). An aminergic code for stimulation of thyrotropin (TSH) in experimental animals seems to involve norepinephrine (NE), and a code for inhibition of TSH in animals and humans involves dopamine (81). Whether these catecholamines act primarily through TSH-releasing hormone (TRH) or directly on thyrotrophs to alter TSH secretion is not yet determined. Another question is whether brain centers controlling the thyroid axis and those controlling sympathetic outflow to the rest of the body operate separately or are linked to one another.

THYROIDAL SECRETION AS INFLUENCED BY THE SYMPATHETIC SYSTEM

Sympathetic innervation of thyroid follicles has provided investigators with a morphologic basis for sympathetic influence on the thyroid other than through an autonomic effect on thyroid blood flow. Melander (64, 65) has reviewed a number of experiments which examine manipulation of catecholaminergic influence at the level of the thyroid in the absence of TSH, utilizing different markers of thyroidal secretion. These experiments indicate that NE and epinephrine can stimulate thyroidal secretion via α- and β-receptors and an adenyl cyclase-dependent mechanism. Melander (64) found significant elevation of blood T_4 and T_3 in patients being treated for psychiatric illness with amphetamines, but there is little evidence that endogenous or exogenous catecholamines can sustain altered peripheral levels of thyroid hormones in the presence of an intact TSH mechanism. Although propranolol can reduce serum T_3 levels (46, 58, 80), this is probably a result of inhibition of peripheral conversion of T_4 to T_3. Since atenolol produced the same antithyrotoxic effects as did propranolol but without a fall in serum T_3, reduction in T_3 does not account for the symptomatic relief obtained with propranolol in thyrotoxic patients (67). Further, propranolol administration reportedly did not alter thyroid iodine release, T_4 turnover, or the TSH response to TRH in patients with thyrotoxicosis (92).

SYMPATHETIC TONE ALTERED BY THYROID HORMONES

The opposite question of whether primary alterations in thyroid hormone levels can alter sympathetic activity leads to the surprising conclusion that dysthyroid signs and symptoms are not a reflection of parallel changes in sympathetic tone. Cardiac NE turnover in rats, a good indicator of sympathetic activity, is accelerated in experimental hypothyroidism and is normal or slowed in thyrotoxicosis (53). Urinary excretion of catecholamines (5, 51, 102), plasma concentration of NE (15, 16, 35, 68, 88), and plasma dopamine-β-hydroxylase (DBH) activity (69, 70) may be normal or reduced in patients with hyperthyroidism and are definitely elevated in hypothyroidism. These findings apparently apply only to the noradrenergic system, since indices of epinephrine function, including secretion rate, (17) have not been altered in hyper- or hypothyroidism. Thus, in this model of primary thyroidal disease, there is good evidence for an inverse relationship between thyroidal function and the activity of the extraadrenal sympathetic nervous system.

END ORGAN SENSITIVITY

It is important to consider whether excess thyroid hormone can produce effects by increasing sensitivity to the corresponding level of catecholamines. The majority of studies in this area pertain to the cardiovascular system, which shows obvious alterations in thyroid disease. These studies have been reviewed by Harrison (39) and Landsberg (52, 53) and Skelton and Sonnenblick (83). Contrary to the results of older reports (39), recent studies have not been able to document an increase in cardiovascular sensitivity to catecholamines induced by thyroid hormone in experimental animals or in humans. Thyroid hormones have no consistent effect on the adenyl cyclase system of the heart, nor do they potentiate the effects of catecholamines on a variety of hemodynamic parameters *in vivo* or *in vitro*. The results of Buccino *et al.* (10), using isolated cat papillary muscles, provide an instructive example. The dysthyroid state profoundly affected intrinsic myocardial contractile properties, unrelated to load effects produced by changes in peripheral tissues. However, neither hyper- nor hypothyroidism changed ventricular NE stores, and the intrinsic contractile state of the myocardium induced by hyperthyroidism was not affected by reserpine-induced depletion of NE stores. Finally, the combined effect of thyroid hormones, administered before excision of the papillary muscle, and NE, in the preparation medium, seemed to project a constant maximum. Hypothyroidism allowed a greater effect of NE, and hyperthyroidism allowed little effect of NE before reaching a tension ceiling in response to graded doses of NE.

In conformity with Harrison's (39) suggestion that thyroid hormones and catecholamines have independent actions on effector systems, recent evidence thus suggests that catecholamines and thyroid hormones appear additive through different mechanisms on the final common pathways of contractility and heart rate (20). Thyrotoxicosis in humans does not augment the hemodynamic responses to NE or epinephrine (1, 2). The general thesis that sympathetic and thyroid actions are independent may not be universally applicable to the entire cardiovascular system, because of reports of a blunted pressure response to NE in hypothyroid animals and humans (see Harrison (39)). Interestingly, developments in the field of catecholamine receptors, reviewed by Landsberg (52) and Sterling (87) are discordant: cardiac catecholamine receptors were increased in experimental hyperthyroidism (94) and β-adrenergic receptors on human mononuclear cells were unaltered in spontaneous hyperthyroidism (95), but were more concentrated in normal subjects given triiodothyronine exogenously (36). The meaning of the receptor observations awaits clarification.

Although less extensively investigated, the metabolic and calorigenic effects of catecholamines do seem to be influenced by thyroid hormones (34, 52). Thyrotoxicosis increases and hypothyroidism decreases the positive effect of epinephrine on glycogenolysis and oxygen consumption (see Harrison (39)). Catecholamine-induced lipolysis is positively influenced by thyroid hormones in rats (9, 44) and humans (75–78). Krishna *et al.* (50) have reported increased adenylate cyclase activity in adipose tissue of thyroid-treated rats, and Malbon *et al.* (59) reported that catecholamine-induced lipolysis and accumulation of cyclic adenosine monophosphate in isolated rat fat cells varied directly with thyroid status. Surprisingly, the latter authors found that the number of β-adrenergic receptors did not change with thyroid status. Thyroxine enhanced the calorigenic effects of catecholamines in pigs (47) and rats (55, 86). It is apparent that thyroid hormones and catecholamines have similar but predominantly independent effects on the cardiovascular system, and they even appear to interact with respect to substrate mobilization and calorigenesis. However, as discussed above, sympathetic outflow varies inversely with a primary alteration in thyroid function.

Nonthyroidal Illness

LOW T_3 SYNDROME

Another setting in which thyroid hormones are known to be altered is severe nonthyroidal illness which may produce the "low T_3 syndrome." Patients accidentally burned over more than 29% of body surface had suppressed total triiodothyronine (T_3) concentrations (74), as did patients with diabetes mellitus (73). Faber *et al.* (33) found that after acute myocardial infarction there was a fall in serum T_3 levels and a rise in the inactive metabolite, reverse T_3 (rT_3), but thyroxine (T_4), T_3 resin uptake, and TSH did not change. One factor influencing the analysis of thyroid hormones in severe illness may be altered thyroid hormone binding in plasma (14). Although there is disagreement about changes in production rates of thyroid hormones in such conditions as infection and liver and kidney disease, there seems to be a consensus that total and free T_3 levels fall and levels of rT_3 rise (12, 38). The fall in serum T_3 is thought to result at least in part from decreased peripheral conversion of T_4 to T_3 (56, 73).

Recent evidence indicates that besides the reduction of circulating T_3, there can also be a fall in total and free T_4 in malnutrition (42) and in severe nonthyroidal illness (79, 84). Whether true hypothyroidism results from severe illness is still not clear. TSH concentrations do not demonstrate classical primary hypothyroidism because either basal TSH has not been elevated (56, 84, 85) or it has been suppressed (79) in severely ill patients. There was a blunted and delayed rise of TSH after an injection of TRH in uremic patients (56) and a blunted TSH response to TRH in fasting obese subjects (11). The TSH response to TRH in severe illness has been reviewed and examined by Maturlo *et al.* (61). Although the TSH response often appears normal in such patients, these authors caution against interpreting that as a sign of euthyroidism, because the patients may have abnormal regulation of TSH. Further reduction of T_3 by iodide administration failed to produce the expected augmentation of the TSH response to TRH. In the uremic low T_3 syndrome, metabolic rate and other indices of thyroid hormone effects were not perceptibly diminished (85). However, Lim *et al.* (57) argue that tissue hypothyroidism is a possibility, because in uremic rats, both liver T_3 content and thyroid hormone-responsive hepatic enzymes were reduced.

STARVATION AND FASTING CONTRASTED WITH PRIMARY THYROID DISEASE

Fasting and starvation lower T_3, and some would imply that food deprivation represents the same low T_3 syndrome seen in severe illness (12, 26). In the fasting model, it is clear that thyroid hormones and the sympathetic nervous system vary directly (19, 44–46, 54, 93, 104): underfeeding reduces thyroid hormone concentrations and sympathetic activity, and overfeeding produces the opposite effect on both systems. This result is envisioned as part of an adaptive servomechanism modulating metabolic rate in order to conserve fuel during periods of relative catabolism or food shortage and to burn the extra fuel during periods of caloric affluence. In this model, in contradistinction to the findings in primary thyroid disease, the noradrenergic sympathetic nervous system and thyroid hormones change in the same direction. This may serve to optimize the balance between output and input of energy.

To explain the apparent disparity of adrenergic thyroid relationships in primary thyroid disease and fasting, Dratman (21, 22) and colleagues (23) propose T_4 and T_3 as amino acid analogs of tyrosine, which enter the biosynthetic pathway of NE as alternate precursors resulting in alternate (thyrogenic) transmitters capable of producing local effects by stimulating β-receptors. In conformity with this proposal, uptake of thyroid hormones appears to take place in nerve endings localized to certain parts of the brain (28, 29), as well as in peripheral nerve endings (26, 30). Experimental hypothyroidism in the rat markedly elevates brain capacity to convert T_4 to T_3, increasing the proportion of its T_3 derived locally, and hyperthyroidism has the opposite effect (25–27). Thus, in primary hypothyroidism, elevated peripheral sympathetic tone might result from the interaction of thyroid hormones with central noradrenergic nerves which help control sympathetic outflow. Alternatively, there could be a change in transmitter substrate in peripheral nerve endings, such that the lesser contribution of thyroid hormones in hypothyroidism might allow tyrosine and its product, NE, to become more dominant. In thyrotoxicosis, the opposite might occur, with more dominant thyronergic alternate transmitters at sympathetic terminals, accounting for the otherwise engimatic benefit of β blockade in this condition.

In the fasting model, restricted caloric intake is a central signal for reduction in sympathetic outflow which would dictate a diminution of the neuronal pathway from T_4 to T_3 or other potential thyronergic alternate transmitters. In the fasted rat, diminished conversion of radioactive T_4 to T_3 was observed in innervated tissues, and radiolabeled T_3 was retained longer, consistent with diminished thyronergic transmitter turnover (24). After feeding, the expected resumption of T_4 to T_3 conversion (26) and increase of NE turnover (105) were both prevented by the ganglionic blocker chlorisondamine. Dratman's thyronergic autonomic theory affords a unification of thyroid-sympathetic interactions, but there remain several gaps to be filled. These include verification of the β-acting thyronergic transmitter, identification of the precise steps in its biosynthetic pathway, determination of the role of central nervous system thyronergic function relative to TSH secretion and sympathetic outflow, and assessment of the magnitude of the thyronergic process in nerve endings in relation to the total hepatic T_4-degradative response to fasting or primary thyroid disease.

SEVERE INJURY

So far, we have considered two categories of altered thyroid and sympathetic function, primary thyroid disease and fasting. Although fasting has been considered a model for the "low T_3 syndrome" observed in severe illness, it is becoming apparent that the hormonal response to some forms of severe illness or injury may define a separate entity quite different from that observed in fasting. The diminished thyroid activity reported with fasting is associated with a reduction of glucose flow, metabolic rate, and sympathetic activity. Severe injury, conversely, is associated with increased glucose flow, increased metabolic rate, hyperthermia, and elevated sympathetic activity (18, 31, 96–99). This category may also include other forms of febrile illness which have been associated with a decrease of both T_3 and the free T_3 index (48) but elevated excretion of catecholamines (60) and hypermetabolism (49).

Unlike other forms of severe injury, such as myocardial infarction or multiple fractures, burn injury can be more precisely quantitated as simply the total burn size (TBS) expressed as percent of body surface burned. Metabolic rate and catecholamine excretion can be correlated with TBS (40, 98). Additional interest in the burn model stems from the observation that the patients often appear thyrotoxic (7), with tremor, sweating, tachycardia, increased pulse pressure, high cardiac output, and increased metabolic rate (MR) However, Becker *et al.* (7) have found that T_4 and T_3 levels are suppressed in those burn patients who are critically ill. The circulating level of T_3 in burn patients correlated inversely with plasma concentration of NE (6), as in primary thyroid disease but unlike the pattern in starvation.

PATIENTS AND METHODS

Relationships among Basal Hormone Values

In order to assess thyroid-sympathetic relationships after burn injury, we measured thyroid hormones, catecholamines and MR in 36 men with widely distributed total burn size (TBS 2 to 93% of body surface) and who were of comparable age (17 to 23 years). Because of the possibility that it would be beneficial for wound healing (106) to reverse the previously observed chemical hypothyroidism in such patients, a prospective study of T_3 treatment was performed. Patients with TBS equal to or greater than 18% were randomly allocated in double-blind fashion to a treatment group receiving placebo or T_3, 200 μg/day orally or by nasogastric tube, until death or until their wounds were judged clinically healed by their attending physicians. Our previous experience had indicated that in critically ill burn patients, this dose of T_3 was required to produce physiologic T_3 concentrations.

Eight patients with TBS of 2 to 7.5% were studied as controls, since this small degree of injury had not previously resulted in a detectable sustained increase in MR (98). The controls remained on the ward for a variable duration, being discharged between postburn day (PBD) 6 and 26. The patients and relatives gave informed consent, and the investigation conformed to the principles of the Declaration of Helsinki.

Among the patients randomized to placebo and treated groups, four died in each group. Thus, in addition to the control (CONT) group, there were four additional groups for consideration: survivors (SURV) and nonsurvivors (NSURV) given placebo, survivors treated with T_3 (SURV-TX), and nonsurvivors treated with T_3 (NSURV-TX). Table 30.1 gives the characteristics of each group.

Beginning on postburn day (PBD) 3 (PBD 5 for CONT), a blood sample was obtained approximately thrice weekly between 0500 and 0700 hours, before the patients arose or underwent other manipulations as part of their daily care. Thyroid hormones and thyrotropin (TSH) were measured by radioimmunoassay, T_3 uptake (T_3U) by radioassay, and NE, epinephrine, dopamine, and DBH by enzymatic radioassay. In the surviving patients, resting metabolic rate was measured weekly at an ambient temperature of 31°C, using a clear plastic head canopy with controlled air flow and continuous measurement of inflow and outflow O_2, CO_2, and H_2O (4). Metabolic rate was thus calculated by indirect calorimetry as O_2 consumption corrected to standard temperature and pressure and for body surface area. No caloric intake occurred for at least 8 hours prior to measurement of MR.

Indices of free thyroid hormone concentration (FT_4I and FT_3I) were calculated as the product of total T_4 (μg/dl) or T_3 (ng/dl) and the T_3U divided by the normal calibrator T_3U provided in the kit. In order to validate FT_4I and FT_3I as a measure of the free hormone concentration over the range of thyroid hormone values observed in burn patients, 100 samples were sent to the Nichols Institute for determination of the dialysable fraction of T_4 and T_3, and subsequent calculation of free hormone concentrations. Comparison of these values with respective indices yielded excellent

Table 30.1
Characteristics of Groups

	N	%TBS (mean)	%TBS (range)	Begin Placebo or TX (PBD)	End Placebo or TX (PBD)
CONT	8	4.5	2–7.5		
SURV	10	44.3	18–22	3	31–104
NSURV	4	68.4	55–93	3	6–54
SURV-TX	10	45.3	28–75	3	26–83
NSURV-TX	4	72.9	62–85	3	12–22

correlation coefficients of 0.93 (FT_4I) and 0.98 (FT_3I) (Fig. 30.1).

The period of PBD 3 to 26 was chosen for analysis because the major decrement in catecholamines and MR occurred by PBD 26; the control patients were available for varying periods up to this time; and all survivors received placebo or T_3 treatment during this time (Table 30.1). All values sampled within 24 hours of dopamine or glucocorticoid administration were discarded from analysis. In one approach to the data, the variables were considered as the mean value for each patient. But, since major changes in most variables took place during this interval, the time factor was accounted in separate analyses using individual values of variables in a standard stepwise multiple linear regression program (BMDP, UCLA) performed on a PDP 1140 computer. To allow maximum ability to account for dependent

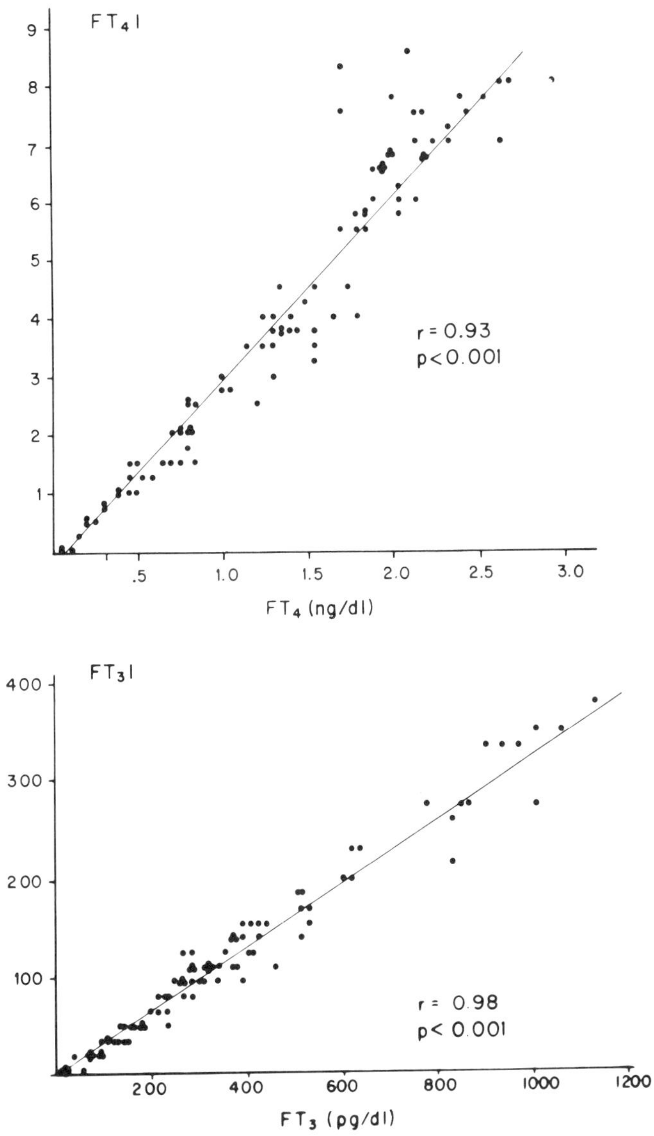

Figure 30.1. Correlation of the free indices (FT_4I and FT_3I) with respective free T_4 and free T_3 concentrations by dialysis in unselected burn patients. Though the indices do not have actual units, their values may be viewed as a correction of the respective total concentrations of T_4 (μg/dl) and T_3 (ng/dl).

variable variability related to TBS and PBD, these independent variables were entered also as TBS^2 and PBD^2 into the stepwise multiple regression analysis. Other possible independent variables were added to see if they would provide a better correlation with the dependent variable than its relationship with TBS or PBD. To assess the impact of nonsurvival on a dependent variable, an independent variable was generated by assigning survival a value of 0 and nonsurvival a value of 1. When assessing the effect of T_3 treatment, treatment group (T_3 or placebo) was entered as an additional independent variable while restricting the analysis to the whole period during which treatment was actually given.

TRH Stimulation

In a separate study, five nonburned healthy control subjects, five survivors of burn injury, five nonsurvivors, and five nonsurvivors receiving dopamine infusion for cardiovascular support underwent a TRH test. Between the 10th and 20th postburn day, a single injection of TRH, 250 μg i.v. bolus, was given after a baseline sample was taken. Additional samples were taken at 15, 30, 45, 60, and 90 minutes for TSH assay, and the TSH curve integral was computed.

Thyroid Hormone Kinetics

In a separate study, $[^{131}I]T_4$ (100 μCi) and $[^{125}I]T_3$ (100 μCi) were injected together in two healthy control subjects and in six burn patients between the 10th and 20th postburn days (mean TBS 51%). Serum was sampled at frequent intervals for 1 week after isotope injection. All calculations were made assuming a single compartment model for thyroid hormone kinetics (66, 72). Production rates were determined using the respective basal thyroid hormone concentrations.

RESULTS

Relationships among Basal Hormone Values

Figure 30.2 shows the time course of FT_4I, FT_3I, and TSH in the five groups. In the untreated patients, the FT_4I ($p < 0.001$) and the FT_3I ($p < 0.001$) were reduced in proportion to TBS^2 (FT_4I) and to TBS (FT_3I). The nonsurvivors, who had larger burns, had lower FT_4I and FT_3I. After accounting for variability associated with TBS and PBD in survivors, T_3 treatment suppressed FT_4I and TSH (both $p < 0.001$) and was successful ($p < 0.001$) in raising the FT_3I. Suppression of FT_4I and TSH, however, was not significant in T_3-treated nonsurvivors compared to untreated nonsurvivors, though those treated with T_3 had the expected rise of FT_3I in response to T_3 treatment. Not shown are the reverse T_3 (rT_3) concentrations which were suppressed ($p < 0.001$) by treatment in the survivors but not in the nonsurvivors, after accounting for variability associated with TBS and PBD. In untreated patients, rT_3 was elevated in proportion to TBS ($p < 0.01$). With survival or nonsurvival as an independent variable along with TBS and PBD but without other hormonal variables or treated patients, FT_4I and FT_3I and TSH were excessively low ($p < 0.001$ for each) in the dying patients compared to expectations based only on TBS or PBD. Using FT_4I and survival as independent variables in all groups combined, rT_3 was positively correlated with FT_4I ($p < 0.001$), but excessively elevated ($p < 0.001$) in nonsurvivors compared to the expected variation with FT_4I.

The mean values of each nontreated patient from PBD 3 to 26 (Fig. 30.3), show the negative relationship between FT_4I or FT_3I and TBS. In both cases, this relationship ($p < 0.005$) was present by PBD 3 or 5. TSH was unrelated to burn size. Figure 30.4 shows the relationships between thyroid hormones and TSH. For FT_4I *vs.* TSH, location nearer the origin indicates suppression of the pituitary-thyroid axis compared to locations farther from the origin. All treated patients are located on the origin side of the dashed line separating them from untreated survivors. Nonsurvivors are the only untreated patients in this area and thus possess a relatively suppressed TSH-thyroid axis, similar to that of the other patients whose axes were suppressed by T_3 treatment. On PBD 3, before the treated groups received T_3 treatment, the nonsurvivors analyzed separately had significant positive regressions of FT_4I and FT_3I on TSH, indicating that the TSH-thyroid axis was suppressed and failed to increase TSH in response to low thyroid hormone levels.

The *bottom panel* of Figure 30.5 shows that rT_3 in the untreated groups was higher with greater TBS. Elevated FT_3I and suppressed FT_4I with T_3 treatment were associated with lower rT_3, consistent with less substrate for conversion of T_4 to T_3. Interestingly, all the nonsurvivors except one had higher rT_3 than expected from the level of FT_4I, falling outside

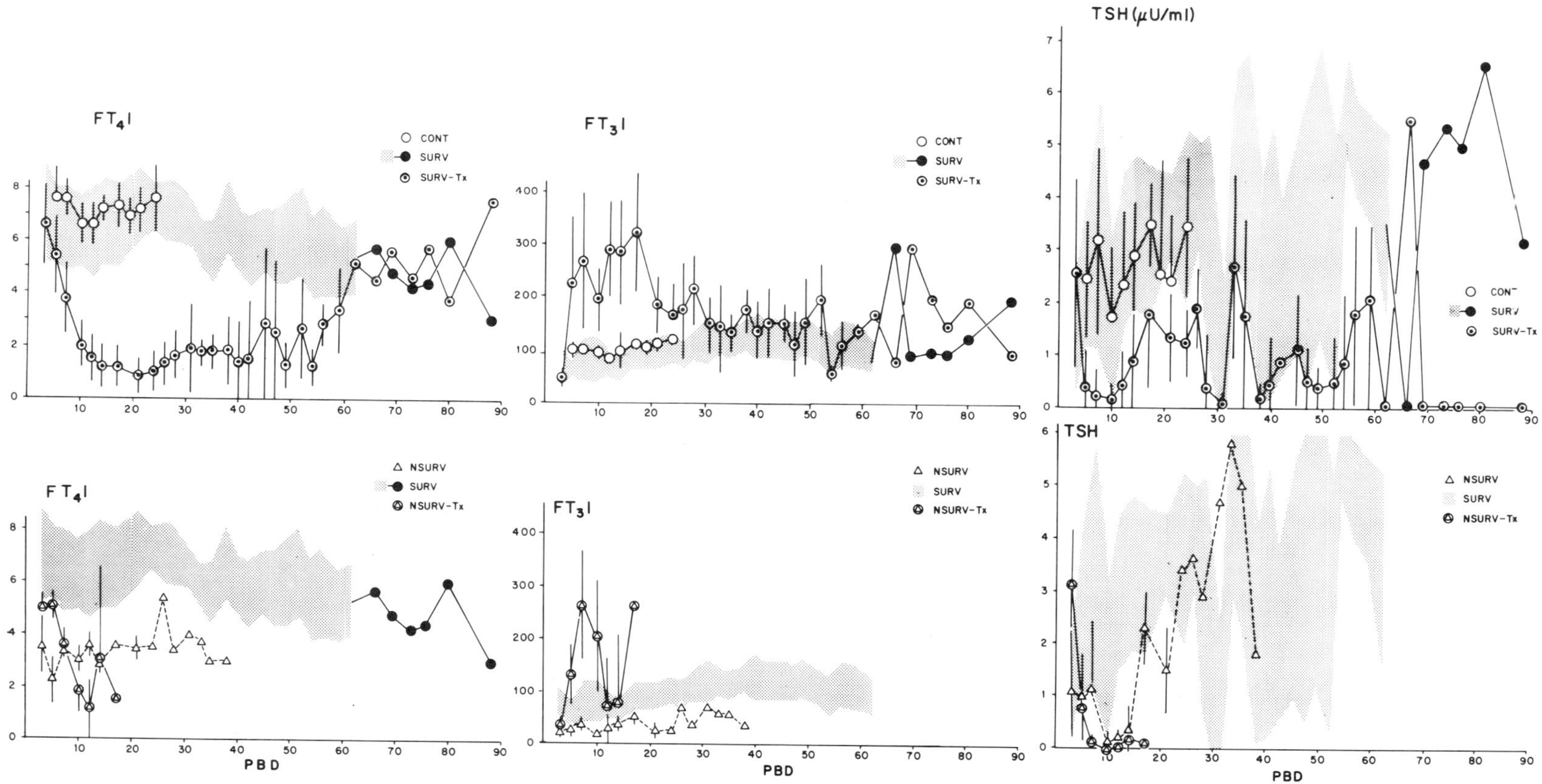

Figure 30.2. Time course of FT$_4$I, FT$_3$I, and TSH (mean ± SD) in burn patients. The SURV group (*stippled areas*) is represented in both the *top* and *bottom panels* for each hormone. *CONT*, controls; *SURV*, survivors; *SURV-Tx*, survivors treated with T$_3$; *NSURV*, nonsurvivors; *NSURV-Tx*, nonsurvivors treated with T$_3$; *PBD*, postburn day.

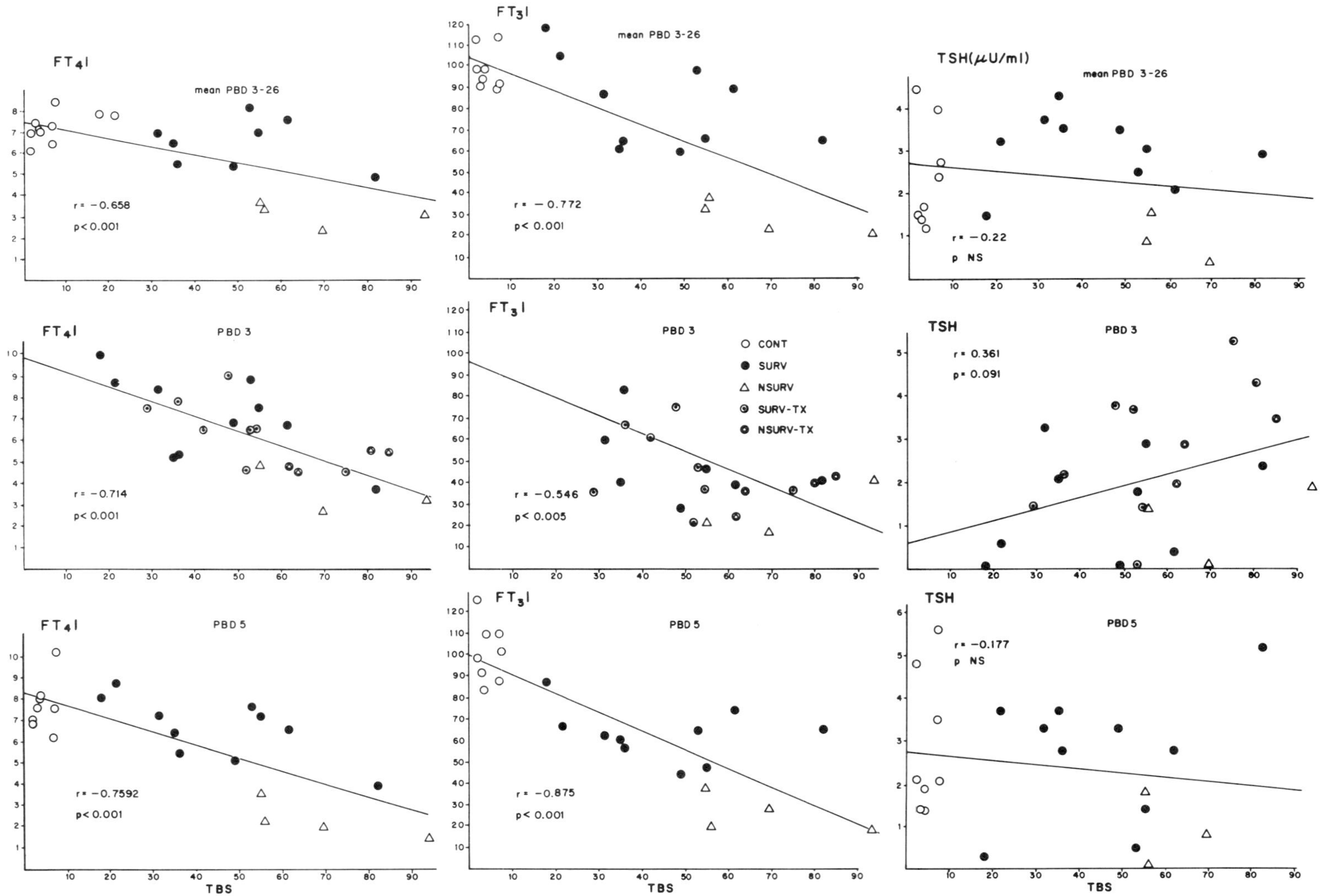

Figure 30.3. TH concentrations as functions of total burn size (*TBS*), based on mean values for each patient over PBD 3–26, or for PBD 3 or 5. On PBD 3, the treated groups had not yet received T_3 treatment. Group names in this and subsequent figures are explained in Figure 30.2.

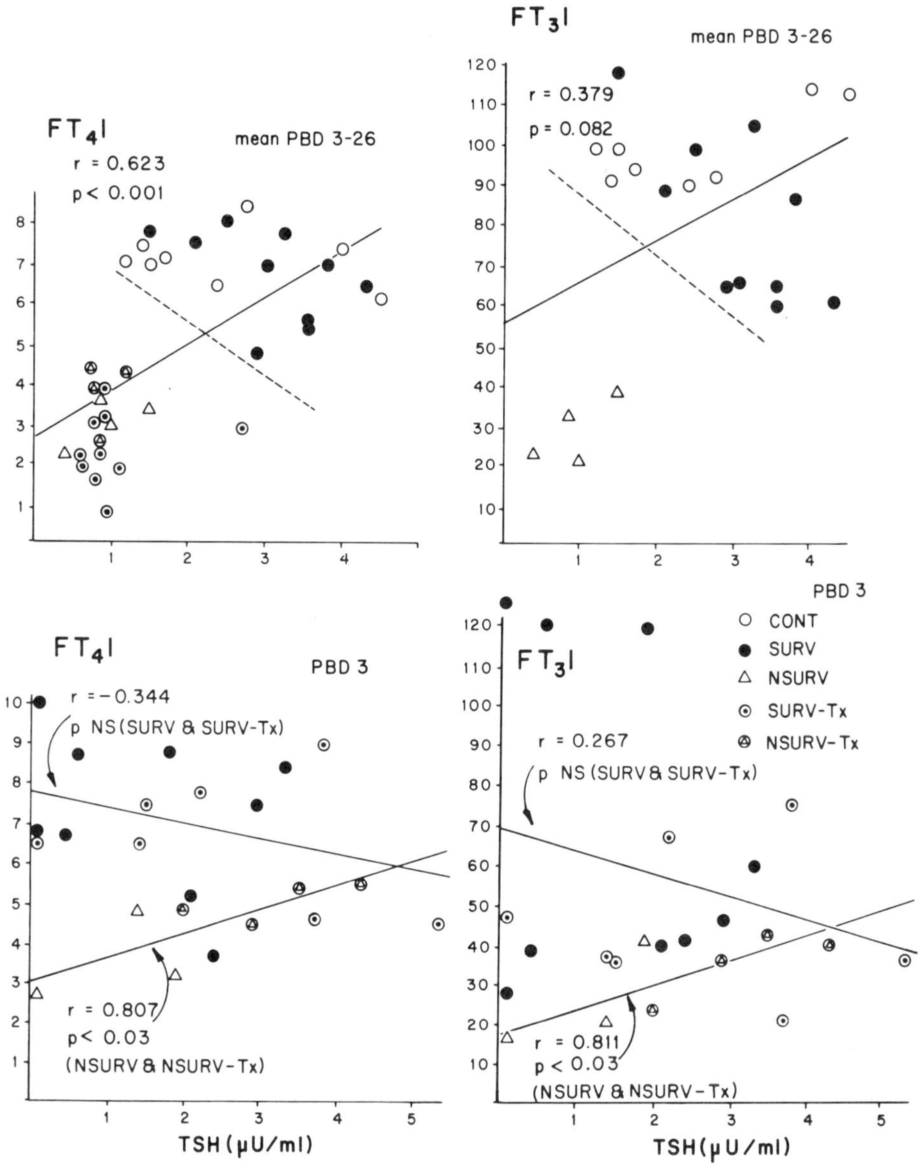

Figure 30.4. FT_4I and FT_3I as functions of TSH. For mean values over PBD 3–26, the *dashed line* separates all the treated patients and nonsurvivors from the rest.

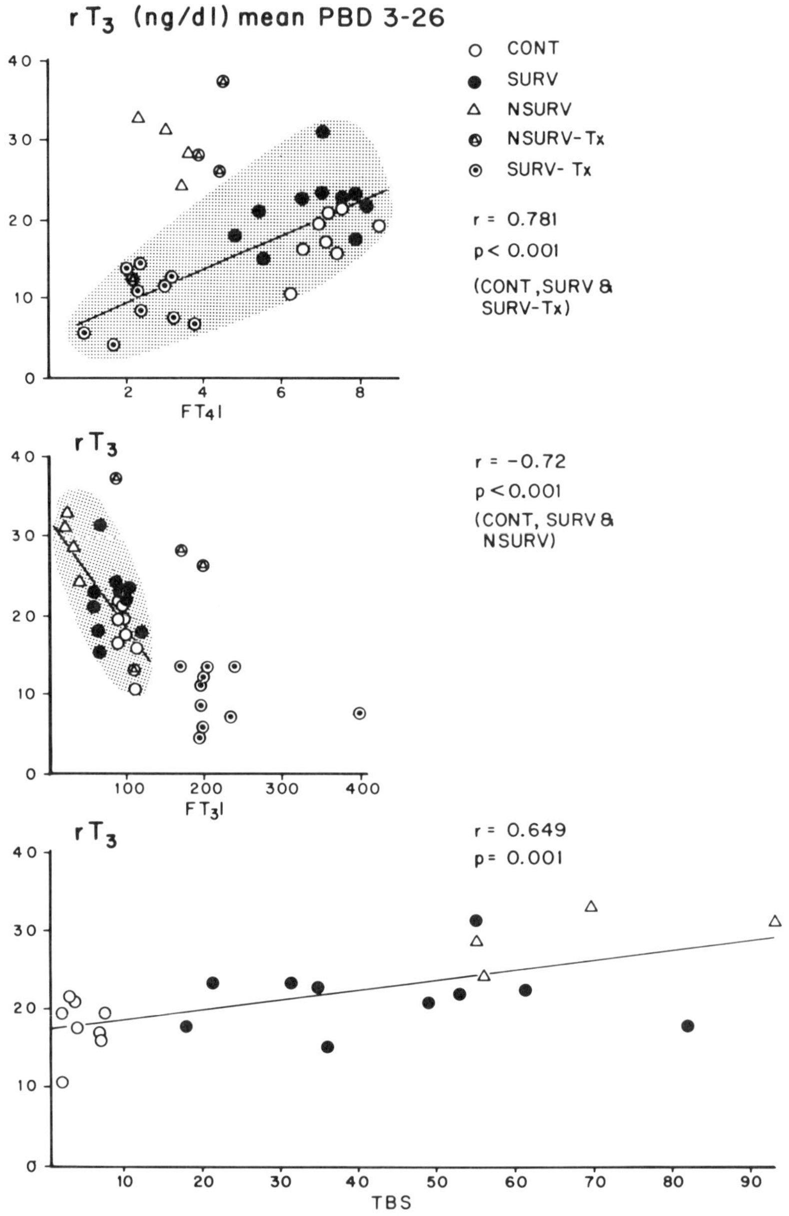

Figure 30.5. Mean rT_3 over PBD 3–26 as a function of FT_4I, FT_3I, or TBS. The *stippled areas* include all the data points used in the respective regressions.

the range for all the other patients. This is compatible with suppression of rT_3 degradation or greater rT_3 production in the nonsurvivors.

Epinephrine, NE, and MR were elevated ($p < 0.001$) in proportion to burn size (Fig. 30.6). The nonsurvivors did not have altered NE or epinephrine concentration out of proportion to TBS. Treatment with T_3 lowered NE levels as T_3 treated survivors had lower values for NE than untreated survivors for any given TBS and PBD ($p < 0.05$), but there was no detectable effect of T_3 treatment on epinephrine or metabolic rate. Plasma dopamine was elevated ($p < 0.001$) in proportion to TBS^2. Treatment with T_3 did not suppress dopamine, which was elevated out of proportion to TBS in nonsurvivors compared to survivors ($p < 0.01$).

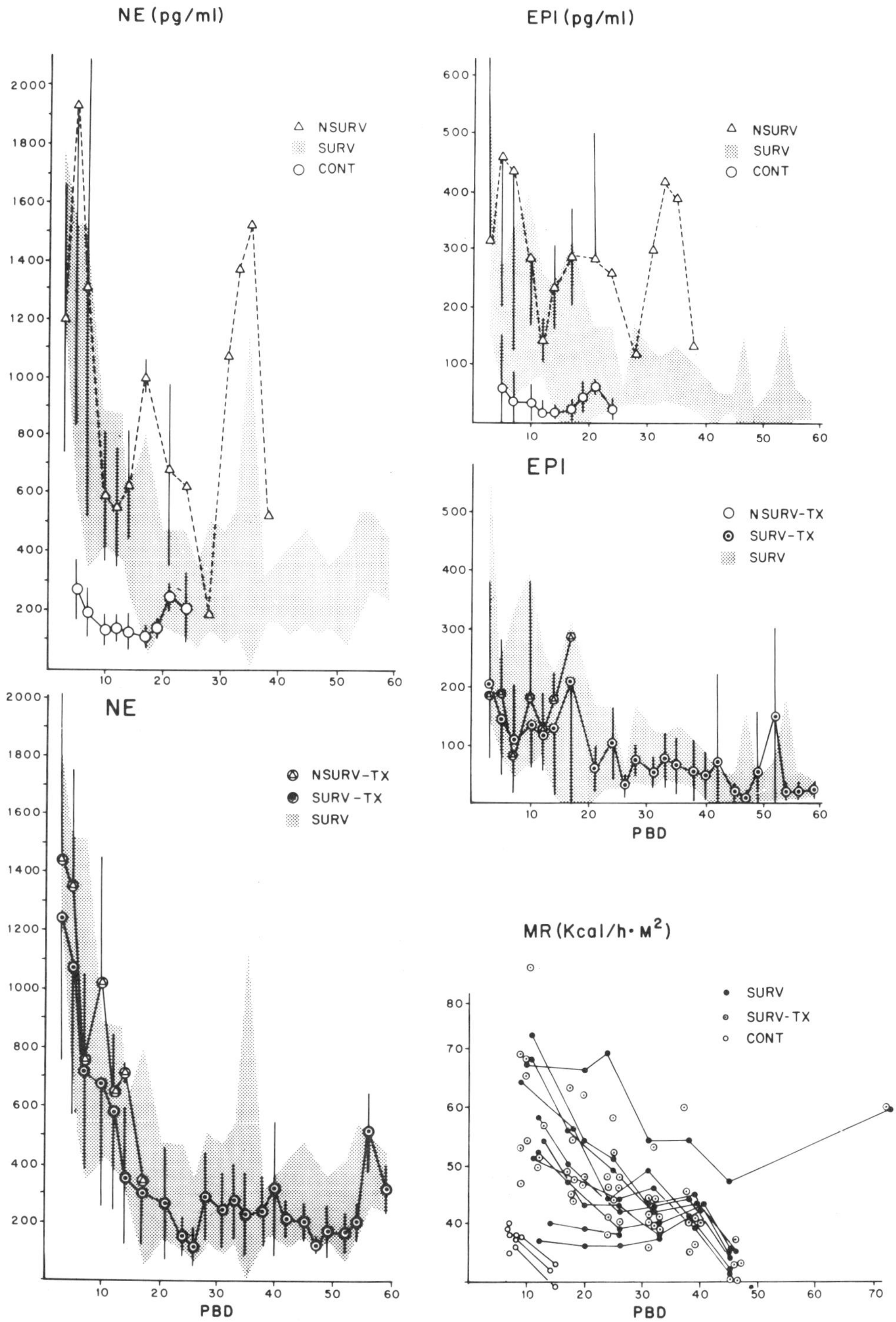

Figure 30.6. Time course of norepinephrine (*NE*), epinephrine (*EPI*) and metabolic rate (*MR*) (mean ± SD) in burn patients. The SURV group (*stippled areas*) is represented in both the *upper* and *lower panels* for NE and EPI.

Patients with more extensive burns had higher NE levels. They also had low T_3 and T_4 levels which were inversely related to NE levels (Fig. 30.7). Metabolic rate increased with larger burns and was closely related to NE levels but inversely related to FT_3I (Fig. 30.8). Catecholamines may drive the hypermetabolism of burn patients, but thyroid hormones clearly do not.

We performed multiple regression analyses on thyroid hormones, catecholamines, dopamine-β-hydroxylase corrected for total serum protein (DBH/P), and MR as dependent variables. When a thyroid hormone was a depend-

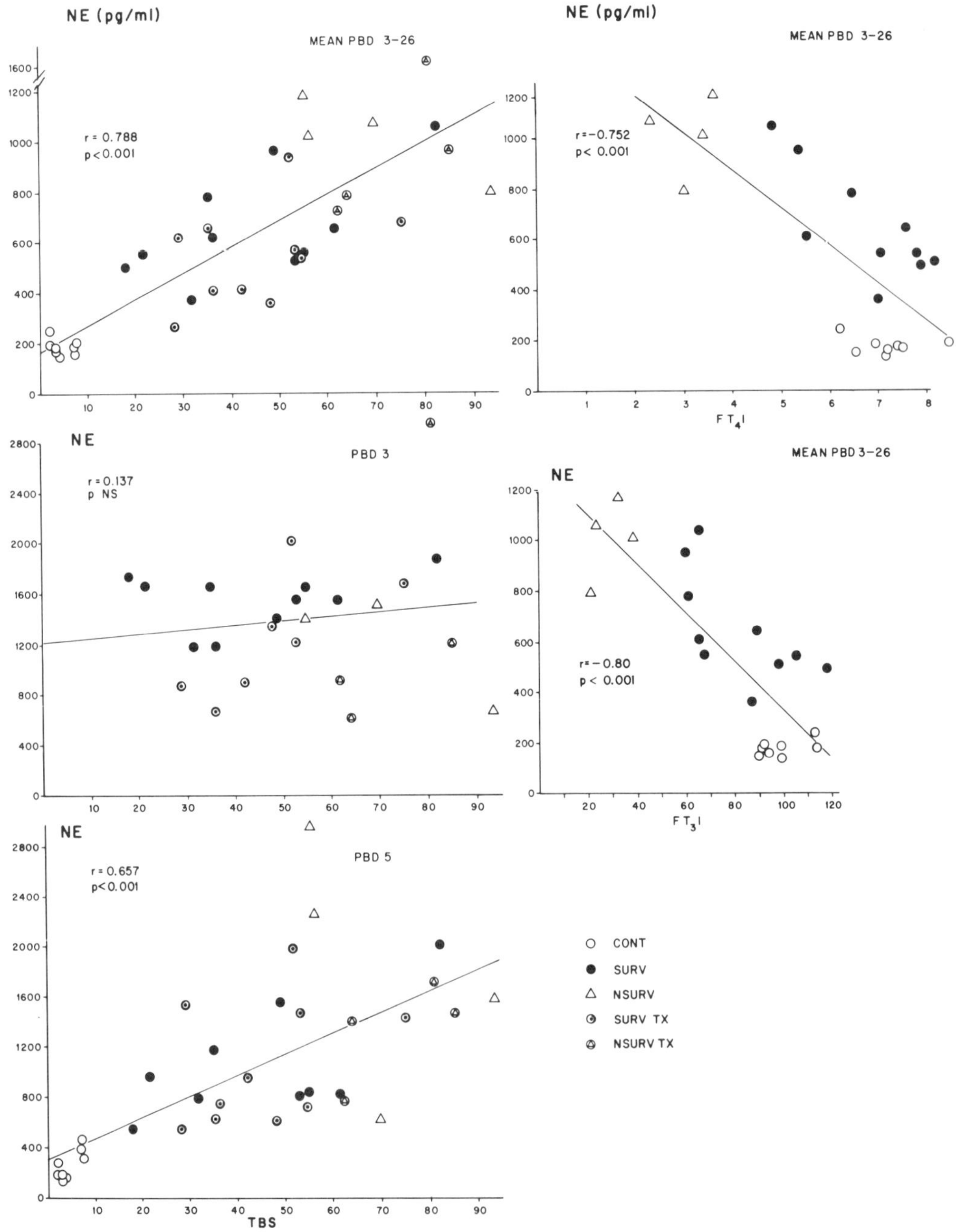

Figure 30.7. Correlation between NE and TBS (*left*) and between NE and FT_4I or FT_3I (*right*).

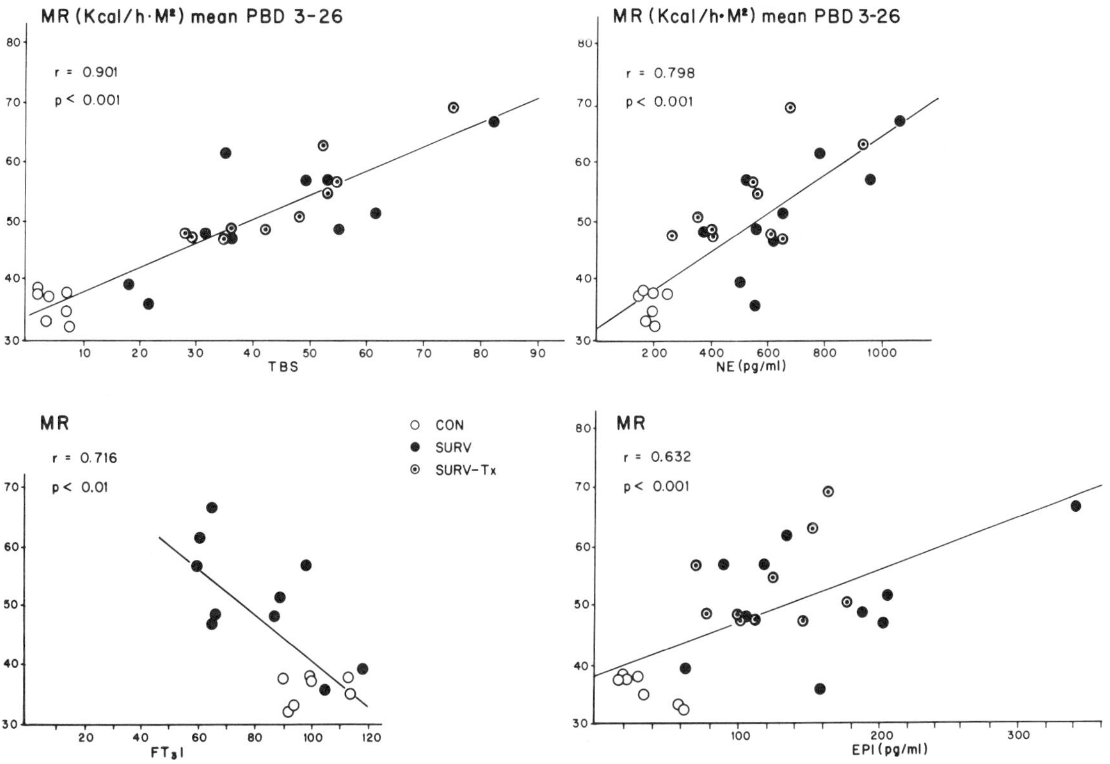

Figure 30.8. Correlation of MR with TBS, FT_3I, NE, and EPI, using mean values over PBD 3–26.

ent variable, the other thyroid hormones were not included as independent variables, and the same was true for catecholamines. TSH was excluded from the independent variables only when it was a dependent variable. The program chose only those independent variables which significantly ($p < 0.05$) reduced the residual variance of the dependent variable to a greater extent than that achieved together by the other independent variables chosen. The results are as follows in untreated patients:

Variables, intercept and coefficients	n	r^2
$FT_4I = 7.34 - 0.0003\ TBS^2 - 0.002\ dopamine + 0.001\ PBD^2$	143	0.344
$FT_3I = 98.6 - 0.568\ TBS + 0.046\ PBD^2 - 0.035\ dopamine$	143	0.417
$rT_3 = 44.2 - 3.75\ PBD + 0.094\ PBD^2 + 0.255\ TBS - 0.002\ TBS^2$	143	0.540
$TSH = 1.69 + 0.085\ PBD$	141	0.129
$NE = 1425 + 22.7\ TBS - 122\ PBD - 0.186\ TBS^2 + 2.92\ PBD^2 - 58.2\ FT_4I$	142	0.639
$Epinephrine = 143 + 3.0\ TBS - 8.75\ PBD$	142	0.397
$Dopamine = 208 + 0.026\ TBS^2 - 1.23\ FT_3I$	142	0.290
$DBH/P = 90.1 - 0.306\ FT_3I$	141	0.037
$MR = 35.1 + 0.243\ TBS + 0.017\ NE - 1.74\ TSH + 0.041\ DBH/P$	36	0.827

While DBH/P and TSH have only weak predictive value for other variables, thyroid hormones, catecholamines, burn size, and time since the burn are closely interrelated.

TRH Stimulation

Patients with large burns had low FT_4I and a low FT_3I. Those with fatal burns had a low TSH response to TRH stimulation (Table 30.2 and Fig. 30.9), even though a low thyroid hormone level is ordinarily a potent stimulus to TSH release. Even the survivors, with four of their five FT_3I below the lowest control value,

Table 30.2
Basal Thyroid Hormone Levels and TSH Curve Integrals in the Patients Who Received a 250-mg i.v. Bolus of TRH[a]

Group	Age (yr)	N/Sex	%TBS (mean)	Days Prior to Death	FT_4I	FT_3I	TSH Integral (U min/ml)
CONT	31–40	5 M			7.9 ±0.35	153 ±9.5	1245 ±208
SURV	19–54	5 M	50–68 (58)		5.9 ±0.7[b]	95.0 ±21.0[c]	1326 ±216
NSURV	18–63	4 M 1 F	28–68 (47)	4–7	2.9 ±0.6[c]	25.0 ±6.0[c]	579 ±109[c]
NSURV-DA[d]	18–61	5 M	63–80 (71)	1–3	2.78 ±0.73	25.6 ±8.9	204 ±47

[a] For survivors (SURV), comparison group is nonburned controls (CONT); for nonsurvivors (NSURV), comparison group is SURV. Error terms are SEM.
[b] $p < 0.05$, Newman-Keuls test.
[c] $p < 0.01$, Newman-Keuls test.
[d] All NSURV-DA patients were receiving dopamine, and one was receiving hydrocortisone.

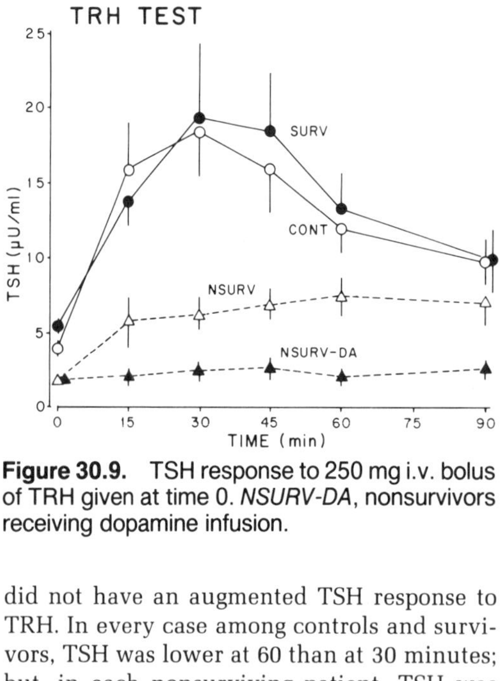

Figure 30.9. TSH response to 250 mg i.v. bolus of TRH given at time 0. *NSURV-DA*, nonsurvivors receiving dopamine infusion.

did not have an augmented TSH response to TRH. In every case among controls and survivors, TSH was lower at 60 than at 30 minutes; but, in each nonsurviving patient, TSH was higher at 60 than at 30 minutes. It is not clear whether dopamine infusion, greater proximity to death, or greater burn size contributed to reduced TSH response in dopamine-treated nonsurvivors.

Thyroid Hormone Kinetics

The PR ratio (T_3/T_4) was 0.51 ± 0.01 (0.49, 0.52) in controls and 0.24 ± 0.1 (0.07–0.71) in the burn patients (Table 30.3). Although the number of controls is small, their results agree with published normal results (13, 72, 91). The data are consistent with greater clearance of T_4 and T_3, and reduced production of T_3, but not T_4, in burn patients. Since rT_3 levels are elevated in burn patients, T_4 is metabolized to rT_3 at the expense of T_3 production, and/or rT_3 is metabolized more slowly.

DISCUSSION

Thyroid hormone and catecholamine levels in burn patients varied with the size of their burn and with the time since the burn occurred. Suppression of thyroid hormone levels was especially profound in nonsurvivors. Other studies have found thyroid suppression in burned patients (6, 7) and in patients who died in a medical intensive care unit (84). The level of TSH was not a very good predictor of circulating thyroid hormone in our surviving patients. Patients with low thyroid hormones should normally have an elevation in TSH, but these patients did not have augmented basal or TRH stimulated TSH levels. In fact, nonsurvivors who had the lowest levels of thyroid hormones also had the lowest basal TSH levels with a suppressed and delayed response to TRH. These results indicate a central failure in the pituitary-thyroid axis, which might be related to stress-induced release of dopamine (81) or cortisol (63, 90). In the untreated patients, TSH did not correlate with plasma dopamine, and it is possible that the TSH deficiency is partially a result of altered brain function as reflected by the delayed response to TRH in nonsurvivors.

Table 30.3
Results of the Kinetic Study in Two Normal Controls and Six Critically Ill Burn Patients (Mean TBS 51%), Using a Single-Compartment Model after Injection of [^{131}I]T_4 and [^{125}I]T_3[a]

	T_4		T_3	
	Controls	Patients	Controls	Patients
Serum concentration[b]	9.3, 7.0 (8.2 ± 1.1)	1.2–6.3 (4.3 ± 0.8)	136, 170 (153 ± 17)	29–75 (50 ± 7)
$T_{1/2}$ (day)	4.0, 5.5 (4.8 ± 0.7)	0.33–3.5 (2.2 ± 0.5)	0.81, 1.0 (0.91 ± 0.1)	0.41–0.75 (0.52 ± 0.06)
K (d^{-1})	0.13, 0.17 (0.15)	0.18–0.89 (0.38)	0.68, 0.85 (0.77)	0.92–1.76 (1.40)
MCR (L/day)	1.3, 1.07 (1.2 ± 0.1)	0.18–6.6 (3.8 ± 0.8)	28, 36 (32.3 ± 4)	27–96 (52.6 ± 11.1)
V_D(L)	6.5, 10.2 (8.4 ± 1.8)	3.7–25.4 (12.5 ± 3.1)	42, 43 (42.1 ± 0.1)	33–80 (54.3 ± 7.1)
PR (μg/day)	79, 118 (98.5 ± 19)	59–277 (144.6 ± 30.5)	39, 62 (50.1 ± 11.6)	12–64 (26.7 ± 7.9)

[a] $T_{1/2}$, serum disappearance half-time; K, fractional removal rate; MCR, metabolic clearance rate; V_D, volume of distribution; PR, production rate. Parentheses indicate mean ± SEM. Other numbers indicate the range of values.
[b] T_4, μg/dl; T_3, ng/dl.

It seems reasonable that therapy with T_3 might be beneficial in severely burned patients who have very low levels of T_3, but the number of patients involved in this study is probably too small to evaluate this question. There were four deaths in both the T_3-treated and nontreated groups. The brain appears to accumulate more of its T_3 from locally converted T_4 than from circulating T_3 when compared to other tissues (71), especially in hypothyroidism (26). Therapy with T_4 thus might be useful in attempts to restore normal brain function, and such a trial is underway.

Plasma catecholamines and metabolic rate were elevated in relation to burn size in these patients. However, low thyroid hormone levels were associated with higher NE levels beyond that predicted by burn size, and metabolic rate was also elevated in relation to high NE and dopamine-β-hydroxylase but low TSH. These relationships suggest a central nervous system response to injury which results in stimulation of sympathetic and metabolic activity and inhibition of thyroid axis, both as components of the efferent limb of the same neuroendocrine reflex. Previous studies (8, 37, 40, 98) have shown elevated urinary catecholamines in burn patients, and Harrison *et al.* (40) found a correlation between metabolic rate and both urinary NE and epinephrine after burns. Wilmore and colleagues (98) additionally noted suppression of postburn hypermetabolism by propranolol administration and proposed catecholamines as the mediator of the hypermetabolic response to thermal injury. Using plasma values, we found a correlation of MR with NE, but a considerably weaker relationship between MR and epinephrine. In studies on hypermetabolism induced by a different kind of stimulus, cold exposure, Jessen (43) found a 300% rise in plasma NE but no change in plasma epinephrine. Our studies also indicate that NE may be more important in driving hypermetabolism. Burn patients treated with T_3 had mild suppression of their plasma NE but no effect on epinephrine or MR in this study or a prior study (6).

Starvation can lower thyroid hormone and catecholamine levels in normal subjects and increased glucose intake has been correlated with greater urinary NE excretion in severely injured patients (31). When we compared the intake of carbohydrate, protein, fat, or total calories with thyroid hormone or NE levels, we found no significant relationship on multiple regression analysis, suggesting that nutrition was not an important determinant of the thyroid hormone and NE changes in these patients.

When normal subjects are starved, both thyroid hormone and NE levels decrease as MR diminishes in an appropriate response to conserve energy. In burn patients who are hypermetabolic, the thyroid axis responds with decreased thyroid stimulation of MR, but sympathetic tone increases markedly. There is accelerated removal of thyroid hormone in burned patients, but there is also diminished TSH stimulation of thyroid secretion. In primary thyroid disease, T_3 levels decrease and NE levels increase in response to the diminished thyroid hormone levels. However, in burn patients, the elevated NE levels cannot be fully attributed to diminished T_3, because replacement suppresses very little of the excess NE concentration. Thus, it is clear that the neuroendocrine response to injury differs significantly from that occurring as a result of hypothyroidism or starvation.

CONCLUSION

Patients with severe burns have suppression of their thyroid hormone levels without a compensatory rise in TSH and without an augmented TSH response to TRH stimulation. Patients with severe burns who did not survive had greater suppression of TSH levels and a reduced and delayed response to TRH. This central hypothyroidism in burn patients may have been accentuated by increased metabolism of thyroid hormones. Levels of thyroid hormones correlated inversely with burn size and metabolic rate (MR), but there was a strong correlation of plasma NE with MR, suggesting that noradrenergic sympathetic activity may account at least partially for postburn hypermetabolism. Whether elevated glucagon (3, 82, 100) or cortisol (101) contributes to this metabolic state remains to be determined.

Burn injury is a type of low T_3 state that probably includes other hyperpyrexic states (32, 48, 49, 60) but is markedly different from the low T_3 syndrome seen in fasting because it is associated both with increased sympathetic activity and hypermetabolism. After burn injury, hypermetabolism and elevated NE levels are both closely related to a suppressed pituitary-thyroid axis. This is compatible with functional linkage between brain centers controlling the thyroid axis and those controlling sympathetic outflow.

HYPOTHESIS

Severe injury can produce a hypermetabolic low T_3 syndrome in which the rate of metabolism is independent of thyroid stimulation and is driven by some other mechanisms. Severe injury elicits a stereotyped response from the central nervous system to increase sympathetic drive and adrenal cortical function (97). This leads to higher circulating levels of cortisol, catecholamines, and catecholamine-stimulated glucagon (100), providing a higher rate of glucose formation. Hyperglycemia develops with relatively low levels of insulin, although an added glucose load will stimulate insulin output (96, 99). Glucagon may stimulate calorigenesis independent of glucose utilization (3), and lipolysis and fatty acid oxidation may lead to further energy consumption.

In sympathetic nerve endings, tyrosine is more rapidly converted into norepinephrine, and thyronergic pathways in sympathetic nerve endings (21) may accelerate the rate of thyroid hormone degradation. The failure of a normal TSH response to low thyroid hormone levels could result from altered modulation of TRH by the brain as part of the central response to injury, which might include accelerated brain thyronergic pathways that would mimic thyrotoxicosis locally. TSH could also be inhibited by high circulating levels of dopamine and cortisol.

Elevated catecholamine levels might stimulate the thyroid and lead to a higher rate of production of T_4 than would be predicted from the low TSH levels (103) in these patients. Blockade of possible catecholamine-stimulated thyroid hormone release with a drug such as propranolol might diminish thyroid hormone levels as well as thyronergic sympathetic activity at synapses. Drugs that act centrally or at the ganglia to block sympathetic nervous activity might block the elevated turnover of thyroid hormone, however, and return blood levels toward normal. Pharmacologic blockade of dopamine might restore the TSH response to low thyroid hormone levels or to TRH stimulation and thus also tend to normalize thyroid hormone levels. Evaluation of these possibilities will likely await development of an animal injury model which has the same thyroidal and sympathetic characteristics as do burn patients.

References

1. Aoki VS, Wilson WR, Theilen EO, Lukensmeyer WW, Leaverton PE: The effects of triiodothyronine on hemodynamic responses to epinephrine and norepinephrine in man. *J Pharmacol Exp Ther* 157:62–68, 1967.
2. Aoki VS, Wilson WR, Theilen EO: Studies of the reputed augmentation of the cardiovascular effects of catecholamines in patients with spontaneous hyperthyroidism. *J Pharmacol Exp Ther* 181:362–368, 1972.

3. Aulick LH, Wilmore DW, Mason AD Jr: Mechanism of glucagon calorigenesis. *Fed Proc* 35:401, 1976.
4. Aulick LH, Hander DH, Wilmore DW, Mason AD Jr, Pruitt BA Jr: The relative significance of thermal and metabolic demands on burn hypermetabolism. *J Trauma* 19:559–566, 1979.
5. Bayliss RIS, Edwards OM: Urinary excretion of free catecholamines in Grave's disease. *J Endocrinol* 49:167–173, 1971.
6. Becker RA, Vaughan GM, Goodwin CW Jr, Ziegler MG, Harrison TS, Mason AD Jr, Pruitt BA Jr: Plasma norepinephrine, epinephrine, and thyroid hormone interactions in severely burned patients. *Arch Surg* 115:439–443, 1980.
7. Becker RA, Wilmore DW, Goodwin CW Jr, Zitzka CA, Wartofsky L, Burman KD, Mason AD Jr, Pruitt BA Jr: Free T_4, free T_3, and reverse T_3 in critically ill, thermally injured patients. *J Trauma* 20:713–721, 1980.
8. Black JA, Harris F, Miller RWS, Robinson S, Carroll J: Catecholamine excretion in burns in children. *Burns* 3:218–224, 1977.
9. Brodie BB, Davies JI, Hynie S, Krishna G, Weiss BP: Interrelationship of catecholamines with other endocrine systems. *Pharmacol Rev* 18:273–289, 1966.
10. Buccino RA, Spann JF Jr, Pool PE, Sonnenblick EH, Braunwald E: Influence of the thyroid state on the intrinsic contractile properties and energy stores of the myocardium. *J Clin Invest* 46:1669–1682, 1967.
11. Burman KD, Smallridge RC, Osburne R, Dimond RC, Whorton NE, Kesler P, Wartofsky L: Nature of suppressed TSH secretion during undernutrition: Effect of fasting and refeeding on TSH responses to prolonged TRH infusions. *Metabolism* 29:46–52, 1980.
12. Cavalieri RR, Rapoport B: Impaired peripheral conversion of thyroxine to triiodothyronine. *Ann Rev Med* 28:57–65, 1977.
13. Cavalieri RR, Steinberg M, Searle GL: Metabolic clearance rate of L-triiodothyronine in man: A comparison of results by single-injection and constant infusion methods. *J Clin Endocrinol* 33:624–629, 1971.
14. Chopra IJ, Teco GNC, Nuyen AH, Solomon DH: In search of an inhibitor of thyroid hormone binding to serum in nonthyroid illnesses. *J Clin Endocrinol Metab* 49:63–68, 1979.
15. Christensen NJ: Increased levels of plasma noradrenaline in hypothyroidism. *J Clin Endocrinol Metab* 35:359–363, 1972.
16. Christensen NJ: Plasma noradrenaline and adrenaline in patients with thyrotoxicosis and myxoedema. *Clin Sci Mol Med* 45:163–171, 1973.
17. Coulombe P, Dussault JH, Letarte J, Simard SJ: Catecholamine metabolism in thyroid diseases. I. Epinephrine secretion rate in hyperthyroidism and hypothyroidism. *J Clin Endocrinol Metab* 42:125–131, 1976.
18. Cuthbertson DP: Surgical metabolism: Historical and evolutionary aspects. In Wilkinson AW, Cuthbertson DP: *Metabolism and the Response to Injury*. Chicago, Year Book Medical Publishers, 1976, pp 1–34.
19. Danforth E Jr, Horton ES, O'Connell M, Sims EAH, Burger AG, Ingbar SH, Braverman L, Vagenakis A: Dietary-induced alterations in thyroid hormone metabolism during overnutrition. *J Clin Invest* 64:1336–1347, 1979.
20. deGroot WJ, Leonard JJ: Hyperthyroidism as a high cardiac output state. *Am Heart J* 79:265–275, 1970.
21. Dratman Mary B: On the mechanism of action of thyroxin, an amino acid analog of tyrosine. *J Theor Biol* 46:255–270, 1974.
22. Dratman MB: The mechanism of thyroxine action. In Li CH: *Hormonal Proteins and Peptides*. New York, Academic Press, 1978, vol 6, pp 205–271.
23. Dratman MB, Crutchfield FL, Marsh E, Axelrod J, Sterling FH: Iodothyronine metabolism and catecholamine enzymes: Effects of alpha-methylparatyrosine in thyroidectomized euthyroid rats. In *Thyroid Research*, Excerpta Medica International Congress Series No. 378. Proceedings of the Seventh International Thyroid Conference, Boston, Massachusetts, June 9–13, 1975. Amsterdam, Excerpta Medica, 1975, pp 248–250.
24. Dratman MB, Crutchfield FL: Conservation of tissue iodocompounds during fasting. 61st Annual Meeting of the Endocrine Society, Anaheim, California, June, 1979. Bethesda, Md, The Endocrine Society, 1979, p 92.
25. Dratman MB, Crutchfield FL: Contrasting effects of hyperthyroidism on formation of labelled 3′,3,5-triiodothyronine (T_3*) in rat brain and liver after I.V. ^{125}I-thyroxine (T_4*). *Endocrinology* 107 (Suppl): T-17, 1980.
26. Dratman MB, Crutchfield FL: Interactions of adrenergic and thyronergic systems in the development of the low T_3 syndrome. In Hesch R-D: Serono Symposium No. 40, *The 'Low T_3 Syndrome*. New York, Academic Press, 1981.
27. Dratman MB, Crutchfield FL: Normally high levels of T_3 generated from T_4 in brain markedly increased by hypothyroidism: contrastingly minimal responses in other tissues. 62nd Annual Meeting of the Endocrine Society, Washington, DC, June 18–20, 1980. Bethesda, Md, The Endocrine Society, 1980, p 223.
28. Dratman MB, Crutchfield FL: Synaptosomal (^{125}I) triiodothyronine after intravenous (^{125}I) thyroxine. *Am J Physiol* 235:E638–E647, 1978.
29. Dratman MD, Crutchfield FL, Axelrod J, Colburn RW, Thoa N: Localization of triiodothyronine in nerve ending fractions of rat brain. *Proc Natl Acad Sci USA* 73:941–944, 1976.
30. Dratman MB, Crutchfield FL, Axelrod J, Utiger RD, Marsh E: Triiodothyronine in rat submaxillary salivary gland: Effects of altered adrenergic function. *Pharmacology* 15:193–207, 1977.
31. Elwyn DH, Kinney JM, Jeevanandam M, Gump FE, Broell JR: Influence of increasing carbohydrate intake on glucose kinetics in injured patients. *Ann Surg* 190:117–127, 1979.
32. Epstein Y, Udassin R, Sack J: Serum 3, 5, 3′-triiodothyronine and 3, 3′, 5′-triiodothyronine concentrations during acute heat load. *J Clin Endocrinol Metab* 49:677–678, 1979.
33. Faber J, Kirkegaard C, Lumholtz IB, Siersbaek-

Nielsen K, Friis T: Variations in serum T_3, rT_3, 3, 3′-diiodothyronine and 3′, 5′-diiodothyronine induced by acute myocardial infarction and propranolol. *Acta Endocrinol* 94:341–345, 1980.
34. Fregly MJ, Field FP, Katovich MJ, Barney CC: Catecholamine-thyroid hormone interaction in cold-acclimated rats. *Fed Proc* 38:2162–2169, 1979.
35. Ghione S, Pellegrini M, Buzzigoli G, Carpi A, Valori C, Donato L: Plasma and urinary catecholamine levels and thyroid activity in relation to cardiovascular changes in hyper- and hypothyroids. *Horm Metab Res* 6:93, 1974.
36. Ginsberg AM, Blutter WE, Shah SD, Cryer PE: Triiodothyronine-induced thyrotoxicosis increases mononuclear leukocyte beta-adrenergic receptor density in man. *J Clin Invest* 67:1785–1791, 1981.
37. Goodall McC: Sympatho-adrenal medullary response to thermal burn. *Ann NY Acad Sci* 150:685–689, 1968.
38. Gregerman RI, Davis PJ: Effects of intrinsic and extrinsic variables on thyroid hormone economy. In Werner SC, Ingbar SH: *The Thyroid.* Hagerstown, Md, Harper & Row, 1978, pp 223–246.
39. Harrison TS: Adrenal medullary and thyroid relationships. *Phys Rev* 44:161–185, 1964.
40. Harrison TS, Seaton JF, Feller I: Relationship of increased oxygen consumption to catecholamine excretion in thermal burns. *Ann Surg* 165:169–172, 1967.
41. Ichikawa A, Matsumoto H, Sakato N, Tomita K: Effect of thyroid hormones on epinephrine-induced lipolysis in adipose tissue of rats. *J Biochem* 69:1055–1064, 1971.
42. Ingenbleek Y, De Nayer P, De Visscher M: Total and free thyroxine in protein-calorie malnutrition. *Acta Paediatr Belg* 29:145–152, 1976.
43. Jessen K: An assessment of human regulatory nonshivering thermogenesis. *Acta Anesth Scand* 24:138–143, 1980.
44. Jung RT, Shetty PS, James WPT: Nutritional effects on thyroid and catecholamine metabolism. *Clin Sci* 58:183–191, 1980.
45. Jung RT, Liddell F, Shetty PS, James WPT: The effect of levodopa with carbidopa (Sinemet[R]) on the metabolic response to semistarvation. *Int J Obes* 4:221–226, 1980.
46. Jung RT, Shetty PS, James WPT: The effect of refeeding after semistarvation on catecholamine and thyroid metabolism. *Int J Obes* 4:95–100, 1980.
47. Kaciuba-Uščiłko, H: The effect of previous thyroxine administration on the metabolic response to adrenaline in newborn pigs. *Biol Neonate* 19:220–226, 1971.
48. Kallner G, Ljunggren J: The role of endogenous cortisol in patients with non-thyroidal illness and decreased T_3 levels. *Acta Med Scand* 206:459–461, 1979.
49. Kinney JM: Surgical hypermetabolism and nitrogen metabolism. In Wilkinson AW, Cuthbertson D: *Metabolism and the Response to Injury.* Chicago, Year Book Publishers, 1976, pp 237–252.
50. Krishna G, Hynie S, Brodie BB: Effects of thyroid hormones on adenyl cyclase in adipose tissue and on free fatty acid mobilization. *Proc Natl Acad Sci USA* 59:884–889, 1968.
51. Kuschke HJ, Wernze H, Becker G: Sympathoadrenal activity in thyrotoxicosis. *Br Med J* 2:1656, 1960.
52. Landsberg L: Catecholamines and hyperthyroidism. *Clin Endocrinol Metab* 6:697–718, 1977.
53. Landsberg L: Catecholamines and the sympathoadrenal system. In Werner SC, Ingbar SH: *The Thyroid.* Hagerstown, Md, Harper & Row, 1978, pp 216–222.
54. Landsberg L, Young JB: Fasting, feeding and regulation of the sympathetic nervous system. *N Engl J Med* 298:1295–1301, 1978.
55. Leblanc J, Villemarie A: Thyroxine and noradrenaline on noradrenaline sensitivity, cold resistance, and brown fat. *Am J Physiol* 218:1742–1745, 1970.
56. Lim VS, Fang VS, Katz AI, Refetoff S: Thyroid dysfunction in chronic renal failure. *J Clin Invest* 60:522–534, 1977.
57. Lim VS, Henriquez C, Seo H, Refetoff S, Martino E: Thyroid function in a uremic rat model. *J Clin Invest* 66:946–954, 1980.
58. Lotti G, Delitala G, Devilla L, Alagna S, Masala A: Reduction of plasma triiodothyronine (T_3) induced by propranolol. *Clin Endocrinol* 6:405–410, 1977.
59. Malbon CC, Moreno FJ, Cabelli RJ, Fain JN: Fat cell adenylate cyclase and β-adrenergic receptors in altered thyroid states. *J Biol Chem* 253:671–677, 1978.
60. Mason JW, Buescher EL, Belfer ML, Artenstein MS, Mougey EH: A prospective study of corticosteroid and catecholamine levels in relation to viral respiratory illness. *J Human Stress* 5:18–28, 1979.
61. Maturlo SJ, Rosenbaum RL, Pan C, Surks MI: Variable thyrotropin response to thyrotropin-releasing hormone after small decreases in plasma free thyroid hormone concentrations in patients with nonthyroidal diseases. *J Clin Invest* 66:451–456, 1980.
62. McCann SM, Drulick L, Ojeda SR, Negro-Vilar A, Vijayan E: Neurotransmitters in the control of anterior pituitary function. In Fuxe K, Hökfelt T, Luft R: *Central Regulation of the Endocrine System.* New York, Plenum Press, 1979, pp 329–347.
63. McKenzie JM: Stress and thyroid function. In Tolis G, Labrie F, Martin J, Naftolin F: *Clinical Neuroendocrinology: A Pathophysiological Approach.* New York, Raven Press, 1979, pp 319–328.
64. Melander A: Sympathetic nervous-adrenal medullary system. In Werner SC, Ingbar SH: *The Thyroid.* Hagerstown, Md, Harper & Row, 1978, pp 216–222.
65. Melander A, Westgren U, Ericson LE, Sundler F: Influence of the sympathetic nervous system on the secretion and metabolism of thyroid hormone. *Endocrinology* 101:1228–1237, 1977.
66. Nicoloff JT: Thyroid hormone transport and metabolism: Pathophysiologic implications. In

Werner SC, Ingbar SH: *The Thyroid.* Hagerstown, Md, Harper & Row, 1978, pp 88–99.
67. Nilsson OR, Karlberg BE, Kågedal B, Tegler L, Almqvist S: Non-selective and selective β-1-adrenoceptor blocking agents in the treatment of hyperthyroidism. *Acta Med Scand* 206:21–25, 1979.
68. Nilsson OR, Karlberg BE, Söderberg A: Plasma catecholamines and cardiovascular responses to hypoglycemia in hyperthryoidism before and during treatment with metoprolol or propranolol. *J Clin Endocrinol Metab* 50:906–911, 1980.
69. Nishizawa Y, Hamada N, Jujii S, Morii H, Okuda K, Wada M: Serum dopamine-beta-hydroxylase activity in thyroid disorders. *J Clin Endocrinol Metab* 39:599–602, 1974.
70. Noth RH, Spaulding SW: Decreased serum dopamine-beta-hydroxylase in hyperthyroidism. *J Clin Endocrinol Metab* 39:614–617, 1974.
71. Obregon MJ, Roelfsema F, DeEscobar GM, Del Rey FE, Querido A: Exchange of triiodothyronine derived from thyroxine with circulating triiodothyronine as studied in the rat. *Clin Endocrinol* 10:305–315, 1979.
72. Oppenheimer JH, Schwartz HL, Surks MI: Determination of common parameters of iodothyronine metabolism and distribution in man by noncompartmental analysis. *J Clin Endocrinol* 33:624–629, 1971.
73. Pittman CS: The effects of diabetes mellitus on thyroid physiology. *Thyroid Today* IV:1–4, 1981.
74. Popp MB, Srivastava LS, Knowles HC Jr, MacMillan BG: Anterior pituitary function in thermally injured male children and young adults. *Surg Gynecol Obstet* 145:517–524, 1977.
75. Rosenqvist U: Adrenergic receptor response in hypothyroidism: An *in vitro* study on human adipose tissue and rabbit aorta. *Acta Med Scand (Suppl)* 532:5–28, 1972.
76. Rosenqvist U: Inhibition of noradrenaline-induced lipolysis in hypothyroid subjects by increased β-adrenergic responsiveness. *Acta Med Scand* 192:353–359, 1972.
77. Rosenqvist U: Noradrenaline-induced lipolysis in subcutaneous adipose tissue from hypothyroid subjects. *Acta Med Scand* 192:361–369, 1972.
78. Rosenqvist U, Boreus LO: Enhancement of the alpha adrenergic response in aorta from hypothyroid rabbits. *Life Sci* 11:595–604, 1972.
79. Rudman D, Fleischer AS, Kutner MH, Raggio JF: Suprahypophyseal hypogonadism and hypothyroidism during prolonged coma after head trauma. *J Clin Endocrinol Metab* 45:747–754, 1977.
80. Saunders J, Hall SEH, Crowther A, Sönksen PH: The effect of propranolol on thyroid hormones and oxygen consumption in thyrotoxicosis. *Clin Endocrinol* 9:67–72, 1978.
81. Scanlon MF, Lewis M, Weightman DR, Chan V, Hall R: The neuroregulation of human thyrotropin secretion. *Front Neuroendocrinol* 6:333–380, 1980.
82. Shuck JM, Eaton RP, Shuch LW, Wachtel TL, Schade DS: Dynamics of insulin and glucagon secretions in severely burned patients. *J Trauma* 17:706–713, 1977.
83. Skelton CL, Sonnenblick EH: Cardiovascular system. In Werner SC, Ingbar SH: *The Thyroid.* Hagerstown, Md, Harper & Row, 1978, pp 688–697.
84. Slag MF, Morley JE, Elson MK, Crowson TW, Nuttail FQ, Shafter RB: Hypothyroxinemia in critically ill patients as a predictor of high mortality. *JAMA* 245:43–45, 1981.
85. Spector DA, Davis PJ, Helderman JH, Bell B, Utiger RD: Thyroid function and metabolic state in chronic renal failure. *Ann Intern Med* 85:724–730, 1976.
86. Steele RE, Wekstein DR: Effects of thyroxine on calorigenic response of the newborn rat to norepinephrine. *Am J Physiol* 224:979–984, 1973.
87. Sterling K: Thyroid hormone action at the cell level. *N Engl J Med* 300:173–177, 1979.
88. Stoffer SS, Nai-Siang J, Gorman CA, Pikler GM: Plasma catecholamines in hypothyroidism and hyperthyroidism. *J Clin Endocrinol Metab* 36:587–589, 1973.
89. Utiger RD: Tests of the hypothalamic-pituitary-thyroid axis. In Werner SC, Ingbar SH: *The Thyroid.* Hagerstown, Md, Harper & Row, 1978, pp 367–374.
90. Vagenakis AG: Regulation of TSH secretion. In Tolis G, Labrie F, Martin JB, Naftolin F: *Clinical Neuroendocrinology: A Pathophysiological Approach.* New York, Raven Press, 1979, pp 329–343.
91. VanMiddlesworth L: Metabolism and excretion of thyroid hormones. In *Handbook of Physiology,* Sect 7 (Endocrinology), Vol III, (Thyroid). Bethesda, Md, American Physiological Society, 1974, pp 214–231.
92. Wartofsky L, Dimond RC, Noel GL, Frantz AG, Earll JM: Failure of propranolol to alter thyroid iodine release, thyroxine turnover, or the TSH and PRL responses to thyrotropin-releasing hormone in patients with thyrotoxicosis. *J Clin Endocrinol Metab* 41:485–490, 1975.
93. Welle S, Lilavivathana U, Campbell RG: Increased plasma norepinephrine concentrations and metabolic rates following glucose ingestion in man. *Metabolism* 29:806–809, 1980.
94. Williams LT, Lefkowitz RJ, Watanabe AM, Hathaway DR, Besch RH Jr: Thyroid hormone regulation of beta-adrenergic receptor number. *J Biol Chem* 252:2787–2789, 1977.
95. Williams RS, Guthrow EC, Lefkowitz RJ: Beta-adrenergic receptors of human lymphocytes are unaltered by hyperthyroidism. *J Clin Endocrinol Metab* 48:503–505, 1979.
96. Wilmore DW: Carbohydrate metabolism in trauma. *Clin Endocrinol Metab* 5:731–745, 1976.
97. Wilmore DW, Long JM, Mason AD Jr, Pruitt BA Jr: Stress in surgical patients as a neurophysiologic reflex response. *Surg Gynecol Obstet* 142:257–269, 1976.
98. Wilmore DW, Long JM, Mason AD Jr, Skreen RW, Pruitt BA Jr: Catecholamines: mediator of the hypermetabolic response to thermal injury. *Ann Surg* 180:653–669, 1974.

99. Wilmore DW, Mason AD Jr, Pruitt BA Jr: Insulin response to glucose in hypermetabolic burn patients. *Ann Surg* 183:314–320, 1976.
100. Wilmore DW, Moylan JA, Pruitt BA Jr, Lindsey CA, Faloona GR, Unger RH: Hyperglucagonaemia after burns. *Lancet* 1:73–75, 1974.
101. Wise L, Margraf HW, Ballinger WF: Adrenal cortical function in severe burns. *Arch Surg* 105:213–220, 1972.
102. Wiswell JG, Hurwitz GE, Corohno V, Bing OHL, Child DL: Urinary catecholamines and their metabolites in hyperthyroidism and hypothyroidism. *J Clin Endocrinol Metab* 23:1102–1106, 1963.
103. Young JB, Landsberg L: Catecholamines and regulation of hormone secretion. *Clin Endocrinol Metab* 6:657–695, 1977.
104. Young JB, Landsberg L: Effect of concomitant fasting and cold exposure on sympathoadrenal activity in rats. *Am J Physiol* 240:E314–E319, 1981.
105. Young JB, Landsberg L: Stimulation of the sympathetic nervous system during sucrose feeding. *Nature* 269:615–617, 1977.
106. Zamick P, Mehregan AH: Effect of 1-tri-iodothyronine on marginal scars of skin grafted burns in rats. *Plast Reconstr Surg* 51:71–75, 1973.

CHAPTER THIRTY-ONE

The Sympathoadrenal System in Human Glucose Counterregulation and Diabetes Mellitus

PHILLIP E. CRYER, M.D.*

INTRODUCTION

Norepinephrine and epinephrine, the major products of the sympathoadrenal system, exert a variety of metabolic effects (7, 48) including direct and indirect effects on carbohydrate, lipid, and protein metabolism. Catecholamines stimulate hepatic glycogenolysis and gluconeogenesis, increase glucose release into the circulation, and decrease glucose clearance from the circulation (37), causing hyperglycemia. Catecholamine-stimulated glycogenolysis and glycolysis in tissues such as muscle result in lactate and pyruvate release, while catecholamine-stimulated lipolysis in fat causes glycerol and fatty acid release. They also stimulate hepatic ketogenesis, resulting in increased circulating levels of acetoacetate and β-hydroxybutyrate, and inhibit amino acid release from muscle (20), causing suppression of plasma amino acid levels (45).

Despite these well-documented metabolic effects of the catecholamines, the precise roles of the sympathoadrenal system in normal metabolic physiology remain to be clearly defined. Further, aside from the metabolic aspects of the predominantly cardiovascular syndrome caused by catecholamine secreting pheochromocytomas, a role of the sympathoadrenal system in the pathophysiology of metabolic disease remains to be firmly established. This chapter focuses on the role of catecholamines in human glucose counterregulation, an example of metabolic physiology, and the potential role of catecholamines in the pathophysiology of human diabetes mellitus.

METABOLIC EFFECTS OF CATECHOLAMINES

The net biologic actions of sympathoadrenal activation result from the effects of catecholamines *per se* and from hormonal, neural, and autoregulatory systems that antagonize, or synergize with, the effects of catecholamines. For example, insulin exerts effects on carbohydrate and lipid metabolism that are the opposite of the effects produced by catecholamines. Insulin suppresses hepatic glycogenolysis, gluconeogenesis, and glucose produc-

*The author's work cited in this review was supported, in part, by U.S.P.H.S. grants AM20579, AM27085, and RR00036 and by grants from the Children's Diabetic Welfare Association/American Diabetes Association, Greater St. Louis Affiliate.

The author acknowledges the collaboration of Mr. Suresh Shah and Drs. Dennis Bier, William Clarke, William Clutter, Alan Garber, John Gerich, Morey Haymond, Steven Leveston, Robert Rizza, Julio Santiago, Alan Silverberg, and Jack Tohmeh in various aspects of the work cited. I also am grateful to the staff of the Washington University General Clinical Research and to Ms. Doris Evans for her help in the preparation of this manuscript.

tion, and stimulates glucose utilization, which results in hypoglycemia. Unlike the catecholamines, insulin suppresses lipolysis and ketogenesis. In contrast, the suppressive effects of insulin on circulating amino acid levels are like those of the catecholamines. Other hormonal systems produce effects more consistently like those of the catecholamines. For example, glucagon, cortisol, and growth hormone also cause hyperglycemia. Recent studies (18) indicate that the hyperglycemic effects of such hormones are more than additive to those of catecholamines, *i.e.*, that there are synergistic interactions among these hyperglycemic factors.

The metabolic effects of catecholamines are the result of both direct and indirect actions. The indirect actions include modulation of the secretion of various hormones, *i.e.*, catecholamine suppression of insulin secretion (32–34). Thus, the hyperglycemic effect of sympathoadrenal activation results both from direct catecholamine effects on glucose production and clearance, as well as from catecholamine-mediated limitation of insulin secretion.

Interestingly, suppression of insulin secretion by catecholamines is mediated through α-adrenergic receptors (32), whereas direct effects of catecholamines on glucose production and clearance appear to be mediated largely through β-adrenergic receptors in human subjects (36). Catecholamine-induced hyperglycemia is the result of activation of both α-adrenergic and β-adrenergic mechanisms in man.

Clearly, the major determinant of metabolic actions of catecholamines is the tightly regulated rate of catecholamine release. The sympathoadrenal system is a rapid communication neuroendocrine system. The neural signals that result in catecholamine release are very rapid; the released catecholamines produce their effects quickly and are swiftly cleared from the extracellular fluid. Thus, the sympathoadrenal system can be activated and inactivated rapidly. Large catecholamine storage pools that are promptly replenished also permit the release of large quantities of catecholamines and, if necessary, repeated release at frequent intervals.

Nonetheless, catecholamines must be transported to their target cells, interact with membrane-bound adrenergic receptors and trigger a series of intramembrane and intracellular events in order to produce their biologic actions. Recent evidence indicates that at least two of these steps—clearance and receptor interaction—are modulated in human subjects and are therefore potentially regulated steps in catecholamine action.

Catecholamines accelerate their own clearance from plasma (7), an effect that appears to be controlled by β-adrenergic mechanisms (13). Mononuclear leukocyte β-adrenergic receptor densities may vary widely among individuals and in the same individual; this variation is three times greater than analytical variation and largely represents biologic variation (25). Mononuclear leukocyte β-adrenergic receptor densities exhibit a rapid biphasic response—an initial increase followed by subsequent decrease—to β-adrenergic agonists *in vivo* (50) and an increase during administration of triiodothyronine *in vivo* (25). These and other observations show that the biologic actions of released catecholamines are modulated by regulatory systems, indicating that catecholamine action is not only a simple function of catecholamine release.

From the foregoing examples it should be clear that much remains to be learned concerning the *in vivo* physiology of the sympathoadrenal system and the biologic results of its activation. This ignorance should be kept in mind as we attempt to interpret experimental data. Operationally, it is appropriate to reduce experimental data to the simplest model compatible with those data; I will do so in the following paragraphs. However, it is also appropriate to recall Albert Einstein's admonition, "Everything should be made as simple as possible, but not simpler."

Catecholamines as Neurotransmitters and Hormones

Outside of the central nervous system, epinephrine is normally synthesized in the adrenal medullae which are the major source of epinephrine in plasma. Epinephrine is a hormone in the traditional sense—it is secreted from the adrenal medullae and transported via the circulation to its target cells. Measurements of plasma epinephrine concentrations can be interpreted, as we interpret measurements of other hormones, as those of a biologic substance en route to its target cells and/or to metabolic degradation and elimination. Dose-response studies in human subjects (7) indicate that plasma epinephrine concentrations sufficient to cause metabolic and hemodynamic effects occur commonly under physiologic conditions (14). Thus, epinephrine may well be a physiologically important hormone in the nonstressed, as well as the stressed, individual.

In contrast to epinephrine, norepinephrine is released primarily from axon terminals of sympathetic postganglionic neurons. Since these axon terminals lie in direct relation to receptor sites on target cells, transport via the circulation is not necessary for the biologic actions of norepinephrine. Indeed, dose-response studies in human subjects indicate that plasma norepinephrine concentrations must be raised nearly 10-fold in order to produce measurable metabolic or hemodynamic effects (48). Plasma norepinephrine concentrations seldom achieve such levels under physiologic conditions (10), so the biologic actions of norepinephrine are generally attributable to its sympathetic neurotransmitter function. Under some physiologic conditions, such as vigorous exercise, and in a variety of pathophysiologic states, plasma norepinephrine concentrations rise to biologically effective levels (10). Norepinephrine may also act as a hormone under these conditions.

The physiology of adrenergic axon terminals and the adrenal medullae has practical relevance to the interpretation of measurements of the plasma norepinephrine concentration. Norepinephrine released from adrenergic axon terminals is largely dissipated locally by reuptake into the axon terminal or by local metabolic degradation. Norepinephrine that escapes into the circulation is only a fraction of that released, and measurements of the plasma norepinephrine concentration provide only an index of sympathetic neural activity. Some norepinephrine is released from the adrenal medullae, although epinephrine is the quantitatively predominant secretory product (8). But when the adrenal medullae are stimulated, during hypoglycemia for example, the adrenal medullae may make an appreciable contribution to the circulating norepinephrine pool.

HYPOGLYCEMIC GLUCOSE COUNTERREGULATION IN NORMAL HUMAN SUBJECTS

Potentially Important Glucose Counterregulatory Factors

The central nervous system requires a continuous supply of glucose, its primary metabolic fuel under most physiologic conditions, and cannot extract glucose from the circulation against a concentration gradient. Consequently, maintenance of the plasma glucose concentration is crucial to survival. It has long been appreciated that insulin is the glucose regulatory hormone; it suppresses glucose production and accelerates glucose utilization, lowering the plasma glucose concentration. However, despite a plethora of candidates, the precise physiologic mechanisms that raise plasma glucose concentration have not been clearly defined. What are the mechanisms that promote recovery from hypoglycemia and that prevent the occurrence of hypoglycemia?

Theoretically, glucose counterregulation could be accomplished by hormonal signals, neural signals, glucose autoregulation, or a combination of these mechanisms. Hormonal signals could include a decrement in insulin secretion or increments in the secretion of glucose counterregulatory hormones such as glucagon, epinephrine, cortisol, growth hormone, or other newly discovered hormones with as yet unknown metabolic action. Neural signals could include norepinephrine released from sympathetic postganglionic neurons among other neural products. Evidence that hepatic glucose release is an inverse function of the circulating glucose concentration independent of hormonal and perhaps neural signals (30, 39, 40, 46) raises the possibility that glucose autoregulation may play a role in glucose counterregulation.

Recovery from Insulin-Induced Hypoglycemia

The rise in plasma glucose following hypoglycemia produced by the intravenous injection of insulin has been the model used most extensively to dissect the mechanisms of acute glucose counterregulation. Changes in the plasma glucose concentration and the rates of glucose production, glucose utilization, and glucose formation from alanine produced by the rapid intravenous injection of regular insulin (0.04 units/kg) into normal human subjects are illustrated in Figure 31.1. Following insulin injection, plasma glucose concentrations decline, reaching a nadir at 30 minutes, and then return toward baseline. The decrease in plasma glucose is the result of both insulin-stimulated glucose utilization and insulin-suppressed glucose production. The subsequent increase in plasma glucose is due to a reduc-

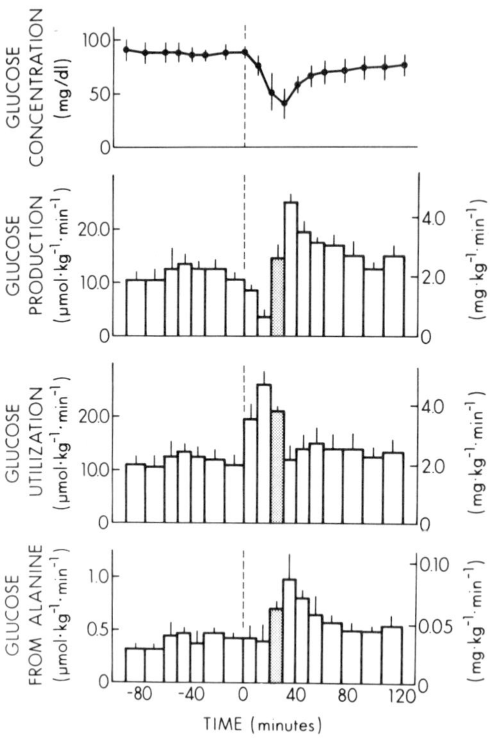

Figure 31.1. Mean (±SE) plasma glucose concentrations, glucose production rates, glucose utilization rates, and rates of glucose formation from alanine in five normal subjects before and after the rapid intravenous injection of regular insulin (0.04 units/kg) at zero minutes. The *stippled columns* mark the onset of glucose counterregulation. Glucose production and utilization rates were determined from infused [3-^{3}H]glucose, and alanine kinetics were determined from infused [U-^{14}C]alanine as detailed in reference 6. Alanine production and utilization rates (not shown) did not change. These data, and those shown in Figures 31.2 to 31.4, include data published in reference 6.

tion of glucose utilization to the baseline rate and a rise in glucose production above the baseline rate. This compensatory increase in glucose production is primarily the result of accelerated hepatic glycogenolysis. However, the corresponding increase in glucose formation from alanine shown in Figure 31.1 indicates that accelerated hepatic gluconeogenesis may contribute to this increase in glucose production, although its quantitative contribution is unknown. The interested reader may wish to compare these patterns with those produced by a larger dose of insulin (19).

As can be seen in Figure 31.1, glucose counterregulation begins 20 to 30 minutes after insulin injection in this model of insulin-induced hypoglycemia. Even though the plasma glucose concentration is still falling (glucose utilization still exceeds glucose production), glucose utilization has begun to decline toward the baseline rate, and glucose production has risen sharply from its nadir. Thus, any model of the mechanisms of glucose recovery from hypoglycemia must include a counterregulatory factor or factors capable of rapidly activating the enzymatic systems responsible for these changes in glucose utilization and production.

As shown in Figure 31.2, mean plasma insulin concentrations are 4- to 10-fold above baseline levels at the onset of glucose counterregulation. This point was even more dramatically made in earlier studies employing a larger dose of insulin (19). Since the onset of glucose counterregulation clearly can occur while plasma insulin concentrations are substantially elevated, a decrease in plasma insulin cannot be the sole explanation for glucose recovery. Further, this observation indicates that glucose recovery can occur in the presence of elevated insulin levels.

Hypoglycemia stimulates the release of epi-

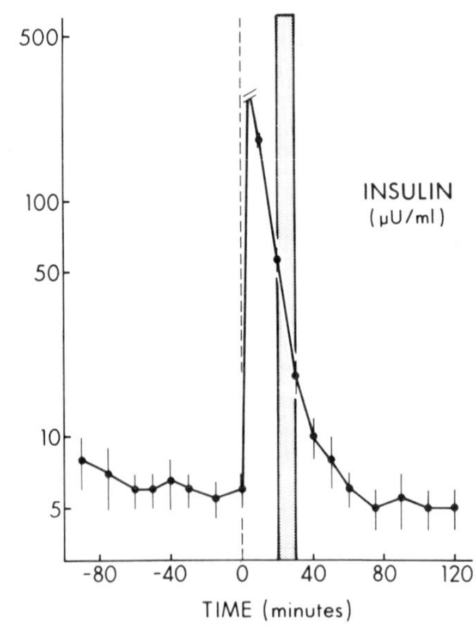

Figure 31.2. Mean (±SE) plasma insulin concentrations in five normal subjects before and after the rapid intravenous injection of regular insulin (0.04 units/kg) at zero minutes. The *stippled column* marks the onset of glucose counterregulation (see Fig. 31.1).

nephrine, norepinephrine, and glucagon (Fig. 31.3) and of cortisol and growth hormone (Fig. 31.4). Although maximum plasma levels of cortisol and growth hormone occur somewhat later than those of epinephrine, norepinephrine, and glucagon, the levels of all of these potentially important glucose counterregulatory factors increase at the onset of glucose counterregulation. Consequently, their relative physiologic roles cannot be determined simply from examination of the temporal patterns of their release in relation to glucose recovery from hypoglycemia.

Based upon this realization, we designed a series of experiments to determine the relative physiologic importance of epinephrine, norepinephrine, glucagon, cortisol, and growth hormone in recovery from insulin-induced hypoglycemia in normal human subjects (6, 23, 35) (for review, see refs. 11 and 22). Because of their rapid onset of action, epinephrine, norepinephrine, and glucagon were considered more likely candidates than the slower acting hormones, cortisol and growth hormone. Although our major focus was on catecholamines and glucagon, the studies nonetheless permit certain conclusions concerning the roles of cortisol and growth hormone as well as those of neural signals and glucose autoregulation.

This series of studies (6, 23, 35) is summarized in Figure 31.5. Glucose recovery from

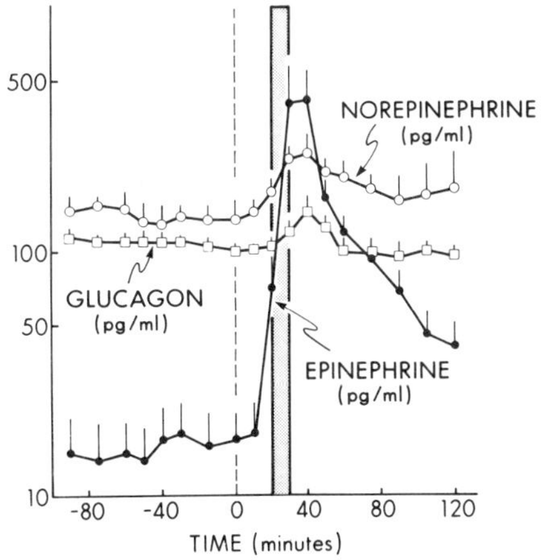

Figure 31.3. Mean (±SE) plasma epinephrine, norepinephrine, and glucagon concentrations in five normal subjects before and after the rapid intravenous injection of regular insulin (0.04 units/kg) at zero minutes. The *stippled column* marks the onset of glucose counterregulation (see Fig. 31.1).

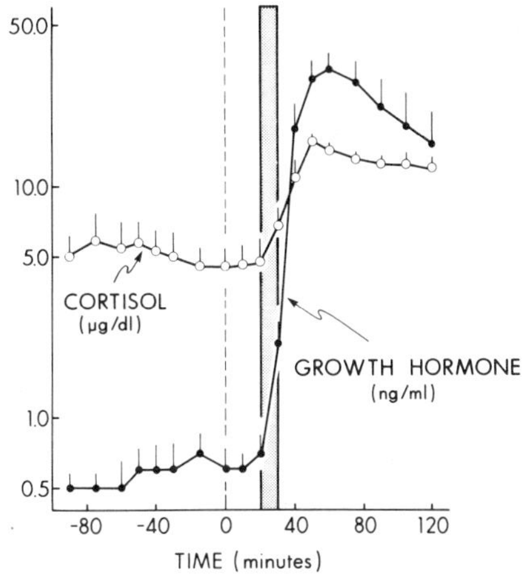

Figure 31.4. Mean (±SE) plasma cortisol and growth hormone concentrations in five normal subjects before and after the rapid intravenous injection of regular insulin (0.04 units/kg) at zero minutes. The *stippled column* marks the onset of glucose counterregulation (see Fig. 31.1).

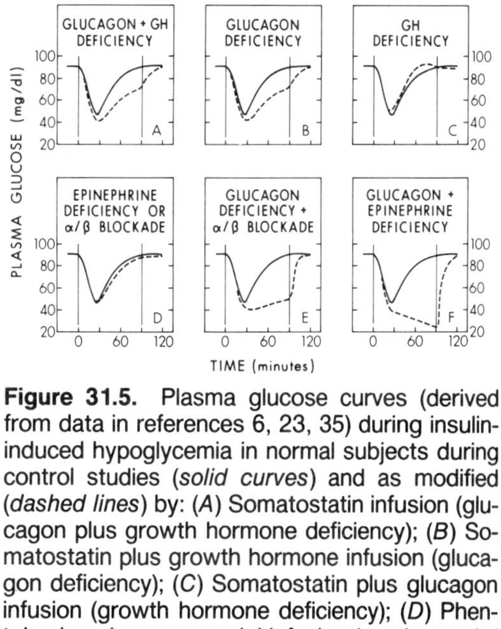

Figure 31.5. Plasma glucose curves (derived from data in references 6, 23, 35) during insulin-induced hypoglycemia in normal subjects during control studies (*solid curves*) and as modified (*dashed lines*) by: (*A*) Somatostatin infusion (glucagon plus growth hormone deficiency); (*B*) Somatostatin plus growth hormone infusion (glucagon deficiency); (*C*) Somatostatin plus glucagon infusion (growth hormone deficiency); (*D*) Phentolamine plus propranolol infusion (α-adrenergic/β-adrenergic blockade) or studies in bilaterally adrenalectomized, glucocorticoid- and mineralocorticoid-replaced patients (epinephrine deficiency); (*E*) Somatostatin plus phentolamine plus propranolol infusions (glucagon deficiency plus α-adrenergic/β-adrenergic blockade); (*F*) Somatostatin infusion in bilaterally adrenalectomized patients (glucagon deficiency plus epinephrine deficiency). (Reproduced with permission from P. E. Cryer (11).)

insulin-induced hypoglycemia is partially impaired (by approximately 40%) during the infusion of somatostatin which suppresses glucagon and growth hormone secretion (*panel A*, Fig. 31.5). Since this effect is not corrected by simultaneous infusion of growth hormone (*panel B*, Fig. 31.5) but is reversed by the simultaneous infusion of glucagon (*panel C*, Fig. 31.5), it is best attributed to deficient glucagon secretion.

In our experience, glucose recovery from insulin-induced hypoglycemia is impaired little, if at all, by adrenergic blockade with the β-adrenergic antagonist propranolol and the α-adrenergic antagonist phentolamine, separately or in combination (*panel D*, Fig. 31.5). As discussed elsewhere (11), several other groups of investigators have found no effect of β-adrenergic blockade on recovery from insulin-induced hypoglycemia, while others have noted partial impairment (averaging approximately 25%) of glucose recovery during β-adrenergic blockade. This apparent discrepancy may be explicable on the basis of the use of generally larger doses of insulin and the production of greater degrees of hypoglycemia in the latter studies. Although plausible, this latter hypothesis is based upon review of the published literature (11) and has not been rigorously tested.

The critical point, in my judgement, is that recovery from insulin-induced hypoglycemia is only partially impaired in the absence of glucagon secretion and is either unimpaired or only minimally impaired during pharmacologic blockade of the hyperglycemic actions of the catecholamines. Stated conversely, major recovery from hypoglycemia occurs in the isolated absence of glucagon secretion and in the isolated absence of catecholamine action. In contrast, glucose recovery is markedly impaired during combined suppression of glucagon secretion and adrenergic blockade (*panel E*, Fig. 31.5). Taken together, these findings suggest that glucagon normally plays a primary role in recovery from insulin-induced hypoglycemia and that adrenergic mechanisms rather effectively compensate for deficient glucagon secretion.

Theoretically, these compensatory adrenergic mechanisms could be mediated by norepinephrine released from sympathetic neurons within the tissues or by epinephrine released into the circulation from the adrenal medullae. (The plasma norepinephrine levels achieved are not sufficient to alter carbohydrate metabolism (48)). To clarify this, studies were performed in bilaterally adrenalectomized (glucocorticoid- and mineralocorticoid-replaced) individuals who are epinephrine deficient. As shown in *panel F* of Figure 31.5, recovery from insulin-induced hypoglycemia was not discernibly abnormal in epinephrine-deficient subjects with intact glucagon secretion. However, glucose recovery failed to occur in the absence of both glucagon and epinephrine secretion.

These studies permit the following conclusions concerning hypoglycemic glucose counterregulation in human subjects. First, glucagon plays a primary role in glucose recovery; glucose recovery is essentially normal when glucagon secretion is intact. Second, enhanced adrenomedullary epinephrine secretion rather effectively compensates for deficient glucagon secretion. Glucose recovery fails to occur only in the absence of both glucagon and epinephrine. Third, neither the acute release of cortisol nor that of growth hormone is critical to acute glucose recovery. Fourth, neither the release of sympathetic neural norepinephrine nor glucose autoregulation is sufficiently potent to promote glucose recovery in the absence of glucagon and epinephrine secretion; neither neural norepinephrine release nor autoregulation need be invoked to explain glucose recovery from insulin-induced hypoglycemia.

NONHYPOGLYCEMIC GLUCOSE COUNTERREGULATION AND THE PREVENTION OF HYPOGLYCEMIA IN NORMAL HUMAN SUBJECTS

The precise mechanisms of nonhypoglycemic glucose counterregulation, *i.e.*, those which reverse physiologic decrements in the plasma glucose concentration and prevent hypoglycemia, have not been determined. It seems likely that suppression of insulin secretion is a factor.

Several groups of investigators (17, 41, 42) have begun to examine the effect of more physiologic nonhypoglycemic decrements in plasma glucose, again produced by intravenous insulin, on the plasma concentrations of potentially important glucose counterregulatory factors. Our experience (42) in human subjects is summarized in Figure 31.6. Controlled reductions of the plasma glucose concentration from 95 mg/dl to 60 mg/dl stimulated significant early increments in plasma epinephrine, norepinephrine, and glucagon, and later increments in plasma cortisol and growth hormone. Interestingly, plasma glucose reductions from 200 mg/dl to 100 mg/dl were also associated with small increments in circulating epinephrine, norepinephrine, glucagon, and growth hormone.

These findings (17, 41, 42) indicate that there is no absolute plasma glucose concentration threshold for activation of glucose counterregulatory systems and that neither the absolute glucose decrement nor the rate of glucose decline are primary determinants of the magnitude of the counterregulatory response. Rather, the magnitude of the counterregulatory response is primarily an inverse function of the absolute plasma glucose concentration. Thus, glucose decrements to hypoglycemic levels trigger a large counterregulatory response, decrements from high to low physiologic glucose levels trigger an intermediate response, and decrements from hyperglycemic to normal levels trigger a small response.

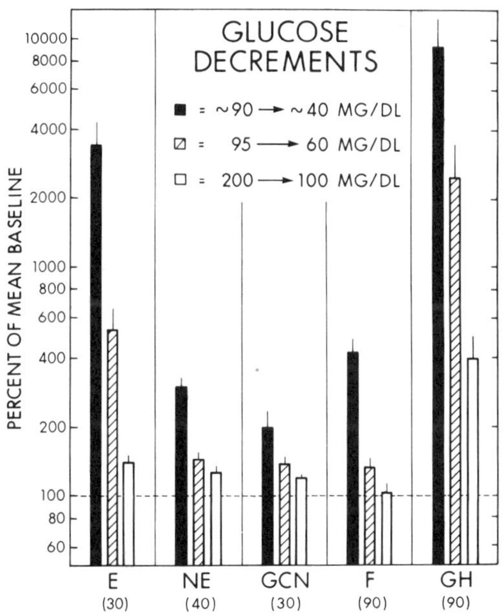

Figure 31.6. Mean (±SE) maximal plasma concentrations of epinephrine (*E*), norepinephrine (*NE*), glucagon (*GCN*), cortisol (*F*), and growth hormone (*GH*) resulting from insulin-induced hypoglycemia (*solid columns*), controlled plasma glucose reductions from 95 mg/dl to 60 mg/dl (*cross-hatched columns*), and controlled plasma glucose reductions from 200 mg/dl to 100 mg/dl (*open columns*). The numbers in parentheses indicate the time, in minutes, after initiation of the change in plasma glucose. These data were derived from those in references 19 and 42. (Reproduced with permission from P. E. Cryer (11).)

These studies, employing a broad range of nadir glucose concentrations (42) (Fig. 31.6), indicate that the depth of the absolute plasma glucose concentration achieved is the primary determinant of the magnitude of the counterregulatory response. However, the absolute glucose decrement, the rate of glucose decline, or both, may well be secondary determinants. It is only reasonable to expect that very small or very slow plasma glucose decrements would not be associated with measurable increments in the circulating concentrations of counterregulatory factors. Whether the absence of significant increments in the plasma concentrations of these factors in studies employing smaller and/or slower plasma glucose decrements (39, 41) reflects no stimulation of release or release too small to significantly elevate the mean absolute plasma levels, the magnitude of the counterregulatory response would appear to be less than that which occurs when similar plasma glucose levels are achieved more rapidly from higher baseline levels. Thus, although this hypothesis remains to be rigorously tested, it is reasonable to conclude that the absolute glucose decrement and the rate of glucose decline are also determinants of the magnitude of the counterregulatory response.

Although clearly important to an understanding of the physiologic mechanisms of activation of the counterregulatory systems, the relevance of these findings to nonhypoglycemic glucose counterregulation and the prevention of hypoglycemia remains to be established. The plasma epinephrine concentrations achieved in association with the 95 mg/dl to 60 mg/dl plasma glucose decrements in our study (42) averaged 230 pg/ml, clearly high enough to produce biologic effects (7). Particularly in view of the experience with hypoglycemic glucose counterregulation discussed earlier, it is reasonable to postulate a biologic role for the increase, albeit small, in glucagon secretion associated with this more physiologic decrement in plasma glucose. Indeed, the recent report of Gauthier *et al.* (21) that the small decrement in plasma glucose caused by phlorizin-induced glycosuria in dogs is markedly enhanced by concomitant suppression of glucagon secretion, indicates that glucagon may well play a physiologic role in nonhypoglycemic glucose counterregulation.

To conclude, hormonal signals, specifically glucagon and epinephrine, are the critical factors in recovery from insulin-induced hypoglycemia. Available evidence is compatible with a central role for these hormones in nonhypoglycemic glucose counterregulation (and the prevention of hypoglycemia) as well, although the latter remains to be fully defined. Obviously, this does not exclude permissive roles for other hormones, such as cortisol and growth hormone, nor does it exclude facilitative roles for neural mechanisms, autoregulatory mechanisms, or both. It should be emphasized that these conclusions apply to short-term glucose counterregulation. The quantitative contribution of each of these mechanisms to maintenance of the postabsorptive plasma glucose concentration remains to be determined.

THE SYMPATHOADRENAL SYSTEM IN PATIENTS WITH DIABETES MELLITUS

Plasma Catecholamine Concentrations

Plasma catecholamine concentrations are markedly elevated in patients with diabetic ketoacidosis (5) and exaggerated plasma catecholamine increments in response to exercise occur in patients with poorly controlled diabetes (5, 49). The levels return toward normal with therapy of ketoacidosis (5), and the response to exercise becomes normal during relatively long-term, effective insulin therapy (49). The mechanism(s) of these plasma catecholamine elevations have not been systematically defined. It is reasonable to suspect that intravascular volume contraction is a hemodynamic stimulus to sympathoadrenal activation, at least in ketoacidosis. In view of the recognition that glucose decrements from supraphysiologic to physiologic levels trigger counterregulatory systems (17, 42), it may be that plasma glucose decrements caused by early fluid and insulin therapy contribute to the elevated plasma catecholamine levels in such patients. With respect to serum growth hormone in diabetic ketoacidosis, this phenomenon was recognized a decade ago (12). Clearly, other factors may well be operative.

The status of sympathoadrenal activity in patients with stable, nonketotic diabetes mellitus is less clear. Christensen (4) found that mean basal and posturally stimulated plasma catecholamine concentrations were not different from normal in a small group of nonneuropathic patients with insulin-dependent diabetes mellitus. In contrast, Robertson *et al.* (38) found elevated mean total plasma catecholamine concentrations in a small group of untreated, noninsulin-dependent diabetic patients. This observation was confirmed by the same group of investigators in a larger group of 16 patients (27). Mean total plasma catecholamine concentrations were significantly higher in the diabetic patients; the levels were more than 2 SDs above the normal mean in eight patients. However, on follow-up of the diabetic patients an average of 6 months later, mean total plasma catecholamine concentrations were indistinguishable from those of nondiabetic controls; this was independent of therapy with insulin or oral hypoglycemic agents.

We measured basal and posturally stimulated plasma catecholamine concentrations in an unselected series of 100 stable, nonketotic, predominantly insulin-treated diabetic patients (15). Mean plasma norepinephrine and epinephrine concentrations—measured with the subject in the supine, resting position and after 2, 5, and 10 minutes of standing—were not significantly different from those of nondiabetic controls. Thus, we concluded that generalized sympathoadrenal activation is not a feature of diabetes mellitus *per se*. That is not to say that sympathoadrenal abnormalities were not detected in individual patients. Postural hypotension with a blunted plasma norepinephrine response to standing, hypoadrenergic postural hypotension attributable to diabetic adrenergic neuropathy, was present in 7%. An additional 11% had exaggerated plasma norepinephrine responses to standing; these included three patients with (hyperadrenergic) postural hypotension. We chose to emphasize and pursue (15, 29, 51) the noradrenergic abnormalities.

An analysis of the primary epinephrine data from our study of 100 patients with diabetes (15) is summarized in Table 31.1. Basal plasma epinephrine concentrations, obtained with the

Table 31.1
Elevated Plasma Epinephrine Concentrations in Patients with Diabetes Mellitus[a]

Definition of Group				Elevated Plasma Epinephrine					
					Standing				
Group	Standing Plasma Norepinephrine	Postural Hypotension[b]	Number in Group	Supine	Any Time Point	2-min Point	5-min Point	10-min Point	Supine → Standing Δ
I	↑[c]		11	0	6	1	2	4	5
1A	↑	No	8	0	4	0	2	3	3
IB	↑	Yes	3	0	2	1	0	1	2
II	N[d]	Yes	8	0	3	2	1	2	2
III	↓[e]	Yes	7	0	1	0	1	1	1
IV	N	No	74	7	14	6	7	9	11
Totals			100	7	24	9	11	16	19
Range of maximum values (pg/ml)				79–302	108–2020	97–521	119–878	108–2020	79–2010
Upper limit of normal[f]				68		83	116	105	74

[a] From the primary data for reference 15.
[b] Postural hypotension, a supine to standing decrement in mean blood pressure of 20 mm Hg or greater.
[c] ↑, increased (more than 2 SDs above the mean for normal subjects).
[d] N, not increased (less than 2 SDs above the mean for normal subjects).
[e] ↓, decreased (increment less than that observed in any normal subject).
[f] Two standard deviations above the mean for normal subjects.

subjects in the supine, resting position for 30 minutes after the insertion of an intravenous sampling line, were elevated in only 7% of the diabetic patients. This is a frequency only slightly greater than that of chance alone. None of the patients with recognized noradrenergic abnormalities (elevated standing plasma norepinephrine concentrations and/or postural hypotension) had elevated basal plasma epinephrine concentrations. On the other hand, plasma epinephrine concentrations obtained while the subjects were standing were elevated at least at one sampling point in 24% of the diabetic patients; the supine to standing increments in plasma epinephrine were increased in 19%. These patients tended to be those with recognized noradrenergic abnormalities. For example, 6 (54%) of the 11 patients with elevated standing norepinephrine concentrations also had elevated standing plasma epinephrine concentrations. Only 14 (19%) of the 74 patients without recognized noradrenergic abnormalities had elevated plasma epinephrine levels at at least one sampling point during standing (8% at 2 minutes, 9% at 5 minutes, and 12% at 10 minutes); 11 (15%) of these 74 patients exhibited an increased supine to standing increment in plasma epinephrine.

Thus, on the basis of our experience in 100 patients with relatively stable, nonketotic diabetes, I would draw the following conclusions: (a) Although elevated basal plasma concentrations of norepinephrine (9%) and epinephrine (7%) and posturally stimulated plasma concentrations of norepinephrine (11%) and epinephrine (24%) occur in some stable nonketotic patients, they do not occur commonly and need to be explained on some basis other than the presence of diabetes *per se*. (b) Increased plasma catecholamine concentrations and, therefore, generalized increments in sympathoadrenal activity, are not a feature of diabetes mellitus *per se*.

These conclusions need to be considered in the perspective of subsequent studies. They are quite consistent with the findings of Weidmann *et al.* (52), who studied 17 hypertensive diabetic patients, but would appear to be at variance with the findings of Gustafson and Kalkhoff (26). The latter investigators found exaggerated plasma epinephrine (but not norepinephrine) responses to standing in 10 patients with insulin-dependent diabetes compared with 10 matched controls. Using a fluorometric method, Bolli *et al.* (3) confirmed that ketotic diabetic patients have elevated plasma catecholamine concentrations and found elevated mean plasma norepinephrine concentrations in 16 insulin-treated patients that they judged to be poorly controlled. Mean plasma norepinephrine and epinephrine concentrations were normal in 11 insulin-treated patients judged to have good control and in five patients with glucose intolerance ("chemical diabetes"). With these data, Bolli *et al.* concluded that sympathoadrenal activation is a direct function of poor metabolic control, implying that it is not a feature of diabetes *per se* but rather a feature of the degree of the metabolic abnormality. Their conclusion was buttressed by the findings of significant correlations between the plasma glucose and catecholamine concentrations in their patients. However, these data may well have been skewed by the inclusion of ketotic patients in the analysis. In our study of 100 nonketotic diabetics (15), there were no significant correlations between the fasting plasma glucose concentrations and the plasma norepinephrine or epinephrine concentrations measured in the basal state and in response to postural stress.

More recently, Zadik *et al.* (53) used a continuous withdrawal pump to obtain 24-hour plasma pools from 11 insulin-dependent diabetic patients. As the investigators pointed out, the patients were not well controlled; their mean (±SD) integrated plasma glucose concentration was 299 ± 81 mg/dl, and their mean glucose excretion was 96 ± 79 gm/24 hour. Nonetheless, they were not ketonuric at the time of study. The mean integrated plasma concentrations of norepinephrine (240 ± 81 pg/ml) and of epinephrine (38 ± 7 pg/ml) were significantly higher in the diabetic patients than in nondiabetic controls (148 ± 74 pg/ml and 26 ± 13 pg/ml, respectively). Mean norepinephrine values were more than 2 SDs above the normal mean in 2 of 11 patients; mean epinephrine values were not more than 2 SDs above the normal mean in any of the patients. These data suggest that a rather subtle increase in the plasma concentrations of norepinephrine and epinephrine occurs in patients with poorly controlled, but nonketotic, diabetes mellitus. The finding of significantly higher mean plasma norepinephrine and epinephrine values may be due to their technique which minimizes variation in the measured values; alternatively, since other investigators have generally measured morning values, it is conceivable that there are differences in diurnal plasma catecholamine patterns between diabetic and nondiabetic patients.

In summary, plasma catecholamine concentrations are elevated, at times markedly so, in patients with diabetic ketoacidosis (5). Plasma catecholamine responses to exercise are exaggerated in patients with poorly controlled diabetes (5, 49). Although major increments in circulating catecholamines do not occur commonly in patients with nonketotic diabetes (4, 15, 27, 38), rather subtle increments may occur in poorly controlled, but nonketotic, patients (3, 26, 52). The extent to which these elevations reflect increased sympathoadrenal catecholamine release or decreased catecholamine clearance and whether or not they occur in patients with better metabolic control remain to be determined. Whether such elevations that do occur are a primary feature of diabetes or are secondary to the metabolic and/or hemodynamic abnormalities of diabetes remains to be determined with certainty. I favor the latter possibility.

Responsiveness to Catecholamines

Shamoon and his colleagues (44, 47) have demonstrated that insulin-dependent diabetic patients who are infused with sufficient insulin to lower the fasting plasma glucose concentration to normal levels exhibit an exaggerated hyperglycemic response to the prolonged intravenous infusion of rather large doses of epinephrine (1.2 $\mu g/m^2/min$). The exaggerated hyperglycemic response is due to a heightened and sustained epinephrine-stimulated increase in (hepatic) glucose production in the patients with diabetes, as opposed to the normal smaller and transient increase in glucose production that occurs in nondiabetic subjects infused with epinephrine.

The mechanism of this altered glucose response is unknown. As the investigators pointed out, it is conceivable that, despite the continued infusion of insulin at basal rates, the fact that the diabetic patients were receiving insulin peripherally and were unable to release endogenous insulin could provide an explanation. Although sufficient insulin was infused peripherally to normalize the plasma glucose concentration prior to epinephrine infusion, relative portal venous (and, thus, hepatic) insulin deficiency may well have existed. Further, as epinephrine-induced hyperglycemia developed, endogenous insulin could not be released. It could be reasoned that this latter point is not relevant, since these investigators observed no significant increase in plasma insulin during epinephrine infusions in normal subjects. This observation is, however, in contrast to the data of others who have shown a significant increase (after a transient decrease) in plasma insulin concentrations during the infusion of epinephrine into nondiabetic subjects (7, 37). The authors cite the data of Rizza *et al.* (37), dealing with the effect of somatostatin on the response to infused epinephrine in nondiabetics, to discount the possible role of diminished insulin release in their diabetic patients. However, inspection of the latter paper (37) reveals that inhibition of insulin secretion with somatostatin in normal subjects converted the glucose production response to epinephrine to a more sustained pattern despite the coexistent (somatostatin-induced) suppression of glucagon secretion.

Regardless of its mechanism, the observation that insulin-dependent diabetic patients treated with peripherally infused insulin exhibit an exaggerated hyperglycemic response to epinephrine has practical implications, since we currently treat such patients with peripheral insulin administration.

Possible Pathophysiologic Roles

DIABETIC KETOACIDOSIS

It is reasonable to conclude that the markedly elevated catecholamine levels observed in diabetic ketoacidosis (5) play a role in the perpetuation, if not the initiation, of that syndrome (43). As mentioned earlier, accelerated glucose production and diminished glucose clearance with resultant hyperglycemia, and accelerated lipolysis and ketogenesis are characteristic features of catecholamine action; they are also characteristic features of diabetic ketoacidosis.

It would appear, however, that enhanced sympathoadrenal activity may not be critical to the development of diabetic ketosis. Alberti *et al.* (1) examined the effects of 41 hours of insulin deprivation on four patients with insulin-dependent diabetes. Two of these patients became ketotic with no apparent increase in their plasma norepinephrine or epinephrine concentrations. In fairness, however, the catecholamines were measured only four times (twice on a control day and twice near the end of the insulin deprivation period) in these patients. Although these data question a critical role for sympathoadrenal activation in the initiation of diabetic ketoacidosis, they do not, in my judgment, convincingly establish that point. Clearly, they do not preclude a perpetuating role for the catecholamines (quite likely in concert with other hormonal abnormalities including relative hypoinsulinemia, hyperglucagonemia, and elevated growth hormone and cortisol levels) in the pathophysiology of diabetic ketoacidosis.

POSTURAL HYPOTENSION

Postural hypotension can be due to relative or absolute intravascular volume contraction, vascular resistance to the pressor action of norepinephrine, or a defect in the sympathetic neural reflex arc (9). Postural activation of the baroreceptor-initiated, CNS-mediated sympathetic reflex that results in release of norepinephrine from axon terminals within the vasculature (and thus, maintenance of the standing blood pressure) is reflected in a sharp rise in the plasma norepinephrine concentration (14). Thus, the integrity of the reflex can be readily assessed in human subjects by measurement of the plasma norepinephrine concentration before and after standing.

A blunted plasma norepinephrine response to standing is the biochemical common denominator of patients with postural hypotension due to a defect in the sympathetic reflex arc (16). Such hypoadrenergic postural hypotension occurs not infrequently as a component of diabetic autonomic neuropathy (15); we have attributed it to a postganglionic axonal lesion in such patients (29).

Patients with postural hypotension due to intravascular volume contraction or vascular resistance to the pressor effect of norepinephrine have a compensatory increase in their sympathetic response to standing. This is reflected in an exaggerated plasma norepinephrine response to standing. They have hyperadrenergic postural hypotension.

As mentioned earlier, we (15) recognized some diabetic patients with hyperadrenergic postural hypotension. We speculated that they might have diminished pressor responses to norepinephrine, diminished plasma volumes, or both, possibly due to hypoaldosteronism. None of these speculations was correct. Detailed studies of four such patients disclosed that they all had sharply reduced red blood cell masses (51), a finding not suggested by their peripheral venous hematocrits. Thus, decreased intravascular volume due to anemia appears to be an important factor in the pathogenesis of hyperadrenergic postural hypotension in some patients with diabetes. A coexisting neuropathic limitation of their compensatory sympathetic response seems likely. Nonetheless, the finding of postural hypotension in a patient with diabetes should not be quickly attributed to diabetic adrenergic neuropathy alone since other, potentially reversible, factors may be playing a pathogenic role.

METABOLIC COMPENSATION AND INSTABILITY

The metabolic impact of counterregulatory system activation in general, and sympathoadrenal activation in particular, in patients with diabetes remain speculative. The likely role of these factors in the pathophysiology of diabetic ketoacidosis was mentioned earlier.

The glucose counterregulatory systems are almost assuredly critical to diabetic patients treated with glucose-lowering drugs such as oral hypoglycemic agents or insulin. Such patients are at risk for major hypoglycemia and must have adequate mechanisms for glucose recovery from hypoglycemia. As mentioned earlier, these same mechanisms may well promote nonhypoglycemic glucose counterregulation and, thus, prevent hypoglycemia. The efficacy of the counterregulatory systems may well define the dosage limits of these drugs and their maximum therapeutic benefit in terms of glucose lowering. Any dose that overwhelms the counterregulatory systems and causes hypoglycemia at any time during its course of action is excessive, even if that dose does not adequately lower the plasma glucose concentration at other times. Indeed, the effectiveness of the counterregulatory systems may explain why a less than perfect insulin dose, delivered with recently popularized open loop insulin delivery systems, can result in near normalization of the metabolic abnormalities in some patients with diabetes.

As discussed earlier, glucagon and epinephrine play critical roles in normal hypoglycemic glucose counterregulation and may play similar roles in normal nonhypoglycemic glucose counterregulation. It is well established that many patients with diabetes have deficient glucagon responses to hypoglycemia (2, 24). We recently demonstrated that diabetic patients have blunted glucagon responses to nonhypoglycemic decrements in plasma glucose as well (42) and may be more dependent upon the glucose counterregulatory actions of epinephrine than are nondiabetic persons. Systematic testing of this hypothesis is currently in progress. Preliminary evidence suggests that some patients with insulin-dependent diabetes mellitus do not recover from hypoglycemia normally and are, indeed, more dependent upon adrenergic mechanisms for glucose recovery than nondiabetic controls (31). Such patients, in contrast to nondiabetics, are at increased risk for hypoglycemia when treated with a β-adrenergic antagonist, such as propranolol for hypertension, angina etc.

Although beneficial to the diabetic patient with respect to the correction of, and possibly the prevention of, hypoglycemia, activation of systems that raise the plasma glucose concentration would be expected to cause metabolic instability in patients incapable of releasing endogenous insulin. Posthypoglycemic hyperglycemia (the Somogyi phenomenon) is a likely example. Once counterregulatory systems are activated to promote recovery from hypoglycemia, the rising plasma glucose concentration cannot be limited by secreted insulin; hyperglycemia results. Our finding that propranolol significantly reduces the hyperglycemia that follows insulin-induced hypoglycemia in patients with diabetes (31) indicates that catecholamines, probably adrenomedullary epinephrine, play an important role in the pathogenesis of posthypoglycemic hyperglycemia. It is conceivable that similar patterns follow nonhypoglycemic glucose decrements and contribute to the fluctuations in plasma glucose that typify insulin-dependent diabetes mellitus.

Clearly, factors that promote hyperglycemia, such as epinephrine, are released in response to stimuli other than hypoglycemia. These stimuli include a variety of nonspecific stresses such as those imposed by intercurrent illnesses which often cause metabolic deterioration, and may precipitate ketoacidosis in patients with diabetes. The observation that small doses of intravenous epinephrine, which raise plasma epinephrine concentrations only to 75 to 80 pg/ml, result in impaired glucose tolerance in nondiabetic subjects (28) underscores the concept that physiologic fluctuations in sympathoadrenal activity may well contribute to metabolic instability in patients with diabetes.

It is not difficult to envision a trade-off concept with respect to the sympathoadrenal system in patients with diabetes mellitus. The catecholamines may well play an important role in the prevention and correction of hypoglycemia but may also contribute to diabetic instability and even precipitate frank metabolic decompensation.

SUMMARY

The precise roles of the sympathoadrenal system, and its major biologically active products norepinephrine and epinephrine, in human metabolic physiology are only beginning to emerge. Much remains to be learned about the biology of the sympathoadrenal system *per se* and about its integrative role in human metabolism.

In concert with glucagon, adrenomedullary epinephrine appears to play an important role in recovery from hypoglycemia; recovery from hypoglycemia fails to occur only in the absence of both glucagon and epinephrine. In contrast, sympathetic neural norepinephrine release does not appear to be critical to recovery from hypoglycemia. Available evidence is consistent with a physiologic role for epinephrine and glucagon in nonhypoglycemic glucose counterregulation and the prevention of hypoglycemia as well, although this remains to be established. If this is the case, this knowledge will provide a rational basis for systematic study of the pathophysiology of human hypoglycemic states.

Deficient sympathetic neural activity, the result of diabetic autonomic neuropathy, causes postural hypotension in some patients with diabetes. Assessment of the integrity of the sympathetic neural reflex by measurement of the plasma norepinephrine response to standing permits identification of such patients and of patients in whom other, potentially reversible factors such as intravascular volume contraction are contributing to the pathogenesis of their postural hypotension. It is likely that markedly enhanced sympathoadrenal activity plays a role in the perpetuation, if not the initiation, of diabetic ketoacidosis and that stress-induced increments in sympathoadrenal activity are a factor in the associated metabolic deterioration that occurs in patients with diabetes. The possibility that rather subtle increments in sympathoadrenal activity may occur in nonstressed, nonketotic diabetics has been raised by recent observations, but their pathophysiologic significance is as yet unclear. It is conceivable that increased catecholamine release might be both beneficial in so far as it serves to prevent hypoglycemia, and detrimental, in so far as it promotes metabolic instability, in patients with diabetes. Clearly, much also remains to be learned about the pathophysiology of the sympathoadrenal system in diabetes as well as in other human diseases.

References

1. Alberti KGMM, Christensen NJ, Iversen J, Orskow H: Role of glucagon and other hormones in development of diabetic ketoacidosis. *Lancet* 1:1307–1311, 1975.
2. Benson JW Jr, Johnson DG, Palmer JP, Werner PL, Ensinck JW: Glucagon and catecholamine secretion during hypoglycemia in normal and diabetic man. *J Clin Endocrinol Metab* 44:459–464, 1977.
3. Bolli G, Cartechini MG, Compagnucci P, Malvicini S, DeFeo P, Santeusanio F, Angeletti G, Brunetti P: Effect of metabolic control on urinary excretion and plasma levels of catecholamines in diabetics. *Horm Metab Res* 11:493–497, 1979.
4. Christensen NJ: Plasma catecholamines in long-term diabetics with and without neuropathy and in hypophysectomized subjects. *J Clin Invest* 51:779–787, 1972.
5. Christensen NJ: Plasma norepinephrine and epinephrine in untreated diabetics, during fasting and after insulin administration. *Diabetes* 23:1–8, 1974.
6. Clarke WL, Santiago JV, Thomas L, Haymond MW, Ben-Galim E, Cryer PE: The role of adrenergic mechanisms in recovery from hypoglycemia in man: Studies with adrenergic blockade. *Am J Physiol* 236:E147–E152, 1979.
7. Clutter WE, Shah SD, Bier DM, Cryer PE: Epinephrine plasma metabolic clearance rates and physiologic thresholds for metabolic and hemodynamic actions in man. *J Clin Invest* 66:94–101, 1980.
8. Cryer PE: Isotope derivative measurements of plasma norepinephrine and epinephrine in man. *Diabetes* 25:1071–1082, 1976.
9. Cryer PE: Mechanisms and management of postural hypotension. In Cohen MP, Foa PP: *Special Topics in Endocrinology and Metabolism.* New York, Alan R. Liss, 1979, vol 1, pp 79–91.
10. Cryer PE: Physiology and pathophysiology of the human sympathoadrenal neuroendocrine system. *N Engl J Med* 303:436–444, 1980.
11. Cryer PE: Glucose counterregulation in man. *Diabetes* 30:261–264, 1981.

12. Cryer PE, Daughaday WH: Diabetic ketosis: Serial plasma growth hormone concentrations during therapy. *Diabetes* 19:519–523, 1970.
13. Cryer PE, Rizza RA, Haymond MW, Gerich JE: Epinephrine and norepinephrine are cleared through beta-adrenergic, but not alpha-adrenergic, mechanisms in man. *Metabolism* 29:1114–1118, 1980.
14. Cryer PE, Santiago JV, Shah SD: Measurement of norepinephrine and epinephrine in small volumes of human plasma by a single isotope derivative method: Response to the upright posture. *J Clin Endocrinol Metab* 39:1025–1029, 1974.
15. Cryer PE, Silverberg AB, Santiago JV, Shah SD: Plasma catecholamines in diabetes: The syndromes of hypoadrenergic and hyperadrenergic postural hypotension. *Am J Med* 64:407–416, 1978.
16. Cryer PE, Weiss S: Reduced plasma norepinephrine response to standing in autonomic dysfunction. *Arch Neurol* 33:275–277, 1976.
17. DeFronzo RA, Hendler R, Christensen NJ: Stimulation of counter-regulatory hormonal responses in diabetic man by a fall in glucose concentration. *Diabetes* 29:125–131, 1980.
18. Eigler N, Sacca L, Sherwin RS: Synergistic interactions of physiologic increments of glucagon, epinephrine and cortisol in the dog: A model for stress induced hyperglycemia. *J Clin Invest* 63:114–123, 1979.
19. Garber AJ, Cryer PE, Santiago JV, Haymond MW, Pagliara AS, Kipnis DM: The role of adrenergic mechanisms in the substrate and hormonal response to insulin-induced hypoglycemia in man. *J Clin Invest* 58:7–15, 1976.
20. Garber AJ, Karl IE, Kipnis DM: Alanine and glutamine synthesis and release from skeletal muscle. IV. β-Adrenergic inhibition of amino acid release. *J Biol Chem* 251:1851–1857, 1976.
21. Gauthier C, Vranic M, Hetenyi G Jr: Importance of glucagon in regulatory rather than emergency responses to hypoglycemia. *Am J Physiol* 238:E131–E140, 1980.
22. Gerich JE, Cryer PE, Rizza RA: Hormonal mechanisms in acute glucose counterregulation: The relative roles of glucagon, epinephrine, norepinephrine, growth hormone and cortisol. *Metabolism* 29:1164–1175, 1980.
23. Gerich J, Davis J, Lorenzi M, Rizza R, Bohannon N, Karam J, Lewis S, Kaplan S, Schultz T, Cryer P: Hormonal mechanisms of recovery from insulin-induced hypoglycemia in man. *Am J Physiol* 235:E380–E385, 1979.
24. Gerich JE, Langlois M, Nozocco C, Karam JH, Forsham PH: Lack of glucagon response to hypoglycemia in diabetes: Evidence for an intrinsic pancreatic alpha-cell defect. *Science* 182:171–173, 1973.
25. Ginsberg AM, Clutter WE, Shah SD, Cryer PE: Triiodothyronine induced thyrotoxicosis increases mononuclear leukocyte β-adrenergic receptor density in man. *J Clin Invest* 67:1785–1791, 1981.
26. Gustafson AB, Kalkhoff RK: Plasma epinephrine disturbances in diabetic subjects during isometric exercise (abstr). *Diabetes* 29(Suppl 2):40A, 1980.
27. Halter J, Porte D Jr: Increased adrenergic activity in diabetes mellitus. Response to therapy and pharmacologic stimulation (abstr). *Clin Res* 25:160A, 1977.
28. Hamburg S, Hendler R, Sherwin RS: Epinephrine: Exquisite sensitivity to its diabetogenic effects in normal man (abstr). *Clin Res* 27:252A, 1979.
29. Leveston SA, Shah SD, Cryer PE: Cholinergic stimulation of norepinephrine release in man: Evidence of a sympathetic postganglionic axonal lesion in diabetic adrenergic neuropathy. *J Clin Invest* 64:374–380, 1979.
30. Liljenquist JE, Mueller GL, Cherrington AD, Perry JM, Rabinowitz D: Hyperglycemia per se (insulin and glucagon withdrawn) can inhibit hepatic glucose production in man. *J Clin Endocrinol Metab* 48:171–174, 1979.
31. Popp DA, Shah SD, Cryer PE: The role of epinephrine mediated β-adrenergic mechanisms in hypoglycemic glucose counterregulation and posthypoglycemic hyperglycemia in insulin dependent diabetes mellitus. *J Clin Invest* 69:315–326, 1982.
32. Porte D Jr: Sympathetic regulation of insulin secretion. *Arch Intern Med* 123:252–260, 1969.
33. Porte D Jr, Garber AL, Kuzuya T, Williams RH: The effect of epinephrine on immunoreactive insulin levels in man. *J Clin Invest* 45:228–236, 1966.
34. Porte D Jr, Williams RH: Inhibition of insulin release by norepinephrine in man. *Science* 152:1248–1250, 1966.
35. Rizza RA, Cryer PE, Gerich JE: Role of glucagon, epinephrine and growth hormone in human glucose counterregulation: Effects of somatostatin and adrenergic blockade on plasma glucose recovery and glucose flux rates following insulin-induced hypoglycemia. *J Clin Invest* 64:62–71, 1979.
36. Rizza RA, Cryer PE, Haymond MW, Gerich JE: Adrenergic mechanisms for the effects of epinephrine on glucose production and clearance in man. *J Clin Invest* 65:682–689, 1980.
37. Rizza RA, Haymond MW, Cryer PE, Gerich JE: Differential effects of epinephrine on glucose production and disposal in man. *Am J Physiol* 237:E356–E362, 1979.
38. Robertson RP, Halter JB, Porte D Jr: A role for alpha-adrenergic receptors in abnormal insulin secretion in diabetes mellitus. *J Clin Invest* 57:791–795, 1976.
39. Sacca L, Cryer PE, Sherwin RS: Blood glucose regulates the effects of insulin and counterregulatory hormones on glucose production in vivo. *Diabetes* 28:533–536, 1979.
40. Sacca L, Hendler R, Sherwin RS: Hyperglycemia inhibits glucose production in man independent of changes in glucoregulatory hormones. *J Clin Endocrinol Metab* 47:1160–1163, 1978.
41. Sacca L, Sherwin R, Hendler R, Felig P: Influence of continuous physiologic hyperinsulinemia on glucose kinetics and counterregulatory hormones in normal and diabetic humans. *J Clin Invest* 63:849–857, 1979.
42. Santiago JV, Clarke WL, Shah SD, Cryer PE: Epinephrine, norepinephrine, glucagon and

growth hormone release in association with physiological decrements in the plasma glucose concentration in normal and diabetic man. *J Clin Endocrinol Metab* 51:877–883, 1980.

43. Schade DS, Eaton RP: Pathogenesis of diabetic ketoacidosis: A reappraisal. *Diabetes Care* 2:296–306, 1979.
44. Shamoon H, Hendler R, Sherwin RS: Altered responsiveness to cortisol, epinephrine and glucagon in insulin infused juvenile-onset diabetics: A mechanism for diabetic instability. *Diabetes* 29:284–291, 1980.
45. Shamoon H, Jacobs R, Sherwin RS: Epinephrine-induced hypoaminoacidemia in normal and diabetic humans: Effect of β-blockade. *Diabetes* 29:875–881, 1980.
46. Shulman GI, Liljenquist JE, Williams PE, Lacy WW, Cherrington AD: Glucose disposal during insulinopenia in somatostatin-treated dogs. The roles of glucose and glucagon. *J Clin Invest* 62:487–491, 1978.
47. Sherwin RS, Shamoon H, Hendler R, Sacca L, Eigler N, Walesky M: Epinephrine and the regulation of glucose metabolism: Effect of diabetes and hormonal interactions. *Metabolism* 29:1146–1154, 1980.
48. Silverberg AB, Shah SD, Haymond MW, Cryer PE: Norepinephrine: Hormone and neurotransmitter in man. *Am J Physiol* 234:E252–E256, 1978.
49. Tamborlane WV, Sherwin RS, Koivisto V, Hendler R, Genel M, Felig P: Normalization of the growth hormone and catecholamine response to exercise in juvenile onset diabetic subjects treated with a portable insulin infusion pump. *Diabetes* 28:785–788, 1979.
50. Tohmeh JF, Cryer PE: Biphasic adrenergic modulation of β-adrenergic receptors in man: Agonist induced early increment and late decrement in β-adrenergic receptor number. *J Clin Invest* 65:836–840, 1980.
51. Tohmeh JF, Shah SD, Cryer PE: The pathogenesis of hyperadrenergic postural hypotension in diabetic patients. *Am J Med* 67:772–778, 1979.
52. Weidman P, Beretta-Piccoli C, Keusch G, Gluck Z, Mujagic M, Grimm M, Meier A, Ziegler WH: Sodium-volume factor, cardiovascular reactivity and hypotensive mechanism of diuretic therapy in mild hypertension associated with diabetes mellitus. *Am J Med* 67:779–784, 1979.
53. Zadik Z, Kayne R, Kappy M, Plotnick LP, Kowarski AA: Increased integrated concentration of norepinephrine, epinephrine, aldosterone and growth hormone in patients with uncontrolled juvenile diabetes mellitus. *Diabetes* 29:655–658, 1980.

CHAPTER THIRTY-TWO

Pheochromocytoma: Clinical Considerations

CHRIS K. FINTON, M.D., BART CHERNOW, M.D., AND HARRY R. KEISER, M.D.

INTRODUCTION—DEFINITION OF THE TUMOR

Pheochromocytoma is a tumor comprised of chromaffin cells of neuroectodermal origin. The name stems from its staining characteristics with chromium salts (*i.e.*, from the Greek, phaiso = dusky + chroma = color); the catecholamine (CA)-containing granules stain brown as a result of oxidation. During fetal development, chromaffin cells are found at the carotid bifurcation, aortic arch, along the posterior sympathetic chain, the organ of Zuckerkandl, and, of course, the adrenal medulla. This wide distribution accounts for the 10% incidence of extraadrenal locations of pheochromocytomas (49).

FREQUENCY OF PHEOCHROMOCYTOMA

The first pheochromocytoma was described in 1886, but it was not until 1922 that it was linked to paroxysms of hypertension. Mayo (37) pioneered the technique for excision of a pheochromocytoma in the mid 1920s. It was in 1937 that the associated hypertension was found to be caused by hormones, CA, released from the tumor. Pheochromocytomas were found in 15 of 15,934 consecutive autopsies performed at the Mayo Clinic, suggesting its occurrence in 0.1% of the population (40). It is estimated that it is the cause of hypertension in 0.6% of hypertensive patients (33). In adults there is an equal incidence of this tumor in men and women. The tumor is bilateral in 10% of the cases and malignant in 3 to 13%, depending upon the reported series.

CLINICAL CHARACTERISTICS

Pheochromocytomas are often present with both paroxysmal and persistent signs and symptoms (Table 32.1). The three most common symptoms are headache, diaphoresis, and palpitations (24). The symptoms usually occur concurrent with paroxysms of hypertension and may be precipitated by emotion, specific movements of the body, or anything that activates the sympathetic nervous system (SNS). This is especially typical of headaches which are most often sudden in onset, and frequently severe, but clinically indistinguishable from other types of headaches. Other symptoms include: anxiety, nausea, weight loss, Raynaud's phenomenon, dizziness, chest pain, and abdominal pain (24). For an excellent in-depth review of this subject, see the monograph by Manges and Gifford (36).

Postural hypotension occurs and may be related to several mechanisms: decreased plasma volume (56); norepinephrine (NE) metabolites acting as false neurotransmitters; decreased adrenergic receptor sensitivity; or release of vasodilator substances (16). Other clinical findings include glucose intolerance which may be related to CA-induced suppression of pancreatic β-cell function (27). The glucose intolerance is reversible by pharmacologic blockade of α-adrenergic receptors (61). Patients with epinephrine (E)-secreting pheochromocytomas may have an increased basal metabolic rate and many eventually suffer weight loss. In some, thyrotoxicosis develops, possibly related to augmented β- and α-adre-

Table 32.1
Signs and Symptoms of Pheochromocytoma[a]

Signs		Symptoms	
Hypertension	++++	Headaches	++++
Tachycardia or reflex bradycardia	+++	Palpitations	+++
Postural hypotension	+++	Sweating	+++
Hypertension, paroxysmal	++	Anxiety/nervousness	++
Weight loss	++	Tremulousness	++
Pallor of face or trunk	++	Nausea/vomiting	++
Hypermetabolism	++	Pain in chest/abdomen	++
Fasting hyperglycemia	++	Weakness/fatigue/prostration	++
Tremor	++	Dizziness	+
Increased respiratory rate	++	Heat intolerance	+
Decreased GI motility	++	Paresthesias	+
Psychosis (rare)	+	Constipation	+
Flushing, paroxysmal (rare)	+		

[a] ++++, 76 to 100%; +++, 51 to 75%; ++, 26 to 50%; +, 1 to 25%.

nergic stimulation of the thyroid gland (39). Severe cardiomyopathies are reported. Recent data (15) suggest that hypertrophic cardiomyopathy tends to occur with NE-secreting tumors, whereas congestive cardiomyopathy tends to occur with E-secreting tumors.

In children, although the signs and symptoms of pheochromocytoma are often similar in adults, the incidence of extraadrenal and multiple tumors is higher, about 30% (57).

ASSOCIATED DISORDERS

In 1961, Sipple (54) reported the association of pheochromocytoma and thyroid carcinoma, subsequently identified as medullary thyroid carcinoma (63). In 1968, the terminology multiple endocrine neoplasia (MEN), type II (Sipple's syndrome) was applied to patients with pheochromocytoma, medullary carcinoma of the thyroid, and hyperparathyroidism (Table 32.2) (57). The pheochromocytoma in these patients is frequently multicentric and bilateral and may be preceded by nodular and diffuse hyperplasia of the adrenal gland (9). A variant of MEN-II is described with hypercalcitonemia, C-cell hyperplasia, and pheochromocytoma (48). Another report describes a pheochromocytoma that secreted calcinonin and was paradoxically associated with hypercalcemia which resolved after removal of the pheochromocytoma (30). MEN-IIB (or MEN type III) is used to describe patients with pheochromocytoma, medullary thyroid carcinoma, hyperparathyroidism, and mucosal neuromas (5). There is also an association between neurofibromatosis and the development of pheochromocytoma (50, 63). A similar association with other neuroectodermal syndromes (von Hippel-Lindau, tuberous sclerosis and Sturge-Weber syndromes) and pheochromocytomas also exists (26, 42) but occurs less frequently. A number of other tumors are reported to be associated with pheochromocytoma. These include Burkitt's lymphoma, neuroblastoma, gastrin-producing islet cell tumors of the pancreas, and pituitary adenomas (1, 4, 9, 21). Familial occurrences of pheochromocytoma have an autosomal-dominant pattern of inheritance (41). In this entity, the tumors are usually multiple and bilateral.

Table 32.2
Multiple Endocrine Neoplasia Syndromes

Type I (Wermer's syndrome)	Hyperparathyroidism Pancreatic tumors secreting gastrin, insulin, glucagon, or VIP Pituitary tumors
Type IIA (Sipple's syndrome)	Hyperparathyroidism Pheochromocytoma Medullary carcinoma of the thyroid
Type IIB or III	Pheochromocytoma Medullary carcinoma of the thyroid Mucosal neuromas Ganglioneuromas Marfanoid habitus

DIAGNOSTIC CONSIDERATIONS

The diagnosis of pheochromocytoma is dependent on the demonstration of elevations in endogenous CA or their metabolites in plasma or urine. This is usually accomplished by the collection of a plasma or urine specimen for CA assay. However, the results may be definitive only if the specimen is collected while the patient is hypertensive. Normal values for tests obtained while the patient is normotensive do not necessarily exclude a pheochromocytoma. Provocative tests, because of their propensity for precipitating a hypertensive crisis, should be reserved for only the most selective cases and performed with close monitoring.

Several biochemical assays are available for the potential diagnosis of pheochromocytoma (Table 32.3). Plasma and urinary CA can be measured directly. E and NE are metabolized by monoamine oxidase (MAO) and catechol-O-methyltransferase (COMT) to vanillylmandelic acid (VMA) which can be measured in the urine (Fig. 32.1). Other metabolites, metanephrine (M) and normetanephrine (NM), are also measurable in the urine. The advantages and disadvantages of each test will be discussed below.

PLASMA DIAGNOSTIC TESTS

Pheochromocytomas secrete CA continuously as well as paroxysmally (19, 47). Plasma CA can be measured by fluorometric methods (34), by radioenzymatic methods (35), and by high-pressure liquid chromatography (HPLC) coupled with electrochemical detection (see Chapter 3). The normal basal range for plasma

Table 32.3
Biochemical Tests for Pheochromocytoma

		Normal[a]
Plasma	Norepinephrine	<390 pg/ml
	Epinephrine	<70 pg/ml
	Total catecholamines	<1000 pg/ml
Urine	VMA	<7.0 mg/24 hr
	Metanephrine and normetanephrine	<1.3 mg/24 hr
	NE	<75 μg/24 hr
	E	<25 μg/24 hr

[a] Since normal values may vary greatly in different laboratories, these ranges are only approximate.

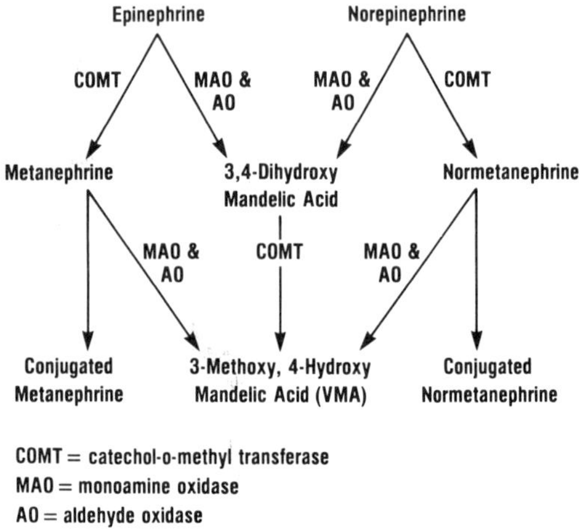

Figure 32.1. The metabolic pathway of epinephrine and norepinephrine, showing the derivation of urine metabolites.

E is 10 to 70 pg/ml and varies between labs by 100% for NE (see Chapter 1), although it has been given as 66 to 390 pg/ml (13). Plasma E and NE increase in response to standing and stress (e.g., hypoglycemia, exercise, surgery, ketoacidosis, and myocardial infarction) (13). To be meaningful and reproducible, plasma specimens for CA determination must not be collected during stress (see Chapter 1). Cigarette smoking and drugs containing sympathomimetic agents (as do many nasal decongestants) can elevate plasma CA levels and therefore must be avoided before testing.

Plasma CA determinations are not widely available and are expensive to obtain in some areas. Bravo *et al.* (6) find total plasma CA levels above 2000 pg/ml to be diagnostic for pheochromocytoma. Levels below 1000 pg/ml exclude the diagnosis, and levels between 1000 pg/ml and 2000 pg/ml are equivocal. They contend that this test, in their hands, is more reliable than urinary VMA or M determinations. Interestingly, significant elevations are found in plasma CA concentrations in patients with pheochromocytoma, even when the blood pressure (BP) is normal at the time of venous sampling, and this opinion is shared by others (3, 14). On the other hand, some (16, 24, 25) report plasma CA to be elevated only in those with persistently "functioning" pheochromocytomas. These latter investigators frequently note normal CA levels in patients with paroxysmal signs and symptoms if the venous sampling is done between paroxysms.

URINARY DIAGNOSTIC STUDIES

Urinary free CA can be measured by a trihydroxyindole fluorometric technique. The measurement of free urinary CA is preferable to that of total CA, as the latter also includes conjugated CA from dietary sources which may lead to a false-positive result (43). Urinary free CA represent about 1 to 3% of the total CA secreted into the circulation (14). False-positive tests are found in patients taking either CA-containing medications or tetracycline, quinine, levo-dopa, prazosin, or α-methyldopa. False-negative tests are extremely rare but are possible in extremely dilute urine (15). In patients with persistently "functioning" pheochromocytomas, the elevations in urinary free CA concentration are diagnostic (16, 17, 24). Because of the wide variation in minute to minute secretion of plasma CA, the collection of 24-hour urine specimens for free CA measurements has been advocated. However, "spot" urine collections for urinary free CA are useful when they are obtained immediately after a paroxysm of symptoms and/or hypertension. The measurement of urinary NE in first voided morning specimens can also be helpful in diagnosing pheochromocytoma (23). The false-negative rate with this latter test occurs most frequently in patients with paroxysmal symptoms and ranges from 0 to 25% (2, 53).

Urinary VMA satisfactorily reflects plasma CA concentrations (Fig. 32.1). This appears to be unaltered in chronic renal failure (32). Urinary VMA may be assayed by oxidizing it to vanillin and then measuring the latter by ultraviolet spectrophotometry (46). Both gas and HPLC methods are available (22). Urinary M can be measured by extraction followed by oxidation to vanillin (45). Urinary VMA and M determinations are the most widely used screening tests in patients with suspected pheochromocytomas. Both tests in some labs appear to be diagnostic in persistently "functioning" pheochromocytomas in greater than 95% of the cases (17, 47). Others report false-negative rates as high as 50% with VMA and M screening (2, 6, 16, 53, 56). These results may be influenced by obtaining samples between paroxysms in patients with intermittently secreting tumors. Diagnostic testing by urinary M measurements is somewhat more accurate than with VMA determinations (6, 49). In general, false-positive tests are more common than false-negative tests, and in situations where the accuracy of laboratory tests is in doubt, assays of M, VMA, and total CA should all be done on the same urines. Collections of urine for less than 24 hours are valuable in diagnosing pheochromocytoma when expressed in terms of urinary VMA or M per gram of urine creatinine (60). "Spot" urinary M concentrations seem to correlate well with 24-hour collections (31). In children, different normal values for both urinary M and VMA must be used to make the diagnosis. These are usually expressed per milligram of creatinine.

PROVOCATIVE DIAGNOSTIC STUDIES

Tyramine, by releasing CA from storage sites (18), histamine, probably by the same mechanism, and glucagon, by an unknown mechanism, all can be used as provocative agents in testing for pheochromocytomas (2). Phentolamine, an α-adrenergic receptor antagonist, is used in doses of 1 to 5 mg via rapid intravenous infusion as both a diagnostic test and for rapid BP control. Phenoxybenzamine, another α-adrenergic blocking agent, has similar effects in pheochromocytoma and is available orally; an i.v. preparation is available for experimental or emergency use (16, 55). Provocative tests should be reserved for patients with equivocal plasma or urine tests who are suspected of having an intermittently secreting pheochromocytoma and in whom spontaneous "attacks" are too infrequent to allow testing during one. Specific details for provocative testing (other than the clonidine suppression test) are summarized in Table 32.4. A recently described test that seems both safe and discriminating is the clonidine suppression test (7). In this test the plasma NE level is measured before and 3 hours after a single oral dose of 0.3 mg of clonidine (Fig. 32.2). A significant decrease in plasma CA at 3 hours is strong evidence against a pheochromocytoma. This test is only of value in patients in whom basal levels of plasma CA are elevated.

PREOPERATIVE LOCALIZATION OF THE TUMOR

Once biochemical evidence for a pheochromocytoma has been found, its size and location should be determined before surgery is undertaken. Computerized axial tomography ("CAT" scan) and ultrasonography are the initial methods of choice. With "CAT" scanning,

Table 32.4
Pharmacologic Tests for Pheochromocytoma (in Order of Clinical Usefulness)

Agent	Amount Given	Results[a]	Comments
Phentolamine (Regitine)	5 mg i.v.	Decrease in BP 35/35 torr from baseline	False-positive tests are common. False-negative tests are rare. Patient should be hypertensive at the time.
Tyramine	1 mg i.v.	Increase in BP 20 torr	10–30% false-negative rate.
Glucagon	0.5–1.0 mg i.v.	Pressor response	May yield both false-positive and negative results.
Histamine	0.025–0.05 mg i.v.	Secondary rise in BP following drop in BP	False-positive and negative results common. Test is of limited utility and very dangerous.

[a] Results which are suggestive of pheochromocytoma. The results of any provocative test are greatly enhanced if a blood or urine sample is obtained during any induced hypertension and a significant increase in catecholamines is demonstrated.

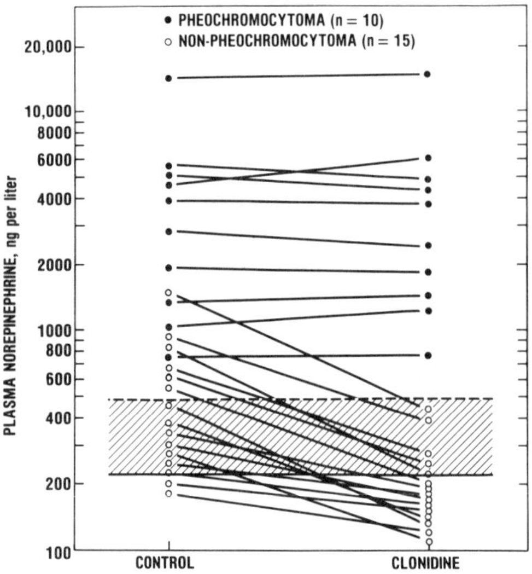

Figure 32.2. Plasma norepinephrine values in individual patients before and 3 hours after a single oral dose of clonidine (0.3 mg). The *hatched area* represents the mean of values obtained from 60 adult healthy subjects. (Reproduced with permission from E. L. Bravo *et al.* (7) and the *New England Journal of Medicine.*)

up to 90% of adrenal tumors can be located noninvasively (59), and adrenal tumors as small as 1.5 cm can be found (23). Ultrasonography is used to localize adrenal tumors (52), but a high degree of expertise is needed to provide reliable, reproducible results (20). The technique formerly used most often was intravenous pyelography with tomography. However, this method has frequently failed to demonstrate the site of the tumor (38). Selective adrenal vein and vena caval sampling for plasma CA has been successfully employed for localization of tumors, but is poor in estimating size (17, 29); such information is needed only in selected cases. Until recently, angiography was the definitive procedure for preoperative localization of the tumor. It is excellent for delineating location and size but has inherent, potential complications, including precipitation of a hypertensive crisis (51). If angiography is to be done, the patient should be prepared as for surgery and then only midstream aortic injection made initially.

A recent study (55) reported success in localizing pheochromocytomas using [^{131}I]-labeled meta-iodobenzylguanidine. This method was successful in identifying both intra- and extraadrenal tumors, including four tumors which were not localized by computerized axial tomography. Further use of this technique will require better and more widespread demonstration of its validity and reliability.

MEDICAL THERAPY: ANESTHESIA AND SURGERY

The significant morbidity and mortality of surgical resection of a pheochromocytoma is related to pre- and intraoperative hypertension (particularly during anesthetic induction and tumor manipulation), postoperative hypotension, and arrhythmias (11). Careful preoperative preparation and close intraoperative monitoring are essential. α-Adrenergic blockade is usually accomplished with phentolamine or phenoxybenzamine. β-Adrenergic blockade is needed only if CA-induced arrhythmias or a severe tachycardia are present. A tyrosine hydroxylase (TH) inhibitor, α-methyl-p-tyrosine (AMPT) (metyrosine-Demser) is available for preoperative management and for chronic use when surgery is not possible (19). Some groups use only α- and β-blockers to prepare pheochromocytoma patients for surgery. Other groups, namely those trained at the NIH, where the use of this drug in patients with pheochromocytoma was developed, also use AMPT for routine preoperative preparation. These groups report that AMPT greatly reduces BP and the incidence of tachyarrhythmias during anesthetic induction and surgery. Halogenated hydrocarbon anesthetics such as methoxyflurane or ethrane are usually the mainstay anesthetics because they do not sensitize the myocardium to CA-induced arrhythmias (12). Intraoperative tachyarrhythmias can usually be controlled with intravenous lidocaine and/or small doses of propranolol. Shock after removal of the tumor is best treated by volume replacement and, only as a last resort, by pressor agents (8).

Inoperable pheochromocytomas can be treated with a combination of α- and β-adrenergic antagonists and with AMPT. Streptozotocin has been used without success in the treatment of malignant pheochromocytoma (25).

SUMMARY

Pheochromocytoma can be a challenging diagnostic and therapeutic problem for the clinician. Improvements in CA assay techniques have allowed for a more accurate diagnostic approach to suspected patients. The newly described clonidine suppression test may help eliminate patients with nonpheochromocytoma hyperadrenergic syndromes. Surgery remains the treatment of choice, and the availability of adrenergic antagonists and agents capable of specific blockage of CA synthesis make this operation safer to perform.

References

1. Alberts WM, McMeekin JO, George JM: Mixed multiple endocrine neoplasia syndromes. *JAMA* 244:1236–1237, 1980.
2. Amery A, Conway J: A critical review of diagnostic tests for pheochromocytoma. *Am Heart J* 73:129–133, 1967.
3. Aronoff SL, Passamani E, Borowsky BA, Weiss AN, Roberts R, Cryer PE: Norepinephrine and epinephrine secretion from a clinically epinephrine-secreting pheochromocytoma. *Am J Med* 69:321–324, 1980.
4. Berman BW, Binder RA, Cornfield DB: Burkitt lymphoma in a patient with neurofibromatosis and pheochromocytoma. *JAMA* 237:993–994, 1977.
5. Block MB, Roberts JP, Kadair RG, Seyfer AE, Hall SF, Nofeldt FD: Multiple endocrine adenomatosis type IIb, diagnosis and treatment. *JAMA* 234:710–714, 1975.
6. Bravo EL, Tarazi RC, Gifford RW, Stewart BH: Circulating and urinary catecholamines in pheochromocytoma, diagnostic and pathophysiologic implications. *N Engl J Med* 301:682–686, 1979.
7. Bravo EL, Tarazi RC, Fouad FM, Vidt DG, Gifford RW: Clonidine suppression test—a useful aid in the diagnosis of pheochromocytoma. *N Engl J Med* 305:623–626, 1981.
8. Brunjes S, Johns V, Crane M: Pheochromocytoma: Postoperative shock and blood volume. *N Engl J Med* 262:393–396, 1960.
9. Carney JA, Co VLW, Gordon H, Northcutt RC, Pearse AGE, Sheps SG: Familial pheochromocytoma and islet cell tumor of the pancreas. *Am J Med* 68:515–521, 1980.
10. Carney JA, Sizemore GW, Sheps SG: Adrenal medullary disease in multiple endocrine neoplasia, type II, pheochromocytoma and its precursors. *Am J Clin Pathol* 66:279–290, 1976.
11. Cooperman LH, Engleman K, Mann PEG: Anesthetic management of pheochromocytoma employing halothane and beta adrenergic blockade. *Anesthesiology* 28:575–582, 1967.
12. Crout J, Brown B: Anesthetic management of pheochromocytoma: The value of phenoxybenzamine and methoxfluorane. *Anesthesiology* 30:29–33, 1969.
13. Cryer P: Physiology and pathophysiology of the human sympathoadrenal neuroendocrine system. *N Engl J Med* 303:436–444, 1980.
14. Cryer PE, Avioli LV: Pheochromocytoma and autonomic dysfunction. *Arch Intern Med* 137:783–787, 1977.
15. Darsee JR, Nutter DO: Two distinct catecholamine cardiomyopathies from norepinephrine and epinephrine secreting pheochromocytomas (abstr). *Circulation* 62:317, 1980.
16. DeQuattro V, Campese VM: Pheochromocytoma: Diagnosis and therapy. In DeGroot L: *Endocrinology*. New York, Grune & Stratton, 1979, pp 1279–1288.
17. Engleman K: Pheochromocytoma. *Clin Endocrinol Metab* 6:769–797, 1977.
18. Engelman K, Horwitz D, Ambrose I, et al: Further evaluation of the tyramine test for pheochromocytoma. *N Engl J Med* 278:705–709, 1968.
19. Engleman K, Horowitz D, Jequier E, Sjoerdsma A: Biochemical and pharmacologic effects of methyltyrosine in man. *J Clin Invest* 47:577–594, 1968.
20. Epstein AJ, Patel SK, Petasnick JP: Computerized tomography of the adrenal gland. *JAMA* 242:2791–2794, 1979.
21. Fairchild RS, Kyner JL, Hermreck A, Schimke N: Neuroblastoma, pheochromocytoma and renal cell carcinoma, occurrence in a single pa-

tient. *JAMA* 242:2210–2211, 1979.

22. Felice LJ, Kissinger PT: A modification of the Pisano method for vanillylmandelic acid using high pressure liquid chromatography. *Clin Chim Acta* 76:317–320, 1977.
23. Ganguly A, Henry DP, Yune HY, Pratt JH, Grim CE, Donohue JP, Weinberger MH: Diagnosis and localization of pheochromocytoma, detection by measurement of urinary norepinephrine excretion during sleep, plasma norepinephrine concentration and computerized axial tomography. *Am J Med* 67:21–26, 1979.
24. Gifford RW, Kvale WF, Maher FT, Roth GM, Priestly JT: Clinical features, diagnosis and treatment of pheochromocytoma: A review of 76 cases. *Mayo Clin Proc* 39:281–302, 1964.
25. Gitlow SE, Mendlowitz M, Wilk EK, Wolf RL, Bertani LM: Excretion of catecholamine catabolites by normal children. *J Lab Clin Med* 72:612–620, 1968.
26. Glusheim AS, Mansuy MM, Litmann DS: Pheochromocytoma: Its relationship to the neuroectodermal syndromes. *Am J Med* 15:318–337, 1953.
27. Hamaji M: Pancreatic α and β cell function in pheochromocytoma. *J Clin Endocrinol Metab* 49:322–325, 1979.
28. Hamilton BPM, Cheich JE, Rivera LE: Attempted treatment of inoperable pheochromocytoma with streptozocin. *Arch Intern Med* 137:762–765, 1977.
29. Harrison TS, Seaton JF, Cerny JC, Bookstein JJ, Bartlett J: Localization of pheochromocytoma by caval catheterization. *Arch Surg* 95:339–343, 1967.
30. Heath H, Edis AJ: Pheochromocytoma associated with hypercalcemia and ectopic secretion of calcitonin. *Ann Intern Med* 91:208–210, 1979.
31. Kaplan NM, Kramer NJ, Holland OB, Sheps SG, Gomez-Sanchez C: Single-voided urine metanephrine assays in screening for pheochromocytoma. *Arch Intern Med* 137:190–193, 1977.
32. Kobayashi K, Miura Y, Tomioka H, Sakuma H, Adachi M, Furuyama T, Yoshinaga K, Ishizaki M, Sekino H: Plasma catecholamines and dopamine-beta-hydroxylase activity in chronic renal failure. *Clin Chim Acta* 95:317–323, 1979.
33. Kvale WF, Roth GM, Manger WM, Priestly JT: Present-day diagnosis and treatment of pheochromocytoma: A review of fifty-one cases. *JAMA* 164:854–861, 1957.
34. Lake CR: Renin-aldosterone profiling and plasma norepinephrine and dopamine-β-hydroxylase, pp 603–605 in Mitchell JR (moderator): Renin-aldosterone profiling on hypertension. *Ann Intern Med* 87:596–612, 1977.
35. Lake CR, Ziegler MG, Kopin IJ: Use of plasma norepinephrine for evaluation of sympathetic neuronal function in man. *Life Sci* 18:1315–1326, 1976.
36. Manges WM, Gifford RW: *Pheochromocytoma.* New York, Springer-Verlag, 1977.
37. Mayo C: Paroxysmal hypertension with tumor of retroperitoneal nerve. *JAMA* 139:1047–1050, 1927.
38. Meaney TF, Buonocore E: Selective arteriography as a localizing and provocative test in the diagnosis of pheochromocytoma. *Radiology* 87:309–314, 1966.
39. Melander A, Ranklev E, Sundler F, et al: Beta-adrenergic stimulation of thyroid hormone secretion. *Endocrinology* 97:332, 1975.
40. Minno AM, Bennett WA, Kvale WF: Pheochromocytoma: A study of 15 cases diagnosed at autopsy. *N Engl J Med* 251:959–965, 1954.
41. Moorhead EL, Brenner MJ, Caldwell JE, et al: Pheochromocytoma: A familial tumor. A study of 11 families. *Henry Ford Hosp Med Bull* 13:467–478, 1965.
42. Mulholland SG, Atuk NO, Walzak MP: Familial pheochromocytoma associated with cerebellar hemangioblastoma. A case history and review of the literature. *JAMA* 207:1709–1711, 1969.
43. Page LB, Copeland RB: Pheochromocytoma. *Dis Month* 1:1–40, 1968.
44. Page LB, Raker JW, Berberich FR: Pheochromocytoma with predominant epinephrine secretion. *Am J Med* 47:648–652, 1969.
45. Pisano JJ: A simple analysis for normetanephrine and metanephrine in urine. *Clin Chim Acta* 5:406–409, 1960.
46. Pisano JJ, Cout R, Abraham D: Determination of 3-methoxy-4-hydroxymandelic acid in urine. *Clin Chim Acta* 7:285–291, 1962.
47. Plouin PF, Doclos JM, Menard J, Comoy E, Bohuon C, Alexandre JM: Biochemical tests for diagnosis of pheochromocytoma: Urinary versus plasma determinations. *Br Med J* 282:853–854, 1981.
48. Ram MD, Kottapalli NR, Brown L: Hypercalcitoninemia, pheochromocytoma, and C-cell hyperplasia, a new variant of Sipple's syndrome. *JAMA* 239:2155–2156, 1978.
49. Remine WH, Chong CG, Van Heerden JA, Sheps SG, Harrison EG: Current management of pheochromocytoma. *Ann Surg* 179:740–748, 1974.
50. Riccardi VM: Von Recklinghausen neurofibromatosis. *N Engl J Med* 305:1617–1626, 1981.
51. Rossi P, Young TS, Panke WF: Technique, usefulness, and hazards of arteriography of pheochromocytoma. *JAMA* 205:547–553, 1968.
52. Sample WF, Sarti DA: Computed tomography and gray scale ultrasonography of the adrenal gland: A comparative study. *Radiology* 128:377–383, 1978.
53. Sheps SG, Tyce GM, Flock EV, Maher FT: Current experience in the diagnosis of pheochromocytoma. *Circulation* 34:473–483, 1966.
54. Sipple J: The association of pheochromocytoma with carcinoma of the thyroid gland. *Am J Med* 31:163–166, 1961.
55. Sisson JC, Frager MS, Valk TW, Gross MD, Swanson DP, Wieland DM, Tobes MC, Beierwaltes WH, Thompson NW: Scientigraphic localization of pheochromocytoma. *N Engl J Med* 305:12–17, 1981.
56. Sjoerdsma A, Engelman K, Waldmann TA, Cooperman IG, Hammond WG: Pheochromocytoma: Current concepts of diagnosis and treatment. *Ann Intern Med* 65:1302–1326, 1966.
57. Stackpole R, Meyer M, Uson A: Pheochromocytoma in children. *J Pediatr* 63:315–330, 1963.

58. Steiner AL, Goodman AD, Powers SR: Study of a kindred with pheochromocytoma medullary thyroid carcinoma, hyperparathyroidism and Cushing's disease: Multiple endocrine neoplasia, type 2. *Medicine* 47:371–409, 1968.
59. Stewart BH, Bravo EL, Haaga J, Meaney TF, Tarazi R: Localization of pheochromocytoma by computed tomography. *N Engl J Med* 299:460–461, 1978.
60. Sullivan JM, Solomon HS: The diagnosis of pheochromocytoma—overnight excretion of catecholamine metabolites. *JAMA* 231:618–619, 1975.
61. Turnbull DM, Johnston DG, Albert KGMM, Hall R: Hormonal and metabolic studies in a patient with a pheochromocytoma. *J Clin Endocrinol Metab* 51:930–933, 1980.
62. Wallace JM, Gill DP: Prazosin in the diagnosis and treatment of pheochromocytoma. *JAMA* 240:2752–2753, 1978.
63. Williams E, Pollock D: Multiple mucosal neuromata with endocrine tumors: A syndrome allied to von Recklinghausen's disease. *J Pathol Bacteriol* 91:71–80, 1966.

Index

Page numbers in italics denote figures; those followed by t or f denote tables and footnotes, respectively.

B

E

F

O